Recursion-Theoretic Hierarchies

Since their inception, the Perspectives in Logic and Lecture Notes in Logic series have published seminal works by leading logicians. Many of the original books in the series have been unavailable for years, but they are now in print once again.

The theory set out in this volume, the 9th publication in the Perspectives in Logic series, is the result of the meeting and common development of two currents of mathematical research: Descriptive Set Theory and Recursion Theory. Both are concerned with notions of definability and with the classification of mathematical objects according to their complexity. These are the common themes which run through the topics discussed here. The author develops a general theory from which the results of both areas can be derived, making these common threads clear.

PETER G. HINMAN works in the Department of Mathematics at the University of Michigan, Ann Arbor.

PERSPECTIVES IN LOGIC

The *Perspectives in Logic* series publishes substantial, high-quality books whose central theme lies in any area or aspect of logic. Books that present new material not now available in book form are particularly welcome. The series ranges from introductory texts suitable for beginning graduate courses to specialized monographs at the frontiers of research. Each book offers an illuminating perspective for its intended audience.

The series has its origins in the old *Perspectives in Mathematical Logic* series edited by the Ω-Group for "Mathematische Logik" of the Heidelberger Akademie der Wissenschaften, whose beginnings date back to the 1960s. The Association for Symbolic Logic has assumed editorial responsibility for the series and changed its name to reflect its interest in books that span the full range of disciplines in which logic plays an important role.

For more information, see www.aslonline.org/books_perspectives.html

PERSPECTIVES IN LOGIC

Recursion-Theoretic Hierarchies

PETER G. HINMAN

University of Michigan

ASSOCIATION FOR SYMBOLIC LOGIC

CAMBRIDGE
UNIVERSITY PRESS

CAMBRIDGE
UNIVERSITY PRESS

University Printing House, Cambridge CB2 8BS, United Kingdom

One Liberty Plaza, 20th Floor, New York, NY 10006, USA

477 Williamstown Road, Port Melbourne, VIC 3207, Australia

4843/24, 2nd Floor, Ansari Road, Daryaganj, Delhi – 110002, India

79 Anson Road, #06–04/06, Singapore 079906

Cambridge University Press is part of the University of Cambridge.

It furthers the University's mission by disseminating knowledge in the pursuit of education, learning, and research at the highest international levels of excellence.

www.cambridge.org
Information on this title: www.cambridge.org/9781107168244
10.1017/9781316717110

Association for Symbolic Logic
Richard A. Shore, Publisher
Department of Mathematics, Cornell University, Ithaca, NY 14853
http://www.aslonline.org

A catalogue record for this publication is available from the British Library.

ISBN 978-1-107-16824-4 Hardback

For *M* and *m*

who make life fun

Preface to the Series

On Perspectives. *Mathematical logic arose from a concern with the nature and the limits of rational or mathematical thought, and from a desire to systematise the modes of its expression. The pioneering investigations were diverse and largely autonomous. As time passed, and more particularly in the last two decades, interconnections between different lines of research and links with other branches of mathematics proliferated. The subject is now both rich and varied. It is the aim of the series to provide, as it were, maps or guides to this complex terrain. We shall not aim at encyclopaedic coverage; nor do we wish to prescribe, like Euclid, a definitive version of the elements of the subject. We are not committed to any particular philosophical programme. Nevertheless we have tried by critical discussion to ensure that each book represents a coherent line of thought; and that, by developing certain themes, it will be of greater interest than a mere assemblage of results and techniques.*

The books in the series differ in level: some are introductory some highly specialised. They also differ in scope: some offer a wide view of an area, others present a single line of thought. Each book is, at its own level, reasonably self-contained. Although no book depends on another as prerequisite, we have encouraged authors to fit their book in with other planned volumes, sometimes deliberately seeking coverage of the same material from different points of view. We have tried to attain a reasonable degree of uniformity of notation and arrangement. However, the books in the series are written by individual authors, not by the group. Plans for books are discussed and argued about at length. Later, encouragement is given and revisions suggested. But it is the authors who do the work; if, as we hope, the series proves of value, the credit will be theirs.

History of the Ω-Group. *During 1968 the idea of an integrated series of monographs on mathematical logic was first mooted. Various discussions led to a meeting at Oberwolfach in the spring of 1969. Here the founding members of the group (R. O. Gandy, A. Levy, G. H. Müller, G. Sacks, D. S. Scott) discussed the project in earnest and decided to go ahead with it. Professor F. K. Schmidt and Professor Hans Hermes gave us encouragement and support. Later Hans Hermes joined the group. To begin with all was fluid. How ambitious should we be? Should we write the books ourselves? How long would it take? Plans for authorless books were promoted, savaged and scrapped. Gradually there emerged a form and a method. At the end of an infinite discussion we found our name, and that of the series. We established our centre in Heidelberg. We agreed to meet twice a year together with authors, consultants and assistants, generally in Oberwolfach. We soon found the value of collaboration: on the one hand the permanence of the founding group gave coherence to the over-all plans; on the other hand the stimulus of new contributors kept the project alive and flexible. Above all, we found how intensive discussion could modify the authors' ideas and our own. Often the battle ended with a detailed plan for a better book which the author was keen to write and which would indeed contribute a perspective.*

Acknowledgements. *The confidence and support of Professor Martin Barner of the Mathematisches Forschungsinstitut at Oberwolfach and of Dr. Klaus Peters of Springer-Verlag made possible the first meeting and the preparation of a provisional plan. Encouraged by the Deutsche Forschungsgemeinschaft and the Heidelberger Akademie der Wissenschaften we submitted this plan to the Stiftung Volkswagenwerk where Dipl. Ing. Penschuck vetted our proposal; after careful investigation he became our adviser and advocate. We thank the Stiftung Volkswagenwerk for a generous grant (1970–73) which made our existence and our meetings possible.*

Since 1974 the work of the group has been supported by funds from the Heidelberg Academy; this was made possible by a special grant from the Kultusministerium von Baden-Württemberg (where Regierungsdirektor R. Goll was our counsellor). The success of the negotiations for this was largely due to the enthusiastic support of the former President of the Academy, Professor Wilhelm Doerr. We thank all those concerned.

Finally we thank the Oberwolfach Institute, which provides just the right atmosphere for our meetings, Drs. Ulrich Felgner and Klaus Gloede for all their help, and our indefatigable secretary Elfriede Ihrig.

Oberwolfach
September 1975

R. O. Gandy *H. Hermes*
A. Levy *G. H. Müller*
G. Sacks *D. S. Scott*

Author's Preface

At a recent meeting of logicians, one speaker complained — mainly, but perhaps not wholly, in jest — that logic is tightly controlled by a small group of people (the cabal) who exercise careful control over the release of new ideas to the general public (especially students) and indeed suppress some material completely. The situation is surely not so grim as this, but any potential reader of this book must have felt at some time that there is at least a minor conspiracy to keep new ideas inaccessible until the "insiders" have worked them over thoroughly. In particular he might well feel this way about the whole subject of Generalized Recursion Theory, which developed in the second half of the 1960s. The basic definitions and results on recursion involving functionals of higher type appeared in the monumental but extremely difficult paper Kleene [1959] and [1963]. Gandy [1967] gave another presentation *ab initio*, but the planned part II of this paper, as well as several other major advances in the subject, never appeared in print. For the theory of recursion on ordinals, the situation was even worse. Much of the basic material had appeared only in the *abstracts* Kripke [1964, 1964a], and although certain parts of the theory had been worked out in papers such as Kreisel-Sacks [1965] and Sacks [1967], there was no reasonably complete account of the basic facts of the subject in print.

When I first contemplated doing something about this situation in the spring of 1971, I planned to write a short monograph on recursion relative to type-2 functionals with enough background on ordinary Recursion Theory to show how the theories fit together. Before I had done much about it, however, the invitation of the Ω-Group to write a volume for this series stimulated me to think in more ambitious terms and my plan expanded gradually to include functionals of types 3 and higher, ordinal recursion, and a more thorough presentation of the material on definability (Chapters III–V). The constant encouragement of the Ω-Group, collective and individual, was essential to the completion of the task.

The original plan arose from a course I gave at the University of Michigan in the Fall Term of 1970. Thanks to Jens-Erik Fenstad and the University of Oslo I had the opportunity to lecture on much of the material during the academic year 1971–1972. Other occasions to lecture on parts of the material were provided by the University of Michigan in 1972–73 and the Winter Term of 1975, the Warsaw

Logic Semester in May, 1973, and the Michigan–Ohio Logic Seminar. The majority of the actual writing was done in the summers of 1973–75 under grants from the National Science Foundation.

Of my many teachers, formal and informal, who have personally helped me to form my conception of this subject, I want especially to mention John Addison, Jens-Erik Fenstad, Robin Gandy, Yiannis Moschovakis, and Joe Shoenfield. Andreas Blass read much of the first draft and made many helpful comments. Mm. Bocuse and Haberlin provided inspiring models of excellence. The boldness of the section and subsection headings in the first third of the book is due to the careful work of Monica Scott and her brown crayon. Barbara Perkel did a superb job of typing. Finally, the person to whom the reader should be most grateful is Anne Zalc. In reading carefully the entire final draft she caught hundreds of errors, serious and minor. More importantly, she was an unrelenting enemy of that peculiar brand of obfuscation which results from an author's implicit assumption that the reader has perfectly understood and remembered every detail of what has preceded any given point. Without her the book would be a denser jungle.

January 30, 1978 Peter G. Hinman
Ann Arbor

Table of Contents

Introduction

The theory set out in this book is the result of the confluence and common development of two currents of mathematical research, Descriptive Set Theory and Recursion Theory. Both are concerned with notions of definability and, more broadly, with the classification of mathematical objects according to various measures of their complexity. These are the common themes which run through the topics discussed in this book.

Descriptive Set Theory arose around the turn of the century as a reaction among some of the mathematical analysts of the day against the free-wheeling methods of Cantorian set theory. People such as Baire, Borel, and Lebesgue felt uneasy with constructions which required the Axiom of Choice or the set of *all* countable ordinals and began to investigate what part of analysis could be carried out by more explicit and constructive means. Needless to say, there was vigorous disagreement over the meaning of these terms. Some of the landmarks in the early days of Descriptive Set Theory are the construction of the Borel sets, Suslin's Theorem that a set of real numbers is Borel just in case both it and its complement are analytic ($\mathbf{\Sigma}^1_1$), and the discovery that analytic sets have many pleasant properties — they are Lebesgue measurable, have the Baire property, and satisfy the Continuum Hypothesis.

A natural concomitant of this interest in the means necessary to effect mathematical constructions is the notion of hierarchy. Roughly speaking, a hierarchy is a classification of a collection of mathematical objects into levels, usually indexed by ordinal numbers. Objects appearing in levels indexed by larger ordinals are in some way more complex that those at lower levels and the index of the first level at which an object appears is thus a measure of the complexity of the object. Such a classification serves both to deepen our understanding of the objects classified and as a valuable technical tool for establishing their properties.

A familiar example, and one which was an important model in the development of the theory, is the hierarchy of Borel sets of real numbers. This class is most simply characterized as the smallest class of sets containing all intervals and closed under the operations of complementation and countable union. In the ρ-th level of the hierarchy are put sets which require a sequence of ρ

applications of complementation and countable union to families of intervals for their construction. The open and closed sets make up the first level, the F_σ (countable union of closed) and G_δ (countable intersection of open) sets the second, etc. This yields an analysis of the class of Borel sets into a strictly increasing sequence of $\aleph_1$ levels (see § V.3).

Recursion Theory developed in the 1930's as an attempt to give a rigorous meaning to the notion of a mechanically or algorithmically calculable function. Such a function is, in an obvious sense, more constructive and less complex than an arbitrary function. The first great success of Recursion Theory was Gödel's application of it in his incompleteness theorems in 1931. The diverse characterizations of the class of recursive functions by Church, Kleene, and Turing suggested strongly that this is a natural class of functions. Other related notions of (relative) complexity developed in the 1940's and 1950's — the notion of one function being recursive in another, the arithmetical and analytical hierarchies of Kleene and Mostowski, various sorts of definability in formal languages, and inductive definability.

The two theories developed essentially independently until the middle 1950's when, largely through the work of Addison, it was realized that to a great extent they are both special cases of a single general theory of definability. The results and methods of Recursion Theory are based on a more restrictive notion of constructivity and in many instances may in hindsight be viewed as refinements of their counterparts in Descriptive Set Theory. Since this is a book of mathematics rather than history, I shall develop here the general theory from which the results of both areas can be derived. As a result, many of the earlier parts of Descriptive Set Theory appear to depend on recursion-theoretic techniques. This dependence is explained by the fact that many of these techniques were known in some form before the advent of Recursion Theory. In a few cases (e.g. the Borel hierarchy) where the older theory is much simpler and more elegant than its recursion-theoretic refinement, I adopt a historical approach and present the classical version first. Roughly half of the material of the book comes from the period when the theories were separate; the other half is a product of the marriage.

In accord with the aims of this series, this book is a perspective on Recursion and Hierarchy Theory and not an encyclopedic treatment. Certain approaches are stressed heavily and other equally valid ones are omitted entirely. My definition of recursive function(al) in § II.2 is non-standard and is chosen for the ease with which it can be adapted to the definition of various classes of generalized recursive function(al)s in Chapters VI–VIII. Inductive definability is portrayed (correctly!) as the cornerstone of almost every aspect of the theory. Many topics closely related to those included are omitted. Some of the most notable of these are degree theory (in both ordinary and generalized recursion theory), abstract recursion theory (over other than the "natural" structures), axiomatic recursion theory, and subrecursive hierarchies. Most of these will be treated in other books in this series.

The book is intended for a variety of audiences. As a whole, it is aimed at a student with some general background in abstract mathematics — at least a smattering of topology, measure theory, and set theory — who has finished a course in logic covering the completeness and incompleteness theorems. In general, those parts which are more recursion-theoretic rely more heavily on a background in logic, while the more descriptive set-theoretic parts use more topology. Of course, as always, lack of formal experience in any area is compensated for by that elusive "mathematical maturity". The book may be used for a variety of formal courses of study under titles such as (Generalized) Recursion Theory, Descriptive Set Theory, or Theory of Definability. Students with sufficient background to skim Chapters I and II quickly can cover most of the book in a full-year course, but otherwise some judicious pruning will be required.

In general, the sections of the book depend on each other as indicated in the following diagram; some individual results may presuppose more or less background.

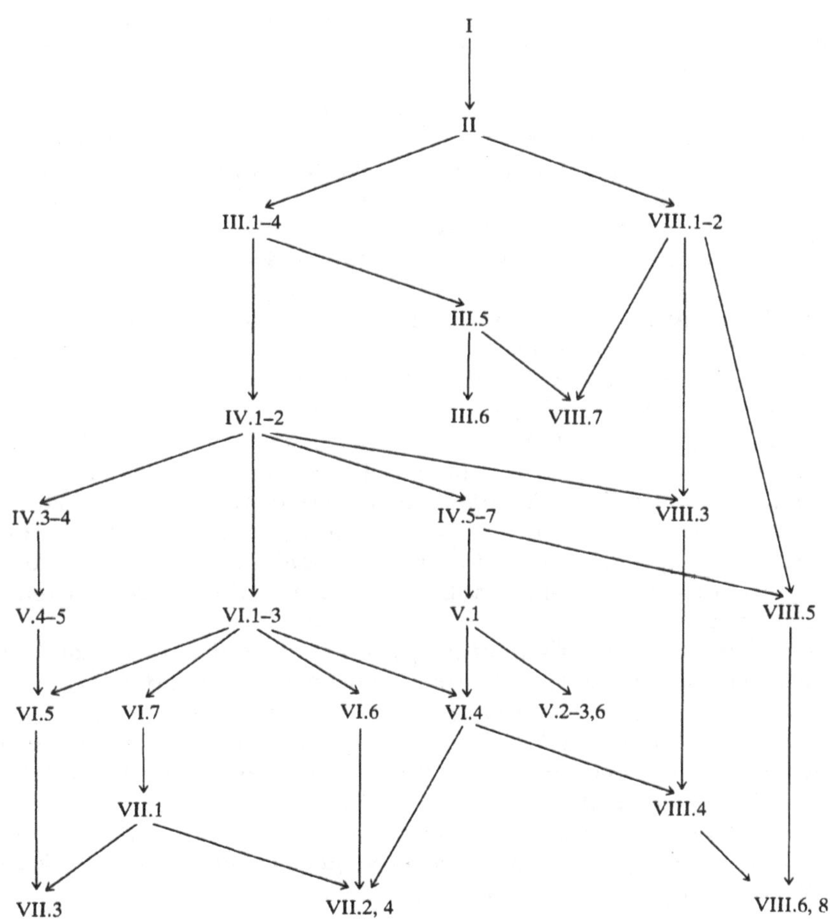

In designing the proofs of results in this book, I have in general made the assumption that the reader has access to pen and paper and will not mind working out a few points for him/herself, but in the main I have tried to give sufficient detail so that anyone who has mastered the prerequisite sections will find this an easy exercise. The subject is full of proofs which require a rather intricate construction followed by a tedious but straightforward inductive verification that the object constructed does the job it was designed for. Furthermore, one often has need in later parts of the theory for a construction quite similar to an earlier one but with one or two additional twists. Indeed, I have often arranged a sequence of lemmas and theorems exactly so that the constructions become increasingly complex in stages in preference to giving at once the most general case. In presenting these I have tried to strike a middle ground between putting the reader to sleep by constant repetition and overly free use of that attractive term "obviously". In general, the construction, or the modification of an earlier construction, is given in full detail, but a good part of the verification is left to the reader. I suspect that the average reader will usually be content to know that the rest of the proof runs "similarly", while the devoted reader (if any!) who seeks a firm grasp on the methods of the subject as well as the results will benefit from the process of working out the details.

The exercises are of two main types. The more routine among them are designed to give the student some experience in handling the methods and techniques of the text and usually require few new ideas. These occasionally include a proof of a lemma from the section. Many of the exercises, however, present results which might well have been included in a larger or more specialized book and constitute a do-it-yourself supplement to the book. (Indeed, one mathematician was somewhat offended that his favorite theorem achieved only the status of an exercise!) I have provided hints and suggestions where they seemed necessary, but many of these exercises will be quite challenging even to the experienced student. The more casual reader should at least glance over the exercises for statements of results.

With respect to the history of the subject, I have taken a middle course between suppressing it altogether and trying to document and credit each minute advance. The primary purpose of the Notes at the end of most sections is to give some idea how, when, and by whom the subject (was) developed, but in the interest of brevity I have omitted mention of many significant contributors. I apologize to those slighted and hope that they will recognize that their sacrifice is for a good cause.

The References similarly contain only a fraction of the articles and books in which our subject matured. Many older references are in a style so different from the current one that they are of little practical use to the working mathematician. I have included a few of these for their historical importance, but in the main the works cited are ones I feel might be of interest to the serious student. In some cases they contain material beyond that of the text, in others they will provide further insight into origins and motivations. In the Epilogue I discuss the current literature and give some guidance for reading which goes beyond the confines of this book.

Part A

Basic Notions of Definability

Chapter I

Groundwork

In this introductory chapter we review some of the prerequisites to the theory to be studied in this book. In the first section we discuss the kinds of objects we shall study — functions and relations with natural numbers and functions of natural numbers as arguments. The second section outlines the application of topological and measure-theoretic notions to these objects. In the third section we discuss inductive definability, a notion which plays a dual role in our theory. Most of our fundamental definitions are given inductively, but in addition we shall *study* inductive definability as a means of classifying sets and relations.

The reader need not master all of Chapter I before going on to the theory proper. Subsections 1.1–1.5, 2.1–2.4, and 3.1–3.11 will suffice for a reading of Chapter II and most of Chapter III, and the other subsections may be used for reference. A list of all of the global notational conventions is on page 467.

1. Logic and Set Theory

1.1 Functions and Sequences. A function φ is a set of ordered pairs (x, y). The *domain* of φ is the set $\mathrm{Dm}\,\varphi = \{x\colon \text{for some } y\,(x, y) \in \varphi\}$, and the *image* of φ is the set $\mathrm{Im}\,\varphi = \{y\colon \text{for some } x\,(x, y) \in \varphi\}$. We often write $\varphi(x)\downarrow$ or say $\varphi(x)$ *is defined* to mean that $x \in \mathrm{Dm}\,\varphi$. Similarly, $\varphi(x)\uparrow$ means $x \notin \mathrm{Dm}\,\varphi$ and is read $\varphi(x)$ *is undefined.* If φ and ψ are two functions, we write $\varphi(x) \simeq \psi(x')$ to mean that *either* both $\varphi(x)$ and $\psi(x')$ are undefined *or* both are defined and have the same value ($\varphi(x) = \psi(x')$). In particular, $\varphi(x) \simeq y$ means that $\varphi(x)$ is defined and has value y — that is, $(x, y) \in \varphi$. We write $\varphi(x) = y$ only in contexts where it is clear that $\varphi(x)$ is defined. If $\mathrm{Dm}\,\varphi$ and $\mathrm{Dm}\,\psi$ are both subsets of a set X, then the statement (for all $x \in X$) $\varphi(x) \simeq \psi(x)$ means simply that φ and ψ denote the same function. If the set X is clear from context, this may be written simply $\varphi(x) \simeq \psi(x)$. The *restriction of* φ *to* X is the function $\varphi \restriction X = \{(x, y)\colon x \in X \text{ and } \varphi(x) \simeq y\}$. The *image of* X *under* φ is the set $\varphi'' X = \mathrm{Im}(\varphi \restriction X)$.

We write $\varphi\colon X \to Y$ to mean φ is a function, $\mathrm{Dm}\,\varphi \subseteq X$, and $\mathrm{Im}\,\varphi \subseteq Y$. If $\mathrm{Dm}\,\varphi = X$ we say φ is *total*; otherwise, φ is *partial.* The set of all total functions $\varphi\colon X \to Y$ is denoted by ${}^{X}Y$.

If y_x denotes an element of Y whenever $x \in Z \subseteq X$, we use any of the expressions $x \mapsto y_x$, $\lambda x . y_x$ and $\langle y_x : x \in Z \rangle$ to denote the function $\{(x, y_x): x \in Z\}$.

The *natural number* m is the set $\{0, 1, \ldots, m-1\}$ of all smaller natural numbers. The set of all natural numbers is denoted by ω.

For any set X, a *finite sequence* from X is a function $\mathbf{x}$ with domain a natural number k called the *length of* $\mathbf{x}$ ($\lg(\mathbf{x})$) and image a subset of X. Hence $\mathbf{x} \in {}^kX$. For $i < \lg(\mathbf{x})$, $\mathbf{x}(i)$ is called the i-th *component* of $\mathbf{x}$ and is usually denoted by x_i. To exhibit all the components we write $(x_0, x_1, \ldots, x_{k-1})$ for $\mathbf{x}$. Note that the *empty sequence* $\varnothing$ is the unique sequence of length 0. We make no distinction in general between X and ${}^1X$. Note that if $\mathbf{x} = (x_0, \ldots, x_{k-1})$ and $\mathbf{y} = (y_0, \ldots, y_{l-1})$ are two finite sequences from X, then $\mathbf{x} \subseteq \mathbf{y}$ just in case $\mathbf{y}$ extends $\mathbf{x}$; that is, $k \leq l$ and for all $i < k$, $x_i = y_i$. The operation $\mathbf{x} * \mathbf{y}$ produces the sequence $(x_0, \ldots, x_{k-1}, y_0, \ldots, y_{l-1})$. If $\varphi \in {}^\omega X$, then $\mathbf{x} * \varphi$ is the function $\psi \in {}^\omega X$ such that for $i < \lg(\mathbf{x})$, $\psi(i) = x_i$, and for $i \geq \lg(\mathbf{x})$, $\psi(i) = \varphi(i - \lg(\mathbf{x}))$. If $Z \subseteq X$, we sometimes write $\mathbf{x} \in Z$ to mean that for all $i < \lg(\mathbf{x})$, $x_i \in Z$. Similarly, $\mathbf{m} < n$ means that for all $i < \lg(\mathbf{m})$, $m_i < n$. If $\varphi: X \to Y$ and $\mathbf{x} \in {}^kX$, then $\varphi(\mathbf{x})$ denotes the sequence $(\varphi(x_0), \ldots, \varphi(x_{k-1}))$.

1.2 Functionals and Relations. For $k, l \in \omega$ we set ${}^{k,l}\omega = {}^k\omega \times {}^l({}^\omega\omega)$. A function $\mathsf{F}: {}^{k,l}\omega \to \omega$ is called a *functional of rank* (k, l). A functional of rank $(k, 0)$ is also called a *function of rank* k and is identified with the corresponding function $F: {}^k\omega \to \omega$. Elements of ${}^\omega\omega$ are thus total functions of rank 1.

Elements of ${}^{k,l}\omega$ are ordered pairs of the form $(\mathbf{m}, \boldsymbol{\alpha})$. However, if F is a functional of rank (k, l), we write $\mathsf{F}(\mathbf{m}, \boldsymbol{\alpha})$ instead of $\mathsf{F}((\mathbf{m}, \boldsymbol{\alpha}))$ and think of $\mathbf{m}, \boldsymbol{\alpha}$ as a *list* of arguments $m_0, \ldots, m_{k-1}, \alpha_0, \ldots, \alpha_{l-1}$. Thus, for example, we write $\mathsf{F}(m_0, \ldots, m_{k-1}, \boldsymbol{\alpha})$ instead of $\mathsf{F}((m_0, \ldots, m_{k-1}), \boldsymbol{\alpha})$ and $\mathsf{F}(p, \mathbf{m}, \boldsymbol{\alpha}, \beta, \gamma)$ instead of of $\mathsf{F}((p) * \mathbf{m}, \boldsymbol{\alpha} * (\beta, \gamma))$. If F is a total functional of rank $(k+1, l)$, then F may also be thought of as a function from ${}^{k,l}\omega$ into ${}^\omega\omega$ whose values are given by:

$$\mathsf{F}[\mathbf{m}, \boldsymbol{\alpha}] = \langle \mathsf{F}(p, \mathbf{m}, \boldsymbol{\alpha}): p \in \omega \rangle = \lambda p . \mathsf{F}(p, \mathbf{m}, \boldsymbol{\alpha}).$$

A subset R of ${}^{k,l}\omega$ is called a *relation of rank* (k, l). We usually write $\mathsf{R}(\mathbf{m}, \boldsymbol{\alpha})$ instead of $(\mathbf{m}, \boldsymbol{\alpha}) \in \mathsf{R}$. A relation of rank $(k, 0)$ is called a *relation* (*on numbers*) *of rank* k and is identified with the corresponding subset R of ${}^k\omega$. In accord with the list notation for functionals, we write, for example, $\mathsf{R}(\mathbf{m}, \alpha_0, \ldots, \alpha_{l-1})$ for $\mathsf{R}(\mathbf{m}, (\alpha_0, \ldots, \alpha_{l-1}))$ and $\mathsf{R}(p, q, \mathbf{m}, r, \beta, \boldsymbol{\alpha})$ for $\mathsf{R}((p, q) * \mathbf{m} * (r), (\beta) * \boldsymbol{\alpha})$. For $\mathsf{R} \subseteq {}^{k,l}\omega$, the *complement* of R (with respect to ${}^{k,l}\omega$) is the relation $\sim\mathsf{R} = \{(\mathbf{m}, \boldsymbol{\alpha}): (\mathbf{m}, \boldsymbol{\alpha}) \in {}^{k,l}\omega \text{ and } (\mathbf{m}, \boldsymbol{\alpha}) \notin \mathsf{R}\}$.

With each relation R of rank (k, l) is associated its *characteristic functional* of rank (k, l) defined by

$$\mathsf{K}_\mathsf{R}(\mathbf{m}, \boldsymbol{\alpha}) = \begin{cases} 0, & \text{if } \mathsf{R}(\mathbf{m}, \boldsymbol{\alpha}); \\ 1, & \text{otherwise.} \end{cases}$$

Conversely, to each functional F of rank (k, l) corresponds it graph Gr_{F} (occasionally $\mathrm{Gr}(\mathsf{F})$), a relation of rank $(k+1, l)$ defined by

$$\mathrm{Gr}_{\mathsf{F}}(n, \mathbf{m}, \boldsymbol{\alpha}) \quad \text{iff} \quad \mathsf{F}(\mathbf{m}, \boldsymbol{\alpha}) \simeq n.$$

By the identification of functionals of rank (1, 0) with functions from ω into ω and of relations of rank (1, 0) with subsets of ω, if $A \subseteq \omega$, then $\mathsf{K}_A \in {}^{\omega}\omega$. We write, for example, $\mathsf{F}(\mathbf{m}, A, \boldsymbol{\alpha}, B)$ instead of $\mathsf{F}(\mathbf{m}, \mathsf{K}_A, \boldsymbol{\alpha}, \mathsf{K}_B)$ and thus extend functionals to admit subsets of ω as arguments.

Compositions of partial functionals and relations are taken to be defined whenever possible. For example, $\mathsf{F}(\mathsf{G}(\mathbf{m}, \boldsymbol{\alpha}), \mathbf{m}, \boldsymbol{\alpha}) \simeq n$ just in case for some $p \in \omega$, $\mathsf{G}(\mathbf{m}, \boldsymbol{\alpha}) \simeq p$ and $\mathsf{F}(p, \mathbf{m}, \boldsymbol{\alpha}) \simeq n$. Similarly, $\mathsf{F}(\mathbf{m}, \boldsymbol{\alpha}, \lambda p . \mathsf{H}(p, \mathbf{m}, \boldsymbol{\alpha})) \simeq n$ just in case for some $\beta \in {}^{\omega}\omega$, $\mathsf{H}(p, \mathbf{m}, \boldsymbol{\alpha}) \simeq \beta(p)$ for all $p \in \omega$, and $\mathsf{F}(\mathbf{m}, \boldsymbol{\alpha}, \beta) \simeq n$. Note that $\mathsf{F}(\mathbf{m}, \boldsymbol{\alpha}, \lambda p . \mathsf{H}(p, \mathbf{m}, \boldsymbol{\alpha}))$ is undefined for any $\mathbf{m}$ and $\boldsymbol{\alpha}$ for which $\lambda p . \mathsf{H}(p, \mathbf{m}, \boldsymbol{\alpha})$ is not total. For relations, we have, for example, $\mathsf{R}(\mathbf{m}, \mathsf{G}(\mathbf{m}, \boldsymbol{\alpha}), \boldsymbol{\alpha})$ is true iff for some $p \in \omega$, $\mathsf{G}(\mathbf{m}, \boldsymbol{\alpha}) \simeq p$ and $\mathsf{R}(\mathbf{m}, p, \boldsymbol{\alpha})$, and false otherwise (not undefined).

Natural numbers are said to be objects of *type* 0. Functions from ${}^{k}\omega$ into ω and subsets of ${}^{k}\omega$ are objects of *type* 1; functionals from ${}^{k,l}\omega$ into ω and subsets of ${}^{k,l}\omega$ $(l > 0)$ are objects of *type* 2. In general a function with natural number values or a relation is of *type* $n+1$ iff its arguments are objects of types at most n. In practice, the arguments of types >0 will almost always be total unary functions. Thus the objects of type 3 discussed in § VI.7 and Chapter VII are functions and relations on ${}^{k,l,l'}\omega = {}^{k}\omega \times {}^{l}({}^{\omega}\omega) \times {}^{l'}({}^{{}^{\omega}\omega}\omega)$. Elements of ${}^{k,l,l'}\omega$ are written $(\mathbf{m}, \boldsymbol{\alpha}, \mathbf{I})$, where $\mathbf{I} = (\mathrm{I}_1, \ldots \mathrm{I}_{l'-1})$. Functionals of type 3 are denoted by letters $\mathbb{F}, \mathbb{G}, \mathbb{H}, \ldots$ and relations of type 3 by $\mathbb{R}, \mathbb{S}, \mathbb{T}, \ldots$.

1.3 Logical Notation. We shall use the logical symbols $\wedge$, $\vee$, $\neg$, $\rightarrow$, and $\leftrightarrow$ as abbreviations for the expressions 'and', 'or', 'not', 'implies', and 'if and only if', respectively. Although we are not, for the most part, dealing with formalized languages, these connectives are to be understood in their usual truth-functional sense. Thus, for example, an expression of the form ——— $\rightarrow$ .... is true just in case ——— is false, or .... is true (or both).

The symbols $\exists$ and $\forall$ will be used as abbreviations for 'there exists' and 'for all', respectively. In most cases the range of the quantifier will be indicated by the type of variable following it in accord with the conventions listed on page 467. For example, an expression of the form $\exists m\ [— m —]$ is true just in case — m — is true for some natural number m. Similarly, the condition for equality of partial functionals is written

$$\mathsf{F} = \mathsf{G} \leftrightarrow \forall \mathbf{m} \forall \boldsymbol{\alpha}\, [\mathsf{F}(\mathbf{m}, \boldsymbol{\alpha}) \simeq \mathsf{G}(\mathbf{m}, \boldsymbol{\alpha})].$$

Further restrictions on the range of a quantifier may be indicated by use of a *bounded quantifier*. For example, $(\exists m < p)$—m— is true iff —m— is true for some m among $0, 1, \ldots, p-1$; $(\forall \gamma \in \mathrm{W}) \ldots . \gamma \ldots .$ is true iff $\ldots . \gamma \ldots .$ is true for all γ belonging to W. We write $\exists! m[$—m—$]$ to mean that —m— is true for exactly one m.

Parentheses, (and), and brackets, [and], are used interchangeably in sufficient quantity to ensure unique readability of expressions. In addition, a single dot . may be used to set off two parts of an expression — for example, $(\forall p < q) . R(p, \mathbf{m})$ or $\lambda p . f(p, \mathbf{m}) + g(p)$.

1.4 Sequence Coding. For each k, ${}^k\omega$ is a countable set and may thus be put in one-one correspondence with a subset of ω. Similarly for $l > 0$, ${}^l({}^\omega\omega)$ may be put in one-one correspondence with a subset of ${}^\omega\omega$. We define here some particularly simple such correspondences which we call *coding functions*.

Temporarily we let p_i denote the i-th prime number: $p_0 = 2, p_1 = 3, \ldots$. For each k we define a total function $\langle \cdot \rangle^k$ of rank k by:

$$\langle \ \rangle^0 = 1, \text{ and for any } k > 0 \text{ and any } \mathbf{m} \in {}^k\omega,$$

$$\langle \mathbf{m} \rangle^k = \langle m_0, \ldots, m_{k-1} \rangle^k = p_0^{m_0+1} \cdot p_1^{m_1+1} \cdot \ldots \cdot p_{k-1}^{m_{k-1}+1}.$$

The unique factorization theorem of arithmetic ensures that if $\langle \mathbf{m} \rangle^k = \langle \mathbf{n} \rangle^l$, then $k = l$ and $\mathbf{m} = \mathbf{n}$. As the superscript is usually clear from the context, we shall usually omit it.

For any s, t, and $i \in \omega$, let

$$(s)_i = \text{least } m \ [m < s \wedge p_i^{m+2} \text{ does not divide } s];$$

$$\mathrm{lg}(s) = \text{least } k \ [k < s \wedge p_k \text{ does not divide } s];$$

$$s * t = s \cdot t', \text{ where } t' \text{ arises from } t \text{ by replacing each}$$

$$\text{factor } p_i^n \text{ in the prime decomposion of } t \text{ by } p_{\mathrm{lg}(s)+i}^n.$$

Then it is an arithmetical exercise to verify that for all k, all $\mathbf{m} \in {}^k\omega$, all $\mathbf{n}$, and all $i < k$, $(\langle \mathbf{m} \rangle)_i = m_i$, $\mathrm{lg}(\langle \mathbf{m} \rangle) = k$, and $\langle \mathbf{m} \rangle * \langle \mathbf{n} \rangle = \langle \mathbf{m} * \mathbf{n} \rangle$. We denote by Sq the set of all s such that $s = \langle \mathbf{m} \rangle$ for some $\mathbf{m}$. Note that for any k and s, s may be regarded as coding a sequence of length k, namely $((s)_0, \ldots, (s)_{k-1})$. We often regard $\mathbf{m}$ and $\langle \mathbf{m} \rangle$ as interchangeable and write, for example, $\mathbf{p} \subseteq \langle \mathbf{m} \rangle$ instead of $\mathbf{p} \subseteq \mathbf{m}$. In particular, $s \subseteq t$ iff for some $\mathbf{m}$ and $\mathbf{n}$, $s = \langle \mathbf{m} \rangle$, $t = \langle \mathbf{n} \rangle$, and $\mathbf{m} \subseteq \mathbf{n}$. For any $\beta \in {}^\omega\omega$, $\bar{\beta}(k)$ denotes the code for the sequence $\beta \restriction k$ — that is, $\bar{\beta}(k) = \langle \beta(0), \ldots, \beta(k-1) \rangle$.

We next define coding functions from ${}^l({}^\omega\omega)$ into ${}^\omega\omega$:

$$\langle \ \rangle^0 = \lambda m . 1, \text{ and for any } l > 0 \text{ and any } \boldsymbol{\alpha} \in {}^l({}^\omega\omega),$$

$$\langle \boldsymbol{\alpha} \rangle^l = \langle \alpha_0, \ldots, \alpha_{l-1} \rangle^l = \lambda m . \langle \alpha_0(m), \ldots, \alpha_{l-1}(m) \rangle.$$

Again it is obvious that if $\langle \boldsymbol{\alpha} \rangle^l = \langle \boldsymbol{\beta} \rangle^k$, then $l = k$ and $\boldsymbol{\alpha} = \boldsymbol{\beta}$, and that we may omit the superscript without ambiguity.

For any γ and $\delta \in {}^\omega\omega$ and any $j \in \omega$, let

$$(\gamma)_j = \lambda m\,.\,(\gamma(m))_j;$$

$$\lg(\gamma) = \lg(\gamma(0));$$

$$\gamma * \delta = \lambda m\,.\,\gamma(m) * \delta(m).$$

Then for all l, all $\boldsymbol{\alpha} \in {}^l({}^\omega\omega)$, all $\boldsymbol{\beta}$, and all $j < l$, $(\langle \boldsymbol{\alpha} \rangle)_i = \alpha_i$, $\lg(\langle \boldsymbol{\alpha} \rangle) = l$, and $\langle \boldsymbol{\alpha} \rangle * \langle \boldsymbol{\beta} \rangle = \langle \boldsymbol{\alpha} * \boldsymbol{\beta} \rangle$. We denote by Sq_1 the set of all γ such that $\gamma = \langle \boldsymbol{\alpha} \rangle$ for some $\boldsymbol{\alpha}$. For any l and γ, γ may be regarded as coding the sequence $((\gamma)_0, \ldots, (\gamma)_{l-1})$.

It will also be useful occasionally to code ω-sequences of functions. We set

$$\langle \alpha_0, \alpha_1, \ldots, \alpha_n, \ldots \rangle = \lambda m\,.\,\alpha_{(m)_0}((m)_1), \quad \text{and}$$

$$(\gamma)^n(m) = \gamma(\langle n, m \rangle).$$

Then clearly $(\langle \alpha_0, \alpha_1, \ldots, \alpha_n, \ldots \rangle)^n(m) = \alpha_n(m)$.

1.5 Set Theory. Except where we specify otherwise, the results of this book are all theorems of ZFC, Zermelo-Fraenkel Set Theory with the Axiom of Choice. The (Generalized) Continuum Hypothesis is not assumed. We shall occasionally want to replace the full Axiom of Choice (AC) by the weaker Axiom of Dependent Choices:

(DC) $\quad \forall x \, \exists y\,.\,(x, y) \in X \rightarrow \exists \varphi \, \forall m\,.\,(\varphi(m), \varphi(m+1)) \in X.$

We recall that this implies the principle of choice for countable families of non-empty sets:

(AC_ω) $\quad \forall m\,.\,Y_m \neq \varnothing \rightarrow \exists \psi \, \forall m\,.\,\psi(m) \in Y_m.$

Most of our set-theoretic conventions are standard and we refer the reader to (for example) Lévy [1978] for further background. A set x is transitive iff $\forall y(y \in x \rightarrow y \subseteq x)$. x is an *ordinal* (*number*) iff x and all of its elements are transitive. For ordinals π and ρ, $\pi < \rho$ iff $\pi \in \rho$; the relation $\leqslant$ is a well-ordering on any set of ordinals. For any ordinal π, $\pi + 1$ is the set $\pi \cup \{\pi\}$, the *ordinal successor* of π. ρ is a *successor ordinal* iff $\rho = \pi + 1$ for some π; ρ is a *limit ordinal* iff $\forall \pi(\pi < \rho \rightarrow \pi + 1 < \rho)$. Every ordinal is either 0, a limit, or a successor. The natural numbers are exactly the finite ordinals and ω is the smallest limit ordinal. Or is the class of all ordinals.

For any set X of ordinals we denote by $\inf X$ the $\leqslant$-least element of X.

Although X need not have a $\leqslant$-greatest element, there is always an ordinal greater than or equal to all elements of X and we denote by $\sup X$ the least such ordinal. In fact, $\sup X$ is exactly the union of the members of X. We set also $\sup^+ X = \sup\{\pi + 1 : \pi \in X\}$. Then $\sup^+ X$ is the least ordinal strictly greater than all elements of X and is the same as $\sup X$ if X has no greatest element; otherwise, $\sup X$ is the greatest element and $\sup^+ X = \sup X + 1$. If φ is a function from ordinals to ordinals and $X \subseteq \mathrm{Dm}\,\varphi$, then

$$\inf_{\pi \in X} \varphi(\pi) = \inf\{\varphi(\pi) : \pi \in X\}, \quad \text{and}$$

$$\sup^{(+)}_{\pi \in X} \varphi(\pi) = \sup^{(+)}\{\varphi(\pi) : \pi \in X\}.$$

An ordinal ρ is a *limit of members of* $X \subseteq \mathrm{Or}$ iff $(\forall \pi < \rho)(\exists \sigma \in X) . \pi < \sigma < \rho$. Any limit of members of any set X is a limit ordinal. If also $\rho \in X$, then ρ is called a *limit point of* X. We denote by $\mathrm{Lim}\, X$ the set of limit points of X. A subset $Y \subseteq X$ is *cofinal in* X iff $(\forall \sigma \in X)(\exists \tau \in Y) \sigma \leqslant \tau$.

If $\mathfrak{A}$ is a proposition which may be true ($\mathfrak{A}(\sigma)$) or false ($\neg \mathfrak{A}(\sigma)$) of each ordinal σ, then to prove $\forall \sigma\, \mathfrak{A}(\sigma)$ we may use the method of proof by *transfinite induction*: if $\forall \sigma ([(\forall \tau < \sigma) \mathfrak{A}(\tau)] \rightarrow \mathfrak{A}(\sigma))$, then $\forall \sigma\, \mathfrak{A}(\sigma)$. We use frequently also the parallel method of *definition by transfinite recursion*: for any total $k+2$-place function ψ, there exists a $k+1$-place function φ such that for all ρ and $\mathbf{x}$,

$$\varphi(\rho, \mathbf{x}) = \psi(\varphi \restriction_{\mathbf{x}} \rho, \rho, \mathbf{x})$$

where

$$\varphi \restriction_{\mathbf{x}} \rho = \{((\pi, \mathbf{x}), z) : \pi < \rho \wedge \varphi(\pi, \mathbf{x}) = z\}.$$

φ is not unique, but any other function φ' which satisfies this equation has $\varphi'(\sigma, \mathbf{x}) = \varphi(\sigma, \mathbf{x})$ for all σ and $\mathbf{x}$.

Since any set X of ordinals is well ordered by the relation $\leqslant$, it is uniquely order-isomorphic to an ordinal which we denote by $\|X\|$, the *order-type* of X. The function φ_X which realizes this isomorphism is recursively defined by:

$$\varphi_X(\rho) = \sup^+\{\varphi_X(\pi) : \pi < \rho \wedge \pi \in X\}.$$

We list here some elementary properties of φ_X which will be needed in Chapter VIII:

(1) $\quad \pi, \rho \in X \wedge \pi < \rho \rightarrow \varphi_X(\pi) < \varphi_X(\rho);$

(2) $\quad \rho \in X \wedge \sigma < \varphi_X(\rho) \rightarrow \exists \pi [\pi \in X \wedge \pi < \rho \wedge \sigma = \varphi_X(\pi)];$

(3) $\quad \rho \in X \rightarrow \varphi_X(\rho + 1) = \varphi_X(\rho) + 1;$

(4) $\rho \subseteq X \to \varphi_X(\rho) = \rho$;

(5) X is an ordinal $\leftrightarrow (\forall \rho \in X) . \varphi_X(\rho) = \rho$.

An ordinal κ is an *initial ordinal* or *cardinal* (*number*) iff there is no one-one correspondence between κ and any $\tau < \kappa$. From the axiom of choice it follows that for every set X there is a unique cardinal κ such that there exists a one-one correspondence between X and κ. We denote this κ by Card(X), the *cardinal of* X. Then there is a one-one correspondence between two sets X and Y just in case $\text{Card}(X) = \text{Card}(Y)$. The natural numbers are exactly the finite cardinals, and ω is the least infinite cardinal. A set X is *countable* iff $\text{Card}(X) \leqslant \omega$ and denumerable or countably infinite iff $\text{Card}(X) = \omega$. The infinite cardinals are enumerated by the function $\aleph$ defined by:

$$\aleph_0 = \omega;$$

$$\aleph_{\sigma+1} = \{\rho : \text{Card}(\rho) \leqslant \aleph_\sigma\};$$

$$\aleph_\sigma = \bigcup \{\aleph_\pi : \pi < \sigma\}, \quad \text{for limit } \sigma.$$

In particular, $\aleph_1$ is the set of countable ordinals.

For any X, $\mathbf{P}(X)$ denotes the power set of X, the set of all subsets of X. If $\text{Card}(X) = \kappa$, then Card $\mathbf{P}(X)$ is denoted by 2^κ. If X is infinite, then $\text{Card}({}^X 2) = \text{Card}({}^X\omega) = 2^\kappa$. In particular, $\text{Card}({}^\omega\omega) = \text{Card}({}^{k,l}\omega) = 2^{\aleph_0}$ for all k and all $l > 0$. By *Cantor's Theorem*, $\kappa < 2^\kappa$ for all cardinals κ. The *Continuum Hypothesis* is the statement that $2^{\aleph_0} = \aleph_1$.

1.6 Ordering Relations. For any set X and any $Z \subseteq {}^2X$, the *field* of Z is the set

$$\text{Fld}(Z) = \{x : \exists y\, [(x, y) \in Z \vee (y, x) \in Z]\}.$$

Z is a *pre-partial-ordering* iff

(1) $(\forall x \in \text{Fld}(Z))[(x, x) \in Z]$, ($Z$ is *reflexive*),

and

(2) $\forall x\, \forall y\, \forall z\, [(x, y) \in Z \wedge (y, z) \in Z \to (x, z) \in Z]$ (Z is *transitive*).

Z is a *pre-linear-ordering* iff (1), (2), and

(3) $\forall x\, \forall y\, [x, y \in \text{Fld}(Z) \wedge x \neq y \to (x, y) \in Z \vee (y, x) \in Z]$ (Z is *connected*).

Z is a *pre-wellordering* iff (1), (2), (3), and

(4) $\forall Y(Y \subseteq \mathrm{Fld}(Z) \wedge Y \neq \varnothing \rightarrow (\exists x \in Y)(\forall y \in Y)[(y,x) \in Z \rightarrow (x,y) \in Z])$

(Z is *well founded*).

Z is a *partial-* (*linear-*, *well-*) *ordering* iff Z is a pre-partial- (linear-, well-) ordering and

(5) $\forall x \forall y\,[(x,y) \in Z \wedge (y,x) \in Z \rightarrow x = y]$ (Z is *antisymmetric*).

From the Axiom of Dependent Choice (DC) it follows that (4) is equivalent to

(4′) $\forall \varphi\,[\forall m.(\varphi(m+1), \varphi(m)) \in Z \rightarrow \exists m.(\varphi(m), \varphi(m+1)) \in Z]$.

If Z is a pre-wellordering, then there is a unique function $|\cdot|_Z$, the *norm associated with* Z, from $\mathrm{Fld}(Z)$ onto an ordinal such that for all $x, y \in \mathrm{Fld}(Z)$,

(6) $(x,y) \in Z \leftrightarrow |x|_Z \leqslant |y|_Z$.

In fact, for any $y \in \mathrm{Fld}(Z)$,

$$|y|_Z = \sup{}^+\{|x|_Z : (x,y) \in Z \wedge (y,x) \notin Z\}.$$

Conversely, if $|\ \ |$ is any function from a set Y into the ordinals, the relation $Z_{|\ |}$ defined by

(7) $(x,y) \in Z_{|\ |} \leftrightarrow |x| \leqslant |y|$

is a pre-wellordering. If the image of $|\ \ |$ is an ordinal, then $|\ \ |$ is the norm associated with $Z_{|\ |}$.

Z is a well-ordering just in case $|\ \ |_Z$ is injective (one-one). The image of $|\ \ |_Z$ is called the (*pre-*)*order-type* of Z and is denoted by $\|Z\|$. Clearly $\|Z\| < \kappa$, where κ is the least cardinal greater than $\mathrm{Card}(X)$. In the context of of set theory without the Axiom of Choice, a useful measure of the size of a set X in terms of ordinals is $o(X) = \sup^+\{\|Z\|: Z$ is a pre-wellordering and $\mathrm{Fld}(Z) \subseteq X\}$.

Orderings will generally be denoted by symbols $\leqslant$ or $\preccurlyeq$ with various sub- and superscripts. In any such context, the symbols $<$ or $\prec$ always denote the associated *strict ordering* defined by:

$$x < y \leftrightarrow x \leqslant y \wedge y \nleqslant x.$$

With any $\gamma \in {}^\omega\omega$ we associate a binary relation $\leqslant_\gamma$ by

$$m \leqslant_\gamma n \leftrightarrow \gamma(\langle m, n\rangle) = 0.$$

We shall mainly be interested in γ such that $\leq_\gamma$ is a partial ordering. We then set

$$\mathrm{Fld}(\gamma) = \mathrm{Fld}(\leq_\gamma);$$

$$\mathsf{W} = \{\gamma: \leq_\gamma \text{ is a well-ordering}\};$$

$$\|\gamma\| = \begin{cases} \|\leq_\gamma\|, & \text{if } \gamma \in \mathsf{W}; \\ \aleph_1, & \text{otherwise}; \end{cases}$$

$$(\gamma \restriction p)(\langle m, n\rangle) = \begin{cases} 0, & \text{if } m \leq_\gamma n \wedge n <_\gamma p; \\ 1, & \text{otherwise}; \end{cases}$$

$$|p|_\gamma = \|\gamma \restriction p\|.$$

The following facts are easily verified: for any $\gamma \in \mathsf{W}$ and any p,

(8) $\gamma \restriction p \in \mathsf{W}$, and if $\|\gamma\| > 0$, then $\|\gamma \restriction p\| < \|\gamma\|$;

(9) for any $\sigma < \|\gamma\|$, there is a unique $p \in \mathrm{Fld}(\gamma)$ such that $\|\gamma \restriction p\| = \sigma$;

(10) $\|\gamma \restriction p\| = \sup^+\{\|\gamma \restriction q\|: q <_\gamma p\}$;

(11) $\aleph_1 = \{\|\gamma\|: \gamma \in \mathsf{W}\}$.

1.7 Notes. The idea of coding finite sequences of natural numbers by prime powers goes back (at least) to Gödel [1931]. For readers less familiar with set theory we recommend Lévy [1978], Zuckerman [1974], or the handiest recent text.

2. *Topology and Measure*

We begin our study of the spaces ${}^{k,l}\omega$ by defining a natural topology and measure theory for them. We define first a topology based on viewing ${}^{k,l}\omega$ as a product of copies of ω, show that with this topology ${}^\omega\omega$ is homeomorphic to the set of binary irrational numbers between 0 and 1 with the topology induced from the reals, and using this homeomorphism, carry Lebesgue measure over to ${}^\omega\omega$.

The set ${}^\omega\omega$ may be viewed as a product $\omega \times \omega \times \cdots \times \omega \times \cdots$ of denumerably many copies of ω. To ω we assign the *discrete topology*: all sets are open (and hence all are also closed). Then to ${}^\omega\omega$ we assign the induced *product topology*: a set $\mathsf{A} \subseteq {}^\omega\omega$ is a *basic open set* iff for some n and some (open) subsets $B_0, \ldots, B_{n-1}$ of ω,

$$\mathsf{A} = B_0 \times B_1 \times \cdots \times B_{n-1} \times \omega \times \cdots \times \omega \times \cdots.$$

In other words, for all α,

$$\alpha \in \mathsf{A} \leftrightarrow (\alpha(0) \in B_0) \wedge (\alpha(1) \in B_1) \wedge \cdots \wedge (\alpha(n-1) \in B_{n-1}).$$

The *open subsets* of ${}^\omega\omega$ are then, of course, arbitrary unions of basic open sets. Finally to ${}^{k,l}\omega$ we again assign the product topology: $\mathsf{R} \subseteq {}^{k,l}\omega$ is a *basic open relation* iff for some $A_0, \ldots, A_{k-1} \subseteq \omega$ and some open sets $\mathsf{A}_0, \ldots, \mathsf{A}_{l-1} \subseteq {}^\omega\omega$,

$$\mathsf{R} = (A_0 \times \cdots \times A_{k-1}) \times (\mathsf{A}_0 \times \cdots \times \mathsf{A}_{l-1}).$$

For any finite sequence $\mathbf{m} = (m_0, \ldots, m_{n-1})$, the *interval* $[\mathbf{m}]$ is defined by:

$$\alpha \in [\mathbf{m}] \leftrightarrow (\alpha(0) = m_0) \wedge \cdots \wedge (\alpha(n-1) = m_{n-1}).$$

If $\mathbf{m} \subseteq \mathbf{n}$, then $[\mathbf{n}] \subseteq [\mathbf{m}]$ is a *subinterval* of $[\mathbf{m}]$. Clearly each interval is a basic open subset of ${}^\omega\omega$. Conversely, if A is the basic open set determined by $B_0, \ldots, B_{n-1}$, then

$$\mathsf{A} = \bigcup \{[\mathbf{m}]: m_0 \in B_0 \wedge \cdots \wedge m_{n-1} \in B_{n-1}\}.$$

Thus the set of intervals is also a base for the topology on ${}^\omega\omega$.

A partial functional is *partial continuous* iff for all n,

$$\mathsf{F}^{-1}(\{n\}) = \{(\mathbf{m}, \boldsymbol{\alpha}): \mathsf{F}(\mathbf{m}, \boldsymbol{\alpha}) \simeq n\}$$

is open. This is equivalent to the more usual condition that $\mathsf{F}^{-1}(B)$ be open for any open set $B \subseteq \omega$. F is *continuous* iff it is partial continuous and total.

2.1 Lemma. *For any* $\mathsf{R} \subseteq {}^{k,l}\omega$,

(i) R *is open iff* R *is the domain of some partial continuous functional*;

(ii) R *is closed-open iff* K_R *is continuous.*

Proof. If F is partial continuous, then $\mathrm{Dm}\,\mathsf{F} = \bigcup \{\mathsf{F}^{-1}(\{n\}): n \in \omega\}$ is a union of open sets and hence is open. Conversely, if R is open, let

$$\mathsf{F}(\mathbf{m}, \boldsymbol{\alpha}) \simeq \begin{cases} 0, & \text{if } \mathsf{R}(\mathbf{m}, \boldsymbol{\alpha}); \\ \text{undefined, otherwise.} \end{cases}$$

Clearly F is partial continuous and $\mathsf{R} = \mathrm{Dm}\,\mathsf{F}$. (ii) follows immediately from the definitions. □

A set $\mathsf{A} \subseteq {}^\omega\omega$ is *dense in* an interval $[\mathbf{m}]$ iff for every subinterval $[\mathbf{m} * \mathbf{n}] \subseteq [\mathbf{m}]$, $\mathsf{A} \cap [\mathbf{m} * \mathbf{n}] \neq \varnothing$. A is *dense* iff it is dense in the interval $[\varnothing] = {}^\omega\omega$. A is *nowhere dense* iff it is dense in no interval. A is *meager* (*first category*) iff it is a countable union of nowhere dense sets. A is *non-meager* (*second category*) iff it is not meager. A is *comeager* (*residual*) iff $\sim$ A is meager. An element α of A is *isolated in* A iff there exists a neighborhood $[\alpha \restriction k]$ of α such that $[\alpha \restriction k] \cap \mathsf{A} = \{\alpha\}$. A is called *perfect* iff it is closed, non-empty, and has no isolated elements.

We mention first some simple direct consequences of these definitions. A singleton is nowhere dense so any countable set is meager. A countable union of meager sets is meager. A subset of a meager set is meager. A set is nowhere dense iff its closure includes no interval. The complement of an open dense set is nowhere dense. A perfect set has power $2^{\aleph_0}$.

2.2 Baire Category Theorem. *No non-empty open set is meager; no comeager set is meager.*

Proof. Both statements follow from the assertion that no interval is meager. Suppose to the contrary that for some $\mathbf{p}$ and some nowhere dense sets A_n, $[\mathbf{p}] = \bigcup\{\mathsf{A}_n : n \in \omega\}$. A_0 is not dense in $[\mathbf{p}]$, so for some sequence $\mathbf{m}^0$, $\mathsf{A}_0 \cap [\mathbf{p} * \mathbf{m}^0] = \varnothing$. A_1 is not dense in $[\mathbf{p} * \mathbf{m}^0]$, so for some $\mathbf{m}^1$, $\mathsf{A}_1 \cap [\mathbf{p} * \mathbf{m}^0 * \mathbf{m}^1] = \varnothing$. In this way we construct $\mathbf{m}^n$ such that for all n,

$$(\mathsf{A}_0 \cup \cdots \cup \mathsf{A}_n) \cap [\mathbf{p} * \mathbf{m}^0 * \cdots * \mathbf{m}^n] = \varnothing.$$

There is a function $\alpha \in \bigcap\{[\mathbf{p} * \mathbf{m}^0 * \cdots * \mathbf{m}^n] : n \in \omega\}$ and $\alpha \in [\mathbf{p}]$ but $\alpha \notin \mathsf{A}_n$ for all n, a contradiction. ☐

We shall have occasion to consider the subspace ${}^{\omega}2$ consisting of all α which assume only the values 0 and 1. ${}^{\omega}2$ is just the set of characteristic functions of subsets of ω and thus in a natural one-one correspondence with $\mathbf{P}(\omega)$. The interval $[\mathbf{m}]$ has a non-empty intersection with ${}^{\omega}2$ iff $\mathbf{m}$ is a *binary sequence* — all m_i are either 0 or 1. If X is a set of finite sequences we say X is *closed downward* iff whenever $\mathbf{n} \subseteq \mathbf{m} \in X$, also $\mathbf{n} \in X$.

2.3 Infinity Lemma. *For any set X of binary sequences which is closed downward, if X is infinite, then X contains an infinite branch — that is, for some* $\alpha \in {}^{\omega}2$, $\alpha \restriction k \in X$ *for all* k.

Proof. Let X satisfy the hypotheses and consider the set

$$Y = \{\mathbf{m} : \mathbf{m} \in X \text{ and } \{\mathbf{n} : \mathbf{m} \subseteq \mathbf{n} \wedge \mathbf{n} \in X\} \text{ is infinite}\}.$$

By hypothesis $\varnothing \in Y$. For any $\mathbf{m}$ and any $\mathbf{n} \neq \mathbf{m}$,

$$\mathbf{m} \subseteq \mathbf{n} \leftrightarrow (\mathbf{m} * (0) \subseteq \mathbf{n})) \vee (\mathbf{m} * (1) \subseteq \mathbf{n}).$$

Hence if $\mathbf{m} \in Y$, then at least one of $\mathbf{m} * (0)$ and $\mathbf{m} * (1)$ also belongs to Y. Thus there exists a unique function α such that for all k,

$$\alpha(k) = \text{least}\, p.(\alpha \restriction k) * (p) \in Y.$$

Since $Y \subseteq X$, this α satisfies the conclusion of the lemma. □

2.4 Theorem. (i) ${}^{\omega}2$ *is a compact subspace of* ${}^{\omega}\omega$;
(ii) *for any* k *and* l, $({}^{k}2) \times {}^{l}({}^{\omega}2)$ *is a compact subspace of* ${}^{k,l}\omega$.

Proof. We prove (i) by showing that any open cover $\mathcal{F}$ of ${}^{\omega}2$ has a finite subcover. Let X be the set of all finite binary sequences $\mathbf{m}$ such that $[\mathbf{m}]$ is included in no member of $\mathcal{F}$. Clearly X is closed downward; suppose X is infinite. Then by the Infinity Lemma, X contains an infinite branch α. Since $\mathcal{F}$ is a cover, $\alpha \in \mathsf{A}$ for some $\mathsf{A} \in \mathcal{F}$. As A is open, for some k, $[\alpha \restriction k] \subseteq \mathsf{A}$, a contradiction. Hence X is finite, so for some k all members of X have length less than k. Let $\mathbf{m}^0, \ldots, \mathbf{m}^{2^k-1}$ be a list of all binary sequences of length k. For each $i < 2^k$ we may choose an $\mathsf{A}_i \in \mathcal{F}$ such that $[\mathbf{m}^i] \subseteq \mathsf{A}_i$. Then $\mathcal{F}_0 = \{\mathsf{A}_i : i < 2^k\}$ is the required finite subcover. The proof of (ii) is similar. □

Note that the proof of Theorem 2.4 depends only on the fact that for any $\mathbf{m} \in X$, $\{p : \mathbf{m} * (p) \in X\}$ is finite. Hence, for example, ${}^{\omega}q$ is a compact subspace of ${}^{\omega}\omega$ for any $q \in \omega$.

The original aim of Descriptive Set Theory was the study and classification of sets of real numbers and their properties which are of interest for mathematical analysis. It was early discovered that little is lost and much is gained in simplicity and elegance if one studies sets of irrational numbers. Indeed, for most properties of interest to analysis — measurability, having the power of the continuum, being meager, etc. — the exclusion of a countable set of points (the rationals) has no effect. On the other hand, there are important topological differences between the reals and the irrationals which simplify the theory of sets of irrationals: the irrationals are of topological dimension 0, there is a base for the topology on the irrationals which consists of closed-open sets, and the irrationals are homeomorphic to their own Cartesian powers. Further simplification was obtained by the discovery that the space of irrationals is homeomorphic to ${}^{\omega}\omega$ with the topology described above. Thus many results concerning ${}^{\omega}\omega$ and the product spaces ${}^{k,l}\omega$ have immediate consequences for the spaces of irrational and real numbers (cf. end of § IV.3).

Temporarily, let ${}^{\omega}\omega$ denote (ambiguously) the topological space described above (as well as its underlying set). Let Ir denote similarly the set of irrational numbers x such that $0 < x < 1$ together with the topology induced by the standard topology on the set of real numbers: $Y \subseteq \mathrm{Ir}$ is *open* iff $Y = \mathrm{Ir} \cap Z$ for some open subset Z of the real interval $(0, 1)$. Then the fact we mentioned is: ${}^{\omega}\omega$ and Ir are homeomorphic. We leave the proof of this to Exercise 2.8 and construct here instead a homeomorphism of ${}^{\omega}\omega$ with another subspace of $(0, 1)$, the space BIr of binary irrationals. This correspondence will serve just as well in transferring results from ${}^{\omega}\omega$ to $(0, 1)$ and is somewhat more natural.

A *finite binary decimal* is a representation of a real number in the form:

$$.r_1r_2\ldots r_l = r_1(2^{-1}) + r_2(2^{-2}) + \cdots + r_l(2^{-l})$$

where each $r_i = 0$ or 1. The real numbers which have finite binary representations are exactly those which can be written as a quotient p/q of natural numbers such that q is a power of 2. Clearly such numbers are dense in (0, 1). An *infinite binary decimal* is a representation

$$.r_1r_2\ldots r_l\ldots = \sum_{i=1}^{\infty} r_i(2^{-i})$$

where each $r_i = 0$ or 1. Any such series converges to a real number between 0 and 1 and every such real number has an infinite binary representation. Two infinite binary decimals represent the same real number iff they are of the forms

$$.r_1r_2\ldots r_l\,1\,0\,0\ldots 0\ldots,$$

and

$$.r_1r_2\ldots r_l\,0\,1\,1\ldots 1\ldots.$$

A *binary irrational* is a real number between 0 and 1 that does not have a finite binary representation. BIr is the topological space consisting of the binary irrationals with the topology induced from (0, 1).

2.5 Theorem. *$^\omega\omega$ and* BIr *are homeomorphic.*

Proof. For any $\alpha \in {}^\omega\omega$, let $\theta(\alpha)$ be the infinite binary decimal:

$$\theta(\alpha) = .\underbrace{11\ldots1}_{\alpha(0)+1}0\,\underbrace{00\ldots0}_{\alpha(1)+1}1\,\underbrace{11\ldots1}_{\alpha(2)+1}0\,\underbrace{00\ldots0}_{\alpha(3)+1}1\ldots.$$

From the preceding remarks it is obvious that θ maps $^\omega\omega$ one-one onto BIr. θ is continuous because if $\alpha \restriction k = \beta \restriction k$, then $|\theta(\alpha) - \theta(\beta)| < 2^{-k}$. For the continuity of θ^{-1}, suppose that $x \in \text{BIr}$ and k are given. To insure that $\theta^{-1}(x)\restriction k = \theta^{-1}(y)\restriction k$ it suffices to take $|x - y| < 2^n$, where $n = \theta^{-1}(x)(0) + \cdots + \theta^{-1}(x)(k-1) + k$. Hence θ is a homeomorphism. □

To compute the image of a given interval **[m]** in $^\omega\omega$, note that $\theta(\alpha)$ has the following representation:

$$\begin{aligned}\theta(\alpha) = {} & .\underbrace{1\,1\ldots1\,1}_{\alpha(0)+1} \\ & -(.\underbrace{0\,0\ldots0\,0}_{\alpha(0)+1}\,\underbrace{1\,1\ldots1\,1}_{\alpha(1)+1}) \\ & +(.\underbrace{0\,0\ldots0\,0}_{\alpha(0)+1}\,\underbrace{0\,0\ldots0\,0}_{\alpha(1)+1}\,\underbrace{1\,1\ldots1\,1}_{\alpha(2)+1}) \\ & -\cdots.\end{aligned}$$

Thus θ induces the following correspondence between intervals of ${}^{\omega}\omega$ and of BIr:

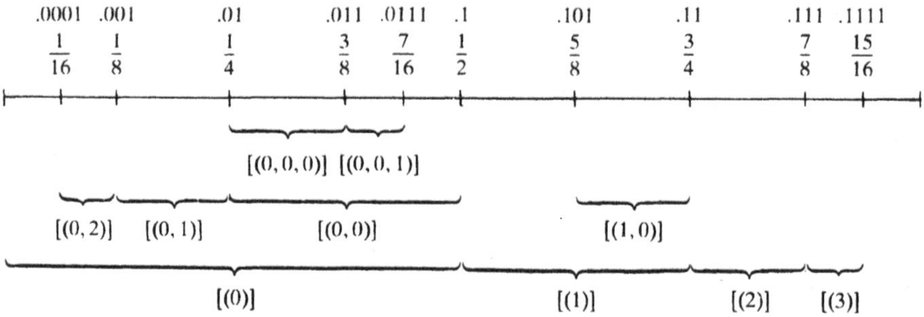

To extend these results to the spaces ${}^{k,l}\omega$, it suffices to show that for each k and l, ${}^{k,l}\omega$ is homeomorphic to ${}^{\omega}\omega$. For this we need sequence coding functions which are *onto* ω. Set

$$\langle\langle m, n\rangle\rangle^2 = \tfrac{1}{2}(m^2 + 2mn + n^2 + 3m + n)$$

and, recursively, for $l > 1$,

$$\langle\langle m_0, \ldots, m_l\rangle\rangle^{l+1} = \langle\langle\, \langle\langle m_0, \ldots, m_{l-1}\rangle\rangle^l, m_l\rangle\rangle.$$

We leave it as an exercise (2.9) to check that the map $\theta^{k,l}$ defined by:

$$\theta^{k,l}(\mathbf{m}, \boldsymbol{\alpha}) = (\mathbf{m}) * \lambda p\,.\langle\langle \alpha_0(p), \ldots, \alpha_{l-1}(p)\rangle\rangle^l$$

is the desired homeomorphism.

The homeomorphism θ induces a natural measure on ${}^{\omega}\omega$. Let mes_{Lb} denote Lebesgue measure restricted to BIr. For $\mathsf{A} \subseteq {}^{\omega}\omega$, we set

$$\text{mes}(\mathsf{A}) = \text{mes}_{\text{Lb}}\{\theta(\alpha)\colon \alpha \in \mathsf{A}\}$$

and say A is *measurable* just in case its θ-image is Lebesgue measurable. Because θ is a homeomorphism, all open and closed sets are measurable. The measure is clearly countably additive and has the property that all subsets of a set of measure 0 are measurable (completeness). This measure may also be described as the *product measure* on ${}^{\omega}\omega$ generated by the measure on ω which assigns $\{n\}$ the measure $2^{-(n+1)}$. Thus for any sequence $\mathbf{m} = (m_0, \ldots, m_{k-1})$,

$$\text{mes}([\mathbf{m}]) = 2^{-(m_0+1)} \cdot \ldots \cdot 2^{-(m_{k-1}+1)} = 2^{-(m_0+\cdots+m_{k-1}+k)}.$$

Similarly, we may define a measure on ${}^{k,l}\omega$ either via the homeomorphism $\theta^{k,l}$ or directly by setting

$$\begin{aligned}\text{mes}(\{m_0\} \times \cdots \times \{m_{k-1}\} \times [p^0] \times \cdots \times [p^{l-1}])\\ = 2^{-(m_0+1)} \cdot \ldots \cdot 2^{-(m_{k-1}+1)} \cdot \text{mes}([p^0]) \cdot \ldots \cdot \text{mes}([p^{l-1}]).\end{aligned}$$

2.6–2.13 Exercises

2.6. Let α^* range over ${}^{\omega}2$. Show that for any set X of finite sequences of 0's and 1's,

$$\forall \alpha^* \exists p . \alpha^* \restriction p \in X \leftrightarrow \exists n \forall \alpha^* (\exists p < n) . \alpha^* \restriction p \in X.$$

2.7. For any $\mathsf{A} \subseteq {}^{\omega}\omega$, let $s * \mathsf{A} = \{s * \alpha \colon \alpha \in \mathsf{A}\}$ and $\mathsf{A}^{(s)} = \{\alpha \colon s * \alpha \in \mathsf{A}\}$. The class Ka of *Kalmar sets* is the smallest class X of subsets of ${}^{\omega}\omega$ such that $\varnothing, {}^{\omega}\omega \in X$ and if for all $n, \mathsf{A}_n \in X$, then $\bigcup \{\langle n \rangle * \mathsf{A}_n \colon n \in \omega\} \in X$. Show for all A and s,

(i) $\mathsf{A} \in \text{Ka} \rightarrow \mathsf{A}^{(s)} \in \text{Ka}$;
(ii) $\mathsf{A} \in \text{Ka} \leftrightarrow s * \mathsf{A} \in \text{Ka}$;
(iii) $\forall n . \mathsf{A}^{(\langle n \rangle)} \in \text{Ka} \rightarrow \mathsf{A} \in \text{Ka}$;
(iv) $\forall \beta \exists n . \mathsf{A}^{(\bar{\beta}(n))} \in \text{Ka} \rightarrow \mathsf{A} \in \text{Ka}$;
(v) $\mathsf{A} \in \text{Ka} \leftrightarrow \mathsf{A}$ is closed-open.

2.8. Prove that the topological spaces ${}^{\omega}\omega$ and Ir are homeomorphic. (Since the sets of rationals and binary rationals are each countable and dense in (0, 1), there is a one-one order-preserving correspondence between them. This may be extended in a unique way to a homeomorphism of (0, 1) with itself. The restriction of this homeomorphism to Ir is a homeomorphism of Ir with BIr.)

2.9. Show that $\theta^{k,l}$ is a homeomorphism of ${}^{k,l}\omega$ onto ${}^{\omega}\omega$.

2.10. (The Zero-One Law). Show that for any measurable set $\mathsf{A} \subseteq {}^{\omega}\omega$, if for all s,

$$\text{mes}(\mathsf{A} \cap [s]) = \text{mes}(\mathsf{A}) \cdot \text{mes}([s]),$$

then mes(A) is either 0 or 1. (Show that this equation holds with $[s]$ replaced by any measurable set.)

2.11. Show that in the usual topology on the real interval (0, 1), Ir and BIr are G_δ sets (countable intersection of open sets) but not F_σ sets (countable union of closed sets).

2.12. Show that a relation $\mathsf{R} \subseteq {}^{k,l}\omega$ is open iff for some $S \subseteq {}^{k+1}\omega$,

$$\mathsf{R}(\mathbf{m}, \boldsymbol{\alpha}) \leftrightarrow \exists p\, S(\mathbf{m}, \bar{\alpha}_0(p), \ldots, \bar{\alpha}_{l-1}(p)).$$

2.13. Show that a homeomorphism between ${}^\omega\omega$ and Ir may be constructed directly. (To each finite sequence $\mathbf{m}$ assign a rational number $\theta(\mathbf{m})$ recursively by the rules:

$$\theta(\varnothing) = 0 \quad \text{and} \quad \theta((p, \mathbf{m})) = \frac{1}{p + 1 + \theta(\mathbf{m})}.$$

For all α, $\theta(\alpha \restriction k)$ converges to a limit $\theta^*(\alpha)$.)

3. *Inductive Definitions*

Let X be any fixed set. A function Γ from the power set of X into itself is called an *operator over* X. Γ is said to be *inclusive* iff for all $Y \subseteq X$, $Y \subseteq \Gamma(Y)$, *monotone* iff for all $Y \subseteq Z \subseteq X$, $\Gamma(Y) \subseteq \Gamma(Z)$, and *inductive* iff Γ is either inclusive or monotone. An operator Γ *defines inductively* a subset $\bar{\Gamma}$ of X as follows. We define by transfinite recursion the sequence Γ^σ by $\Gamma^\sigma = \Gamma(\bigcup\{\Gamma^\tau : \tau < \sigma\})$ and set $\bar{\Gamma} = \bigcup\{\Gamma^\sigma : \sigma \in \mathrm{Or}\}$. We write $\Gamma^{(\sigma)}$ for $\bigcup\{\Gamma^\tau : \tau < \sigma\}$ so that $\Gamma^\sigma = \Gamma(\Gamma^{(\sigma)})$.

We think of the set $\bar{\Gamma}$ as being "built up" in stages. Starting from the empty set we get successively $\Gamma(\varnothing), \Gamma(\Gamma(\varnothing)), \ldots$. Γ^σ is called the σ-th *stage* or *level*.

3.1. Lemma. *For any inductive operator Γ and any ordinal σ,*

(i) $\Gamma^{(\sigma)} \subseteq \Gamma^\sigma$;

(ii) $\Gamma^{\sigma+1} = \Gamma(\Gamma^\sigma)$;

(iii) $\Gamma^{(\sigma)} = \Gamma^\sigma \to \Gamma^\tau = \Gamma^\sigma = \bar{\Gamma}$ *for all* $\tau \geqslant \sigma$;

(iv) $\Gamma^{(\sigma)} = \Gamma^\sigma$ *for some σ such that* $\mathrm{Card}(\sigma) \leqslant \mathrm{Card}(X)$.

Proof. For inclusive Γ, (i) is immediate from the definitions; for monotone Γ it follows from the obvious fact that for $\tau \leqslant \sigma$, $\Gamma^{(\tau)} \subseteq \Gamma^{(\sigma)}$. (ii) is immediate from the observation that by (i), $\Gamma^{(\sigma+1)} = \Gamma^\sigma$. (iii) is proved by induction on τ: for $\tau = \sigma$, clearly $\Gamma^\tau = \Gamma^\sigma$; for $\tau > \sigma$, the induction hypothesis yields $\Gamma^{(\tau)} = \Gamma^\sigma$ and we have $\Gamma^\tau = \Gamma(\Gamma^{(\tau)}) = \Gamma(\Gamma^\sigma) = \Gamma(\Gamma^{(\sigma)}) = \Gamma^\sigma$. For (iv), suppose that for each σ with $\mathrm{Card}(\sigma) \leqslant \mathrm{Card}\, X$, $\Gamma^{(\sigma)} \subsetneqq \Gamma^\sigma$, and let x_σ be an element of $\Gamma^\sigma \sim \Gamma^{(\sigma)}$. If $\tau \neq \sigma$, also

$x_\tau \neq x_\sigma$, and this defines an injection of $\{\sigma\colon \text{Card}(\sigma) \leq \text{Card}(X)\}$ into X. But this set is exactly the least cardinal *larger* than Card X, so this is impossible. □

We denote by $|\Gamma|$ the least σ such that $\Gamma^{(\sigma)} = \Gamma^\sigma$, the *closure ordinal* of Γ. Then

3.2 Corollary. *For any inductive operator Γ over X,* $\text{Card}(|\Gamma|) \leq \text{Card}(X)$ *and* $\bar{\Gamma} = \Gamma^{(|\Gamma|)}$. □

Thus we need not think of the sequence Γ^σ as extended over all ordinals but only over those less than $|\Gamma|$. In particular, if $X = \omega$ we need only consider countable ordinals.

Note that for any inductive Γ, $\Gamma(\bar{\Gamma}) = \bar{\Gamma}$; but for $\sigma < |\Gamma|$, $\Gamma(\Gamma^{(\sigma)}) \neq \Gamma^{(\sigma)}$. In other words, $\bar{\Gamma}$ is the first *fixed point* of Γ in the sequence Γ^σ.

3.3 Theorem. *For any monotone operator Γ over a set X, $\bar{\Gamma}$ is the smallest fixed point of Γ — that is,*

$$\bar{\Gamma} = \bigcap \{Z\colon Z \subseteq X \wedge \Gamma(Z) = Z\}.$$

Proof. By the preceding remark, $\bar{\Gamma} \in \{Z\colon Z \subseteq X \wedge \Gamma(Z) = Z\}$ so that the intersection of this set is included in $\bar{\Gamma}$. Conversely, let Z be any subset of X such that $\Gamma(Z) = Z$; we prove by induction on σ that for all σ, $\Gamma^\sigma \subseteq Z$. Assume as induction hypothesis that this holds for all $\tau < \sigma$ so $\Gamma^{(\sigma)} \subseteq Z$. Then by monotonicity, $\Gamma^\sigma = \Gamma(\Gamma^{(\sigma)}) \subseteq \Gamma(Z) = Z$. □

Note that the proof yields also that for monotone Γ,

$$\bar{\Gamma} = \bigcap \{Z\colon Z \subseteq X \wedge \Gamma(Z) \subseteq Z\}.$$

These results have two distinct aspects. First, they give a characterization of $\bar{\Gamma}$ which does not involve ordinals. Second, they provide a very convenient way of proving that all $x \in \bar{\Gamma}$ have some property: one shows that the set Z of all $x \in X$ which have the property satisfies $\Gamma(Z) \subseteq Z$. In applying this method we say that the proof is by *Γ-induction* or by *induction over $\bar{\Gamma}$*.

In many contexts where we are defining inductively a particular set Y it will be convenient to avoid direct reference to the inductive operator involved. Thus if Y is defined as $\bar{\Gamma}$, we may write Y^σ and $Y^{(\sigma)}$ instead of Γ^σ and $\Gamma^{(\sigma)}$ and describe proofs by Γ-induction as proofs by *induction over* Y.

In the remainder of this section we consider the properties of two special classes of inductive definitions. Let Y be a subset of X and $\mathcal{F}$ a family of finitary functions on X — that is, for each $\varphi \in \mathcal{F}$, there is a natural number $k(\varphi)$ such that $\text{Dm}\,\varphi = {}^{k(\varphi)}X$ and $\text{Im}\,\varphi \subseteq X$. For each such pair $(Y, \mathcal{F})$, we define an inductive operator $\Gamma_{Y,\mathcal{F}}$ by:

$$\Gamma_{Y,\mathscr{F}}(Z) = Y \cup \{\varphi(\mathbf{z}) \colon \varphi \in \mathscr{F} \wedge \mathbf{z} \in {}^{k(\varphi)}Z\}.$$

The resulting set $\bar{\Gamma}_{Y,\mathscr{F}}$ is called the *closure of* Y *under* $\mathscr{F}$. Since $\Gamma_{Y,\mathscr{F}}$ is clearly monotone, $\bar{\Gamma}_{Y,\mathscr{F}}$ is also the *smallest set including* Y *and closed under* $\mathscr{F}$.

3.4 Lemma. *For any* Y *and* $\mathscr{F}$ *as above,* $|\Gamma_{Y,\mathscr{F}}| \leq \omega$.

Proof. Let Y and $\mathscr{F}$ be fixed and write Γ for $\Gamma_{Y,\mathscr{F}}$. By 3.1 (i) it suffices to show $\Gamma^{\omega} \subseteq \Gamma^{(\omega)}$. Let x be any element of $\Gamma^{\omega} = \Gamma(\Gamma^{(\omega)})$. If $x \in Y$, then $x \in \Gamma^0 \subseteq \Gamma^{(\omega)}$. Otherwise, for some $\varphi \in \mathscr{F}$ and some $\mathbf{z} \in {}^{k(\varphi)}(\Gamma^{(\omega)})$, $x = \varphi(\mathbf{z})$. For each $i < k(\varphi)$, let r_i be the least natural number such that $z_i \in \Gamma^{r_i}$, and set $r = \max\{r_i \colon i < k(\varphi)\}$. Then $\mathbf{z} \in {}^{k(\varphi)}\Gamma^r$ so $x \in \Gamma(\Gamma^r) = \Gamma^{r+1} \subseteq \Gamma^{(\omega)}$. □

The method of inductive definition is a generalization of the definition of the set ω of natural numbers in set theory: ω is the smallest set including $\{0\}$ and closed under the successor function, $\mathrm{Sc}(x) = x \cup \{x\}$. Many of the fundamental notions of elementary logic are most naturally defined inductively, often by operators of the form $\Gamma_{Y,\mathscr{F}}$. For example, the set of formulas of a (finitary) first-order formal language is the closure of the set of atomic formulas under functions corresponding to the propositional connectives and quantifiers (cf. § III.5). The set of formal theorems of an axiomatic theory is the closure of the set of axioms under functions corresponding to the rules of inference. An example which is not a closure under finitary functions is the class of formulas of the infinitary language $L_{\omega_1\omega}$ (cf. Keisler [1971]).

We shall also need a generalization of the method of definition by recursion. Roughly speaking, for any set X^* we may define a function $\theta \colon \omega \to X^*$ by specifying a value $\theta(0)$ and a method for calculating $\theta(m+1)$ from $\theta(m)$. The corresponding generalization will allow us to define a function $\theta \colon \bar{\Gamma}_{Y,\mathscr{F}} \to X^*$ by specifying the values $\theta(y)$ for $y \in Y$ and methods for calculating $\theta(\varphi(\mathbf{x}))$ from $\theta(x_0), \ldots, \theta(x_{k(\varphi)-1})$ for all $\varphi \in \mathscr{F}$ (the x_i should be thought of as the immediate predecessors of $\varphi(\mathbf{x})$). In the case of ω, m is uniquely determined by $m+1$, but for arbitrary Y and $\mathscr{F}$ it may happen that $\varphi(\mathbf{x}) = \varphi'(\mathbf{x}')$ or $\varphi(\mathbf{x}) \in Y$ so that the rules would not determine a unique value for $\theta(\varphi(\mathbf{x}))$.

We call the pair $(Y, \mathscr{F})$ *monomorphic* iff all $\varphi \in \mathscr{F}$ are one-one and the sets Y and $\{\mathrm{Im}\,\varphi \colon \varphi \in \mathscr{F}\}$ are pairwise disjoint. The inductive definitions of ω and the class of formulas of a first-order language are monomorphic whereas that of the class of formal theorems is not.

3.5 Theorem (Definition by Recursion). *For any monomorphic pair* $(Y, \mathscr{F})$ *and any set* X^*, *suppose that* $\psi \colon X^* \to X^*$ *and for each* $\varphi \in \mathscr{F}$, $\varphi^* \colon {}^{k(\varphi)}X^* \to X^*$. *Then there exists a unique function* $\theta \colon \bar{\Gamma}_{Y,\mathscr{F}} \to X^*$ *such that*

(i) *for all* $y \in Y$, $\theta(y) = \psi(y)$;

(ii) *for all* $\varphi \in \mathscr{F}$ *and all* $\mathbf{x} \in {}^{k(\varphi)}\bar{\Gamma}_{Y,\mathscr{F}}$,

$$\theta(\varphi(\mathbf{x})) = \varphi^*(\theta(x_0), \ldots, \theta(x_{k(\varphi)-1})).$$

Proof. Let $\Gamma = \Gamma_{Y,\mathcal{F}}$, X^*, ψ, and φ^* be given as in the hypothesis. We define functions $\theta_r: \Gamma^r \to X^*$ by ordinary recursion as follows. $\theta_0 = \psi$. Suppose θ_r is defined and $x \in \Gamma^{r+1}$. If $x \in \Gamma^r$ we set $\theta_{r+1}(x) = \theta_r(x)$. If $x \in \Gamma^{r+1} \sim \Gamma^r$, then by the assumption that $(Y, \mathcal{F})$ is monomorphic there exist unique $\varphi \in \mathcal{F}$ and $\mathbf{z} \in {}^{k(\varphi)}\Gamma^r$ such that $x = \varphi(\mathbf{z})$. We then set $\theta_{r+1}(x) = \varphi^*(\theta_r(z_0), \ldots, \theta_r(z_{k(\varphi)-1}))$. Finally, $\theta = \bigcup \{\theta_r : r \in \omega\}$. We leave to the reader the easy verification that θ satisfies conditions (i) and (ii) (Exercise 3.11). □

Our second special class consists of operators over a product space $X \times Y$. Operators of this type will be used in defining subsets of ${}^{k,l}\omega$.

3.6 Definition. An operator Γ over $X \times Y$ is *decomposable* iff there exists a family of operators Γ_y over X, indexed by $y \in Y$, such that for any $Z \subseteq X \times Y$,

$$\Gamma(Z) = \{(x, y) : x \in \Gamma_y(\{z : (z, y) \in Z\})\}.$$

3.7 Lemma. *For any decomposable operator Γ over $X \times Y$,*
(i) $\bar{\Gamma} = \{(x, y) : y \in Y \wedge x \in \bar{\Gamma}_y\}$;
(ii) $|\Gamma| = \sup\{|\Gamma_y| : y \in Y\}$.

Proof. Both parts follows easily from the assertion that for all σ

$$\Gamma^\sigma = \{(x, y) : y \in Y \wedge x \in \Gamma_y^\sigma\}.$$

To establish this by induction, suppose that it holds for all $\tau < \sigma$. Then

$$\Gamma^{(\sigma)} = \bigcup_{\tau<\sigma} \{(x, y) : y \in Y \wedge x \in \Gamma_y^\tau\} = \{(x, y) : y \in Y \wedge x \in \Gamma_y^{(\sigma)}\},$$

and

$$\Gamma^\sigma = \Gamma(\Gamma^{(\sigma)}) = \{(x, y) : y \in Y \wedge x \in \Gamma_y(\Gamma_y^{(\sigma)})\} = \{(x, y) : y \in Y \wedge x \in \Gamma_y^\sigma\}. \quad \square$$

Decomposable inductive operators over ${}^{k,l}\omega$ are given by families of operators Γ_α over ${}^k\omega$. By Corollary 3.2 each $|\Gamma_\alpha|$ is countable and thus $|\Gamma| \leq \aleph_1$, whereas the closure ordinal of an arbitrary operator over ${}^{k,l}\omega$ is bounded only by the least cardinal greater than $2^{\aleph_0}$. This fact will play an important role in § III.3.

3.8–3.13 Exercises

3.8. Show that any monotone operator over a set X has a largest fixed point $\bar{\Gamma}$. In fact, $\bar{\Gamma} = \sim\Gamma^\circ$, where Γ° is a monotone operator defined by $\Gamma^\circ(Y) = \sim\Gamma(\sim Y)$.

3.9. Construct an example of a non-monotone inductive operator which has no smallest fixed point.

3.10. An operator Γ is called κ*-compact* (for any cardinal κ) iff whenever $x \in \Gamma(Y)$, also $x \in \Gamma(Z)$ for some $Z \subseteq Y$ with $\text{Card}(Z) < \kappa$. Note that any $\Gamma_{Y,\mathcal{F}}$ is ω-compact. What can be said in general about the closure ordinal of a κ-compact inductive operator?

3.11. Complete the proof of Theorem 3.5. Sketch an alternative proof in which θ is defined inductively as the smallest set of pairs (x, x^*) such that ....

3.12. Suppose $Y \subseteq X$ and $\mathcal{F}$ is a family of finitary functions on X such that $(Y, \mathcal{F})$ is monomorphic. For each $x \in \bar{\Gamma}_{Y,\mathcal{F}}$, define $\text{Sp}(x)$, the *support* of x, recursively by:

$$\text{Sp}(y) = \varnothing, \quad \text{for} \quad y \in Y;$$

$$\text{Sp}(\varphi(\mathbf{x})) = \bigcup \{\text{Sp}(x_i) \colon i < k(\varphi)\} \cup \{x_i \colon i < k(\varphi)\}.$$

Establish the following *principle of proof by course-of-values induction* over $\bar{\Gamma}_{Y,\mathcal{F}}$: for any $Z \subseteq \bar{\Gamma}_{Y,\mathcal{F}}$, if $(\forall x \in \bar{\Gamma}_{Y,\mathcal{F}})[\text{Sp}(x) \subseteq Z \to x \in Z]$, then $Z = \bar{\Gamma}_{Y,\mathcal{F}}$.

3.13. There is also a natural notion of definition by course-of-values recursion that says roughly that we may define a function $\theta \colon \bar{\Gamma}_{Y,\mathcal{F}} \to X^*$ by specifying the values $\theta(y)$ for $y \in Y$ and methods for calculating $\theta(x)$ from values $\theta(z)$ for $z \in \text{Sp}(x)$. Formulate precisely a principle of this kind as general as possible and prove that it is valid.

3.14 Notes. Inductive definitions have long played a fundamental role in many areas of mathematics but have been studied as objects only much more recently. Definitions in Algebra of the subgroup, subring, etc. generated by certain elements are all inductive. The class of Borel sets of a topological space is inductively defined as is (the complement of) the perfect kernel of a set of reals (cf. Exercise 3.8). The principal objects of study of Logic and Recursion Theory are all inductively defined. The general study of inductive definability begins explicitly with Spector [1961] but it is close to the surface in many earlier papers of Kleene, especially [1955] and [1955a]. Moschovakis [1974, pp. 3–4] gives a more extended history of the subject.

Chapter II
Ordinary Recursion Theory

The notion of a recursive function resulted from an attempt in the 1930's to provide a precise mathematical characterization of the concept of a mechanically or algorithmically calculable function from ${}^k\omega$ into ω. One way to understand this concept is to imagine an idealized digital computer not subject to error or limitations of memory or storage space. Then a partial function F is mechanically calculable just in case there is a finite program (or algorithm) for this computer which directs it to accept inputs of the form $\mathbf{m}$ and carry out a computation with two possible results: if $\mathbf{m} \in \mathrm{Dm}\, F$, the computation terminates after finitely many steps with the correct value $F(\mathbf{m})$ as output; if $\mathbf{m} \notin \mathrm{Dm}\, F$, the computation does not terminate.

As this is an intuitive concept, however, it cannot be described completely except by convention. Not only is any attempt subject to legitimate disagreement on the basis of current knowledge, but also the possibility remains open that in the future a new means of calculation will be discovered which will be agreed by mathematicians to be mechanical but will not fall under the proposed description. Still, from a practical point of view, the notion seems to be a viable one: most people with a thorough understanding of the concepts involved will agree on the question of whether or not a given method of calculation is mechanical.

In particular, although we cannot give a rigorous proof that every recursive function is mechanically calculable, our justification of this assertion in § 2 below should be convincing to almost everyone. The converse proposition, known as Church's Thesis, that all mechanically calculable functions are recursive, is somewhat more problematic. Without a precise delineation of the class of mechanically calculable functions, we are in no position to *prove* that all of its members are recursive. We are forced, therefore, to rely on what might be called circumstantial evidence. Most importantly, no one has exhibited a function which is agreed to be mechanically calculable but is not recursive. In a similar vein, every known procedure which produces from calculable functions another calculable function also produces a recursive function from recursive functions.

Another kind of evidence is given by the variety of ways that the class of recursive functions can be characterized. Although these characterizations have quite different intuitive content (based on different conceptions of mechanical calculability) they all describe exactly the class of recursive functions. This shows that this is a very natural class and is at least intimately connected with the notion of mechanical calculability.

A discussion of these diverse characterizations and a more detailed examination of the evidence for Church's Thesis may be found in Kleene [1952].

As we shall be discussing functionals as well as functions, we shall want, for comparison, also a notion of mechanical calculability for partial functions from ${}^{k,l}\omega$ into ω. At first glance there seems to be no way for our idealized computer to accept inputs of the form $(\mathbf{m}, \boldsymbol{\alpha})$. Even if we allow the computer to have infinite memory facilities sufficient to store all the values of an argument α, it would seemingly take infinitely long just to "read in" these values. Hence to preserve the finiteness of computations we say that the computer receives an input $(\mathbf{m}, \boldsymbol{\alpha})$ when it is connected to an infinite memory device in which have previously been stored $m_0, \ldots, m_{k-1}$ and the complete graphs of $\alpha_0, \ldots, \alpha_{l-1}$. The computer may then refer to this device at any point in the computation to transfer to its working "registers" either an m_i or a value $\alpha_j(p)$. Since the computation of a value $\mathsf{F}(\mathbf{m}, \boldsymbol{\alpha})$ must be finite, only finitely many values of each argument are actually used. Thus mechanically calculable functionals are continuous.

1. *Primitive Recursion*

We examine first the class of primitive recursive functionals. We shall show that this class includes many familiar functionals but fails to exhaust the class of mechanically calculable functionals. Although in this section we are concerned only with total functionals, we state some of the definitions with '$\simeq$' rather than '$=$' for future application to partial functionals.

1.1 Definition. For any k, l, and n, any $i < k$ and $j < l$, and any $(\mathbf{m}, \boldsymbol{\alpha}) \in {}^{k,l}\omega$,

(i) (*the initial functionals*)

$$\mathsf{Cs}_n^{k,l}(\mathbf{m}, \boldsymbol{\alpha}) = n, \qquad \mathsf{Pr}_i^{k,l}(\mathbf{m}, \boldsymbol{\alpha}) = m_i,$$

$$\mathsf{Sc}_i^{k,l}(\mathbf{m}, \boldsymbol{\alpha}) = m_i + 1, \quad \text{and} \quad \mathsf{Ap}_{i,j}^{k,l}(\mathbf{m}, \boldsymbol{\alpha}) = \alpha_j(m_i);$$

(ii) (*functional composition*) for any k' and any functionals $\mathsf{G}, \mathsf{H}_0, \ldots, \mathsf{H}_{k'-1}$, $\mathsf{FCmp}_{k'}^{k,l}(\mathsf{G}, \mathsf{H}_0, \ldots, \mathsf{H}_{k'-1})$ is the functional F of rank (k, l) such that
(a) if G is of rank (k', l) and $\mathsf{H}_0, \ldots, \mathsf{H}_{k'-1}$ are all of rank (k, l), then

$$\mathsf{F}(\mathbf{m}, \boldsymbol{\alpha}) \simeq \mathsf{G}(\mathsf{H}_0(\mathbf{m}, \boldsymbol{\alpha}), \ldots, \mathsf{H}_{k'-1}(\mathbf{m}, \boldsymbol{\alpha}), \boldsymbol{\alpha});$$

(b) otherwise, $\mathsf{F}(\mathbf{m}, \boldsymbol{\alpha}) \simeq 0$;

(iii) (*primitive recursion*) for any functionals G and H, $\mathrm{Rec}^{k+1,l}(\mathsf{G},\mathsf{H})$ is the functional F of rank $(k+1, l)$ such that

(a) if G is of rank (k, l) and H is of rank $(k+2, l)$, then $\mathsf{F}(0, \mathbf{m}, \boldsymbol{\alpha}) \simeq \mathsf{G}(\mathbf{m}, \boldsymbol{\alpha})$, and for all p,

$$\mathsf{F}(p+1, \mathbf{m}, \boldsymbol{\alpha}) \simeq \mathsf{H}(\mathsf{F}(p, \mathbf{m}, \boldsymbol{\alpha}), p, \mathbf{m}, \boldsymbol{\alpha});$$

(b) otherwise, $\mathsf{F}(p, \mathbf{m}, \boldsymbol{\alpha}) \simeq 0$.

1.2 Definition. The class Prf of *primitive recursive functionals* is the smallest class of total functionals which contains the initial functionals and is closed under functional composition and primitive recursion.

Note that Prf is inductively defined by closure under finitary functions. By induction over Prf it follows that every primitive recursive functional is total and mechanically calculable: clearly this is true for the initial functionals and these properties are preserved by functional composition and primitive recursion.

1.3 Examples. The *addition* function $(+)$ is defined by the equations $0 + m = n$ and $(p+1) + m = (p+m) + 1$. A simple calculation shows that

$$+ = \mathrm{Rec}^{2,0}(\mathsf{Pr}_0^{1,0}, \mathsf{Sc}_0^{3,0})$$

and is thus primitive recursive. *Multiplication* satisfies $0 \cdot m = 0$ and $(p+1) \cdot m = p \cdot m + m$, so that

$$\cdot = \mathrm{Rec}^{2,0}(\mathsf{Cs}_0^{1,0}, \mathrm{FCmp}_2^{3,0}(+, \mathsf{Pr}_0^{3,0}, \mathsf{Pr}_2^{3,0}))$$

and is thus primitive recursive. The *exponential* function $\exp(p, m) = m^p$ satisfies $\exp(0, m) = 1$ and $\exp(p+1, m) = \exp(p, m) \cdot m$ and is similarly shown to be primitive recursive. The *factorial* function (!) satisfies $0! = 1$ and $(p+1)! = p!(p+1)$ and is primitive recursive. Let

$$\mathrm{sg}^+(p) = \begin{cases} 0, & \text{if } p = 0; \\ 1, & \text{if } p > 0; \end{cases} \quad \text{and} \quad \mathrm{sg}^-(p) = \begin{cases} 1, & \text{if } p = 0; \\ 0, & \text{if } p > 0. \end{cases}$$

Then

$$\mathrm{sg}^+ = \mathrm{Rec}^{1,0}(\mathsf{Cs}_0^{0,0}, \mathsf{Cs}_1^{2,0}) \quad \text{and} \quad \mathrm{sg}^- = \mathrm{Rec}^{1,0}(\mathsf{Cs}_1^{0,0}, \mathsf{Cs}_0^{2,0}),$$

so both are primitive recursive. Let f be the primitive recursive function $\mathrm{Rec}^{1,0}(\mathsf{Cs}_0^{0,0}, \mathsf{Pr}_1^{2,0})$ so that $f(0) = 0$ and $f(p+1) = p$ (the *predecessor* function). Then if we set

$$g = \mathrm{Rec}^{2,0}(\mathsf{Pr}_1^{1,0}, \mathrm{FCmp}^{3,0}(f, \mathsf{Pr}_0^{3,0})),$$

it is straightforward to check that

$$g(p, m) = \begin{cases} m - p, & \text{if} \quad m \geq p; \\ 0, & \text{otherwise.} \end{cases}$$

$g(p, m)$ is usually written $m \dot{-} p$.

We call a relation R *primitive recursive* just in case its characteristic functional K_R is primitive recursive. Then $\mathsf{K}_\leq(m,p) = \mathrm{sg}^+(m \dot{-} p)$, $\mathsf{K}_\geq(m,p) = \mathrm{sg}^+(p \dot{-} m)$, $\mathsf{K}_<(m,p) = \mathrm{sg}^-(p \dot{-} m)$, $\mathsf{K}_>(m,p) = \mathrm{sg}^-(m \dot{-} p)$, and $\mathsf{K}_=(m,p) = \mathrm{sg}^+(\mathsf{K}_\leq(m,p) + \mathsf{K}_\geq(m,p))$ so these relations are all primitive recursive. Furthermore, if R and S are primitive recursive relations of the same rank, then

$$\mathsf{K}_{\mathsf{R}\cup\mathsf{S}}(\mathbf{m}, \boldsymbol{\alpha}) = \mathsf{K}_\mathsf{R}(\mathbf{m}, \boldsymbol{\alpha}) \cdot \mathsf{K}_\mathsf{S}(\mathbf{m}, \boldsymbol{\alpha}),$$

$$\mathsf{K}_{\mathsf{R}\cap\mathsf{S}}(\mathbf{m}, \boldsymbol{\alpha}) = \mathrm{sg}^+(\mathsf{K}_\mathsf{R}(\mathbf{m}, \boldsymbol{\alpha}) + \mathsf{K}_\mathsf{S}(\mathbf{m}, \boldsymbol{\alpha})), \quad \text{and}$$

$$\mathsf{K}_{\sim\mathsf{R}}(\mathbf{m}, \boldsymbol{\alpha}) = 1 \dot{-} \mathsf{K}_\mathsf{R}(\mathbf{m}, \boldsymbol{\alpha})$$

so that the class of primitive recursive relations of a given rank forms a Boolean algebra.

Our next aim is to show that the sequence coding and decoding functions of §I.1.4 are primitive recursive. To this end we establish some further closure properties of the classes of primitive recursive functionals and relations.

1.4 Definition. For any k, l, k', and l', any functionals $\mathsf{G}, \mathsf{G}_0, \ldots, \mathsf{G}_{k'}$, and any relations $\mathsf{R}_0, \ldots, \mathsf{R}_{k'-1}$, and S,

(i) (*expansion*) if G has rank (k, l), then $\mathrm{Ex}_{k',l'}(\mathsf{G})$ is the functional F of rank $(k + k', l + l')$ such that

$$\mathsf{F}(\mathbf{m}, \mathbf{n}, \boldsymbol{\alpha}, \boldsymbol{\beta}) \simeq \mathsf{G}(\mathbf{m}, \boldsymbol{\alpha});$$

(ii) (*bounded search*) if G has rank $(k + 2, l)$, then $\mathrm{Bs}(\mathsf{G})$ is the functional F of rank $(k + 1, l)$ such that

$$\mathsf{F}(p, \mathbf{m}, \boldsymbol{\alpha}) \simeq \begin{cases} q, & \text{if} \quad q < p,\ \mathsf{G}(q, \mathbf{m}, \boldsymbol{\alpha}) \simeq 0, \quad \text{and} \\ & \qquad (\forall r < q)(\exists n > 0).\, \mathsf{G}(r, \mathbf{m}, \boldsymbol{\alpha}) \simeq n; \\ p, & \text{if} \quad (\forall q < p)(\exists n > 0).\, \mathsf{G}(q, \mathbf{m}, \boldsymbol{\alpha}) \simeq n; \end{cases}$$

we write

$$\mathsf{F}(p, \mathbf{m}, \boldsymbol{\alpha}) \simeq \text{“least”} \quad q < p.\, \mathsf{G}(q, \mathbf{m}, \boldsymbol{\alpha}) \simeq 0;$$

(iii) (*definition by cases*) if $\mathsf{G}_0, \ldots, \mathsf{G}_{k'}, \mathsf{R}_0, \ldots, \mathsf{R}_{k'-1}$ all have rank (k, l) and for any $(\mathbf{m}, \boldsymbol{\alpha}) \in {}^{k,l}\omega$ there is at most one $i < k'$ such that $\mathsf{R}_i(\mathbf{m}, \boldsymbol{\alpha})$, then $\mathrm{Cases}_{k'}(\mathsf{G}_0, \ldots, \mathsf{G}_{k'}, \mathsf{R}_0, \ldots, \mathsf{R}_{k'-1})$ is the functional F of rank (k, l) such that

$$\mathsf{F}(\mathbf{m}, \boldsymbol{\alpha}) \simeq \begin{cases} \mathsf{G}_0(\mathbf{m}, \boldsymbol{\alpha}), & \text{if } \mathsf{R}_0(\mathbf{m}, \boldsymbol{\alpha}); \\ \quad\vdots & \\ \mathsf{G}_{k'-1}(\mathbf{m}, \boldsymbol{\alpha}), & \text{if } \mathsf{R}_{k'-1}(\mathbf{m}, \boldsymbol{\alpha}); \\ \mathsf{G}_{k'}(\mathbf{m}, \boldsymbol{\alpha}), & \text{otherwise } (\forall i < k' \sim \mathsf{R}_i(\mathbf{m}, \boldsymbol{\alpha})); \end{cases}$$

(iv) (*relational composition*) if $\mathsf{G}_0, \ldots, \mathsf{G}_{k'-1}$ have rank (k, l) and S has rank (k', l), then $\mathrm{RCmp}_{k'}(\mathsf{S}, \mathsf{G}_0, \ldots, \mathsf{G}_{k'-1})$ is the relation R of rank (k, l) such that

$$\mathsf{R}(\mathbf{m}, \boldsymbol{\alpha}) \leftrightarrow \mathsf{S}(\mathsf{G}_0(\mathbf{m}, \boldsymbol{\alpha}), \ldots, \mathsf{G}_{k'-1}(\mathbf{m}, \boldsymbol{\alpha}), \boldsymbol{\alpha});$$

(v) (*bounded quantification*) if S is of rank $(k+2, l)$, then $\exists^0_<(\mathsf{S})$ and $\forall^0_<(\mathsf{S})$ are the relations P and Q of rank $(k+1, l)$ such that

$$\mathsf{P}(p, \mathbf{m}, \boldsymbol{\alpha}) \leftrightarrow (\exists q < p)\mathsf{S}(q, \mathbf{m}, \boldsymbol{\alpha}), \quad \text{and}$$
$$\mathsf{Q}(p, \mathbf{m}, \boldsymbol{\alpha}) \leftrightarrow (\forall q < p)\mathsf{S}(q, \mathbf{m}, \boldsymbol{\alpha}).$$

1.5 Theorem. *The class of primitive recursive functionals and relations is closed under expansion, bounded search, definition by cases, relational composition, and bounded quantification.*

Proof. (i) Clearly any expansion of an initial functional is still an initial functional. Any expansion of $\mathrm{FCmp}(\mathsf{G}, \mathsf{H}_0, \ldots, \mathsf{H}_{k'-1})$ is $\mathrm{FCmp}(\mathsf{G}', \mathsf{H}'_0, \ldots, \mathsf{H}'_{k'-1})$ for suitable expansions G' and H'_i of G and H_i. Similarly, any expansion of $\mathrm{Rec}(\mathsf{G}, \mathsf{H})$ is $\mathrm{Rec}(\mathsf{G}', \mathsf{H}')$ for suitable G' and H'. Hence by induction the expansion of any primitive recursive functional is primitive recursive.

(ii) If G is primitive recursive, then so is F defined by:

$$\mathsf{F}(0, \mathbf{m}, \boldsymbol{\alpha}) = 0;$$
$$\mathsf{F}(p+1, \mathbf{m}, \boldsymbol{\alpha}) = \mathsf{F}(p, \mathbf{m}, \boldsymbol{\alpha}) + \mathrm{sg}^+(\mathsf{G}(\mathsf{F}(p, \mathbf{m}, \boldsymbol{\alpha}), \mathbf{m}, \boldsymbol{\alpha})).$$

We leave to the reader the amusing verification that F is $\mathrm{Bs}(\mathsf{G})$.

(iii) Suppose that $\mathsf{G}_0, \ldots, \mathsf{G}_{k'}$ and $\mathsf{R}_0, \ldots, \mathsf{R}_{k'-1}$ satisfy the hypothesis for definition by cases. Then the F defined there is also given by

$$\mathsf{F}(\mathbf{m}, \boldsymbol{\alpha}) = [\mathsf{G}_0(\mathbf{m}, \boldsymbol{\alpha}) \cdot \mathrm{sg}^-(\mathsf{K}_{\mathsf{R}_0}(\mathbf{m}, \boldsymbol{\alpha}))] + \cdots + [\mathsf{G}_{k'-1}(\mathbf{m}, \boldsymbol{\alpha}) \cdot \mathrm{sg}^-(\mathsf{K}_{\mathsf{R}_{k'-1}}(\mathbf{m}, \boldsymbol{\alpha}))]$$
$$+ [\mathsf{G}_{k'}(\mathbf{m}, \boldsymbol{\alpha}) \cdot \mathsf{K}_{\mathsf{R}_0}(\mathbf{m}, \boldsymbol{\alpha}) \cdot \cdots \cdot \mathsf{K}_{\mathsf{R}_{k'-1}}(\mathbf{m}, \boldsymbol{\alpha})],$$

and is thus seen to be primitive recursive.

(iv) If $\mathsf{R} = \mathrm{RCmp}(\mathsf{S}, \mathsf{G}_0, \ldots, \mathsf{G}_{k'-1})$, then $\mathsf{K}_\mathsf{R} = \mathrm{FCmp}(\mathsf{K}_\mathsf{S}, \mathsf{G}_0, \ldots, \mathsf{G}_{k'-1})$; so if $\mathsf{S}, \mathsf{G}_0, \ldots,$ and $\mathsf{G}_{k'-1}$ are all primitive recursive, so is R.

(v) Let P, Q, and S be as in the definition. Then

$$\mathsf{P}(p, \mathbf{m}, \boldsymbol{\alpha}) \leftrightarrow [\text{“least”}\ q < p\,.\,\mathsf{S}(q, \mathbf{m}, \boldsymbol{\alpha})] < p,$$

and $\mathsf{Q} = \sim\exists^0_<(\sim\mathsf{S})$. □

These results will be used, usually without reference, to justify the claim that some explicitly defined functional or relation is primitive recursive. For example, if G, H, I, R, S, and T are primitive recursive, and F is defined by:

$$\mathsf{F}(\mathbf{m}, \boldsymbol{\alpha}) = \begin{cases} m_0 + \alpha_2(m_1), & \text{if } (\exists q < s)\mathsf{R}(q, \mathsf{G}(q, \mathbf{m}, \boldsymbol{\alpha}), \alpha_4); \\ \alpha_1(\text{“least”}\ q < \mathsf{H}(m_3, \alpha_0, \alpha_2) \\ \qquad [q \cdot m_1 \geqslant \mathsf{I}(m_2, \alpha_3)]), & \text{if } \mathsf{S}(\mathbf{m}, \boldsymbol{\alpha}) \wedge \neg\mathsf{T}(m_0, \boldsymbol{\alpha}); \\ 0, & \text{otherwise}; \end{cases}$$

then a succession of applications of the clauses of Theorem 1.5 together with the remarks preceding it shows that F is primitive recursive.

As a first application, we obtain the primitve recursiveness of the relation “m divides p” and the function $\langle p_m : m \in \omega\rangle$ which enumerates the prime numbers:

$$m \text{ divides } p \leftrightarrow (\exists q < p + 1)(q \cdot m = p);$$

$$p_0 = 2, \quad \text{and}$$

$$p_{m+1} = (\text{“least”}\ q < p_m! + 2)[p_m < q \wedge (\neg\exists r < q)(1 < r \wedge r \text{ divides } q)].$$

1.6 Corollary. *The sequence coding and decoding functions and the set* Sq *of* §I.1.4 *are all primitive recursive.*

Proof. For the functions, this is immediate from their definitions. Also $s \in \mathrm{Sq} \leftrightarrow (\forall i < s)[p_i \text{ divides } s \rightarrow i < \lg(s)]$. □

For any functional F of rank $(k+1, l)$ we set

$$\bar{\mathsf{F}}(0, \mathbf{m}, \boldsymbol{\alpha}) = \langle\ \rangle \quad \text{and}$$

$$\bar{\mathsf{F}}(p + 1, \mathbf{m}, \boldsymbol{\alpha}) = \bar{\mathsf{F}}(p, \mathbf{m}, \boldsymbol{\alpha}) * \langle \mathsf{F}(p, \mathbf{m}, \boldsymbol{\alpha})\rangle.$$

Thus $\bar{\mathsf{F}}(p, \mathbf{m}, \boldsymbol{\alpha}) \simeq \langle \mathsf{F}(0, \mathbf{m}, \boldsymbol{\alpha}), \ldots, \mathsf{F}(p-1, \mathbf{m}, \boldsymbol{\alpha})\rangle$. From the definition and the preceding Corollary it is clear that if F is primitive recursive, so is $\bar{\mathsf{F}}$. Furthermore, for any $q > p$, $\mathsf{F}(p, \mathbf{m}, \boldsymbol{\alpha}) \simeq (\bar{\mathsf{F}}(q, \mathbf{m}, \boldsymbol{\alpha}))_p$ so the primitive recursiveness of $\bar{\mathsf{F}}$ implies that of F. The same argument shows that the functional $\mathsf{H}(p, \alpha) = \bar{\alpha}(p)$ is primitive recursive.

1.7 Definition (*course-of-values recursion*). For any k and l, and any functional G of rank $(k+2, l)$, CvRec(G) is the functional F of rank $(k+1, l)$ such that

$$\mathsf{F}(p, \mathbf{m}, \boldsymbol{\alpha}) \simeq \mathsf{G}(\bar{\mathsf{F}}(p, \mathbf{m}, \boldsymbol{\alpha}), p, \mathbf{m}, \boldsymbol{\alpha}).$$

1.8 Theorem. *The class of primitive recursive functionals is closed under course-of-values recursion.*

Proof. If G is of rank $(k+2, l)$ and $\mathsf{F} = \text{CvRec}(\mathsf{G})$, then

$$\bar{\mathsf{F}}(p+1, \mathbf{m}, \boldsymbol{\alpha}) = \bar{\mathsf{F}}(p, \mathbf{m}, \boldsymbol{\alpha}) * \langle \mathsf{G}(\bar{\mathsf{F}}(p, \mathbf{m}, \boldsymbol{\alpha}), p, \mathbf{m}, \boldsymbol{\alpha})\rangle.$$

Hence, if G is primitive recursive, so is $\bar{\mathsf{F}}$. But then by the preceding remarks, also F is primitive recursive. □

In applying these theorems to show that a particular relation R is primitive recursive, we must formally work with K_R and show this to be a primitive recursive functional. Usually, however, it is more perspicuous to describe directly recursive conditions on R. For example, the condition

$$R(p) \leftrightarrow p = 0 \vee (\exists q < p)[q + 7 = p \wedge R(q)]$$

is equivalent to

$$\mathsf{K}_R(p) = \begin{cases} 0, & \text{if} \quad p = 0; \\ F(p, \bar{\mathsf{K}}_R(p)), & \text{if} \quad p > 0; \end{cases}$$

where

$$F(p, s) = \begin{cases} 0, & \text{if} \quad (\exists q < p)[q + 7 = p \wedge (s)_q = 0]; \\ 1, & \text{otherwise.} \end{cases}$$

In such cases we shall leave to the reader the translation of the conditions on R to conditions on K_R.

Another technique we shall use frequently is to give definitions of the form

$$\mathsf{F}(i, j, \langle \mathbf{m}\rangle, \langle \boldsymbol{\alpha}\rangle) = \mathsf{G}(\alpha_j(m_i), \langle \mathbf{m}\rangle, \langle \boldsymbol{\alpha}\rangle).$$

This should be taken as an abbreviation for

$$\mathsf{F}(i, j, s, \gamma) = \mathsf{G}(((\gamma((s)_i))_j, s, \gamma)$$

so that F is defined for all arguments, not only those of the form $\langle \mathbf{m} \rangle$ and $\langle \boldsymbol{\alpha} \rangle$. Since the decoding functions are primitive recursive, if G is primitive recursive, so is F.

We turn now to the assertion that not all mechanically calculable functionals are primitive recursive. Our method derives from the description of mechanical computability in terms of an idealized computer. We shall in effect specify a particular computer and a "programming language" which suffices to write programs for computing all primitive recursive functionals. We can then exhibit a functional which is mechanically calculable but cannot be "programmed" in this language and hence is not primitive recursive.

The "language" is simply a set $\mathrm{Pri} \subseteq \omega$, members of which we call *primitive recursive indices.* To each $a \in \mathrm{Pri}$ is assigned a primitive recursive functional $[a]$ by interpreting a as a program for an idealized computer whose basic operations correspond to the clauses of Definition 1.1.

1.9 Definition. Pri is the smallest subset of ω such that for all k, l, and n, all $i < k$, and all $j < l$,
(0) $\langle 0, k, l, 0, n\rangle$, $\langle 0, k, l, 1, i\rangle$, $\langle 0, k, l, 2, i\rangle$, and $\langle 0, k, l, 3, i, j\rangle$ all belong to Pri;
(1) for any k' and any $b, c_0, \ldots, c_{k'-1} \in \mathrm{Pri}$, $\langle 1, k, l, b, c_0, \ldots, c_{k'-1}\rangle \in \mathrm{Pri}$;
(2) for any $b, c \in \mathrm{Pri}$, $\langle 2, k+1, l, b, c\rangle \in \mathrm{Pri}$.

This is clearly a monomorphic inductive definition, so by Theorem I.3.5 there exists a unique map $[\cdot]$ from Pri into the class of functionals such that

(0) $[\langle 0, k, l, 0, n\rangle] = \mathsf{Cs}_n^{k,l}$;

$[\langle 0, k, l, 1, i\rangle] = \mathsf{Pr}_i^{k,l}$;

$[\langle 0, k, l, 2, i\rangle] = \mathsf{Sc}_i^{k,l}$;

$[\langle 0, k, l, 3, i, j\rangle] = \mathsf{Ap}_i^{k,l}$;

(1) $[\langle 1, k, l, b, c_0, \ldots, c_{k'-1}\rangle] = \mathrm{FCmp}_{k'}^{k,l}([b], [c_0], \ldots, [c_{k'-1}])$;
(2) $[\langle 2, k+1, l, b, c\rangle] = \mathrm{Rec}^{k,l}([b], [c])$.

1.10 Theorem. $\mathrm{Prf} = \{[a] \colon a \in \mathrm{Pri}\}$.

Proof. For the inclusion ($\subseteq$) we observe that $\{[a] \colon a \in \mathrm{Pri}\}$ clearly contains the initial functions and is closed under composition and primitive recursion. For ($\supseteq$) consider $\{a \colon a \in \mathrm{Pri} \wedge [a] \text{ is primitive recursive}\}$. This set satisfies clauses (0)–(2) of Definition 1.9 and thus includes Pri. □

For each k and l, set

$$\mathsf{Ev}^{k,l}(a, \mathbf{m}, \boldsymbol{\alpha}) = \begin{cases} [a](\mathbf{m}, \boldsymbol{\alpha}), & \text{if} \quad a \in \text{Pri} \wedge (a)_1 = k \wedge (a)_2 = l; \\ 0, & \text{otherwise.} \end{cases}$$

$\mathsf{Ev}^{k,l}$ is called an *evaluation function.*

1.11 Theorem. *for all* $k > 0$ *and all* l, $\mathsf{Ev}^{k,l}$ *is mechanically calculable but not primitive recursive.*

Proof. Suppose first that $\mathsf{Ev}^{k+1,l}$ were primitive recursive. Then if

$$\mathsf{F}(a, \mathbf{m}, \boldsymbol{\alpha}) = \mathsf{Ev}^{k+1,l}(a, a, \mathbf{m}, \boldsymbol{\alpha}) + 1$$

also F is primitive recursive. Hence by Theorem 1.10, $\mathsf{F} = [b]$ for some $b \in \text{Pri}$. But then

$$\mathsf{F}(b, \mathbf{m}, \boldsymbol{\alpha}) = \mathsf{Ev}^{k+1,l}(b, b, \mathbf{m}, \boldsymbol{\alpha}) + 1 = [b](b, \mathbf{m}, \boldsymbol{\alpha}) + 1 = \mathsf{F}(b, \mathbf{m}, \boldsymbol{\alpha}) + 1,$$

a contradiction.

To see that $\mathsf{Ev}^{k,l}$ is mechanically calculable we examine the notion of a *computation tree.* Such a tree may be thought of as a schematic representation of the action of the idealized computer in calculating a value of a given functional. Each computation tree has a top node x_0. Each node x has 0 or more *immediate predecessors* which lie just below x. Each node x is labeled with a triple $(a, \mathbf{m}, \boldsymbol{\alpha})$. x is said to be *evaluated* when $[a](\mathbf{m}, \boldsymbol{\alpha})$ is computed. If a is an index for one of the initial functions, then x has no immediate predecessors. If $a = \langle 1, k, l, b, c_0, \ldots, c_{k'-1} \rangle$, then x has $k' + 1$ immediate predecessors labeled as follows:

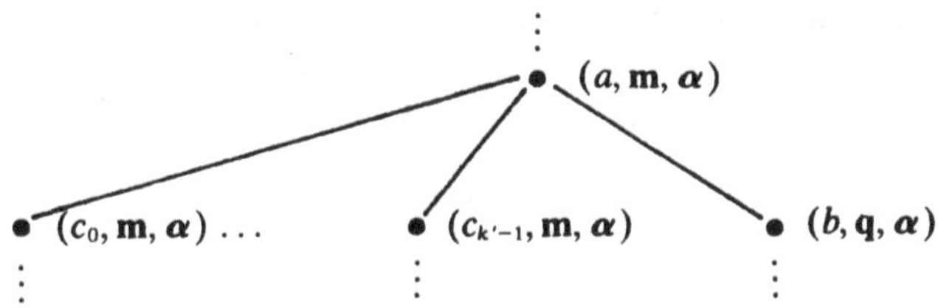

where $\mathbf{q} = (q_0, \ldots, q_{k'-1})$ and for all $i < k'$, $[c_i](\mathbf{m}, \boldsymbol{\alpha}) = q_i$. If $a = \langle 2, k+1, l, b, c \rangle$, then x has either 1 or 2 immediate predecessors:

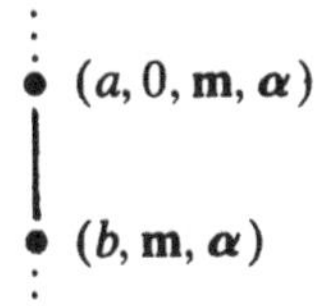

or

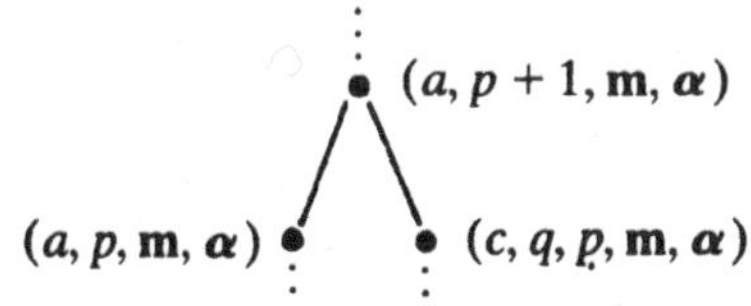

where $[a](p, \mathbf{m}, \boldsymbol{\alpha}) = q$. Thus the labels on the immediate predecessors of x correspond to the *subcomputations* necessary to evaluate the label at x.

For any triple $(a, \mathbf{m}, \boldsymbol{\alpha})$ with $a \in \text{Pri}$, $(a)_1 = k$, and $(a)_2 = l$, we generate and label a computation tree as follows. The top node is labeled $(a, \mathbf{m}, \boldsymbol{\alpha})$. Depending on a, the appropriate number of immediate predecessors of x_0 are constructed and all but possibly the right-most one labeled. The number of immediate predecessors of these may then be determined and so on. If node x lies below node y, then the index at x is not greater than the index at y, and if they are equal, then the first argument at x is strictly less than the first argument at y. Hence each branch terminates with a node labeled with an index for one of the initial functions. This is immediately evaluable. If at some stage all nodes at a given level except the right-most one have been evaluated, then this one may be labeled. When all immediate predecessors of a given node x have been evaluated, then x may be evaluated and has the value of the right-most immediate predecessor as its value. In any application of composition, $k' < a$. Hence by the Infinity Lemma (I.2.3) the tree is finite and this process terminates after a finite number of steps with an evaluation of the top node x_0 and hence with the value $\text{Ev}^{k,l}(a, \mathbf{m}, \boldsymbol{\alpha})$.

A mechanical procedure for calculating $\text{Ev}^{k,l}(a, \mathbf{m}, \boldsymbol{\alpha})$ now goes as follows. Determine first whether or not $a \in \text{Pri}$, $(a)_1 = k$, and $(a)_2 = l$. This is possible by Corollary 1.6 and Exercise 1.16. If not, the value is 0. If so, construct the computation tree as described above and read off the value of the top node. □

1.12–1.18 Exercises

1.12. Find explicitly primitive recursive indices for addition and multiplication.

1.13. Show that any primitive recursive function has infinitely many indices.

1.14. Show that if G is primitive recursive and

$$\mathsf{F}(p, \mathbf{m}, \boldsymbol{\alpha}) = \mathsf{G}(p, \mathbf{m}, \boldsymbol{\alpha}, \lambda q \,.\, \mathsf{F}_p(q, \mathbf{m}, \boldsymbol{\alpha}))$$

where

$$\mathsf{F}_p(q, \mathbf{m}, \boldsymbol{\alpha}) = \begin{cases} \mathsf{F}(q, \mathbf{m}, \boldsymbol{\alpha}), & \text{if } q < p; \\ 0, & \text{otherwise;} \end{cases}$$

then also F is primitive recursive.

1.15. Show that if G and H are both primitive recursive and

$$\mathsf{F}(\mathbf{m}, \boldsymbol{\alpha}) = \mathsf{G}(\mathbf{m}, \boldsymbol{\alpha}, \lambda p \,.\, \mathsf{H}(p, \mathbf{m}, \boldsymbol{\alpha})),$$

then F is primitive recursive.

1.16. Show that the set Pri of primitive recursive indices is a primitive recursive set.

1.17. For each of the closure properties established above, there is a stronger *effective closure property* which asserts that an index for the resulting functional or relation can be computed from indices of the component functionals and relations. For example, Prf is *effectively closed* under bounded search iff there exists a primitive recursive function f such that for all $a \in$ Pri, $[f(a)] = \mathrm{Bs}([a])$. Show that the class of primitive recursive functionals and relations is effectively closed under expansion, bounded search, and course-of-values recursion.

1.18. For any $\boldsymbol{\beta} = (\beta_0, \ldots, \beta_{r-1})$, let Prf$[\boldsymbol{\beta}]$, the class of functionals *primitive recursive in* $\boldsymbol{\beta}$, be the smallest class of total functionals which contains $\beta_0, \ldots, \beta_{r-1}$, and the initial functionals, and is closed under functional composition and primitive recursion. Show that $\mathsf{F} \in \mathrm{Prf}[\boldsymbol{\beta}]$ iff, for some primitive recursive G, $\mathsf{F}(\mathbf{m}, \boldsymbol{\alpha}) = \mathsf{G}(\mathbf{m}, \boldsymbol{\alpha}, \boldsymbol{\beta})$.

1.19 Notes. The primitive recursive functions have played a varying role over the years. Gödel [1931] used "rekursiv" to mean primitive recursive and used only these functions in the incompleteness proof. As various formalized analogues of the class of mechanically calculable functions came to be studied in the 1930's and the class of (general) recursive functions was introduced, the primitive recursive functions were relegated to the secondary role of providing a basic stock of methods for elementary calculations (cf. Exercise 1.17 and Corollary 3.3 below). That some kinds of "recursions" lead beyond the primitive recursive functions was already shown in Ackermann [1928] and by the method of Theorem 1.11 in Péter [1935]. Péter [1967] has a good deal of information on various extensions of the class of primitive recursive functions by means of more complicated recursions.

The method of proof used in Theorem 1.11 is called a *diagonal argument* and is used repeatedly in the remainder of the book. It is an adaptation of Cantor's proof that the set of real numbers is uncountable.

2. *Recursive Functionals and Relations*

If the class of recursive functionals is to include all mechanically calculable functionals, then it must evidently contain each $\mathsf{Ev}^{k,l}$. This suggests that we consider the class Prf_1, the smallest class of total functionals which contains the initial functionals and all $\mathsf{Ev}^{k,l}$ and is closed under functional composition and primitive recursion. Unfortunately, the methods of §1 are easily adapted to show that there is an indexing $[\cdot]_1$ of Prf_1 such that the functionals

$$\mathsf{Ev}_1^{k,l}(a, \mathbf{m}, \boldsymbol{\alpha}) = \begin{cases} [a]_1(\mathbf{m}, \boldsymbol{\alpha}), & \text{if} \quad a \in \mathrm{Pri}_1, \quad (a)_1 = k, \quad \text{and} \quad (a)_2 = l; \\ 0, & \text{otherwise;} \end{cases}$$

are mechanically calculable but not in Prf_1.

It is still instructive to examine the properties of Prf_1 more closely. For any $\mathsf{G} \in \mathrm{Prf}_1$, the functional F defined by

$$\mathsf{F}(\mathbf{m}, \boldsymbol{\alpha}) = \mathsf{Ev}^{k,l}(\mathsf{G}(\mathbf{m}, \boldsymbol{\alpha}), \mathbf{m}, \boldsymbol{\alpha})$$

is also in Prf_1. The computation of $\mathsf{F}(\mathbf{m}, \boldsymbol{\alpha})$ proceeds in two parts: first a value $\mathsf{G}(\mathbf{m}, \boldsymbol{\alpha})$ is determined, then this value is interpreted as a "program" for the rest of the computation. The principle that the results of one part of a computation may be interpreted as coded instructions for another part we call the *self-determination principle.* It is an idealized form of the various devices for program modification which play an essential role in most programming languages for electronic computers.

The class Prf_1 exhibits the self-determination principle only in a limited form. Although the functional G may be any element of Prf_1, the remainder of the computation described above is determined by an index in Pri (not Pri_1) and thus cannot make further use of the functionals $\mathsf{Ev}^{k,l}$. We should like to allow for the possibility that the result $\mathsf{G}(\mathbf{m}, \boldsymbol{\alpha})$ of one computation might be interpreted as an index for any functional in the class under consideration (*full self-determination principle*).

To accomplish this we change slightly our point of view. Instead of fixing in advance a class of indices which code instructions for computing total functionals, we shall regard every natural number a as an index for a *partial* functional $\{a\}$. As before we shall have $\{a\}(\mathbf{m}, \boldsymbol{\alpha}) \simeq n$ just in case a codes a "program" which, when applied to the arguments $(\mathbf{m}, \boldsymbol{\alpha})$ leads to the value n. Now, however, a program may lead to a value for some arguments but not for others. Indeed many a (0, for example) are not of the prescribed form for coding instructions, so that $\{a\}$ is the empty functional. A functional will be called *partial recursive* just in case it is one of the functionals $\{a\}$ $(a \in \omega)$.

Many of the indices will have the same interpretation as before. For example, we shall arrange that

$$\{\langle 1, k, l, b, c_0, \ldots, c_{k'-1}\rangle\}(\mathbf{m}, \boldsymbol{\alpha}) \simeq \{b\}(\{c_0\}(\mathbf{m}, \boldsymbol{\alpha}), \ldots, \{c_{k'-1}\}(\mathbf{m}, \boldsymbol{\alpha}), \boldsymbol{\alpha})$$

so that the partial recursive functionals are closed under composition. The key to obtaining the full self-determination principle is the fact that we shall have

$$\{\langle 2, k+1, l\rangle\}(b, \mathbf{m}, \boldsymbol{\alpha}) \simeq \{b\}(\mathbf{m}, \boldsymbol{\alpha}).$$

The index $\langle 2, k+1, l\rangle$ may be thought of as coding the following program: apply the program coded by the first number argument to the remaining arguments. If $\{b\}(\mathbf{m}, \boldsymbol{\alpha})$ is undefined, so is $\{\langle 2, k+1, l\rangle\}(b, \mathbf{m}, \boldsymbol{\alpha})$; it is for this reason that we are forced to deal with partial functionals. It is now evident that the full self-determination principle holds: if G is partial recursive, so is the functional F defined by:

$$\mathsf{F}(\mathbf{m}, \boldsymbol{\alpha}) \simeq \{\mathsf{G}(\mathbf{m}, \boldsymbol{\alpha})\}(\mathbf{m}, \boldsymbol{\alpha}) \simeq \{\langle 2, k+1, l\rangle\}(\mathsf{G}(\mathbf{m}, \boldsymbol{\alpha}), \mathbf{m}, \boldsymbol{\alpha}).$$

The reader may well be puzzled as to why a diagonal argument does not exhibit an inconsistency. Indeed, if we set $\mathsf{F}(a, \mathbf{m}, \boldsymbol{\alpha}) \simeq \{a\}(a, \mathbf{m}, \boldsymbol{\alpha})+1$, then F is partial recursive and thus coincides with some $\{b\}$. Thus

$$\mathsf{F}(b, \mathbf{m}, \boldsymbol{\alpha}) \simeq \{b\}(b, \mathbf{m}, \boldsymbol{\alpha})+1 \simeq \mathsf{F}(b, \mathbf{m}, \boldsymbol{\alpha})+1,$$

but F is now a *partial* functional and from this we conclude merely that $\mathsf{F}(b, \mathbf{m}, \boldsymbol{\alpha})$ is not defined.

To realize these provisions, we shall define inductively the set Ω of sequences $(a, \mathbf{m}, \boldsymbol{\alpha}, n)$ such that $\{a\}(\mathbf{m}, \boldsymbol{\alpha}) \simeq n$. At each stage $r+1$ we add to Ω^{r+1} sequences corresponding to computations which are intuitively one step more complex than computations corresponding to sequences in Ω^r. The elements of Ω^0 correspond to computations of some initial functionals which we regard as being immediately computable. These are the same as the initial primitive recursive functionals with two additions which serve as a technical device to enable us to derive the closure of the class of partial recursive functionals under primitive recursion without putting this into the definition. The function Sb_0 is defined in Lemma 2.5 below.

2.1 Definition. Ω is the smallest set such that for all k, l, n, p, q, r, and s, all $i<k$ and $j<l$ and all $(\mathbf{m}, \boldsymbol{\alpha}) \in {}^{k,l}\omega$,

(0) $(\langle 0, k, l, 0, n\rangle, \mathbf{m}, \boldsymbol{\alpha}, n) \in \Omega$;
$(\langle 0, k, l, 1, i\rangle, \mathbf{m}, \boldsymbol{\alpha}, m_i) \in \Omega$;
$(\langle 0, k, l, 2, i\rangle, \mathbf{m}, \boldsymbol{\alpha}, m_i+1) \in \Omega$;
$(\langle 0, k, l, 3, i, j\rangle, \mathbf{m}, \boldsymbol{\alpha}, \alpha_j(m_i)) \in \Omega$;
$(\langle 0, k+4, l, 4\rangle, p, q, r, s, \mathbf{m}, \boldsymbol{\alpha}, p) \in \Omega$, if $r=s$;
$(\langle 0, k+4, l, 4\rangle, p, q, r, s, \mathbf{m}, \boldsymbol{\alpha}, q) \in \Omega$, if $r \neq s$;
$(\langle 0, k+2, l, 5\rangle, p, q, \mathbf{m}, \boldsymbol{\alpha}, \mathrm{Sb}_0(p, q)) \in \Omega$;

(1) for any k', b, $c_0, \ldots, c_{k'-1}$, $q_0, \ldots, q_{k'-1}$, if for all $i<k'$ $(c_i, \mathbf{m}, \boldsymbol{\alpha}, q_i) \in \Omega$ and $(b, \mathbf{q}, \boldsymbol{\alpha}, n) \in \Omega$, then

$$(\langle 1, k, l, b, c_0, \ldots, c_{k'-1}\rangle, \mathbf{m}, \boldsymbol{\alpha}, n) \in \Omega;$$

(2) for any b, if $(b, \mathbf{m}, \boldsymbol{\alpha}, n) \in \Omega$, then

$$(\langle 2, k+1, l\rangle, b, \mathbf{m}, \boldsymbol{\alpha}, n) \in \Omega.$$

Although it is not phrased explicitly as such, the definition of Ω is another example of closure under finitary functions. For example, clause (2) requires Ω to be closed under the function $\lambda x.(\langle 2, k+1, l\rangle) * x$. In particular, there is an inductive operator Γ such that $\Omega = \bar{\Gamma} = \Gamma^{(\omega)}$.

2.2 Lemma. *For all a, $\mathbf{m}$, and $\boldsymbol{\alpha}$, there is at most one n such that $(a, \mathbf{m}, \boldsymbol{\alpha}, n) \in \Omega$.*

Proof. With Γ as above, it suffices to prove by induction on r that there is at most one n such that $(a, \mathbf{m}, \boldsymbol{\alpha}, n) \in \Gamma^r$. Assume as induction hypothesis that this holds for all $s < r$ and suppose that both $(a, \mathbf{m}, \boldsymbol{\alpha}, n)$ and $(a, \mathbf{m}, \boldsymbol{\alpha}, n')$, with $n \neq n'$, belong to Γ^r. This is clearly impossible if $(a)_0 = 0$ so suppose $(a)_0 = 1$. Then for some k, k', l, b, c, $\mathbf{q}$, and $\mathbf{q}'$, $a = \langle 1, k, l, b, c\rangle$, and for all $i < k'$, $(c_i, \mathbf{m}, \boldsymbol{\alpha}, q_i)$ and $(c_i, \mathbf{m}, \boldsymbol{\alpha}, q'_i)$ belong to $\Gamma^{(r)}$ and $(b, \mathbf{q}, \boldsymbol{\alpha}, n)$ and $(b, \mathbf{q}', \boldsymbol{\alpha}, n')$ belong to $\Gamma^{(r)}$. But then by the induction hypothesis, $\mathbf{q} = \mathbf{q}'$ so $n = n'$. The case $(a)_0 = 2$ is similar. □

Thus if we set

$$\{a\}(\mathbf{m}, \boldsymbol{\alpha}) \simeq n \leftrightarrow (a, \mathbf{m}, \boldsymbol{\alpha}, n) \in \Omega,$$

then for each $a \in \omega$, $\{a\}$ is a partial functional.

2.3 Definition. For all F and R,

(i) F is *partial recursive* iff for some $a \in \omega$, $\mathsf{F} = \{a\}$; a is called an *index for* F;

(ii) F is *recursive* iff F is partial recursive and total;

(iii) R is *recursive* iff K_R is recursive; an index for K_R is also called an *index for* R.

2.4 Remark. There is a somewhat technical point which should be mentioned here. The assertion

$$\{\langle 2, k+1, l\rangle\}(b, \mathbf{m}, \boldsymbol{\alpha}) \simeq \{b\}(\mathbf{m}, \boldsymbol{\alpha})$$

is equivalent to: for all n,

$$\{\langle 2, k+1, l\rangle\}(b, \mathbf{m}, \boldsymbol{\alpha}) \simeq n \leftrightarrow \{b\}(\mathbf{m}, \boldsymbol{\alpha}) \simeq n.$$

The implication $(\leftarrow)$ is built into the definition, but the converse must be proved. The situation is similar with

$$\{\langle 1, k, l, b, c_0, \ldots, c_{k'-1}\rangle\}(\mathbf{m}, \boldsymbol{\alpha}) \simeq \{b\}(\{c_0\}(\mathbf{m}, \boldsymbol{\alpha}), \ldots, \{c_{k'-1}\}(\mathbf{m}, \boldsymbol{\alpha}), \boldsymbol{\alpha}).$$

If the right-hand side is $\simeq n$, then for some $q_0, \ldots, q_{k'-1}$, each $\{c_i\}(\mathbf{m}, \boldsymbol{\alpha}) \simeq q_i$ and $\{b\}(\mathbf{q}, \boldsymbol{\alpha}) \simeq n$ so that also the left-hand side $\simeq n$ by the definition of Ω. The reader should not find great difficulty in supplying the missing proofs. (Exercise 2.14)

From this discussion it is clear that the class of partial recursive functionals is closed under functional composition. To show that it is also closed under primitive recursion, we first prove a very fundamental result known as the Recursion Theorem.

2.5 Lemma. *For every $i \in \omega$, there exists a primitive recursive function* Sb_i *such that for any a, any $k > i$, any l, and any $(\mathbf{m}, \boldsymbol{\alpha}) \in {}^{k,l}\omega$,*

$$\{\mathrm{Sb}_i(a, m_0, \ldots, m_i)\}(m_{i+1}, \ldots, m_{k-1}, \boldsymbol{\alpha}) \simeq \{a\}(\mathbf{m}, \boldsymbol{\alpha}).$$

Proof. Let $k' = (a)_1 - 1$ and $l' = (a)_2$. Then set

$$\mathrm{Sb}_0(a, m_0) = \langle 1, k', l', a, \langle 0, k', l', 0, m_0\rangle, \langle 0, k', l', 1, 0\rangle, \ldots, \langle 0, k', l', 1, k'-1\rangle\rangle;$$

$$\mathrm{Sb}_{i+1}(a, m_0, \ldots, m_{i+1}) = \mathrm{Sb}_0(\mathrm{Sb}_i(a, m_0, \ldots, m_i), m_{i+1}).$$

The functions Sb_i are clearly all primitive recursive. We leave it to the reader to check that they all have the required property. □

2.6 Recursion Theorem. *For any partial recursive* F *there exists an $\bar{e} \in \omega$ such that for all* $(\mathbf{m}, \boldsymbol{\alpha})$,

$$\{\bar{e}\}(\mathbf{m}, \boldsymbol{\alpha}) \simeq \mathsf{F}(\bar{e}, \mathbf{m}, \boldsymbol{\alpha}).$$

Proof. Let F be given and set $\mathsf{G}(a, \mathbf{m}, \boldsymbol{\alpha}) \simeq \mathsf{F}(\mathrm{Sb}_0(a, a), \mathbf{m}, \boldsymbol{\alpha})$. G is partial recursive, so let b be an index for G and take $\bar{e} = \mathrm{Sb}_0(b, b)$. Then

$$\{\bar{e}\}(\mathbf{m}, \boldsymbol{\alpha}) \simeq \{b\}(b, \mathbf{m}, \boldsymbol{\alpha}) \simeq \mathsf{G}(b, \mathbf{m}, \boldsymbol{\alpha}) \simeq \mathsf{F}(\mathrm{Sb}_0(b, b), \mathbf{m}, \boldsymbol{\alpha}) \simeq \mathsf{F}(\bar{e}, \mathbf{m}, \boldsymbol{\alpha}). \quad \square$$

We observe that the Recursion Theorem holds also for primitive recursion. If F is primitive recursive, then so is G in the preceding proof. If b is a primitive recursive index for G, then $\bar{e} = \mathrm{Sb}_0(b, b)$ is also a primitive recursive index and clearly

$$[\bar{e}](m, \alpha) = \mathsf{F}(\bar{e}, m, \alpha).$$

We shall refer to this fact as the *Primitive Recursion Theorem.*

The following is a property of the partial recursive functionals which is not shared by the primitive recursive functionals.

2.7 Definition (*unbounded search*). For any functional G of rank $(k+1, l)$, $\mathrm{Se}(\mathsf{G})$ is the functional F of rank (k, l) such that

$$\mathsf{F}(\mathbf{m}, \boldsymbol{\alpha}) \simeq \begin{cases} q, \text{ if } \mathsf{G}(q, \mathbf{m}, \boldsymbol{\alpha}) \simeq 0, \text{ but for all } r < q, \\ \quad \text{there exists } n > 0 \text{ such that } \mathsf{G}(r, \mathbf{m}, \boldsymbol{\alpha}) \simeq n; \\ \\ \text{undefined, if there is no such } q. \end{cases}$$

(We write $\mathsf{F}(\mathbf{m}, \boldsymbol{\alpha}) \simeq$ "least" $q \,.\, \mathsf{G}(q, \mathbf{m}, \boldsymbol{\alpha}) \simeq 0$. Note that if $\mathsf{G}(r, \mathbf{m}, \boldsymbol{\alpha})$ is undefined for some r less than the smallest q such that $\mathsf{G}(q, \mathbf{m}, \boldsymbol{\alpha}) \simeq 0$ (if any), then $\mathsf{F}(\mathbf{m}, \boldsymbol{\alpha})$ is undefined. If this cannot happen, say, because G is total, we write $\mathsf{F}(\mathbf{m}, \boldsymbol{\alpha}) \simeq$ least $q \,.\, \mathsf{G}(q, \mathbf{m}, \boldsymbol{\alpha}) \simeq 0$. In particular, we write least $p \,.\, \mathsf{R}(p, \mathbf{m}, \boldsymbol{\alpha})$ for least $p \,.\, \mathsf{K}_{\mathsf{R}}(p, \mathbf{m}, \boldsymbol{\alpha}) = 0$.)

2.8 Theorem. *The class of partial recursive functionals is closed under unbounded search.*

Proof. Suppose G is partial recursive and $\mathsf{F} = \mathrm{Se}(\mathsf{G})$. Let b and c be indices such that for all $e, q, \mathbf{m}$, and $\boldsymbol{\alpha}$,

$$\{b\}(e, q, \mathbf{m}, \boldsymbol{\alpha}) \simeq 0, \quad \text{and} \quad \{c\}(e, q, \mathbf{m}, \boldsymbol{\alpha}) \simeq \{e\}(q+1, \mathbf{m}, \boldsymbol{\alpha}) + 1,$$

and set

$$\mathsf{H}(e, q, \mathbf{m}, \boldsymbol{\alpha}) \simeq \{\{\langle 0, 4, 0, 4\rangle\}(b, c, \mathsf{G}(q, \mathbf{m}, \boldsymbol{\alpha}), 0)\}(e, q, \mathbf{m}, \boldsymbol{\alpha}).$$

By the Recursion Theorem there exists an $\bar{e}$ such that

$$\{\bar{e}\}(q, \mathbf{m}, \boldsymbol{\alpha}) \simeq \mathsf{H}(\bar{e}, q, \mathbf{m}, \boldsymbol{\alpha}).$$

Let $\mathsf{I} = \{\bar{e}\}$.

Suppose first that $\mathsf{G}(q, \mathbf{m}, \boldsymbol{\alpha}) \simeq 0$. Then

$$\{\langle 0, 4, 0, 4\rangle\}(b, c, \mathsf{G}(q, \mathbf{m}, \boldsymbol{\alpha}), 0) \simeq b$$

so

$$\mathsf{I}(q, \mathbf{m}, \boldsymbol{\alpha}) \simeq \mathsf{H}(\bar{e}, q, \mathbf{m}, \boldsymbol{\alpha}) \simeq \{b\}(\bar{e}, q, \mathbf{m}, \boldsymbol{\alpha}) \simeq 0.$$

On the other hand, if $\mathsf{G}(q, \mathbf{m}, \boldsymbol{\alpha}) \simeq n > 0$, then

$$\mathsf{I}(q, \mathbf{m}, \boldsymbol{\alpha}) \simeq \{c\}(\bar{e}, q, \mathbf{m}, \boldsymbol{\alpha}) \simeq \mathsf{I}(q+1, \mathbf{m}, \boldsymbol{\alpha}) + 1.$$

If $\mathsf{G}(q, \mathbf{m}, \boldsymbol{\alpha})$ is undefined, so is $\mathsf{I}(q, \mathbf{m}, \boldsymbol{\alpha})$. From these facts it follows easily that $\mathsf{F}(\mathbf{m}, \boldsymbol{\alpha}) \simeq \mathsf{I}(0, \mathbf{m}, \boldsymbol{\alpha})$. □

2.9 Corollary. *The predecessor function $p \dot{-} 1$ is partial recursive.*

Proof. Let b and c be indices such that for all p,

$$\{b\}(p) \simeq 0 \quad \text{and} \quad \{c\}(p) \simeq \text{“least”}\, q . \{\langle 0,4,0,4\rangle\}(0,1,q+1,p) \simeq 0.$$

Then

$$p \dot{-} 1 \simeq \{\{\langle 0,4,0,4\rangle\}(b,c,p,0)\}(p). \quad \square$$

2.10 Theorem. *The class of partial recursive functionals is closed under primitive recursion and course-of-values recursion.*

Proof. Suppose G and H are partial recursive and F = Rec(G, H). Let G′ and H′ be defined by:

$$\mathsf{G}'(e,p,\mathbf{m},\boldsymbol{\alpha}) \simeq \mathsf{G}(\mathbf{m},\boldsymbol{\alpha});$$

$$\mathsf{H}'(e,p,\mathbf{m},\boldsymbol{\alpha}) \simeq \mathsf{H}(\{e\}(p \dot{-} 1,\mathbf{m},\boldsymbol{\alpha}), p \dot{-} 1,\mathbf{m},\boldsymbol{\alpha}).$$

By the corollary, G′ and H′ are partial recursive. Let b' and c' be indices for G′ and H′, respectively, and set

$$\mathsf{I}(e,p,\mathbf{m},\boldsymbol{\alpha}) \simeq \{\{\langle 0,4,0,4\rangle\}(b',c',p,0)\}(e,p,\mathbf{m},\boldsymbol{\alpha}).$$

Thus

$$\mathsf{I}(e,0,\mathbf{m},\boldsymbol{\alpha}) \simeq \mathsf{G}(\mathbf{m},\boldsymbol{\alpha})$$

and

$$\mathsf{I}(e,p+1,\mathbf{m},\boldsymbol{\alpha}) \simeq \mathsf{H}(\{e\}(p,\mathbf{m},\boldsymbol{\alpha}),p,\mathbf{m},\boldsymbol{\alpha}).$$

Let $\bar{e}$ be an index provided by the Recursion Theorem such that

$$\{\bar{e}\}(p,\mathbf{m},\boldsymbol{\alpha}) \simeq \mathsf{I}(\bar{e},p,\mathbf{m},\boldsymbol{\alpha}).$$

Then

$$\{\bar{e}\}(0,\mathbf{m},\boldsymbol{\alpha}) \simeq \mathsf{G}(\mathbf{m},\boldsymbol{\alpha})$$

and

$$\{\bar{e}\}(p+1,\mathbf{m},\boldsymbol{\alpha}) \simeq \mathsf{H}(\{\bar{e}\}(p,\mathbf{m},\boldsymbol{\alpha}),p,\mathbf{m},\boldsymbol{\alpha}).$$

So $\mathsf{F} = \{\bar{e}\}$ and thus F is partial recursive. □

2.11 Corollary. *Every primitive recursive functional is recursive.* □

We leave it to the reader to check the obvious fact that the primitive recursive functionals form a proper subset of the recursive functionals (Exercise 2.17).

Computations of partial recursive functionals may be arranged in labeled trees similar to those for primitive recursions. There is one additional type of node:

$$\bullet\ (\langle 2, k+1, l\rangle, b, \mathbf{m}, \boldsymbol{\alpha})$$

$$\bullet\ (b, \mathbf{m}, \boldsymbol{\alpha})$$

Branching in the tree is still finite, so that a tree either is finite or has an infinite branch. The latter case corresponds to an undefined computation. The procedure for computing on the tree is the same as before and we conclude that the functional $\lambda\, a\, \mathbf{m}\, \boldsymbol{\alpha}\,.\{a\}(\mathbf{m}, \boldsymbol{\alpha})$ is mechanically computable, and hence so is every partial recursive functional.

We want now to consider other closure conditions satisfied by the (partial) recursive functionals and the recursive relations. It follows as in 1.3 that the recursive relations form a Boolean algebra. If we restrict ourselves to recursive (total) functionals, then we can establish closure under bounded search, definition by cases, relational composition, and bounded quantification with exactly the same proofs as for Theorem 1.5. The proof for expansion given there depends on the inductive characterization of the class Prf and does not generalize directly; the result is proved as Corollary 3.4.

What happens to these properties when the functionals are allowed to be partial? Closure under relational composition is no longer true (Exercise 4.26). On the other hand it is easy to verify that the proof for bounded search still works: if G is partial recursive so is $\mathrm{Bs}(\mathsf{G})$. The proof for definition by cases, however, does not work if some of $\mathsf{G}_0, \dots, \mathsf{G}_{k'}$ are not total functionals. If F is as defined in 1.5(iii), $\mathsf{F}(\mathbf{m}, \boldsymbol{\alpha})$ is undefined unless *all* of $\mathsf{G}_0(\mathbf{m}, \boldsymbol{\alpha}), \dots, \mathsf{G}_{k'}(\mathbf{m}, \boldsymbol{\alpha})$ are defined, whereas $\mathrm{Cases}_{k'}(\mathsf{G}, \mathsf{R})$ is defined for any $(\mathbf{m}, \boldsymbol{\alpha})$ such that for some i, $\mathsf{R}_i(\mathbf{m}, \boldsymbol{\alpha})$ and $\mathsf{G}_i(\mathbf{m}, \boldsymbol{\alpha})$ is defined. The result is nevertheless true.

2.12 Theorem. *The class of partial recursive functionals and recursive relations is closed under definition by cases.*

Proof. Suppose $\mathsf{R}_0, \dots, \mathsf{R}_{k'-1}$ are recursive relations and $\mathsf{G}_0, \dots, \mathsf{G}_{k'}$ are partial recursive functionals which satisfy the hypothesis of 1.4(iii). Let b_i be an index for G_i and let H be the recursive functional defined by:

$$\mathsf{H}(\mathbf{m}, \boldsymbol{\alpha}) = \begin{cases} b_0, & \text{if } \mathsf{R}_0(\mathbf{m}, \boldsymbol{\alpha}); \\ \vdots & \\ b_{k'-1}, & \text{if } \mathsf{R}_{k'-1}(\mathbf{m}, \boldsymbol{\alpha}); \\ b_{k'}, & \text{otherwise.} \end{cases}$$

Then $\text{Cases}_{k'}(\mathsf{G}_0, \ldots, \mathsf{G}_{k'}, \mathsf{R}_0, \ldots, \mathsf{R}_{k'-1})$ is the partial recursive functional

$$\mathsf{F}(\mathbf{m}, \boldsymbol{\alpha}) \simeq \{\mathsf{H}(\mathbf{m}, \boldsymbol{\alpha})\}(\mathbf{m}, \boldsymbol{\alpha}). \quad \square$$

2.13–2.20 Exercises

2.13. Show that there exists a primitive recursive functional F such that if for each $\boldsymbol{\alpha}$ we define

$$\Gamma_{\boldsymbol{\alpha}}(A) = \{u : \exists p\, (\forall i < p) . \mathsf{F}(p, i, u, \langle \boldsymbol{\alpha} \rangle) \in A \cup \{0\}\},$$

then

$$\{a\}(\mathbf{m}, \boldsymbol{\alpha}) \simeq n \leftrightarrow \langle a, \mathbf{m}, n \rangle \in \bar{\Gamma}_{\boldsymbol{\alpha}}.$$

2.14. Supply the missing proofs discussed in Remark 2.4.

2.15. Prove the *Effective Recursion Theorem*: there exists a primitive recursive function g such that for all $a, \mathbf{m}$, and $\boldsymbol{\alpha}$,

$$\{g(a)\}(\mathbf{m}, \boldsymbol{\alpha}) \simeq \{a\}(g(a), \mathbf{m}, \boldsymbol{\alpha}).$$

2.16. With each sequence $(a, \mathbf{m}, \boldsymbol{\alpha}, n)$ such that $\{a\}(\mathbf{m}, \boldsymbol{\alpha}) \simeq n$, we associate the natural number $|a, \mathbf{m}, \boldsymbol{\alpha}|$, the least number r such that $(a, \mathbf{m}, \boldsymbol{\alpha}, n) \in \Omega^r$. Show that $\bar{e}$ in the Recursion Theorem may be chosen so that whenever $\mathsf{F}(\bar{e}, \mathbf{m}, \boldsymbol{\alpha}) \simeq n$, then $|a, \bar{e}, \mathbf{m}, \boldsymbol{\alpha}| < |\bar{e}, \mathbf{m}, \boldsymbol{\alpha}|$ (a is an index for F).

2.17. Check that the class of primitive recursive functionals is a proper subclass of the class of recursive functionals.

2.18. Prove an effective version of Corollary 2.11: there exists a primitive recursive function f such that for all $a \in \text{Pri}$, $[a] = \{f(a)\}$.

2.19. Suppose that F is defined from G and H by the following recursion:

$$\mathsf{F}(0, q, \mathbf{m}, \boldsymbol{\alpha}) \simeq \mathsf{G}(q, \mathbf{m}, \boldsymbol{\alpha}),$$
$$\mathsf{F}(p+1, q, \mathbf{m}, \boldsymbol{\alpha}) \simeq \mathsf{H}(p, q, \mathbf{m}, \boldsymbol{\alpha}, \lambda r . \mathsf{F}(p, r, \mathbf{m}, \boldsymbol{\alpha})).$$

If G and H are partial recursive, does it follow that F is partial recursive?

2.20. Is it possible to define recursive functions Sb_j^l such that for all a, all $l > j, k$, and $(\mathbf{m}, \boldsymbol{\alpha}) \in {}^{k,l}\omega$,

$$\{\mathrm{Sb}_j^l(a, \alpha_0, \ldots, \alpha_j)\}(\mathbf{m}, \alpha_{j+1}, \ldots, \alpha_{l-1}) \simeq \{a\}(\mathbf{m}, \boldsymbol{\alpha})?$$

2.21 Notes. The definition we have given for the class of partial recursive functionals via indexed schemata is an adaptation of that of Kleene [1959]. Historically, the most important characterizations of this class are those of Kleene [1935/36] and Turing [1936]. A good informal discussion of these may be found in Rogers [1967, § 1.5]. We have chosen the Kleene [1959] definition for two reasons. First the basic properties of the partial recursive functionals are easier to derive because the indexing is built in from the beginning. More importantly, it is this definition which is most easily generalized to functionals of higher types and functions on ordinals. Of course, there are many minor variations possible. For example, the schemata in clause (0) for equality and sequence coding may be omitted in favor of a schema for primitive recursion. The sequence coding functions may be replaced by + and $\cdot$.

The fact that recursiveness for functionals can be defined exactly as for functions was apparently first noted in print in Kleene [1952, § 55] and developed in Kleene [1955 b].

3. Normal Forms

Much of the remaining theory of partial recursive functionals depends on the fact that the relation $\{a\}(\mathbf{m}, \boldsymbol{\alpha}) \simeq n$ can be represented in a simple way in terms of primitive recursive relations, namely, as $\exists u\, \mathsf{P}(u, a, \mathbf{m}, n, \boldsymbol{\alpha})$ with P primitive recursive. The key idea is that the number u codes the sequence of steps which are necessary to carry out the computation.

3.1 Theorem. *There exists a primitive recursive relation* $\mathsf{T} \subseteq {}^{3,1}\omega$ *such that for all* $a, \mathbf{m}, \boldsymbol{\alpha}$, *and* n,

$$\{a\}(\mathbf{m}, \boldsymbol{\alpha}) \simeq n \leftrightarrow \exists u\, [\mathsf{T}(a, \langle \mathbf{m} \rangle, u, \langle \boldsymbol{\alpha} \rangle) \wedge (u)_0 = n].$$

Proof. We first define a primitive recursive relation Cmt—$\mathrm{Cmt}(u, \langle \boldsymbol{\alpha} \rangle)$ is to mean that u codes some computation relative to $\boldsymbol{\alpha}$. Each such u will be of the form $u = \langle n, a, \mathbf{m}, v_0, \ldots, v_{r-1} \rangle$. $v_0, \ldots, v_{r-1}$ code in turn the immediate subcomputations of u, that is the computations indicated by the labels on the immediate predecessors of the node labeled $(a, \mathbf{m}, \boldsymbol{\alpha})$ in a computation tree. We shall write $u = \langle n, a, \mathbf{m}, \ldots \rangle$ as an abbreviation for: $\mathrm{Sq}(u)$ and $\mathrm{lg}(u) \geqslant (a)_1 + 2$ and $(u)_0 = n$ and $(u)_1 = a$ and $(\forall i < (a)_1)(u)_{i+2} = m_i$.

For any l and any $\alpha \in {}^{l}({}^{\omega}\omega)$, $\mathsf{Cmt}(u, \langle\alpha\rangle)$ iff for some $k, m_0, \ldots, m_{k-1}, n, p, q, r$, and s all less than u, one of the following holds:

(0) $u = \langle n, \langle 0, k, l, 0, n\rangle, \mathbf{m}\rangle$;

$u = \langle m_i, \langle 0, k, l, 1, i\rangle, \mathbf{m}\rangle$ for some $i < k$;

$u = \langle m_i + 1, \langle 0, k, l, 2, i\rangle, \mathbf{m}\rangle$ for some $i < k$;

$u = \langle \alpha_j(m_i), \langle 0, k, l, 3, i, j\rangle, \mathbf{m}\rangle$ for some $i < k$ and $j < l$;

$u = \langle p, \langle 0, k+4, l, 4\rangle, p, q, r, s, \mathbf{m}\rangle$ and $r = s$;

$u = \langle q, \langle 0, k+4, l, 4\rangle, p, q, r, s, \mathbf{m}\rangle$ and $r \neq s$;

$u = \langle \mathsf{Sb}_0(p, q), \langle 0, k+2, l, 5\rangle, p, q, \mathbf{m}\rangle$;

(1) for some $k', b, c_0, \ldots, c_{k'-1}, q_0, \ldots, q_{k'-1}, v_0, \ldots, v_{k'} < u$,

$$u = \langle n, \langle 1, k, l, b, \mathbf{c}\rangle, \mathbf{m}, \mathbf{v}\rangle, \text{ for all } i \leq k', \ \mathsf{Cmt}(v_i, \langle\alpha\rangle),$$

$$\text{for all } i < k', \ v_i = \langle q_i, c_i, \mathbf{m}, \ldots\rangle, \text{ and } v_{k'} = \langle n, b, \mathbf{q}, \ldots\rangle;$$

(2) for some b and $v < u$,

$$u = \langle n, \langle 2, k+1, l\rangle, b, \mathbf{m}, v\rangle, \quad \mathsf{Cmt}(v, \langle\alpha\rangle), \quad \text{and} \quad v = \langle n, b, \mathbf{m}, \ldots\rangle.$$

All quantifiers in the definition of Cmt are bounded to u and thus Cmt may be seen to be primitive recursive by the techniques discussed following Theorem 1.8. We now set

$$\mathsf{T}(a, \langle\mathbf{m}\rangle, u, \langle\alpha\rangle) \leftrightarrow \mathsf{Cmt}(u, \langle\alpha\rangle) \wedge u = \langle (u)_0, a, \mathbf{m}, \ldots\rangle.$$

Then T is clearly primitive recursive and it suffices to show that for all $a, \mathbf{m}, \alpha$, and n,

(i) $\{a\}(\mathbf{m}, \alpha) \simeq n \rightarrow \exists u\,[\mathsf{T}(a, \langle\mathbf{m}\rangle, u, \langle\alpha\rangle) \wedge (u)_0 = n]$;

(ii) $\forall u\,[\mathsf{T}(a, \langle\mathbf{m}\rangle, u, \langle\alpha\rangle) \rightarrow \{a\}(\mathbf{m}, \alpha) \simeq (u)_0]$.

The proof of (i) is by induction over Ω — that is, we show that the set

$$Z = \{(a, \mathbf{m}, \alpha, n) \colon \exists u\,[\mathsf{T}(a, \langle\mathbf{m}\rangle, u, \langle\alpha\rangle) \wedge (u)_0 = n]\}$$

is closed under the conditions which define Ω. If $(a)_0 = 0$ and $(a, \mathbf{m}, \alpha, n) \in \Omega$, then the condition for membership in Z is satisfied with $u = \langle n, a, \mathbf{m}\rangle$. Suppose $a = \langle 1, k, l, b, \mathbf{c}\rangle$ and for some $\mathbf{q}$, $(\forall i < k')(c_i, \mathbf{m}, \alpha, q_i) \in Z$ and $(b, \mathbf{q}, \alpha, n) \in Z$. Then there exist $v_0, \ldots, v_{k'}$ such that $(\forall i < k')[\mathsf{T}(c_i, \langle\mathbf{m}\rangle, v_i, \langle\alpha\rangle) \wedge (v_i)_0 = q_i]$ and $[\mathsf{T}(b, \langle\mathbf{q}\rangle, v_{k'}, \langle\alpha\rangle) \wedge (v_{k'})_0 = n]$. Then by clause (1) in the definition of Cmt, if $u = \langle n, a, \mathbf{m}, \mathbf{v}\rangle$, $\mathsf{Cmt}(u, \langle\alpha\rangle)$ and hence $\mathsf{T}(a, \langle\mathbf{m}\rangle, u, \langle\alpha\rangle)$. Since also $(u)_0 = n$, $(a, \mathbf{m}, \alpha, n) \in Z$. The proof for clause (2) is easier.

The proof of (ii) is by ordinary induction on u. We assume as induction hypothesis that the matrix of (ii) is true for all $v < u$ and that $\mathsf{T}(a, \langle\mathbf{m}\rangle, u, \langle\alpha\rangle)$. If u falls under clause (0), the result is obvious. Suppose u falls under clause (1). Then since $v_0, \ldots, v_{k'} < u$, and clearly $\mathsf{T}(c_i, \langle\mathbf{m}\rangle, v_i, \langle\alpha\rangle)$ and $T(b, \langle\mathbf{q}\rangle, v_{k'}, \langle\alpha\rangle)$, the induction hypothesis guarantees that $\{c_i\}(\mathbf{m}, \boldsymbol{\alpha}) \simeq q_i$ and $\{b\}(\mathbf{q}, \boldsymbol{\alpha}) \simeq n$. Hence $\{a\}(\mathbf{m}, \boldsymbol{\alpha}) \simeq n = (u)_0$. Case (2) is similar. □

3.2 Corollary (First Normal Form). *For any partial recursive functional* F, *there exists an a such that*

$$\mathsf{F}(\mathbf{m}, \boldsymbol{\alpha}) \simeq (\text{least } u \,.\, \mathsf{T}(a, \langle \mathbf{m} \rangle, u, \langle \boldsymbol{\alpha} \rangle))_0.$$

Proof. If a is any index for F, then clearly $\mathsf{F}(\mathbf{m}, \boldsymbol{\alpha})$ is defined just in case $\exists u\, \mathsf{T}(a, \langle \mathbf{m} \rangle, u, \langle \boldsymbol{\alpha} \rangle)$ and if so, then $\mathsf{F}(\mathbf{m}, \boldsymbol{\alpha}) \simeq (u)_0$ for any such u. □

As an immediate consequence of the First Normal Form we obtain direct inductive characterizations of the classes of partial recursive and recursive functionals:

3.3 Corollary. (i) *The class of partial recursive functionals is the smallest class of partial functionals which includes the class of primitive recursive functionals and is closed under functional composition and unbounded search*;

(ii) *the class of recursive functionals is the smallest class of total functionals which includes the class of primitive recursive functionals and is closed under functional composition and unbounded search restricted to functionals* G *such that* $\forall \mathbf{m} \forall \boldsymbol{\alpha} \exists q \,.\, \mathsf{G}(q, \mathbf{m}, \boldsymbol{\alpha}) = 0$.

Proof. Let X be the class of partial recursive functionals and Y the other class described in (i). Then $Y \subseteq X$ by the results of § 2 and $X \subseteq Y$ by Corollary 3.2. The proof of (ii) is similar. □

3.4 Corollary. *The class of partial recursive functionals is closed under expansion.*

Proof. Suppose $\mathsf{F}(\mathbf{m}, \mathbf{n}, \boldsymbol{\alpha}, \boldsymbol{\beta}) \simeq \mathsf{G}(\mathbf{m}, \boldsymbol{\alpha})$ and let b be an index for G. By expansion for primitive recursive functionals there is a primitive recursive relation R such that

$$\mathsf{R}(u, \mathbf{m}, \mathbf{n}, \boldsymbol{\alpha}, \boldsymbol{\beta}) \leftrightarrow \mathsf{T}(b, \langle \mathbf{m} \rangle, u, \langle \boldsymbol{\alpha} \rangle).$$

Then

$$\mathsf{F}(\mathbf{m}, \mathbf{n}, \boldsymbol{\alpha}, \boldsymbol{\beta}) \simeq (\text{least } u \,.\, \mathsf{R}(u, \mathbf{m}, \mathbf{n}, \boldsymbol{\alpha}, \boldsymbol{\beta}))_0$$

and is thus partial recursive by Theorem 2.11. □

In the introduction to this chapter we observed that all mechanically calculable functionals are continuous. The fact that all partial recursive functionals are partial continuous plays an important role in the theory of the later chapters and we shall find useful a second normal form which makes this explicit.

3.5 Theorem. *There exists a primitive recursive relation* $T \subseteq {}^4\omega$ *such that for all* $a, \mathbf{m}, \boldsymbol{\alpha}$, *and* n,

$$\{a\}(\mathbf{m}, \boldsymbol{\alpha}) \simeq n \leftrightarrow \exists u\,[T(a, \langle \mathbf{m} \rangle, u, \langle \bar{\alpha}_0(u), \dots, \bar{\alpha}_{l-1}(u) \rangle) \wedge (u)_0 = n].$$

Proof. Let Cmt be the relation defined in the proof of Theorem 3.1 and set

$$\mathsf{Cmt}(u, s) \leftrightarrow \exists \boldsymbol{\beta}\,[s = \langle \bar{\beta}_0(u), \dots, \bar{\beta}_{l-1}(u) \rangle \wedge \mathsf{Cmt}(u, \langle \boldsymbol{\beta} \rangle)],$$

and

$$T(a, \langle \mathbf{m} \rangle, u, s) \leftrightarrow \mathrm{Cmt}(u, s) \wedge u = \langle (u)_0, a, \mathbf{m}, \dots \rangle.$$

Suppose first that $\{a\}(\mathbf{m}, \boldsymbol{\alpha}) \simeq n$ and by Theorem 3.1 choose u such that $\mathsf{T}(a, \langle \mathbf{m} \rangle, u, \langle \boldsymbol{\alpha} \rangle)$ and $(u)_0 = n$. Since $\mathsf{Cmt}(u, \langle \boldsymbol{\alpha} \rangle)$ holds we have $\mathsf{Cmt}(u, \langle \bar{\alpha}_0(u), \dots, \bar{\alpha}_{l-1}(u) \rangle)$ and thus $T(a, \langle \mathbf{m} \rangle, u, \langle \bar{\alpha}_0(u), \dots, \bar{\alpha}_{l-1}(u) \rangle)$. For the implication ($\leftarrow$) we first note that it is easy to prove by induction on u that

$$\mathsf{Cmt}(u, \langle \boldsymbol{\alpha} \rangle) \wedge (\forall j < l)\bar{\alpha}_j(u) = \bar{\beta}_j(u) \rightarrow \mathsf{Cmt}(u, \langle \boldsymbol{\beta} \rangle).$$

Intuitively, the computation u "uses" values of the functions only for arguments less than u. Then if $T(a, \langle \mathbf{m} \rangle, u, \langle \bar{\alpha}_0(u), \dots, \bar{\alpha}_{l-1}(u) \rangle)$ holds, it follows that also $\mathsf{T}(a, \langle \mathbf{m} \rangle, u, \langle \boldsymbol{\alpha} \rangle)$ and the result is immediate from Theorem 3.1.

Of course, as we have defined them above it is not at all evident that Cmt and T are primitive recursive. In fact, however, Cmt may also be defined exactly as was Cmt except for the replacement of the clause

$$u = \langle \alpha_j(m_i), \dots \rangle \quad \text{by} \quad u = \langle ((s)_j)_{m_i}, \dots \rangle$$

and the addition of the conditions: $\lg(s) = l$ and $(\forall j < l)\lg((s)_j) = u$. □

3.6 Corollary (Second Normal Form). *For any partial recursive functional* F, *there exists an a such that*

$$\mathsf{F}(\mathbf{m}, \boldsymbol{\alpha}) \simeq (\text{least } u\,.\, T(a, \langle \mathbf{m} \rangle, u, \langle \bar{\alpha}_0(u), \dots, \bar{\alpha}_{l-1}(u) \rangle))_0. \quad \square$$

3.7 Corollary. *Every partial recursive functional is partial continuous. Every recursive relation is closed-open.*

Proof. If F is partial recursive and $(\mathbf{m}, \boldsymbol{\alpha}) \in \mathsf{F}^{-1}(\{n\})$, then for some u, $T(a, \langle \mathbf{m} \rangle, u, \langle \bar{\alpha}_0(u), \dots, \bar{\alpha}_{l-1}(u) \rangle)$ and $(u)_0 = n$. Then for any $(\mathbf{p}, \boldsymbol{\beta})$ belonging to the neighborhood $\{m_0\} \times \dots \times \{m_{k-1}\} \times [\alpha_0 \restriction u] \times \dots \times [\alpha_{l-1} \restriction u]$ of $(\mathbf{m}, \boldsymbol{\alpha})$ also $\mathsf{F}(\mathbf{p}, \boldsymbol{\beta}) \simeq n$ so $(\mathbf{p}, \boldsymbol{\beta}) \in \mathsf{F}^{-1}(\{n\})$. Hence $\mathsf{F}^{-1}(\{n\})$ is open and F is partial continuous. If R is recursive, K_R is recursive, hence continuous so R is closed-open by Lemma I.2.1. □

The recursive functionals are sometimes called *effectively continuous* and the recursive relations *effectively closed-open.*

We conclude this section by applying the Second Normal Form to establish another closure property of the class of partial recursive functionals.

3.8 Definition. (i) (*functional substitution*) For any functionals G and H of ranks $(k, l+1)$ and $(k+1, l)$, respectively, FSub(G, H) is the functional F of rank (k, l) such that for all $\mathbf{m}$ and $\boldsymbol{\alpha}$,

$$\mathsf{F}(\mathbf{m}, \boldsymbol{\alpha}) \simeq \mathsf{G}(\mathbf{m}, \boldsymbol{\alpha}, \lambda p . \mathsf{H}(p, \mathbf{m}, \boldsymbol{\alpha}));$$

(ii) (*relational substitution*) for any relation S of rank $(k, l+1)$ and any functional H of rank $(k+1, l)$, RSub(S, H) is the relation R of rank (k, l) such that for all $\mathbf{m}$ and $\boldsymbol{\alpha}$,

$$\mathsf{R}(\mathbf{m}, \boldsymbol{\alpha}) \leftrightarrow \mathsf{S}(\mathbf{m}, \boldsymbol{\alpha}, \lambda p . \mathsf{H}(p, \mathbf{m}, \boldsymbol{\alpha})).$$

At first consideration, it seems natural that the class of partial recursive functionals should be closed under functional substitution. If G and H are mechanically calculable then so should be F — whenever in the computation of a value $\mathsf{F}(\mathbf{m}, \boldsymbol{\alpha})$ a value of $\lambda p . \mathsf{H}(p, \mathbf{m}, \boldsymbol{\alpha})$ is called for, it could be computed by a "subroutine". This procedure leads to an answer as long as $\mathsf{H}(p, \mathbf{m}, \boldsymbol{\alpha})$ is defined for all those finitely many p for which a value is required during the computation. The right-hand side, however, is defined at most when $\lambda p . \mathsf{H}(p, \mathbf{m}, \boldsymbol{\alpha})$ is a *total* function, as we have admitted only total functions as arguments. Thus the procedure described actually calculates an extension of F.

This difficulty could be removed by extending our functionals to allow partial functions as arguments, but this leads to other problems which we prefer to avoid. Of course, this argument establishes only that the proposed procedure may fail to compute F, but in fact the partial recursive functionals are *not* closed under functional substitution (Exercise 4.27). We have instead the following restricted version.

3.9 Theorem. *For any partial recursive functionals* G *and* H *there exists a partial recursive functional* F *such that for all* $\mathbf{m}$ *and* $\boldsymbol{\alpha}$ *such that* $\lambda p . \mathsf{H}(p, \mathbf{m}, \boldsymbol{\alpha})$ *is total,*

$$\mathsf{F}(\mathbf{m}, \boldsymbol{\alpha}) \simeq \mathsf{G}(\mathbf{m}, \boldsymbol{\alpha}, \lambda p . \mathsf{H}(p, \mathbf{m}, \boldsymbol{\alpha})).$$

Proof. To minimize notational confusion we suppose $k = l = 1$. Let b and c be indices for G and H, respectively. By Theorem 3.5 we have for any m, α, and β,

$$\mathsf{G}(m, \alpha, \beta) \simeq n \leftrightarrow \exists u\, [T(b, \langle m \rangle, u, \langle \bar{\alpha}(u), \bar{\beta}(u) \rangle) \wedge (u)_0 = n]$$
$$\leftrightarrow \exists s\, \exists u\, [s = \bar{\beta}(u) \wedge T(b, \langle m \rangle, u, \langle \bar{\alpha}(u), s \rangle) \wedge (u)_0 = n].$$

Then if $\lambda p\,.\,\mathsf{H}(p, m, \alpha)$ is total,

$$\mathsf{G}(m, \alpha, \lambda p\,.\,\mathsf{H}(p, m, \alpha)) \simeq n \leftrightarrow \exists s\, \exists u\, \exists v\, [(\forall p < u)(\mathsf{T}(c, \langle p, m \rangle, (v)_p, \langle \alpha \rangle))$$
$$\wedge ((v)_p)_0 = (s)_p) \wedge T(b, \langle m \rangle, u, \langle \bar{\alpha}(u), s \rangle) \wedge (u)_0 = n].$$

If we abbreviate the right-hand side of this equivalence by $\exists s\, \exists u\, \exists v$ $\mathsf{P}(s, u, v, m, \alpha)$, then P is recursive and

$$\mathsf{G}(m, \alpha, \lambda p\,.\,\mathsf{H}(p, m, \alpha)) \simeq ((\text{least } w\,.\,\mathsf{P}((w)_0, (w)_1, (w)_2, m, \alpha))_1)_0.$$

Hence if we take F to be the functional defined by the right-hand expression, F has the desired property. ☐

3.10 Corollary. *The class of recursive functionals is closed under functional substitution. The class of recursive relations and recursive functionals is closed under relational substitution.*

Proof. If H is recursive, then the hypothesis is automatically satisfied. ☐

3.11–3.13 Exercises

3.11. Show that the class of partial recursive functionals is the smallest class of functionals containing the functionals of clause (0) of Definition 2.1 and closed under functional composition and unbounded search.

3.12. Prove the following effective version of Theorem 3.9: there exists a primitive recursive function f such that for any $a, d, \mathbf{m}$, and $\boldsymbol{\alpha}$ such that $\lambda p\,.\,\{d\}(p, \mathbf{m}, \boldsymbol{\alpha})$ is total,

$$\{f(a, d)\}(\mathbf{m}, \boldsymbol{\alpha}) \simeq \{a\}(\mathbf{m}, \boldsymbol{\alpha}, \lambda p\,.\,\{d\}(p, \mathbf{m}, \boldsymbol{\alpha})).$$

3.13. The definition schema of Exercise 1.14 makes sense if we repace the relation $<$ on ω by any relation $<^*$ such that $\leqslant^*$ is a well-ordering of ω. Show that if $\leqslant^*$ is recursive, then the class of recursive functionals is closed under this schema.

3.14. Notes. The Normal Forms are due to Kleene [1935/36] and are characteristic of Ordinary Recursion Theory as opposed to recursion relative to a functional of type 2 or higher. To appreciate their value, the reader might glance ahead at the work required to prove Corollary VI.2.11, the analogue of Theorem 3.9 (functional substitution) for functionals partial recursive in a type-2 functional.

4. Semi-Recursive Relations

We noted following 3.7 that the recursive relations might be viewed as the effectively closed-open relations. In this section we study *effectively open* (semi-recursive) and *effectively closed* (co-semi-recursive) relations.

4.1 Definition. For any R,

(i) R is *semi-recursive* iff R = Dm G for some partial recursive functional G; an index for G is called a *semi-index* for R;

(ii) R is *co-semi-recursive* iff ~R is semi-recursive.

The semi-recursive relations are, of course, a formal counterpart of the *mechanically semi-calculable* relations. R is semi-calculable iff there is a finite program which directs an idealized computer of the sort discussed in the introduction to this chapter to accept inputs of the form $(\mathbf{m}, \boldsymbol{\alpha})$ and carry out a computation which terminates iff $\mathsf{R}(\mathbf{m}, \boldsymbol{\alpha})$.

4.2 Lemma. *Every recursive relation is semi-recursive but there exist semi-recursive subsets of ω which are not recursive.*

Proof. For any recursive R, define F and G by

$$F(m) = \begin{cases} \text{an index for } \lambda p.0, & \text{if } m = 0; \\ 0, & \text{otherwise}; \end{cases}$$

$$\mathsf{G}(\mathbf{m}, \boldsymbol{\alpha}) \simeq \{F(\mathsf{K}_\mathsf{R}(\mathbf{m}, \boldsymbol{\alpha}))\}(0).$$

Then F and G are partial recursive and R = Dm G.

For the second part we use a diagonal argument. Let $g(a) \simeq \{a\}(a)$ and $A = \mathrm{Dm}\, g$. Clearly g is partial recursive so A is semi-recursive. Suppose, for a contradiction that A were in fact recursive. Then by Theorem 2.9 the function f defined by

$$f(a) = \begin{cases} g(a) + 1, & \text{if } a \in A; \\ 0, & \text{otherwise} \end{cases}$$

is recursive. Let b be an index for f. Since $\{b\}$ is total, $b \in A$ and thus

$$f(b) = g(b) + 1 = \{b\}(b) + 1 = f(b) + 1,$$

a contradiction. □

Note the similarity of this proof with that of Theorem 1.11. Essentially the same argument shows that the set of indices of total functions is not a recursive set (Exercise 4.23).

We begin now to study the properties of the class of semi-recursive relations. The first lemma follows directly from the definition.

4.3 Lemma. (i) *The class of semi-recursive relations is closed under finite intersection and bounded universal quantification*;

(ii) *the class of semi-recursive relations and partial recursive functionals is closed under relational composition.*

Proof. If $\mathsf{R} = \mathrm{Dm}\,\mathsf{F}$ and $\mathsf{S} = \mathrm{Dm}\,\mathsf{G}$ with F and G partial recursive, then $\mathsf{R} \cap \mathsf{S}$ is the domain of $\lambda \mathbf{m}\boldsymbol{\alpha}.\,\mathsf{F}(\mathbf{m}, \boldsymbol{\alpha}) + \mathsf{G}(\mathbf{m}, \boldsymbol{\alpha})$. If

$$\mathsf{Q}(p, \mathbf{m}, \boldsymbol{\alpha}) \leftrightarrow (\forall q < p)\mathsf{S}(q, \mathbf{m}, \boldsymbol{\alpha}),$$

define a partial recursive functional H by

$$\mathsf{H}(0, \mathbf{m}, \boldsymbol{\alpha}) \simeq 0;$$

$$\mathsf{H}(p+1, \mathbf{m}, \boldsymbol{\alpha}) \simeq \mathsf{H}(p, \mathbf{m}, \boldsymbol{\alpha}) + \mathsf{G}(p, \mathbf{m}, \boldsymbol{\alpha}).$$

Then $\mathsf{Q} = \mathrm{Dm}\,\mathsf{H}$ and thus is semi-recursive.

For (ii), if $\mathsf{S} = \mathrm{Dm}\,\mathsf{I}$ and

$$\mathsf{R}(\mathbf{m}, \boldsymbol{\alpha}) \leftrightarrow \mathsf{S}(\mathsf{G}_0(\mathbf{m}, \boldsymbol{\alpha}), \ldots, \mathsf{G}_{k'-1}(\mathbf{m}, \boldsymbol{\alpha}), \boldsymbol{\alpha}),$$

then R is the domain of the partial recursive functional $\mathrm{FCmp}(\mathsf{I}, \mathsf{G})$. □

Note, however, that the *recursive* relations and partial recursive functionals are *not* closed under relational composition (Exercise 4.26).

The remainder of the theory of semi-recursive relations depends on the normal forms of § 3. First, it is an immediate consequence of Theorem 3.9 that the class of semi-recursive relations and recursive functionals is closed under relational substitution.

4.4 Selection Theorem. *There exists a partial recursive functional* Sel *such that for all* $a, \mathbf{m}$, *and* $\boldsymbol{\alpha}$,

(i) $\exists p.\{a\}(p, \mathbf{m}, \boldsymbol{\alpha})\downarrow$ *iff* $\mathrm{Sel}(a, \langle \mathbf{m} \rangle, \langle \boldsymbol{\alpha} \rangle)\downarrow$, *and*

(ii) *if* $\exists p.\{a\}(p, \mathbf{m}, \boldsymbol{\alpha})\downarrow$, *then* $\{a\}(\mathrm{Sel}(a, \langle \mathbf{m} \rangle, \langle \boldsymbol{\alpha} \rangle), \mathbf{m}, \boldsymbol{\alpha})\downarrow$.

Proof. With T as in Theorem 3.1 we take

$$\mathrm{Sel}(a, \langle \mathbf{m} \rangle, \langle \boldsymbol{\alpha} \rangle) \simeq (\text{least } q \,.\, \mathsf{T}(a, \langle (q)_0, \mathbf{m} \rangle, (q)_1, \langle \boldsymbol{\alpha} \rangle))_0. \quad \square$$

4.5 Corollary. *For any semi-recursive relation* R, *there exists a partial recursive functional* $\mathrm{Sel}_{\mathsf{R}}$ *such that for all* **m** *and* $\boldsymbol{\alpha}$,

$$\exists p \,.\, \mathsf{R}(p, \mathbf{m}, \boldsymbol{\alpha}) \leftrightarrow \mathsf{R}(\mathrm{Sel}_{\mathsf{R}}(\mathbf{m}, \boldsymbol{\alpha}), \mathbf{m}, \boldsymbol{\alpha}) \leftrightarrow \mathrm{Sel}_{\mathsf{R}}(\mathbf{m}, \boldsymbol{\alpha}) \downarrow .$$

Proof. If a is any semi-index for R, take

$$\mathrm{Sel}_{\mathsf{R}}(\mathbf{m}, \boldsymbol{\alpha}) \simeq \mathrm{Sel}(a, \langle \mathbf{m} \rangle, \langle \boldsymbol{\alpha} \rangle).$$

$\mathrm{Sel}_{\mathsf{R}}$ is partial recursive by Theorem 3.9 and clearly has the required property. □

By analogy with the notation "least" we shall sometimes write $\mathrm{Sel}\, p \,.\, \mathsf{R}(p, \mathbf{m}, \boldsymbol{\alpha})$ for $\mathrm{Sel}_{\mathsf{R}}(\mathbf{m}, \boldsymbol{\alpha})$.

4.6 Corollary. *The class of semi-recursive relations of a given rank is closed under finite union.*

Proof. Suppose $\mathsf{R} = \mathrm{Dm}\{a\}$ and $\mathsf{S} = \mathrm{Dm}\{b\}$ are semi-recursive. Let F be the recursive function defined by

$$F(p) = \begin{cases} a, & \text{if } p = 0; \\ b, & \text{if } p > 0. \end{cases}$$

Set

$$\mathsf{H}(p, \mathbf{m}, \boldsymbol{\alpha}) \simeq \{F(p)\}(\mathbf{m}, \boldsymbol{\alpha}).$$

Then $(\mathsf{R} \cup \mathsf{S})(\mathbf{m}, \boldsymbol{\alpha}) \leftrightarrow \exists p \,.\, \mathsf{H}(p, \mathbf{m}, \boldsymbol{\alpha}) \downarrow$ so that $\mathsf{R} \cup \mathsf{S} = \mathrm{Dm}(\mathrm{Sel}_{\mathrm{Dm}\,\mathsf{H}})$ and is thus semi-recursive. □

4.7 Definition ((*unbounded*) *number quantification*). If R is any relation of rank $(k+1, l)$ then $\exists^0\mathsf{R}$ and $\forall^0\mathsf{R}$ are the relations P and Q of rank (k, l) such that

$$\mathsf{P}(\mathbf{m}, \boldsymbol{\alpha}) \leftrightarrow \exists p \,.\, \mathsf{R}(p, \mathbf{m}, \boldsymbol{\alpha}),$$

and

$$\mathsf{Q}(\mathbf{m}, \boldsymbol{\alpha}) \leftrightarrow \forall p \,.\, \mathsf{R}(p, \mathbf{m}, \boldsymbol{\alpha}).$$

4.8 Corollary. *The class of semi-recursive relations is closed under existential number quantification* ($\exists^0$), *hence also under bounded existential quantification* ($\exists^0_<$).

Proof. Immediate from Corollary 4.5. □

It follows, of course, that the co-semi-recursive relations are closed under $\forall^0$ and $\forall^0_<$. The semi-recursive relations are *not*, however, closed under $\forall^0$ (Exercise 4.24).

To extend definition by cases to semi-recursive relations we must modify the definition (cf. Exercise 4.28). We say that F is *defined by positive cases* from $\mathsf{G}_0, \ldots, \mathsf{G}_{k'-1}$ and $\mathsf{R}_0, \ldots, \mathsf{R}_{k'-1}$ if all are of rank (k, l), for any $(\mathbf{m}, \boldsymbol{\alpha}) \in {}^{k,l}\omega$ there is at most one $i < k'$ such that $\mathsf{R}_i(\mathbf{m}, \boldsymbol{\alpha})$, and

$$\mathsf{F}(\mathbf{m}, \boldsymbol{\alpha}) \simeq \begin{cases} \mathsf{G}_0(\mathbf{m}, \boldsymbol{\alpha}), & \text{if } \mathsf{R}_0(\mathbf{m}, \boldsymbol{\alpha}); \\ \vdots & \\ \mathsf{G}_{k'-1}(\mathbf{m}, \boldsymbol{\alpha}), & \text{if } \mathsf{R}_{k'-1}(\mathbf{m}, \boldsymbol{\alpha}); \\ \text{undefined, otherwise.} \end{cases}$$

4.9 Theorem. *The class of partial recursive functionals and semi-recursive relations is closed under definition by positive cases.*

Proof. Let f and g be recursive functions defined by

$$f(i) = \begin{cases} \text{a semi-index for } \mathsf{R}_i, & \text{if } i < k'; \\ 0, & \text{otherwise}; \end{cases}$$

$$g(i) = \begin{cases} \text{an index for } \mathsf{G}_i, & \text{if } i < k'; \\ 0, & \text{otherwise.} \end{cases}$$

Then if

$$\mathsf{H}(i, \mathbf{m}, \boldsymbol{\alpha}) \simeq \{f(i)\}(\mathbf{m}, \boldsymbol{\alpha}),$$

it is easy to check that the desired F is given by

$$\mathsf{F}(\mathbf{m}, \boldsymbol{\alpha}) \simeq \{g(\mathrm{Sel}_{\mathrm{Dm}\,\mathsf{H}}(\mathbf{m}, \boldsymbol{\alpha}))\}(\mathbf{m}, \boldsymbol{\alpha}). \quad \square$$

4.10 Corollary. *A relation is recursive iff it is both semi-recursive and co-semi-recursive.*

Proof. If R is recursive so is $\sim\mathsf{R}$ and thus R is both semi-recursive and co-semi-recursive by Lemma 4.2. Conversely, if both R and $\sim\mathsf{R}$ are semi-recursive, then

$$\mathsf{K}_{\mathsf{R}}(\mathbf{m},\boldsymbol{\alpha}) \simeq \begin{cases} 0, & \text{if } \mathsf{R}(\mathbf{m},\boldsymbol{\alpha}); \\ 1, & \text{if } \sim\mathsf{R}(\mathbf{m},\boldsymbol{\alpha}); \end{cases}$$

is a definition by positive semi-recursive cases and thus K_{R} is recursive. □

4.11 Corollary. *For any partial functional* F, F *is partial recursive iff its graph* Gr_{F} *is semi-recursive, and* F *is recursive iff* F *is total and* Gr_{F} *is recursive.*

Proof. If F is partial recursive, then since = is recursive and

$$\mathrm{Gr}_{\mathsf{F}}(n,\mathbf{m},\boldsymbol{\alpha}) \leftrightarrow \mathsf{F}(\mathbf{m},\boldsymbol{\alpha}) = n,$$

Gr_{F} is semi-recursive by 4.3(ii). For the converse we have

$$\mathsf{F}(\mathbf{m},\boldsymbol{\alpha}) \simeq \mathrm{Sel}_{\mathrm{Gr}(\mathsf{F})}(\mathbf{m},\boldsymbol{\alpha}).$$

The second part follows from the observation that if F is total, $\mathsf{F}(\mathbf{m},\boldsymbol{\alpha}) \neq n$ iff $(\exists n' \neq n)\mathsf{F}(\mathbf{m},\boldsymbol{\alpha}) \simeq n'$. □

To this point we have used the normal forms only to prove the Selection Theorem. Now we use them to obtain some new characterizations of the class of semi-recursive relations.

4.12 Theorem. *For any relation* R, *the following are equivalent:*

(i) R *is semi-recursive;*

(ii) *for some (primitive) recursive relation* S, $\mathsf{R}(\mathbf{m},\boldsymbol{\alpha}) \leftrightarrow \exists p\, \mathsf{S}(p,\mathbf{m},\boldsymbol{\alpha})$;

(iii) *for some (primitive) recursive relation S,*

$$\mathsf{R}(\mathbf{m},\boldsymbol{\alpha}) \leftrightarrow \exists p\, S(\mathbf{m},\bar{\alpha}_0(p),\ldots,\bar{\alpha}_{l-1}(p)) \text{ (only when } l>0).$$

Proof. Suppose R is semi-recursive, say with semi-index a. Then

$$\mathsf{R}(\mathbf{m},\boldsymbol{\alpha}) \leftrightarrow \exists n.\{a\}(\mathbf{m},\boldsymbol{\alpha}) \simeq n \leftrightarrow \exists u\, \mathsf{T}(a,\langle\mathbf{m}\rangle,u,\langle\boldsymbol{\alpha}\rangle)$$
$$\leftrightarrow \exists u\, T(a,\langle\mathbf{m}\rangle,u,\langle\bar{\alpha}_0(u),\ldots,\bar{\alpha}_{l-1}(u)\rangle)$$

by Theorems 3.1 and 3.5. Hence (ii) and (iii) follow if we set

$$\mathsf{S}(p,\mathbf{m},\boldsymbol{\alpha}) \leftrightarrow \mathsf{T}(a,\langle\mathbf{m}\rangle,p,\langle\boldsymbol{\alpha}\rangle)$$

and

$$S(\mathbf{m},s_0,\ldots,s_{l-1}) \leftrightarrow T(a,\langle\mathbf{m}\rangle,\mathrm{lg}(s_0),\langle s_0,\ldots,s_{l-1}\rangle).$$

On the other hand, if (ii) holds (in either form) then R is the domain of the function $\lambda \mathbf{m}\boldsymbol{\alpha}.\operatorname{least} p.\mathsf{S}(p,\mathbf{m},\boldsymbol{\alpha})$ and is thus semi-recursive. The implication (iii)→(i) is similar. □

Of course, combinations of forms (ii) and (iii) are possible. For example, every semi-recursive R may also be expressed in the form

$$\mathsf{R}(\mathbf{m},\boldsymbol{\alpha}) \leftrightarrow \exists p\ \mathsf{S}(\bar{\alpha}_0(p),\mathbf{m},\bar{\alpha}_1(p),\alpha_2,\ldots,\alpha_{l-1}).$$

4.13 Definition (*function quantification*). For any relation R of rank $(k, l+1)$, $\exists^1\mathsf{R}$ and $\forall^1\mathsf{R}$ are the relations P and Q of rank (k,l) such that

$$\mathsf{P}(\mathbf{m},\boldsymbol{\alpha}) \leftrightarrow \exists\beta\ \mathsf{R}(\mathbf{m},\boldsymbol{\alpha},\beta),$$

and

$$\mathsf{Q}(\mathbf{m},\boldsymbol{\alpha}) \leftrightarrow \forall\beta\ \mathsf{R}(\mathbf{m},\boldsymbol{\alpha},\beta).$$

4.14 Theorem. *The class of semi-recursive relations is closed under existential function quantification* ($\exists^1$).

Proof. Suppose R is semi-recursive. Then by 4.12 there exists a recursive S such that $\mathsf{R}(\mathbf{m},\boldsymbol{\alpha},\beta) \leftrightarrow \exists p.\mathsf{S}(\bar{\beta}(p),\mathbf{m},\boldsymbol{\alpha})$. Then

$$\begin{aligned}\exists\beta\ \mathsf{R}(\mathbf{m},\boldsymbol{\alpha},\beta) &\leftrightarrow \exists\beta\ \exists p\ \mathsf{S}(\bar{\beta}(p),\mathbf{m},\boldsymbol{\alpha})\\ &\leftrightarrow \exists s\ \mathsf{S}(s,\mathbf{m},\boldsymbol{\alpha}).\end{aligned}$$

Thus $\exists^1\mathsf{R} = \exists^0\mathsf{S}$ and is thus semi-recursive by 4.12. □

4.15. Theorem. *For any non-empty set $A \subseteq \omega$, the following are equivalent*:

(i) *A is semi-recursive*;
(ii) *$A = \operatorname{Im} F$, for some recursive function F*;
(iii) *$A = \operatorname{Im} \mathsf{F}$, for some recursive functional F*;
(iv) *$A = \operatorname{Im} \mathsf{F}$, for some partial recursive functional F.*

Proof. Suppose first that A is semi-recursive, so by Theorem 4.12 there exists a recursive relation S such that

$$n \in A \leftrightarrow \exists p.S(p,n).$$

Then if $\bar{n}$ is some fixed element of A, A is the image of the recursive function F defined by

$$F(m) = \begin{cases} (m)_1, & \text{if } S((m)_0, (m)_1); \\ \bar{n}, & \text{otherwise.} \end{cases}$$

The implications (ii)→(iii)→(iv) are trivial. Suppose $A = \operatorname{Im} \mathsf{F}$ for a partial recursive F. Then

$$n \in A \leftrightarrow \exists \mathbf{m} \exists \boldsymbol{\alpha} . \mathrm{Gr}_{\mathsf{F}}(n, \mathbf{m}, \boldsymbol{\alpha}).$$

Gr_{F} is semi-recursive by Corollary 4.11 and thus so is A by 4.8 and 4.14. □

Because of this theorem, semi-recursive subsets of ω are often called *recursively enumerable.*

We conclude this section with a discussion of two notions which originated in descriptive set theory and have played a significant role in recursion theory as well.

4.16 Definition. A class X of relations has the *reduction property* iff for any R and S belonging to X, there exist R^* and S^* in X such that:

(i) $\mathsf{R}^* \subseteq \mathsf{R}$ and $\mathsf{S}^* \subseteq \mathsf{S}$;

(ii) $\mathsf{R}^* \cap \mathsf{S}^* = \emptyset$;

(iii) $\mathsf{R}^* \cup \mathsf{S}^* = \mathsf{R} \cup \mathsf{S}$.

The pair $(\mathsf{R}^*, \mathsf{S}^*)$ is said to *reduce* (R, S).

4.17 Theorem. *The class of semi-recursive relations has the reduction property.*

Proof. Let R and S be any two semi-recursive relations. If R and S are not of the same rank, then they are disjoint and it suffices to take $\mathsf{R}^* = \mathsf{R}$ and $\mathsf{S}^* = \mathsf{S}$. Otherwise by Theorem 4.12 there exist recursive relations P and Q such that $\mathsf{R} = \exists^0 \mathsf{P}$ and $\mathsf{S} = \exists^0 \mathsf{Q}$. We set

$$\mathsf{R}^*(\mathbf{m}, \boldsymbol{\alpha}) \leftrightarrow \exists p \, [\mathsf{P}(p, \mathbf{m}, \boldsymbol{\alpha}) \wedge (\forall q < p) \sim \mathsf{Q}(q, \mathbf{m}, \boldsymbol{\alpha})];$$

$$\mathsf{S}^*(\mathbf{m}, \boldsymbol{\alpha}) \leftrightarrow \exists q \, [\mathsf{Q}(q, \mathbf{m}, \boldsymbol{\alpha}) \wedge (\forall p \leq q) \sim \mathsf{P}(p, \mathbf{m}, \boldsymbol{\alpha})].$$

Again by Theorem 4.12, R^* and S^* are semi-recursive, and it is straightforward to check that conditions (i)–(iii) are satisfied. □

4.18 Definition. A class X of relations has the *separation property* iff for all R and S belonging to X, if R and S are of the same rank and $\mathsf{R} \cap \mathsf{S} = \emptyset$, then there exists a relation P such that both P and $\sim\mathsf{P}$ belong to X and $\mathsf{R} \subseteq \mathsf{P} \subseteq \sim\mathsf{S}$. P is said to *separate* R and S.

These two properties are closely related. First we have

4.19 Lemma. *For any class X of relations, if X has the reduction property, then* $cX = \{\sim\mathsf{R}\colon \mathsf{R} \in X\}$ *has the separation property.*

Proof. Suppose X has the reduction property, R and S belong to X and have the same rank (k, l), and $(\sim\mathsf{R}) \cap (\sim\mathsf{S}) = \varnothing$. Then $\mathsf{R} \cup \mathsf{S} = {}^{k,l}\omega$. Let $(\mathsf{R}^*, \mathsf{S}^*)$ reduce (R, S). Since also $\mathsf{R}^* \cup \mathsf{S}^* = {}^{k,l}\omega$ and $\mathsf{R}^* \cap \mathsf{S}^* = \varnothing$, $\mathsf{R}^* = \sim\mathsf{S}^*$ so both S^* and $\sim\mathsf{S}^*$ belong to X. Also, $\sim\mathsf{R} \subseteq \sim\mathsf{R}^* = \mathsf{S}^* \subseteq \mathsf{S} = \sim(\sim\mathsf{S})$, so S^* separates $\sim\mathsf{R}$ and $\sim\mathsf{S}$. □

4.20 Definition. A class X of relations is *indexable* iff there exists a relation $\mathsf{U} \in X$ such that for every $\mathsf{R} \in X$, there exists a number a such that

$$\mathsf{R}(\mathbf{m}, \boldsymbol{\alpha}) \leftrightarrow \mathsf{U}(a, \langle \mathbf{m} \rangle, \langle \boldsymbol{\alpha} \rangle).$$

U is said to be *universal* for X.

4.21 Lemma. *For any indexable class X of relations which is closed under composition with recursive functionals, if X has the reduction property, then X does not have the separation property.*

Proof. Let U be universal for X and set

$$R(m) \leftrightarrow \mathsf{U}((m)_0, \langle m \rangle, \langle\ \rangle) \quad \text{and} \quad S(m) \leftrightarrow \mathsf{U}((m)_1, \langle m \rangle, \langle\ \rangle).$$

R and S both belong to X so there exists (R^*, S^*) which reduces (R, S). Suppose that X has the separation property and that P separates R^* and S^*. Since both P and $\sim P \in X$, there exist a and b such that

$$P(m) \leftrightarrow \mathsf{U}(a, \langle m \rangle, \langle\ \rangle) \quad \text{and} \quad \sim P(m) \leftrightarrow \mathsf{U}(b, \langle m \rangle, \langle\ \rangle).$$

Consider the number $m = \langle b, a \rangle$. We have

(1) $$P(\langle b, a \rangle) \leftrightarrow \mathsf{U}(a, \langle\langle b, a \rangle\rangle, \langle\ \rangle) \leftrightarrow S(\langle b, a \rangle),$$

and

(2) $$\sim P(\langle b, a \rangle) \leftrightarrow \mathsf{U}(b, \langle\langle b, a \rangle\rangle, \langle\ \rangle) \leftrightarrow R(\langle b, a \rangle).$$

Suppose that $P(\langle b, a \rangle)$. Since $P \cap S^* = \varnothing$, it follows from (1) that $\langle b, a \rangle \in S \sim S^*$. Then $R^*(\langle b, a \rangle)$ so $R(\langle b, a \rangle)$ and hence by (2), $\sim P(\langle b, a \rangle)$, a contradiction. A similar argument obtains a contradiction from the hypothesis $\sim P(\langle b, a \rangle)$. □

4.22 Corollary. (i) *The class of semi-recursive relations has the reduction property but not the separation property;*

(ii) *the class of co-semi-recursive relations has the separation property but not the reduction property.*

Proof. The class of semi-recursive relations has the universal relation

$$\mathsf{U}(a, \langle \mathbf{m} \rangle, \langle \boldsymbol{\alpha} \rangle) \leftrightarrow \{a\}(\mathbf{m}, \boldsymbol{\alpha})\downarrow. \quad \square$$

Note that in view of Corollary 4.10, the separation property for the class of (co-) semi-recursive relations asserts separability by a *recursive* relation. The second part of 4.22(i) is often referred to as establishing the existence of a pair of *recursively inseparable* semi-recursive relations.

4.23–4.34 Exercises

4.23. Show that the set of indices a such that $\{a\}$ is a total unary function is not recursive or even semi-recursive.

4.24. Show that the class of semi-recursive relations is not closed under $\forall^0$.

4.25. Let $\mathrm{Sel}'_{\mathsf{R}}(\mathbf{m}, \boldsymbol{\alpha}) \simeq \text{least } p\,.\,\mathsf{R}(p, \mathbf{m}, \boldsymbol{\alpha})$. Show that for some semi-recursive relations R, $\mathrm{Sel}'_{\mathsf{R}}$ is not partial recursive.

4.26. Show that the class of recursive relations and partial recursive functionals is *not* closed under relational composition.

4.27. Show that the class of partial recursive functionals is *not* closed under functional substitution (cf. Theorem 3.9).

4.28. Show that the class of partial recursive functionals and semi-recursive relations is *not* closed under definition by (not necessarily positive) cases.

4.29. Prove or refute the following analogue of Corollary 4.11: for any total functional F, F is primitive recursive iff Gr_{F} is primitive recursive.

4.30. Show that a non-empty subset A of ω is semi-recursive iff A is the image of a primitive recursive function f. Show that an infinite subset A of ω is semi-recursive iff A is the image of a one-one recursive function F.

4.31. For any partial function f of rank 1, let $\Omega[f]$ be defined as is Ω with the following additional clause:

$$\text{if} \quad f(p) \simeq n, \quad \text{then} \quad (\langle 0, k+1, l, 6 \rangle, p, \mathbf{m}, \boldsymbol{\alpha}, n) \in \Omega[f].$$

We write

$$\{a\}(\mathbf{m}, \boldsymbol{\alpha}, f) \simeq n \leftrightarrow (a, \mathbf{m}, \boldsymbol{\alpha}, n) \in \Omega[f].$$

We have thus defined the class of partial recursive functionals with arguments of type $(\mathbf{m}, \boldsymbol{\alpha}, f)$. Show that for any such partial recursive functional F

(i) there exists an ordinary partial recursive functional G such that

$$\mathsf{G}(e, \mathbf{m}, \boldsymbol{\alpha}) \simeq \mathsf{F}(\mathbf{m}, \boldsymbol{\alpha}, \{e\});$$

(ii) (*First Recursion Theorem*) there exists a partial recursive function $\bar{f}$ such that for all p, $\mathsf{F}(p, \bar{f}) \simeq \bar{f}(p)$ and for any h, if also for all p, $\mathsf{F}(p, h) \simeq h(p)$, then $\bar{f} \subseteq h$.

(iii) Formulate and prove a version of (ii) which allows for the presence of parameters.

Hint. For (i) use an analogue of the Second Normal Form of § 3. For (ii), let $g_0 = \varnothing$ and $g_{r+1}(p) \simeq \mathsf{F}(p, g_r)$. Take $\bar{f}(p) \simeq n \leftrightarrow \exists r \,.\, g_r(p) \simeq n$.

4.32. Use the techniques of Exercise I.2.7 to show that the class of closed subsets of ${}^{\omega}\omega$ has the separation property. (Show that for any A and B, if $\forall n \,.\, \mathsf{A}^{(\langle n \rangle)}$ and $\mathsf{B}^{(\langle n \rangle)}$ can be separated by a Kalmar set, then also A and B can be separated by a Kalmar set.)

4.33. Give an alternate proof using the Selection Theorem that the class of semi-recursive relations has the reduction property.

4.34. A class of relations X is said to have the *second separation property* iff for any $\mathsf{R}, \mathsf{S} \in X$ of the same rank, there exist R^* and S^* such that $\sim\mathsf{R}^*$ and $\sim\mathsf{S}^* \in X$ and $\mathsf{R} \sim \mathsf{S} \subseteq \mathsf{R}^* \subseteq \sim\mathsf{S}^* \subseteq \sim(\mathsf{S} \sim \mathsf{R})$. Show that if $\mathrm{c}X$ has the reduction property, then X has the second separation property.

4.35 Notes. The remarks made in the Notes to the preceding section apply also to the Selection Theorem and its consequences — all of § VI.3 is devoted to a technical lemma needed to prove the Selection Theorem for functionals partial recursive in a type-2 functional (VI.4.1).

Theorem 4.12 does not depend on any of the earlier results of this section and could be used to give alternate proofs for many of them. For example, if R and R' are semi-recursive and S and S' are recursive relations such that $\mathsf{R} = \exists^0\mathsf{S}$ and $\mathsf{R}' = \exists^0\mathsf{S}'$, then $\mathsf{R} \cup \mathsf{R}' = \exists^0(\mathsf{S} \cup \mathsf{S}')$ and is thus also semi-recursive. Similarly, if $\mathsf{R}(q, \mathbf{m}, \boldsymbol{\alpha}) \leftrightarrow \exists p\, \mathsf{S}(p, q, \mathbf{m}, \boldsymbol{\alpha})$, $\exists q\, \mathsf{R}(q, \mathbf{m}, \boldsymbol{\alpha}) \leftrightarrow \exists r\, \mathsf{S}((r)_0, (r)_1, \mathbf{m}, \boldsymbol{\alpha})$, so that if R is semi-recursive, so is $\exists^0\mathsf{R}$. We have given the proofs as we have in order to emphasize the parallel with the generalized recursion theories of Chapters VI and VII where the analogues of 4.2–4.11 hold, but that of 4.12 does not.

The term "semi-recursive" is relatively new and many authors still use "recursively enumerable" (r.e.) even for subsets of ${}^{k,l}\omega$ (which are not enumerable at all!). The need for separate terms, even for sets of numbers, becomes especially clear in § VI.1 where the class of "recursively-in-E enumerable" sets of numbers (images of functions partial recursive in E) is properly included in the class of sets semi-recursive in E (domains of functions partial recursive in E). To complete the confusion, at least one author has used "semi-recursive" in a totally different sense from ours.

5. Relativization

In our description of mechanical calculability, the input to the idealized computer consisted of an infinite memory device which contained the arguments and could be connected to the computer. In this section we consider the notion of calculability which arises from considering a second such memory device with fixed content permanently connected to the computer. If this device contains the graphs of a sequence of functions $\boldsymbol{\beta}$, then we say that a functional computed by the machine is *calculable in* (or *relative to*) $\boldsymbol{\beta}$.

5.1 Definition. For any F, R, and $\boldsymbol{\beta}$,

(i) F is *partial recursive in* $\boldsymbol{\beta}$ iff there exists a partial recursive G such that $\mathsf{F}(\mathbf{m}, \boldsymbol{\alpha}) \simeq \mathsf{G}(\mathbf{m}, \boldsymbol{\alpha}, \boldsymbol{\beta})$; an index for G is also called an *index for* F *from* $\boldsymbol{\beta}$;

(ii) F is *recursive in* $\boldsymbol{\beta}$ iff F is partial recursive in $\boldsymbol{\beta}$ and total;

(iii) R is *recursive in* $\boldsymbol{\beta}$ iff K_R is recursive in $\boldsymbol{\beta}$; an index for K_R from $\boldsymbol{\beta}$ is also called an *index for* R *from* $\boldsymbol{\beta}$;

(iv) R is *semi-recursive in* $\boldsymbol{\beta}$ iff R is the domain of some functional G partial recursive in $\boldsymbol{\beta}$; an index for G from $\boldsymbol{\beta}$ is also called a *semi-index for* R *from* $\boldsymbol{\beta}$.

In accord with the conventions of § I.1 we also say that F is partial recursive in $(A_0, \ldots, A_r)$ iff F if partial recursive in $(\mathsf{K}_{A_0}, \ldots, \mathsf{K}_{A_r})$, F is partial recursive in R iff F is partial recursive in $\{\langle\mathbf{m}\rangle: R(\mathbf{m})\}$, etc.

Note that we did *not* define F to be recursive in $\boldsymbol{\beta}$ just in case $\mathsf{F}(\mathbf{m}, \boldsymbol{\alpha}) = \mathsf{G}(\mathbf{m}, \boldsymbol{\alpha}, \boldsymbol{\beta})$ for some *recursive* (total) G. This is a more restrictive notion (cf. Exercises 5.8–9). From (iv) it is clear that R is semi-recursive in $\boldsymbol{\beta}$ iff $\mathsf{R}(\mathbf{m}, \boldsymbol{\alpha}) \leftrightarrow \mathsf{S}(\mathbf{m}, \boldsymbol{\alpha}, \boldsymbol{\beta})$ for some semi-recursive S, but again it is false that R is recursive in $\boldsymbol{\beta}$ iff this equivalence holds for a recursive S.

If we replace "primitive recursive" by "primitive recursive in $\boldsymbol{\beta}$" (cf. Exercise 1.18), "partial recursive" by "partial recursive in $\boldsymbol{\beta}$", etc., the theory of §§ 1–4 all goes through with essentially the same proofs. We shall refer to these results as the *relativized versions* of the various theorems.

5.2 Lemma (Transitivity). *If all of* $\beta_0, \ldots, \beta_r$ *are recursive* (*in* γ), *and* F *is partial recursive in* $\boldsymbol{\beta}$, *then* F *is partial recursive* (*in* γ).

Proof. We prove the version with γ. By assumption there exist partial recursive $\mathsf{G}, \mathsf{H}_0, \ldots, \mathsf{H}_r$ such that

$$\mathsf{F}(\mathbf{m}, \boldsymbol{\alpha}) \simeq \mathsf{G}(\mathbf{m}, \boldsymbol{\alpha}, \boldsymbol{\beta}), \quad \text{and for all} \quad i \leq r, \quad \beta_i(p) = \mathsf{H}_i(p, \boldsymbol{\gamma}).$$

By Theorem 3.9 there exists a partial recursive functional I such that for all $\boldsymbol{\gamma}$ such that $\lambda p . \mathsf{H}_i(p, \boldsymbol{\gamma})$ is total for all $i \leq r$,

$$\mathsf{I}(\mathbf{m}, \boldsymbol{\alpha}, \boldsymbol{\gamma}) \simeq \mathsf{G}(\mathbf{m}, \boldsymbol{\alpha}, \lambda p . \mathsf{H}_0(p, \boldsymbol{\gamma}), \ldots, \lambda p . \mathsf{H}_r(p, \boldsymbol{\gamma})).$$

In particular this is true for the fixed $\boldsymbol{\gamma}$ under consideration and thus $\mathsf{F}(\mathbf{m}, \boldsymbol{\alpha}) \simeq \mathsf{I}(\mathbf{m}, \boldsymbol{\alpha}, \boldsymbol{\gamma})$. Hence F is partial recursive in $\boldsymbol{\gamma}$. □

It follows that the relation "recursive in" is transitive on members of ${}^\omega\omega$. For any α, the *degree* of α is defined by

$$\mathrm{dg}(\alpha) = \{\beta : \alpha \text{ is recursive in } \beta \text{ and } \beta \text{ is recursive in } \alpha\}.$$

The degrees inherit a partial ordering — we say

$$\mathrm{dg}(\alpha) \leq \mathrm{dg}(\beta) \leftrightarrow \alpha \text{ is recursive in } \beta;$$

$$\mathrm{dg}(\alpha) < \mathrm{dg}(\beta) \leftrightarrow \mathrm{dg}(\alpha) \leq \mathrm{dg}(\beta) \quad \text{and} \quad \mathrm{dg}(\beta) \not\leq \mathrm{dg}(\alpha).$$

Because of transitivity these notions do not depend on the choice of representative. This is far from being a linear ordering. Because there are only countably many partial recursive functionals, $\{\beta : \beta \text{ is recursive in } \alpha\}$ is always countable.

5.3 Theorem. *For any non-recursive α, $\{\beta : \alpha$ is recursive in $\beta\}$ is meager.*

Proof. Let α be a fixed non-recursive function, $\mathsf{A} = \{\beta : \alpha \text{ is recursive in } \beta\}$, and for each a, $\mathsf{A}_a = \{\beta : \alpha \text{ is recursive in } \beta \text{ with index } a\}$. If A is not meager, then for some a and some p, A_a is dense in $[p]$. We claim, however, that from this follows that

$$\alpha(m) = n \leftrightarrow \exists u\, \exists s\, [(p \subseteq s \vee s \subseteq p) \wedge \lg(s) = u \wedge T(a, \langle m \rangle, u, \langle s \rangle) \wedge (u)_0 = n].$$

This implies, by Corollary 4.11, that α is recursive, contrary to assumption.

Suppose first that $\alpha(m) = n$. Since A_a is dense in $[p]$ there exists a $\beta \in [p]$ such that $\{a\}(m, \beta) \simeq n$. Then $\exists u\, [T(a, \langle m \rangle, u, \langle \bar{\beta}(u) \rangle) \wedge (u)_0 = n]$ and clearly $\bar{\beta}(u) \subseteq p$ or $p \subseteq \bar{\beta}(u)$, and $\lg(\bar{\beta}(u)) = u$.

Conversely suppose there are u and s such that $(p \subseteq s \vee s \subseteq p)$, $\lg(s) = u$, $T(a, \langle m \rangle, u, \langle s \rangle)$, and $(u)_0 = n$. Let β be any element of A_a and the smaller of $[p]$

and $[\bar{\beta}(u)]$ so that $s=\bar{\beta}(u)$. Thus $T(a,\langle m\rangle,u,\langle\bar{\beta}(u)\rangle)$ and $(u)_0=n$ so $\{a\}(m,\beta)\simeq n$ and hence also $\alpha(m)=n$. □

5.4 Corollary. *For any non-recursive α, there are uncountably many β such that neither α is recursive in β nor β is recursive in α.*

Proof. By 5.3 and the Baire Category Theorem (I.2.2). □

One of the main themes of this book is the notion of complexity for functionals and relations. The simplest objects beyond the finite that we have discussed are the primitive recursive ones. Recursive functionals and relations are more complex in that an additional technique of calculation (the self-determination principle or unbounded search) is needed to compute them. Partial recursive functionals and semi-recursive relations are still more complex in that they are not completely calculable — if $(\mathbf{m},\boldsymbol{\alpha})\in \mathrm{Dm}\,\mathsf{F}$, that fact will become known at the conclusion of the computation of $\mathsf{F}(\mathbf{m},\boldsymbol{\alpha})$, but if $(\mathbf{m},\boldsymbol{\alpha})\notin \mathrm{Dm}\,\mathsf{F}$, the attempt to compute $\mathsf{F}(\mathbf{m},\boldsymbol{\alpha})$ will go on forever (cf. Lemma 4.2).

The relationship "recursive in" is a measure of *relative* complexity among elements of ${}^{\omega}\omega$ and $\mathbf{P}(\omega)$. If α is recursive in β, then "given" the graph of β, α is mechanically calculable. Thus in some sense β is at least as complex as α and more complex if also β is not recursive in α. According to this measure, the recursive functions are the simplest ones, since any recursive function is recursive in every β. Of course, by the preceding corollary, the complexity of two functions may be incomparable.

For any A and $B\subseteq\omega$, if A is recursive and B is semi-recursive but not recursive, then clearly A is recursive in B but B is not recursive in A. Thus in this sense also semi-recursive sets are more complex than recursive sets. It is natural to ask if all non-recursive semi-recursive sets are equally complex. This is answered by:

Friedberg-Mučnik Theorem. There exist semi-recursive sets A and B such that neither A is recursive in B nor B is recursive in A. This result belongs to the theory of degrees, which will be treated in a separate volume in this series Lerman [198?], and we shall not give the proof.

We mentioned in the Introduction the fact that many of the ideas of hierarchy theory arose independently in Descriptive Set Theory and Recursion Theory, and that the fruitful methods of investigation in the two fields are very similar. The following theorem, which characterizes the topology of ${}^{k,l}\omega$ in recursion-theoretic terms, is the basis for this similarity.

5.5 Theorem. *For any* F *and* R,

(i) F *is partial continuous iff* F *is partial recursive in some* β;

(ii) R *is open iff* R *is semi-recursive in some* β;

(iii) R *is closed iff* R *co-semi-recursive in some* β;
(iv) R *is closed-open iff* R *is recursive in some* β.

Proof. Clauses (ii)–(iv) are immediate from (i). The implication $(\leftarrow)$ of (i) is the relativized version of Corollary 3.7. Suppose that F is partial continuous; for simplicity take F to be of rank (1,1). Define β by

$$\beta(\langle m, p\rangle) = \begin{cases} n+1, & \text{if } \mathsf{F}(m,\alpha) \simeq n \quad \text{for all} \quad \alpha \in [p]; \\ 0, & \text{if there is no such } n. \end{cases}$$

The continuity of F implies that for all m and α,

$$\mathsf{F}(m,\alpha)\downarrow \leftrightarrow \exists k\,.\,\beta(\langle m, \bar{\alpha}(k)\rangle) > 0.$$

Hence

$$\mathsf{F}(m,\alpha) \simeq \beta(\langle m, \bar{\alpha}(\text{least } k\,.\,\beta(\langle m, \bar{\alpha}(k)\rangle) > 0)\rangle) \dot{-} 1$$

and is thus partial recursive in β. □

The relationship "recursive in" is by no means the only one which compares the complexity of two objects. The next chapter will provide many examples, but one is appropriate here. A relation R is called (*many-one*) *reducible* to a set A $(\mathsf{R} \leqslant A)$ iff for some recursive functional F,

$$\mathsf{R}(\mathbf{m}, \boldsymbol{\alpha}) \leftrightarrow \mathsf{F}(\mathbf{m}, \boldsymbol{\alpha}) \in A.$$

From the composability of partial recursive functionals it follows that if R is many-one reducible to A, then R is recursive in A and if A is (semi-) recursive, so is R (cf. Exercise 5.13).

Among the sets semi-recursive in a given set A there is one which under comparison by many-one reducibility is most complex:

5.6 Definition. For any $A \subseteq \omega$, the (*ordinary*) *jump* of A is the set

$$A^{\text{oJ}} = \{\langle a, \mathbf{m}\rangle : \{a\}(\mathbf{m}, A)\downarrow\}.$$

Clearly A^{oJ} is semi-recursive in A but not recursive in A — in fact,

5.7 Theorem. *For any* $A \subseteq \omega$, *any* k, *and any* $R \subseteq {}^k\omega$,

$$R \text{ is semi-recursive in } A \leftrightarrow R \leqslant A^{\text{oJ}}.$$

Proof. Suppose that R is semi-recursive in A with semi-index a from A. Then

$$R(\mathbf{m}) \leftrightarrow \{a\}(\mathbf{m}, A)\downarrow \leftrightarrow \langle a, \mathbf{m}\rangle \in A^{\omega J}$$

and thus $R \leqslant A^{\omega J}$. On the other hand, $A^{\omega J}$ is semi-recursive in A by its definition and hence so is any R many-one reducible to it. □

The jump may also be applied to functions:

$$\beta^{\omega J}(\langle a, \mathbf{m}\rangle) = \begin{cases} 0, & \text{if } \{a\}(\mathbf{m}, \beta)\downarrow; \\ 1, & \text{otherwise;} \end{cases}$$

and $\beta^{\omega J}(p) = 0$ if p is not of the form $\langle a, \mathbf{m}\rangle$. Corresponding to Theorem 5.7 we have only the implication:

$$R \text{ is semi-recursive in } \beta \rightarrow R \text{ is recursive in } \beta^{\omega J}.$$

In each of these results the restriction to relations on numbers is a necessary one. Consider the relation (set) $\mathsf{R} = \{\alpha : \exists m . \alpha(m) = 0\}$. R is easily seen to be semi-recursive (and hence open) but not closed. Then by Theorem 5.5, R is not recursive in *any* β.

This restriction is closely related to the fact that we have defined here only the notion "F is partial recursive in β" rather than the more general "F is partial recursive in I". This latter notion is the subject of Chapter VI, and we shall have there a version of Theorem 5.7 for arbitrary relations (VI.1.11).

5.8–5.14 Exercises

5.8. A is called *truth-table reducible* to B iff there exist recursive F and D such that

$$m \in A \leftrightarrow \bar{\mathsf{K}}_B(F(m)) \in D.$$

Show that A is truth-table reducible to B iff there exists a (total) recursive functional G such that $\mathsf{K}_A(m) = \mathsf{G}(m, \mathsf{K}_B)$. (Suppose such a G exists and R is a recursive relation such that

$$\mathsf{G}(m, \alpha) = q \leftrightarrow \exists p . R(\bar{\alpha}(p), m, q).$$

If α^* ranges over ${}^{\omega}2$, we have

$$\forall m\, \forall \alpha^*\, \exists p\, \exists q\, R(\bar{\alpha}^*(p), m, q).$$

Apply the result of Exercise I.2.6.)

5.9. Show that there exist A and B such that A is recursive in B but A is not truth-table reducible to B. (Let

$$a \in A \leftrightarrow \{a\}(a) \quad \text{is defined,}$$

choose by Exercise 4.30 a one-one recursive function f which enumerates A and set

$$n \in B \leftrightarrow (\exists p > n) . f(p) < f(n).$$

Show first that $\sim B$ is infinite and

$$\text{if} \quad n \notin B \quad \text{and} \quad f(n) > a, \quad \text{then} \quad a \in A \leftrightarrow (\exists p < n)[a = F(p)],$$

and conclude that A is recursive in B. Suppose that A were truth-table reducible to B, say F and D are recursive and

$$a \in A \leftrightarrow \bar{\mathsf{K}}_B(F(a)) \in D.$$

Let G be a recursive function such that

$$\{G(m)\}(a) \quad \text{is defined iff} \quad \langle \mathsf{K}_B(0), \ldots, \mathsf{K}_B(m-1), 0, \ldots, 0 \rangle \notin D,$$

where the sequence on the right has length $F(a)$. Show that for every m, there is an $n \notin B$ such that $m \leqslant n < F(G(m))$ and conclude that A is recursive, a contradiction.)

5.10. Show that
(i) the class of open relations has the reduction property but not the separation property;
(ii) the class of closed relations has the separation property but not the reduction property (cf. Corollary 4.22 and Exercise 4.32).

5.11. Parallel to Definition 4.20, call a class X of relations *parametrizable* iff there exists a relation $\mathsf{U} \in X$ such that for every $\mathsf{R} \in X$, there exists a function β such that

$$\mathsf{R}(\mathbf{m}, \boldsymbol{\alpha}) \leftrightarrow \mathsf{U}(\langle \mathbf{m} \rangle, \langle \boldsymbol{\alpha} \rangle, \beta).$$

U is said to be *universal* for X. Show that the class of open relations is parametrizable.

5.12. Prove that the class of partial continuous functionals is the smallest class of partial functionals which includes ${}^{\omega}\omega$ and the class of primitive recursive functionals and is closed under expansion, functional composition, and unbounded search.

5.13. Show that if R is reducible to A, then R is recursive in A and if A is (semi-) recursive, then also R is (semi-) recursive. Give an example of R recursive in A but not reducible to A.

5.14. Show that there exists a primitive recursive function f such that if A is recursive in B with index a, then $A^{\circ J}$ is recursive in $B^{\circ J}$ with index $f(a)$.

5.15 Notes. The theory of degrees is one of the most developed areas of recursion theory. Shoenfield [1971a] is an excellent introduction. Many-one reducibility and other related notions are discussed in detail in Rogers [1967, Chapters 6–10].

Historically, the notion of relative recursiveness for functions on numbers preceded and led to the notion of recursive functional (see the first few pages of Kleene [1955b]). In our framework this would go as follows. First we would define for a fixed β a set $\Omega[\beta]$ of sequences of the form $(a, \mathbf{m}, n)$. The definition would be exactly like Definition 2.1 except that $\boldsymbol{\alpha}$ and the parameter l would be omitted and the clause introducing the initial functionals $\mathsf{Ap}^{k,l}$ would be replaced by one introducing β as an initial function:

$$(\langle 0, k, 3, i\rangle, \mathbf{m}, \beta(m_i)) \in \Omega[\beta].$$

We would write $\{a\}^{\beta}(\mathbf{m}) \simeq n$ iff $(a, \mathbf{m}, n) \in \Omega[\beta]$ and say F is partial recursive in β iff $F = \{a\}^{\beta}$ for some $a \in \omega$. Next we would extend this in the obvious way to $\Omega[\boldsymbol{\beta}]$, etc. and finally define F to be a partial recursive functional iff for some a, $\mathsf{F}(\mathbf{m}, \boldsymbol{\beta}) \simeq \{a\}^{\boldsymbol{\beta}}(\mathbf{m})$. Of course, in hindsight, this is a trivial variant, but it was not always so obvious.

Chapter III
Hierarchies and Definability

In the preceding chapter we saw that the semi-recursive relations are exactly those which arise from the recursive relations by existential number quantification (II.4.12). In this chapter we study the relations which arise from the recursive relations by all kinds of quantification: existential, universal, number, and function. After classifying these relations according to the number and type of quantifiers used and establishing the simplest combinatorial properties of this classification in §§ 1 and 2, we relate it to other notions of definability. In § 3 we compare the complexity of definition of an inductive operator Γ with that of the set $\bar{\Gamma}$. In § 4, we investigate the relationship between the complexity of a subset A of ${}^{\omega}\omega$ and that of its elements. In § 5 we show that the relations we are considering are exactly those definable in certain natural first- and second-order formal languages. Finally in § 6 we introduce the method of forcing to extend and complete some earlier results.

1. The Arithmetical Hierarchy

1.1 Definition. The class of *arithmetical relations* is the smallest class of relations containing the recursive relations and closed under number quantification ($\exists^0$ and $\forall^0$).

We next define a classification of the arithmetical relations based on the number of quantifiers needed to define a relation.

1.2 Definition (The Arithmetical Hierarchy). For all r,

(i) $\Sigma^0_0 = \Pi^0_0 =$ the class of recursive relations;

(ii) $\Sigma^0_{r+1} = \{\exists^0 \mathsf{P}\colon \mathsf{P} \in \Pi^0_r\}$;

(iii) $\Pi^0_{r+1} = \{\forall^0 \mathsf{P}\colon \mathsf{P} \in \Sigma^0_r\}$;

(iv) $\Delta^0_r = \Sigma^0_r \cap \Pi^0_r$;

(v) $\Delta^0_{(\omega)} = \bigcup \{\Sigma^0_r \cup \Pi^0_r\colon r \in \omega\}$.

It is immediate by induction on r that all of the classes Σ^0_r and Π^0_r, and hence $\Delta^0_{(\omega)}$, are included in the class of arithmetical relations. The converse inclusion is immediate from Theorem 1.5 below.

1.3 Examples. Note first that by Theorem II.4.12, Σ^0_1 is exactly the class of semi-recursive relations and Π^0_1 is the class of co-semi-recursive relations. Hence by Corollary II.4.10, $\Delta^0_0 = \Delta^0_1 =$ the class of recursive relations.

Consider the set $\mathsf{A} = \{\alpha : \alpha \text{ is recursive}\}$. We have

$$\begin{aligned}\alpha \in \mathsf{A} &\leftrightarrow \exists a\, \forall m\, [\{a\}(m) \simeq \alpha(m)]\\ &\leftrightarrow \exists a\, \forall m\, \exists u\, [\mathsf{T}(a, \langle m\rangle, u, \langle\ \ \rangle) \wedge (u)_0 = \alpha(m)]\end{aligned}$$

and thus $\mathsf{A} \in \Sigma^0_3$. The set B of primitive recursive functions is Σ^0_2:

$$\alpha \in \mathsf{B} \leftrightarrow \exists a\, \forall m\, [a \in \mathrm{Pri} \wedge \mathsf{Ev}^{1,0}(a, m) = \alpha(m)].$$

Let C be the set of γ such that the relation $\leqslant_\gamma = \{(p, q) : \gamma(\langle p, q\rangle) = 0\}$ is a linear ordering. Then an easy computation shows $\mathsf{C} \in \Pi^0_1$ (try this now without using Theorem 1.5).

Let $\mathscr{L}$ be a first-order formal language with (for simplicity) only one binary relation symbol and no function symbols, and suppose Gödel numbers have been assigned to the symbols, formulas, and sequences of formulas in some standard way (as in Shoenfield [1967], for example). We henceforth identify these objects with their Gödel numbers. An analysis of the notion of formal proof shows that the relation

$$\mathsf{R}(p, m, A) \leftrightarrow p \text{ is a proof of } m \text{ from } A$$

is recursive. Hence, by the Completeness Theorem,

$$\mathsf{S}(m, A) \leftrightarrow m \text{ is a logical consequence of } A$$

is Σ^0_1. The set $\{A : A \text{ is consistent}\}$ is Π^0_1 and $\{A : A \text{ is complete}\}$ is Π^0_2. Every denumerable structure for $\mathscr{L}$ is isomorphic to one of the form (ω, R) with $R \subseteq {}^2\omega$. Analysis of any standard proof of the Completeness Theorem shows that if A is recursive and has infinite models, then A has a model (ω, R) with $R \in \Delta^0_2$.

We write $\mathsf{F} \in \Sigma^0_r\ (\Pi^0_r, \Delta^0_r)$ to mean $\mathsf{Gr}_\mathsf{F} \in \Sigma^0_r\ (\Pi^0_r, \Delta^0_r)$. Note that $\mathsf{R} \in \Sigma^0_r$ does not in general imply $\mathsf{K}_\mathsf{R} \in \Sigma^0_r$. We often use the terms Σ^0_r, etc. as adjectives and write, for example, "for any Σ^0_r relation R" instead of "for any $\mathsf{R} \in \Sigma^0_r$".

1.4 Lemma. *For all r and* R,

$$\mathsf{R} \in \Sigma^0_r \leftrightarrow \sim\mathsf{R} \in \Pi^0_r \quad \textit{and} \quad \mathsf{R} \in \Pi^0_r \leftrightarrow \sim\mathsf{R} \in \Sigma^0_r.$$

Proof. The case $r = 0$ is immediate. Suppose the result holds for r and suppose

$\mathsf{R} \in \Sigma^0_{r+1}$. Then $\mathsf{R} = \exists^0\mathsf{P}$ for some $\mathsf{P} \in \Pi^0_r$ and $\sim\mathsf{R} = \forall^0{\sim}\mathsf{P}$. Then $\sim\mathsf{P} \in \Sigma^0_r$ so $\sim\mathsf{R} \in \Pi^0_{r+1}$ by definition. The case for $\mathsf{R} \in \Pi^0_{r+1}$ is similar. □

1.5 Theorem. *The classes of the arithmetical hierarchy have the following closure properties for all r:*

	Σ^0_r	Π^0_r	Δ^0_r	$\Delta^0_{(\omega)}$
Composition and substitution with recursive functions	√	√	√	√
Finite union and intersection	√	√	√	√
Expansion	√	√	√	√
Complementation			√	√
Bounded quantification ($\exists^0_<$ and $\forall^0_<$)	√	√	√	√
Existential number quantification ($\exists^0$)	√($r>0$)			√
Universal number quantification ($\forall^0$)		√($r>0$)		√

Proof. In the proofs we use the following equivalences and their duals (obtained by negating both sides):

$$\exists p\, \mathsf{P}(p, \mathbf{m}, \boldsymbol{\alpha}) \vee \exists q\, \mathsf{Q}(q, \mathbf{m}, \boldsymbol{\alpha}) \leftrightarrow \exists p\, [\mathsf{P}(p, \mathbf{m}, \boldsymbol{\alpha}) \vee \mathsf{Q}(p, \mathbf{m}, \boldsymbol{\alpha})];$$

$$\exists p\, \mathsf{P}(p, \mathbf{m}, \boldsymbol{\alpha}) \wedge \exists q\, \mathsf{Q}(q, \mathbf{m}, \boldsymbol{\alpha}) \leftrightarrow \exists p\, [\mathsf{P}((p)_0, \mathbf{m}, \boldsymbol{\alpha}) \wedge \mathsf{Q}((p)_1, \mathbf{m}, \boldsymbol{\alpha})];$$

$$(\exists q < s) \exists p\, \mathsf{P}(p, q, s, \mathbf{m}, \boldsymbol{\alpha}) \leftrightarrow \exists p\, [(p)_1 < s \wedge \mathsf{P}((p)_0, (p)_1, s, \mathbf{m}, \boldsymbol{\alpha})];$$

$$(\forall q < s) \exists p\, \mathsf{P}(p, q, s, \mathbf{m}, \boldsymbol{\alpha}) \leftrightarrow \exists p\, (\forall q < s)\, \mathsf{P}((p)_q, q, s, \mathbf{m}, \boldsymbol{\alpha});$$

$$\exists p\, \exists q\, \mathsf{P}(p, q, \mathbf{m}, \boldsymbol{\alpha}) \leftrightarrow \exists p\, \mathsf{P}((p)_0, (p)_1, \mathbf{m}, \boldsymbol{\alpha}).$$

For example, we prove by induction on r that Σ^0_r and Π^0_r are closed under bounded quantification. For $r = 0$, this is known (§ II.2). Suppose it holds for r and R is any Σ^0_{r+1} relation, say

$$\mathsf{R}(q, s, \mathbf{m}, \boldsymbol{\alpha}) \leftrightarrow \exists p\, \mathsf{P}(p, q, s, \mathbf{m}, \boldsymbol{\alpha})$$

with $\mathsf{P} \in \Pi^0_r$. Using the third and fourth equivalences and the induction hypothesis, it follows that the relations defined by

$$(\exists q < s)\, \mathsf{R}(q, s, \mathbf{m}, \boldsymbol{\alpha}) \quad \text{and} \quad (\forall q < s)\, \mathsf{R}(q, s, \mathbf{m}, \boldsymbol{\alpha})$$

are also Σ^0_{r+1}. The other proofs are similar. □

1.6 Corollary. *For all r, $\Sigma^0_r \cup \Pi^0_r \subseteq \Delta^0_{r+1}$.*

Proof. We observed in 1.3 that this holds for $r = 0$ and we proceed by induction. Given the conclusion for r, suppose $\mathsf{R} \in \Sigma^0_{r+1}$ and let S be defined by

$$\mathsf{S}(p, \mathbf{m}, \boldsymbol{\alpha}) \leftrightarrow \mathsf{R}(\mathbf{m}, \boldsymbol{\alpha}).$$

Then $\mathsf{R} = \exists^0\mathsf{S} = \forall^0\mathsf{S}$ and $\mathsf{S} \in \Sigma^0_{r+1}$ so immediately $\mathsf{R} \in \Pi^0_{r+2}$. But $\mathsf{S} = \exists^0\mathsf{P}$ for some $\mathsf{P} \in \Pi^0_r \subseteq \Pi^0_{r+1}$ so also $\mathsf{R} \in \Sigma^0_{r+2}$ and thus $\mathsf{R} \in \Delta^0_{r+2}$. The proof for $\mathsf{R} \in \Pi^0_{r+1}$ is similar. □

These last two results are often used together to "compute" where a given arithmetical relation falls in the hierarchy. For example, if

$$\mathsf{R}(\mathbf{m}, \boldsymbol{\alpha}) \leftrightarrow \exists p\, \exists q\, \forall r\, [\exists s\, \forall t\, \mathsf{P}(p, G(q, s, \mathbf{m}), t, \mathbf{m}, \boldsymbol{\alpha}) \vee \exists u\, \mathsf{Q}(p, u, r, \mathbf{m}, \boldsymbol{\alpha})],$$

with G, P, and Q recursive, then the relation described inside the brackets is the union of a Σ^0_2 and a Σ^0_1 relation, hence is Σ^0_2. Then by applying successively $\forall^0$, $\exists^0$, and $\exists^0$ we conclude $\mathsf{R} \in \Sigma^0_4$. Of course, such a computation does not always yield an optimal classification — in fact, in this example we have also

$$\mathsf{R}(\mathbf{m}, \boldsymbol{\alpha}) \leftrightarrow \exists p\, \exists q\, [\exists s\, \forall t\, \mathsf{P}(p, G(q, s, \mathbf{m}), t, \mathbf{m}, \boldsymbol{\alpha}) \vee \forall r\, \exists u\, \mathsf{Q}(p, u, r, \mathbf{m}, \boldsymbol{\alpha})]$$

from which we obtain $\mathsf{R} \in \Sigma^0_3$. We consider later in the chapter methods for showing that a relation does *not* belong to some class Σ^0_r or Π^0_r.

1.7 Corollary. *For any $r > 0$, $\mathsf{F}\colon {}^{k,l}\omega \to \omega$, $F\colon {}^{k}\omega \to \omega$, and $\mathsf{R} \subseteq {}^{k,l}\omega$,*
(i) *if $\mathsf{F} \in \Sigma^0_r$ and F is total, then also $\mathsf{F} \in \Delta^0_r$;*
(ii) *if $\mathsf{F} \in \Sigma^0_r$ then $\mathrm{Dm}\,\mathsf{F} \in \Sigma^0_r$;*
(iii) *if $F \in \Sigma^0_r$ then $\mathrm{Im}\,F \in \Sigma^0_r$;*
(iv) *if $\mathsf{R} \in \Sigma^0_r \cup \Pi^0_r$, then $\mathsf{K}_\mathsf{R} \in \Delta^0_{r+1}$;*
(v) *Σ^0_r and Π^0_r are closed under composition with total Δ^0_r functionals.*

Proof. For (i), if F is total, then

$$\mathsf{F}(\mathbf{m}, \boldsymbol{\alpha}) = n \leftrightarrow \forall n'[n' \neq n \to \mathsf{F}(\mathbf{m}, \boldsymbol{\alpha}) \neq n'].$$

Hence if Gr_F is Σ^0_r, it is also Π^0_r. (ii) and (iii) are immediate from the closure of Σ^0_r under $\exists^0$. For (iv) we have

$$\mathsf{K}_\mathsf{R}(\mathbf{m}, \boldsymbol{\alpha}) = n \leftrightarrow [\mathsf{R}(\mathbf{m}, \boldsymbol{\alpha}) \wedge n = 0] \vee [\sim\mathsf{R}(\mathbf{m}, \boldsymbol{\alpha}) \wedge n = 1]$$

so $\mathrm{Gr}_{\mathrm{K_R}}$ is the union of a Σ_r^0 and a Π_r^0 relation, hence is Δ_{r+1}^0. For (v) we have for total G,

$$\mathrm{R}(\mathrm{G}(\mathbf{m}, \boldsymbol{\alpha}), \mathbf{m}, \boldsymbol{\alpha}) \leftrightarrow \exists n\, [\mathrm{G}(\mathbf{m}, \boldsymbol{\alpha}) = n \wedge \mathrm{R}(n, \mathbf{m}, \boldsymbol{\alpha})]$$
$$\leftrightarrow \forall n\, [\mathrm{G}(\mathbf{m}, \boldsymbol{\alpha}) = n \rightarrow \mathrm{R}(n, \mathbf{m}, \boldsymbol{\alpha})].$$

The first equivalence serves if $\mathrm{R} \in \Sigma_r^0$ and the second if $\mathrm{R} \in \Pi_r^0$. □

The following diagram exhibits the inclusions of Corollary 1.6:

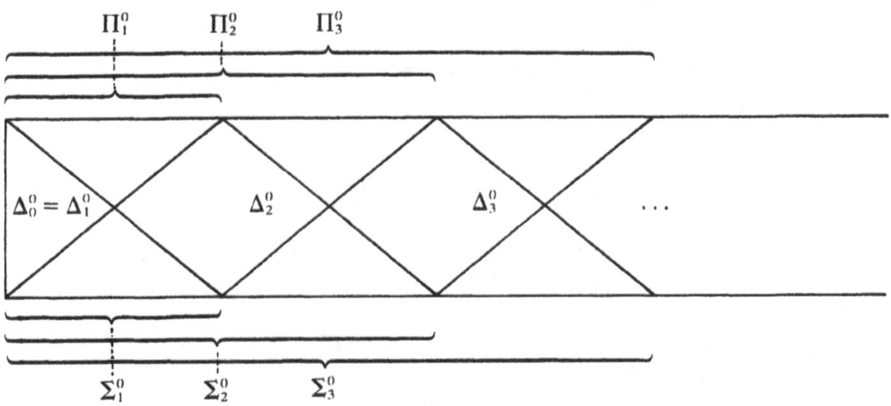

To show that each of these inclusions is proper — that each space in the diagram represents a non-empty class of relations, we use a diagonal argument. For $r \geq 1$, let

$$\mathrm{U}_1^0(a, \langle \mathbf{m} \rangle, \langle \boldsymbol{\alpha} \rangle) \leftrightarrow \exists u\ \mathrm{T}(a, \langle \mathbf{m} \rangle, u, \langle \boldsymbol{\alpha} \rangle);$$

$$\mathrm{U}_{r+1}^0(a, \langle \mathbf{m} \rangle, \langle \boldsymbol{\alpha} \rangle) \leftrightarrow \exists p \sim \mathrm{U}_r^0(a, \langle p, \mathbf{m} \rangle, \langle \boldsymbol{\alpha} \rangle);$$

$$U_r^0(a, \langle \mathbf{m} \rangle) \leftrightarrow \mathrm{U}_r^0(a, \langle \mathbf{m} \rangle, \langle\ \rangle).$$

Recall (II.4.20) that U is universal for a class X iff $\mathrm{U} \in X$ and for every $\mathrm{R} \in X$ there exists an $a \in \omega$ such that

$$\mathrm{R}(\mathbf{m}, \boldsymbol{\alpha}) \leftrightarrow \mathrm{U}(a, \langle \mathbf{m} \rangle, \langle \boldsymbol{\alpha} \rangle).$$

1.8 Arithmetical Indexing Theorem. *For all* $r > 0$,

(i) U_r^0 *is universal for* Σ_r^0;
(ii) $\sim\mathrm{U}_r^0$ *is universal for* Π_r^0;
(iii) U_r^0 *is universal for* $\{R : R \in \Sigma_r^0\}$;
(iv) $\sim U_r^0$ *is universal for* $\{R : R \in \Pi_r^0\}$.

Proof. A straightforward induction on r shows that U^0_r and U^0_r are Σ^0_r and $\sim\mathsf{U}^0_r$ and $\sim U^0_r$ are Π^0_r. For $r = 1$, (i)–(iv) are clear from the proof of Theorem II.4.12. Given (i)–(iv) for r, suppose $\mathsf{R} \in \Sigma^0_{r+1}$, say $\mathsf{R} = \exists^0\mathsf{P}$ with $\mathsf{P} \in \Pi^0_r$. Then for some a,

$$\mathsf{P}(p, \mathbf{m}, \boldsymbol{\alpha}) \leftrightarrow \sim \mathsf{U}^0_r(a, \langle p, \mathbf{m}\rangle, \langle \boldsymbol{\alpha}\rangle),$$

which implies

$$\mathsf{R}(\mathbf{m}, \boldsymbol{\alpha}) \leftrightarrow \exists p \sim \mathsf{U}^0_r(a, \langle p, \mathbf{m}\rangle, \langle \boldsymbol{\alpha}\rangle) \leftrightarrow \mathsf{U}^0_{r+1}(a, \langle \mathbf{m}\rangle, \langle \boldsymbol{\alpha}\rangle).$$

(ii)–(iv) are proved similarly. □

1.9 Arithmetical Hierarchy Theorem. *For all* $r > 0$

(i) $\Sigma^0_r \not\subseteq \Delta^0_r$ *and* $\Pi^0_r \not\subseteq \Delta^0_r$;

(ii) $\Delta^0_{r+1} \not\subseteq \Sigma^0_r \cup \Pi^0_r$.

Proof. Since $U^0_r \in \Sigma^0_r$, it will suffice for (i) to show $U^0_r \notin \Delta^0_r$. Suppose the contrary and let

$$A = \{a : U^0_r(a, \langle a\rangle)\}.$$

Then also $A \in \Delta^0_r$ and in particular $\sim A \in \Sigma^0_r$. Since U^0_r is universal, there is some $b \in \omega$ such that for all m,

$$m \notin A \leftrightarrow U^0_r(b, \langle m\rangle).$$

In particular,

$$b \notin A \leftrightarrow U^0_r(b, \langle b\rangle) \leftrightarrow b \in A,$$

a contradiction.

For (ii), let

$$B = \{2m : m \in A\} \cup \{2m + 1 : m \notin A\}.$$

Clearly B is a union of a Σ^0_r set and a Π^0_r set, so $B \in \Delta^0_{r+1}$. Suppose, however, that $B \in \Sigma^0_r$. Then

$$m \notin A \leftrightarrow 2m + 1 \in B$$

and $\sim A \in \Sigma^0_r$ which was seen to be impossible. If $B \in \Pi^0_r$, then

$$m \in A \leftrightarrow 2m \in B$$

which implies $A \in \Pi^0_r$, hence $\sim A \in \Sigma^0_r$. Therefore $B \notin \Sigma^0_r \cup \Pi^0_r$. □

One application of the Hierarchy Theorem is to obtain precise classifications in the arithmetical hierarchy. Recall that a relation R is (*many-one*) *reducible* to a set A (in symbols, $\mathsf{R} \leqslant A$) iff for some recursive functional F,

$$\mathsf{R}(\mathbf{m}, \boldsymbol{\alpha}) \leftrightarrow \mathsf{F}(\mathbf{m}, \boldsymbol{\alpha}) \in A.$$

Clearly if $A \in \Sigma_r^0$ or Π_r^0, so does R. Suppose that A is such that *all* Σ_r^0 relations on numbers are reducible to A (such an A is called Σ_r^0-*complete*). In particular, if R is a relation which is Σ_r^0 but not Π_r^0, R is reducible to A. Hence $A \notin \Pi_r^0$. For example, let

$$\mathrm{Tot} = \{a : \{a\} \text{ is a total function of rank } 1\}.$$

Since for any a,

$$a \in \mathrm{Tot} \leftrightarrow \forall m\, \exists u\, \mathsf{T}(a, \langle m \rangle, u, \langle\ \rangle),$$

$\mathrm{Tot} \in \Pi_2^0$. In fact, this is an optimal estimate of the complexity of Tot — i.e., $\mathrm{Tot} \notin \Sigma_2^0$. To establish this we show that Tot is Π_2^0-complete.

If R is any Π_2^0 relation and P is a recursive relation such that $\forall p\, \exists q\, P(p, q, \mathbf{m}) \leftrightarrow R(\mathbf{m})$, let f be the partial recursive function defined by

$$f(\mathbf{m}, p) \simeq \text{least } q\,.\, P(p, q, \mathbf{m}).$$

Clearly, $R(\mathbf{m}) \leftrightarrow \forall p\,.\, f(\mathbf{m}, p)\downarrow$. Let b be an index for f. Then by Lemma II.2.5,

$$R(\mathbf{m}) \leftrightarrow Sb_{k-1}(b, \mathbf{m}) \in \mathrm{Tot}$$

and $R \leqslant \mathrm{Tot}$. Some other results of this type are given in the exercises.

To this point it appears that the properties of the classes Σ_r^0, Π_r^0, and Δ_r^0 and the relationships among them strongly resemble the corresponding properties and relationships for Σ_1^0, Π_1^0, and Δ_1^0. In the remainder of this section we shall examine how well this analogy holds up.

First the reduction and separation properties hold for $r > 1$ just as they do for $r = 1$:

1.10 Theorem. *For all* $r > 0$,

(i) Σ_r^0 *has the reduction property but not the separation property*;

(ii) Π_r^0 *has the separation property but not the reduction property.*

Proof. That Σ_r^0 has the reduction property is proved exactly as in Theorem II.4.17 using the closure properties of Theorem 1.5. The other results now follow from Lemmas II.4.19 and II.4.21. □

One flaw in the analogy appears in connection with function quantification. Σ_1^0 is closed under existential function quantification (II.4.14), but Σ_{r+2}^0 is not so closed for any r. Indeed, if R is any relation in $\Pi^0_{r+2} \sim \Delta^0_{r+2}$, say $\mathsf{R}(\mathbf{m}, \boldsymbol{\alpha}) \leftrightarrow \forall p\, \exists q\, \mathsf{S}(p, q, \mathbf{m}, \boldsymbol{\alpha})$ with $\mathsf{S} \in \Pi^0_r$, then

$$\mathsf{R}(\mathbf{m}, \boldsymbol{\alpha}) \leftrightarrow \exists \beta\, \forall p\, \mathsf{S}(p, \beta(p), \mathbf{m}, \boldsymbol{\alpha})$$

so that $\mathsf{R} = \exists^1 \mathsf{P}$ for a relation $\mathsf{P} \in \Pi^0_r \subseteq \Sigma^0_{r+2}$.

For relations on numbers, we shall see that the Σ^0_{r+1} relations are exactly those which are semi-recursive in a certain Σ^0_r set. To investigate this situation and for later use we prove first a general result on substitution of arithmetical functionals in arithmetical relations.

1.11 Arithmetical Substitution Theorem. *For any r and s, any $\mathsf{S} \in \Sigma^0_{r+1}$ (Π^0_{r+1}), and any total functionals $\mathsf{H}_0, \ldots, \mathsf{H}_n \in \Delta^0_s$, if*

$$\mathsf{R}(\mathbf{m}, \boldsymbol{\alpha}) \leftrightarrow \mathsf{S}(\mathbf{m}, \boldsymbol{\alpha}, \lambda p\,.\, \mathsf{H}_0(p, \mathbf{m}, \boldsymbol{\alpha}), \ldots, \lambda p\,.\, \mathsf{H}_n(p, \mathbf{m}, \boldsymbol{\alpha})),$$

then $\mathsf{R} \in \Sigma^0_{r+s}$ (Π^0_{r+s}).

Proof. We proceed by induction on r and take $n = 0$ for simplicity. Suppose first $\mathsf{S} \in \Sigma^0_1$. By Theorem II.4.12 there exists a recursive relation P such that

$$\mathsf{S}(\mathbf{m}, \boldsymbol{\alpha}, \beta) \leftrightarrow \exists p\, \mathsf{P}(\bar{\beta}(p), \mathbf{m}, \boldsymbol{\alpha}).$$

Then

$$\mathsf{R}(\mathbf{m}, \boldsymbol{\alpha}) \leftrightarrow \exists p\, \exists s\, [s = \bar{\mathsf{H}}_0(p, \mathbf{m}, \boldsymbol{\alpha}) \wedge \mathsf{P}(s, \mathbf{m}, \boldsymbol{\alpha})].$$

Since

$$s = \bar{\mathsf{H}}_0(p, \mathbf{m}, \boldsymbol{\alpha}) \leftrightarrow \mathrm{Sq}(s) \wedge \lg(s) = p \wedge (\forall i < p)((s)_i = \mathsf{H}(i, \mathbf{m}, \boldsymbol{\alpha})),$$

the relation inside the brackets is Δ^0_s, so $\mathsf{R} \in \Sigma^0_s$.

If $\mathsf{S} \in \Pi^0_1$, then for some recursive Q,

$$\mathsf{S}(\mathbf{m}, \boldsymbol{\alpha}, \beta) \leftrightarrow \forall p\, \mathsf{Q}(\bar{\beta}(p), \mathbf{m}, \boldsymbol{\alpha})$$

and we use the equivalence

$$\mathsf{R}(\mathbf{m}, \boldsymbol{\alpha}) \leftrightarrow \forall p\, \forall s\, [s = \bar{\mathsf{H}}_0(p, \mathbf{m}, \boldsymbol{\alpha}) \rightarrow \mathsf{Q}(s, \mathbf{m}, \boldsymbol{\alpha})].$$

The induction step is straightforward and is left to the reader. □

The next two results are known jointly as *Post's Theorem.*

1.12 Theorem. *For all r and all $R \subseteq {}^k\omega$,*

(i) *$R \in \Sigma^0_{r+1} \leftrightarrow R$ is semi-recursive in some set $A \in \Sigma^0_r$ (Π^0_r);*

(ii) *$R \in \Delta^0_{r+1} \leftrightarrow R$ is recursive in some set $A \in \Sigma^0_r$ (Π^0_r).*

Proof. The implications $(\leftarrow)$ are immediate from the preceding theorem together with Corollary 1.7(iv). Suppose now that $R = \exists^0 P$ with $P \in \Pi^0_r$. Let $A = \{\langle p, \mathbf{m}\rangle : \sim P(p, \mathbf{m})\}$. Then $A \in \Sigma^0_r$ and $R(\mathbf{m}) \leftrightarrow \exists p\, [\langle p, \mathbf{m}\rangle \notin A]$, so R is semi-recursive in A.

For (ii) ($\to$), if $R \in \Delta^0_{r+1}$ then both R, $\sim R \in \Sigma^0_{r+1}$ so by (i) there are $A, B \in \Sigma^0_r$ such that R is semi-recursive in A and $\sim R$ is semi-recursive in B. Let

$$C = \{2m: m \in A\} \cup \{2m+1: m \in B\}.$$

Then both R and $\sim R$ are semi-recursive in C, hence R is recursive in C (by the relativized version of Corollary II.4.10) and $C \in \Sigma^0_r$. Because a set is recursive in its complement, we may take as well $A \in \Pi^0_r$. □

Let $D_0 = \{0\}$ and $D_{r+1} = (D_r)^{\omega J}$.

1.13 Theorem. *For all r and all $R \subseteq {}^k\omega$,*

(i) $R \in \Sigma^0_r \leftrightarrow R \leqslant D_r$ *(in particular, $D_r \in \Sigma^0_r$);*

(ii) $R \in \Sigma^0_{r+1} \leftrightarrow R$ *is semi-recursive in D_r;*

(iii) $R \in \Delta^0_{r+1} \leftrightarrow R$ *is recursive in D_r.*

Proof. We prove (i)–(iii) simultaneously by induction on r. For each r, (ii) and (iii) follow from (i) by Theorem 1.12. For $r = 0$, D_0 is recursive and hence so is any R which is many-one reducible to it. If $R \in \Sigma^0_0$, R is recursive and $R(\mathbf{m}) \leftrightarrow \mathsf{K}_R(\mathbf{m}) \in D_0$, so $R \leqslant D_0$.

Suppose (i)–(iii) hold for r. Then (i) for $r+1$ is immediate from (ii) for r and Theorem II.5.7. □

There are two natural definitions for the relativized arithmetical hierarchy. We may either set:

$\Sigma^0_0[\boldsymbol{\beta}] = \Pi^0_0[\boldsymbol{\beta}] =$ the class of relations recursive in β;

$\Sigma^0_{r+1}[\boldsymbol{\beta}] = \{\exists^0\mathsf{P}: \mathsf{P} \in \Pi^0_r[\boldsymbol{\beta}]\}$;

etc, as in Definition 1.2,

or we may define, for $r > 0$, $\mathsf{R} \in \Sigma^0_r[\boldsymbol{\beta}]$ ($\Pi^0_r[\boldsymbol{\beta}]$) iff $\mathsf{R}(\mathbf{m}, \boldsymbol{\alpha}) \leftrightarrow \mathsf{S}(\mathbf{m}, \boldsymbol{\alpha}, \boldsymbol{\beta})$ for some $\mathsf{S} \in \Sigma^0_r$ (Π^0_r). Fortunately, these two definitions are equivalent (Exercise 1.24). Note that $\mathsf{R} \in \Delta^0_r[\boldsymbol{\beta}]$ is *not* in general equivalent to the condition that $\mathsf{R}(\mathbf{m}, \boldsymbol{\alpha}) \leftrightarrow \mathsf{S}(\mathbf{m}, \boldsymbol{\alpha}, \boldsymbol{\beta})$ for some $\mathsf{S} \in \Delta^0_r$ (cf. remarks following II.5.1). It is easy to check that R is (semi-) recursive in $\boldsymbol{\beta}$ just in case $\mathsf{R} \in \Delta^0_1[\boldsymbol{\beta}]$ ($\Sigma^0_1[\boldsymbol{\beta}]$). □

With appropriate changes, the results of this section all hold for the relativized arithmetical hierarchy with essentially the same proofs. In particular, there exists for each $r > 0$ a relation $\mathsf{U}^0_r[\boldsymbol{\beta}]$ universal for $\Sigma^0_r[\boldsymbol{\beta}]$ and thus not in $\Delta^0_r[\boldsymbol{\beta}]$. (*Relativized Arithmetical Hierarchy Theorem*)

Since α is recursive in β just in case $\alpha \in \Delta^0_1[\beta]$, it is natural to enquire the properties of the relations '$\alpha \in \Delta^0_r[\beta]$'. Surprisingly they are not transitive for $r > 1$ (Exercise 1.23). The best one can say in general is

1.14 Theorem. *For any r and s, if $\alpha \in \Delta^0_{r+1}[\beta]$ and $\beta \in \Delta^0_s[\gamma]$, then $\alpha \in \Delta^0_{r+s}[\gamma]$.*

Proof. Immediate from the relativized version of Theorem 1.11. □

Finally, motivated by Theorem II.5.5, we define

1.15 Definition. For all $r > 0$,

(i) $\mathbf{\Sigma}_r^0 = \bigcup\{\Sigma_r^0[\beta]: \beta \in {}^\omega\omega\}$;

(ii) $\mathbf{\Pi}_r^0 = \bigcup\{\Pi_r^0[\beta]: \beta \in {}^\omega\omega\}$;

(iii) $\mathbf{\Delta}_r^0 = \mathbf{\Sigma}_r^0 \cap \mathbf{\Pi}_r^0$.

Thus $\mathbf{\Sigma}_1^0$ is the class of open relations, $\mathbf{\Pi}_1^0$ is the class of closed relations, and $\mathbf{\Delta}_1^0$ is the class of closed-open relations.

1.16 Theorem. *For all r and* R, $\mathsf{R} \in \mathbf{\Sigma}_{r+1}^0 \leftrightarrow \mathsf{R}$ *is the union of countably many* $\mathbf{\Pi}_r^0$ *relations of the same rank.*

Proof. Suppose first that $\mathsf{R} \in \mathbf{\Sigma}_{r+1}^0$, so for some β and some $\mathsf{P} \in \mathbf{\Pi}_r^0[\beta]$, $\mathsf{R} = \exists^0\mathsf{P}$. If for each $p \in \omega$,

$$\mathsf{P}_p(\mathbf{m}, \boldsymbol{\alpha}) \leftrightarrow \mathsf{P}(p, \mathbf{m}, \boldsymbol{\alpha}),$$

then $\mathsf{P}_p \in \Pi_r^0[\beta] \subseteq \mathbf{\Pi}_r^0$ and $\mathsf{R} = \bigcup\{\mathsf{P}_p: p \in \omega\}$.

For the converse, let $\mathsf{R} = \bigcup\{\mathsf{P}_p: p \in \omega\}$ with each $\mathsf{P}_p \in \mathbf{\Pi}_r^0$. For each p choose β_p such that $\mathsf{P}_p \in \Pi_r^0[\beta_p]$. Then for each p there exists a_p such that

$$\mathsf{P}_p(\mathbf{m}, \boldsymbol{\alpha}) \leftrightarrow \sim\mathsf{U}_r^0((\gamma(\langle p, 0\rangle))_0, \langle\mathbf{m}\rangle, \langle\boldsymbol{\alpha}, (\gamma)_1^p\rangle)$$

Let γ be a function such that for all p and q, $\gamma(\langle p, q\rangle) = \langle a_p, \beta_p(q)\rangle$. Then for all p,

$$\mathsf{P}_p(\mathbf{m}, \boldsymbol{\alpha}) \leftrightarrow \sim\mathsf{U}_r^0(\gamma(\langle p, 0\rangle)_0, \langle\mathbf{m}\rangle, \langle\boldsymbol{\alpha}, (\gamma)_1^p\rangle)$$

and thus for some relation $\mathsf{Q} \in \Pi_r^0[\gamma]$,

$$\mathsf{P}_p(\mathbf{m}, \boldsymbol{\alpha}) \leftrightarrow \mathsf{Q}(p, \mathbf{m}, \boldsymbol{\alpha}).$$

Then $\mathsf{R} = \exists^0\mathsf{Q} \in \Sigma_{r+1}^0[\gamma] \subseteq \mathbf{\Sigma}_{r+1}^0$. □

Thus, for example, $\mathbf{\Sigma}_2^0$ is the class of countable unions of closed relations, commonly called F_σ in analysis, $\mathbf{\Pi}_2^0$ is the class G_δ, $\mathbf{\Sigma}_3^0 = G_{\delta\sigma}$, etc. These comprise the finite levels of the *Borel Hierarchy* which will be studied further in the next section and in § IV.3. (See also Exercises 1.25–30.)

1.17–1.30 Exercises

1.17. Show that the following relations are arithmetical and estimate the level at which they occur in the arithmetical hierarchy:

$\mathsf{P}_1(p, \gamma) \leftrightarrow \leqslant_\gamma$ is a linear ordering and p is the $\leqslant_\gamma$-least element of $\mathsf{Fld}(\gamma)$;

$P_2(\gamma) \leftrightarrow \leqslant_\gamma$ is a discrete linear ordering (every element of Fld(γ) has an immediate $\leqslant_\gamma$-successor);
$P_3(\gamma) \leftrightarrow \leqslant_\gamma$ is a well-ordering of type ω;
$P_4(n, \gamma) \leftrightarrow \leqslant_\gamma$ is a well-ordering of type $\leqslant \omega \cdot n$.

1.18. Show that for all r, $\{\alpha : \alpha \in \Delta_r^0\}$ is arithmetical and estimate its level in the arithmetical hierarchy.

1.19. Write out the proofs for two of the other parts of Theorem 1.5.

1.20. Let α be a function such that $\{\alpha\}$ is a Π_1^0 set (α is *implicitly* Π_1^0). Show that if Im α is bounded, then α is recursive, but otherwise α may be non-recursive.

1.21. Let $X_0 = \Sigma_1^0$, $\bar{X}_0 = \Pi_1^0$, and for all r,

$$X_{r+1} = \{R \cup S\colon R \in X_r, S \in \bar{X}_r, \text{ R and S of the same rank}\}$$

and

$$\bar{X}_{r+1} = \{R \cap S\colon R \in X_r, S \in \bar{X}_r, \text{ R and S of the same rank}\}.$$

Show that for all r,
(i) $R \in X_r \leftrightarrow {\sim} R \in \bar{X}_r$;
(ii) $X_r \cup \bar{X}_r \subseteq X_{r+1} \cap \bar{X}_{r+1}$;
(iii) $X_r \not\subseteq \bar{X}_r$ and $\bar{X}_r \not\subseteq X_r$;
(iv) $\bigcup \{X_r : r \in \omega\}$ is a proper subclass of Δ_2^0.

1.22. From Theorem 1.11 we see that if $\beta \in \Delta_s^0$, then $\Sigma_{r+1}^0[\beta] \subseteq \Sigma_{r+s}^0$. Is this ever an equality?

1.23. Let $\alpha \leqslant_r^0 \beta \leftrightarrow \alpha \in \Delta_r^0[\beta]$. Show that for $r > 1$, $\leqslant_r^0$ is not transitive.

1.24. Show that the two characterizations of $\boldsymbol{\Sigma}_{r+1}^0[\boldsymbol{\beta}]$ are equivalent.

1.25. Verify that
(i) R is semi-recursive in $\boldsymbol{\beta} \leftrightarrow R \in \Sigma_1^0[\boldsymbol{\beta}]$;
(ii) $\boldsymbol{\Sigma}_{r+1}^0 = \{\exists^0 P\colon P \in \boldsymbol{\Pi}_r^0\}$;
(iii) $\boldsymbol{\Delta}_r^0 = \bigcup \{\Delta_r^0[\beta]\colon \beta \in {}^\omega\omega\}$.

1.26. Show that for all $r \geqslant 1$, $\boldsymbol{\Sigma}_r^0$ and $\boldsymbol{\Pi}_r^0$ are parametrizable (cf. Exercise II.5.11).

1.27. Show that for all $r > 0$ and all relations R,
(i) $R \in \boldsymbol{\Sigma}_r^0 \leftrightarrow {\sim} R \in \boldsymbol{\Pi}_r^0$;

(ii) $\mathbf{\Sigma}_r^0 \cup \mathbf{\Pi}_r^0 \subseteq \mathbf{\Delta}_{r+1}^0$;
(iii) $\mathbf{\Sigma}_r^0 \not\subseteq \mathbf{\Delta}_r^0$ and $\mathbf{\Pi}_r^0 \not\subseteq \mathbf{\Delta}_r^0$;
(iv) $\mathbf{\Delta}_{r+1}^0 \not\subseteq \mathbf{\Sigma}_r^0 \cup \mathbf{\Pi}_r^0$.

1.28. Show that the Arithmetical Hierarchy Theorem may be improved to: for all $r > 0$, $\Sigma_r^0 \not\subseteq \mathbf{\Delta}_r^0$ and $\Pi_r^0 \not\subseteq \mathbf{\Delta}_r^0$.

1.29. Show that for all r,
(i) $\mathbf{\Sigma}_r^0$ has the reduction property but not the separation property;
(ii) $\mathbf{\Pi}_r^0$ has the separation property but not the reduction property.

1.30. Let $\mathsf{V}_1^0 = \mathsf{U}_1^0$, and for $r > 0$,

$$\mathsf{V}_{r+1}^0(a, \langle \mathbf{m} \rangle, \langle \boldsymbol{\alpha} \rangle) \leftrightarrow \exists p \sim \mathsf{V}_r^0(\{a\}(p), \langle \mathbf{m} \rangle, \langle \boldsymbol{\alpha} \rangle).$$

Set

$$\mathsf{P}_a^r(\mathbf{m}, \boldsymbol{\alpha}) \leftrightarrow \mathsf{V}_r^0(a, \langle \mathbf{m} \rangle, \langle \boldsymbol{\alpha} \rangle).$$

Show that for all $r > 0$,

$$\Sigma_r^0 = \{\mathsf{P}_a^r\colon a \in \omega\}.$$

(Note: P_a^{r+1} is the *recursive union* of the relations $\sim\mathsf{P}_{\{a\}(p)}^r$ so that Σ_{r+1}^0 consists exactly of recursive unions of Π_r^0 relations. The arithmetical hierarchy is thus sometimes called the (finite) *effective Borel hierarchy*. Cf. § IV.4.)

2. The Analytical Hierarchy

We take up next the relations which are obtained from the recursive relations by application of both number and function quantifiers. The basic structure we develop in this section is parallel in most points to that of § 1, but we shall see in Part B that these are indeed much more complex relations. Proofs will be omitted when they are very similar to corresponding proofs in § 1.

2.1 Definition. The class of *analytical relations* is the smallest class of relations containing the arithmetical relations and closed under function quantification ($\exists^1$ and $\forall^1$).

2.2 Definition (The Analytical Hierarchy). For all r,
(i) $\Sigma_0^1 = \Pi_0^1 =$ the class of arithmetical relations;

(ii) $\Sigma^1_{r+1} = \{\exists^1 \mathsf{P}: \mathsf{P} \in \Pi^1_r\}$;
(iii) $\Pi^1_{r+1} = \{\forall^1 \mathsf{P}: \mathsf{P} \in \Sigma^1_r\}$;
(iv) $\Delta^1_r = \Sigma^1_r \cap \Pi^1_r$;
(v) $\Delta^1_{(\omega)} = \bigcup \{\Sigma^1_r \cup \Pi^1_r: r \in \omega\}$.

Clearly every relation in $\Delta^1_{(\omega)}$ is analytical; the converse inclusion follows from Theorem 2.6 below. As before, we write $\mathsf{F} \in \Sigma^1_r$, etc., to mean $\mathrm{Gr}_{\mathsf{F}} \in \Sigma^1_r$, etc., and use the terms Σ^1_r, Π^1_r, and Δ^1_r as adjectives.

2.3 Examples. Recall that

$$\mathsf{W} = \{\gamma: \leqslant_\gamma \text{ is a well-ordering}\}.$$

Then using (4′) of I.1.6,

$$\begin{aligned} \gamma \in \mathsf{W} \leftrightarrow {\leqslant_\gamma} & \text{ is a linear ordering} \wedge \\ & \forall\alpha\,(\forall m\,[\alpha(m+1) \leqslant_\gamma \alpha(m)] \rightarrow \exists m\,.\,\alpha(m) \leqslant_\gamma \alpha(m+1)) \\ & \leftrightarrow \forall\alpha\,[\leqslant_\gamma \text{ is a linear ordering} \wedge (\forall m\,.\,\gamma(\langle\alpha(m+1), \alpha(m)\rangle) = 0 \\ & \qquad \rightarrow \exists m\,.\,\gamma(\langle\alpha(m), \alpha(m+1)\rangle) = 0)]. \end{aligned}$$

The part inside the brackets is easily seen to be arithmetical so that $\mathsf{W} \in \Pi^1_1$.

Let

$$\begin{aligned} \gamma \preccurlyeq \delta \leftrightarrow {\leqslant_\gamma} & \text{ and } \leqslant_\delta \text{ are linear orderings and} \\ & \leqslant_\gamma \text{ is isomorphic to a subordering of } \leqslant_\delta. \end{aligned}$$

Then the second conjunct is equivalent to

$$\exists\alpha\,\forall p\,\forall q\,[\alpha \text{ is 1–1 on } \mathrm{Fld}(\gamma) \wedge (p \leqslant_\gamma q \rightarrow \alpha(p) \leqslant_\delta \alpha(q))]$$

so that the relation $\preccurlyeq$ is Σ^1_1. Note that if $\gamma, \delta \in \mathsf{W}$,

$$\gamma \preccurlyeq \delta \leftrightarrow \|\gamma\| \leqslant \|\delta\|.$$

A similar argument shows that

$$\{(R, S): \text{the relational structures } (\omega, R) \text{ and } (\omega, S) \text{ are isomorphic}\}$$

is Σ^1_1.

Suppose A is an arithmetical set of functions with just one element, α. Then α is Δ^1_1 since for any m and n

$$\alpha(m)=n \leftrightarrow \exists\beta\,[\beta\in\mathsf{A} \wedge \beta(m)=n]$$
$$\leftrightarrow \forall\beta\,[\beta\in\mathsf{A}\rightarrow \beta(m)=n].$$

Let Γ be a monotone operator over ω such that the relation '$m \in \Gamma(\{n\colon \alpha(n)=0\})$' is arithmetical. By Theorem I.3.3,

$$m\in\bar{\Gamma}\leftrightarrow\forall B\,[\Gamma(B)\subseteq B\rightarrow m\in B]$$
$$\leftrightarrow\forall\alpha\,[\forall p\,(p\in\Gamma(\{n\colon\alpha(n)=0\})\rightarrow\alpha(p)=0)\rightarrow\alpha(m)=0].$$

Hence $\bar{\Gamma}\in\Pi^1_1$.

Two other examples which will be treated in § V.2 but may serve to orient the reader familiar with other parts of logic are:

$\{(m, R)$: m is the Gödel number of a formula valid in $(\omega, R)\}$ is Δ^1_1;

$\{\alpha$: α is constructible (in the sense of Gödel)$\}$ is Σ^1_2.

2.4 Lemma. *For all r and* R,

$$\mathsf{R}\in\Sigma^1_r\leftrightarrow\sim\mathsf{R}\in\Pi^1_r\quad\textit{and}\quad\mathsf{R}\in\Pi^1_r\leftrightarrow\sim\mathsf{R}\in\Sigma^1_r.\quad\square$$

2.5 Theorem. *The classes of the analytical hierarchy have the following closure properties for all r:*

	Σ^1_r	Π^1_r	Δ^1_r	$\Delta^1_{(\omega)}$
Composition and substitution with recursive functionals	$\surd$	$\surd$	$\surd$	$\surd$
Finite union and intersection	$\surd$	$\surd$	$\surd$	$\surd$
Expansion	$\surd$	$\surd$	$\surd$	$\surd$
Complementation			$\surd$	$\surd$
Bounded quantification ($\exists^0_<$ and $\forall^0_<$)	$\surd$	$\surd$	$\surd$	$\surd$
Existential number quantification ($\exists^0$)	$\surd$	$\surd$	$\surd$	$\surd$
Universal number quantification ($\forall^0$)	$\surd$	$\surd$	$\surd$	$\surd$
Existential function quantification ($\exists^1$)	$\surd(r>0)$			$\surd$
Universal function quantification ($\forall^1$)		$\surd(r>0)$		$\surd$

Proof. In the proofs we use the following equivalences and their duals (obtained by negating both sides):

$$\exists\beta\, \mathsf{P}(\mathbf{m}, \boldsymbol{\alpha}, \beta) \vee \exists\gamma\, \mathsf{Q}(\mathbf{m}, \boldsymbol{\alpha}, \gamma) \leftrightarrow \exists\beta\, [\mathsf{P}(\mathbf{m}, \boldsymbol{\alpha}, \beta) \vee \mathsf{Q}(\mathbf{m}, \boldsymbol{\alpha}, \beta)];$$

$$\exists\beta\, \mathsf{P}(\mathbf{m}, \boldsymbol{\alpha}, \beta) \wedge \exists\gamma\, \mathsf{Q}(\mathbf{m}, \boldsymbol{\alpha}, \gamma) \leftrightarrow \exists\beta\, [\mathsf{P}(\mathbf{m}, \boldsymbol{\alpha}, (\beta)_0) \wedge \mathsf{Q}(\mathbf{m}, \boldsymbol{\alpha}, (\beta)_1)];$$

$$(\exists q < r)\exists\beta\, \mathsf{P}(q, r, \mathbf{m}, \boldsymbol{\alpha}, \beta) \leftrightarrow \exists\beta\, (\exists q < r)\mathsf{P}(q, r, \mathbf{m}, \boldsymbol{\alpha}, \beta);$$

$$(\forall q < r)\exists\beta\, \mathsf{P}(q, r, \mathbf{m}, \boldsymbol{\alpha}, \beta) \leftrightarrow \exists\beta\, (\forall q < r)\mathsf{P}(q, r, \mathbf{m}, \boldsymbol{\alpha}, (\beta)_q);$$

$$\exists p\, \exists\beta\, \mathsf{P}(p, \mathbf{m}, \boldsymbol{\alpha}, \beta) \leftrightarrow \exists\beta\, \mathsf{P}(\beta(0), \mathbf{m}, \boldsymbol{\alpha}, \lambda q \,.\, \beta(q+1));$$

$$\forall p\, \exists\beta\, \mathsf{P}(p, \mathbf{m}, \boldsymbol{\alpha}, \beta) \leftrightarrow \exists\beta\, \forall p\, \mathsf{P}(p, \mathbf{m}, \boldsymbol{\alpha}, (\beta)^p);$$

$$\exists\beta\, \exists\gamma\, \mathsf{P}(\mathbf{m}, \boldsymbol{\alpha}, \beta, \gamma) \leftrightarrow \exists\beta\, \mathsf{P}(\mathbf{m}, \boldsymbol{\alpha}, (\beta)_0, (\beta)_1).$$

For example, we prove by induction on r that Σ^1_r and Π^1_r are closed under $\exists^0$ and $\forall^0$. For $r = 0$ this is contained in Theorem 1.5. Suppose it holds for r and R is any Σ^1_{r+1} relation, say

$$\mathsf{R}(p, \mathbf{m}, \boldsymbol{\alpha}) \leftrightarrow \exists\beta\, \mathsf{P}(p, \mathbf{m}, \boldsymbol{\alpha}, \beta)$$

with $\beta \in \Pi^1_r$. Using the fifth and sixth equivalences and the induction hypothesis, it follows that the relations $\exists^0\mathsf{R}$ and $\forall^0\mathsf{R}$ are also Σ^1_r. The result for Π^1_r follows by dualization. □

2.6 Corollary. *For all r, $\Sigma^1_r \cup \Pi^1_r \subseteq \Delta^1_{r+1}$.* □

2.7 Corollary. *For any $r > 0$, $\mathsf{F}\colon {}^{k,l}\omega \to \omega$, $F\colon {}^{k}\omega \to \omega$, $\alpha \in {}^{\omega}\omega$, and $\mathsf{R} \subseteq {}^{k,l}\omega$*

(i) *if $\mathsf{F} \in \Sigma^1_r \cup \Pi^1_r$ and F is total, then also $\mathsf{F} \in \Delta^1_r$;*
(ii) *if $\mathsf{F} \in \Sigma^1_r$ (Π^1_r), then* $\operatorname{Dm} \mathsf{F} \in \Sigma^1_r$ (Π^1_r);
(iii) *if $F \in \Sigma^1_r$ (Π^1_r), then* $\operatorname{Im} F \in \Sigma^1_r$ (Π^1_r);
(iv) *if $\mathsf{F} \in \Sigma^1_r$, then* $\operatorname{Im} \mathsf{F} \in \Sigma^1_r$;
(v) *if $\mathsf{R} \in \Sigma^1_r \cup \Pi^1_r$, then $\mathsf{K}_{\mathsf{R}} \in \Delta^1_{r+1}$;*
(vi) *Σ^1_r (Π^1_r) is closed under composition with Σ^1_r (Π^1_r) functionals;*
(vii) *$\alpha \in \Delta^1_r$ iff $\{\alpha\} \in \Delta^1_r$ iff $\{\alpha\} \in \Sigma^1_r$.*

Proof. The proofs are in general like those for Corollary 1.8 and we leave them to Exercise 2.21. □

We may picture the arithmetical and analytical hierarchies together as follows:

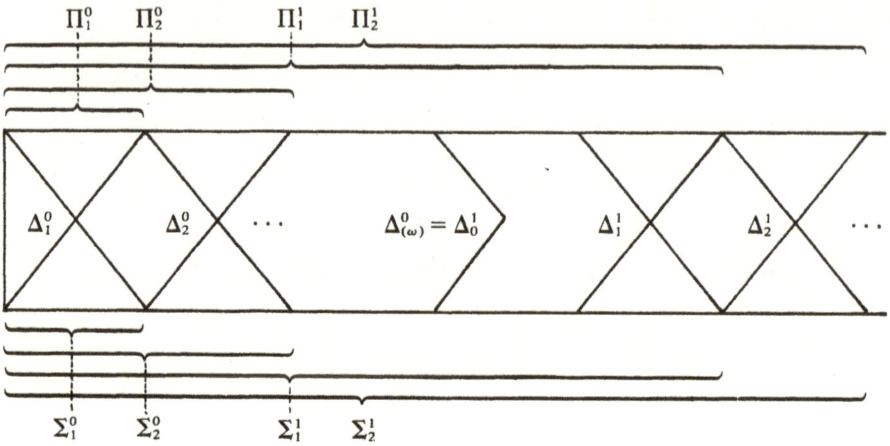

We want next to show that the indicated inclusions are proper. We postpone to the next section (3.8) the proof that $\Delta^0_{(\omega)} \neq \Delta^1_1$. For the others we proceed as for the arithmetical hierarchy to define relations universal for Σ^1_r and Π^1_r. However, we shall need one preliminary result.

2.8 Lemma. $\Sigma^1_1 = \{\exists^1 \mathsf{P} : \mathsf{P} \in \Pi^0_1\}$.

Proof. The inclusion ($\supseteq$) is immediate from the definitions. On the other hand, it follows from the last three equivalences used in the proof of Theorem 2.5 that $\{\exists^1 \mathsf{P} : \mathsf{P} \in \Pi^0_1\}$ is closed under $\exists^0$, $\forall^0$, and $\exists^1$. Hence this set includes all arithmetical relations and thus also Σ^1_1. □

We now set, for $r \geq 1$,

$$\mathsf{U}^1_1(a, \langle \mathbf{m} \rangle, \langle \boldsymbol{\alpha} \rangle) \leftrightarrow \exists \beta \sim \mathsf{U}^0_1(a, \langle \mathbf{m} \rangle, \langle \boldsymbol{\alpha}, \beta \rangle);$$

$$\mathsf{U}^1_{r+1}(a, \langle \mathbf{m} \rangle, \langle \boldsymbol{\alpha} \rangle) \leftrightarrow \exists \beta \sim \mathsf{U}^1_r(a, \langle \mathbf{m} \rangle, \langle \boldsymbol{\alpha}, \beta \rangle);$$

$$U^1_r(a, \langle \mathbf{m} \rangle) \leftrightarrow U^1_r(a, \langle \mathbf{m} \rangle, \langle\ \rangle).$$

2.9 Analytical Indexing Theorem. *For all $r > 0$,*

(i) U^1_r *is universal for* Σ^1_r;
(ii) $\sim\mathsf{U}^1_r$ *is universal for* Π^1_r;
(iii) U^1_r *is universal for* $\{R : R \in \Sigma^1_r\}$;
(iv) $\sim U^1_r$ *is universal for* $\{R : R \in \Pi^1_r\}$.

Proof. We first prove by induction that for all $r > 0$, $\mathsf{U}^1_r \in \Sigma^1_r$ and $\sim\mathsf{U}^1_r \in \Pi^1_r$. By the convention discussed following Theorem II.1.8, the definition of U^1_1 given above is an abbreviation for:

$$\mathsf{U}^1_1(a, s, \gamma) \leftrightarrow \exists \beta \sim \mathsf{U}^0_1(a, s, \lambda m(\gamma(m) * (\beta(m)))).$$

The part following the quantifier is Π_1^0 by the Arithmetical Substitution Theorem (1.11) and thus $\mathsf{U}_1^1 \in \Sigma_1^1$. Similarly,

$$\mathsf{U}_{r+1}^1(a, s, \gamma) \leftrightarrow \exists\beta \sim \mathsf{U}_r^1(a, s, \lambda m(\gamma(m) * (\beta(m)))).$$

Thus under the induction hypothesis that $\sim\mathsf{U}_r^1 \in \Pi_r^1$ we have, using Lemma 2.5, that $\mathsf{U}_{r+1}^1 \in \Sigma_{r+1}^1$.

If R is any Σ_1^1 relation, then by Lemma 2.8, $\mathsf{R} = \exists^1\mathsf{P}$ for some $\mathsf{P} \in \Pi_1^0$. Since $\sim\mathsf{U}_1^0$ is universal for Π_1^0, there is a number a such that

$$\mathsf{P}(\mathbf{m}, \boldsymbol{\alpha}, \beta) \leftrightarrow \sim\mathsf{U}_1^0(a, \langle\mathbf{m}\rangle, \langle\boldsymbol{\alpha}, \beta\rangle).$$

Then

$$\mathsf{R}(\mathbf{m}, \boldsymbol{\alpha}) \leftrightarrow \mathsf{U}_1^1(a, \langle\mathbf{m}\rangle, \langle\boldsymbol{\alpha}\rangle).$$

The remainder of the proof is by induction on r just as in the arithmetical case. □

2.10 Analytical Hierarchy Theorem. *For all* $r > 0$

(i) $\Sigma_r^1 \not\subseteq \Delta_r^1$ *and* $\Pi_r^1 \not\subseteq \Delta_r^1$;

(ii) $\Delta_{r+1}^1 \not\subseteq \Sigma_r^1 \cup \Pi_r^1$.

Proof. Just as for the arithmetical hierarchy. □

From this point on the theory of the analytical hierarchy begins to diverge from that of the arithmetical hierarchy. The question of which of the classes Σ_r^1 and Π_r^1 have the reduction and separation properties is much more complicated. Results for $r = 1$ and $r = 2$ will be obtained in §§ IV.1 and V.1, respectively, but for $r \geqslant 3$ these questions cannot be decided on the basis of the usual axioms for set theory (cf. end of § V.3).

2.11 Analytical Substitution Theorem. *For any* r, *any* $\mathsf{S} \in \Sigma_r^1$ (Π_r^1), *and any partial functional* $\mathsf{H} \in \Sigma_r^1$ (Π_r^1), *if*

$$\mathsf{R}(\mathbf{m}, \boldsymbol{\alpha}) \leftrightarrow \mathsf{S}(\mathbf{m}, \boldsymbol{\alpha}, \lambda p\,.\,\mathsf{H}(p, \mathbf{m}, \boldsymbol{\alpha})),$$

then also $\mathsf{R} \in \Sigma_r^1$ (Π_r^1).

Proof. For $r = 0$, suppose S and H are both arithmetical. H may be extended to a total arithmetical functional I:

$$\mathsf{I}(p, \mathbf{m}, \boldsymbol{\alpha}) \simeq n \leftrightarrow \mathsf{H}(p, \mathbf{m}, \boldsymbol{\alpha}) \simeq n \vee (\neg\exists q\,.\,\mathsf{H}(p, \mathbf{m}, \boldsymbol{\alpha}) \simeq q \wedge n = 0).$$

Then

$$\mathsf{R}(\mathbf{m}, \boldsymbol{\alpha}) \leftrightarrow \forall p\, \exists n\,.\, \mathsf{H}(p, \mathbf{m}, \boldsymbol{\alpha}) \simeq n \wedge \mathsf{S}(\mathbf{m}, \boldsymbol{\alpha}, \lambda p\,.\, \mathsf{I}(p, \mathbf{m}, \boldsymbol{\alpha})),$$

and R is arithmetical by the Arithmetical Substitution Theorem 1.11.

Suppose $r > 0$ and $\mathsf{H}, \mathsf{S} \in \Sigma^1_r$. Then the result follows from the equivalence:

$$\mathsf{R}(\mathbf{m}, \boldsymbol{\alpha}) \leftrightarrow \exists \beta\, (\forall p\, [\mathsf{H}(p, \mathbf{m}, \boldsymbol{\alpha}) = \beta(p)] \wedge \mathsf{S}(\mathbf{m}, \boldsymbol{\alpha}, \beta)).$$

If $\mathsf{H}, \mathsf{S} \in \Pi^1_r$ we have

$$\mathsf{R}(\mathbf{m}, \boldsymbol{\alpha}) \leftrightarrow \forall p\, \exists n\, [\mathsf{H}(p, \mathbf{m}, \boldsymbol{\alpha}) \simeq n] \wedge$$
$$\forall \beta\, (\forall p\, n [\mathsf{H}(p, \mathbf{m}, \boldsymbol{\alpha}) = n \rightarrow \beta(p) = n] \rightarrow \mathsf{S}(\mathbf{m}, \boldsymbol{\alpha}, \beta)). \quad \square$$

The relativized analytical hierarchy may be defined in two equivalent ways:

$\Sigma^1_0[\boldsymbol{\beta}] = \Pi^1_0[\boldsymbol{\beta}] =$ the class of relations arithmetical in $\boldsymbol{\beta}$;

$\Sigma^1_{r+1}[\boldsymbol{\beta}] = \{\exists^1 \mathsf{P} \colon \mathsf{P} \in \Pi^1_r[\boldsymbol{\beta}]\}$;

etc. as in Definition 2.2;

or, for $r > 0$, $\mathsf{R} \in \Sigma^1_r[\boldsymbol{\beta}]$ ($\Pi^1_r[\boldsymbol{\beta}]$) iff $\mathsf{R}(\mathbf{m}, \boldsymbol{\alpha}) \leftrightarrow \mathsf{S}(\mathbf{m}, \boldsymbol{\alpha}, \boldsymbol{\beta})$ for some $\mathsf{S} \in \Sigma^1_r$ (Π^1_r). The proof that these are equivalent is the same as in the arithmetical case (Exercise 1.24).

Again there is no problem in extending all the results of this section to the relativized hierarchy.

2.12 Corollary. *For all r and β, if $\beta_0, \ldots, \beta_n \in \Delta^1_r$, then*

$$\Sigma^1_r[\boldsymbol{\beta}] = \Sigma^1_r, \quad \Pi^1_r[\boldsymbol{\beta}] = \Pi^1_r, \quad \textit{and} \quad \Delta^1_r[\boldsymbol{\beta}] = \Delta^1_r.$$

Proof. Immediate from Theorem 2.11. $\square$

2.13 Corollary. *For all r, α, β, and γ, if $\alpha \in \Delta^1_r[\beta]$ and $\beta \in \Delta^1_r[\gamma]$, then $\alpha \in \Delta^1_r[\gamma]$.* $\square$

Thus the relation "Δ^1_r in" is transitive and one may consider Δ^1_r degrees analogous to the (ordinary) degrees of § II.5:

$$\Delta^1_r\text{-dg}(\alpha) = \{\beta \colon \alpha \in \Delta^1_r[\beta] \text{ and } \beta \in \Delta^1_r[\alpha]\}.$$

The Δ^1_1 degrees are called *hyperdegrees* and are considered further in § IV.2.

There are several ways one might hope to extend Post's Theorem to the analytical hierarchy. The most direct generalization is

(1) $R \in \Delta^1_{r+1} \overset{?}{\leftrightarrow} R$ recursive in some set $A \in \Sigma^1_r$.

It could be argued that a relationship stronger than "recursive in" is needed here. Since "recursive in" coincides with "Δ^0_1 in", a natural choice is "Δ^1_1 in" or even "Δ^1_r in":

(2) $R \in \Delta^1_{r+1} \overset{?}{\leftrightarrow} R \in \Delta^1_1[A]$ for some $A \in \Sigma^1_r$;

(3) $R \in \Delta^1_{r+1} \overset{?}{\leftrightarrow} R \in \Delta^1_r[A]$ for some $A \in \Sigma^1_r$.

The implication ($\leftarrow$) of (3), and hence of (1) and (2) is immediate from Corollary 2.12: if $A \in \Sigma^1_r$, $\mathsf{K}_A \in \Delta^1_{r+1}$ and $R \in \Delta^1_r[A] \subseteq \Delta^1_{r+1}[A]$, so $R \in \Delta^1_{r+1}$. However the implication ($\rightarrow$) of (3), and hence of (1) and (2), is false for all $r > 0$ (also for $r = 0$ but for a different reason — see Corollary 3.8 below). By the relativized analytical hierarchy theorem there exists a relation $R \in \Sigma^1_r[U^1_r] \sim \Delta^1_r[U^1_r]$. Since every $A \in \Sigma^1_r$ is recursive in U^1_r, $R \notin \Delta^1_r[A]$ for any $A \in \Sigma^1_r$. However, $\Sigma^1_r[U^1_r] \subseteq \Delta^1_{r+1}[U^1_r]$ and $U^1_r \in \Sigma^1_r \subseteq \Delta^1_{r+1}$, so by Corollary 2.12, $R \in \Delta^1_{r+1}$. Thus there is no analogue of Post's Theorem for the analytical hierarchy.

2.14 Definition. For all $r > 0$,
(i) $\boldsymbol{\Sigma}^1_r = \bigcup \{\Sigma^1_r[\beta]: \beta \in {}^\omega\omega\}$;
(ii) $\boldsymbol{\Pi}^1_r = \bigcup \{\Pi^1_r[\beta]: \beta \in {}^\omega\omega\}$;
(iii) $\boldsymbol{\Delta}^1_r = \boldsymbol{\Sigma}^1_r \cap \boldsymbol{\Pi}^1_r$.

The classes $\boldsymbol{\Sigma}^1_r$ and $\boldsymbol{\Pi}^1_r$ comprise what is known as the *projective hierarchy*, and were known and studied long before the invention of recursion theory. It follows easily from Lemma 2.8 that $\boldsymbol{\Sigma}^1_1 = \{\exists^1 \mathsf{P}: \mathsf{P}$ is closed$\}$. That is, the $\boldsymbol{\Sigma}^1_1$ relations are exactly the *projections* (with respect to a function coordinate) of closed relations. Similarly the $\boldsymbol{\Sigma}^1_{r+1}$ relations are exactly the projections of $\boldsymbol{\Pi}^1_r$ relations. The class of *projective relations* ($= \boldsymbol{\Delta}^1_{(\omega)}$) is the smallest class containing the closed relations and closed under projection ($\exists^1$) and complementation.

Contrasting with Theorem 1.16 we have

2.15 Theorem. *For all $r > 0$, $\boldsymbol{\Sigma}^1_r$ and $\boldsymbol{\Pi}^1_r$ are closed under countable unions and intersections of relations of the same rank.*

Proof. Suppose $\mathsf{P}_p \in \boldsymbol{\Sigma}^1_r$ for all $p \in \omega$ and let

$$\mathsf{Q}(p, \mathbf{m}, \boldsymbol{\alpha}) \leftrightarrow \mathsf{P}_p(\mathbf{m}, \boldsymbol{\alpha}).$$

Then as in the proof of Theorem 1.16, $\mathsf{Q} \in \boldsymbol{\Sigma}_r^1$. Since

$$\bigcup \{\mathsf{P}_p : p \in \omega\} = \exists^0 \mathsf{Q} \quad \text{and} \quad \bigcap \{\mathsf{P}_p : p \in \omega\} = \forall^0 \mathsf{Q},$$

these are both in $\boldsymbol{\Sigma}_r^1$ by the relativized version of Theorem 2.6. □

The class Bo of *Borel relations* is the smallest class containing the open relations and closed under countable unions and intersections of relations of the same rank.

2.16 Corollary. $\mathrm{Bo} \subseteq \boldsymbol{\Delta}_1^1$. □

We will prove in § IV.3 that in fact $\mathrm{Bo} = \boldsymbol{\Delta}_1^1$. This classical result of descriptive set theory is paradigmatic for many of the other results of later chapters.

2.17–2.23 Exercises

2.17. Show that $\{\alpha : \alpha \in \Delta_{(\omega)}^0\} \in \Delta_1^1$ (cf. Exercise 1.18 and Corollary 4.21 below).

2.18. We might have defined the analytical hierarchy by quantifying sets rather than functions. Let $\dot{\exists}^1 \mathsf{P}$ be the relation $\exists B\, \mathsf{P}(\mathbf{m}, \boldsymbol{\alpha}, B)$ and define analogously $\dot{\forall}^1$, $\dot{\Sigma}_r^1$, etc. Show that

$$\dot{\Sigma}_r^1 = \Sigma_r^1, \text{ but that corresponding to Lemma 2.8 we have}$$
$$\{\dot{\exists}^1 \mathsf{P} : \mathsf{P} \in \Pi_1^0\} \subsetneqq \{\dot{\exists}^1 \mathsf{P} : \mathsf{P} \in \Pi_2^0\} = \dot{\Sigma}_1^1.$$

2.19. For any countable indexed family of relations $\langle \mathsf{P}_s : s \in \omega \rangle$, let $\mathscr{A}\langle \mathsf{P}_s : s \in \omega \rangle = \bigcup \{ \bigcap \{\mathsf{P}_{\bar{\beta}(p)} : p \in \omega\} : \beta \in {}^\omega\omega\}$. Show that for any R, the following are equivalent:

(i) $\mathsf{R} \in \boldsymbol{\Sigma}_1^1$;
(ii) $\mathsf{R} = \mathscr{A}\langle \mathsf{P}_s : s \in \omega \rangle$ for some $\mathsf{P}_s \in \boldsymbol{\Delta}_1^0$;
(iii) $\mathsf{R} = \exists^1 \mathsf{P}$ for some $\mathsf{P} \in \boldsymbol{\Delta}_1^1$.

2.20. The operation $\mathscr{A}$ of the preceding exercise may also be regarded as a quantifier:

$$(\mathscr{A}\mathsf{P})(\mathbf{m}, \boldsymbol{\alpha}) \leftrightarrow \exists \beta\, \forall p\, \mathsf{P}(\bar{\beta}(p), \mathbf{m}, \boldsymbol{\alpha}).$$

Show that for all r and P

(i) if $r \geqslant 1$ and $\mathsf{P} \in \Sigma_r^1$ ($\boldsymbol{\Sigma}_r^1$), then $\mathscr{A}\mathsf{P} \in \Sigma_r^1$ ($\boldsymbol{\Sigma}_r^1$);
(ii) if $r \geqslant 2$ and $\mathsf{P} \in \Pi_r^1$ ($\boldsymbol{\Pi}_r^1$), then $\mathscr{A}\mathsf{P} \in \Pi_r^1$ ($\boldsymbol{\Pi}_r^1$).

2.21. Prove Corollary 2.7.

2.22. Prove that the class of Borel relations is closed under complementation.

2.23. Prove the following *Strong Hierarchy Theorem*: for all $r>0$, $\Sigma_r^1 \not\subseteq \Delta_r^1$.

2.24 Notes. The arithmetical hierarchy was developed independently by Kleene [1943] and Mostowski [1946] (because of the Second World War Mostowski in Warsaw did not see Kleene's paper until his manuscript was finished). The analytical hierarchy was first studied in Kleene [1955b]. Kleene's approach was motivated by Gödel's results on incompleteness, whereas Mostowski saw the arithmetical hierarchy as analogous to the projective hierarchy, with "recursive" corresponding to "Borel" and existential number quantification ($\exists^0$) corresponding to projection ($\exists^1$). When Kleene [1950] showed that Σ_1^0 does not have the separation property, Mostowski pointed out that this is a flaw in the analogy, since Σ_1^1 does have the separation property. Addison, in his thesis (announced in Addison [1955]) proposed that a better correspondence is that between "recursive" and "closed-open" with $\exists^0$ corresponding to countable union. The unified approach to these hierarchies grew out of this analogy and is also largely due to Addison.

The technique of proof of Theorem 2.11 is due to Shoenfield [1962].

3. *Inductive Definability*

To this point we have used inductive definitions mainly as a tool. We begin now to consider their use as a measure of complexity. We shall be interested here in the relationship between the complexity of Γ and that of $\bar{\Gamma}$ as measured by their classifications in the arithmetical and analytical hierarchies.

For each $\alpha \in {}^\omega\omega$ we put $Z_\alpha = \{p\colon \alpha(p)=0\}$. Then for any inductive operator Γ over ω we define a relation P_Γ by

$$\mathsf{P}_\Gamma(m,\alpha) \leftrightarrow m \in \Gamma(Z_\alpha).$$

We write $\Gamma \in \Sigma_r^i$, etc. to mean $\mathsf{P}_\Gamma \in \Sigma_r^i$, etc. We showed in Example 2.3 that for monotone $\Gamma \in \Delta_{(\omega)}^0$, $\bar{\Gamma} \in \Pi_1^1$. The same proof establishes

3.1 Theorem. *For any $r>0$ and any monotone operator Γ over ω,*

$$\Gamma \in \Pi_r^1 \to \bar{\Gamma} \in \Pi_r^1.$$

Proof. If Γ is monotone, we have

$$m \in \bar{\Gamma} \leftrightarrow \forall B\,[\Gamma(B) \subseteq B \rightarrow m \in B]$$
$$\leftrightarrow \forall \alpha\,[\forall p\,(\mathsf{P}_\Gamma(p, \alpha) \rightarrow \alpha(p) = 0) \rightarrow \alpha(m) = 0].$$

An easy calculation based on the techniques of the preceding section shows that if $\mathsf{P}_\Gamma \in \Pi^1_r$, so is $\bar{\Gamma}$. □

Much of the rest of this section concerns the question of when a strengthening of the hypothesis of Theorem 3.1 allows a strengthening of the conclusion. We begin with the case $r = 1$ and find that limiting Γ to lie in Π^0_1 permits no stronger conclusion, whereas if $\Gamma \in \Sigma^0_1$ (and Γ is monotone), then $\bar{\Gamma} \in \Sigma^0_1$.

3.2 Theorem. *For every set $A \in \Pi^1_1$, there exists a monotone operator $\Gamma \in \Pi^0_1$ such that A is many-one reducible to $\bar{\Gamma}$.*

Proof. Let A be an arbitrary Π^1_1 set. By Lemma 2.9, $A = \forall^1 \mathsf{P}$ for some $\mathsf{P} \in \Sigma^0_1$. Thus by Theorem II.4.12 there exists a recursive relation R such that

$$m \in A \leftrightarrow \forall \beta\, \exists p\, R(m, \bar{\beta}(p)).$$

Let Γ be the operator defined by

$$\langle m, s\rangle \in \Gamma(B) \leftrightarrow R(m, s) \vee \forall n\,(\langle m, s * \langle n\rangle\rangle \in B).$$

Clearly Γ is monotone and $\Gamma \in \Pi^0_1$. We claim that

$$\langle m, s\rangle \in \bar{\Gamma} \leftrightarrow \forall \beta\, \exists p\, R(m, s * \bar{\beta}(p)).$$

Once this is established we have

$$m \in A \leftrightarrow \langle m, \langle\ \rangle\rangle \in \bar{\Gamma}$$

so that A is many-one reducible to $\bar{\Gamma}$.

To establish the claim, let

$$C = \{\langle m, s\rangle : \forall \beta\, \exists p\, R(m, s * \bar{\beta}(p))\}.$$

We must show $C = \bar{\Gamma}$. As usual, to show $\bar{\Gamma} \subseteq C$ it suffices to show $\Gamma(C) \subseteq C$. Suppose $\langle m, s\rangle \in \Gamma(C)$. Then either

(1) $R(m, s)$ or (2) $\forall n.\langle m, s * \langle n\rangle\rangle \in C$.

In case (1), any β satisfies $R(m, s * \bar{\beta}(0))$ so that $\langle m, s\rangle \in C$. In case (2), for all n, $\forall \beta\, \exists p\, R(m, s * \langle n\rangle * \bar{\beta}(p))$, so also $\langle m, s\rangle \in C$.

For the converse inclusion we assume $\langle m, s\rangle \notin \bar{\Gamma}$ and construct a function β such that $\forall p \sim R(m, s * \bar{\beta}(p))$. Let $D = \{t: \langle m, s * t\rangle \notin \bar{\Gamma}\}$. By assumption, $\langle\ \rangle \in D$. Since $\Gamma(\bar{\Gamma}) = \bar{\Gamma}$ we have

$$t \in D \rightarrow \langle m, s * t\rangle \notin \Gamma(\bar{\Gamma}) \rightarrow \exists n\, (t * \langle n\rangle \in D).$$

Hence there is a unique function β such that for all p,

$$\beta(p) = \text{least } n[\bar{\beta}(p) * \langle n\rangle \in D].$$

For this β, $\forall p\, (\langle m, s * \bar{\beta}(p)\rangle \notin \bar{\Gamma})$ so in particular, $\forall p \sim R(m, s * \bar{\beta}(p))$. Hence $\langle m, s\rangle \notin C$. $\square$

Combining 3.1 and 3.2, a set A is Π^1_1 iff A is many-one reducible to $\bar{\Gamma}$ for some monotone operator $\Gamma \in \Pi^0_1$. The same proof works also for relations R on numbers.

Since by the Analytical Hierarchy Theorem there exist sets A wich are Π^1_1 but not Δ^1_1, there exist Π^0_1 monotone operators such that $\bar{\Gamma}$ is Π^1_1 but not Δ^1_1. A natural question to ask in conjunction with this theorem is whether or not every Π^1_1 set is equal to $\bar{\Gamma}$ for some $\Gamma \in \Pi^0_1$. We shall show in § 6 that this is false if Γ is required to be monotone. The answer is unknown if non-monotone Γ are admitted.

We turn now to Σ^0_1 operators, which turn out to be much weaker.

3.3 Lemma. *For any inductive operator $\Gamma \in \Sigma^0_1$, $|\Gamma| \leqslant \omega$.*

Proof. Suppose $\Gamma \in \Sigma^0_1$ and let R be a recursive relation such that $\mathsf{P}_\Gamma(m, \alpha) \leftrightarrow \exists n R(m, \bar{\alpha}(n))$. $|\Gamma|$ is the least ordinal σ such that $\Gamma(\Gamma^{(\sigma)}) \subseteq \Gamma^{(\sigma)}$, so to show $|\Gamma| \leqslant \omega$ it suffices to show $\Gamma(\Gamma^{(\omega)}) \subseteq \Gamma^{(\omega)}$. Suppose $m \in \Gamma(\Gamma^{(\omega)})$, so for some n, $R(m, \bar{\mathsf{K}}_{\Gamma^{(\omega)}}(n))$. For each $i < n$ let

$$s_i = \begin{cases} \text{least } s \,.\, i \in \Gamma^s, & \text{if } \ i \in \Gamma^{(\omega)}; \\ 0, & \text{otherwise;} \end{cases}$$

and

$$t = \max\{s_i : i < n\}.$$

Then $\bar{\mathsf{K}}_{\Gamma^{(\omega)}}(n) = \bar{\mathsf{K}}_{\Gamma^t}(n)$ so $R(m, \bar{\mathsf{K}}_{\Gamma^t}(n))$ and thus $m \in \Gamma(\Gamma^t) = \Gamma^{t+1} \subseteq \Gamma^{(\omega)}$. $\square$

As a special notation for the following lemma and theorem only, we write

$$s < A \leftrightarrow (\forall i < \lg(s))((s)_i \leqslant 1 \wedge [(s)_i = 0 \rightarrow i \in A]).$$

Thus if $s = \bar{\mathsf{K}}_B(n)$, then $s < A \leftrightarrow B \cap n \subseteq A \cap n$.

The import of the following lemma is that for monotone Σ^0_1 operators Γ, the relation $m \in \Gamma(A)$ depends only on *positive* information about A — that is, information concerning membership in A but not concerning *non*-membership in A.

3.4 Lemma. *For any monotone operator $\Gamma \in \Sigma^0_1$, there exists a recursive relation R such that for all m and A,*

$$m \in \Gamma(A) \leftrightarrow \exists s\,[s < A \wedge R(m,s)].$$

Proof. If $\Gamma \in \Sigma^0_1$ there exists by definition a recursive R such that

$$\begin{aligned} m \in \Gamma(A) &\leftrightarrow \exists n\, R(m, \bar{K}_A(n)) \\ &\leftrightarrow \exists s\, \exists n\, [s = \bar{K}_A(n) \wedge R(m,s)]. \end{aligned}$$

Thus for this R,

$$m \in \Gamma(A) \rightarrow \exists s\,[s < A \wedge R(m,s)].$$

The converse implication holds also. Suppose $s < A$ and $R(m,s)$. Let

$$B = \{i\colon [i < \lg(s) \wedge (s)_i = 0] \vee [i \geqslant \lg(s) \wedge i \in A]\}.$$

Then $B \subseteq A$ and $s = \bar{K}_B(\lg(s))$. Hence $m \in \Gamma(B)$ and by the monotonicity of Γ, $\Gamma(B) \subseteq \Gamma(A)$ so also $m \in \Gamma(A)$. □

3.5 Theorem. *For any monotone operator Γ,*

$$\Gamma \in \Sigma^0_1 \rightarrow \bar{\Gamma} \in \Sigma^0_1.$$

Proof. Let Γ be a Σ^0_1 monotone operator. We shall define a primitive recursive function f such that for all r, $f(r+1)$ is a semi-index for the semi-recursive set Γ^r. Once this is done, we have by Lemma 3.3

$$m \in \bar{\Gamma} \leftrightarrow \exists r\,.\, m \in \Gamma^r \leftrightarrow \exists r\, \exists u\, \mathrm{T}(f(r+1), \langle m \rangle, u, \langle\ \rangle)$$

and thus $\bar{\Gamma} \in \Sigma^0_1$.

We let $f(0) = 0$ and $f(r+1) = \mathrm{Sb}_0(c, f(r))$, where c is an index chosen as follows. Let R be as in Lemma 3.4, so that

$$m \in \Gamma^r \leftrightarrow \exists s\,[s < \Gamma^{(r)} \wedge R(m,s)].$$

Under the induction hypothesis that $\Gamma^{(r)} = \mathrm{Dm}\{f(r)\}$ (which is valid for $r = 0$) we have

$$m \in \Gamma^r \leftrightarrow \exists s\,(\forall i < \lg(s)[(s)_i \leq 1 \wedge ((s)_i = 0 \rightarrow \exists u\, \mathrm{T}(f(r), \langle i \rangle, u, \langle\ \rangle))] \wedge R(m, s))$$
$$\leftrightarrow \exists p\, S(p, f(r), m)$$

for an appropriate recursive relation S. Choose c such that

$$(n, m) \in \mathrm{Dm}\{c\} \leftrightarrow \exists p\, S(p, n, m).$$

Then

$$m \in \Gamma^r \leftrightarrow (f(r), m) \in \mathrm{Dm}\{c\} \leftrightarrow m \in \mathrm{Dm}\{\mathrm{Sb}_0(c, f(r))\} = \mathrm{Dm}\{f(r+1)\}. \quad \square$$

If Γ is not monotone, then $\bar{\Gamma}$ need not even be arithmetical. Let

$$U^0_{(\omega)} = \{\langle r, a, \mathbf{m} \rangle \colon U^0_r(a, \langle \mathbf{m} \rangle)\}.$$

It is clear that for any arithmetical relation R, there exist r and a such that

$$R(\mathbf{m}) \leftrightarrow \langle r, a, \mathbf{m} \rangle \in U^0_{(\omega)}.$$

Hence $U^0_{(\omega)}$ is not itself arithmetical as if it were, say, Σ^0_r, then so would be every arithmetical relation contrary to the Arithmetical Hierarchy Theorem. On the other hand,

3.6 Theorem. *There exists a Σ^0_1 inductive operator Γ such that $\bar{\Gamma} = U^0_{(\omega)}$.*

Proof. Let

$$s \in \Gamma(A) \leftrightarrow \exists a\, \exists \mathbf{m}\, [s = \langle 1, a, \mathbf{m} \rangle \wedge U^0_1(a, \langle \mathbf{m} \rangle)]$$
$$\vee\, \exists r\, \exists a\, \exists \mathbf{m}\, [\exists b\, \exists \mathbf{n}\, \langle r, b, \mathbf{n} \rangle \in A \wedge s = \langle r+1, a, \mathbf{m} \rangle \wedge$$
$$\exists p\, (\langle r, a, p, \mathbf{m} \rangle \notin A)] \vee (s \in A).$$

Clearly $\Gamma \in \Sigma^0_1$ and it is easy to show by induction on r that

$$\Gamma^r = \{\langle t, a, \mathbf{m} \rangle \colon 0 < t \leq r+1 \wedge U^0_t(a, \langle \mathbf{m} \rangle)\}.$$

Hence $\bar{\Gamma} = U^0_{(\omega)}$. $\square$

Note that Γ is non-monotone because of the condition in its definition that something *not* belong to A. On the positive side,

3.7 Theorem. *For any inductive operator* Γ,

$$\Gamma \in \Sigma_1^0 \rightarrow \bar{\Gamma} \in \Delta_1^1.$$

Proof. Let Γ be a Σ_1^0 inductive operator and let β^* be the function such that for all r and m,

$$\beta^*(\langle r+1, m\rangle) = 0 \leftrightarrow m \in \Gamma^r,$$

and $\beta^*(t) = 1$, otherwise. Consider the following arithmetical relation S:

$$\begin{aligned} \mathsf{S}(\beta) \leftrightarrow \forall t\,[\beta(t) \leqslant 1] \wedge \\ \wedge \forall t\,[\sim \mathrm{Sq}(t) \vee \lg(t) \neq 2 \vee (t)_0 = 0 \rightarrow \beta(t) = 1] \\ \wedge \forall r\, \forall m\,[\beta(\langle r+1, m\rangle) = 0 \leftrightarrow \mathsf{P}_\Gamma(m, \lambda n\,.\,\beta(\langle r, n\rangle))]. \end{aligned}$$

Note first that $\mathsf{S}(\beta^*)$. Furthermore, for any β such that $\mathsf{S}(\beta)$, $\beta(t) = 1 = \beta^*(t)$ for all t not of the form $\langle r+1, m\rangle$ and it is easy to show by induction on r that for all m

$$\beta(\langle r+1, m\rangle) = 0 \leftrightarrow m \in \Gamma^r.$$

Thus β^* is the unique member of S and we have

$$\begin{aligned} m \in \bar{\Gamma} \leftrightarrow \exists r\,.\, m \in \Gamma^r \leftrightarrow \exists \beta\,[\mathsf{S}(\beta) \wedge \exists r\,.\,\beta(\langle r+1, m\rangle) = 0] \\ \leftrightarrow \forall \beta\,[\mathsf{S}(\beta) \rightarrow \exists r\,.\,\beta(\langle r+1, m\rangle) = 0]. \end{aligned}$$

Thus $\bar{\Gamma} \in \Delta_1^1$. □

3.8 Corollary. $U^0_{(\omega)} \in \Delta_1^1 \sim \Delta^0_{(\omega)}$.

Proof. By 3.6 and 3.7. □

The proof of 3.7 actually establishes a stronger result:

$$\Gamma \in \Delta_1^1 \wedge |\Gamma| \leqslant \omega \rightarrow \bar{\Gamma} \in \Delta_1^1.$$

Since by Theorem 3.2 there exist monotone $\Gamma \in \Pi_1^0$ such that $\bar{\Gamma} \notin \Delta_1^1$, it follows that for some such $\Gamma, |\Gamma| > \omega$. It will soon be clear that Π_1^0 operators may have quite large closure ordinal; an upperbound for these is obtained in § IV.2.

For the remainder of this section we exploit the fact (Corollary I.3.2) that all inductive operators over ω have countable closure ordinal. Thus for any such Γ we have

$$m \in \bar{\Gamma} \leftrightarrow (\exists \sigma < \aleph_1) m \in \Gamma^\sigma$$
$$\leftrightarrow (\forall \sigma < \aleph_1)[\Gamma^\sigma \subseteq \Gamma^{(\sigma)} \rightarrow m \in \Gamma^\sigma].$$

Recall from the end of § I.1 that W denotes the set of functions γ such that $\leqslant_\gamma$ is well-ordering of order type $\|\gamma\| < \aleph_1$. Since every countable ordinal is $\|\gamma\|$ for some $\gamma \in \mathsf{W}$, we have

$$m \in \bar{\Gamma} \leftrightarrow (\exists \gamma \in \mathsf{W}) m \in \Gamma^{\|\gamma\|}$$
$$\leftrightarrow (\forall \gamma \in \mathsf{W})[\Gamma^{\|\gamma\|} \subseteq \Gamma^{(\|\gamma\|)} \rightarrow m \in \Gamma^{\|\gamma\|}].$$

Thus we can classify $\bar{\Gamma}$ in the analytical hierarchy if we can classify the relations $m \in \Gamma^{(\|\gamma\|)}$ and $m \in \Gamma^{\|\gamma\|}$.

3.9 Theorem. *For any $r > 0$ and any inductive operator $\Gamma \in \Delta^1_r$, there exist relations $\mathsf{V}^{()}_\Sigma$ and $\mathsf{V}_\Sigma \in \Sigma^1_r$ and $\mathsf{V}^{()}_\Pi$ and $\mathsf{V}_\Pi \in \Pi^1_r$ such that for any $\gamma \in \mathsf{W}$ and any m,*

(i) $m \in \Gamma^{(\|\gamma\|)} \leftrightarrow \mathsf{V}^{()}_\Sigma(m, \gamma) \leftrightarrow \mathsf{V}^{()}_\Pi(m, \gamma)$;
(ii) $m \in \Gamma^{\|\gamma\|} \leftrightarrow \mathsf{V}_\Sigma(m, \gamma) \leftrightarrow \mathsf{V}_\Pi(m, \gamma)$.

Proof. The technique is an elaboration of that used for Theorem 3.7 in which we index the stages of Γ by $p \in \mathrm{Fld}(\gamma)$ rather than by $r \in \omega$. Suppose $\Gamma \in \Delta^1_r$ with $r > 0$. For each $\gamma \in \mathsf{W}$ we define α_γ and β_γ by:

$$\alpha_\gamma(\langle p, m \rangle) = 0 \leftrightarrow p \in \mathrm{Fld}(\gamma) \wedge m \in \Gamma^{(|p|_\gamma)};$$

$$\beta_\gamma(\langle p, m \rangle) = 0 \leftrightarrow p \in \mathrm{Fld}(\gamma) \wedge m \in \Gamma^{|p|_\gamma};$$

and $\alpha_\gamma(t) = \beta_\gamma(t) = 1$, otherwise. Let S be the relation defined as follows:

$$\begin{aligned} \mathsf{S}(\alpha, \beta, \gamma) \leftrightarrow & \forall t [\alpha(t) \leqslant 1 \wedge \beta(t) \leqslant 1] \\ & \wedge \forall t [\sim \mathrm{Sq}(t) \vee \mathrm{lg}(t) \neq 2 \vee (t)_0 \notin \mathrm{Fld}(\gamma) \rightarrow \alpha(t) = \beta(t) = 1] \\ & \wedge \forall p \, \forall m \, (p \in \mathrm{Fld}(\gamma) \rightarrow [\alpha(\langle p, m \rangle) = 0 \leftrightarrow (\exists q <_\gamma p) \beta(\langle q, m \rangle) = 0] \\ & \wedge [\beta(\langle p, m \rangle) = 0 \leftrightarrow \mathsf{P}_\Gamma(m, \lambda n \, . \, \alpha(\langle p, n \rangle))]). \end{aligned}$$

As $\mathsf{P}_\Gamma \in \Delta^1_r$ and everything else is arithmetical, $\mathsf{S} \in \Delta^1_r$. It is routine to check that for any $\gamma \in \mathsf{W}$, $S(\alpha_\gamma, \beta_\gamma, \gamma)$. We claim that for any $\gamma \in \mathsf{W}$ and any α and β such that $S(\alpha, \beta, \gamma)$, $\alpha = \alpha_\gamma$ and $\beta = \beta_\gamma$. For any such α, β, and γ, let

$$Z_{\alpha, p} = \{m : \alpha(\langle p, m \rangle) = 0\}$$

and similarly $Z_{\beta, p}$. It will suffice to show that for all $p \in \mathrm{Fld}(\gamma)$,

$$Z_{\alpha,p} = \Gamma^{(|p|_\gamma)} \quad \text{and} \quad Z_{\beta,p} = \Gamma^{|p|_\gamma}.$$

We proceed by induction on $|p|_\gamma$ and assume as induction hypothesis that for all $q <_\gamma p$,

$$Z_{\alpha,q} = \Gamma^{(|q|_\gamma)} \quad \text{and} \quad Z_{\beta,q} = \Gamma^{|q|_\gamma}.$$

Then by the definition of S and formulas (8)–(10) of § I.1,

$$\begin{aligned} m \in Z_{\alpha,p} &\leftrightarrow (\exists q <_\gamma p) m \in Z_{\beta,q} \leftrightarrow (\exists q <_\gamma p) m \in \Gamma^{|q|_\gamma} \\ &\leftrightarrow (\exists \sigma < |p|_\gamma) m \in \Gamma^\sigma \leftrightarrow m \in \Gamma^{(|p|_\gamma)}, \end{aligned}$$

and

$$\begin{aligned} m \in Z_{\beta,p} &\leftrightarrow \mathsf{P}_\Gamma(m, \lambda n \, . \, \alpha(\langle p, n\rangle)) \leftrightarrow m \in \Gamma(Z_{\alpha,p}) \\ &\leftrightarrow m \in \Gamma(\Gamma^{(|p|_\gamma)}) = \Gamma^{|p|_\gamma}. \end{aligned}$$

We now set

$$\mathsf{V}_\Sigma^{()}(m,\gamma) \leftrightarrow \exists \alpha\beta\, [\mathsf{S}(\alpha,\beta,\gamma) \wedge \exists p\, (\beta(\langle p, m\rangle) = 0)];$$
$$\mathsf{V}_\Pi^{()}(m,\gamma) \leftrightarrow \forall \alpha\beta\, [\mathsf{S}(\alpha,\beta,\gamma) \rightarrow \exists p\, (\beta(\langle p, m\rangle) = 0)];$$
$$\mathsf{V}_\Sigma(m,\gamma) \leftrightarrow \exists \alpha\beta\delta\, [\mathsf{S}(\alpha,\beta,\gamma) \wedge \forall n\, (\delta(n) = 0 \leftrightarrow \exists p\, [\beta(\langle p, m\rangle) = 0]) \wedge \mathsf{P}_\Gamma(m,\delta)];$$
$$\mathsf{V}_\Pi(m,\gamma) \leftrightarrow \forall \alpha\beta\delta\, [\mathsf{S}(\alpha,\beta,\gamma) \wedge \forall n\, (\delta(n) = 0 \leftrightarrow \exists p\, [\beta(\langle p, m\rangle) = 0]) \rightarrow \mathsf{P}_\Gamma(m,\delta)].$$

We leave to the reader the straightforward verification that these relations satisfy the conditions of the theorem. □

3.10 Theorem. *For any $r \geqslant 2$ and any inductive operator Γ,*

$$\Gamma \in \Delta_r^1 \rightarrow \bar{\Gamma} \in \Delta_r^1.$$

Proof. With the notation of the preceding theorem and the remarks before it we have

$$\begin{aligned} m \in \bar{\Gamma} &\leftrightarrow \exists \gamma\, [\gamma \in \mathsf{W} \wedge \mathsf{V}_\Sigma(m,\gamma)] \\ &\leftrightarrow \forall \gamma\, [\gamma \in \mathsf{W} \wedge \forall p\, (\mathsf{V}_\Pi(p,\gamma) \rightarrow \mathsf{V}_\Sigma^{()}(p,\gamma)) \rightarrow \mathsf{V}_\Pi(m,\gamma)]. \end{aligned}$$

Since $r \geqslant 2$, $\mathsf{W} \in \Pi_1^1 \subseteq \Delta_r^1$ and an easy computation shows that the first formula gives a Σ_r^1 definition for $\bar{\Gamma}$, the second a Π_r^1 definition. □

By Theorem 3.2, the result fails for $r = 1$, but we can get some information about Δ_1^1 operators.

3.11 Definition. An ordinal σ is *recursive* iff $\sigma = \|\gamma\|$ for some recursive $\gamma \in \mathsf{W}$. The least non-recursive ordinal is denoted by ω_1.

As there are only countably many recursive functions, ω_1 is a countable ordinal. Furthermore it is easy to see that any ordinal less than a recursive ordinal is also recursive so that σ is recursive iff $\sigma < \omega_1$. Other properties of ω_1 are indicated in Exercise 3.27.

3.12 Corollary. *For any inductive operator Γ,*

$$\Gamma \in \Delta_1^1 \wedge |\Gamma| < \omega_1 \to \bar{\Gamma} \in \Delta_1^1.$$

Proof. If $|\Gamma| < \omega_1$, then there exists a recursive function $\gamma \in \mathsf{W}$ such that $\|\gamma\| = |\Gamma|$. Then by Theorem 3.9,

$$m \in \bar{\Gamma} \leftrightarrow m \in \Gamma^{\|\gamma\|} \leftrightarrow \mathsf{V}_\Sigma(m, \gamma) \leftrightarrow \mathsf{V}_\Pi(m, \gamma).$$

The conclusion follows from Theorem 2.5. □

The proof of Corollary 3.12 establishes an apparently stronger result. We say an ordinal σ is Δ_r^1 iff $\sigma = \|\gamma\|$ for some $\gamma \in \Delta_r^1 \cap \mathsf{W}$; δ_r^1 denotes the least non-Δ_r^1 ordinal. Then using the Analytical Substitution Theorem (2.11) instead of Theorem 2.5 we have by the same argument,

$$\Gamma \in \Delta_1^1 \wedge |\Gamma| < \delta_1^1 \to \bar{\Gamma} \in \Delta_1^1.$$

However, we shall see in § IV.2 that $\delta_1^1 = \omega_1$ so that this is no improvement. We can also now extend the reasoning following Corollary 3.8 to conclude that for some Π_1^0 monotone operators Γ we have $|\Gamma| \geq \omega_1$. In §IV.2 we shall also prove that for any Π_1^0 operator $\Gamma, |\Gamma| \leq \omega_1$ and $\bar{\Gamma} \in \Pi_1^1 \sim \Delta_1^1$ just in case $|\Gamma| = \omega_1$.

For arbitrary inductive operators $\Gamma \in \Sigma_r^1$, the best possible classification of $\bar{\Gamma}$ in the analytical hierarchy is that given by Theorem 3.10: $\bar{\Gamma} \in \Delta_{r+1}^1$. For monotone Γ, however, a refinement of the proof of Theorem 3.10 yields a result parallel to Theorem 3.1.

3.13 Theorem. *For any $r > 0$ and any monotone operator $\Gamma \in \Sigma_r^1$ (Π_r^1), there exist relations $\mathsf{V}^{()}$ and $\mathsf{V} \in \Sigma_r^1$ (Π_r^1) such that for any $\gamma \in \mathsf{W}$ and any m,*

(i) $m \in \Gamma^{(\|\gamma\|)} \leftrightarrow \mathsf{V}^{()}(m, \gamma)$;

(ii) $m \in \Gamma^{\|\gamma\|} \leftrightarrow \mathsf{V}(m, \gamma)$.

Proof. We follow closely the proof of Theorem 3.9 and only indicate the necessary modifications. Suppose first that $\Gamma \in \Sigma_r^1$ and is monotone. Define α_γ and β_γ for $\gamma \in \mathsf{W}$ as before and let S' be the relation defined as S except that the clause

$$[\beta(\langle p, m\rangle) = 0 \leftrightarrow \mathsf{P}_\Gamma(m, \lambda n . \alpha(\langle p, n\rangle))]$$

is replaced by

$$[\beta(\langle p, m\rangle) = 0 \rightarrow \mathsf{P}_\Gamma(m, \lambda n . \alpha(\langle p, n\rangle))].$$

Clearly $\mathsf{S}' \in \Sigma^1_r$ and for any $\gamma \in \mathsf{W}$, $\mathsf{S}'(\alpha_\gamma, \beta_\gamma, \gamma)$, but α_γ and β_γ are no longer the only functions for which this is true. Rather we prove by induction on $|p|_\gamma$ that for all $p \in \mathrm{Fld}(\gamma)$, if $\mathsf{S}'(\alpha, \beta, \gamma)$, then

$$Z_{\alpha, p} \subseteq \Gamma^{(|p|_\gamma)} \quad \text{and} \quad Z_{\beta, p} \subseteq \Gamma^{|p|_\gamma}.$$

Assuming as induction hypothesis that this holds for all $q <_\gamma p$, we have

$$\begin{aligned} m \in Z_{\alpha, p} &\leftrightarrow (\exists q <_\gamma p) m \in Z_{\beta, q} \rightarrow (\exists q <_\gamma p) m \in \Gamma^{|q|_\gamma} \\ &\leftrightarrow (\exists \sigma < |p|_\gamma) m \in \Gamma^\sigma \leftrightarrow m \in \Gamma^{(|p|_\gamma)}, \end{aligned}$$

and

$$\begin{aligned} m \in Z_{\beta, p} &\rightarrow \mathsf{P}_\Gamma(m, \lambda n . \alpha(\langle p, n\rangle)) \leftrightarrow m \in \Gamma(Z_{\alpha, p}) \\ &\rightarrow m \in \Gamma(\Gamma^{(|p|_\gamma)}) = \Gamma^{|p|_\gamma}. \end{aligned}$$

The last implication is the only place where the monotonicity of Γ is used.

Finally, define $\mathsf{V}^{(\,)}$ and V from S' just as $\mathsf{V}^{(\,)}_\Sigma$ and V_Σ are defined from S. Clearly $\mathsf{V}^{(\,)}$ and V are Σ^1_r. If $\gamma \in \mathsf{W}$ and $m \in \Gamma^{(\|\gamma\|)}$, then for some $p \in \mathrm{Fld}(\gamma)$, $m \in \Gamma^{|p|_\gamma}$ and thus $\beta_\gamma(\langle p, m\rangle) = 0$. Since $\mathsf{S}'(\alpha_\gamma, \beta_\gamma, \gamma)$, also $\mathsf{V}^{(\,)}(m, \gamma)$. Conversely, suppose $\mathsf{V}^{(\,)}(m, \gamma)$ holds, say $\mathsf{S}'(\alpha, \beta, \gamma)$ and $\beta(\langle p, m\rangle) = 0$. Then $p \in \mathrm{Fld}(\gamma)$ and $m \in Z_{\beta, p} \subseteq \Gamma^{|p|_\gamma}$ so $m \in \Gamma^{(\|\gamma\|)}$. We leave the similar verification of (ii) to the reader.

In case Γ is a monotone Π^1_r operator, we define S'' by replacing the last clause in the definition of S by

$$[\mathsf{P}_\Gamma(m, \lambda n . \alpha(\langle p, n\rangle)) \rightarrow \beta(\langle p, m\rangle) = 0].$$

Then $\mathsf{S}'' \in \Sigma^1_r$ (!), $\mathsf{S}''(\alpha_\gamma, \beta_\gamma, \gamma)$ holds for any $\gamma \in \mathsf{W}$, and for all $\gamma \in \mathsf{W}$, α and β such that $\mathsf{S}''(\alpha, \beta, \gamma)$ and $p \in \mathrm{Fld}(\gamma)$:

$$\Gamma^{(|p|_\gamma)} \subseteq Z_{\alpha, p} \quad \text{and} \quad \Gamma^{|p|_\gamma} \subseteq Z_{\beta, p}.$$

Then if $\mathsf{V}^{(\,)}$ and V are defined from S'' as $\mathsf{V}^{(\,)}_\Pi$ and V_Π are defined from S in 3.9, we have $\mathsf{V}^{(\,)}$ and V in Π^1_r and an easy computation verifies (i) and (ii). □

3.14 Theorem. *For any $r \geqslant 2$ and any monotone operator Γ,*

$$\Gamma \in \Sigma^1_r \to \bar{\Gamma} \in \Sigma^1_r.$$

Proof. With the notation of the preceding theorem, we have

$$m \in \bar{\Gamma} \leftrightarrow \exists\gamma\,[\gamma \in \mathrm{W} \wedge m \in \Gamma^{\|\gamma\|}] \leftrightarrow \exists\gamma\,[\gamma \in \mathrm{W} \wedge \mathrm{V}(m, \gamma)]. \quad \square$$

At this point the strongest result we have for $\Gamma \in \Sigma^1_1$ is that $\bar{\Gamma} \in \Delta^1_2$. Although this is clearly the best estimate possible in terms of the analytical hierarchy, we shall return to this question in § VI.6 and find better bounds in terms of other measures of complexity.

So far in this section we have treated only inductive operators over ω. There is little difficulty in extending our analysis to operators over ${}^k\omega$. For any α let

$$Z^k_\alpha = \{\mathbf{p}\colon \mathbf{p} \in {}^k\omega \wedge \alpha(\langle\mathbf{p}\rangle) = 0\},$$

and for any operator Γ over ${}^k\omega$ set

$$\mathrm{P}_\Gamma(\mathbf{m}, \alpha) \leftrightarrow \mathbf{m} \in \Gamma(Z^k_\alpha).$$

Then we classify Γ as Σ^i_r, etc. ($i = 0, 1$), according as $\mathrm{P}_\Gamma \in \Sigma^i_r$, etc. and all the results of this section hold with only minor changes in the proofs.

If we attempt to do the same for operators over ${}^{k,l}\omega$ or even over ${}^\omega\omega$, however, we encounter an immediate difficulty. In place of Z^k_α we should have to use

$$\mathrm{Z}^{k,l}_{\mathrm{F}} = \{(\mathbf{p}, \boldsymbol{\beta})\colon (\mathbf{p}, \boldsymbol{\beta}) \in {}^{k,l}\omega \wedge \mathrm{F}(\mathbf{p}, \boldsymbol{\beta}) = 0\}$$

and attempt to classify Γ by means of the relation

$$\mathbb{P}_\Gamma(\mathbf{m}, \boldsymbol{\alpha}, \mathrm{F}) \leftrightarrow (\mathbf{m}, \boldsymbol{\alpha}) \in \Gamma(\mathrm{Z}^{k,l}_{\mathrm{F}}).$$

Relations with functionals as arguments are not included in our present system so we have no way to assess the complexity of such a $\mathbb{P}_\Gamma$. We shall develop such means in Chapter VI and return to this question in § VI.7.

We can, however, with our current machinery, treat inductive operators over ${}^{k,l}\omega$ which are decomposable in the sense of Definition I.3.6. Recall that such an operator Γ is defined by a family $\langle\Gamma_{\boldsymbol{\alpha}}\colon \boldsymbol{\alpha} \in {}^l({}^\omega\omega)\rangle$ of operators over ${}^k\omega$ by

$$(\mathbf{m}, \boldsymbol{\alpha}) \in \Gamma(\mathrm{R}) \leftrightarrow \mathbf{m} \in \Gamma_{\boldsymbol{\alpha}}(\mathrm{R}_{\boldsymbol{\alpha}})$$

where $\mathrm{R}_{\boldsymbol{\alpha}}(\mathbf{m}) \leftrightarrow \mathrm{R}(\mathbf{m}, \boldsymbol{\alpha})$. For such an operator we define

$$\mathrm{P}_\Gamma(\mathbf{m}, \boldsymbol{\alpha}, \delta) \leftrightarrow \mathbf{m} \in \Gamma_{\boldsymbol{\alpha}}(Z^k_\delta)$$

and say that Γ is Σ^i_r, etc. according as $\mathsf{P}_\Gamma \in \Sigma^i_r$, etc. Note that with these conventions,

$$(\mathbf{m}, \boldsymbol{\alpha}) \in \Gamma(\mathsf{R}) \leftrightarrow \mathsf{P}_\Gamma(\mathbf{m}, \boldsymbol{\alpha}, \lambda p \,.\, \mathsf{K}_\mathsf{R}((p)_0, \ldots, (p)_{k-1}, \boldsymbol{\alpha})).$$

The crucial restriction is that the arguments $\boldsymbol{\alpha}$ enter only as *parameters*: whether or not $(\mathbf{m}, \boldsymbol{\alpha}) \in \Gamma(\mathsf{R})$ depends only on the membership or non-membership of other sequences $(\mathbf{p}, \boldsymbol{\alpha})$ in R not on that of sequences $(\mathbf{p}, \boldsymbol{\beta})$ for any $\boldsymbol{\beta} \neq \boldsymbol{\alpha}$. One sometimes says that the operators $\Gamma_{\boldsymbol{\alpha}}$ are Σ^i_r, etc. *uniformly in* $\boldsymbol{\alpha}$.

At first glance, the class of decomposable operators over ${}^{k,l}\omega$ may appear very limited. It turns out, however, that not only can all of the results of this section be extended to this class, but also that these extensions are just what is needed for many applications in later chapters. We shall state some of these but relegate most of the details of the proofs to the exercises.

In what follows we always assume that the decomposable operator Γ is defined by the family $\langle \Gamma_{\boldsymbol{\alpha}} \colon \boldsymbol{\alpha} \in {}^l({}^\omega\omega)\rangle$. Note that Γ is monotone just in case each $\Gamma_{\boldsymbol{\alpha}}$ is monotone.

3.15 Theorem. *For any $r > 0$ and any decomposable monotone operator Γ over ${}^{k,l}\omega$,*

$$\Gamma \in \Pi^1_r \rightarrow \bar{\Gamma} \in \Pi^1_r.$$

Proof. As in the proof of Theorem 3.1, by Lemma I.3.7,

$$(\mathbf{m}, \boldsymbol{\alpha}) \in \bar{\Gamma} \leftrightarrow \mathbf{m} \in \bar{\Gamma}_{\boldsymbol{\alpha}} \leftrightarrow \forall \delta\, [\forall \mathbf{p}\, (\mathsf{P}_\Gamma(\mathbf{p}, \boldsymbol{\alpha}, \delta) \rightarrow \delta(\langle \mathbf{p} \rangle) = 0) \rightarrow \delta(\langle \mathbf{m} \rangle) = 0]. \quad \square$$

3.16 Theorem. *For every $\mathsf{R} \in \Pi^1_1$, there exists a decomposable monotone operator $\Gamma \in \Pi^0_1$ and a recursive function f such that*

$$\mathsf{R}(\mathbf{m}, \boldsymbol{\alpha}) \leftrightarrow (f(\mathbf{m}), \boldsymbol{\alpha}) \in \bar{\Gamma}.$$

Proof. For any $\mathsf{R} \in \Pi^1_1$ there exists a recursive relation S such that

$$\mathsf{R}(\mathbf{m}, \boldsymbol{\alpha}) \leftrightarrow \forall \beta\, \exists p\, \mathsf{S}(\bar{\beta}(p), \mathbf{m}, \boldsymbol{\alpha}).$$

Let

$$\langle \mathbf{m}, s \rangle \in \Gamma_{\boldsymbol{\alpha}}(B) \leftrightarrow \mathsf{S}(s, \mathbf{m}, \boldsymbol{\alpha}) \vee \forall n\, (\langle \mathbf{m}, s * \langle n \rangle \rangle \in B).$$

Then $\Gamma \in \Pi^0_1$, Γ is monotone, and as in the proof of Theorem 3.2,

$$\mathsf{R}(\mathbf{m}, \boldsymbol{\alpha}) \leftrightarrow \langle \mathbf{m}, \langle\ \rangle \rangle \in \bar{\Gamma}_{\boldsymbol{\alpha}} \leftrightarrow (\langle \mathbf{m}, \langle\ \rangle \rangle, \boldsymbol{\alpha}) \in \bar{\Gamma}. \quad \square$$

3.17 Theorem. *For any decomposable operator Γ over* ${}^{k,l}\omega$,

(i) $\Gamma \in \Sigma_1^0 \wedge \Gamma$ *monotone* $\rightarrow \bar{\Gamma} \in \Sigma_1^0$;

(ii) $\Gamma \in \Sigma_1^0 \rightarrow \bar{\Gamma} \in \Delta_1^1$.

Proof. We leave the proof of (i) to Exercise 3.31. Suppose $\Gamma \in \Sigma_1^0$. For each $\boldsymbol{\alpha}$, let

$$\beta_{\boldsymbol{\alpha}}(\langle r+1, \mathbf{m}\rangle) = 0 \leftrightarrow \mathbf{m} \in \Gamma_{\boldsymbol{\alpha}}^{r},$$

and $\beta_{\boldsymbol{\alpha}}(t) = 1$, otherwise. Set

$$\begin{aligned} \mathsf{S}(\boldsymbol{\alpha}, \beta) \leftrightarrow \forall t\, [\beta(t) \leqslant 1] \\ \wedge \forall t\, [\sim \mathrm{Sq}(t) \vee \mathrm{lg}(t) \neq k+1 \vee (t)_0 = 0 \rightarrow \beta(t) = 1] \\ \wedge \forall r\, \forall \mathbf{m}\, [\beta(\langle r+1, \mathbf{m}\rangle) = 0 \leftrightarrow \mathsf{P}_\Gamma(\mathbf{m}, \boldsymbol{\alpha}, \lambda n\,.\, \beta(\langle r, n\rangle))]. \end{aligned}$$

As in the proof of Theorem 3.7, for each $\boldsymbol{\alpha}$, $\beta_{\boldsymbol{\alpha}}$ is the unique β such that $S(\boldsymbol{\alpha}, \beta)$ and we have

$$\begin{aligned} (\mathbf{m}, \boldsymbol{\alpha}) \in \bar{\Gamma} &\leftrightarrow \exists \beta\, [\mathsf{S}(\boldsymbol{\alpha}, \beta) \wedge \exists r\,.\, \beta(\langle r+1, \mathbf{m}\rangle) = 0] \\ &\leftrightarrow \forall \beta\, [\mathsf{S}(\boldsymbol{\alpha}, \beta) \rightarrow \exists r\,.\, \beta(\langle r+1, \mathbf{m}\rangle) = 0]. \quad \square \end{aligned}$$

3.18 Theorem. *For any $r \geqslant 2$ and any decomposable inductive operator Γ over* ${}^{k,l}\omega$,

(i) $\Gamma \in \Delta_r^1 \rightarrow \bar{\Gamma} \in \Delta_r^1$;

(ii) Γ *monotone* $\wedge\, \Gamma \in \Sigma_r^1 \rightarrow \bar{\Gamma} \in \Sigma_r^1$.

Proof. By modifications of the proof of Theorem 3.9 similar to those in the preceding proof we obtain for $\Gamma \in \Delta_r^1$ relations $\mathsf{V}_\Sigma^{(\ )}$ and $\mathsf{V}_\Sigma \in \Sigma_r^1$ and $\mathsf{V}_\Pi^{(\ )}$ and $\mathsf{V}_\Pi \in \Pi_r^1$ such that for all $\gamma \in \mathsf{W}$,

$$\mathbf{m} \in \Gamma_{\boldsymbol{\alpha}}^{(\|\gamma\|)} \leftrightarrow \mathsf{V}_\Sigma^{(\ )}(\mathbf{m}, \boldsymbol{\alpha}, \gamma) \leftrightarrow \mathsf{V}_\Pi^{(\ )}(\mathbf{m}, \boldsymbol{\alpha}, \gamma)$$

and

$$\mathbf{m} \in \Gamma_{\boldsymbol{\alpha}}^{\|\gamma\|} \leftrightarrow \mathsf{V}_\Sigma(\mathbf{m}, \boldsymbol{\alpha}, \gamma) \leftrightarrow \mathsf{V}_\Pi(\mathbf{m}, \boldsymbol{\alpha}, \gamma).$$

Then the proof may be completed as above. The construction for (ii) is similar. $\square$

As usual the results of this section may be relativized to any $\beta \in {}^\omega\omega$. In particular we shall later have occasion to use the "boldface" versions of Theorems 3.15 and 3.18: for any $r > 0$ and any decomposable operator Γ over ${}^{k,l}\omega$,

$$\Gamma \text{ monotone} \wedge \Gamma \in \mathbf{\Pi}^1_r \rightarrow \bar{\Gamma} \in \mathbf{\Pi}^1_r;$$

and for $r \geqslant 2$,

$$\Gamma \in \mathbf{\Delta}^1_r \rightarrow \bar{\Gamma} \in \mathbf{\Delta}^1_r$$

and

$$\Gamma \text{ monotone} \wedge \Gamma \in \mathbf{\Sigma}^1_r \rightarrow \bar{\Gamma} \in \mathbf{\Sigma}^1_r.$$

For reference, we summarize in a table the results concerning the relative complexity of Γ and $\bar{\Gamma}$:

Operator Γ	Σ^0_1 monotone	Σ^0_1	Π^0_1 monotone	Arithmetical monotone	Π^1_r monotone $(r \geq 1)$	Σ^1_r monotone $(r \geq 2)$	Δ^1_r $(r \geq 2)$
Closure $\bar{\Gamma}$	Σ^0_1	Δ^1_1	Π^1_1	Π^1_1	Π^1_r	Σ^1_r	Δ^1_r

3.19–3.34 Exercises

3.19. Let D be the smallest subset of ω such that $0 \in D$ and for all a, if $\forall p.\{a\}(p) \in D$, then $a \in D$. Show that $D \in \Pi^1_1$ and every Π^1_1 relation on numbers is reducible to D (D is Π^1_1-*complete*).

3.20. Let $\mathscr{L}$ be a countable first-order language and $\mathscr{T}$ a theory of $\mathscr{L}$ which under some standard Gödel numbering is recursively axiomatizable. Sketch a proof that both $\mathrm{Fm} = \{n: n \text{ is the Gödel number of some formula of } \mathscr{L}\}$ and $\mathrm{Th} = \{n: n \text{ is the Gödel number of a theorem of } \mathscr{T}\}$ are both of the form $\bar{\Gamma}$ for monotone Σ^0_1 operators Γ. What difference in these inductive definitions accounts for the fact that Fm is recursive, whereas in general Th is only Σ^0_1?

3.21. Show that for any monotone Σ^0_1 operator Γ, there exists a partial recursive function f such that

(i) for all e and m,

$$f(e, m) \simeq \begin{cases} 0, & \text{if } \ m \in \Gamma(\{p: \{e\}(p) \simeq 0\}); \\ \text{undefined, otherwise;} \end{cases}$$

(ii) there exists an index $\bar{e}$ such that for all m,

$$f(\bar{e}, m) \simeq \{\bar{e}\}(m) \quad \text{and} \quad \{\bar{e}\}(m) \simeq 0 \leftrightarrow m \in \bar{\Gamma}.$$

3.22 (Moschovakis [1972]). The Π^1_1 relations are exactly those expressible in the form $\forall \beta\, \exists n\, \mathsf{P}(\bar{\beta}(n), \mathbf{m}, \boldsymbol{\alpha})$ with P recursive. The function quantifier may be thought of as an infinite string of number quantifiers thus:

$$\forall p_0 \forall p_1 \forall p_2 \ldots \exists n\, \mathsf{P}(\langle p_0, \ldots, p_{n-1}\rangle, \mathbf{m}, \boldsymbol{\alpha}).$$

The purpose of this exercise is to establish that the class of relations expressible in the form

$$(*) \qquad \forall p_0 \exists p_1 \forall p_2 \exists p_3 \forall p_4 \ldots \exists n\, \mathsf{P}(\langle p_0, \ldots, p_{n-1}\rangle, \mathbf{m}, \boldsymbol{\alpha})$$

is also exactly Π^1_1. The expression (*) may be interpreted in several equivalent ways. We shall take it here to mean

$$\exists \gamma\, \forall p_0 \forall p_2 \forall p_4 \ldots \exists n\, \mathsf{P}(\langle p_0, \gamma(\langle p_0\rangle), p_2, \gamma(\langle p_0, p_2\rangle), \ldots\rangle, \mathbf{m}, \boldsymbol{\alpha})$$

where the second ... terminates with p_{n-1} if n is odd and with $\gamma(\langle p_0, \ldots, p_{n-2}\rangle)$ if n is even. (The function γ may be thought of either as a Skolem Function or as a strategy for a certain infinite game — cf. V.3 and the discussion preceding Definition V.4.7.)

It is easy to see that every Π^1_1 relation is expressible in the form (*) with P recursive. For the converse, show first that the same class of relations is expressible in the form

$$\exists p_0 \forall p_1 \exists p_2 \forall p_3 \exists p_4 \ldots \exists n\, \mathsf{P}(\langle p_0, \ldots, p_{n-1}\rangle, \mathbf{m}, \boldsymbol{\alpha}),$$

that is,

$$\exists \delta\, \forall p_1 \forall p_3 \forall p_5 \ldots \exists n\, \mathsf{P}(\langle \delta(\langle\ \rangle), p_1, \delta(\langle p_1\rangle), p_3, \ldots\rangle, \mathbf{m}, \boldsymbol{\alpha}).$$

Then imitate the proof of Theorem 3.2 to show that every relation (*) is reducible to the closure of a Δ^0_2 monotone operator.

3.23. Extend the reasoning of the preceding exercise to show that every relation definable in the form

$$\forall \beta_0 \exists p_0 \forall \beta_1 \exists p_1 \ldots \exists n\, \mathsf{P}(\langle \bar{\beta}_0(p_0), \ldots, \bar{\beta}_{n-1}(p_{n-1})\rangle, \mathbf{m}, \boldsymbol{\alpha})$$

with P recursive is Π^1_1.

3.24. Show that every Σ^0_1 set is reducible to $\bar{\Gamma}$ for some $\Gamma \in \Delta^0_1$.

3.25 (Richter). Show that for each $n \geqslant 1$, there exists a Π^0_1 inductive operator Γ such that $|\Gamma| = \omega^n$. (For any two operators Γ_0 and Γ_1, let

$$[\Gamma_0, \Gamma_1](A) = \begin{cases} \Gamma_0(A), & \text{if } \Gamma_0 \not\subseteq A; \\ \Gamma_1(A), & \text{otherwise.} \end{cases}$$

Choose appropriate Γ_0 and Γ_1 such that $\Gamma = [\Gamma_0, \Gamma_1]$ is Π_1^0 and has closure ordinal ω^2.)

3.26. Show that there exist (non-monotone) inductive operators $\Gamma \in \Pi_2^0$ such that $\bar{\Gamma} \notin \Pi_1^1$. (Choose Γ_0 such that $\bar{\Gamma}_0 \in \Pi_1^1 \sim \Delta_1^1$, set

$$\Gamma_1(A) = \{\langle 0, m\rangle: m \in \Gamma_0(\{p: \langle 0, p\rangle \in A\})\}$$

and

$$\Gamma(A) = \Gamma_1(A) \cup \{\langle 1, m\rangle: \Gamma_1(A) \subseteq A \wedge m \notin A\}.)$$

3.27. Show that the recursive ordinals are closed under ordinal addition and multiplication.

3.28. Show that for X any of Σ_r^i or Π_r^i ($i = 0, 1$; $r > 0$), there exists an "X-universal" operator Γ_* — that is, $\Gamma_* \in X$ and for any $\Gamma \in X$ there exists an a such that

$$m \in \bar{\Gamma} \leftrightarrow \langle a, m\rangle \in \bar{\Gamma}_*.$$

3.29. Find a monotone operator $\Gamma \in \Sigma_1^1$ such that $\bar{\Gamma} \notin \Sigma_1^1 \cup \Pi_1^1$.

3.30. Show that not every Δ_2^1 subset of ω is reducible to $\bar{\Gamma}$ for some monotone Σ_1^1 operator Γ.

3.31. Complete the proofs of Theorems 3.17 and 3.18.

3.32. Let X be any of the classes Σ_r^i or Π_r^i ($i = 0$ or 1, $r \geq 0$) and $\bar{X}$ the class of relations reducible to some $\bar{\Gamma}$ with $\Gamma \in X$. Show that

(i) $\bar{X}$ is closed under $\exists^0$;

(ii) if X is not Σ_0^0, Π_0^0, or Σ_1^0, then $\bar{X}$ is closed under $\forall^0$ and indeed under the Suslin quantifier $\mathscr{A}$, where

$$(\mathscr{A} R)(\mathbf{m}) \leftrightarrow \forall \beta\, \exists n\, R(\bar{\beta}(n), \mathbf{m}).$$

3.33 (Aczel). For any monotone operator Γ over ω, let

$$|m|_\Gamma = \begin{cases} \text{least } \sigma . m \in \Gamma^\sigma, & \text{if } \quad m \in \bar{\Gamma}; \\ |\Gamma|, & \text{otherwise}; \end{cases}$$

$$m \leqslant_\Gamma n \leftrightarrow m \in \bar{\Gamma} \wedge |m|_\Gamma \leqslant |n|_\Gamma;$$

$$m <_\Gamma n \leftrightarrow m \in \bar{\Gamma} \wedge |m|_\Gamma < |n|_\Gamma;$$

$$\Gamma^\circ(A) = \sim \Gamma(\sim A).$$

(i) Show that for any $\tau < |\Gamma|$ and any $m \in \Gamma^\tau$

$$|m|_\Gamma < |n|_\Gamma \leftrightarrow n \in \Gamma^\circ(\{q : |m|_\Gamma \leqslant |q|_\Gamma\})$$

and

$$|m|_\Gamma \leqslant |q|_\Gamma \leftrightarrow m \in \Gamma(\{p : p \in \Gamma^{(\tau)} \wedge |p|_\Gamma < |q|_\Gamma\}).$$

(ii) Let Δ be the operator defined by:

$$\Delta(A) = \{\langle m, n\rangle : n \in \Gamma^\circ(\{q : m \in \Gamma(\{p : \langle p, q\rangle \in A\})\})\}.$$

Show that for all m and n and all $\tau < |\Gamma|$,

$$m \in \Gamma^\tau \wedge |m|_\Gamma < |n|_\Gamma \leftrightarrow \langle m, n\rangle \in \Delta^\tau,$$

and

$$m <_\Gamma n \leftrightarrow \langle m, n\rangle \in \bar{\Delta}.$$

(iii) Similarly, show

$$n \leqslant_\Gamma m \leftrightarrow \langle m, n\rangle \in \bar{\Delta}^\circ.$$

3.34. Use the results of the preceding exercise together with Theorem 3.2 to show that:

(i) the class of Π^1_1 relations on numbers has the reduction property;

(ii) for any monotone arithmetical operator Γ, $|\Gamma| \leqslant \delta^1_1$, the least ordinal not the order-type of a Δ^1_1 well-ordering of ω; $\bar{\Gamma} \in \Delta^1_1$ iff $|\Gamma| < \delta^1_1$;

(iii) (Cenzer [1974a]) for any $r \geqslant 2$ and any monotone operator $\Gamma \in \Delta^1_r$, $|\Gamma| < \delta^1_r$.

3.35 Notes. Theorem 3.2 is essentially due to Kleene [1955a], but the simple direct proof given here seems to appear first in Lorenzen-Myhill [1959]. Theorem 3.10 is from Putnam [1964], although it is stated there in very different terms.

4. Implicit Definability and Bases

If $A \subseteq {}^{\omega}\omega$ is a set of given complexity, what can one say about the complexity of the elements of A? Since ${}^{\omega}\omega$ is recursive and has elements of all complexities, we shall at best be able to prove that a simple set has some simple elements. This leads to the notion of basis:

4.1 Definition. For any class X of relations and any $B \subseteq {}^{\omega}\omega$, B is a *basis* for X iff for all $A \in X$,

$$\exists\alpha\,.\,\alpha \in A \rightarrow (\exists\alpha \in B)\,.\,\alpha \in A.$$

We shall present in this section a number of positive results concerning bases along with counterexamples to indicate that these are optimal in the sense that they fail if B is reduced or X enlarged. Among the classes X to be considered are classes of *singletons*. If $\{\beta\} \in \Delta^i_r$ (etc.) we say that β is *implicitly* Δ^i_r. Note that B is a basis for the class of Δ^i_r singletons iff every implicitly Δ^i_r function belongs to B.

4.2 Theorem. *Δ^0_1 is a basis for Σ^0_1.*

Proof. The set of recursive (Δ^0_1) functions is dense and hence intersects every open (Σ^0_1) set. □

4.3 Theorem. *$\Delta^0_{(\omega)}$ is not a basis for the class of Π^0_2 singletons.*

Proof. Let Γ be the inductive operator of Theorem 3.6 and β^* and S be as in the proof of Theorem 3.7 for this Γ. Since

$$\beta^*(\langle r+1, \langle t, a, \mathbf{m}\rangle\rangle) = 0 \leftrightarrow \langle t, a, \mathbf{m}\rangle \in \Gamma^r$$
$$\leftrightarrow 0 < t \leq r+1 \wedge U^0_t(a, \langle\mathbf{m}\rangle)$$

β^* is non-arithmetical. On the other hand, β^* is the unique member of S and it is routine to check that $S \in \Pi^0_2$. □

4.4 Lemma. *For any β, if β is implicitly arithmetical, then β is recursive in some function γ which is implicitly Π^0_1.*

Proof. We shall prove the conclusion for $\{\beta\} \in \Pi^0_r$ by induction on r. If $r = 1$, there is nothing to prove. Suppose $\{\beta\} \in \Pi^0_{r+1}$ with $r > 0$ and let R be a Δ^0_r relation such that

$$\forall\alpha\,[\alpha = \beta \leftrightarrow \forall m\,\exists n\,R(m, n, \alpha)].$$

In particular, $\forall m\, \exists n\, \mathsf{R}(m, n, \beta)$ so define

$$\delta(m) = \langle \beta(m), \text{least } n\,.\, \mathsf{R}(m, n, \beta)\rangle.$$

Since for all m, $\beta(m) = (\delta(m))_0$, β is recursive in δ. We claim that

$$(*) \qquad \begin{aligned} \forall \alpha\, (\alpha = \delta \leftrightarrow \mathrm{Sq}_1(\alpha) \wedge \mathrm{lg}(\alpha) = 2 \wedge \forall m\, \mathsf{R}(m, (\alpha(m))_1, (\alpha)_0) \\ \wedge\, \forall m\, \forall p\, [p < (\alpha(m))_1 \rightarrow \sim \mathsf{R}(m, p, (\alpha)_0)]), \end{aligned}$$

and thus that $\{\delta\} \in \Pi^0_r$. Then by the induction hypothesis, δ is recursive in some implicitly Π^0_1 function γ and thus so is β.

The implication $(\rightarrow)$ of $(*)$ is obvious from the definition of δ. Conversely, if α satisfies the right-hand side, then from the third clause follows that $\forall m\, \exists n\, \mathsf{R}(m, n, (\alpha)_0)$ and thus that $(\alpha)_0 = \beta$. Then from the fourth clause it is immediate that $(\alpha(m))_1 = \text{least } n\,.\, \mathsf{R}(m, n, \beta)$ and thus that $\alpha = \delta$. □

4.5 Corollary. *$\Delta^0_{(\omega)}$ is not a basis for the class of Π^0_1 singletons.*

Proof. Immediate from 4.3 and 4.4. □

In the positive direction, we have from the examples of 2.3:

4.6 Lemma. *For all $r > 0$, Δ^1_r is a basis for the class of Σ^1_r singletons.* □

Hence in particular, Δ^1_1 is a basis for the class of Π^0_1 singletons. To settle the question for arbitrary Π^0_1 sets we shall need a result from the next chapter:

$$\text{(IV.2.6)} \qquad \{\alpha : \alpha \in \Delta^1_1\} \in \Pi^1_1$$

together with the following Lemma.

4.7 Lemma. *For any $r > 0$ and any set B which is closed under "recursive in", if B is a basis for Π^0_1 (Π^1_r), then B is also a basis for Σ^1_1 (Σ^1_{r+1}).*

Proof. Suppose $\alpha \in \mathsf{A} \leftrightarrow \exists \beta\, \mathsf{R}(\alpha, \beta)$, $\mathsf{R} \in \Pi^0_1$, and B is a basis for Π^0_1. Let $\mathsf{C} = \{\gamma : \mathsf{R}((\gamma)_0, (\gamma)_1)\}$. If $\mathsf{A} \neq \varnothing$, also $\mathsf{C} \neq \varnothing$ so there exists $\gamma \in \mathsf{C} \cap \mathsf{B}$. Then $(\gamma)_0 \in \mathsf{A}$ and as $(\gamma)_0$ is recursive in γ, also $(\gamma)_0 \in \mathsf{B}$, so $\mathsf{A} \cap \mathsf{B} \neq \varnothing$. The proof for $\mathsf{R} \in \Pi^1_r$ is identical. □

4.8 Corollary. *Δ^1_1 is not a basis for Π^0_1.*

Proof. Let $\mathsf{A} = \{\alpha : \alpha \notin \Delta^1_1\}$. By IV.2.6, $\mathsf{A} \in \Sigma^1_1$, but clearly A has no Δ^1_1 element. Hence Δ^1_1 is not a basis for Σ^1_1, so by Lemma 4.7 also not for Π^0_1. □

Note that since a function α is Σ^1_1 or Π^1_1 just in case it is Δ^1_1, neither Σ^1_1 nor Π^1_1 is a basis for Π^0_1. However,

4.9 Kleene Basis Theorem. $\{\alpha\colon \alpha$ *is recursive in some* $B \in \Sigma^1_1\}$ *is a basis for* Π^0_1, *hence also for* Σ^1_1.

Proof. Let A be a non-empty Π^0_1 set. Then for some recursive set $A \subseteq \omega$,

$$\alpha \in \mathsf{A} \leftrightarrow \forall m\, [\bar{\alpha}(m) \in A].$$

Let

$$B = \{s\colon \exists \alpha\, \forall m\, [s * \bar{\alpha}(m) \in A]\}.$$

Then $B \in \Sigma^1_1$ and from the assumption $\mathsf{A} \neq \emptyset$ it follows that $\langle\ \rangle \in B$. Furthermore, $(\forall s \in B) \exists n.\, s * \langle n \rangle \in B$. Thus there is a unique function β such that for all m,

$$\beta(m) = \text{least } n[\bar{\beta}(m) * \langle n \rangle \in B].$$

Clearly β is recursive in B. Since $B \subseteq A$ and for all m, $\bar{\beta}(m) \in B$, also $\beta \in \mathsf{A}$. $\square$

In the remainder of this section we investigate the basis properties of various special arithmetical classes. We have not yet developed techniques to deal successfully with most basis questions in the analytical hierarchy, but for comparison we mention some results from later chapters:

(V.6.1) $\{\alpha\colon \alpha$ is recursive in some $B \in \Sigma^1_1\}$ is not a basis for the class of Π^1_1 singletons;

(IV.7.9) Δ^1_2 is a basis for Π^1_1, hence for Σ^1_2;

(IV.2.20) every $R \in \Delta^1_1$ is recursive in some implicitly Π^0_1 function γ;

but

(III.6.11) there exist $\alpha \in \Delta^1_1$ such that $\{\alpha\} \notin \Delta^0_{(\omega)}$.

The existence of bases for Π^1_2 and higher classes in the analytical hierarchy is independent of the axioms of set theory. We shall show in § V.2 that it is consistent that for all $r \geq 2$, Δ^1_r be a basis for Σ^1_r. However it is also known to be consistent that $\Delta^1_{(\omega)}$ not be a basis for Π^1_2 (Lévy [1965a]).

4.10 Kreisel Basis Theorem. Δ^0_2 *is a basis for* $\{\mathsf{A}\colon \mathsf{A} \in \Sigma^0_2 \wedge \mathsf{A} \cap {}^\omega 2 \neq \emptyset\}$.

Proof. Suppose that $\mathsf{A}=\{\alpha: \exists p\,\forall m\,R(p,\bar{\alpha}(m))\}$ with R recursive and $\mathsf{A}\cap {}^\omega 2\neq\varnothing$. Then for some p_0 there exists a function $\alpha^*\in{}^\omega 2$ such that $\forall m\,R(p_0,\bar{\alpha}^*(m))$. Let

$$B=\{s: (\exists\alpha^*\in{}^\omega 2)\forall m\,R(p_0, s*\bar{\alpha}^*(m))\}.$$

As in the proof of Theorem 4.9 there exists a function $\beta\in\mathsf{A}$ which is recursive in B. In this case, however, we shall show $B\in\Pi_1^0$ which by Post's Theorem (1.12) implies $\beta\in\Delta_2^0$.

By Exercise I.2.6,

$$s\in B\leftrightarrow\forall n\,(\exists\alpha^*\in{}^\omega 2)(\forall m<n)R(p_0, s*\bar{\alpha}^*(m)).$$

Let δ be the primitive recursive function defined by: $\delta(0)=1$ and $\delta(n+1)=\delta(n)*\langle 1\rangle$. Any code for a sequence of 0's and 1's of length at most n is less than $\delta(n)$. Hence

$$s\in B\leftrightarrow\forall n\,(\exists t\leqslant\delta(n))(\forall m<n)[(t)_m\leqslant 1\wedge R(p_0, s*\langle(t)_0,\ldots,(t)_{m-1}\rangle)]$$

which implies $B\in\Pi_1^0$. □

The following shows that this is the best possible result.

4.11 Theorem. (i) *There exists a non-empty* Π_1^0 *set* $A\subseteq{}^\omega 2$ *which contains no characteristic function of a* Σ_1^0 *or* Π_1^0 *set;*

(ii) *there exists a non-empty* Π_2^0 *set* $\mathsf{B}\subseteq{}^\omega 2$ *which contains no* Δ_1^1 *function.*

Proof. Corollary II.4.22 asserts that the class of semi-recursive relations does not have the separation property. The proof actually shows that there exist semi-recursive *sets* $A, B\subseteq\omega$ such that no recursive set C separates A and B. For any set D and all $i<2$, let $D_i=\{m: \langle i,m\rangle\in D\}$. Set

$$\mathsf{A}=\{\mathsf{K}_D: A\subseteq D_0\subseteq{\sim}B\wedge D_0={\sim}D_1\}.$$

Since $A, B\in\Sigma_1^0$, it follows directly that $\mathsf{A}\in\Pi_1^0$. Suppose that there is a $D\in\Sigma_1^0$ such that $\mathsf{K}_D\in\mathsf{A}$. Then $D_0, D_1\in\Sigma_1^0$ so D_0 is a recursive set which separates A and B, contrary to assumption.

We established in Corollary 4.8 that there exists a non-empty Π_1^0 set with no Δ_1^1 elements. If C is such a set, it suffices for (ii) to take $\mathsf{B}=\{\mathsf{K}_{\mathrm{Gr}(\alpha)}: \alpha\in\mathsf{C}\}$ (Exercise 4.26). □

Note that without using 4.8 (which depends on IV.2.6) we have already in the proof of Theorem 4.3 a Π_2^0 subset of ${}^\omega 2$ with no $\Delta_{(\omega)}^0$ element.

The theme of the remainder of this section is that a set of a given complexity which is in some sense "large" is more likely to contain some simple elements than a "small" set of the same complexity. In particular, this is true when we interpret "large" by "non-meager" or "of positive measure".

4.12 Lemma. *Δ_1^0 is a basis for the class of Π_2^0 sets which are dense in some interval.*

Proof. Suppose $\mathsf{A} = \{\alpha : \forall m\, \exists n\, R(m, \bar{\alpha}(n))\}$ with R recursive, and A is dense in $[s_0]$. Then for any s which extends s_0 and any m, there is a sequence t compatible with s such that $R(m, t)$ holds (cf. proof of II.5.3). Hence if we set

$$s_{m+1} = \text{least } t[\mathrm{Sq}(t) \wedge s_m \subsetneq t \wedge (\exists u \subseteq t) R(m, u)],$$

then s_m is defined for all m and $s_0 \subseteq s_1 \subseteq \cdots \subseteq s_m \subseteq \cdots$. Hence there is a unique function α which is the "limit" of the s_m, given by $\alpha(m) = (s_m)_m$. Clearly α is recursive and belongs to A.

4.13 Corollary. *Δ_1^0 is a basis for the class of non-meager Σ_3^0 sets.*

Proof. A Σ_3^0 set is a countable union of Π_2^0 sets. If it is non-meager, one of these must be dense in some interval and hence contain a recursive function. □

An interesting by-product is the classification of the class of recursive functions:

4.14 Corollary. $\{\alpha : \alpha \in \Delta_1^0\} \in \Sigma_3^0 \sim \Delta_3^0$.

Proof. It was shown as one of the Examples 1.3 that the set A of recursive functions is Σ_3^0. Suppose also $\mathsf{A} \in \Pi_3^0$ so that $\sim\mathsf{A} \in \Sigma_3^0$. As A is denumerable it is meager, so by the Baire Category Theorem (I.2.2), $\sim$A is non-meager. But then by Corollary 4.13, $\sim$A contains a recursive element, a contradiction. □

Note that it also follows from this argument that Δ_1^0 is not a basis for the class of non-meager or even co-meager Π_3^0 sets. Also, since the class of primitive recursive functions is Σ_2^0, this class does not form a basis for the class of non-meager Π_2^0 sets.

We shall extend these results to all levels of the arithmetical hierarchy in § 6. If we replace "non-meager" by "of positive measure" the results have a similar flavor but are weaker in the sense that the bases are larger. First, in contrast with 4.13,

4.15 Theorem. *Δ_1^0 is not a basis for either of*

(i) *the class of Π_1^0 sets of positive measure, or*

(ii) *the class of Σ_2^0 sets of measure* 1.

Proof. For each n and a let

$$A_{n,a} = \{\alpha : (\forall m < n+a+1)\{a\}(m) \simeq \alpha(m)\}.$$

Each $A_{n,a}$ is either empty or is an interval of length $n+a+1$. In either case, since the measure of an interval of length k is at most 2^{-k}, $\text{mes}(A_{n,a}) \leq 2^{-n} \cdot 2^{-(a+1)}$. Hence for each n, $A_n = \bigcup\{A_{n,a} : a \in \omega\}$ has measure at most $\Sigma_{a=0}^{\infty} 2^{-n} \cdot 2^{-(a+1)} = 2^{-n}$. Furthermore all recursive functions belong to every A_n, so that each $\sim A_n$ is a Π_1^0 set of positive measure with no recursive elements, and $\sim \bigcap\{A_n : n \in \omega\}$ is a Σ_2^0 set of measure 1 with no recursive elements. □

On the other hand, we have

4.16 Theorem. *Δ_1^0 is a basis for the class of Π_2^0 sets of measure* 1.

Proof. A Π_2^0 set A of measure 1 is a countable intersection of Σ_1^0 sets A_n each also of measure 1. Clearly each A_n is open and dense so that $\sim A_n$ is nowhere dense. Thus $\sim A = \bigcup\{\sim A_n : n \in \omega\}$ is meager and by the Baire Category Theorem, A is non-meager so has a recursive element by Corollary 4.13. □

In the remainder of this section we compute for all r bases for the class of Σ_r^0 sets of positive measure. For the next lemma only, if F is any total functional we set for any R,

$$R_{\text{mes}}^{F}(\mathbf{m}, \boldsymbol{\alpha}) \leftrightarrow \text{mes}\{\beta : R(\mathbf{m}, \boldsymbol{\alpha}, \beta)\} \leq \frac{1}{F(\mathbf{m}, \boldsymbol{\alpha})}.$$

4.17 Lemma. *For all $r > 0$, all ε, any recursive functional* F *and any* R, *if* $R \in \Sigma_r^0[\varepsilon]$, *then* $R_{\text{mes}}^{F} \in \Pi_r^0[\varepsilon]$.

Proof. We proceed by induction on r and omit reference to ε. Suppose first that $R \in \Sigma_1^0$, say

$$R(\mathbf{m}, \boldsymbol{\alpha}, \beta) \leftrightarrow \exists n\, S(\bar{\beta}(n), \mathbf{m}, \boldsymbol{\alpha})$$

for a recursive S. Let θ be the following function from ${}^{k+1,l}\omega$ into the rational interval [0, 1]:

$$\theta(s, \mathbf{m}, \boldsymbol{\alpha}) = \begin{cases} \text{mes}[s], & \text{if } \; s \in \text{Sq} \wedge S(s, \mathbf{m}, \boldsymbol{\alpha}) \wedge (\forall t \subsetneq s) \sim S(t, \mathbf{m}, \boldsymbol{\alpha}); \\ 0, & \text{otherwise.} \end{cases}$$

Since $\{\beta : R(\mathbf{m}, \boldsymbol{\alpha}, \beta)\}$ is the disjoint union of the intervals $[s]$ for which $\theta(s, \mathbf{m}, \boldsymbol{\alpha}) > 0$, we have

$$\begin{aligned} R_{\text{mes}}^{F}(\mathbf{m}, \boldsymbol{\alpha}) &\leftrightarrow \sum_{s=0}^{\infty} \theta(s, \mathbf{m}, \boldsymbol{\alpha}) \leq \frac{1}{F(\mathbf{m}, \boldsymbol{\alpha})} \\ &\leftrightarrow \forall q \left(\sum_{s=0}^{q} \theta(s, \mathbf{m}, \boldsymbol{\alpha}) \leq \frac{1}{F(\mathbf{m}, \boldsymbol{\alpha})} \right). \end{aligned}$$

Since if $s \in \mathrm{Sq}$ and $\lg(s) = k$, $\mathrm{mes}[s] = 2^{-((s)_0 + \cdots + (s)_{k-1} + k)}$ (cf. end of § I.2), a simple calculation shows that there exist recursive H and I such that

$$\sum_{s=0}^{q} \theta(s, \mathbf{m}, \boldsymbol{\alpha}) = \frac{\mathsf{H}(q, \mathbf{m}, \boldsymbol{\alpha})}{\mathsf{I}(q, \mathbf{m}, \boldsymbol{\alpha})}.$$

Thus

$$\mathsf{R}^{\mathsf{F}}_{\mathrm{mes}}(\mathbf{m}, \boldsymbol{\alpha}) \leftrightarrow \forall q\, [\mathsf{F}(\mathbf{m}, \boldsymbol{\alpha}) \cdot \mathsf{H}(q, \mathbf{m}, \boldsymbol{\alpha}) \leqslant \mathsf{I}(q, \mathbf{m}, \boldsymbol{\alpha})].$$

Assume now as induction hypothesis that the result holds for $1, \ldots, r$. We consider first $\mathsf{R} \in \Pi^0_r$, say $\mathsf{R}(\mathbf{m}, \boldsymbol{\alpha}, \beta) \leftrightarrow \forall p\, \mathsf{S}(p, \mathbf{m}, \boldsymbol{\alpha}, \beta)$, with $\mathsf{S} \in \Sigma^0_{r-1}$. Then

$$\mathsf{R}^{\mathsf{F}}_{\mathrm{mes}}(\mathbf{m}, \boldsymbol{\alpha}) \leftrightarrow \forall n\, \exists q \left[\mathrm{mes}\{\beta : (\forall p \leqslant q)\mathsf{S}(p, \mathbf{m}, \boldsymbol{\alpha}, \beta)\} \leqslant \frac{1}{\mathsf{F}(\mathbf{m}, \boldsymbol{\alpha})} + \frac{1}{n}\right]$$

so by the induction hypothesis $\mathsf{R}^{\mathsf{F}}_{\mathrm{mes}} \in \Pi^0_{r+1}$.

Now if $\mathsf{R} \in \Sigma^0_{r-1}$, say $\mathsf{R}(\mathbf{m}, \boldsymbol{\alpha}, \beta) \leftrightarrow \exists p\, \mathsf{P}(p, \mathbf{m}, \boldsymbol{\alpha}, \beta)$ with $\mathsf{P} \in \Pi^0_r$ we have

$$\mathsf{R}^{\mathsf{F}}_{\mathrm{mes}}(\mathbf{m}, \boldsymbol{\alpha}) \leftrightarrow \forall q \left[\mathrm{mes}\{\beta : (\exists p \leqslant q)\mathsf{P}(p, \mathbf{m}, \boldsymbol{\alpha}, \beta)\} \leqslant \frac{1}{\mathsf{F}(\mathbf{m}, \boldsymbol{\alpha})}\right]$$

and again $\mathsf{R}^{\mathsf{F}}_{\mathrm{mes}} \in \Pi^0_{r+1}$. □

4.18 Lemma. *For all r and ε, and any* $\mathsf{R} \in \Pi^0_r[\varepsilon]$, *if* $\mathrm{mes}\{\alpha : \forall p\, \exists q\, \mathsf{R}(p, q, \alpha)\} > 0$, *then there exists a* $\beta \in \Delta^0_{r+2}[\varepsilon]$ *such that* $\mathrm{mes}\{\alpha : \forall p\, (\exists q \leqslant \beta(p))\mathsf{R}(p, q, \alpha)\} > 0$.

Proof. Let $\mathsf{A} = \{\alpha : \forall p\, \exists q\, \mathsf{R}(p, q, \alpha)\}$, $\mathsf{A}_p = \{\alpha : \exists q\, \mathsf{R}(p, q, \alpha)\}$ and $\mathsf{A}_{p,n} = \{\alpha : (\exists q \leqslant n)\mathsf{R}(p, q, \alpha)\}$. Suppose $\mathrm{mes}\, \mathsf{A} > 1/2^m$ and define β by

$$\beta(p) = \text{least } n \left[\mathrm{mes}(\mathsf{A}_p \sim \mathsf{A}_{p,n}) \leqslant \frac{1}{2^{m+p+2}}\right].$$

As $\mathsf{A}_p \sim \mathsf{A}_{p,n} \in \Sigma^0_{r+1}[\varepsilon]$, it follows from Lemma 4.15 that $\beta \in \Delta^0_{r+2}[\varepsilon]$.

Let $\mathsf{B} = \{\alpha : \forall p\, (\exists q \leqslant \beta(p))\mathsf{R}(p, q, \alpha)\}$. Then since

$$\mathsf{A} = \bigcap \{\mathsf{A}_p : p \in \omega\} \subseteq B \cup \bigcup \{\mathsf{A}_p \sim \mathsf{A}_{p,\beta(p)} : p \in \omega\},$$

we have

$$\frac{1}{2^m} < \mathrm{mes}\, \mathsf{A} \leqslant \mathrm{mes}\, \mathsf{B} + \sum_{p=0}^{\infty} \frac{1}{2^{m+p+2}} = \mathrm{mes}\, \mathsf{B} + \frac{1}{2^{m+1}}$$

which implies $\text{mes}\,\mathsf{B} > 1/2^{m+1} > 0$. □

4.19 Lemma. *For all r, β, ε, and all $\mathsf{R} \in \Pi^0_r[\varepsilon]$, if $\mathsf{S}(\mathbf{m}, \boldsymbol{\alpha}) \leftrightarrow \forall p\,(\exists q \leqslant \beta(p))\mathsf{R}(p, q, \mathbf{m}, \boldsymbol{\alpha})$, then $\mathsf{S} \in \Pi^0_{\max(1,r)}[\beta, \varepsilon]$.*

Proof. We leave this straightforward calculation to the reader. □

Now let $f(r)$ denote the greatest integer not exceeding $r^2/4 + 1$.

4.20 Lemma. *For any r and ε, and any $\mathsf{A} \in \Pi^0_r[\varepsilon]$ such that* $\text{mes}\,\mathsf{A} > 0$, *there exists $\gamma \in \Delta^0_{f(r)}[\varepsilon]$ and $\mathsf{C} \in \Pi^0_1[\gamma]$ such that $\mathsf{C} \subseteq \mathsf{A}$ and also* $\text{mes}\,\mathsf{C} > 0$.

Proof. We proceed by induction on r. For $r = 0$ or 1 the result is obvious. Suppose $\mathsf{A} \in \Pi^0_{r+2}[\varepsilon]$. By the two preceding lemmas there exist $\mathsf{B} \subseteq \mathsf{A}$ and $\beta \in \Delta^0_{r+2}[\varepsilon]$ such that $\mathsf{B} \in \Pi^0_{\max(1,r)}[\beta, \varepsilon]$ and $\text{mes}\,\mathsf{B} > 0$. By the induction hypothesis there exist $\gamma \in \Delta^0_{f(r)}[\beta, \varepsilon]$ and $\mathsf{C} \in \Pi^0_1[\gamma]$ such that $\mathsf{C} \subseteq \mathsf{B}$ and $\text{mes}\,\mathsf{C} > 0$. Then by the Arithmetical Substitution Theorem, $\gamma \in \Delta^0_{f(r)+r+1}[\varepsilon] = \Delta^0_{f(r+2)}[\varepsilon]$ as required. □

4.21 Theorem. *For all r, $\Delta^0_{f(r)+2}$ is a basis for the class of Π^0_r sets of positive measure.*

Proof. Let A be a Π^0_r set of positive measure and let γ and C be as in Lemma 4.20. Then there exists a unique function δ such that for all m,

$$\delta(m) = \text{least } \mathrm{n}\,.\,\text{mes}(\mathsf{C} \cap [\bar{\delta}(m) * \langle n \rangle]) > 0.$$

Since $\mathsf{C} \in \Pi^0_{f(r)}$, by Lemma 4.17 $\delta \in \Delta_{f(r)+2}$. That $\delta \in \mathsf{C}$ is immediate from the fact that C is closed. □

4.22 Corollary. *For all r, $\Delta^0_{f(r)+2}$ is a basis for the class of Σ^0_{r+1} sets of positive measure.* □

4.23 Corollary. $\{\alpha : \alpha \in \Delta^0_{(\omega)}\} \notin \Delta^0_{(\omega)}$.

Proof. Since $\{\alpha : \alpha \notin \Delta^0_{(\omega)}\}$ is the complement of a denumerable set it has measure 1. Hence if it were arithmetical it would have an arithmetical element, which is absurd. □

4.24–4.27 Exercises

4.24. Show that for $r \geqslant 2$, if β is implicitly Π^0_r and β and γ are each recursive in the other, then also γ is implicitly Π^0_r. (Exercise 6.17 shows that this is false for $r = 1$.)

4.25. Show that if β is implicitly Π^0_2, then so is $\beta^{\circ J}$. (Let F be a recursive functional such that for all β, $\beta = \lambda m\,.\,\mathsf{F}(m, \beta^{\circ J})$. Then $\gamma = \beta^{\circ J} \leftrightarrow \lambda m\,.\,\mathsf{F}(m, \gamma) = \beta$ and $\gamma = (\lambda m\,.\,\mathsf{F}(m, \gamma))^{\circ J}$.)

4.26. Show that the set B in the proof of Theorem 4.11 (ii) is in fact Π^0_2.

4.27. Is $\{\alpha : \alpha$ is primitive recursive$\}$ a basis for the class of non-meager Σ^0_2 sets?

4.28 Notes. Corollary 4.14 is from Shoenfield [1958] and Theorem 4.15 is from Tanaka [1970a]. 4.17–4.22 are due to Sacks [1969] and independently to Tanaka [1967], although neither formulated the sharp versions given here. Corollary 4.23 was originally proved by Addison using forcing methods (cf. Corollary 6.10 and Notes to § 6). Extensions of many of the results of this section and § 6 may be found in Kechris [1973].

5. Definability in Formal Languages for Arithmetic

The reader familiar with formal languages will certainly have noticed a similarity between the arithmetical and analytical hierarchies and classifications of the formulas of a formal language by the complexity of their structure. In this section we show that the classes Σ^i_r and Π^i_r $(i = 0, 1)$ consist exactly of the relations formally definable over the standard model of arithmetic by certain classes $\exists^i_r$ and $\forall^i_r$ $(i = 0, 1)$ of formulas. At the end we sketch briefly how these results may be applied to derive the undecidability and incompleteness theorems for axiomatic theories of arithmetic and discuss how much of the theory of this book could be developed in such systems.

We shall assume in this section some familiarity with formal languages and their interpretations and we shall omit many standard details, all of which may be found in Shoenfield [1967] or Enderton [1972]. Some of our notational conventions are suspended for this section as indicated below.

By the *standard model for arithmetic* we mean the structure

$$\mathfrak{N} = (\omega, {}^{\omega}\omega, <, +, \cdot, ', 0)$$

where $'$ denotes the successor function and $<$, $+$, and $\cdot$ have their usual arithmetical meanings. With $\mathfrak{N}$ we associate the second-order language $\mathscr{L}$ described as follows. The *symbols* of $\mathscr{L}$ are:

$$\neg, \vee, \exists, \ominus, \otimes, \oplus, \odot, {}^{\circledcirc}, \bar{0}, x_0, x_1, \ldots, \phi_0, \phi_1, \ldots.$$

The set of *terms* is defined inductively as the smallest class such that

(i) $\bar{0}, x_0, x_1, \ldots$ are terms;

(ii) if σ and τ are terms, then so are $\sigma^{\textcircled{1}}$, $\sigma \oplus \tau$, $\sigma \odot \tau$, and $\phi_j(\sigma)$ for $j = 0, 1, \ldots$.

The *atomic formulas* of $\mathscr{L}$ are the expressions $\sigma \circleddash \tau$ and $\sigma \circledS \tau$ for any terms σ and τ. The set of *formulas* is the smallest class such that

(i) atomic formulas are formulas;

(ii) if $\mathfrak{A}$ and $\mathfrak{B}$ are formulas, then so are $\neg\mathfrak{A}$, $\mathfrak{A} \vee \mathfrak{B}$, $\exists x_i\, \mathfrak{A}$, and $\exists \phi_j\, \mathfrak{A}$ for $i, j = 0, 1, \ldots$.

We denote by $\mathfrak{A}(\sigma/x_i)$ the result of substituting the term σ for all free occurrences of the variable x_i in the formula $\mathfrak{A}$ and assume that before the substitution all bound variables of $\mathfrak{A}$ which occur in σ are changed to the first unused variables of the proper type. $\mathfrak{A}(\sigma_0, \ldots, \sigma_{k-1})$ is an abbreviation for $\mathfrak{A}(\sigma_0/x_0)\ldots(\sigma_{k-1}/x_{k-1})$. Similarly, $\mathfrak{A}(\phi_j/\phi_i)$ denotes the result of substituting ϕ_j for all free occurrences of ϕ_i in $\mathfrak{A}$. We denote by $\bar{n}$ the term $\bar{0}^{\textcircled{1}\cdots\textcircled{1}}$ with n successor symbols. The symbols $\wedge, \rightarrow, \leftrightarrow$, and $\forall$ are used as abbreviations in the usual way. We write $(\exists x_i < \sigma)\mathfrak{A}$ for $\exists x_i\,(x_i < \sigma \wedge \mathfrak{A})$ and $(\forall x_i < \sigma)\mathfrak{A}$ for $\forall x_i\,(x_i < \sigma \rightarrow \mathfrak{A})$.

5.1 Definition. For any k and l, all $(\mathbf{m}, \boldsymbol{\alpha}) \in {}^{k,l}\omega$, and any term σ whose variables are included among $x_0, \ldots, x_{k-1}, \phi_0, \ldots, \phi_{l-1}$, we define $\sigma[\mathbf{m}, \boldsymbol{\alpha}]$, the *value of* σ *at* $(\mathbf{m}, \boldsymbol{\alpha})$, recursively by:

(i) $\bar{0}[\mathbf{m}, \boldsymbol{\alpha}] = 0$;

(ii) $x_i[\mathbf{m}, \boldsymbol{\alpha}] = m_i$;

(iii) $\sigma^{\textcircled{1}}[\mathbf{m}, \boldsymbol{\alpha}] = \sigma[\mathbf{m}, \boldsymbol{\alpha}] + 1$;

(iv) $(\sigma \oplus \tau)[\mathbf{m}, \boldsymbol{\alpha}] = \sigma[\mathbf{m}, \boldsymbol{\alpha}] + \tau[\mathbf{m}, \boldsymbol{\alpha}]$;

(v) $(\sigma \odot \tau)[\mathbf{m}, \boldsymbol{\alpha}] = \sigma[\mathbf{m}, \boldsymbol{\alpha}] \cdot \tau[\mathbf{m}, \boldsymbol{\alpha}]$;

(vi) $\phi_j(\sigma)[\mathbf{m}, \boldsymbol{\alpha}] = \alpha_j(\sigma[\mathbf{m}, \boldsymbol{\alpha}])$.

Note that this definition and the next one rely on Theorem I.3.5 for their justification.

5.2 Definition. For any k and l, all $(\mathbf{m}, \boldsymbol{\alpha}) \in {}^{k,l}\omega$, and any formula $\mathfrak{A}$ whose free variables are included among $x_0, \ldots, x_{k-1}, \phi_0, \ldots, \phi_{l-1}$, we define $\models \mathfrak{A}[\mathbf{m}, \boldsymbol{\alpha}]$, $\mathfrak{A}$ *is true at* $(\mathbf{m}, \boldsymbol{\alpha})$, recursively by:

(i) $\models (\sigma \circleddash \tau)[\mathbf{m}, \boldsymbol{\alpha}]$ iff $\sigma[\mathbf{m}, \boldsymbol{\alpha}] = \tau[\mathbf{m}, \boldsymbol{\alpha}]$;

(ii) $\models (\sigma \circledS \tau)[\mathbf{m}, \boldsymbol{\alpha}]$ iff $\sigma[\mathbf{m}, \boldsymbol{\alpha}] < \tau[\mathbf{m}, \boldsymbol{\alpha}]$;

(iii) $\models (\neg\mathfrak{A})[\mathbf{m}, \boldsymbol{\alpha}]$ iff not $\models \mathfrak{A}[\mathbf{m}, \boldsymbol{\alpha}]$;

(iv) $\models (\mathfrak{A} \vee \mathfrak{B})[\mathbf{m}, \boldsymbol{\alpha}]$ iff $\models \mathfrak{A}[\mathbf{m}, \boldsymbol{\alpha}]$ or $\models \mathfrak{B}[\mathbf{m}, \boldsymbol{\alpha}]$;

(v) $\models \exists x_i\, \mathfrak{A}[\mathbf{m}, \boldsymbol{\alpha}]$ iff $\exists n\, (\models \mathfrak{A}'(x_k/x_i)[\mathbf{m}, n, \boldsymbol{\alpha}])$;

(vi) $\models \exists \phi_j\, \mathfrak{A}[\mathbf{m}, \boldsymbol{\alpha}]$ iff $\exists \beta\, (\models \mathfrak{A}'(\phi_l/\phi_j)[\mathbf{m}, \boldsymbol{\alpha}, \beta])$;

where $\mathfrak{A}'$ is a variant of $\mathfrak{A}$ in which the variables x_k and ϕ_l do not occur.

Note that for any $\mathfrak{A}$,

$$\models \mathfrak{A}[\mathbf{m}, \mathbf{n}, \boldsymbol{\alpha}] \quad \text{iff} \quad \models \mathfrak{A}(\bar{n}_0/x_k)\cdots(\bar{n}_{k'-1}/x_{k+k'-1})[\mathbf{m}, \boldsymbol{\alpha}].$$

5.3 Definition. For all r,

(i) $\exists_0^0 = \forall_0^0 =$ the smallest class of formulas such that

(a) all atomic formulas belong to $\exists_0^0$;

(b) if $\mathfrak{A}$ and $\mathfrak{B}$ belong to $\exists_0^0$, then so do $\neg\mathfrak{A}$, $\mathfrak{A} \vee \mathfrak{B}$, and $(\exists x_i < \sigma)\mathfrak{A}$ for $i = 0, 1, \ldots$ and σ any term in which x_i does not occur;

(ii) $\exists_{r+1}^0 = \{\exists x_i \mathfrak{A}: \mathfrak{A} \in \forall_r^0 \wedge i \in \omega\}$;

(iii) $\forall_{r+1}^0 = \{\neg\mathfrak{A}: \mathfrak{A} \in \exists_{r+1}^0\}$;

(iv) $\exists_0^1 = \forall_0^1 = \bigcup\{\exists_r^0: r \in \omega\} \cup \bigcup\{\forall_r^0: r \in \omega\}$;

(v) $\exists_{r+1}^1 = \{\exists \phi_j \mathfrak{A}: \mathfrak{A} \in \forall_r^1 \wedge j \in \omega\}$;

(vi) $\forall_{r+1}^1 = \{\neg\mathfrak{A}: \mathfrak{A} \in \exists_{r+1}^1\}$.

5.4 Definition. For any k and l, any $\mathsf{R} \subseteq {}^{k,l}\omega$, any r, and any $i < 2$, R is $\exists_r^i$ ($\forall_r^i$)-*definable* (in the standard model) iff for some formula $\mathfrak{A} \in \exists_r^i$ ($\forall_r^i$) with free variables included among $x_0, \ldots, x_{k-1}, \phi_0, \ldots, \phi_{l-1}$,

$$\mathsf{R}(\mathbf{m}, \boldsymbol{\alpha}) \leftrightarrow \models \mathfrak{A}[\mathbf{m}, \boldsymbol{\alpha}].$$

A functional is $\exists_r^i$ ($\forall_r^i$)-definable just in case its graph is. We also use $\exists_r^i$ and $\forall_r^i$ to denote the corresponding classes of relations.

5.5 Lemma. *For all r and all $i < 2$, $\exists_r^i \subseteq \Sigma_r^i$ and $\forall_r^i \subseteq \Pi_r^i$.*

Proof. Relations defined by atomic formulas are recursive. Since the class of recursive relations is closed under complementation, union, and bounded quantification, all relations in $\exists_0^0$ are recursive. The remainder of the proof is a straightforward induction based on the fact that $\exists_{r+1}^i = \{\exists^i\mathsf{R}: \mathsf{R} \in \forall_r^i\}$. □

The main result of this section is that for i and r not both 0, the converse inclusions hold also. To this end we first establish some closure properties of $\exists_0^0$ and $\exists_1^0$. First it is clear from the definitions that the class of $\exists_0^0$ relations is a Boolean algebra and is closed under bounded quantification.

5.6 Lemma. *The class of $\exists_1^0$ relations is closed under binary union and intersection, bounded quantification ($\exists_<^0$ and $\forall_<^0$) and existential number quantification ($\exists^0$).*

Proof. For any $\exists_1^0$ relations R and S, there exist $\exists_0^0$ relations P and Q such that $\mathsf{R} = \exists^0\mathsf{P}$ and $\mathsf{S} = \exists^0\mathsf{Q}$. Then (cf. the proof of Theorem 1.6) the lemma follows from the following equivalences:

$$\exists p\, \mathsf{P}(p, \mathbf{m}, \boldsymbol{\alpha}) \vee \exists q\, \mathsf{Q}(q, \mathbf{m}, \boldsymbol{\alpha}) \leftrightarrow \exists p\, [\mathsf{P}(p, \mathbf{m}, \boldsymbol{\alpha}) \vee \mathsf{Q}(p, \mathbf{m}, \boldsymbol{\alpha})];$$

$$\exists p\, \mathsf{P}(p, \mathbf{m}, \boldsymbol{\alpha}) \wedge \exists q\, \mathsf{Q}(q, \mathbf{m}, \boldsymbol{\alpha}) \leftrightarrow \exists r\, [(\exists p < r)\mathsf{P}(p, \mathbf{m}, \boldsymbol{\alpha}) \wedge (\exists q < r)\mathsf{Q}(q, \mathbf{m}, \boldsymbol{\alpha})];$$

$$(\exists p < r)\exists q\, \mathsf{P}(p, q, r, \mathbf{m}, \boldsymbol{\alpha}) \leftrightarrow \exists q\, (\exists p < r)\mathsf{P}(p, q, r, \mathbf{m}, \boldsymbol{\alpha});$$

$$(\forall p < r)\exists q\, \mathsf{P}(p, q, r, \mathbf{m}, \boldsymbol{\alpha}) \leftrightarrow \exists s\, (\forall p < r)(\exists q < s)\mathsf{P}(p, q, r, \mathbf{m}, \boldsymbol{\alpha});$$

$$\exists p\, \exists q\, \mathsf{P}(p, q, \mathbf{m}, \boldsymbol{\alpha}) \leftrightarrow \exists r\, (\exists p < r)(\exists q < r)\mathsf{P}(p, q, \mathbf{m}, \boldsymbol{\alpha}). \quad \square$$

The key to the fact that all primitive recursive relations are $\exists_1^0$-definable is that there is an $\exists_0^0$-definable function which codes finite sequences.

5.7 Lemma. *There exists an $\exists_0^0$-definable function h such that for any $p, q_0, \ldots, q_{p-1}$, there exist s and t such that for all $i < p$, $h(s, t, i) = q_i$ and $q_i < s$.*

Proof. Let $h(s, t, i)$ be the remainder when s is divided by $1 + t(i+1)$ — that is,

$$h(s, t, i) = n \leftrightarrow n < 1 + t(i+1) \wedge (\exists u < s)\,.\, s = u(1 + t(i+1)) + n.$$

It is obvious that h is $\exists_0^0$-definable. Given p and $q_0, \ldots, q_{p-1}$, let $t = (\max\{p, q_0, \ldots, q_{p-1}\} + 1)!$. It is easy to check that the numbers $1 + t(i+1)$ for $i < p$ are pairwise relatively prime and greater than q_i. The Chinese Remainder Theorem of number theory asserts that in this situation there exists a number s such that for all $i < p$, $s \equiv q_i \pmod{1 + t(i+1)}$ as required. $\square$

5.8 Theorem. *All primitive recursive functionals are $\exists_1^0$-definable.*

Proof. We show that the class of $\exists_1^0$-definable functions contains the initial functionals and is closed under composition and primitive recursion. First, we have for the initial functionals:

$$\mathsf{Cs}_p^{k,l}(\mathbf{m}, \boldsymbol{\alpha}) = n \leftrightarrow \models (x_k \circeq \bar{p})[\mathbf{m}, n, \boldsymbol{\alpha}];$$

$$\mathsf{Pr}_i^{k,l}(\mathbf{m}, \boldsymbol{\alpha}) = n \leftrightarrow \models (x_k \circeq x_i)[\mathbf{m}, n, \boldsymbol{\alpha}];$$

$$\mathsf{Sc}_i^{k,l}(\mathbf{m}, \boldsymbol{\alpha}) = n \leftrightarrow \models (x_k \circeq x_i^{\oplus})[\mathbf{m}, n, \boldsymbol{\alpha}];$$

$$\mathsf{Ap}_{i,j}^{k,l}(\mathbf{m}, \boldsymbol{\alpha}) = n \leftrightarrow \models (x_k \circeq \phi_j(x_i))[\mathbf{m}, n, \boldsymbol{\alpha}].$$

If $\mathsf{G}, \mathsf{H}_0, \ldots, \mathsf{H}_{k'-1}$ are all $\exists_1^0$-definable and $\mathsf{F} = \mathrm{FCmp}_{k'}^{k,l}(\mathsf{G}, \mathsf{H}_0, \ldots, \mathsf{H}_{k'-1})$, then

$$\begin{aligned}\mathsf{F}(\mathbf{m}, \boldsymbol{\alpha}) = n \leftrightarrow \exists q_0 \ldots \exists q_{k'-1} [\mathsf{H}_0(\mathbf{m}, \boldsymbol{\alpha}) = q_0 \wedge \ldots \\ \wedge\, \mathsf{H}_{k'-1}(\mathbf{m}, \boldsymbol{\alpha}) = q_{k'-1} \wedge \mathsf{G}(q_0, \ldots, q_{k'-1}, \boldsymbol{\alpha}) = n],\end{aligned}$$

so F is $\exists_1^0$-definable by Lemma 5.6.

If G and H are $\exists_1^0$-definable and $\mathsf{F} = \mathrm{Rec}^{k+1,l}(\mathsf{G}, \mathsf{H})$, then

$$\mathsf{F}(p, \mathbf{m}, \boldsymbol{\alpha}) = n \leftrightarrow \exists s\, \exists t\, (h(s,t,0) = \mathsf{G}(\mathbf{m}, \boldsymbol{\alpha}) \wedge h(s,t,p) = n \wedge$$
$$\wedge (\forall i < p)(\exists q < s)[h(s,t,i) = q \wedge h(s,t,i+1) = \mathsf{H}(q, i, \mathbf{m}, \boldsymbol{\alpha})]),$$

where h is the function of Lemma 5.7. Again by Lemma 5.6, F is $\exists_1^0$-definable. □

5.9 Theorem. *For all* $r > 0$, $\exists_r^0 = \Sigma_r^0$ *and* $\forall_r^0 = \Pi_r^0$; *for all* r, $\exists_r^1 = \Sigma_r^1$ *and* $\forall_r^1 = \Pi_r^1$.

Proof. Let R be any Σ_1^0 relation. By Theorem II.4.12, $\mathsf{R} = \exists^0\mathsf{S}$ for some primitive recursive S. By the preceding Theorem, $\mathsf{K_S}$ is definable by some $\exists_1^0$ formula $\mathfrak{A}$. Then

$$\mathsf{S}(p, \mathbf{m}, \boldsymbol{\alpha}) \leftrightarrow \models \mathfrak{A}(\bar{0}/x_k)[p, \mathbf{m}, \boldsymbol{\alpha}],$$

so $\mathsf{S} \in \exists_1^0$, and thus also $\mathsf{R} \in \exists_1^0$ by Lemma 5.6. The remainder of the proof is a straightforward induction. □

In the rest of this section we shall give a brief survey of some facts concerning axiomatic theories of arithmetic. We do not intend to give a complete treatment of these topics but only to point out some of the ways they are related to the main themes of this book.

A formal theory $\mathcal{T}$ in the language $\mathcal{L}$ consists of a set of sentences (formulas without free variables) called the *axioms* of $\mathcal{T}$ and some *rules of inference* for deducing theorems from the axioms. We shall always assume that among the axioms and rules of $\mathcal{T}$ are a complete set of logical rules and axioms (for example, as in Shoenfield [1967, § 2.6]). We write $\mathcal{T} \vdash \mathfrak{A}$ to mean that $\mathfrak{A}$ is a theorem of $\mathcal{T}$.

We first consider *first-order* theories, that is, theories in which the variables ϕ_i do not occur and the rules of inference are only the usual ones. Let $\mathcal{T}_0$ be the theory with the following nine non-logical axioms. To make the formulas more readable we shall write $x, y, z, \ldots$ instead of $x_0, x_1, x_2, \ldots$.

(1) $\forall x \neg (x^{\circledS} \circleddash \bar{0})$;
(2) $\forall x \forall y (x^{\circledS} \circleddash y^{\circledS} \to x \circleddash y)$;
(3) $\forall x (x \oplus \bar{0} \circleddash x)$;
(4) $\forall x \forall y (x \oplus y^{\circledS} \circleddash (x + y)^{\circledS})$;
(5) $\forall x (x \odot \bar{0} \circleddash \bar{0})$;
(6) $\forall x \forall y (x \odot y^{\circledS} \circleddash (x \odot y) \oplus x)$;
(7) $\forall x \neg (x \circledless \bar{0})$;
(8) $\forall x \forall y (x \circledless y^{\circledS} \to x \circledless y \vee x \circleddash y)$;
(9) $\forall x \forall y (x \circledless y \vee x \circleddash y \vee y \circledless x)$.

$\mathcal{T}_0$ is a very weak theory. For example, it is easy to see that the commutativity of addition is not a theorem of $\mathcal{T}_0$ (Exercise 5.18). However it is just strong enough to carry through the undecidability and incompleteness results of Gödel and we sketch these next. Of course, the axioms of $\mathcal{T}_0$ are true and the rules of inference preserve truth, so whenever $\mathcal{T}_0 \vdash \mathfrak{A}$, also $\models \mathfrak{A}$. The converse is false (Corollary 5.15) but at least we have:

5.10 Lemma. *For all $\exists^0_0$-sentences $\mathfrak{A}$, if $\models \mathfrak{A}$, then $\mathcal{T}_0 \vdash \mathfrak{A}$.*

Proof. Consider the set X of all formulas $\mathfrak{A}$ such that for any k such that the free variables of $\mathfrak{A}$ are included among $x_0, \ldots, x_{k-1}$ and any $\mathbf{m} \in {}^k\omega$, if $\models \mathfrak{A}(\bar{\mathbf{m}})$, then $\mathcal{T}_0 \vdash \mathfrak{A}(\bar{\mathbf{m}})$. It is straightforward to show that X is closed under the clauses of the inductive definition 5.3(i) of $\exists^0_0$ and hence contains all $\exists^0_0$ formulas. □

5.11 Lemma. *For any two disjoint semi-recursive relations R and $S \subseteq {}^k\omega$, there exists an $\exists^0_1$ formula $\mathfrak{C}$ such that for all $\mathbf{m} \in {}^k\omega$,*

(i) $R(\mathbf{m}) \rightarrow \mathcal{T}_0 \vdash \mathfrak{C}(\bar{\mathbf{m}})$;

(ii) $S(\mathbf{m}) \rightarrow \mathcal{T}_0 \vdash \neg\mathfrak{C}(\bar{\mathbf{m}})$.

Proof. By Theorem 5.9 there exist $\exists^0_0$ formulas $\mathfrak{A}$ and $\mathfrak{B}$ such that for all $\mathbf{m} \in {}^k\omega$,

$$R(\mathbf{m}) \leftrightarrow \models \exists x\, \mathfrak{A}(\bar{\mathbf{m}}) \quad \text{and} \quad S(\mathbf{m}) \leftrightarrow \models \exists y\, \mathfrak{B}(\bar{\mathbf{m}}).$$

Let $\mathfrak{C}$ be the formula

$$\exists x\, (\mathfrak{A} \wedge (\forall y \circledS x) \neg \mathfrak{B}).$$

Then (i) and (ii) follow easily by use of Lemma 5.10 and axioms (8) and (9). □

For any theory $\mathcal{T}$, a formula $\mathfrak{A}$ with free variables among $x_0, \ldots, x_{k-1}$ *$\mathcal{T}$-represents* a relation R iff for all $\mathbf{m} \in {}^k\omega$,

$$R(\mathbf{m}) \rightarrow \mathcal{T} \vdash \mathfrak{A}(\bar{\mathbf{m}}) \quad \text{and} \quad \sim R(\mathbf{m}) \rightarrow \mathcal{T} \vdash \neg \mathfrak{A}(\bar{\mathbf{m}}).$$

R is *$\mathcal{T}$-representable* iff R is $\mathcal{T}$-represented by some formula. It follows immdiately from Lemma 5.11 that all recursive relations are $\mathcal{T}_0$-representable (take $S = \sim R$).

5.12 Lemma. *For any $\mathcal{T}$ which extends $\mathcal{T}_0$, if $\mathcal{T}$ is consistent, then there exists a formula $\mathfrak{C}$ with only x_0 free such that $\{m : \mathcal{T} \vdash \mathfrak{C}(\bar{m})\}$ is not recursive.*

Proof. By Theorem 1.10(i), there exist disjoint semi-recursive sets A and B such that there is no recursive set C such that $A \subseteq C \subseteq \sim B$. Take A and B for R and S in Lemma 5.11 and let $C = \{m : \mathcal{T} \vdash \mathfrak{C}(\bar{m})\}$. By (i) of 5.11, $A \subseteq C$, and by (ii) together with the consistency of $\mathcal{T}$, $C \subseteq \sim B$. Hence C is not recursive. □

Informally, we say that $\mathcal{T}$ is decidable iff there is an algorithm for deciding among the sentences of $\mathcal{L}$ which are theorems of $\mathcal{T}$ and which are not. Thus on the intuitive level, Lemma 5.12 implies that any consistent $\mathcal{T}$ which extends $\mathcal{T}_0$ is undecidable, as otherwise we could effectively decide for each m whether or not $\mathcal{T} \vdash \mathfrak{C}(\bar{m})$. These intuitions are made precise through the technique of Gödel numbering. To each expression (symbol, term, or formula) z of $\mathcal{L}$ we assign a natural number $\ulcorner z \urcorner$ much as we did to each description of a primitive recursive functional in § II.1. The square quotes are to suggest that the assigned number be thought of as a name for z.

We begin by assigning successive odd numbers $1, 3, \ldots$ to the symbols of $\mathcal{L}$. The terms are (recursively) assigned numbers by: $\ulcorner \sigma^{\odot} \urcorner = \langle \ulcorner \odot \urcorner, \ulcorner \sigma \urcorner \rangle$, $\ulcorner \sigma \oplus \tau \urcorner = \langle \ulcorner \oplus \urcorner, \ulcorner \sigma \urcorner, \ulcorner \tau \urcorner \rangle$, etc. Similarly for formulas, $\ulcorner \sigma \circeq \tau \urcorner = \langle \ulcorner \circeq \urcorner, \ulcorner \sigma \urcorner, \ulcorner \tau \urcorner \rangle$, $\ulcorner \mathfrak{A} \vee \mathfrak{B} \urcorner = \langle \ulcorner \vee \urcorner, \ulcorner \mathfrak{A} \urcorner, \ulcorner \mathfrak{B} \urcorner \rangle$, and so on. It is an elementary exercise to show that the sets of Gödel numbers corresponding to various syntactic classes of expressions are recursive — for example, $\{\ulcorner \sigma \urcorner : \sigma \text{ is a term}\}$, $\{\ulcorner \mathfrak{A} \urcorner : \mathfrak{A} \text{ is a formula}\}$ and $\{\ulcorner \mathfrak{A} \urcorner : \mathfrak{A}$ is an $\exists^0_{29}$-formula with at most x_{12} and ϕ_{47} free$\}$. Similarly, there is a recursive function f such that for all $\mathfrak{A}$ and all $\mathbf{m} \in {}^k\omega$, $f(\ulcorner \mathfrak{A} \urcorner, \mathbf{m}) = \ulcorner \mathfrak{A}(\bar{\mathbf{m}}) \urcorner$. Now we call $\mathcal{T}$ *decidable* iff $\{\ulcorner \mathfrak{A} \urcorner : \mathcal{T} \vdash \mathfrak{A}\}$ is recursive.

5.13 Corollary. *Any consistent theory $\mathcal{T}$ which extends $\mathcal{T}_0$ is undecidable.*

Proof. Immediate from Lemma 5.12 and the above discussion. □

A theory $\mathcal{T}$ is called *(first-order) complete* iff for every (first-order) sentence $\mathfrak{A}$, either $\mathfrak{A}$ or $\neg\mathfrak{A}$ is a theorem of $\mathcal{T}$. $\mathcal{T}$ is *recursively axiomatizable* iff there exists a theory $\mathcal{T}'$ such that $\{\ulcorner \mathfrak{A} \urcorner : \mathfrak{A} \text{ is an axiom of } \mathcal{T}'\}$ is recursive and for all $\mathfrak{A}$, $\mathcal{T} \vdash \mathfrak{A}$ iff $\mathcal{T}' \vdash \mathfrak{A}$. $\mathcal{T}$ is *semi-decidable* iff $\{\ulcorner \mathfrak{A} \urcorner : \mathcal{T} \vdash \mathfrak{A}\}$ is semi-recursive.

5.14 Lemma. *For any $\mathcal{T}$,*

(i) *if $\mathcal{T}$ is recursively axiomatizable, then $\mathcal{T}$ is semi-decidable;*

(ii) *if $\mathcal{T}$ is recursively axiomatizable and complete, then $\mathcal{T}$ is decidable.*

Proof. Let $\Gamma_{\mathcal{T}}$ be the monotone operator defined by:

$$\Gamma_{\mathcal{T}}(A) = \{m : m \in A \text{ or } m \text{ is the Gödel number of an axiom of } \mathcal{T} \text{ or } m \text{ is the Gödel number of a formula which follows from formulas with Gödel numbers in } A \text{ by a single rule of inference}\}.$$

It is easy to see that $\Gamma_{\mathcal{T}}$ is Σ^0_1 and that $\bar{\Gamma}_{\mathcal{T}} = \{\ulcorner \mathfrak{A} \urcorner : \mathcal{T} \vdash \mathfrak{A}\}$. Hence (i) follows from Theorem 3.5. Suppose now $\mathcal{T}$ is recursively axiomatizable and complete. If $\mathcal{T}$ is inconsistent all formulas are provable so $\mathcal{T}$ is decidable. Otherwise the sets $\{\ulcorner \mathfrak{A} \urcorner : \mathfrak{A} \text{ is a sentence and } \mathcal{T} \vdash \mathfrak{A}\}$ and $\{m : m$ is not the Gödel number of a sentence or m is the Gödel number of a sentence $\mathfrak{A}$ such that $\mathcal{T} \vdash \neg\mathfrak{A}\}$ are complementary. By (i), both are semi-recursive, hence both are recursive and $\mathcal{T}$ is decidable. □

5.15 Corollary. *Any consistent recursively axiomatizable theory which extends $\mathcal{T}_0$ is incomplete.*

Proof. Immediate from 5.13 and 5.14. □

As a consequence, the complete theory $\mathcal{T}_{\mathfrak{N}}$ whose axioms are all first-order formulas valid in the standard model is not recursively axiomatizable. It also follows from 5.14(i) that for any consistent recursively axiomatizable theory $\mathcal{T}$ which extends $\mathcal{T}_0$, a relation R is $\mathcal{T}$-representable iff R is recursive.

That $\mathcal{T}_{\mathfrak{N}}$ is not recursively axiomatizable can also be seen by a slightly different argument. From Theorem 5.9 and the Arithmetical Hierarchy Theorem it follows that $\{\ulcorner\mathfrak{A}\urcorner : \models \mathfrak{A}\}$ is not arithmetical, in particular not semi-recursive, and thus is not the set of theorems of any recursively axiomatizable theory by 5.14(i).

$\mathcal{T}_{\mathfrak{N}}$ can, however, be recursively axiomatized if the system is expanded by adding a rule of inference to which the proof of 5.14(i) does not apply — that is, one such that the operator $\Gamma_{\mathcal{T}}$ defined there is not Σ_1^0. For any theory $\mathcal{T}$, we write $\mathcal{T} \vdash^{\omega} \mathfrak{A}$ to mean that $\mathfrak{A}$ is derivable from the rules and axioms of $\mathcal{T}$ together with the *ω-rule*: from the (infinitely many) premises $\mathfrak{A}(\bar{0}), \mathfrak{A}(\bar{1}), \ldots$, infer $\forall x_0\mathfrak{A}$.

5.16 Theorem. (i) *For all first-order formulas $\mathfrak{A}$, $\mathcal{T}_0 \vdash^{\omega} \mathfrak{A}$ iff $\models \mathfrak{A}$;*
(ii) $\{\ulcorner\mathfrak{A}\urcorner : \mathcal{T}_0 \vdash^{\omega} \mathfrak{A}\}$ *is* Π_1^1.

Proof. (i) follows from Lemma 5.10 by an easy induction. (ii) follows from Theorem 3.1 and the observation that the operator $\Gamma_{\mathcal{T},\omega}$ associated with $\mathcal{T}$ and the ω-rule is arithmetical and monotone. □

We turn now to the full (second-order) language $\mathcal{L}$. In the first part of this section we considered only the standard interpretation for $\mathcal{L}$, but now we shall need a more general notion. A (general) *structure* for $\mathcal{L}$ is sequence $\mathfrak{U} = (U, \Phi, <_{\mathfrak{U}}, +_{\mathfrak{U}}, \cdot_{\mathfrak{U}}, '_{\mathfrak{U}}, 0_{\mathfrak{U}})$ such that U is a set, Φ is a set of unary total functions $U \to U$, $<_{\mathfrak{U}}$ is a binary relation on U, $+_{\mathfrak{U}}$ and $\cdot_{\mathfrak{U}}$ are binary total functions on U, $'_{\mathfrak{U}}$ is a unary total function on U and $0_{\mathfrak{U}} \in U$. For $\mathbf{u} \in {}^kU$, $\varphi \in {}^l\Phi$, and σ a term of $\mathcal{L}$ with free variables among $x_0, \ldots, x_{k-1}$, $\phi_0, \ldots, \phi_{l-1}$, we define $\sigma_{\mathfrak{U}}[\mathbf{u}, \varphi]$, the *value of σ in $\mathfrak{U}$ at* $(\mathbf{u}, \varphi)$ just as in Definition 5.1 except that 0 is replaced by $0_{\mathfrak{U}}$, + by $+_{\mathfrak{U}}$, etc. Similarly, the relation $\mathfrak{U} \models \mathfrak{A}[\mathbf{u}, \varphi]$ is defined as in Definition 5.2 with $<$ replaced by $<_{\mathfrak{U}}$, $\exists n$ replaced by $(\exists v \in U)$, and $\exists\beta$ replaced by $(\exists\psi \in \Phi)$. $\mathfrak{U}$ is a *model* of a theory $\mathcal{T}$ iff $\mathfrak{U} \models \mathfrak{A}$ for all axioms $\mathfrak{A}$ of $\mathcal{T}$. $\mathfrak{U}$ is called an *ω-structure* iff the values in $\mathfrak{U}$ of all the terms $\bar{n}$ ($n \in \omega$) exhaust U. In this case $\mathfrak{U}$ is isomorphic to a structure $\mathcal{A} = (\omega, \mathsf{A}, <_{\mathcal{A}}, +_{\mathcal{A}}, \cdot_{\mathcal{A}}, ', 0)$, where $'$ and 0 have their usual meanings. If in addition $\mathfrak{U}$ (and hence $\mathcal{A}$) is a model of $\mathcal{T}_0$, then $<_{\mathcal{A}}$, $+_{\mathcal{A}}$ and $\cdot_{\mathcal{A}}$ must also coincide with the usual $<$, $+$, and $\cdot$. Hence such a structure is determined by the set $\mathsf{A} \subseteq {}^{\omega}\omega$ and we write simply $\mathsf{A} \models \mathfrak{A}[\mathbf{m}, \boldsymbol{\alpha}]$. In particular, $\models \mathfrak{A}[\mathbf{m}, \boldsymbol{\alpha}]$ iff ${}^{\omega}\omega \models \mathfrak{A}[\mathbf{m}, \boldsymbol{\alpha}]$.

Let $\mathcal{T}_1$ denote the theory obtained from $\mathcal{T}_0$ by extending the logical axioms and rules of $\mathscr{T}_0$ to all formulas of $\mathscr{L}$ and adjoining the following (infinitely many) axioms: the universal closures of

(10) (*Induction*) $\mathfrak{A}(\bar{0}) \wedge \forall x_0[\mathfrak{A} \rightarrow \mathfrak{A}(x_0^{\oplus})] \rightarrow \forall x_0 \mathfrak{A}$ for all formulas $\mathfrak{A}$;

(11) (*Extensionality*) $\forall \phi_0 \forall \phi_1 [\forall x (\phi_0(x) = \phi_1(x)) \rightarrow \phi_0 = \phi_1]$;

(12) (Δ_0^1-*Comprehension*) $\forall x \exists! y\, \mathfrak{A} \rightarrow \exists \phi_0 \forall x\, \mathfrak{A}(\phi_0(x)/y)$ for all $\exists_0^1$-formulas $\mathfrak{A}$.

The completeness theorem of first-order logic has the following natural extension to $\mathscr{L}$:

5.17 Theorem. *For any theory $\mathscr{T}$ which extends $\mathscr{T}_1$ and any sentence $\mathfrak{A}$ of $\mathscr{L}$*
(i) *$\mathscr{T} \vdash \mathfrak{A}$ iff $\mathfrak{U} \models \mathfrak{A}$ for all models $\mathfrak{U}$ of $\mathscr{T}$;*
(ii) *$\mathscr{T} \vdash^{\omega} \mathfrak{A}$ iff $\mathsf{A} \models \mathfrak{A}$ for all ω-models A of $\mathscr{T}$.*

Proof. See Shoenfield [1967, § 8.5]. □

From this and 5.16(i) it is immediate that for all $\forall_1^1$ sentences $\mathfrak{A}$, $\mathscr{T}_1 \vdash^{\omega} \mathfrak{A}$ iff $\models \mathfrak{A}$. From 5.16(ii) it follows that any relation which is $\mathscr{T}_0$-ω-representable is Δ_1^1. The same proof applies to $\mathscr{T}_1$ and from the preceding it can be derived that the $\mathscr{T}_1$-ω-representable relations are exactly the Δ_1^1-relations.

The Δ_0^1-Comprehension axioms guarantee that the range of the variable ϕ_i is not too small. In particular, if A is an ω-model of Δ_0^1-Comprehension, then not only does A contain all arithmetical functions, but because the formula $\mathfrak{A}$ in the Δ_0^1-Comprehension schema may have free variables, A is also closed under the relation "arithmetical in".

As is sketched in Shoenfield [1967, § 8.5], the notion of truth for first-order formulas in $\mathfrak{N}$ may be formalized in the theory $\mathscr{T}_1$ (although the system S there includes the comprehension axioms for all formulas, only Δ_0^1-Comprehension is needed here; indeed, for a fixed n, the truth of $\exists_n^0$ sentences may be defined without use of the comprehension axioms). By virtue of Theorem 5.9, much of the theory of the arithmetical hierarchy may thus be developed in $\mathscr{T}_1$. Relations are denoted in $\mathscr{T}_1$ by the Gödel numbers of their defining formulas and the results of § 1 become first-order arithmetical theorems in $\mathscr{T}_1$. Alternatively, the arithmetical hierarchy of relations on numbers may be developed in terms of characteristic functions.

For each r, the truth of $\exists_r^1$ and $\forall_r^1$ sentences in $\mathfrak{N}$ and the classes of formulas which define Σ_r^1 and Π_r^1 relations can similarly be defined in $\mathscr{T}_1$. However, some of the basic closure properties of Σ_r^1 and Π_r^1 cannot be proved without additional axioms. Furthermore, without additional comprehension axioms the class of functions defined by second-order formulas is not provably larger than the class of first-order definable functions. To guarantee the existence of all characteristic functions of Σ_r^1 relations we need

(13) ($\exists_r^1$-*Comprehension*) $\exists\phi_0 \forall x\, [\phi_0(x) = \bar{0} \leftrightarrow \mathfrak{A}]$ for all $\mathfrak{A} \in \exists_r^1$.

Note that in $\mathcal{T}_1$ this is equivalent to the corresponding $\forall_r^1$-Comprehension schema.

Following the development in § 2, we encounter the first difficulty with the next-to-last formula in the proof of Theorem 2.5. The translation of the implication ($\rightarrow$) into $\mathcal{L}$ need not hold in every model of $\mathcal{T}_1$ (see Exercise 5.20). Hence to prove the closure of Σ_r^1 under number quantification we need to add the axioms

(14) ($\exists_r^1$-*Choice*) $\forall x\, \exists\phi_0\, \mathfrak{A} \rightarrow \exists\phi_1 \forall x\, \mathfrak{A}(\phi_1^x/\phi_0)$ for all $\mathfrak{A} \in \exists_r^1$.

Here $\mathfrak{A}(\phi_1^x/\phi_0)$ denotes the formula obtained from $\mathfrak{A}$ by replacing all occurrences of terms $\phi_0(\sigma)$ by $\phi_1(\langle x, \sigma\rangle)$, where $\langle\cdot,\cdot\rangle$ is a symbol for a pairing function introduced by definition into $\mathcal{T}_1$.

Then 2.6–2.10 follow easily. It is worth noting that there is a different proof of Lemma 2.8 which does not require the $\exists_1^1$-Choice axioms. This is based on the equivalence

$$\exists\phi_0 \ldots \forall x\, \exists y \ldots \mathfrak{A} \leftrightarrow \exists\phi_0 \ldots \exists\phi_1 \forall x \ldots \mathfrak{A}(\phi_1(x)/y)$$

which is provable in $\mathcal{T}_1$ (by use of the Induction and Δ_0^1-Comprehension axioms).

The Analytical Substitution Theorem (2.11) for r follows from the $\exists_r^1$-Comprehension axioms, but here a smaller collection of comprehension axioms will suffice. The second equivalence in the proof of that theorem depends on the fact that if $\forall p\, \mathsf{H}(p, \mathbf{m}, \boldsymbol{\alpha})\downarrow$, then $\exists\beta\, \forall p\, [\mathsf{H}(p, \mathbf{m}, \boldsymbol{\alpha}) \simeq \beta(p)]$. This is provable from the schema:

(15) (Δ_r^1-*Comprehension*) $\forall x\, \exists!\, y\, \mathfrak{A} \rightarrow \exists\phi_0 \forall x\, \mathfrak{A}(\phi_0(x)/y)$ for all $\mathfrak{A} \in \exists_r^1$.

Note that Δ_r^1-Comprehension is derivable from $\exists_r^1$-Choice (Kreisel [1962]). If $\forall x\, \exists!\, y\, \mathfrak{A}$, then $\forall x\, \exists\phi_0\, \mathfrak{A}(\phi_0(x)/y)$ so by $\exists_r^1$-Choice, $\exists\phi_1 \forall x\, \mathfrak{A}(\phi_1^x(x)/y)$. Then if $\phi_2(x) = \phi_1(\langle x, x\rangle)$, $\forall x\, \mathfrak{A}(\phi_2(x)/y)$. The existence of ϕ_2 follows from that of ϕ_1 by Δ_0^1-Comprehension (cf. Exercise 5.21).

In the counterexample following Corollary 2.13 an $\exists_r^1$-Comprehension axiom is essential to guarantee the existence of (the characteristic function of) U_r^1. Some results concerning monotone inductive definitions can be established in $\mathcal{T}_1$ together with some of (13)–(15), but those which involve ordinals are in general beyond the scope of these theories. The closure $\bar{\Gamma}$ must be *defined* as the intersection of sets closed under Γ and for $\Gamma \in \Pi_r^1$ the $\exists_r^1$-Comprehension schema is needed to prove the existence of $\bar{\Gamma}$. We leave it to the interested reader to determine which axioms are needed to prove the remaining results of §§ 3 and 4.

5.18–5.21 Exercises

5.18. Show that the commutativity of $\oplus$ is not provable in $\mathcal{T}_0$.

5.19. Let V be the set of all $\langle \ulcorner A \urcorner, \mathbf{m}\rangle$ such that for some k, $\mathfrak{A}$ is a first-order $\mathcal{L}$-formula with free variables among $x_0, \ldots, x_{k-1}$, $\mathbf{m} \in {}^k\omega$, and $\models \mathfrak{A}[\mathbf{m}]$. Show that V is reducible to $\bar{\Gamma}$ for some Σ^0_1 inductive operator Γ. (Use the result of Exercise 3.32(i)).

5.20. Show that $\{\alpha : \alpha \in \Delta^0_{(\omega)}\}$ is not a model of $\exists^1_1$-Choice. In fact, it is not a model of Δ^1_1-Comprehension.

5.21. Show that $\exists^1_1$-Comprehension implies $\exists^1_1$-Choice.

5.22 Notes. The main ideas through Corollary 5.15 are due to Gödel [1931], but the presentation here benefits from many modifications and improvements due to Rosser, Tarski, Kleene, and others.

The difference in the forms of $\exists^1_r$-Comprehension and Δ^1_r-Comprehension is due to the fact that we are working in a language with function variables rather than set variables. A (total) function which is $\exists^1_r$-definable is is also $\forall^1_r$-definable. The import of $\exists^1_r$-Comprehension (for ω-models) is that (the characteristic functions of) *sets* which are $\exists^1_r$-definable belong to the model.

6. Arithmetical Forcing

The technique of forcing was first developed by Cohen [1963/64] in the context of set theory. We shall develop here a simpler form of this technique due to Feferman [1964/65] which leads to several interesting results in the arithmetical hierarchy.

As motivation, consider an arbitrary open ($\mathbf{\Sigma}^0_1$) subset A of ${}^\omega\omega$. There exists a set $A \subseteq \omega$ such that for all α,

$$(*) \qquad \alpha \in \mathsf{A} \leftrightarrow \exists p \,.\, \bar{\alpha}(p) \in A.$$

That a function α belong to A when it does is "forced" by some initial segment $\bar{\alpha}(p)$ and every other function β such that $\bar{\beta}(p) = \bar{\alpha}(p)$ also belongs to A. Of course for A which are not open the situation is different — no finite initial segment s can "force" all $\beta \in [s]$ to belong to $\{\alpha : \forall m \,.\, \alpha(m) = 0\}$, for example.

The key idea of forcing is that even for more complex A, there exists a set A such that (*) holds for "many" α. The success of the applications turns on the interpretation of "many".

It is convenient to use here a formal language $\mathcal{L}'$ slightly different from the language $\mathcal{L}$ of §5. The *symbols* of $\mathcal{L}'$ are $\neg, \vee, \exists, {}^{\oplus}, \bar{0}, x_0, x_1, \ldots, \phi, \bar{R}_a$ ($a \in$ Pri).

Note that we have removed all the relation and function symbols except ${}^{\oplus}$ from $\mathcal{L}$ in favor of an infinite collection of relation symbols R_a and left only one function variable. The *terms* of $\mathcal{L}'$ are the expressions obtained from $\bar{0}, x_0, x_1, \ldots$ by application of ${}^{\oplus}$ and ϕ. The *atomic formulas* of $\mathcal{L}'$ are all expressions of the form $\bar{R}_a(\sigma_0, \ldots, \sigma_{k-1})$ for $a \in$ Pri and $k = \text{rank}[a]$. The *formulas* of $\mathcal{L}'$ are all those expressions obtained from the atomic formulas by application of $\neg$, $\vee$, and $\exists x_i$ ($i = 0, 1, \ldots$) — the variable ϕ is not quantified.

A term σ with free variables included among $x_0, \ldots, x_{k-1}, \phi$, has *value* $\sigma[\mathbf{m}, \alpha]$ defined recursively by clauses (i)–(iii) and (vi) (without the subscript) of Definition 5.1. If $\mathfrak{A}$ is a formula of $\mathcal{L}'$ with free variables included among $x_0, \ldots, x_{k-1}, \phi$, then $\models \mathfrak{A}[\mathbf{m}, \alpha]$ ($\mathfrak{A}$ is *true at* $(\mathbf{m}, \alpha)$) is defined recursively by clauses (iii)–(v) of Definition 5.2 together with:

$$\models \bar{R}_a(\sigma_0, \ldots, \sigma_{k-1})[\mathbf{m}, \alpha] \quad \text{iff} \quad [a](\sigma_0[\mathbf{m}, \alpha], \ldots, \sigma_{k-1}[\mathbf{m}, \alpha]) = 0.$$

The $\exists_0^0$ and $\forall_0^0$ formulas of $\mathcal{L}'$ are simply the atomic formulas and the classes $\exists_{r+1}^0$ and $\forall_{r+1}^0$ are defined as in 5.3. Relations and functionals of rank $(k, 1)$ are $\exists_r^0$ or $\forall_r^0$ definable as before. Then for $\mathcal{L}'$ as for $\mathcal{L}$ we have:

6.1 Theorem. *For all $r > 0$ and all $\mathsf{R} \subseteq {}^{k,1}\omega$, $\mathsf{R} \in \Sigma_r^0$ iff R is $\exists_r^0$-definable and $\mathsf{R} \in \Pi_r^0$ iff R is $\forall_r^0$-definable.*

Proof. It is immediate from the definition that every primitive recursive relation on numbers is $\exists_0^0$-definable and that the class of $\exists_0^0$-definable relations is a Boolean algebra closed under bounded quantification. Hence the proof of Lemma 5.6 also establishes here that the class of $\exists_1^0$-definable relations is closed under binary union and intersection, bounded quantification, and existential number quantification. The function h of Lemma 5.7 is primitive recursive, hence also $\exists_0^0$-definable here and the result follows as for $\mathcal{L}$. □

For any term σ of $\mathcal{L}'$ the value $\sigma[\mathbf{m}, \alpha]$ clearly depends on only finitely many values of α. Hence we may also define the value of a term relative to a finite sequence s thought of as an initial segemnt of α; if s is too short, this will be undefined.

6.2 Definition. For any k, any $\mathbf{m} \in {}^k\omega$, any $s \in$ Sq, and any term σ of $\mathcal{L}'$ whose variables are included among $x_0, \ldots, x_{k-1}, \phi$, $\sigma[\mathbf{m}, s]$ is defined recursively by:

(i) $\bar{0}[\mathbf{m}, s] \simeq 0$;

(ii) $x_i[\mathbf{m}, s] \simeq m_i$;

(iii) $\sigma^{\oplus}[\mathbf{m}, s] \simeq \sigma[\mathbf{m}, s] + 1$;

(iv) if $\sigma[\mathbf{m}, s] < \text{lg}(s)$, $\phi(\sigma)[\mathbf{m}, s] \simeq (s)_{\sigma[\mathbf{m}, s]}$.

It is easy to check that for sufficiently large n, $\sigma[\mathbf{m}, \bar{\alpha}(n)] \simeq \sigma[\mathbf{m}, \alpha]$. Hence, if $\sigma[\mathbf{m}, s] \simeq n$, then $\sigma[\mathbf{m}, \alpha] = n$ for all $\alpha \in [s]$.

6.3 Definition. For any k, any $\mathbf{m} \in {}^k\omega$, and $s \in \mathrm{Sq}$, and any formula $\mathfrak{A}$ of $\mathscr{L}'$ whose free variables are included among $x_0, \ldots, x_{k-1}, \phi$, we define $\Vdash \mathfrak{A}[\mathbf{m}, s]$, $\mathfrak{A}$ is *forced at* $(\mathbf{m}, s)$, recursively by:

(i) $\Vdash \bar{R}_a(\sigma_0, \ldots, \sigma_{k-1})[\mathbf{m}, s]$ iff $[a](\sigma_0[\mathbf{m}, s], \ldots, \sigma_{k-1}[\mathbf{m}, s]) \simeq 0$;
(ii) $\Vdash (\neg \mathfrak{A})[\mathbf{m}, s]$ iff $\sim \exists t\, (t \supseteq s \wedge \Vdash \mathfrak{A}[\mathbf{m}, t])$;
(iii) $\Vdash (\mathfrak{A} \vee \mathfrak{B})[\mathbf{m}, s]$ iff $\Vdash \mathfrak{A}[\mathbf{m}, s]$ or $\Vdash \mathfrak{B}[\mathbf{m}, s]$;
(iv) $\Vdash \exists x_i\, \mathfrak{A}[\mathbf{m}, s]$ iff $\exists n\, (\Vdash \mathfrak{A}'(x_k/x_i))[\mathbf{m}, n, s]$;

where $\mathfrak{A}'$ is a variant of $\mathfrak{A}$ in which the variable x_k does not appear.

Note that except for the clause (ii), this definition is nearly identical with that of $\models \mathfrak{A}[\mathbf{m}, \alpha]$. In terms of the discussion at the beginning of the section, clause (ii) says that α is "forced" to belong to $\sim$A by its initial segment s just in case no extension t forces membership in A.

6.4 Lemma. *For all* $\mathbf{m}$, s, *and* $\mathfrak{A}$ *as in Definition* 6.3,

(i) *not both* $\Vdash \mathfrak{A}[\mathbf{m}, s]$ *and* $\Vdash \neg \mathfrak{A}[\mathbf{m}, s]$;
(ii) *for any* $t \supseteq s$, *if* $\Vdash \mathfrak{A}[\mathbf{m}, s]$, *then also* $\Vdash \mathfrak{A}[\mathbf{m}, t]$;
(iii) *for some* $t \supseteq s$, *either* $\Vdash \mathfrak{A}[\mathbf{m}, t]$ *or* $\Vdash \neg \mathfrak{A}[\mathbf{m}, t]$.

Proof. (i) and (iii) are immediate from clause (ii) of the definition. We prove (ii) by induction on formulas. For $\mathfrak{A}$ atomic, the statement follows from the fact that if $\sigma[\mathbf{m}, s] \simeq n$ and $s \subseteq t$, then also $\sigma[\mathbf{m}, t] \simeq n$. If $\Vdash \neg \mathfrak{A}[\mathbf{m}, s]$ and $s \subseteq t$, then for any $u \supseteq t$ also $u \supseteq s$ so *not* $\Vdash A[\mathbf{m}, u]$. Hence $\Vdash \neg \mathfrak{A}[\mathbf{m}, t]$. The other two clauses follow similarly. □

We call a formula of $\mathscr{L}'$ *closed* iff none of the variables x_i occur free in $\mathfrak{A}$ — ϕ may occur.

6.5 Definition. For all r and α, α is *r-generic* iff for all closed formulas $\mathfrak{A}$ in $\bigcup \{\exists_i^0 : i \leq r\}$,

$$\exists p\, (\Vdash \mathfrak{A}[\bar{\alpha}(p)] \quad \text{or} \quad \Vdash \neg \mathfrak{A}[\bar{\alpha}(p)]).$$

α is *generic* iff α is r-generic for all $r \in \omega$.

We now obtain the promised generalization of (*). Note that all α are 0-generic.

6.6 Theorem. *For all r, all closed* $\mathfrak{A}$ *in* $\exists_{r+1}^0$, *and all r-generic functions* α,

$$\models \mathfrak{A}[\alpha] \textit{ iff for some } p, \Vdash \mathfrak{A}[\bar{\alpha}(p)].$$

Proof. We proceed by induction on r. Suppose $r = 0$ and $\models \bar{R}_a(\sigma_0, \ldots, \sigma_{k-1})[\alpha]$.

Then if p is chosen sufficiently large so that $\sigma_0[\bar{\alpha}(p)], \ldots, \sigma_{k-1}[\bar{\alpha}(p)]$ are all defined, then $\models \bar{R}_a(\sigma_0, \ldots, \sigma_{k-1})[\bar{\alpha}(p)]$, i.e. $\Vdash \bar{R}_a(\sigma_0, \ldots, \sigma_{k-1})[\bar{\alpha}(p)]$. Conversely, if $\Vdash \bar{R}_a(\sigma_0, \ldots, \sigma_{k-1})[\bar{\alpha}(p)]$, then for all $i < k$, $\sigma_i[\alpha] = \sigma_i[\bar{\alpha}(p)]$ so also $\models \bar{R}_a(\sigma_0, \ldots, \sigma_{k-1})[\alpha]$.

Assume the result for r and suppose α is $r+1$-generic and first that $\mathfrak{A} \in \forall^0_{r+1}$. Then $\mathfrak{A} = \neg \mathfrak{B}$ for some $\mathfrak{B} \in \exists^0_{r+1}$ and we have

$$\models \mathfrak{A}[\alpha] \leftrightarrow \text{not} \models \mathfrak{B}[\alpha] \leftrightarrow \text{not } \exists p\,.\Vdash \mathfrak{B}[\bar{\alpha}(p)].$$

Because α is $r+1$-generic, this is equivalent to $\exists p\,.\Vdash \neg \mathfrak{B}[\bar{\alpha}(p)]$.

Now if $\mathfrak{A} \in \exists^0_{r+2}$, $\mathfrak{A} = \exists x_i\, \mathfrak{C}$ for some $\mathfrak{C} \in \forall^0_{r+1}$ and we have

$$\models \mathfrak{A}[\alpha] \leftrightarrow \exists n\,.\models \mathfrak{C}(\bar{n}/x_i)[\alpha]$$
$$\leftrightarrow \exists n\, \exists p\,.\Vdash \mathfrak{C}(\bar{n}/x_i)[\bar{\alpha}(p)] \leftrightarrow \exists p\,.\Vdash \mathfrak{A}[\bar{\alpha}(p)]. \quad \square$$

For Theorem 6.6 to be useful there must be sufficiently many r-generic functions. First we have

6.7 Lemma. *For any s there are $2^{\aleph_0}$ generic functions in $[s]$.*

Proof. Let $\mathfrak{A}_0, \mathfrak{A}_1, \ldots$ be a list of the denumerably many closed formulas of $\mathscr{L}'$. For any s and any β we define recursively:

$$\gamma(0) = s; \qquad \gamma(2n+1) = \gamma(2n) * \langle \beta(n) \rangle;$$
$$\gamma(2n+2) = \text{least } t[t \in \text{Sq} \wedge \gamma(2n+1) \subseteq t \wedge (\Vdash \mathfrak{A}_n[t] \text{ or } \Vdash \neg \mathfrak{A}_n[t])].$$

γ is a well-defined function by Lemma 6.4(iii). As $\gamma(n) \subseteq \gamma(n+1)$ and $\gamma(2n+1)$ is a proper extension of $\gamma(2n)$, there exists a unique limit function β^* such that for all m, $\beta^*(m) = (\gamma(n))_m$ for all sufficiently large n. Clearly $\beta^* \in [s]$ and is generic. Since if $\beta \neq \delta$, also $\beta^* \neq \delta^*$, there are $2^{\aleph_0}$ distinct such β^*. $\square$

Our next aim is to show that there are some relatively simple r-generic and generic functions. For this we shall use a construction similar to that in the preceding proof together with an assessment of the complexity of the relation $\Vdash \mathfrak{A}[s]$. Let $\ulcorner \mathfrak{A} \urcorner$ and $\ulcorner \sigma \urcorner$ denote the Gödel numbers of $\mathfrak{A}$ and σ in an assignment of numbers to the formulas and terms of $\mathscr{L}'$ similar to that described for $\mathscr{L}$ in § 5. By the discussion there it is clear that for each r, $\{\ulcorner \mathfrak{A} \urcorner : \mathfrak{A}$ *is a closed* $\exists^0_r$-formula$\}$ is recursive. Similarly, there are partial recursive functions f_k such that for all terms σ and all m and s, $f_k(\ulcorner \sigma \urcorner, \mathbf{m}, s) \simeq \sigma[\mathbf{m}, s]$. Let

$$\text{Fo}^+_r(\ulcorner \mathfrak{A} \urcorner, s) \leftrightarrow \mathfrak{A} \text{ is a closed } \exists^0_r\text{-formula and } \Vdash \mathfrak{A}[s];$$
$$\text{Fo}^-_r(\ulcorner \mathfrak{A} \urcorner, s) \leftrightarrow \mathfrak{A} \text{ is a closed } \exists^0_r\text{-formula and } \Vdash \neg \mathfrak{A}[s];$$

$$\mathrm{Fo}^+ = \bigcup \{\mathrm{Fo}_r^+ : r \in \omega\};$$
$$\mathrm{Fo}^- = \bigcup \{\mathrm{Fo}_r^- : r \in \omega\}.$$

6.8 Lemma. *For each* $r > 0$, $\mathrm{Fo}_r^+ \in \Sigma_r^0$ *and* $\mathrm{Fo}_r^- \in \Pi_r^0$; Fo^+ *and* Fo^- *are* Δ_1^1.

Proof. If $\mathfrak{A}$ is a closed $\exists_1^0$-formula, then for some $a \in \mathrm{Pri}$ and some terms $\sigma_0, \ldots, \sigma_{k-1}$, $\mathfrak{A} = \exists x_i \bar{R}_a(\sigma_0, \ldots, \sigma_{k-1})$. Then

$$\begin{aligned}\Vdash \mathfrak{A}[s] &\leftrightarrow \exists n . \Vdash \bar{R}_a(\sigma_0, \ldots, \sigma_{k-1})(\bar{n}/x_i)[s] \\ &\leftrightarrow \exists n . [a](\sigma_0[n, s], \ldots, \sigma_{k-1}[n, s]) = 0 \\ &\leftrightarrow \exists n . [a](f_1(\ulcorner\sigma_0\urcorner, n, s), \ldots, f_1(\ulcorner\sigma_{k-1}\urcorner, n, s)) = 0.\end{aligned}$$

Since the Gödel numbers $\ulcorner\sigma_i\urcorner$ of the terms occuring in $\mathfrak{A}$ can be recursively calculated from $\ulcorner\mathfrak{A}\urcorner$, this shows $\mathrm{Fo}_1^+ \in \Sigma_1^0$. That $\mathrm{Fo}_1^- \in \Pi_1^0$ is immediate from this and the definition of $\Vdash$.

Suppose now that $\mathrm{Fo}_r^+ \in \Sigma_r^0$ and $\mathrm{Fo}_r^- \in \Pi_r^0$. Then for any closed $\exists_{r+1}^0$-formula $\mathfrak{A} = \exists x_i \mathfrak{B}$,

$$\Vdash \mathfrak{A}[s] \leftrightarrow \exists n . \Vdash \mathfrak{B}(\bar{n}/x_i)[s] \leftrightarrow \exists n . \mathrm{Fo}_r^-(g(\ulcorner\mathfrak{A}\urcorner, n), s),$$

where g is a recursive function such that

$$g(\ulcorner\exists x_i \mathfrak{B}\urcorner, n) = \ulcorner\mathfrak{B}(\bar{n}/x_i)\urcorner.$$

Hence $\mathrm{Fo}_{r+1}^+ \in \Sigma_{r+1}^0$. That $\mathrm{Fo}_{r+1}^- \in \Pi_{r+1}^0$ follows immediately.

To evaluate Fo^+ and Fo^-, recall the relation $U_{(\omega)}^0$:

$$U_{(\omega)}^0(r, a, \langle \mathbf{m} \rangle) \leftrightarrow U_r^0(a, \langle \mathbf{m} \rangle).$$

By Theorems 3.6 and 3.7, $U_{(\omega)}^0 \in \Delta_1^1$. The preceding part of the proof may be interpreted as providing instructions for computing recursive functions h^+ and h^- such that for all $r > 0$,

$$\begin{aligned}&\mathrm{Fo}_r^+(m, s) \leftrightarrow U_{(\omega)}^0(r, h^+(r), \langle m, s \rangle), \quad \text{and} \\ &\mathrm{Fo}_r^-(m, s) \leftrightarrow \sim U_{(\omega)}^0(r, h^-(r), \langle m, s \rangle);\end{aligned}$$

that is, $h^+(r)$ is an index of Fo_r^+ as a Σ_r^0 set and correspondingly for h^-. Then

$$\begin{aligned}&\mathrm{Fo}^+(m, s) \leftrightarrow \exists r\, U_{(\omega)}^0(r, h^+(r), \langle m, s \rangle), \quad \text{and} \\ &\mathrm{Fo}^-(m, s) \leftrightarrow \exists r \sim U_{(\omega)}^0(r, h^-(r), \langle m, s \rangle),\end{aligned}$$

so both are Δ_1^1. □

6.9 Theorem. *For all r there exist r-generic functions in Δ^0_{r+1}. There exist generic functions in Δ^1_1.*

Proof. Let f be a recursive function which enumerates the Gödel numbers of closed $\exists^0_r$-formulas and let $\mathfrak{A}_n$ denote the formula with Gödel number n. Set $\gamma(0) = \langle\ \ \rangle$ and

$$\gamma(n+1) = \text{least } t[t \in \mathrm{Sq} \wedge \gamma(n) \subsetneq t \wedge (\Vdash \mathfrak{A}_n[t] \text{ or } \Vdash \neg\mathfrak{A}_n[t])]$$
$$= \text{least } t[t \in \mathrm{Sq} \wedge \gamma(n) \subsetneq t \wedge (\mathrm{Fo}^+_r(f(n), t) \vee \mathrm{Fo}^-_r(f(n), t))].$$

As in the proof of 6.7, if β the limit of the sequences $\gamma(n)$, β is r-generic. Both γ and β are recursive in the relations Fo^+_r and Fo^-_r, hence are Δ^0_{r+1} by Post's Theorem.

The second part is proved similarly using Fo^+ and Fo^-. □

6.10 Corollary. *For all r,* $\{\alpha: \alpha \in \Delta^0_{r+1}\} \notin \Sigma^0_{r+1}$. $\{\alpha: \alpha \in \Delta^0_{(\omega)}\} \notin \Delta^0_{(\omega)}$ (second proof — cf. Corollary 4.23).

Proof. Suppose, contrary to the first assertion, that there exists an $\exists^0_r$-formula $\mathfrak{A}$ of $\mathscr{L}'$ such that

$$\alpha \in \Delta^0_{r+1} \leftrightarrow \vDash \exists x_0 \neg\mathfrak{A}[\alpha].$$

Then for some n and some r-generic β in Δ^0_{r+1}, $\vDash \neg\mathfrak{A}(\bar{n}/x_0)[\beta]$. It follows from Theorem 6.6 that for some p, $\Vdash \neg\mathfrak{A}(\bar{n}/x_0)[\bar{\beta}(p)]$. By Lemma 6.7 there are $2^{\aleph_0}$ r-generic functions γ such that $\bar{\gamma}(p) = \bar{\beta}(p)$. For each of these $\Vdash \neg\mathfrak{A}(\bar{n}/x_0)[\bar{\gamma}(p)]$, hence $\vDash \neg\mathfrak{A}(n/x_0)[\gamma]$, hence $\vDash \exists x_0 \neg\mathfrak{A}[\gamma]$, hence $\gamma \in \Delta^0_{r+1}$. This contradicts the fact that there are only countably many Δ^0_{r+1} functions.

For the second part, suppose that $\mathsf{A} = \{\alpha: \alpha \in \Delta^0_{(\omega)}\}$ were, say, Σ^0_{r+1}. The foregoing proof shows that no denumerable Σ^0_{r+1} set includes $\{\alpha: \alpha \in \Delta^0_{r+1}\}$, and A would be such a set, a contradiction. □

A strengthening of the first part of 6.10 is proved in 6.15.

In the examples of 2.3 we saw that if $\{\alpha\} \in \Delta^0_{(\omega)}$, then $\alpha \in \Delta^1_1$ and in IV.2.22 we shall show that every $\alpha \in \Delta^1_1$ is recursive in some β with $\{\beta\} \in \Pi^0_1$. The natural conjecture that every Δ^1_1 function is implicitly $\Delta^0_{(\omega)}$ is, however, false:

6.11 Corollary. *There exist $\alpha \in \Delta^1_1$ such that $\{\alpha\} \notin \Delta^0_{(\omega)}$.*

Proof. Let α be any Δ^1_1 generic function. Suppose that for some $\mathfrak{A} \in \exists^0_r$, $\mathfrak{A}$ defines $\{\alpha\}$ — that is, for all β,

$$\beta = \alpha \leftrightarrow \vDash \mathfrak{A}[\beta].$$

Since $\models \mathfrak{A}[\alpha]$ and α is generic, $\Vdash \mathfrak{A}[\bar{\alpha}(p)]$ for some p. But then for any generic $\beta \in [\bar{\alpha}(p)]$ also $\Vdash \mathfrak{A}[\bar{\beta}(p)]$, hence $\models \mathfrak{A}[\beta]$. By Lemma 6.7 there are $2^{\aleph_0}$ such β, a contradiction. □

Related to this question is the relationship between Δ_1^1 subsets of ω and sets of the form $\bar{\Gamma}$ for monotone arithmetical operators Γ. It follows from Theorem 3.2 that every Δ_1^1 set is many-one reducible to such a $\bar{\Gamma}$, but the following shows that the reduction cannot in general be omitted.

6.12 Corollary. *There exist $A \in \Delta_1^1$ such that $A \neq \bar{\Gamma}$ for any monotone arithmetical operator Γ.*

Proof. Let α be a Δ_1^1 generic function and $A = \{m : \alpha(m) = 0\} = Z_\alpha$. Suppose that $A = \bar{\Gamma}$ for some monotone arithmetical Γ. Since P_Γ is arithmetical, there exists a formula $\mathfrak{A} \in \exists_0^1$ such that for all β,

$$\forall m\, [\mathsf{P}_\Gamma(m, \beta) \rightarrow \beta(m) = 0] \leftrightarrow \models \mathfrak{A}[\beta].$$

Since Γ is monotone, A is the smallest set whose characteristic function satisfies the left-hand side of this equivalence. Hence $\models \mathfrak{A}[\alpha]$ and for any β, if $\models \mathfrak{A}[\beta]$, then $A \subseteq Z_\beta$. Since α is generic, $\Vdash \mathfrak{A}[\bar{\alpha}(p)]$ for some p. Clearly A is not finite so there exists a $q \geqslant p$ such that $q \in A$. Let β be any generic function in $[\bar{\alpha}(q) * \langle 1 \rangle]$. Then $\Vdash \mathfrak{A}[\bar{\beta}(p)]$ so $\models \mathfrak{A}[\beta]$, but $q \in A \sim Z_\beta$, a contradiction. □

To obtain the promised extensions of 4.12–4.14 we need to measure the size of the set of r-generic functions in yet another way:

6.13 Lemma. *For all r, the set of r-generic functions is comeager.*

Proof. For each closed $\exists_r^0$-formula $\mathfrak{A}$, let

$$\mathsf{A}_{\mathfrak{A}} = \{\alpha : \exists p\, (\Vdash \mathfrak{A}[\bar{\alpha}(p)] \text{ or } \Vdash \neg \mathfrak{A}[\bar{\alpha}(p)])\}.$$

It is immediate from Lemma 6.4 that each $\mathsf{A}_{\mathfrak{A}}$ is open and dense, so that $\sim\mathsf{A}_{\mathfrak{A}}$ is nowhere dense. Since

$$\{\alpha : \alpha \text{ is not } r\text{-generic}\} = \bigcup \{\sim\mathsf{A}_{\mathfrak{A}} : \mathfrak{A} \text{ is a closed } \exists_r^0\text{-formula}\},$$

this set is meager. □

6.14 Theorem. *For all r, Δ_{r+1}^0 is a basis for the class of non-meager Σ_{r+3}^0 sets.*

Proof. Suppose $\mathsf{A} \in \Sigma_{r+3}^0$ is non-meager and let $\mathfrak{B}$ be an $\forall_r^0$-formula such that

$$\alpha \in \mathsf{A} \leftrightarrow \models \exists x_0 \forall x_1 \exists x_2\, \mathfrak{B}[\alpha].$$

A is the countable union of sets A_m, where

$$\alpha \in \mathsf{A}_m \leftrightarrow \models \forall x_1 \exists x_2 \mathfrak{B}[m, \alpha].$$

Let Ge_r denote the set of r-generic functions. By the preceding Lemma, $\mathsf{A} \sim \mathrm{Ge}_r$ is meager so $\mathsf{A} \cap \mathrm{Ge}_r$ is non-meager. Hence for some $\bar{m}$ and some $\bar{s}$, $\mathsf{A}_{\bar{m}} \cap \mathrm{Ge}_r$ is dense in $[\bar{s}]$.

We shall contruct a Δ^0_{r+1} function in $\mathsf{A}_{\bar{m}} \cap \mathrm{Ge}_r$. As in the proof of Theorem 6.9, let f be a recursive function which enumerates $\{\ulcorner \mathfrak{A} \urcorner : \mathfrak{A}$ is a closed $\exists^0_r$ formula$\}$, and let $\mathfrak{A}_n$ be the formula with Gödel number $f(n)$. Set

$$\gamma(0) = \bar{s};$$

$$\gamma(2n+1) = \text{least } t[t \in \mathrm{Sq} \wedge \gamma(2n) \subseteq t \wedge (\Vdash \mathfrak{A}_n[t] \text{ or } \Vdash \neg \mathfrak{A}_n[t])];$$

$$\gamma(2n+2) = (\text{least } u[(u)_1 \in \mathrm{Sq} \wedge \gamma(2n+1) \subsetneqq (u)_1 \wedge \Vdash \mathfrak{B}[\bar{m}, n, (u)_0, (u)_1]])_1.$$

To see that γ is well defined, suppose that $\gamma(2n+1)$ is defined. Since

$$\mathsf{A}_{\bar{m}} \cap \mathrm{Ge}_r \subseteq \{\alpha : \models \exists x_2 \mathfrak{B}[\bar{m}, n, \alpha]\} \cap \mathrm{Ge}_r,$$

this latter set is also dense in $[\bar{s}]$ and thus has a non-empty intersection with $[\gamma(2n+1)]$. If δ is a member of this intersection, then for some p, $\models \mathfrak{B}[\bar{m}, n, p, \delta]$, so since δ is r-generic, $\Vdash \mathfrak{B}[\bar{m}, n, p, \bar{\delta}(q)]$ for some q. We may choose q larger than $\lg(\gamma(2n+1))$. Then $u = \langle p, \bar{\delta}(q)\rangle$ satisfies the condition and $\gamma(2n+2)$ is defined.

Now let β be the limit of the sequences $\gamma(n)$. The odd stages of this construction ensure that β is r-generic and the even stages ensure that

$$\forall n \exists p \exists q . \Vdash \mathfrak{B}[\bar{m}, n, p, \bar{\beta}(q)].$$

Hence

$$\forall n \exists p \models \mathfrak{B}[\bar{m}, n, p, \beta],$$

and thus

$$\models \forall x_1 \exists x_2 \mathfrak{B}[\bar{m}, \beta].$$

Thus $\beta \in \mathsf{A}_{\bar{m}} \cap \mathrm{Ge}_r$. That also $\beta \in \Delta^0_{r+1}$ follows from Lemma 6.8 and Post's Theorem. β is the required element of $\mathsf{A} \cap \Delta^0_{r+1}$. □

6.15 Corollary. *For all* $r > 0$, $\{\alpha : \alpha \in \Delta^0_r\} \in \Sigma^0_{r+2} \sim \Delta^0_{r+2}$.

Proof. The positive half follows from the equivalence

$$\alpha \in \Delta_r^0 \leftrightarrow \exists a\, \forall m\, \forall n\, [\alpha(m) = n \leftrightarrow U_r^0(a, \langle m, n\rangle)].$$

Suppose that also $\{\alpha : \alpha \in \Delta_r^0\} \in \Pi_{r+2}^0$. Then $\{\alpha : \alpha \notin \Delta_r^0\}$ is a comeager, hence non-meager, Σ_{r+2}^0 set, so by the Theorem, it has a Δ_r^0 element, a contradiction. □

6.16–6.17 Exercises

6.16. Show that for all $r \geq 1$ there exist functions β which are implicitly Π_{r+1}^0 but not implicitly Π_r^0.

6.17. Use the result of the preceding Exercise to show that Exercise 4.24 is false for $r = 1$.

6.18 Notes. For readers familiar with forcing in the context of set theory as described (say) in Shoenfield [1971], we note that the r-generic functions are those which meet a certain collection of Δ_{r+1}^0 dense sets, namely those of the form

$$\{s : \Vdash \mathfrak{A}[s] \text{ or } \Vdash \neg \mathfrak{A}[s]\}$$

for $\exists_r^0$ formulas $\mathfrak{A}$. Similarly, the generic functions are those which meet a certain collection of arithmetical dense sets.

Corollary 6.10 is due to Addison [1965], the proof given here is from Hinman [1969a]. 6.14 and 6.15 are also from Hinman [1969a]. The other results are due to Feferman [1964/65].

Part B

The Analytical and Projective Hierarchies

Most of the results of Chapter III have in common that their proofs rely heavily on a direct analysis of the number and kind of quantifiers needed to define certain concepts. With these methods we were able to establish the main structural features of the arithmetical hierarchy, but were much less successful with the corresponding questions concerning the analytical hierarchy. In the two chapters of Part B we shall introduce several new techniques — prewellordering, uniformization, and construction of transfinite hierarchies — which will provide answers for many of the questions left open in the preceding chapter. These by no means exhaust the tools which are useful in studying the analytical and projective hierarchies and we shall return to these questions from time to time in Part C. On the other hand, the results of §§ V.2–3 indicate that some questions about analytical and projective relations cannot be decided on the basis of the current axioms of set theory.

Chapter IV
The First Level

We shall study in this chapter the classes Σ_1^1, Π_1^1, and Δ_1^1 and their relativized and boldface counterparts. Not only are these classes more amenable to analysis than the higher levels of the analytical hierarchy, but they have played a much larger role in the development both of descriptive set theory and of generalized recursion theory.

Among the early objects of study in Descriptive Set Theory were the Borel sets and the projective classes $\mathbf{\Sigma}_r^1$ and $\mathbf{\Pi}_r^1$. Although the Borel sets were easily proved to have many pleasant properties, the operation of projection ($\exists^1$) used in defining the projective classes seemed too non-constructive to allow much to be proved about these sets. In 1915 Suslin discovered that the $\mathbf{\Sigma}_1^1$ (*analytic*) sets could all be constructed by use of the much more explicit operation $\mathscr{A}$ (cf. Exercise III.2.19) and that $\mathbf{\Delta}_1^1$ coincides with the class of Borel sets (cf. Theorem IV.3.3). These results stimulated much of the later development of the theory. In particular, the structure of $\mathbf{\Pi}_1^1$ and $\mathbf{\Delta}_1^1$ is reflected in many other pairs of classes of relations to be studied in this book.

From the side of recursion theory, the notational analogy of Δ_1^1 with Δ_1^0 together with the transitivity of the relation "$\alpha \in \Delta_1^1[\beta]$", leads to the conjecture that the Δ_1^1 relations are in some sense "generalized recursive". This is supported by the results of §2 below which show a great similarity between the properties of Σ_1^0 and Π_1^1 (not Σ_1^1!). One such similarity has already been noted: for any monotone operator Γ, if $\Gamma \in \Sigma_1^0$ (Π_1^1), then also $\bar{\Gamma} \in \Sigma_1^0$ (Π_1^1). The conjecture will be verified in § VI.2, where we show that the (Π_1^1) Δ_1^1 relations are exactly those (semi-) recursive in the type-2 functional E, and in § VIII.3 where we see that the (Π_1^1) Δ_1^1 relations on numbers are exactly those which are ω_1-(semi-) recursive.

1. Π_1^1 *and Well-Orderings*

The basic tool in the study of Π_1^1 relations is a characterization of this class in terms of W, the set of well-ordering functions. Recall that for a total functional F of rank $(k+1, l)$, $\mathsf{F}[\mathbf{m}, \boldsymbol{\alpha}]$ denotes the function $\lambda p.\mathsf{F}(p, \mathbf{m}, \boldsymbol{\alpha})$ so that F serves also as a functional from ${}^{k,l}\omega$ into ${}^{\omega}\omega$. A relation R is called (*many-one*) *reducible* to a set A ($\mathsf{R} \leqslant \mathsf{A}$) iff for some recursive functional F,

$$\mathsf{R}(\mathbf{m}, \boldsymbol{\alpha}) \leftrightarrow \mathsf{F}[\mathbf{m}, \boldsymbol{\alpha}] \in \mathsf{A}.$$

We also say R is *reduced to* A *via* F. Note that by the Analytical Substitution Theorem (III.2.12), if $\mathsf{A} \in \Sigma^1_r$ (Π^1_r) and $\mathsf{R} \leqslant \mathsf{A}$, then also $\mathsf{R} \in \Sigma^1_r$ (Π^1_r).

1.1 Theorem. *For all* R, $\mathsf{R} \in \Pi^1_1 \leftrightarrow \mathsf{R} \leqslant \mathsf{W}$.

Proof. $\mathsf{W} \in \Pi^1_1$ by the examples of III.2.3, so if $\mathsf{R} \leqslant \mathsf{W}$ also $\mathsf{R} \in \Pi^1_1$. Suppose now that $\mathsf{R} \in \Pi^1_1$. Then there exists a recursive relation P such that

$$\mathsf{R}(\mathbf{m}, \boldsymbol{\alpha}) \leftrightarrow \forall\beta\, \exists n\, \mathsf{P}(\bar{\beta}(n), \mathbf{m}, \boldsymbol{\alpha}).$$

We shall associate with each $(\mathbf{m}, \boldsymbol{\alpha})$ a linear ordering $\trianglelefteq^{\mathsf{P}}_{\mathbf{m},\boldsymbol{\alpha}}$ such that

(1) $\mathsf{R}(\mathbf{m}, \boldsymbol{\alpha}) \leftrightarrow \trianglelefteq^{\mathsf{P}}_{\mathbf{m},\boldsymbol{\alpha}}$ is a well-ordering;

(2) the functional F defined by

$$\mathsf{F}(\langle s, t\rangle, \mathbf{m}, \boldsymbol{\alpha}) = \begin{cases} 0, & \text{if } s \trianglelefteq^{\mathsf{P}}_{\mathbf{m},\boldsymbol{\alpha}} t; \\ 1, & \text{otherwise} \end{cases}$$

is recursive.

Then R is reduced to W via F.

First, let $\trianglelefteq$ be the recursive linear ordering defined by:

$$s \trianglelefteq t \leftrightarrow s, t \in \mathrm{Sq} \wedge (t \subseteq s \vee (\exists n < \mathrm{lg}(s))[n < \mathrm{lg}(t) \wedge (\forall i < n)((s)_i = (t)_i) \wedge (s)_n < (t)_n]).$$

The ordering $\trianglelefteq^{\mathsf{P}}_{\mathbf{m},\boldsymbol{\alpha}}$ is a restriction of $\trianglelefteq$. Let

$$\mathsf{P}'(s, \mathbf{m}, \boldsymbol{\alpha}) \leftrightarrow (\exists u \leqslant s)[u \in \mathrm{Sq} \wedge u \subseteq s \wedge \mathsf{P}(u, \mathbf{m}, \boldsymbol{\alpha})].$$

Then also P' is recursive and

$$\mathsf{R}(\mathbf{m}, \boldsymbol{\alpha}) \leftrightarrow \forall\beta\, \exists n\, \mathsf{P}'(\bar{\beta}(n), \mathbf{m}, \boldsymbol{\alpha}).$$

We set

$$s \trianglelefteq^{\mathsf{P}}_{\mathbf{m},\boldsymbol{\alpha}} t \leftrightarrow s \trianglelefteq t \wedge \sim\mathsf{P}'(s, \mathbf{m}, \boldsymbol{\alpha}) \wedge \sim\mathsf{P}'(t, \mathbf{m}, \boldsymbol{\alpha}).$$

Claim (2) is obvious. For the implication ($\leftarrow$) of (1), suppose $\sim\mathsf{R}(\mathbf{m}, \boldsymbol{\alpha})$ so that for some β, $\forall n \sim \mathsf{P}'(\bar{\beta}(n), \mathbf{m}, \boldsymbol{\alpha})$. Then for all n, $\bar{\beta}(n+1) \trianglelefteq^{\mathsf{P}}_{\mathbf{m},\boldsymbol{\alpha}} \bar{\beta}(n)$ and by condition (4′) of I.1.6, $\trianglelefteq^{\mathsf{P}}_{\mathbf{m},\boldsymbol{\alpha}}$ is not a well-ordering.

Suppose now that $\trianglelefteq^{\mathsf{P}}_{\mathbf{m},\boldsymbol{\alpha}}$ is not a well-ordering and let $A = \{s : s \in \mathrm{Sq} \wedge \sim\mathsf{P}'(s, \mathbf{m}, \boldsymbol{\alpha})\}$. Then $A = \mathrm{Fld}(\trianglelefteq^{\mathsf{P}}_{\mathbf{m},\boldsymbol{\alpha}})$ and it suffices to show $\exists\beta\,\forall n\,.\,\bar{\beta}(n) \in A$. Note that by the definition of P', for any s and t,

$$(3) \qquad s \in A \wedge t \subseteq s \rightarrow t \in A.$$

By assumption there exists a non-empty set $B \subseteq A$ with no $\trianglelefteq^{\mathsf{P}}_{\mathbf{m},\boldsymbol{\alpha}}$-least element — that is, $(\forall s \in B)(\exists t \in B)\; t \vartriangleleft^{\mathsf{P}}_{\mathbf{m},\boldsymbol{\alpha}} s$. We may assume that (3) holds also with B in place of A. Let f be a partial function defined by

$$f(n) \simeq \text{least } p((\exists s \in B)[\bar{f}(n) \subsetneq s \wedge (s)_n = p]).$$

Clearly either f is total or for some $\bar{n}$, $f(n)$ is defined exactly for $n < \bar{n}$. In the first case we have that for every n, $\bar{f}(n) \subsetneq s$ for some $s \in B \subseteq A$. Hence by (3), $\forall n\,.\,\bar{f}(n) \in A$ and we have the desired conclusion: $\exists\beta\,\forall n\,.\,\bar{\beta}(n) \in A$. Suppose, on the other hand, that $f(n)$ is defined exactly for $n < \bar{n}$. As B has no $\trianglelefteq^{\mathsf{P}}_{\mathbf{m},\boldsymbol{\alpha}}$-least element, there is some $s \in B$ with $s \vartriangleleft^{\mathsf{P}}_{\mathbf{m},\boldsymbol{\alpha}} \bar{f}(\bar{n})$. If $\bar{f}(\bar{n}) \subsetneq s$, then $f(\bar{n})$ would be defined, contrary to assumption. Hence for some $n < \bar{n}$,

$$(\forall i < n)((s)_i = (\bar{f}(\bar{n}))_i = f(i)) \wedge (s)_n < (\bar{f}(\bar{n}))_n = f(n).$$

But then $\bar{f}(n) \subsetneq s$ so by the definition of f, $f(n) \leq (s)_n$, a contradiction. Thus this case does not arise. □

1.2 Corollary. $\mathsf{W} \notin \Sigma_1^1$.

Proof. If $\mathsf{W} \in \Sigma_1^1$, then by the Theorem every Π_1^1 relation is also Σ_1^1, contrary to the Analytical Hierarchy Theorem. □

From Theorem 1.1 we can already begin to see why Π_1^1 rather than Σ_1^1 plays the role of the class of generalized semi-recursive relations. If P is a recursive relation of rank $(k+1, l)$, let

$$|\mathbf{m}, \boldsymbol{\alpha}|_0^{\mathsf{P}} = \begin{cases} \text{least } p\,.\,\mathsf{P}(p, \mathbf{m}, \boldsymbol{\alpha}), & \text{if any;} \\ \omega, & \text{otherwise;} \end{cases}$$

$$|\mathbf{m}, \boldsymbol{\alpha}|_1^{\mathsf{P}} = \begin{cases} \text{order-type of } \trianglelefteq^{\mathsf{P}}_{\mathbf{m},\boldsymbol{\alpha}}, & \text{if this is a well-ordering;} \\ \aleph_1, & \text{otherwise.} \end{cases}$$

Then

$$\exists p\, \mathsf{P}(p, \mathbf{m}, \boldsymbol{\alpha}) \leftrightarrow |\mathbf{m}, \boldsymbol{\alpha}|_0^{\mathsf{P}} < \omega$$

and

$$\forall \beta\, \exists p\, \mathsf{P}(\bar{\beta}(p), \mathbf{m}, \boldsymbol{\alpha}) \leftrightarrow |\mathbf{m}, \boldsymbol{\alpha}|_1^{\mathsf{P}} < \aleph_1.$$

In these terms, the proof that the class of semi-recursive relations has the reduction property (II.4.17) proceeds by defining

$$\mathsf{R}^*(\mathbf{m}, \boldsymbol{\alpha}) \leftrightarrow \mathsf{R}(\mathbf{m}, \boldsymbol{\alpha}) \wedge |\mathbf{m}, \boldsymbol{\alpha}|_0^{\mathsf{P}} \leq |\mathbf{m}, \boldsymbol{\alpha}|_0^{\mathsf{Q}}, \quad \text{and}$$

$$\mathsf{S}^*(\mathbf{m}, \boldsymbol{\alpha}) \leftrightarrow \mathsf{S}(\mathbf{m}, \boldsymbol{\alpha}) \wedge |\mathbf{m}, \boldsymbol{\alpha}|_0^{\mathsf{Q}} < |\mathbf{m}, \boldsymbol{\alpha}|_0^{\mathsf{P}}.$$

To establish similarly the reduction property for Π_1^1, we need only evaluate the complexity of the relation $|\mathbf{m}, \boldsymbol{\alpha}|_1^{\mathsf{P}} \leq |\mathbf{m}, \boldsymbol{\alpha}|_1^{\mathsf{Q}}$.

We recall from Examples III.2.3 the Σ_1^1 relation $\preccurlyeq$ defined by:

$$\begin{aligned} \gamma \preccurlyeq \delta &\leftrightarrow \leqslant_\gamma \text{ and } \leqslant_\delta \text{ are linear orderings and } \leqslant_\gamma \text{ is isomorphic} \\ &\qquad \text{to a subordering of } \leqslant_\delta \\ &\leftrightarrow \leqslant_\gamma \text{ and } \leqslant_\delta \text{ are linear orderings and} \\ &\qquad \exists \alpha\, \forall p\, \forall q\, [\alpha \text{ is 1–1 on } \mathsf{Fld}(\gamma) \wedge (p \leqslant_\gamma q \rightarrow \alpha(p) \leqslant_\delta \alpha(q))]. \end{aligned}$$

We define also a Σ_1^1 relation $\prec$ by

$$\begin{aligned} \gamma \prec \delta &\leftrightarrow \leqslant_\gamma \text{ and } \leqslant_\delta \text{ are linear orderings and } \leqslant_\gamma \text{ is isomorphic} \\ &\qquad \text{to a subordering of a proper initial segment of } \leqslant_\delta \\ &\leftrightarrow \leqslant_\gamma \text{ and } \leqslant_\delta \text{ are linear orderings and} \\ &\qquad \exists \alpha\, \exists r\, \forall p\, \forall q\, [\alpha \text{ is 1–1 on} \\ &\qquad\qquad \mathsf{Fld}(\gamma) \wedge (p \leqslant_\gamma q \rightarrow \alpha(p) \leqslant_\delta \alpha(q) <_\delta r)]. \end{aligned}$$

If both γ and δ belong to W, then clearly

$$\gamma \preccurlyeq \delta \leftrightarrow \|\gamma\| \leq \|\delta\| \quad \text{and} \quad \gamma \prec \delta \leftrightarrow \|\gamma\| < \|\delta\|.$$

We shall need, however, relations which have this property whenever at least *one* of γ and δ belongs to W.

1.3 Definition. For all γ and δ,

(i) $\gamma \preccurlyeq_\Sigma \delta \leftrightarrow \gamma \preccurlyeq \delta \vee \delta \notin \mathsf{W}$;

(ii) $\gamma \prec_\Sigma \delta \leftrightarrow \gamma \prec \delta \vee \delta \notin \mathsf{W}$;

(iii) $\gamma \preccurlyeq_\Pi \delta \leftrightarrow \neg(\delta \prec_\Sigma \gamma)$;

(iv) $\gamma \prec_\Pi \delta \leftrightarrow \neg(\delta \preccurlyeq_\Sigma \gamma)$.

Note that $\leqslant_\Sigma$ and $<_\Sigma$ are Σ_1^1 relations, whereas $\leqslant_\Pi$ and $<_\Pi$ are Π_1^1 relations. None is Δ_1^1, but part of the import of the following theorem is that each is "Δ_1^1 on W".

1.4 Theorem. *For all γ and δ, if either $\gamma \in \mathsf{W}$ or $\delta \in \mathsf{W}$, then*
(i) $\gamma \leqslant_\Sigma \delta \leftrightarrow [\gamma \in \mathsf{W} \wedge \|\gamma\| \leqslant \|\delta\|] \leftrightarrow \gamma \leqslant_\Pi \delta$;
(ii) $\gamma <_\Sigma \delta \leftrightarrow [\gamma \in \mathsf{W} \wedge \|\gamma\| < \|\delta\|] \leftrightarrow \gamma <_\Pi \delta$.

Proof. Suppose first that $\gamma \leqslant_\Sigma \delta$. If $\gamma \in \mathsf{W}$ then either $\delta \notin \mathsf{W}$, so $\|\gamma\| < \aleph_1 = \|\delta\|$, or $\delta \in \mathsf{W}$ and $\gamma \leqslant \delta$, so $\|\gamma\| \leqslant \|\delta\| < \aleph_1$. If $\delta \in \mathsf{W}$, then $\gamma \leqslant \delta$, so as any subordering of a well-ordering is also a well-ordering, also $\gamma \in \mathsf{W}$ and $\|\gamma\| \leqslant \|\delta\|$.

Suppose now that the middle condition of (i) holds: $\gamma \in \mathsf{W}$ and $\|\gamma\| \leqslant \|\delta\|$. Then either $\delta \notin \mathsf{W}$, or $\delta \in \mathsf{W}$ and $\leqslant_\delta$ is a well-ordering of type at least $\|\gamma\|$. In either case $\neg\, \delta < \gamma$, which together with $\gamma \in \mathsf{W}$ implies $\gamma \leqslant_\Pi \delta$.

Next suppose that $\gamma \leqslant_\Pi \delta$ — i.e., $\neg(\delta < \gamma)$ and $\gamma \in \mathsf{W}$. If $\delta \notin \mathsf{W}$, then $\gamma \leqslant_\Sigma \delta$. If $\delta \in \mathsf{W}$ then by the comparability of well-orderings, $\gamma \leqslant \delta$ and again $\gamma \leqslant_\Sigma \delta$.

The implications (ii) are proved similarly. □

1.5 Theorem. (i) Π_1^1 *has the reduction property but not the separation property*;
(ii) Σ_1^1 *has the separation property but not the reduction property.*

Proof. By Lemmas II.4.19 and II.4.21 it suffices to show that Π_1^1 has the reduction property. Let R and S be any two Π_1^1 relations of the same rank. By Theorem 1.1 there exist recursive functionals F and G such that

$$\mathsf{R}(\mathbf{m}, \boldsymbol{\alpha}) \leftrightarrow \mathsf{F}[\mathbf{m}, \boldsymbol{\alpha}] \in \mathsf{W} \quad \text{and} \quad \mathsf{S}(\mathbf{m}, \boldsymbol{\alpha}) \leftrightarrow \mathsf{G}[\mathbf{m}, \boldsymbol{\alpha}] \in \mathsf{W}.$$

We set

$$\mathsf{R}^*(\mathbf{m}, \boldsymbol{\alpha}) \leftrightarrow \mathsf{R}(\mathbf{m}, \boldsymbol{\alpha}) \wedge \mathsf{F}[\mathbf{m}, \boldsymbol{\alpha}] \leqslant_\Pi \mathsf{G}[\mathbf{m}, \boldsymbol{\alpha}];$$

$$\mathsf{S}^*(\mathbf{m}, \boldsymbol{\alpha}) \leftrightarrow \mathsf{S}(\mathbf{m}, \boldsymbol{\alpha}) \wedge \mathsf{G}[\mathbf{m}, \boldsymbol{\alpha}] <_\Pi \mathsf{F}[\mathbf{m}, \boldsymbol{\alpha}].$$

Observe that if $(\mathbf{m}, \boldsymbol{\alpha}) \in \mathsf{R} \cup \mathsf{S}$, then at least one of $\mathsf{F}[\mathbf{m}, \boldsymbol{\alpha}]$ and $\mathsf{G}[\mathbf{m}, \boldsymbol{\alpha}]$ belongs to W so that either $\mathsf{F}[\mathbf{m}, \boldsymbol{\alpha}] \leqslant_\Pi \mathsf{G}[\mathbf{m}, \boldsymbol{\alpha}]$ or $\mathsf{G}[\mathbf{m}, \boldsymbol{\alpha}] <_\Pi \mathsf{F}[\mathbf{m}, \boldsymbol{\alpha}]$. With this it is straightforward to verify that $(\mathsf{R}^*, \mathsf{S}^*)$ reduces (R, S). □

In the proof of Theorem 1.1, the relation $\trianglelefteq^{P}_{\mathbf{m}, \boldsymbol{\alpha}}$ is a well-ordering recursive in $\boldsymbol{\alpha}$. In particular, if P is a recursive relation and ω_1 denotes the least non-recursive ordinal (Definition III.3.11), then $|\mathbf{m}|_1^P$ is either a recursive ordinal or $\aleph_1$ so that

$$\forall \beta\, \exists p\, P(\bar{\beta}(p), \mathbf{m}) \leftrightarrow |\mathbf{m}|_1^P < \omega_1.$$

Hence if for any ordinal σ we set

$$\mathsf{W}_\sigma = \{\gamma : \gamma \in \mathsf{W} \wedge \|\gamma\| < \sigma\},$$

then in the notation of Theorem 1.1,

$$R(\mathbf{m}) \leftrightarrow \forall\beta\, \exists p\, P(\bar{\beta}(p), \mathbf{m}) \leftrightarrow \mathsf{F}[\mathbf{m}] \in \mathsf{W}_{\omega_1}.$$

Thus we have proved the implication ($\rightarrow$) of

1.6 Theorem. *For all* $R \subseteq {}^k\omega$, $R \in \Pi^1_1 \leftrightarrow R \preccurlyeq \mathsf{W}_{\omega_1}$.

Proof. It suffices to show $\mathsf{W}_{\omega_1} \in \Pi^1_1$. This follows from the equivalences:

$$\begin{aligned} \gamma \in \mathsf{W}_{\omega_1} &\leftrightarrow \exists\delta\, [\delta \in \mathsf{W} \wedge \delta \text{ is recursive} \wedge \|\gamma\| \leqslant \|\delta\|] \\ &\leftrightarrow \exists c\, [\{c\} \text{ is a total function of rank } 1 \wedge \\ &\qquad \{c\} \in \mathsf{W} \wedge \gamma \leqslant_\Pi \{c\}]. \quad \square \end{aligned}$$

The same proof shows that every Π^1_1 relation on numbers is reducible to the denumerable set $\mathsf{W}_{\text{rec}} = \{\gamma : \gamma \in \mathsf{W} \wedge \gamma \text{ is recursive}\}$. It will, however, be more convenient to use the set of indices of recursive well-orderings.

1.7 Definition. $W = \{c : \{c\} \text{ is a total function of rank } 1 \wedge \{c\} \in \mathsf{W}\}$. For $c \in W$, $\|c\| = \|\{c\}\|$.

1.8 Theorem. *For all* $R \subseteq {}^k\omega$, $R \in \Pi^1_1 \leftrightarrow R \preccurlyeq W$.

Proof. That W and hence all R reducible to it are Π^1_1 is immediate from the Analytical Substitution Theorem. If $R \in \Pi^1_1$ and F is a recursive function which reduces R to W, let a be an index such that $\{a\}(\mathbf{m}, p) = F(p, \mathbf{m})$. Then

$$R(\mathbf{m}) \leftrightarrow \lambda p.\{a\}(\mathbf{m}, p) \in \mathsf{W} \leftrightarrow \lambda p.\{\mathrm{Sb}_{k-1}(a, \mathbf{m})\}(p) \in \mathsf{W} \leftrightarrow \mathrm{Sb}_{k-1}(a, \mathbf{m}) \in W. \quad \square$$

1.9 Corollary. *Neither* W_{ω_1} *nor* W *is* Σ^1_1. $\square$

All of the preceding results have relativized and "boldface" extensions which we sketch below. In most cases the proofs are straightforward and are left to the Exercises.

1.10 Definition. For any $\beta, \mathsf{R}, \mathsf{A}$, and A,

(i) $\omega_1[\beta] = \sup^+\{\|\gamma\| : \gamma \in \mathsf{W} \wedge \gamma \text{ is recursive in } \beta\}$;

(ii) $W[\beta] = \{c : \lambda p.\{c\}(p, \beta) \in \mathsf{W}\}$;

(iii) for all $c \in W[\beta]$, $\|c\|_\beta = \|\lambda p.\{c\}(p, \beta)\|$;

(iv) $\mathsf{R} \leqslant_\beta A$ iff for some functional F recursive in β,

$$\mathsf{R}(\mathbf{m}, \boldsymbol{\alpha}) \leftrightarrow \mathsf{F}(\mathbf{m}, \boldsymbol{\alpha}) \in A;$$

(v) $\mathsf{R} \leqslant_\beta \mathsf{A}$ iff for some functional F recursive in β,

$$\mathsf{R}(\mathbf{m}, \boldsymbol{\alpha}) \leftrightarrow \mathsf{F}[\mathbf{m}, \boldsymbol{\alpha}] \in \mathsf{A};$$

(vi) $\mathsf{R} \blacktriangleleft \mathsf{A}$ iff for some continuous functional F,

$$\mathsf{R}(\mathbf{m}, \boldsymbol{\alpha}) \leftrightarrow \mathsf{F}[\mathbf{m}, \boldsymbol{\alpha}] \in \mathsf{A}.$$

1.11 Theorem. *For all* β, R, *and* R,

(i) $\mathsf{R} \in \Pi_1^1[\beta] \leftrightarrow \mathsf{R} \leqslant_\beta \mathsf{W}$;

(ii) $R \in \Pi_1^1[\beta] \leftrightarrow R \leqslant_\beta \mathsf{W}_{\omega_1[\beta]} \leftrightarrow R \leqslant W[\beta]$;

(iii) $\mathsf{R} \in \mathbf{\Pi}_1^1 \leftrightarrow \mathsf{R} \blacktriangleleft \mathsf{W}$;

(iv) *neither* $\mathsf{W}_{\omega_1[\beta]}$ *nor* $W[\beta]$ *is* $\Sigma_1^1[\beta]$;

(v) $\mathsf{W} \notin \mathbf{\Sigma}_1^1$. □

1.12 Corollary. *For all* β

(i) $\Pi_1^1[\beta]$ *and* $\mathbf{\Pi}_1^1$ *have the reduction property but not the separation property*;

(ii) $\Sigma_1^1[\beta]$ *and* $\mathbf{\Sigma}_1^1$ *have the separation property but not the reduction property.* □

1.13–1.25 Exercises

1.13. Show that $\trianglelefteq$ is a dense linear ordering with greatest element but no least element.

1.14. A set $A \subseteq \mathrm{Sq}$ is called a *tree* iff there is no sequence $s_0, s_1, \ldots$ such that for all i, $s_i \in A$ and $s_i \subsetneq s_{i+1}$. Let

$$\mathsf{Tr} = \{A : A \text{ is a tree}\}.$$

Show that for all R, $\mathsf{R} \in \Pi_1^1$ iff $\mathsf{R} \leqslant \mathsf{Tr}$.

1.15. For any tree A and any s, let

$$A_s = \{t : t \in A \wedge s \subsetneq t\}.$$

Set

$$\mathsf{Tr}_{(\sigma)} = \bigcup \{\mathsf{Tr}_\tau : \tau < \sigma\} \cup \{\varnothing\}$$

and

$$A \in \mathsf{Tr}_\sigma \leftrightarrow \forall s\,.\, A_s \in \mathsf{Tr}_{(\sigma)}.$$

Show that $\mathsf{Tr} = \mathsf{Tr}_{(\aleph_1)}$, but that for all $\sigma < \aleph_1$, $\mathsf{Tr} \not\subseteq \mathsf{Tr}_{(\sigma)}$.

1.16. For $A \in \mathsf{Tr}$, let

$$|A| = \text{least } \sigma\,.\, A \in \mathsf{Tr}_\sigma.$$

Show that there exist relations $\preccurlyeq_\Sigma$ and $\preccurlyeq_\Pi$ which satisfy an analogue of Theorem 1.4. It follows that Tr could be used in place of W for all the results of this section.

1.17. For which ordinals σ is W_σ arithmetical? Classify as many W_σ as you can (cf. Exercise III.1.17).

1.18. Show that for every $r \geq 0$, there exists an ordinal $\sigma_r < \omega_1$ such that every Σ_r^0 relation is reducible to W_{σ_r}.

1.19. Show that a relation R is Π_1^1 iff it is reducible to $\{\gamma : \gamma \in \mathsf{W} \wedge \gamma \text{ is primitive recursive}\}$. Conclude that ω_1 is also the least ordinal which is not primitive recursive.

1.20. Show that there exists a recursive function F such that for all m,

$$F(m) \in W \quad \text{and} \quad \|F(m)\| = m.$$

1.21. Let

$$\mathsf{Pl}_W = \{\langle a, b, c\rangle : a, b, c \in W \wedge \|a\| + \|b\| = \|c\|\},$$

and

$$\mathsf{Ti}_W = \{\langle a, b, c\rangle : a, b, c \in W \wedge \|a\| \cdot \|b\| = \|c\|\}.$$

Show that Pl_W and Ti_W are Π_1^1 sets.

1.22. Show that there exists a Π_1^1 set $W^* \subseteq W$ such that for all $\sigma < \omega_1$ there is exactly one $a \in W^*$ such that $\|a\| = \sigma$.

1.23. Prove Theorem 1.11 and Corollary 1.12.

1.24. Show that $\mathsf{R} \in \Pi^1_1$ iff R is *uniformly reducible* to $W[\boldsymbol{\alpha}]$ — that is, there exists a recursive function G such that for all $(\mathbf{m}, \boldsymbol{\alpha})$

$$\mathsf{R}(\mathbf{m}, \boldsymbol{\alpha}) \leftrightarrow G(\mathbf{m}) \in W[\boldsymbol{\alpha}].$$

1.25. Show that every Σ^1_2 relation is expressible in the form $(\exists \gamma \in \mathsf{W})\mathsf{P}(\mathbf{m}, \boldsymbol{\alpha}, \gamma)$ with $\mathsf{P} \in \Delta^1_1$.

1.26 Notes. The representation of Π^1_1 relations in terms of well-orderings is one of the oldest and most fundamental results of Descriptive Set Theory. It essentially originates with Lebesgue in 1905 and is studied further by Luzin, Suslin, and Sierpinski from 1915 on. Luzin [1930] is a good exposition of the state of knowledge then. Kleene [1955a] rediscovered the technique in his proof that the set O of notations for constructive ordinals (Definition 4.16) is a complete Π^1_1 set. Spector [1955] was the first explicit statement that every Π^1_1 set of numbers is reducible to W.

Alternative developments of the theory of this section are sketched in Exercises 1.14–16 and Exercises III.3.33–34.

2. The Boundedness Principle and Other Applications

2.1 Boundedness Theorem. *For any* $\mathsf{A} \subseteq {}^{\omega}\omega$ *and any* $A \subseteq \omega$,

(i) $\mathsf{A} \in \boldsymbol{\Sigma}^1_1 \wedge \mathsf{A} \subseteq \mathsf{W} \rightarrow \sup^+\{\|\gamma\| : \gamma \in \mathsf{A}\} < \aleph_1$;
(ii) $\mathsf{A} \in \Sigma^1_1 \wedge \mathsf{A} \subseteq \mathsf{W} \rightarrow \sup^+\{\|\gamma\| : \gamma \in \mathsf{A}\} < \omega_1$;
(iii) $A \in \Sigma^1_1 \wedge A \subseteq W \rightarrow \sup^+\{\|c\| : c \in A\} < \omega_1$.

Proof. Suppose, contrary to (i), that $\mathsf{A} \in \boldsymbol{\Sigma}^1_1$, $\mathsf{A} \subseteq \mathsf{W}$, but $\{\|\gamma\| : \gamma \in \mathsf{A}\}$ contains arbitrarily large countable ordinals. Thus any δ belongs to W just in case $\|\delta\| \leqslant \|\gamma\|$ for some $\gamma \in \mathsf{A}$. Since $\mathsf{A} \subseteq \mathsf{W}$, this yields

$$\delta \in \mathsf{W} \leftrightarrow \exists \gamma\, [\gamma \in \mathsf{A} \wedge \delta \leqslant_\Sigma \gamma]$$

which implies $\mathsf{W} \in \boldsymbol{\Sigma}^1_1$, contrary to Theorem 1.11.

Next, suppose that A were a counterexample to (iii). Then,

$$\delta \in \mathsf{W}_{\omega_1} \leftrightarrow \exists c\, [c \in A \wedge \delta \leqslant_\Sigma \{c\}],$$

which implies that $\mathsf{W}_{\omega_1} \in \Sigma^1_1$ contrary to Corollary 1.9.

Finally, if A were a counterexample to (ii), then

$$A = \{c : \{c\} \text{ is a total unary function} \wedge \exists \gamma\, [\gamma \in \mathsf{A} \wedge \{c\} \leqslant_\Sigma \gamma]\}$$

would be a counterexample to (iii). □

For $\rho < \omega_1$, we set

$$W_\rho = \{c : c \in W \wedge \|c\| < \rho\}.$$

2.2 Theorem. *For all* R *and all* R,
(i) $\mathsf{R} \in \mathbf{\Delta}^1_1 \leftrightarrow \mathsf{R} \pmb{\ll} \mathsf{W}_\rho$ *for some* $\rho < \aleph_1$;
(ii) $\mathsf{R} \in \Delta^1_1 \leftrightarrow \mathsf{R} \ll \mathsf{W}_\rho$ *for some* $\rho < \omega_1$;
(iii) $R \in \Delta^1_1 \leftrightarrow R \ll W_\rho$ *for some* $\rho < \omega_1$.

Proof. If ρ is any countable ordinal, let δ be an element of W such that $\|\delta\| = \rho$. Then for all γ,

$$\gamma \in \mathsf{W}_\rho \leftrightarrow \gamma <_\Sigma \delta \leftrightarrow \gamma <_\Pi \delta.$$

Hence $\mathsf{W}_\rho \in \mathbf{\Delta}^1_1$, which yields the implication ($\leftarrow$) of (i). If $\rho < \omega_1$, δ may be chosen to be recursive and we thus have ($\leftarrow$) of (ii) and (iii).

For the implications ($\rightarrow$), suppose first that $\mathsf{R} \in \mathbf{\Delta}^1_1$. Since $\mathsf{R} \in \mathbf{\Pi}^1_1$, there exists by Theorem 1.1 a continuous functional F such that $\mathsf{R}(\mathbf{m}, \boldsymbol{\alpha}) \leftrightarrow \mathsf{F}[\mathbf{m}, \boldsymbol{\alpha}] \in \mathsf{W}$. Let

$$\mathsf{A} = \{\gamma : \exists \mathbf{m} \exists \boldsymbol{\alpha} (\mathsf{R}(\mathbf{m}, \boldsymbol{\alpha}) \wedge \gamma = \mathsf{F}[\mathbf{m}, \boldsymbol{\alpha}])\}.$$

Since also $\mathsf{R} \in \mathbf{\Sigma}^1_1$ we have $\mathsf{A} \in \mathbf{\Sigma}^1_1$, and clearly $\mathsf{A} \subseteq \mathsf{W}$. Hence by the Boundedness Theorem there exists a $\rho < \aleph_1$ such that all $\gamma \in \mathsf{A}$ have $\|\gamma\| < \rho$. Hence $\mathsf{R}(\mathbf{m}, \boldsymbol{\alpha}) \leftrightarrow \mathsf{F}[\mathbf{m}, \boldsymbol{\alpha}] \in \mathsf{W}_\rho$ and thus $\mathsf{R} \pmb{\ll} \mathsf{W}_\rho$. The argument for (ii) and (iii) ($\rightarrow$) is nearly identical. $\square$

The import of this theorem is that the Δ^1_1 ($\mathbf{\Delta}^1_1$) relations may be naturally arranged in a sort of hierarchy of length ω_1 ($\aleph_1$). Let $X_\rho = \{\mathsf{R} : (\exists \tau \leqslant \rho) \mathsf{R} \ll \mathsf{W}_\tau\}$. Then if $\rho \leqslant \sigma$, $X_\rho \subseteq X_\sigma$ and $\Delta^1_1 = \bigcup \{X_\rho : \rho < \omega_1\}$. Similarly there are classes $\mathbf{X}_\rho$ such that $\mathbf{\Delta}^1_1 = \bigcup \{\mathbf{X}_\rho : \rho < \aleph_1\}$. In §§ 3–4 we shall construct other hierarchies on Δ^1_1 and $\mathbf{\Delta}^1_1$. See also Exercise 2.24.

We aim next to show that the set of Δ^1_1 functions is Π^1_1 but not Δ^1_1. This will complete the proof of Lemma III.4.8 that Δ^1_1 is not a basis for Π^0_1. We need first some technical lemmas. We extend the relations $\leqslant_\Sigma$, etc. to W in the obvious way:

$$c \leqslant_\Sigma d \leftrightarrow \{c\} \text{ and } \{d\} \text{ are total functions of rank 1 and } \{c\} \leqslant_\Sigma \{d\}$$

and similarly for $\leqslant_\Pi$, $<_\Sigma$, and $<_\Pi$.

2.3 Lemma. *There exist relations* P^Σ *and* P^Π *in* Σ^1_1 *and* Π^1_1 *respectively, such that for any* α, b, *and* c, *if* $c \in W$, *then*

$$\mathrm{Gr}_\alpha \preccurlyeq W_{\|c\|}\ via\{b\} \leftrightarrow \mathsf{P}^\Sigma(b, c, \alpha) \leftrightarrow \mathsf{P}^\Pi(b, c, \alpha).$$

Proof. Let

$$\begin{aligned}\mathsf{P}^\Sigma(b, c, \alpha) \leftrightarrow \{b\} \text{ is a total function of rank } 2 \wedge \\ \wedge \forall m\, \forall n\, [\alpha(m) = n \rightarrow \{b\}(m, n) \prec_\Sigma c] \wedge \\ \wedge \forall m\, \forall n\, [\{b\}(m, n) \prec_\Pi c \rightarrow \alpha(m) = n].\end{aligned}$$

Then for $c \in W$ and $\{b\}$ a total function of rank 2,

$$\mathsf{P}^\Sigma(b, c, \alpha) \leftrightarrow \forall m\, \forall n\, (\alpha(m) = n \leftrightarrow \|\{b\}(m, n)\| < \|c\|)$$

as required. P^Π is obtained by interchanging $\prec_\Sigma$ and $\prec_\Pi$. □

2.4 Lemma. *For every $\rho < \omega_1$, there exists a set $B \in \Delta^1_1$ such that $B \preccurlyeq W_\sigma$ for no $\sigma \leqslant \rho$.*

Proof. Let ρ be any ordinal $< \omega_1$ and $d \in W$ such that $\|d\| = \rho$. Set

$$A = \{\langle a, c\rangle : a \in W \wedge \|a\| < \|c\| \wedge \|c\| \leqslant \|d\|\}.$$

Since for any a and c,

$$\langle a, c\rangle \in A \leftrightarrow a \prec_\Sigma c \wedge c \preccurlyeq_\Sigma d \leftrightarrow a \prec_\Pi c \wedge c \preccurlyeq_\Pi d,$$

$A \in \Delta^1_1$. Hence also $B = A^{\omega J}$, the jump of A, is Δ^1_1. Suppose that for some $\sigma \leqslant \rho$, $B \preccurlyeq W_\sigma$. Then there is a recursive function f and a $c \in W$ such that $\|c\| = \sigma$ and

$$m \in B \leftrightarrow f(m) \in W_{\|c\|} \leftrightarrow \langle f(m), c\rangle \in A.$$

But then $B \preccurlyeq A$, a contradiction, since $A^{\omega J}$ is not recursive in A, hence not reducible to A. □

2.5 Selection Theorem. *For any Π^1_1 relation R, there exists a partial functional Sel_R with Π^1_1 graph such that for all $\mathbf{m}$ and $\boldsymbol{\alpha}$,*

$$\exists p\, \mathsf{R}(p, \mathbf{m}, \boldsymbol{\alpha}) \leftrightarrow \mathsf{R}(\mathrm{Sel}_\mathsf{R}(\mathbf{m}, \boldsymbol{\alpha}), \mathbf{m}, \boldsymbol{\alpha}) \leftrightarrow \mathrm{Sel}_\mathsf{R}(\mathbf{m}, \boldsymbol{\alpha})\downarrow.$$

Proof. Suppose $\mathsf{R} \in \Pi^1_1$ and is reducible to W via the recursive functional F. We define

$$\begin{aligned}\mathrm{Sel}_\mathsf{R}(\mathbf{m}, \boldsymbol{\alpha}) \simeq \text{least } p[\mathsf{R}(p, \mathbf{m}, \boldsymbol{\alpha}) \\ \wedge \|\mathsf{F}[p, \mathbf{m}, \boldsymbol{\alpha}]\| = \min\{\|\mathsf{F}[q, \mathbf{m}, \boldsymbol{\alpha}]\| : q \in \omega \wedge \mathsf{R}(q, \mathbf{m}, \boldsymbol{\alpha})\}].\end{aligned}$$

It is clear that $\mathsf{Sel_R}$ has the required property. That its graph is Π^1_1 follows from the easily proved equivalence:

$$\mathsf{Sel_R}(\mathbf{m}, \boldsymbol{\alpha}) \simeq p \leftrightarrow \mathsf{R}(p, \mathbf{m}, \boldsymbol{\alpha}) \wedge \forall q\, (\mathsf{F}[p, \mathbf{m}, \boldsymbol{\alpha}] \leqslant_\Pi \mathsf{F}[q, \mathbf{m}, \boldsymbol{\alpha}]) \\ \wedge (\forall q < p)(\mathsf{F}[p, \mathbf{m}, \boldsymbol{\alpha}] <_\Pi \mathsf{F}[q, \mathbf{m}, \boldsymbol{\alpha}]). \quad \square$$

2.6 Theorem. $\{\alpha : \alpha \in \Delta^1_1\} \in \Pi^1_1 \sim \Delta^1_1$.

Proof. To show that the set is Π^1_1 we have, with the notation of Lemma 2.3,

$$\begin{aligned} \alpha \in \Delta^1_1 &\leftrightarrow (\exists \sigma < \omega_1)\,.\,\mathrm{Gr}_\alpha \leqslant W_\sigma \\ &\leftrightarrow (\exists c \in W)\,.\,\mathrm{Gr}_\alpha \leqslant W_{\|c\|} \\ &\leftrightarrow \exists b\, \exists c\, [c \in W \wedge \mathsf{P}^\Pi(b, c, \alpha)]. \end{aligned}$$

Now for a contradiction suppose $\{\alpha : \alpha \in \Delta^1_1\} \in \Delta^1_1$. Let R be defined by:

$$\mathsf{R}(c, \alpha) \leftrightarrow c \in W \wedge ([\alpha \in \Delta^1_1 \wedge \exists b\, \mathsf{P}^\Pi(b, c, \alpha)] \\ \vee [\alpha \notin \Delta^1_1 \wedge \forall m\,.\,\{c\}(m) \simeq 1]).$$

Clearly $\mathsf{R} \in \Pi^1_1$, and by Theorem 2.2, $\forall \alpha\, \exists c\, \mathsf{R}(c, \alpha)$. Hence the functional $\mathsf{Sel_R}$ is total and thus by Corollary III.2.7 has Δ^1_1 graph. Let $A = \mathrm{Im}\,\mathsf{Sel_R}$. $A \in \Sigma^1_1$ and $A \subseteq W$ so by the Boundedness Theorem (2.1) there exists a $\rho < \omega_1$ such that for all $c \in A$, $\|c\| < \rho$. But then $\forall \alpha\, \exists c\, [\mathsf{R}(c, \alpha) \wedge \|c\| < \rho]$ from which it follows that for all $\alpha \in \Delta^1_1$, $\mathrm{Gr}_\alpha \leqslant W_\sigma$ for some $\sigma < \rho$. This contradicts Lemma 2.4. $\square$

By Theorem 1.6 any Π^1_1 relation on numbers can be represented in the form $(\exists \sigma < \omega_1)\mathsf{F}[\mathbf{m}] \in \mathsf{W}_\sigma$. Our next theorem gives another "existential" representation of Π^1_1 relations.

2.7 Lemma. *For any γ and $\delta \in \mathsf{W} \cap \Delta^1_1$, if $\mathrm{Fld}(\delta) \neq \omega$, then*

$$\|\gamma\| \leqslant \|\delta\| \leftrightarrow (\exists \beta \in \Delta^1_1)\forall p\, \forall q\, [p <_\gamma q \leftrightarrow \beta(p) <_\delta \beta(q)].$$

Proof. The implication $(\leftarrow)$ is immediate. Suppose $\gamma, \delta \in \mathsf{W} \cap \Delta^1_1$ and $\|\gamma\| \leqslant \|\delta\|$ and suppose $m \notin \mathrm{Fld}(\delta)$. Set

$$\mathsf{A} = \{\beta : \forall p\, \forall q\, [p <_\gamma q \leftrightarrow \beta(p) <_\delta \beta(q)] \wedge \beta \text{ maps } \mathrm{Fld}(\gamma) \\ \text{onto an intial segment of } \mathrm{Fld}(\delta) \wedge (\forall p \notin \mathrm{Fld}(\delta))\beta(p) = m\}.$$

A is arithmetical in γ and δ, hence is Δ^1_1. A has exactly one element β_0. By Corollary III.2.7(vii), $\beta_0 \in \Delta^1_1$. $\square$

2.8 Definition. $\Sigma_1^{1,\mathrm{Hyp}}$ is the class of relations R such that for some arithmetical relation P,

$$\mathsf{R}(\mathbf{m}, \boldsymbol{\alpha}) \leftrightarrow (\exists \beta \in \Delta_1^1[\boldsymbol{\alpha}])\mathsf{P}(\mathbf{m}, \boldsymbol{\alpha}, \beta)$$

("Hyp" stands for "hyperarithmetical").

2.9 Spector–Gandy Theorem. $\Sigma_1^{1,\mathrm{Hyp}} = \Pi_1^1$.

Proof. We shall give the proof only for relations on numbers where the theorem takes the form: $R \in \Pi_1^1$ iff for some arithmetical relation P,

$$R(\mathbf{m}) \leftrightarrow (\exists \beta \in \Delta_1^1)\mathsf{P}(\mathbf{m}, \beta)$$

(cf. Exercise 2.28).

Suppose first that R and P satisfy this equivalence and let

$$S(b, c) \leftrightarrow c \in W \wedge \{b\} \text{ is a total function of rank } 2 \wedge \\ \forall m\, \exists!\, n(\{b\}(m, n) \in W_{\|c\|}).$$

Then $S \in \Pi_1^1$ and

$$R(\mathbf{m}) \leftrightarrow \exists \beta\, \exists c\, (c \in W \wedge \mathrm{Gr}_\beta \leqslant W_{\|c\|} \wedge \mathsf{P}(\mathbf{m}, \beta)) \\ \leftrightarrow \exists b\, \exists c\, (S(b, c) \wedge \forall \beta\, [\mathsf{P}^\Sigma(b, c, \beta) \rightarrow \mathsf{P}(\mathbf{m}, \beta)])$$

with P^Σ from Lemma 2.3. Thus $R \in \Pi_1^1$.

For the converse implication, it suffices to show $W \in \Sigma_1^{1,\mathrm{Hyp}}$. By Theorems 2.6 and 1.1, there exists a recursive functional F such that

$$\alpha \in \Delta_1^1 \leftrightarrow \mathsf{F}[\alpha] \in \mathsf{W}.$$

Since $\{\alpha : \alpha \in \Delta_1^1\}$ is not Δ_1^1, it is not reducible to any W_ρ with $\rho < \omega_1$. Hence the ordinals $\|\mathsf{F}[\alpha]\|$ for $\alpha \in \Delta_1^1$ are unbounded in ω_1 and we have

$$c \in W \leftrightarrow (\exists \alpha \in \Delta_1^1)(\|c\| \leqslant \|\mathsf{F}[\alpha]\|) \\ \leftrightarrow (\exists \alpha \in \Delta_1^1)(\exists \beta \in \Delta_1^1)\forall p\, \forall q\, [p <_{\{c\}} q \leftrightarrow \beta(p) <_{\mathsf{F}[\alpha]} \beta(q)].$$

It is easy to verify that the last expression defines a set in $\Sigma_1^{1,\mathrm{Hyp}}$. □

We say $\mathsf{R} \in \Delta_1^{1,\mathrm{Hyp}}$ just in case both R and $\sim\mathsf{R} \in \Sigma_1^{1,\mathrm{Hyp}}$.

2.10 Corollary. *For all* R, $\mathsf{R} \in \Delta_1^1 \leftrightarrow \mathsf{R} \in \Delta_1^{1,\mathrm{Hyp}}$. □

These last two results have a natural interpretation in terms of the second-order comprehension axioms discussed in § III.5:

$$^\omega\omega \cap \Delta_1^1 \models \Delta_1^1\text{-Comprehension but } ^\omega\omega \cap \Delta_1^1 \not\models \exists_1^1\text{-Comprehension.}$$

Exercise 2.15 also asserts that

$$^\omega\omega \cap \Delta_1^1 \models \exists_1^1\text{-Choice.}$$

Turning now to the well-orderings of ω which occur in the various classes, we have the somewhat surprising

2.11 Theorem. (i) *For any R, if R is a* Σ_1^1 *well-ordering relation, then* $\|R\| < \omega_1$; (ii) *there exists a* Π_1^1 *well-ordering relation R such that* $\|R\| = \omega_1$.

Proof. Suppose, contrary to (i), that R is a Σ_1^1 well-ordering of type $\rho \geqslant \omega_1$. Then

$$\mathsf{W}_\rho = \{\gamma : \exists\alpha\, \forall p\, \forall q\, [p <_\gamma q \rightarrow R(\alpha(p), \alpha(q)) \wedge \alpha(p) \neq \alpha(q)]$$
$$\wedge \leqslant_\gamma \text{ is a linear ordering}\}$$

so that $\mathsf{W}_\rho \in \Sigma_1^1$, which contradicts the Boundedness Theorem.

For (ii), let

$$W^* = \{c : c \in W \wedge (\forall d < c)[\{c\} <_\Pi \{d\} \vee \{d\} <_\Pi \{c\}]\}.$$

Then $W^* \in \Pi_1^1$ and for each $\sigma < \omega_1$, there is a unique $c \in W^*$ such that $\|c\| = \sigma$. Hence if we set

$$R(c, d) \leftrightarrow c, d \in W^* \wedge \{c\} \leqslant_\Pi \{d\},$$

then R is a Π_1^1 well-ordering of type ω_1. □

Note as a consequence that δ_1^1, the least non-Δ_1^1 ordinal, is just ω_1 (cf. Corollary III.3.12).

2.12 Theorem. *For any R, A, and* α,

(i) $A \in \Pi_1^1 \sim \Delta_1^1 \wedge R \in \Pi_1^1 \rightarrow R \in \Delta_1^1[A]$;

(ii) $\omega_1[\alpha] > \omega_1 \wedge R \in \Pi_1^1 \rightarrow R \in \Delta_1^1[\alpha]$.

Proof. For (i), let F be a recursive function such that A is reducible to W_{ω_1} via F. Since $A \notin \Delta_1^1$, A is not reducible to any W_σ with $\sigma < \omega_1$ so the ordinals $\|F[m]\|$ with $m \in A$ are unbounded in ω_1. Hence, for any c,

$$c \in W \leftrightarrow (\exists m \in A)(\{c\} \leqslant_\Sigma F[m]).$$

From this follows $W \in \Sigma_1^1[A]$. If $R \in \Pi_1^1$, then $R \leqslant W$ so also $R \in \Sigma_1^1[A]$. But $\Pi_1^1 \subseteq \Pi_1^1[A]$ so also $R \in \Delta_1^1[A]$.

For (ii), if $\omega_1[\alpha] > \omega_1$, then there is a function δ recursive in α with $\|\delta\| = \omega_1$. Then for any c

$$c \in W \leftrightarrow \{c\} \text{ is a total unary function} \wedge \{c\} <_\Sigma \delta$$

and thus $W \in \Sigma_1^1[\alpha]$. The rest of the proof is as for (i). □

Part (i) of the theorem has a natural interpretation in terms of Δ_1^1-degrees, usually called *hyperdegrees* (cf. III.2.13 and the paragraph following). Let

$$A \leqslant_1^1 B \leftrightarrow A \in \Delta_1^1[B]$$

and

$$A <_1^1 B \leftrightarrow A \leqslant_1^1 B \wedge B \nleqslant_1^1 A:'$$

so that

$$\text{hydg}(A) = \{B : A \leqslant_1^1 B \wedge B \leqslant_1^1 A\}.$$

The hyperdegrees inherit a natural partial ordering also denoted by $\leqslant_1^1$. The hyperdegree $\mathbf{0} = \{A : A \in \Delta_1^1\}$ is the least element of this partial ordering.

2.13 Corollary. *For all A, $B \in \Pi_1^1 \sim \Delta_1^1$,* $\text{hydg}(A) = \text{hydg}(B)$.

Proof. Immediate from 2.12(i). □

This points out an imperfection in the analogy of Δ_1^1 and Π_1^1 with the classes of recursive and semi-recursive relations. The Friedberg–Mučnik Theorem mentioned at the end of § II.5 asserts the existence of semi-recursive sets A and B such that $\text{dg}(A)$ and $\text{dg}(B)$ are not only different but incomparable. We shall return to this point in § VIII.3.

In § V.6 we shall need a relativized version of Theorem 2.12 which we state here but leave the proof to Exercise 2.33.

2.14 Theorem. *For any R, A, α, and β, if $\beta \in \Delta_1^1[A]$ and $\beta \in \Delta_1^1[\alpha]$, then*

(i) $A \in \Pi_1^1[\beta] \sim \Delta_1^1[\beta] \wedge R \in \Pi_1^1[\beta] \rightarrow R \in \Delta_1^1[A]$;

(ii) $\omega_1[\alpha] > \omega_1[\beta] \wedge R \in \Pi_1^1[\beta] \rightarrow R \in \Delta_1^1[\alpha]$;

(iii) $\omega_1[\alpha] > \omega_1[\beta] \rightarrow \beta <_1^1 \alpha$. □

We conclude this section by establishing a precise bound for the closure ordinals of Π_1^1 monotone operators and Π_1^0 inductive operators over ω (cf. Exercise III.3.34).

2.15 Theorem. *For any operator Γ over ω,*

$$\Gamma \text{ monotone} \wedge \Gamma \in \Pi_1^1 \rightarrow |\Gamma| \leqslant \omega_1.$$

Proof. Let Γ be any Π_1^1 monotone operator. It will suffice to show $\Gamma^{\omega_1} \subseteq \Gamma^{(\omega_1)}$. Let $\bar{m}$ be a fixed element of Γ^{ω_1}. We first observe that for any ρ,

$$\text{(1)} \qquad \begin{aligned} \bar{m} \in \Gamma^{\rho} &\leftrightarrow \forall A\,[\Gamma^{(\rho)} \subseteq A \rightarrow \bar{m} \in \Gamma(A)] \\ &\leftrightarrow \forall A\,[\bar{m} \notin \Gamma(A) \rightarrow (\exists \sigma < \rho)\Gamma^{\sigma} \nsubseteq A]. \end{aligned}$$

Hence if we define

$$\varphi(A) = \begin{cases} \text{least } \sigma < \omega_1 \,.\, \Gamma^{\sigma} \nsubseteq A, & \text{if } \quad \bar{m} \notin \Gamma(A); \\ 0, & \text{otherwise;} \end{cases}$$

then φ is a total function defined on all $A \subseteq \omega$. Let

$$\bar{\rho} = \sup{}^{+}\{\varphi(A) : A \subseteq \omega\}.$$

Clearly $\bar{\rho} \leqslant \omega_1$ and by (1), $\bar{m} \in \Gamma^{\bar{\rho}}$; we shall show $\bar{\rho} < \omega_1$.

If $\bar{m} \in \Gamma^0$ we are done. Otherwise, for any σ

$$\sigma < \bar{\rho} \leftrightarrow \exists A\,[\bar{m} \notin \Gamma(A) \wedge (\forall \tau < \omega_1)(\Gamma^{\tau} \nsubseteq A \rightarrow \sigma \leqslant \tau)].$$

In other words, if

$$\mathsf{A} = \{\gamma : \exists A\,[\bar{m} \notin \Gamma(A) \wedge (\forall \tau < \omega_1)(\Gamma^{\tau} \nsubseteq A \rightarrow \|\gamma\| \leqslant \tau)]\},$$

then $\bar{\rho} = \sup^{+}\{\|\gamma\| : \gamma \in \mathsf{A}\}$. Let V be the Π_1^1 "coding relation" whose existence is claimed by Theorem III.3.13. Then

$$\gamma \in \mathsf{A} \leftrightarrow \exists A\,[\bar{m} \notin \Gamma(A) \wedge \forall c\,(c \in W \wedge \exists p\,[\mathsf{V}(p, \{c\}) \wedge p \notin A] \rightarrow \gamma \leqslant_{\Sigma} \{c\})].$$

Thus $\mathsf{A} \in \Sigma_1^1$ so by the Boundedness Theorem, $\bar{\rho} < \omega_1$. $\square$

2.16 Theorem. *For any inductive operator Γ over ω,*

$$\Gamma \in \Pi_1^0 \rightarrow |\Gamma| \leqslant \omega_1.$$

Proof. Suppose $\Gamma \in \Pi_1^0$ and let R be a recursive relation such that

$$\mathsf{P}_\Gamma(m, \alpha) \leftrightarrow \forall n\, R(m, \bar{\alpha}(n)).$$

As in the proof of Theorem 1.1 we may assume that

$$(2) \qquad R(m, s) \wedge t \subseteq s \rightarrow R(m, t).$$

Let $\bar{m}$ be a fixed element of Γ^{ω_1}; we shall prove $\bar{m} \in \Gamma^{(\omega_1)}$.

For all ordinals σ, let (for typographical reaons) f_σ ($f_{(\sigma)}$) be the characteristic function of $\Gamma^\sigma(\Gamma^{(\sigma)})$. We note first that for any n and any limit ordinal ρ

$$(3) \qquad (\exists \sigma < \rho)(\forall \tau < \rho)[\sigma \leqslant \tau \rightarrow \bar{f}_\tau(n) = \bar{f}_{(\rho)}(n)].$$

To see this, let

$$\tau_i = \begin{cases} \text{least } \tau . i \in \Gamma^\tau, & \text{if} \quad i \in \Gamma^{(\rho)}; \\ 0, & \text{otherwise}; \end{cases}$$

and take $\sigma = \max\{\tau_i : i < n\}$.

From the assumption $\bar{m} \in \Gamma^{\omega_1}$, we have $\forall n\, R(\bar{m}, \bar{f}_{(\omega_1)}(n))$ and thus by (3), $\forall n\, (\exists \sigma < \omega_1)\, R(\bar{m}, \bar{f}_\sigma(n))$. Let

$$\varphi(n) = \text{least } \sigma < \omega_1 \,.\, R(\bar{m}, \bar{f}_\sigma(n))$$

and

$$\bar{\rho} = \sup{}^+\{\varphi(n) : n \in \omega\}.$$

As in the proof of the preceding theorem, $\bar{\rho} \leqslant \omega_1$ and $\bar{\rho} = \sup^+\{\|\gamma\| : \gamma \in \mathsf{A}\}$ where

$$\begin{aligned} \gamma \in \mathsf{A} &\leftrightarrow \exists n\, (\forall \tau < \omega_1)[R(\bar{m}, \bar{f}_\tau(n)) \rightarrow \|\gamma\| \leqslant \tau] \\ &\leftrightarrow \exists n\, \forall c\, [c \in W \wedge R(\bar{m}, \bar{f}_{\|c\|}(n)) \rightarrow \gamma \leqslant_\Sigma \{c\}]. \end{aligned}$$

First note that $\mathsf{A} \subseteq \mathsf{W}$ as if $\gamma \in \mathsf{A}$ because the condition is satisfied for n, then $\|\gamma\| \leqslant \varphi(n) < \omega_1$ so $\gamma \in \mathsf{W}$. To see that $\mathsf{A} \in \Sigma_1^1$, let V_Π and V_Σ be respectively the Π_1^1 and Σ_1^1 "coding relations" from Theorem III.3.9. Then for $c \in W$,

$$\begin{aligned} R(\bar{m}, \bar{f}_{\|c\|}(n)) \leftrightarrow \exists s\, (s \in \mathrm{Sq} \wedge \mathrm{lg}(s) = n \wedge (\forall i < n)((s)_i \leqslant 1) \wedge \\ R(\bar{m}, s) \wedge (\forall i < n)[(s)_i = 0 \rightarrow \mathsf{V}_\Pi(i, \{c\}) \\ \wedge \mathsf{V}_\Sigma(i, \{c\}) \rightarrow (s)_i = 0]). \end{aligned}$$

Thus this relation is Π^1_1 from which it easily follows that $\mathsf{A} \in \Sigma^1_1$ and thus by Boundedness that $\bar{\rho} < \omega_1$. To complete the proof we show $\bar{m} \in \Gamma^{\bar{\rho}}$.

We first observe that the function φ is monotone. If $n \leqslant p$, then $R(\bar{m}, \bar{f}_{\varphi(p)}(p))$ so by (2) also $R(\bar{m}, \bar{f}_{\varphi(p)}(n))$. As $\varphi(n)$ is the *least* σ such that $R(\bar{m}, \bar{f}_\sigma(n))$, it follows that $\varphi(n) \leqslant \varphi(p)$.

Case 1. $\bar{\rho}$ is a successor ordinal. Then for some n_0, $\bar{\rho} = \varphi(n_0) + 1$ and for all n, $\varphi(n) \leqslant \varphi(n_0)$. Because φ is monotone, $\varphi(n) = \varphi(n_0)$ for all $n \geqslant n_0$. Hence for all $n \geqslant n_0$, $R(\bar{m}, \bar{f}_{\varphi(n_0)}(n))$. But then it follows from (2) that this holds also for all $n < n_0$ and thus for all n. Thus $\mathsf{P}_\Gamma(\bar{m}, f_{\varphi(n_0)})$ and $\bar{m} \in \Gamma(\Gamma^{\varphi(n_0)}) = \Gamma^{\bar{\rho}}$.

Case 2. $\bar{\rho}$ is a limit ordinal. Let n be a fixed natural number and σ be as in (3) such that $\sigma < \bar{\rho}$ and

$$(4) \qquad (\forall \tau < \bar{\rho})[\sigma \leqslant \tau \rightarrow \bar{f}_\tau(n) = \bar{f}_{(\bar{\rho})}(n)].$$

Since φ is monotone and $\bar{\rho} = \sup^+\{\varphi(p) : p \in \omega\}$, there is an $n_0 \geqslant n$ such that $\sigma \leqslant \varphi(n_0)$. By the definition of φ, $R(\bar{m}, \bar{f}_{\varphi(n_0)}(n_0))$. But then by (2), $R(\bar{m}, \bar{f}_{\varphi(n_0)}(n))$ and by (4), $R(\bar{m}, \bar{f}_{(\bar{\rho})}(n))$. As n was arbitrary, we have $\forall n R(\bar{m}, \bar{f}_{(\bar{\rho})}(n))$ and thus $\bar{m} \in \Gamma(\Gamma^{(\bar{\rho})}) = \Gamma^{\bar{\rho}}$. □

2.17 Corollary. *For any Π^0_1 inductive operator Γ, $\bar{\Gamma} \in \Pi^1_1$.*

Proof. For any Π^0_1 inductive Γ, the following equivalences hold:

$$m \in \bar{\Gamma} \leftrightarrow (\exists \sigma < \omega_1)(m \in \Gamma^\sigma) \leftrightarrow (\exists c \in W)(m \in \Gamma^{\|c\|})$$
$$\leftrightarrow (\exists c \in W)\mathsf{V}_\Pi(m, \{c\}). \quad \square$$

Similar techniques lead also to a kind of boundedness theorem for inductive definitions:

2.18 Lemma. *For any inductive operator Γ over ω and any Σ^1_1 set $A \subseteq \Gamma^{(\omega_1)}$, if either Γ is Δ^1_1 or Γ is Π^1_1 and monotone, then for some $\rho < \omega_1$, $A \subseteq \Gamma^\rho$.*

Proof. Suppose A and Γ satisfy the hypotheses, define $|m|_\Gamma = \text{least } \sigma . m \in \Gamma^\sigma$ for $m \in \Gamma^{(\omega_1)}$, and set $\bar{\rho} = \sup^+\{|m|_\Gamma : m \in A\}$. Clearly $A \subseteq \Gamma^{\bar{\rho}}$ and by definition $\bar{\rho} \leqslant \omega_1$.

If $A = \varnothing$, then $A \subseteq \Gamma^0$. Otherwise $\bar{\rho} = \sup^+\{\|\gamma\| : \gamma \in \mathsf{A}\}$ where

$$\gamma \in \mathsf{A} \leftrightarrow (\exists m \in A)(\forall \tau < \omega_1)[m \in \Gamma^\tau \rightarrow \|\gamma\| \leqslant \tau]$$
$$\leftrightarrow (\exists m \in A)(\forall c \in W)[\mathsf{V}(m, \{c\}) \rightarrow \gamma \leqslant_\Sigma \{c\}],$$

with V as in III.3.13 if Γ is Π^1_1 monotone and $\mathsf{V} = \mathsf{V}_\Pi$ as in III.3.9 if Γ is Δ^1_1. Then $\mathsf{A} \in \Sigma^1_1$ and $\mathsf{A} \subseteq \mathsf{W}$, so by the Boundedness Theorem, $\bar{\rho} < \omega_1$. □

2.19 Corollary. *For any inductive operator Γ over ω and any Σ_1^1 set $A \subseteq \bar{\Gamma}$, if either Γ is Π_1^0 or Γ is Π_1^1 and monotone, then for some $\rho < \omega_1$, $A \subseteq \Gamma^\rho$.* □

2.20 Corollary. *For any inductive operator Γ over ω, if either Γ is Π_1^0 or Γ is Π_1^1 and monotone, then*

(i) *if $\bar{\Gamma} \in \Sigma_1^1$, then $|\Gamma| < \omega_1$;*

(ii) *for any $R \in \Sigma_1^1$, if $R \leqslant \bar{\Gamma}$, then for some $\rho < \omega_1$, $R \leqslant \Gamma^\rho$.* □

2.21. Corollary. *For any inductive operator Γ over ω, if either Γ is Π_1^0 or Γ is Δ_1^1 and monotone, then*

$$\bar{\Gamma} \in \Delta_1^1 \quad \textit{iff} \quad |\Gamma| < \omega_1.$$

Proof. By Corollary 2.20 and Corollary III.3.12. □

2.22 Corollary. *For any $R \in \Delta_1^1$, there exists an implicitly Π_1^0 function δ such that R is recursive in δ.*

Proof. Suppose $R \in \Delta_1^1$. Then by Theorem III.3.2 there exists a Π_1^0 monotone operator Γ such that $R \leqslant \bar{\Gamma}$. By Corollary 2.20, $R \leqslant \Gamma^\rho$ for some $\rho < \omega_1$. Let S be the relation defined in the proof of Theorem III.3.9; as P_Γ is Π_1^0, S is arithmetical. Let γ be a recursive function such that $\|\gamma\| = \rho + 1$ and set $\varepsilon = \langle \alpha_\gamma, \beta_\gamma \rangle$. Then ε is the unique function which satisfies the arithmetical relation $\mathsf{S}((\varepsilon)_0, (\varepsilon)_1, \gamma)$, so ε is implicitly arithmetical. Furthermore, if p is the unique element of $\mathrm{Fld}(\gamma)$ such that $|p|_\gamma = \rho$, then $m \in \Gamma^\rho \leftrightarrow (\varepsilon)_1(\langle p, m \rangle) = 0$ so that Γ^ρ and hence R is recursive in ε. Finally, by Lemma III.4.4, ε is recursive in some δ which is implicitly Π_1^0. □

2.23–2.38 Exercises

2.23. Prove the following *Effective Boundedness Principle*: there exists a primitive recursive function h such that for any a, if a is a Σ_1^1 index for a set $A \subseteq W$ — that is

$$c \in A \leftrightarrow U_1^1(a, \langle c \rangle)$$

— then $h(a) \in W$ and $\|h(a)\| \geqslant \sup^+\{\|c\| : c \in A\}$. Formulate and prove a corresponding result for Σ_1^1 subsets of W.

2.24. The hierarchy of Δ_1^1 relations described following Theorem 2.2 is deficient in that it may happen that for some $\rho < \sigma < \omega_1$, no new relations are reducible to some W_τ $(\tau \leqslant \sigma)$ that are not already reducible to some W_τ $(\tau \leqslant \rho)$ — that is, $X_\rho = X_\sigma$. This can be remedied by omitting superfluous levels in the hierarchy, and for relations on numbers, the resulting hierarchy may be seen still to have length ω_1. Let

$$Z = \{\sigma : \sigma < \omega_1 \wedge W_\sigma \not\approx W_\rho \text{ for any } \rho < \sigma\}$$

and

$$\bar{\omega}_1 = \text{order-type of } Z.$$

Show that $\bar{\omega}_1 = \omega_1$.

Hint. Let

$$\bar{W} = \{c : c \in W \wedge \|c\| \in Z \wedge \neg \exists d\,(d < c \wedge \|c\| = \|d\|)\}$$

and suppose that for some $e \in W$, $\bar{\omega}_1 = \|e\| < \omega_1$. Set

$$\begin{aligned} P(c, d) \leftrightarrow (\|c\| \not< \|e\| \wedge d = 0) \vee (\|c\| < \|e\| \wedge d \in W \wedge \\ \forall \alpha\, [\text{if } \alpha \text{ is an ordinal-preserving map of} \\ \{a : \|a\| \leqslant \|c\|\} \text{ into } \{b : b \in \bar{W} \wedge \|b\| \leqslant \|d\|\}, \\ \text{then } \alpha(c) = d]). \end{aligned}$$

Show that $P \in \Pi^1_1$, there is a function $\beta \in \Delta^1_1$ such that $\forall c\, P(c, \beta(c))$ and

$$d \mathrel{\dot{\in}} \bar{W} \leftrightarrow \exists c\,(\|c\| \leqslant \|e\| \wedge \beta(c) = d).$$

2.25 (Kreisel [1962]). Prove the following two *effective choice principles*: for any Π^1_1 relation R,

(i) if $\forall \mathbf{m} \forall \boldsymbol{\alpha} \exists p\, \mathsf{R}(p, \mathbf{m}, \boldsymbol{\alpha})$, then there exists a Δ^1_1 functional F such that $\forall \mathbf{m} \forall \boldsymbol{\alpha}\, \mathsf{R}(\mathsf{F}(\mathbf{m}, \boldsymbol{\alpha}), \mathbf{m}, \boldsymbol{\alpha})$;

(ii) if $\forall \mathbf{m} \forall \boldsymbol{\alpha}\, (\exists \beta \in \Delta^1_1[\boldsymbol{\alpha}])\mathsf{R}(\mathbf{m}, \boldsymbol{\alpha}, \beta)$, then there exists a Δ^1_1 functional G such that $\forall \mathbf{m} \forall \boldsymbol{\alpha}\, \mathsf{R}(\mathbf{m}, \boldsymbol{\alpha}, \lambda q\,.\,\mathsf{G}(q, \mathbf{m}, \boldsymbol{\alpha}))$. Conclude in particular that $\{\alpha : \alpha \in \Delta^1_1\}$ is a model for $\exists^1_1$-Choice.

2.26. Show that there exists a set $\mathsf{A} \in \Sigma^1_1$ such that $\mathsf{A} \cap \mathsf{W} = \mathsf{W}_{\omega_1}$.

2.27. Show that if a function $\theta : {}^\omega\omega \to {}^\omega\omega$ has Δ^1_1 graph, then $\mathrm{Dm}\, \theta \in \Delta^1_1$. If $\boldsymbol{\theta}$ is one-one, then also $\mathrm{Im}\, \theta \in \Delta^1_1$.

2.28. Prove the general case of the Spector–Gandy Theorem (2.9).

2.29 (Spector [1959]). Show that a relation R is Π_1^1 iff for some Π_1^0 relation P, $\mathsf{R}(\mathbf{m}, \boldsymbol{\alpha}) \leftrightarrow \exists! \beta\, \mathsf{P}(\mathbf{m}, \boldsymbol{\alpha}, \beta)$. (*Hint*: In one direction ($\leftarrow$) use the Spector–Gandy Theorem. For the other, P will have the property that $\forall \mathbf{m} \forall \boldsymbol{\alpha} \exists \beta\, \mathsf{P}(\mathbf{m}, \boldsymbol{\alpha}, \beta)$.)

2.30. Show that for any $A \in \Pi_1^1$, $A \in \Delta_1^1$ iff $\omega_1[A] = \omega_1$.

2.31. Show that if $\gamma \in \Delta_2^1$, then $\omega_1[\gamma]$ is the order-type of a Δ_2^1 well-ordering of ω.

2.32. For any $A \subseteq \omega$, the *hyperjump* of A is the set

$$A^{\mathsf{hJ}} = \{\langle a, \mathbf{m}\rangle : \mathsf{U}_1^1(a, \langle \mathbf{m}\rangle, \langle A\rangle)\}.$$

Show that if $A \leqslant_1^1 B$, then also $A^{\mathsf{hJ}} \leqslant_1^1 B^{\mathsf{hJ}}$ so that the hyperjump is well defined on hyperdegrees.

2.33. Prove Theorem 2.14.

2.34. Show that

(i) for any $\sigma < \omega_1$, there exists a Π_1^0 inductive operator Γ such that $|\Gamma| = \sigma$;

(ii) any $A \in \Delta_1^1$ is reducible to $\bar{\Gamma}$ for some Π_1^0 inductive operator Γ such that $|\Gamma| < \omega_1$.

2.35. Show that for any monotone arithmetical operator Γ, $\{\langle c, m\rangle : c \in W \wedge m \in \Gamma^{|c|}\}$ is implicitly Π_1^1 (cf. proof of Theorem III.3.7). Conclude that every Π_1^1 relation on numbers is reducible to an implicitly Π_1^1 set and hence that Δ_1^1 is not a basis for the class of Π_1^1 singletons.

2.36. Show that for any α, $\alpha \in \Delta_1^1$ iff for some function δ, $\langle \alpha, \delta\rangle$ is implicitly Π_2^0.

2.37. Construct an arithmetical monotone operator Γ such that for all $\sigma < \omega_1$, $\Gamma^\sigma = W_\sigma$.

2.38. The partial functionals with Π_1^1 graph may be indexed as follows. Let

$$\mathsf{V}(n, a, \langle \mathbf{m}\rangle, \langle \boldsymbol{\alpha}\rangle) \leftrightarrow \sim \mathsf{U}_1^1(a, \langle n, \mathbf{m}\rangle, \langle \boldsymbol{\alpha}\rangle)$$

and set

$$\{a\}^*(\mathbf{m}, \boldsymbol{\alpha}) \simeq \mathrm{Sel}_{\mathsf{V}}(a, \langle \mathbf{m}\rangle, \langle \boldsymbol{\alpha}\rangle).$$

Which properties of the class of partial recursive functionals are shared by the class of Π_1^1 functionals with this indexing? Does the Recursion Theorem hold?

2.39 Notes. The boundedness principle 2.1 (i) is also an old result of Descriptive Set Theory — the effective versions 2.1 (ii) and (iii) are implicitly in Spector [1955]. The Spector–Gandy Theorem was conjectured by Mostowski and proved independently by Spector [1959] and Gandy [1960]. The proof here is due to Moschovakis. Theorems 2.12 and 2.14 are due to Spector [1955] and 2.15 to Spector [1959]. Theorem 2.16 was proved by Gandy [unpublished] essentially as done here. Another quite different proof appears in Richter [1971].

3. The Borel Hierarchy

In the Introduction we discussed briefly the notion of a hierarchy as a decomposition of a class of objects into levels indexed by ordinals in such a way that members of higher levels are in some sense more complex than members of lower levels. The main examples to this point have been the arithmetical and analytical hierarchies and their relativized and boldface counterparts. The "hierarchy" discussed following Theorem 2.2 suffers from the fact that some levels with different indices coincide. In this and the next section we shall construct hierarchies for the $\mathbf{\Delta}^1_1$ and Δ^1_1 relations which do not have this failing (see also Exercise 2.24).

The first of these, the Borel hierarchy, is both the simplest and historically the first transfinite hierarchy. Recall that the class Bo of Borel relations is the smallest class containing the open relations (equivalently, the closed-open relations) and closed under countable union and intersection of relations of the same rank. This is an inductive definition and the corresponding hierarchy consists essentially of the stages. The actual definition will differ slightly from this idea as we shall define classes $\mathbf{\Sigma}^0_\rho$ and $\mathbf{\Pi}^0_\rho$ for all $\rho < \aleph_1$ such that $\mathrm{Bo} = \bigcup \{\mathbf{\Sigma}^0_\rho : \rho < \aleph_1\}$. Thus the Borel hierarchy is a natural extension of the boldface arithmetical hierarchy.

Although these results appear in this context to depend heavily on recursion theory, in fact they were first derived long before the development of recursion theory. It is only in hindsight that we see that the hierarchies of recursion theory are refinements of those of Descriptive Set Theory and that many of the techniques of recursion theory were known already in some form to the early descriptive set theorists. This section and the next have been arranged to emphasize the relationship between the classical hierarchy and its effective or recursion-theoretic counterpart. These constructions also serve as paradigms for several others in later parts of the book — those of §§ V.4, 5 which involve operations more complex than countable union and intersection, and those of § VI.5, § VI.6, and § VII.3.

3.1 Lemma. *The class of Borel relations is closed under complementation and composition and substitution of continuous functionals.*

Proof. Let F and G be continuous and consider

$$X = \{\mathsf{R} : \{(\mathbf{m}, \boldsymbol{\alpha}) : \mathsf{R}(\mathsf{F}(\mathbf{m}, \boldsymbol{\alpha}), \mathbf{m}, \boldsymbol{\alpha}, \lambda p . \mathsf{G}(p, \mathbf{m}, \boldsymbol{\alpha}))\} \text{ is Borel}\}.$$

By II.3.9 and II.5.5, X includes the class of closed-open relations and it is trivial to show that X is closed under countable union and intersection. Hence Bo $\subseteq X$. The proof for complementation is similar. □

3.2 Lemma. *For all countable ordinals ρ, W_ρ is Borel.*

Proof. We proceed by induction on ρ. Clearly $\mathsf{W}_0 = \varnothing$ and W_1 are Borel. If ρ is a limit ordinal, then $\mathsf{W}_\rho = \bigcup \{\mathsf{W}_\sigma : \sigma < \rho\}$. If ρ is countable, then this is a countable union so if all W_σ $(\sigma < \rho)$ are Borel, so is W_ρ. If $\rho = \sigma + 1 > 1$, then $\gamma \in \mathsf{W}_\rho$ iff for all p, $\gamma \restriction p \in \mathsf{W}_\sigma$. Since W_σ is Borel, it follows from Lemma 3.1 that also for each p, $\mathsf{A}_p = \{\gamma : \gamma \restriction p \in \mathsf{W}_\sigma\}$ is Borel. Hence so is $\mathsf{W}_\rho = \bigcap \{\mathsf{A}_p : p \in \omega\}$. □

3.3 Theorem (Suslin). Bo $= \boldsymbol{\Delta}^1_1$.

Proof. The inclusion ($\subseteq$) is Corollary III.2.16. If $\mathsf{R} \in \boldsymbol{\Delta}^1_1$, then by Theorem 2.2, $\mathsf{R} \ll \mathsf{W}_\rho$ for some $\rho < \aleph_1$. Hence by Lemmas 3.1 and 3.2, R is Borel. □

3.4 Definition (The Borel Hierarchy). For all $\rho > 0$,

(i) $\boldsymbol{\Sigma}^0_0 = \boldsymbol{\Pi}^0_0 =$ the class of closed-open relations;
(ii) $\boldsymbol{\Sigma}^0_{(\rho)} = \bigcup \{\boldsymbol{\Sigma}^0_\tau : \tau < \rho\}$; $\boldsymbol{\Pi}^0_{(\rho)} = \bigcup \{\boldsymbol{\Pi}^0_\tau : \tau < \rho\}$;
(iii) $\boldsymbol{\Sigma}^0_\rho = \{\bigcup \{\mathsf{P}_p : p \in \omega\}$: all P_p have the same rank and belong to $\boldsymbol{\Pi}^0_{(\rho)}\}$;
(iv) $\boldsymbol{\Pi}^0_\rho = \{\bigcap \{\mathsf{P}_p : p \in \omega\}$: all P_p have the same rank and belong to $\boldsymbol{\Sigma}^0_{(\rho)}\}$;
(v) $\boldsymbol{\Delta}^0_\rho = \boldsymbol{\Sigma}^0_\rho \cap \boldsymbol{\Pi}^0_\rho$; $\boldsymbol{\Delta}^0_{(\rho)} = \bigcup \{\boldsymbol{\Delta}^0_\tau : \tau < \rho\}$.

It is immediate by induction on ρ that all of the classes $\boldsymbol{\Sigma}^0_\rho$ and $\boldsymbol{\Pi}^0_\rho$ are included in the class of Borel relations. The following lemma together with Theorem III.1.16 implies that this definition is consistent with the former definition of $\boldsymbol{\Sigma}^0_\rho$ for $\rho < \omega$ (III.1.15).

3.5 Lemma. *For all $\rho > 0$ and all* R,

(i) $\mathsf{R} \in \boldsymbol{\Sigma}^0_\rho \leftrightarrow \sim\mathsf{R} \in \boldsymbol{\Pi}^0_\rho$;
(ii) $\boldsymbol{\Sigma}^0_{(\rho)} \cup \boldsymbol{\Pi}^0_{(\rho)} \subseteq \boldsymbol{\Delta}^0_\rho$;
(iii) $\boldsymbol{\Sigma}^0_{(\rho+1)} = \boldsymbol{\Sigma}^0_\rho$; $\boldsymbol{\Pi}^0_{(\rho+1)} = \boldsymbol{\Pi}^0_\rho$.

Proof. The proof of (i) is an easy induction on ρ. If $\mathsf{R} \in \boldsymbol{\Sigma}^0_{(\rho)}$, then $\mathsf{R} \in \boldsymbol{\Sigma}^0_\tau$ for some $\tau < \rho$. Hence for some $\mathsf{P}_p \in \boldsymbol{\Pi}^0_{(\tau)}$, $\mathsf{R} = \bigcup \{\mathsf{P}_p : p \in \omega\}$. But $\boldsymbol{\Pi}^0_{(\tau)} \subseteq \boldsymbol{\Pi}^0_{(\rho)}$, so also $\mathsf{R} \in \boldsymbol{\Sigma}^0_\rho$. On the other hand, let $\mathsf{Q}_p = \mathsf{R}$ for all p. Then all $\mathsf{Q}_p \in \boldsymbol{\Sigma}^0_{(\rho)}$ so $\mathsf{R} = \bigcap \{\mathsf{Q}_p : p \in \omega\} \in \boldsymbol{\Pi}^0_\rho$. (ii) now follows from (i) and (iii) follows immediately. □

Many of the properties of the Borel hierarchy mirror those of the arithmetical hierarchy and we state in the following lemma some of the more important of these. The proofs are easy adaptations of previously used techiques and are omitted.

3.6 Lemma. *For all $\rho > 0$,*

(i) $\mathbf{\Sigma}^0_\rho$ *is closed under countable union, finite intersection, expansion, bounded quantification, existential number quantification, and composition and substitution of continuous functionals;*

(ii) $\mathbf{\Pi}^0_\rho$ *is closed under countable intersection, finite union, expansion, bounded quantification, universal number quantification, and composition and substitution of continuous functionals.* □

3.7 Theorem. $\mathrm{Bo} = \mathbf{\Delta}^0_{(\aleph_1)}$.

Proof. It suffices to show that $\mathbf{\Delta}^0_{(\aleph_1)}$ is closed under countable union and intersection. Suppose that for each $p \in \omega$, $\mathsf{P}_p \in \mathbf{\Delta}^0_{(\aleph_1)}$, say $\mathsf{P}_p \in \mathbf{\Delta}^0_{\sigma_p}$ with $\sigma_p < \aleph_1$. Then if $\rho = \sup\{\sigma_p : p \in \omega\}$, $\rho < \aleph_1$ and by Lemma 3.5 for all $p, \mathsf{P}_p \in \mathbf{\Delta}^0_\rho = \mathbf{\Sigma}^0_{(\rho+1)} \cap \mathbf{\Pi}^0_{(\rho+1)}$. Hence $\bigcup\{\mathsf{P}_p : p \in \omega\} \in \mathbf{\Sigma}^0_{\rho+1} \subsetneq \mathbf{\Delta}^0_{\rho+2} \subsetneq \mathbf{\Delta}^0_{(\aleph_1)}$. □

3.8 Corollary. $\mathbf{\Delta}^1_1 = \mathbf{\Delta}^0_{(\aleph_1)}$.

Proof. Immediate from 3.3 and 3.7. □

We want now to show that all of the levels of the Borel hierarchy are distinct. We proceed analogously to Theorem III.1.9 (the Arithmetical Hierarchy) and define relations U^0_ρ which are universal for $\mathbf{\Sigma}^0_\rho$ — that is, such that $\mathsf{U}^0_\rho \in \mathbf{\Sigma}^0_\rho$ and for every $\mathsf{R} \in \mathbf{\Sigma}^0_\rho$ there exists a $\beta \in {}^\omega\omega$ such that

$$\mathsf{R}(\mathbf{m}, \boldsymbol{\alpha}) \leftrightarrow \mathsf{U}^0_\rho(\langle \mathbf{m} \rangle, \langle \boldsymbol{\alpha} \rangle, \beta)$$

(cf. Exercise II.5.11).

We set

$$\mathsf{U}^0_1(\langle \mathbf{m} \rangle, \langle \boldsymbol{\alpha} \rangle, \beta) \leftrightarrow \mathsf{U}^0_1(\beta(0), \langle \mathbf{m} \rangle, \langle \boldsymbol{\alpha}, \lambda p . \beta(p+1) \rangle);$$

for $\rho > 0$,

$$\mathsf{U}^0_{\rho+1}(\langle \mathbf{m} \rangle, \langle \boldsymbol{\alpha} \rangle, \beta) \leftrightarrow \exists p \sim \mathsf{U}^0_\rho(\langle \mathbf{m} \rangle, \langle \boldsymbol{\alpha} \rangle, (\beta)^p);$$

and for limit ρ,

$$\mathsf{U}^0_\rho(\langle \mathbf{m} \rangle, \langle \boldsymbol{\alpha} \rangle, \beta) \leftrightarrow \exists p\, (\exists \sigma < \rho)[\|(\beta)^p_0\| = \sigma \wedge \sim \mathsf{U}^0_\sigma(\langle \mathbf{m} \rangle, \langle \boldsymbol{\alpha} \rangle, (\beta)^p_1)].$$

3.9 Lemma. *For all* $\rho < \aleph_1$, $W_\rho \in \mathbf{\Delta}^0_{\rho+1}$.

Proof. Follow the proof of Lemma 3.2 with this stronger induction hypothesis. This shows in fact that for successor ρ, $W_\rho \in \mathbf{\Pi}^0_\rho$ and for limit ρ, $W_\rho \in \mathbf{\Sigma}^0_\rho$. □

3.10 Borel Indexing Theorem. *For all* ρ *such that* $0 < \rho < \aleph_1$,
(i) U^0_ρ *is universal for* $\mathbf{\Sigma}^0_\rho$;
(ii) $\sim U^0_\rho$ *is universal for* $\mathbf{\Pi}^0_\rho$.

Proof. That $U^0_1 \in \mathbf{\Sigma}^0_1$ is obvious. Suppose that $U^0_\rho \in \mathbf{\Sigma}^0_\rho$. Then $\sim U^0_\rho \in \mathbf{\Pi}^0_\rho$ and

$$U^0_{\rho+1} = \bigcup\{\{(\langle \mathbf{m}\rangle, \langle \boldsymbol{\alpha}\rangle, \beta) : \sim U^0_\rho(\langle \mathbf{m}\rangle, \langle \boldsymbol{\alpha}\rangle, (\beta)^p)\} : p \in \omega\}$$

so that $U^0_{\rho+1} \in \mathbf{\Sigma}^0_{\rho+1}$. If ρ is a limit ordinal, then by the preceding lemma, $\{\gamma : \|\gamma\| = \sigma\} = W_{\sigma+1} \sim W_\sigma \in \mathbf{\Delta}^0_{\sigma+2}$, so that for all $\sigma < \rho$ this set is in $\mathbf{\Pi}^0_{(\rho)}$. As ρ is countable this shows that U^0_ρ is a countable union of $\mathbf{\Pi}^0_{(\rho)}$ relations, hence is $\mathbf{\Sigma}^0_\rho$.

Again it is obvious that U^0_1 is universal for $\mathbf{\Sigma}^0_1$. Suppose that U^0_ρ is universal for $\mathbf{\Sigma}^0_\rho$ and $R \in \mathbf{\Sigma}^0_{\rho+1}$, so for some relations $P_p \in \mathbf{\Pi}^0_\rho$, $R = \bigcup\{P_p : p \in \omega\}$. By hypothesis, there exist functions β_p such that

$$P_p(\mathbf{m}, \boldsymbol{\alpha}) \leftrightarrow \sim U^0_\rho(\langle \mathbf{m}\rangle, \langle \boldsymbol{\alpha}\rangle, \beta_p).$$

Then if β is a function such that for all p, $(\beta)^p = \beta_p$, then

$$R(\mathbf{m}, \boldsymbol{\alpha}) \leftrightarrow U^0_{\rho+1}(\langle \mathbf{m}\rangle, \langle \boldsymbol{\alpha}\rangle, \beta).$$

Finally, suppose that ρ is a limit ordinal, for all $\sigma < \rho$, U^0_σ is universal for $\mathbf{\Sigma}^0_\sigma$, and $R = \bigcup\{P_p : p \in \omega\} \in \mathbf{\Sigma}^0_\rho$. For each p there is an ordinal $\sigma_p < \rho$ such that $P_p \in \mathbf{\Pi}^0_{\sigma p}$ and thus a function β_p such that

$$P_p(\mathbf{m}, \boldsymbol{\alpha}) \leftrightarrow \sim U^0_{\sigma p}(\langle \mathbf{m}\rangle, \langle \boldsymbol{\alpha}\rangle, \beta_p).$$

Then if β is a function such that for all p, $\|(\beta)^p_0\| = \sigma_p$ and $(\beta)^p_1 = \beta_p$, we again have

$$R(\mathbf{m}, \boldsymbol{\alpha}) \leftrightarrow U^0_\rho(\langle \mathbf{m}\rangle, \langle \boldsymbol{\alpha}\rangle, \beta)$$

so that U^0_ρ is universal for $\mathbf{\Sigma}^0_\rho$. □

3.11 Borel Hierarchy Theorem. *For all* ρ *such that* $0 < \rho < \aleph_1$,
(i) $\mathbf{\Sigma}^0_\rho \not\subseteq \mathbf{\Delta}^0_\rho$ *and* $\mathbf{\Pi}^0_\rho \not\subseteq \mathbf{\Delta}^0_\rho$;
(ii) $\mathbf{\Delta}^0_{\rho+1} \not\subseteq \mathbf{\Sigma}^0_\rho \cup \mathbf{\Pi}^0_\rho$.

Proof. It suffices for (i) to show that $\mathsf{U}^0_\rho \notin \mathbf{\Delta}^0_\rho$. Suppose the contrary and let

$$\mathsf{A} = \{\alpha : \mathsf{U}^0_\rho(\langle\ \rangle, \langle\alpha\rangle, \alpha)\}.$$

Then by Lemma 3.6, $\mathsf{A} \in \mathbf{\Delta}^0_\rho$ so in particular, $\sim\mathsf{A} \in \mathbf{\Sigma}^0_\rho$. Since U^0_ρ is universal, there exists a β such that for all α,

$$\alpha \notin \mathsf{A} \leftrightarrow \mathsf{U}^0_\rho(\langle\ \rangle, \langle\alpha\rangle, \beta).$$

In particular,

$$\beta \notin \mathsf{A} \leftrightarrow \mathsf{U}^0_\rho(\langle\ \rangle, \langle\beta\rangle, \beta) \leftrightarrow \beta \in \mathsf{A},$$

a contradiction.

For (ii), let

$$\mathsf{R}(m, \alpha) \leftrightarrow (m = 0 \wedge \alpha \in \mathsf{A}) \vee (m = 1 \wedge \alpha \notin \mathsf{A}).$$

Then as in the proof of Theorem III.1.9, R is $\mathbf{\Delta}^0_{\rho+1}$ but neither $\mathbf{\Sigma}^0_\rho$ nor $\mathbf{\Pi}^0_\rho$. □

By the homeomorphism discussed in § I.2, all of these results apply equally well to the space BIr of binary irrationals in the real interval (0, 1) with the induced topology: $Y \subseteq \text{BIr}$ is open iff $Y = \text{BIr} \cap Z$ for some open subset Z of (0, 1). Of course, it is not in general true that if $\text{BIr} \cap Z$ is open in BIr, then Z is open in (0, 1). In fact, if $Z = \text{BIr}$, then Z is not open or even an F_σ-set in (0, 1) (Exercise I.2.11), but $\text{BIr} \cap Z = \text{BIr}$ which is open in BIr.

Let BIr-$\mathbf{\Sigma}^0_\rho$, etc., denote the Borel hierarchy on BIr and (0, 1)-$\mathbf{\Sigma}^0_\rho$ denote the Borel hierarchy defined similarly over (0, 1), starting with (0, 1)-$\mathbf{\Sigma}^0_1$ = class of open subsets of (0, 1).

3.12 Theorem. *For all $\rho \geq 3$ and all $Z \subseteq (0, 1)$,*

$$\text{BIr} \cap Z \textit{ is } \text{BIr-}\mathbf{\Sigma}^0_\rho\ (\mathbf{\Pi}^0_\rho) \leftrightarrow Z \textit{ is } (0, 1)\text{-}\mathbf{\Sigma}^0_\rho\ (\mathbf{\Pi}^0_\rho).$$

Proof. The implication ($\leftarrow$) is immediate by induction on ρ (for *all* ρ). For ($\rightarrow$), it is first easy to prove by induction that for all $\rho > 0$,

$$X \in \text{BIr-}\mathbf{\Sigma}^0_\rho\ (\mathbf{\Pi}^0_\rho) \rightarrow X = \text{BIr} \cap Y \text{ for some } Y \in (0, 1)\text{-}\mathbf{\Sigma}^0_\rho\ (\mathbf{\Pi}^0_\rho).$$

Then if $\text{BIr} \cap Z$ is BIr-$\mathbf{\Sigma}^0_\rho$, $\text{BIr} \cap Z = \text{BIr} \cap Y$ for some $Y \in (0, 1)$-$\mathbf{\Sigma}^0_\rho$. But then there exist (countable) sets of binary rationals Y_0 and Y_1 such that $Z = (Y \cup Y_0) \sim Y_1$. Since any countable set is F_σ ((0, 1)-$\mathbf{\Sigma}^0_2$), if $\rho \geq 3$, Z is also (0, 1)-$\mathbf{\Sigma}^0_\rho$. □

We can now see that (i) and (ii) of Corollary 3.11 hold also for $(0, 1)\text{-}\boldsymbol{\Sigma}^0_\rho$ for all $\rho \geq 1$. For $\rho = 1$, this is because $\varnothing$ and $(0, 1)$ are the only closed-open subsets of $(0, 1)$. For $\rho = 2$ we have that BIr is $(0, 1)\text{-}\boldsymbol{\Pi}^0_2$ but not $(0, 1)\text{-}\boldsymbol{\Sigma}^0_2$. For $\rho \geq 3$, it follows from Theorem 3.12.

Of course, the theorem applies to relations as well as sets. For any $\mathsf{R} \subsetneq {}^{k+1}(0, 1)$, let $\exists^{(0,1)}\mathsf{R}$ denote the relation $\mathsf{P}(x) \leftrightarrow \exists y\,[y \in (0, 1) \wedge \mathsf{R}(y, x)]$ and $\exists^{\mathrm{BIr}}\mathsf{R}$ the relation $\mathsf{Q}(x) \leftrightarrow \exists y\,[y \in \mathrm{BIr} \wedge \mathsf{R}(y, x)]$. The $(0, 1)\text{-}\boldsymbol{\Sigma}^1_1$ relations are those of the from $\exists^{(0,1)}\mathsf{R}$ for R in $(0, 1)$-Borel and the $\mathrm{BIr}\text{-}\boldsymbol{\Sigma}^1_1$ relations are those of the form $\exists^{\mathrm{BIr}}\mathsf{R}$ for R BIr-Borel. The other classes of the projective hierarchy are defined similarly. Clearly $(0, 1)\text{-}\boldsymbol{\Delta}^1_1$ contains all $(0, 1)$-Borel relations.

3.13 Theorem. *For all $r > 0$ and all $Z \subseteq (0, 1)$,*

$$\mathrm{BIr} \cap Z \textit{ is } \mathrm{BIr}\text{-}\boldsymbol{\Sigma}^1_r\ (\boldsymbol{\Pi}^1_r) \leftrightarrow Z \textit{ is } (0, 1)\text{-}\boldsymbol{\Sigma}^1_r\ (\boldsymbol{\Pi}^1_r).$$

Proof. If $\mathrm{BIr} \cap Z$ is $\mathrm{BIr}\text{-}\boldsymbol{\Sigma}^1_1$, then $\mathrm{BIr} \cap Z = \exists^{\mathrm{BIr}} R$ for some R in BIr-Borel. But since $\mathsf{R} \subseteq {}^2\mathrm{BIr}$, $\exists^{\mathrm{BIr}}\mathsf{R} = \exists^{(0,1)}\mathsf{R}$ and by Theorem 3.12, R is also $(0, 1)$-Borel. Hence $\mathrm{BIr} \cap Z$ is also $(0, 1)\text{-}\boldsymbol{\Sigma}^1_1$, and since Z differs from $\mathrm{BIr} \cap Z$ by a countable set, also Z is $(0, 1)\text{-}\boldsymbol{\Sigma}^1_1$.

Conversely, if Z is $(0, 1)\text{-}\boldsymbol{\Sigma}^1_1$, then for some $(0, 1)$-Borel relation R, $Z = \exists^{(0,1)}\mathsf{R}$. Then

$$\mathrm{BIr} \cap Z = \exists^{\mathrm{BIr}}[{}^2\mathrm{BIr} \cap \mathsf{R}] \cup (\mathrm{BIr} \cap \exists^{\mathrm{BRa}}\mathsf{R})$$

where $\exists^{\mathrm{BRa}}$ means "there exists a binary rational". By Theorem 3.12, ${}^2\mathrm{BIr} \cap \mathsf{R}$ is BIr-Borel so the first term is $\mathrm{BIr}\text{-}\boldsymbol{\Sigma}^1_1$. $\exists^{\mathrm{BRa}}\mathsf{R}$ is a countable union of $(0, 1)$-Borel relations, hence is $(0, 1)$-Borel. Thus the second term is BIr-Borel and $\mathrm{BIr} \cap Z$ is $\mathrm{BIr}\text{-}\boldsymbol{\Sigma}^1_1$.

The extension to larger r is by induction. □

3.14 Corollary. *For all $X \subseteq (0, 1)$, X is $(0, 1)$-Borel iff X is $(0, 1)\text{-}\boldsymbol{\Delta}^1_1$.* □

3.15–3.19 Exercises

3.15. Complete the following outline of an alternative proof that $\boldsymbol{\Delta}^1_1 \subseteq \mathrm{Bo}$ and that $\boldsymbol{\Sigma}^1_1$ has the separation property (cf. Exercises I.2.7 and II.4.32). For any $\mathsf{R} \subseteq {}^\omega\omega \times {}^\omega\omega$, let

$$\mathsf{R}^{(s,t)} = \{(\alpha, \beta) : \mathsf{R}(s * \alpha, t * \beta)\}.$$

Two sets are called *Borel separable* iff they can be separated by a Borel set. Show

(i) if for all m, n, and p, $\exists^1\mathsf{R}^{(\langle m\rangle, \langle n\rangle)}$ and $\exists^1\mathsf{S}^{(\langle m\rangle, \langle p\rangle)}$ are Borel separable, then $\exists^1\mathsf{R}$ and $\exists^1\mathsf{S}$ are Borel separable;

(ii) if for all α, β, and γ, there exists an n such that $\exists^1 \mathrm{R}^{(\bar{\alpha}(n), \bar{\beta}(n))}$ and $\exists^1 \mathrm{S}^{(\bar{\alpha}(n), \bar{\gamma}(n))}$ are Borel separable, then $\exists^1\mathrm{R}$ and $\exists^1\mathrm{S}$ are Borel separable;
(iii) any two disjoint $\boldsymbol{\Sigma}_1^1$ sets are Borel separable;
(iv) every $\boldsymbol{\Delta}_1^1$ set is Borel.

3.16. For any indexed family $\langle \mathrm{P}_p : p \in \omega\rangle$ of relations of the same rank, let

$$\underline{\mathrm{Lim}}\langle \mathrm{P}_p\rangle(\mathbf{m}, \boldsymbol{\alpha}) \leftrightarrow \exists p\, (\forall q \geqslant p)\, \mathrm{P}_q(\mathbf{m}, \boldsymbol{\alpha})$$

and

$$\overline{\mathrm{Lim}}\langle \mathrm{P}_p\rangle(\mathbf{m}, \boldsymbol{\alpha}) \leftrightarrow \forall p\, (\exists q \geqslant p)\, \mathrm{P}_q(\mathbf{m}, \boldsymbol{\alpha}).$$

When $\underline{\mathrm{Lim}}\langle \mathrm{P}_p\rangle = \overline{\mathrm{Lim}}\langle \mathrm{P}_p\rangle$, we denote the common value by $\mathrm{Lim}\langle \mathrm{P}_p\rangle$; otherwise $\mathrm{Lim}\langle \mathrm{P}_p\rangle$ is undefined. For all ρ, let

$$\boldsymbol{\Delta}_0 = \text{the class of closed-open relations;}$$

$$\boldsymbol{\Delta}_{(\rho)} = \bigcup\{\boldsymbol{\Delta}_\tau : \tau < \rho\};$$

$$\boldsymbol{\Delta}_\rho = \{\mathrm{Lim}\langle \mathrm{P}_p\rangle : \forall p\,.\, \mathrm{P}_p \in \boldsymbol{\Delta}_{(\rho)}\}.$$

Prove for all ρ:
(i) $\boldsymbol{\Delta}_\rho \subseteq \boldsymbol{\Delta}^0_{\rho+1}$;
(ii) $\boldsymbol{\Delta}_\rho$ is a Boolean algebra;
(iii) if $\forall p\,.\,\mathrm{P}_p \in \boldsymbol{\Delta}_{(\rho)}$, then both $\bigcup\{\mathrm{P}_p : p \in \omega\}$ and $\bigcap\{\mathrm{P}_p : p \in \omega\}$ belong to $\boldsymbol{\Delta}_\rho$;
(iv) if $\mathrm{R} = \bigcup\{\mathrm{P}_p : p \in \omega\} = \bigcap\{\mathrm{Q}_q : q \in \omega\}$ where for all p and q, P_p and Q_q belong to $\boldsymbol{\Delta}_\rho$, then also $\mathrm{R} \in \boldsymbol{\Delta}_\rho$;
(v) $\boldsymbol{\Delta}_\rho = \boldsymbol{\Delta}^0_{\rho+1}$.

Hint for (iv). Suppose $\mathrm{P}_p = \mathrm{Lim}\langle \mathrm{P}_{p,m} : m \in \omega\rangle$ and $\mathrm{Q}_q = \mathrm{Lim}\langle \mathrm{Q}_{q,n} : n \in \omega\rangle$ with all $\mathrm{P}_{p,m}, \mathrm{Q}_{q,n} \in \boldsymbol{\Delta}_{(\rho)}$. Let

$$\mathrm{S}_r = (\mathrm{P}_{0,r} \cap \mathrm{Q}_{0,r}) \cup (\mathrm{P}_{1,r} \cap \mathrm{Q}_{0,r} \cap \mathrm{Q}_{1,r}) \cup \ldots \cup (\mathrm{P}_{r,r} \cap \mathrm{Q}_{0,r} \cap \ldots \cap \mathrm{Q}_{r,r}).$$

Show that $\mathrm{R} \subseteq \underline{\mathrm{Lim}}\langle \mathrm{S}_r\rangle$ and $\sim \mathrm{R} \subseteq \sim \overline{\mathrm{Lim}}\langle \mathrm{S}_r\rangle$.

3.17. Let S and H be, respectively, a Borel relation and a Borel functional, and

$$\mathrm{Q}(\mathbf{m}, \boldsymbol{\alpha}) \leftrightarrow \mathrm{S}(\mathbf{m}, \boldsymbol{\alpha}, \lambda p\,.\,\mathrm{H}(p, \mathbf{m}, \boldsymbol{\alpha})).$$

If $\mathrm{S} \in \boldsymbol{\Sigma}^0_\rho$ and $\mathrm{H} \in \boldsymbol{\Delta}^0_\sigma$, what can you conclude about the level of Q?

3.18. Show that for all $\rho < \aleph_1$, $\boldsymbol{\Sigma}^0_\rho$ has the reduction property.

3.19. Show that every $(0, 1)$-$\mathbf{\Sigma}_1^1$ set is the projection of a $(0, 1)$-$\mathbf{\Pi}_2^0$ relation.

3.20 Notes. It might fairly be said that Suslin's Thorem (3.3), proved in 1917, was the result that put Descriptive Set Theory on the map. It was a tremendous breakthrough in its time and is still today the model for most results which give equivalent characterizations of a class of relations "from below" and "from above".

4. *The Effective Borel and Hyperarithmetical Hierarchies*

We want now to define a similar extension of the arithmetical hierarchy to obtain classes Σ_ρ^0 and Π_ρ^0 $(\rho < \omega_1)$ such that $\Delta_{(\omega_1)}^0 = \Delta_1^1$. This construction is called the *effective* Borel hierarchy because it is derived from that in the preceding section by replacing the generating operation of countable union by ***recursively enumerable*** union. Roughly, the idea is to attach to each relation P as it is generated an index $\iota(\mathsf{P})$ and admit unions only of those families $\langle \mathsf{P}_p : p \in \omega \rangle$ such that the function $\lambda p \,.\, \iota(\mathsf{P}_p)$ is recursive. The details differ from this sketch in that we shall first define the set of indices and then assign relations to the indices.

In the second part of the section we consider an alternative method for constructing a hierarchy of the Δ_1^1 relations. This is based on a notion of iterating the ordinary jump operator oJ over a set of ordinal notations and should be considered as an extension of the ideas of Theorem III.1.13.

4.1 Definition. For each k and l, $N^{k,l}$ is the smallest subset of ω such that for all a and b,

(i) if $(b)_1 = k$ and $(b)_2 = l$, then $\langle 7, b\rangle \in N^{k,l}$;
(ii) if for all p, $\{a\}(p) \in N^{k,l}$, then $a \in N^{k,l}$.

As usual we denote by $N_{(\rho)}^{k,l}$ and $N_\rho^{k,l}$ the sets $\Gamma^{(\rho)}$ and Γ^ρ, where Γ is the monotone Π_1^0 operator such that $\bar{\Gamma} = N^{k,l}$. We also put $N = \bigcup\{N^{k,l} : k, l \in \omega\}$ and $N_\rho = \bigcup\{N_\rho^{k,l} : k, l \in \omega\}$. By Theorem 2.15, $|\Gamma| \leq \omega_1$, so $N^{k,l} = N_{(\omega_1)}^{k,l}$. It is easy to check that if $(k, l) \neq (k', l')$, then $N^{k,l} \cap N^{k',l'} = \emptyset$. Note that $N_0^{k,l}$ is recursive and if $a \in N^{k,l}$ by virtue of clause (ii), then $(a)_0 \neq 7$ so $a \notin N_0^{k,l}$.

We next assign to every $a \in N^{k,l}$ a relation $\mathsf{P}_a \subseteq {}^{k,l}\omega$. The assignment is recursive over $N^{k,l}$ and may be justified by an extension of the technique used in the proof of Theorem I.3.5.

4.2 Definition. For each k and l and any $a \in N^{k,l}$,

(i) if $a \in N_0^{k,l}$, then $\mathsf{P}_a = \mathrm{Dm}\{(a)_1\}$;
(ii) if $a \notin N_0^{k,l}$, then $\mathsf{P}_a = \bigcup\{\sim \mathsf{P}_{\{a\}(p)} : p \in \omega\}$.

4.3 Definition. For all ρ,
(i) $\Sigma^0_{1+\rho} = \{\mathsf{P}_a : a \in N_\rho\}$;
(ii) $\Pi^0_{1+\rho} = \{\sim\mathsf{P}_a : a \in N_\rho\}$;
(iii) $\Delta^0_\rho = \Sigma^0_\rho \cap \Pi^0_\rho$;
(iv) $\Sigma^0_{(\rho)} = \bigcup\{\Sigma^0_\tau : 0 < \tau < \rho\}$; $\Pi^0_{(\rho)} = \bigcup\{\Pi^0_\tau : 0 < \tau < \rho\}$;

$$\Delta^0_{(\rho)} = \bigcup\{\Delta^0_\tau : 0 < \tau < \rho\};$$

(v) $\mathrm{EfBo} = \Delta^0_{(\omega_1)}$ = the class of *effective Borel relations.*

Note that for $\rho < \omega$, $1 + \rho = \rho + 1$, while for $\rho \geq \omega$, $1 + \rho = \rho$. A relation is in Σ^0_ρ $(\rho \geq 2)$ just in case it is of the form $\bigcup\{\sim\mathsf{P}_{\{a\}(p)} : p \in \omega\}$ where all $\sim\mathsf{P}_{\{a\}(p)}$ are in $\Pi^0_{(\rho)}$. Similarly if $\mathsf{R} = \sim\mathsf{P}_a$ is in Π^0_ρ, then $\mathsf{R} = \bigcap\{\mathsf{P}_{\{a\}(p)} : p \in \omega\}$ and all $\mathsf{P}_{\{a\}(p)}$ are in $\Sigma^0_{(\rho)}$.

We want first to show that the notation here is not in conflict with that of Definition III.1.2 (arithmetical hierarchy) — that is, the classes Σ^0_ρ for $0 < \rho < \omega$ are just those previously defined.

4.4 Lemma. *For all* $\rho > 0$,
(i) $\Sigma^0_{(\rho)} \cup \Pi^0_{(\rho)} \subseteq \Delta^0_\rho$;
(ii) $\Sigma^0_{(\rho+1)} = \Sigma^0_\rho$; $\Pi^0_{(\rho+1)} = \Pi^0_\rho$.

Proof. $\Sigma^0_{(\rho)} \subseteq \Sigma^0_\rho$ by definition as $N_{(\rho)} \subseteq N_\rho$. If $\mathsf{R} \in \Pi^0_{(\rho)}$, then $\mathsf{R} = \sim\mathsf{P}_a$ for some $a \in N_{(\rho)}$. Thus if b is an index such that for all p, $\{b\}(p) = a$, then $b \in N_\rho$ and $\mathsf{R} = \bigcup\{\sim\mathsf{P}_{\{b\}(p)} : p \in \omega\} = \mathsf{P}_b$ which is in Σ^0_ρ. Thus $\Sigma^0_{(\rho)} \cup \Pi^0_{(\rho)} \subseteq \Sigma^0_\rho$ and (i) follows from the fact that $\Sigma^0_{(\rho)} \cup \Pi^0_{(\rho)}$ is closed under complementation. (ii) is then immediate. □

4.5 Lemma. *For all* $\rho > 0$, Σ^0_ρ *is effectively closed under recursively enumerable union and finite intersection — that is, for each* k *and* l *there exist primitive recursive functions* f *and* g *such that for any* $a, \mathbf{m}, \boldsymbol{\alpha}$, *and any* $\rho > 0$,
(i) *if for all* p, $\{a\}(p) \in N^{k,l}_\rho$, *then* $f(a) \in N^{k,l}_\rho$ *and*

$$\mathsf{P}_{f(a)}(\mathbf{m}, \boldsymbol{\alpha}) \leftrightarrow \exists p\,.\,\mathsf{P}_{\{a\}(p)}(\mathbf{m}, \boldsymbol{\alpha});$$

(ii) *if* $a, b \in N^{k,l}_\rho$, *then* $g(a, b) \in N^{k,l}_\rho$ *and*

$$\mathsf{P}_{g(a,b)} = \mathsf{P}_a \cap \mathsf{P}_b.$$

Proof. Let h be a primitive recursive function such that for all a, p, and q,

$$\{h(a, p, q)\}(\mathbf{m}, \boldsymbol{\alpha})\downarrow \text{ iff } \sim\mathsf{T}((\{a\}(p))_1, \langle\mathbf{m}\rangle, q, \langle\boldsymbol{\alpha}\rangle).$$

Then choose f to be a primitive recursive function such that

$$\{f(a)\}(\langle p,q\rangle) \simeq \begin{cases} \langle 7, h(a,p,q)\rangle, & \text{if} \quad \{a\}(p) \in N_0; \\ \{\{a\}(p)\}(q), & \text{otherwise;} \end{cases}$$

and for r not of the form $\langle p,q\rangle, \{f(a)\}(r) \simeq \langle 7, a_0\rangle$, where $\mathrm{Dm}\{a_0\} = {}^{k,l}\omega$.

Suppose that for all $p, \{a\}(p) \in N_p^{k,l}$. For any p such that $\{a\}(p) \in N_0^{k,l}$, clearly also $\{f(a)\}(\langle p,q\rangle) \in N_0^{k,l}$. For other p we have for all $q, \{\{a\}(p)\}(q) \in N_{(p)}^{k,l}$, hence also for all $q, \{f(a)\}(\langle p,q\rangle) \in N_{(p)}^{k,l}$. Thus for all $r, \{f(a)\}(r) \in N_{(p)}^{k,l}$ so $f(a) \in N_p^{k,l}$.

For p such that $\{a\}(p) \in N_0^{k,l}$ we have

$$\begin{aligned} \mathsf{P}_{\{a\}(p)}(\mathbf{m}, \boldsymbol{\alpha}) &\leftrightarrow \{(\{a\}(p))_1\}(\mathbf{m}, \boldsymbol{\alpha}) \downarrow \\ &\leftrightarrow \exists q\, \mathsf{T}((\{a\}(p))_1, \langle \mathbf{m}\rangle, q, \langle \boldsymbol{\alpha}\rangle) \\ &\leftrightarrow \exists q \sim \mathsf{P}_{\{f(a)\}(\langle p,q\rangle)}(\mathbf{m}, \boldsymbol{\alpha}). \end{aligned}$$

For other p,

$$\begin{aligned} \mathsf{P}_{\{a\}(p)}(\mathbf{m}, \boldsymbol{\alpha}) &\leftrightarrow \exists q \sim \mathsf{P}_{\{\{a\}(p)\}(q)}(\mathbf{m}, \boldsymbol{\alpha}) \\ &\leftrightarrow \exists q \sim \mathsf{P}_{\{f(a)\}(\langle p,q\rangle)}(\mathbf{m}, \boldsymbol{\alpha}). \end{aligned}$$

Hence

$$\begin{aligned} \exists p \,.\, \mathsf{P}_{\{a\}(p)}(\mathbf{m}, \boldsymbol{\alpha}) &\leftrightarrow \exists p\, \exists q \sim \mathsf{P}_{\{f(a)\}(\langle p,q\rangle)}(\mathbf{m}, \boldsymbol{\alpha}) \\ &\leftrightarrow \exists r \sim \mathsf{P}_{\{f(a)\}(r)}(\mathbf{m}, \boldsymbol{\alpha}) \\ &\leftrightarrow \mathsf{P}_{f(a)}(\mathbf{m}, \boldsymbol{\alpha}). \end{aligned}$$

We define g by the following four cases:
(1) if $a, b \in N_0$, then $g(a,b) = \langle 7, c\rangle$, where

$$\{c\}(\mathbf{m}, \boldsymbol{\alpha}) \simeq \{(a)_1\}(\mathbf{m}, \boldsymbol{\alpha}) + \{(b)_1\}(\mathbf{m}, \boldsymbol{\alpha});$$

(2) if $a \notin N_0$ but $b \in N_0$, then $g(a,b)$ is an index c calculated from an index of the function f of (i) such that if $\{a\}(p) \in N$,

$$\mathsf{P}_{\{c\}(p)} = \mathsf{P}_{\{a\}(p)} \cup \sim \mathrm{Dm}\{(b)_1\};$$

(3) if $a \in N_0$ but $b \notin N_0$, then similarly

$$\mathsf{P}_{\{g(a,b)\}(p)} = \sim \mathrm{Dm}\{(a)_1\} \cup \mathsf{P}_{\{b\}(p)};$$

(4) if $a, b \notin N_0$, then

$$\mathsf{P}_{\{g(a,b)\}(p)} = \mathsf{P}_{\{a\}(p)} \cup \mathsf{P}_{\{b\}(p)}.$$

The proof that g satisfies (ii) is straightforward. For example, if $a \in N_0^{k,l}$ and $b \in N^{k,l} \sim N_0^{k,l}$,

$$\begin{aligned}\mathsf{P}_a \cap \mathsf{P}_b &= \mathrm{Dm}\{(a)_1\} \cap \bigcup\{\sim \mathsf{P}_{\{b\}(p)} : p \in \omega\}\\ &= \bigcup\{\sim [\sim \mathrm{Dm}\{(a)_1\} \cup \mathsf{P}_{\{b\}(p)}] : p \in \omega\}\\ &= \bigcup\{\sim \mathsf{P}_{\{g(a,b)\}(p)} : p \in \omega\}\\ &= \mathsf{P}_{g(a,b)}. \quad \square\end{aligned}$$

Before proceeding, we make a few remarks on the methods of proof we shall use in the remainder of this section and often in later parts of the book. In establishing closure properties of the classes Σ^0_ρ we shall in most cases need to prove that the classes are *effectively* closed. In the preceding lemma the definition of g depends on the fact that f is primitive recursive, and this would be so even if we were not requiring that g be primitive recursive. In the next and many succeeding lemmas we shall need to define functions by *effective transfinite recursion*. In outline, the method is as follows. Suppose $<$ is a well-founded transitive relation on ω. For any function F and any $u \in \mathrm{Fld}(<)$, let $F \restriction u$ denote the partial function g such that

$$g(v, p) \simeq \begin{cases} F(v, p), & \text{if} \quad v < u;\\ \text{undefined}, & \text{otherwise.}\end{cases}$$

The set-theoretic principle of definition by transfinite recursion asserts that for any function φ there is a function F such that for all $u \in \mathrm{Fld}(<)$,

$$F(u, m) \simeq \varphi(F \restriction u, u, m),$$

and that the values of F for $u \in \mathrm{Fld}(<)$ are uniquely determined. Roughly speaking, the principle of effective transfinite recursion asserts that if φ is partial recursive, then F may also be chosen to be partial recursive.

In practice, when we apply the method φ will be such that there exists a partial recursive function H such that for all e and m and all $u \in \mathrm{Fld}(<)$,

$$\varphi(\{e\} \restriction u, u, m) \simeq H(e, u, m)$$

and we shall apply the Recursion Theorem to obtain an index $\bar{e}$ such that

$$\{\bar{e}\}(u, m) \simeq H(\bar{e}, u, m).$$

Then clearly

$$\{\bar{e}\}(u, m) \simeq \varphi(\{\bar{e}\} \restriction u, u, m)$$

and $F = \{\bar{e}\}$ is partial recursive.

A variant of this method defines a function f such that for $u \in \mathrm{Fld}(<)$, $\{f(u)\}(m) \simeq F(u, m)$. In a typical such situation there will be a primitive recursive function h such that

$$\varphi(\{\{e\} \restriction u\}, u, m) \simeq \{h(e, u)\}(m),$$

where

$$\{\{e\} \restriction u\}(v, m) \simeq \begin{cases} \{\{e\}(v)\}(m), & \text{if} \quad v < u; \\ \text{undefined}, & \text{otherwise.} \end{cases}$$

Then if $\bar{e}$ is chosen by the Primitive Recursion Theorem such that $\{\bar{e}\}(u) \simeq h(\bar{e}, u)$, the function $f = \{\bar{e}\}$ is in fact *primitive* recursive and has the desired property.

This leads also to an extension of the principle in which the value $F(u, m)$ may depend not only on the values $F(v, p)$ for $v < u$ but also on indices for the functions $\lambda p . F(v, p)$. In such a situation we have a function ψ such that $\psi(f \restriction u, u, m)$ depends both on values $f(v)$ for $v < u$ and on the partial functions $\{f(v)\}$ that they index and a primitive recursive h such that

$$\psi(f \restriction u, u, m) \simeq \{h(e, u)\}(m).$$

Again $\bar{e}$ is chosen so that $\{\bar{e}\}(u) \simeq h(\bar{e}, u)$, $f = \{\bar{e}\}$, and $F(u, m) \simeq \{f(u)\}(m)$.

There are many variations on these paradigms and mastery of the method comes only with practice. Rogers [1959] is also helpful.

4.6 Lemma. *There exists a primitive recursive function f such that for any ρ and any $a \in N_\rho^{k+1,l}$, for all p, $f(a, p) \in N_\rho^{k,l}$ and*

$$\mathsf{P}_{f(a,p)}(\mathbf{m}, \boldsymbol{\alpha}) \leftrightarrow \mathsf{P}_a(p, \mathbf{m}, \boldsymbol{\alpha}).$$

Proof. Let h be a primitive recursive function such that

$$h(e, a, p) \simeq \begin{cases} \langle 7, \mathrm{Sb}_0((a)_1 p) \rangle, & \text{if} \quad a \in N_0; \\ \text{an index } c \text{ such that for all } q, & \\ \{c\}(q) \simeq \{e\}(\{a\}(q), p), & \text{otherwise.} \end{cases}$$

By the Primitive Recursion Theorem there exists an index $\bar{e}$ such that $h(\bar{e}, a, p) = \{\bar{e}\}(a, p)$. We take $f = \{\bar{e}\}$. If $a \in N_0$, then

$$\begin{aligned} \mathsf{P}_{f(a,p)}(\mathbf{m}, \boldsymbol{\alpha}) &\leftrightarrow \{\mathrm{Sb}_0((a)_1, p)\}(\mathbf{m}, \boldsymbol{\alpha})\downarrow \\ &\leftrightarrow \{(a)_1\}(p, \mathbf{m}, \boldsymbol{\alpha})\downarrow \\ &\leftrightarrow \mathsf{P}_a(p, \mathbf{m}, \boldsymbol{\alpha}). \end{aligned}$$

If $a \in N \sim N_0$ and we assume as induction hypothesis that for all q,

$$\mathsf{P}_{f(\{a\}(q),p)}(\mathbf{m}, \boldsymbol{\alpha}) \leftrightarrow \mathsf{P}_{\{a\}(p)}(p, \mathbf{m}, \boldsymbol{\alpha}),$$

then

$$\begin{aligned} \mathsf{P}_{f(a,b)}(\mathbf{m}, \boldsymbol{\alpha}) &\leftrightarrow \exists q \sim \mathsf{P}_{f(\{a\}(q),p)}(\mathbf{m}, \boldsymbol{\alpha}) \\ &\leftrightarrow \exists q \sim \mathsf{P}_{\{a\}(q)}(p, \mathbf{m}, \boldsymbol{\alpha}) \\ &\leftrightarrow \mathsf{P}_a(p, \mathbf{m}, \boldsymbol{\alpha}). \quad \square \end{aligned}$$

4.7 Corollary. *For all $\rho > 0$, Σ^0_ρ is effectively closed under existential number quantification ($\exists^0$) and Π^0_ρ is effectively closed under universal number quantification ($\forall^0$) — that is, there exists a primitive recursive function f such that for all ρ and all $a \in N^{k+1,l}_\rho$, $f(a) \in N^{k,l}_\rho$ and*

$$\mathsf{P}_{f(a)}(\mathbf{m}, \boldsymbol{\alpha}) \leftrightarrow \exists p\,.\,\mathsf{P}_a(p, \mathbf{m}, \boldsymbol{\alpha}).$$

Proof. It suffices to set $f(a) = f_5(\mathrm{Sb}_0(b_6, a))$, where f_5 is the function f of Lemma 4.5 and b_6 is an index for the function f of Lemma 4.6. $\square$

For the next corollary, let the finite levels of the effective Borel hierarchy be denoted by $\Sigma^{0'}_r$ and $\Pi^{0'}_r$ to distinguish them from the levels of the arithmetical hierarchy.

4.8 Corollary. *For all $r > 0$, $\Sigma^0_r = \Sigma^{0'}_r$ and $\Pi^0_r = \Pi^{0'}_r$.*

Proof. For each k, l, and r, and all $(\mathbf{m}, \boldsymbol{\alpha}) \in {}^{k,l}\omega$, let

$$\mathsf{V}^{k,l}_r(a, \mathbf{m}, \boldsymbol{\alpha}) \leftrightarrow a \in N^{k,l}_{r-1} \wedge \mathsf{P}_a(\mathbf{m}, \boldsymbol{\alpha}).$$

We prove by induction on r the stronger assertion

(*) $\qquad \Sigma^0_r = \Sigma^{0'}_r,\ N^{k,l}_r \in \Pi^0_1, \quad \text{and} \quad \mathsf{V}^{k,l}_r \in \Sigma^0_r.$

Clearly (*) holds for $r = 1$ as

$$\mathsf{V}_1^{k,l}(a,\mathbf{m},\boldsymbol{\alpha}) \leftrightarrow (a)_0 = 7 \wedge (a)_{1,0} = k \wedge (a)_{1,1} = l \wedge \{(a)_1\}(\mathbf{m},\boldsymbol{\alpha})\downarrow .$$

Suppose (*) holds for r. If $\mathsf{R} \in \Sigma^0_{r+1}$, then $\mathsf{R} = \exists^0\mathsf{S}$ for some $\mathsf{S} \in \Pi^0_r$. Then $\mathsf{S} \in \Pi^{0'}_r \subseteq \Sigma^{0'}_{r+1}$ by the induction hypothesis and Lemma 4.4 so $\mathsf{R} \in \Sigma^{0'}_{r+1}$ by Corollary 4.7. Hence $\Sigma^0_{r+1} \subseteq \Sigma^{0'}_{r+1}$.

Next we observe that $a \in N^{k,l}_{r+1} \leftrightarrow \forall p.\{a\}(p) \in N^{k,l}_r$ so that if $N^{k,l}_r \in \Pi^0_1$, so is $N^{k,l}_{r+1}$. Then

$$\mathsf{V}^{k,l}_{r+1}(a,\mathbf{m},\boldsymbol{\alpha}) \leftrightarrow a \in N^{k,l}_r \wedge \exists p \sim \mathsf{V}^{k,l}_r(\{a\}(p),\mathbf{m},\boldsymbol{\alpha}).$$

Hence $\mathsf{V}^{k,l}_{r+1} \in \Sigma^0_{r+1}$. Since $\mathsf{V}^{k,l}_{r+1}$ is universal for $\Sigma^{0'}_{r+1}$, this yields $\Sigma^{0'}_{r+1} \subseteq \Sigma^0_{r+1}$. □

We aim next to establish the equation $\Delta^1_1 = \Delta^0_{(\omega_1)}$.

4.9 Lemma. $\Delta^0_{(\omega_1)} \subseteq \Delta^1_1$.

Proof. For a fixed pair (k,l), let

$$\mathsf{V}(a,i,\mathbf{m},\boldsymbol{\alpha}) \leftrightarrow a \in N^{k,l} \wedge [i=0 \wedge \mathsf{P}_a(\mathbf{m},\boldsymbol{\alpha})] \vee [i=1 \wedge \sim\mathsf{P}_a(\mathbf{m},\boldsymbol{\alpha})].$$

It will suffice to show $\mathsf{V} \in \Pi^1_1$ as then for all $a \in N^{k,l}$, both P_a and $\sim\mathsf{P}_a$ are Π^1_1 so $\mathsf{P}_a \in \Delta^1_1$. For this we define a decomposable monotone arithmetical operator Γ such that $\bar{\Gamma} = \mathsf{V}$. Γ is essentially simply a combination of definitions 4.1 and 4.2. For any $\mathsf{R} \subseteq {}^{k+2,l}\omega$, all i, and all $(\mathbf{m},\boldsymbol{\alpha}) \in {}^{k,l}\omega$,

(i) if $(b)_1 = k$ and $(b)_2 = l$, then
  (1) if $\{b\}(\mathbf{m},\boldsymbol{\alpha})\downarrow$, then $(\langle 7,b\rangle,0,\mathbf{m},\boldsymbol{\alpha}) \in \Gamma(\mathsf{R})$;
  (2) if $\{b\}(\mathbf{m},\boldsymbol{\alpha})\uparrow$, then $(\langle 7,b\rangle,1,\mathbf{m},\boldsymbol{\alpha}) \in \Gamma(\mathsf{R})$;

(ii) if for all p $(\exists i \leqslant 1)$ $(\{a\}(p),i,\mathbf{m},\boldsymbol{\alpha}) \in \mathsf{R}$, then
  (1) if $\exists p.(\{a\}(p),1,\mathbf{m},\boldsymbol{\alpha}) \in \mathsf{R}$, then $(a,0,\mathbf{m},\boldsymbol{\alpha}) \in \Gamma(\mathsf{R})$;
  (2) if $\forall p.(\{a\}(p),0,\mathbf{m},\boldsymbol{\alpha}) \in \mathsf{R}$, then $(a,1,\mathbf{m},\boldsymbol{\alpha}) \in \Gamma(\mathsf{R})$.

Γ is clearly monotone, arithmetical, and decomposable; we leave it to the reader to verify that $\bar{\Gamma}$ is as claimed. □

See Exercise 4.25 for an alternative proof of Lemma 4.9. Toward the converse inclusion we establish some further closure properties of the effective Borel hierarchy.

4.10 Lemma. *For all $\rho > 0$, Σ^0_ρ and Π^0_ρ are effectively closed under composition and substitution of recursive functionals — that is, for any recursive functionals* F *and* G, *there exist primitive recursive functions f and g such that for all $\rho > 0$ and all a and b,*

(i) *if $a \in N^{k+1,l}_\rho$, then $f(a) \in N^{k,l}_\rho$ and*

$$\mathsf{P}_{f(a)}(\mathbf{m},\boldsymbol{\alpha}) \leftrightarrow \mathsf{P}_a(\mathsf{F}(\mathbf{m},\boldsymbol{\alpha}),\mathbf{m},\boldsymbol{\alpha});$$

(ii) *if* $b \in N_\rho^{k,l+1}$, *then* $g(b) \in N_\rho^{k,l}$ *and*

$$\mathsf{P}_{g(b)}(\mathbf{m}, \boldsymbol{\alpha}) \leftrightarrow \mathsf{P}_b(\mathbf{m}, \boldsymbol{\alpha}, \lambda p \,.\, \mathsf{G}(p, \mathbf{m}, \boldsymbol{\alpha})).$$

Proof. For (i), let h be a primitive recursive function such that

$$h(e, a) = \begin{cases} \langle 7, b\rangle, \text{ where } \{b\}(\mathbf{m}, \boldsymbol{\alpha}) \simeq \{(a)_1\}(\mathsf{F}(\mathbf{m}, \boldsymbol{\alpha}), \mathbf{m}, \boldsymbol{\alpha}), \text{ if } a \in N_0; \\ \text{an index } c \text{ such that for all } p, \{c\}(p) \simeq \{e\}(\{a\}(p)), \\ \qquad\qquad\qquad\qquad\qquad\qquad\qquad\qquad \text{otherwise.} \end{cases}$$

By the Primitive Recursion Theorem choose $\bar{e}$ such that $h(\bar{e}, a) = \{\bar{e}\}(a)$ and set $f = \{\bar{e}\}$. A straightforward induction over $N^{k+1,l}$ shows that f satisfies (i). The construction of g is very similar. □

4.11 Lemma. *For all* $\rho < \omega_1$, W_ρ *is effective Borel* $(\Delta^0_{(\omega_1)})$.

Proof. Let ρ be any recursive ordinal and $\delta \in \mathsf{W}$ a recursive function such that $\|\delta\| = \rho + 1$. We shall show the existence of a primitive recursive function f such that for all $r \in \mathrm{Fld}(\delta)$, $f(r) \in N^{0,1}$ and $\mathsf{W}_{|r|_\delta} = \mathsf{P}_{f(r)}$.

Choose $a_0 \in N^{0,1}$ such that $\mathsf{P}_{a_0} = \varnothing$ and let G be a partial recursive function such that

$$G(e, p, r) \simeq \begin{cases} \{e\}(p), & \text{if } \ p <_\delta r; \\ a_0, & \text{otherwise.} \end{cases}$$

By Lemmas 4.5, 4.7, and 4.10, there is a primitive recursive function h defined as follows. If $|r|_\delta = 0$ or 1, $h(e, r)$ is any index such that $\mathsf{P}_{h(e,r)} = \mathsf{W}_{|r|_\delta}$. For all other r, $h(e, r)$ is such that if for all p, $G(e, p, r) \in N^{0,1}$, then also $h(e, r) \in N^{0,1}$ and

$$\mathsf{P}_{h(e,r)}(\gamma) \leftrightarrow \exists p\, \forall q\, \mathsf{P}_{G(e,p,r)}(\gamma \restriction q).$$

By the Primitive Recursion Theorem choose $\bar{e}$ such that $h(\bar{e}, r) = \{\bar{e}\}(r)$ and set $f = \{\bar{e}\}$. We prove by induction on $|r|_\delta$ that f has the required property. This is clear if $|r|_\delta = 0$ or 1 so suppose $|r|_\delta \geq 2$ and for all $p <_\delta r$, $f(p) \in N^{0,1}$ and $\mathsf{W}_{|p|_\delta} = \mathsf{P}_{f(p)}$. Then for all p, $G(\bar{e}, p, r) \in N^{0,1}$, so $f(r) = h(\bar{e}, r) \in N^{0,1}$, and for all γ,

$$\begin{aligned} \gamma \in \mathsf{W}_{|r|_\delta} &\leftrightarrow (\exists p <_\delta r) \forall q \,.\, \gamma \restriction q \in \mathsf{W}_{|p|_\delta} \\ &\leftrightarrow \exists p\, \forall q\, \mathsf{P}_{G(\bar{e},p,r)}(\gamma \restriction q) \\ &\leftrightarrow \mathsf{P}_{h(\bar{e},r)}(\gamma) \\ &\leftrightarrow \mathsf{P}_{f(r)}(\gamma). \quad \square \end{aligned}$$

4.12 Theorem. $\mathrm{EfBo} = \Delta^1_1$.

Proof. The inclusion ($\subseteq$) is Lemma 4.9. By Theorem 2.2, for any $\mathsf{R} \in \Delta^1_1$ there exists a $\rho < \omega_1$ such that $\mathsf{R} \preccurlyeq \mathsf{W}_\rho$. By the preceding lemma, $\mathsf{W}_\rho \in \mathrm{EfBo}$ and by Lemma 4.10 so is R. $\square$

To complete the picture of the effective Borel hierarchy, we shall show that the levels are distinct. The method is again essentially the same as before — to find universal relations for each level.

4.13 Lemma. *For all $\rho > 0$, Σ^0_ρ and Π^0_ρ are effectively closed under expansion — that is, for all k' and l', there exist primitive recursive functions $f_{k',l'}$ such that for all $(\mathbf{n}, \boldsymbol{\beta}) \in {}^{k',l'}\omega$, all k, l and all $a \in N^{k,l}_\rho$, $f_{k',l'}(a) \in N^{k+k',l+l'}_\rho$ and*

$$\mathsf{P}_{f_{k',l'}(a)}(\mathbf{m}, \mathbf{n}, \boldsymbol{\alpha}, \boldsymbol{\beta}) \leftrightarrow \mathsf{P}_a(\mathbf{m}, \boldsymbol{\alpha}).$$

Proof. For each k' and l', let

$$h_{k',l'}(e, a) = \begin{cases} \langle 7, b\rangle \text{ such that } \{b\}(\mathbf{m}, \mathbf{n}, \boldsymbol{\alpha}, \boldsymbol{\beta}) \simeq \{(a)_1\}(\mathbf{m}, \boldsymbol{\alpha}), \text{ if } a \in N_1; \\ \text{an index } c \text{ such that for all } p, \\ \{c\}(p) \simeq \{e\}(\{a\}(p)), \quad \text{otherwise.} \end{cases}$$

Set $f_{k',l'}(a) = \{\bar{e}\}(a)$, where $\bar{e}$ is chosen by the Primitive Recursion Theorem so that $h_{k',l'}(\bar{e}, a) = \{\bar{e}\}(a)$. $\square$

4.14 Theorem. *For all ρ such that $0 < \rho < \omega_1$ and each (k, l), there exists a relation $\mathsf{U}^0_\rho \in \Sigma^0_\rho$ which is universal for Σ^0_ρ.*

Proof. Let ρ be any recursive ordinal greater than 0 and fix (k, l). We shall construct a special recursive well-ordering δ such that $\|\delta\| > \rho$ and a primitive recursive function f such that for all $u \in \mathrm{Fld}(\delta)$, $f(u) \in N^{k+1,l}_{|u|_\delta}$ and $\mathsf{P}_{f(u)}$ is universal for $\Sigma^0_{1+|u|_\delta}$.

Let γ be any recursive well-ordering such that $\|\gamma\| > \rho$ and define δ by

$$\langle p, r\rangle \leqslant_\delta \langle q, s\rangle \leftrightarrow p \leqslant_\gamma q \wedge (p = q \rightarrow r \leqslant s).$$

Clearly δ is also a recursive well-ordering with $\|\delta\| > \rho$ — in fact, $\|\delta\| = \omega \cdot \|\gamma\|$ (ordinal multiplication). Furthermore, for all $u \in \mathrm{Fld}(\delta)$, $|u|_\delta$ is a limit ordinal or 0 iff $u = \langle p, 0\rangle$ for some $p \in \mathrm{Fld}(\gamma)$ and $|u|_\delta$ is a successor ordinal iff $u = \langle p, r+1\rangle$ for some $p \in \mathrm{Fld}(\gamma)$ and then $|\langle p, r+1\rangle|_\delta = |\langle p, r\rangle|_\delta + 1$.

Before proceeding to the definition of f, we define an auxiliary primitive recursive function g with the property that for all $u \in \mathrm{Fld}(\delta)$, $g(u) \in N^{1,0}_{|u|_\delta+1}$ and $N^{k,l}_{|u|_\delta} = \sim\mathsf{P}_{g(u)}$ — in short, g verifies that $N^{k,l}_\sigma \in \Pi^0_{1+\sigma+1}$. We obtain g via the

Primitive Recursion Theorem and a primitive recursive h defined as follows. If $|u|_\delta = 0$, then $h(e,u)$ is any index in $N_0^{1,0}$ for the recursive set $N_0^{k,l}$. If $u = \langle p, r+1\rangle$, then $h(e,u)$ is chosen by Lemmas 4.5–7, 10 such that whenever $\{e\}(\langle p,r\rangle) \in N_{(\sigma)}^{1,0}$, then $h(e,u) \in N_\sigma^{1,0}$ and

$$\sim \mathsf{P}_{h(e,u)}(a) \leftrightarrow \forall n\, [\{a\}(n)\downarrow \wedge \sim \mathsf{P}_{\{e\}(\langle p,r\rangle)}(\{a\}(n))].$$

If $u = \langle p, 0\rangle$ and $|u|_\delta > 0$, choose $a_1 \in N_0^{1,0}$ such that $\mathsf{P}_{a_1} = \omega$, let

$$G(e,v,u) \simeq \begin{cases} \{e\}(v), & \text{if } v <_\delta u; \\ a_1, & \text{otherwise}; \end{cases}$$

and let $h(e,u)$ be chosen such that whenever for all v, $G(e,v,u) \in N_{(\sigma)}^{1,0}$, then $h(e,u) \in N_\sigma^{1,0}$ and

$$\sim \mathsf{P}_{h(e,u)}(a) \leftrightarrow \forall n\, [\{a\}(n)\downarrow \wedge \exists v \sim \mathsf{P}_{G(e,v,u)}(\{a\}(n))].$$

We leave to the reader the easy proof by induction on $|u|_\delta$ that if $\bar{e}$ is chosen by the Primitive Recursion Theorem such that $h(\bar{e},u) = \{\bar{e}\}(u)$, then $g = \{\bar{e}\}$ is as required.

We now define f via the Recursion Theorem and a primitive recursive function H defined as follows. If $|u|_\delta = 0$, then $H(e,u) = \langle 7, b\rangle$ where $\{b\}(a, \mathbf{m}, \boldsymbol{\alpha}) \simeq \{a\}(\mathbf{m}, \boldsymbol{\alpha})$ for all $(\mathbf{m}, \boldsymbol{\alpha}) \in {}^{k,l}\omega$. If $u = \langle p, r+1\rangle$, then $H(e,u)$ is chosen such that whenever $\{e\}(\langle p,r\rangle) \in N_{(\sigma)}^{k+1,l}$, then $H(e,u) \in N_\sigma^{k+1,l}$ and

$$\mathsf{P}_{H(e,u)}(a, \mathbf{m}, \boldsymbol{\alpha}) \leftrightarrow \forall n.\{a\}(n)\downarrow \wedge \exists n \sim \mathsf{P}_{\{e\}(\langle p,r\rangle)}(\{a\}(n), \mathbf{m}, \boldsymbol{\alpha}).$$

If $u = \langle p, 0\rangle$ and $|u|_\delta > 0$, choose $H(e,u)$ such that whenever for all v, $G(e,v,u) \in N_{(\sigma)}^{k+1,l}$, then $H(e,u) \in N_\sigma^{k+1,l}$ and

$$\mathsf{P}_{H(e,u)}(a, \mathbf{m}, \boldsymbol{\alpha}) \leftrightarrow \forall n.\{a\}(n)\downarrow \wedge \exists n\, \exists v\, [v <_\delta u \wedge \{a\}(n) \in N_{|v|_\delta}^{k,l} \wedge \sim \mathsf{P}_{G(e,v,u)}(\{a\}(n), \mathbf{m}, \boldsymbol{\alpha})].$$

Now if $f = \{\bar{e}\}$, where $\bar{e}$ is chosen by the Primitive Recursion Theorem such that $H(\bar{e},u) = \{\bar{e}\}(u)$, then it is straightforward to prove by induction on $|u|_\delta$ that f is as required. $\square$

4.15 Effective Borel Hierarchy Theorem. *For all ρ such that $0 < \rho < \omega_1$,*
(i) $\Sigma_\rho^0 \not\subseteq \Delta_\rho^0$ *and* $\Pi_\rho^0 \not\subseteq \Delta_\rho^0$;
(ii) $\Delta_{\rho+1}^0 \not\subseteq \Sigma_\rho^0 \cup \Pi_\rho^0$.

Proof. Exactly as for Theorem III.1.9. $\square$

In the remainder of this section we sketch another approach to constructing a hierarchy for Δ_1^1. This method applies only to relations on numbers and has no descriptive-set-theoretic analogue. The idea arises from the fact established in Theorem III.1.13 that $R \in \Delta_{r+1}^0$ iff R is recursive in the set D_r obtained by applying the jump operator r times to a recursive set. We would like to extend this by defining a notion of iterating the jump operator a transfinite number of times and show that $R \in \Delta_1^1$ iff R is recursive in D_ρ for some $\rho < \omega_1$. This is not quite possible: although we may take $D_{\rho+1} = D_\rho^{\mathsf{J}}$, there is no canonical way to define D_ρ for limit ordinals ρ. What is missing is a way to "piece together" the sets D_σ $(\sigma < \rho)$ to form a set D_ρ which contains essentially just the sum of the information contained in the D_σ's. To get around this difficulty we shall index our sets not by ordinals, but by natural numbers which serve as *notations* for ordinals. This will be arranged in such a way that to any notation for a limit ordinal ρ is canonically associated a recursive function whose values are notations for a sequence of smaller ordinals with limit ρ. Although the sets D_u and D_v assigned to two notations u and v for the same ordinal will not coincide, they will be recursive in each other.

4.16 Definition. *$<_O$ is the smallest subset of $\omega \times \omega$ such that for all a, u, and v,*

(i) $1 <_O 2$;

(ii) *if $u <_O v$, then $v <_O 2^v$;*

(iii) *if $\{a\}$ is a total unary function and for all p $\{a\}(p) <_O \{a\}(p+1)$, then for all p, $\{a\}(p) <_O 3^a$;*

(iv) *if $u <_O v$ and $v <_O w$, then $u <_O w$.*

This is clearly a monotone arithmetic definition and thus has at most ω_1 stages (Theorem 2.15). Furthermore, it is easy to check that $<_O$ is a well-founded partial ordering (Exercise 4.27). We denote the field of $<_O$ by O and assign ordinals $|u|_O$ to elements u of O according to their positions in this ordering. Thus, $|1|_O = 0$, $|2^u|_O = |u|_O + 1$, and $|3^a|_O = \sup^+\{|\{a\}(p)|_O : p \in \omega\}$. Except for $\sigma < \omega$, the σ-th stage of the inductive definition of $<_O$ consists of those u such that $|u|_O \leq \sigma$. In the following we write u^+ for 2^u.

4.17 Definition. For all $u \in O$,

(i) $D_1 = \{0\}$;

(ii) $D_{u^+} = D_u^{\mathsf{J}}$;

(iii) if $u = 3^a$, then $D_u = \{\langle m, p\rangle : m \in D_{\{a\}(p)}\}$.

4.18 Lemma. *For all $u \in O$, $D_u \in \Delta_1^1$.*

Proof. We proceed similarly as for Lemma 4.9. Let

$$V(u, i, m) \leftrightarrow u \in O \wedge ([i = 0 \wedge m \in D_u] \vee [i = 1 \wedge m \notin D_u]).$$

We can define a monotone arithmetical operator Γ such that

$$\bar{\Gamma} = \{(u, i, m): V(u, i, m)\} \cup \{(u, 2, v): u <_O v\}$$

by combining definitions 4.16 and 4.17. Once this is done, it follows that V is Π^1_1 and hence that each D_u is Δ^1_1.

The clauses (1)–(4) of the definition of Γ corresponding to 4.16 are straightforward and we leave it to the reader to write them out. Corresponding to 4.17 we have

(5) $(1,0,0) \in \Gamma(R)$; for all $m > 0$, $(1,1,m) \in \Gamma(R)$;

(6) (a) if $(u, w, u^+) \in R$ and $\exists w\, \exists s\, [(\forall n < \lg(s)) . (u, (s)_n, n) \in R \wedge T(a, \langle m \rangle, w, \langle s \rangle)]$ then $(u^+, 0, \langle a, m \rangle) \in \Gamma(R)$;

(b) if $(u, 2, u^+) \in R$ and $\forall w\, \forall s\, [(\forall n < \lg(s))(\forall i . (u, i, n) \in R \rightarrow \rightarrow (s)_n = i) \rightarrow \sim T(a, \langle m \rangle, w, \langle s \rangle)]$ then $(u^+, 1, \langle a, m \rangle) \in \Gamma(R)$;

(c) if n is not of the form $\langle a, m \rangle$, then $(u^+, 1, n) \in \Gamma(R)$;

(7) (a) if for all p, $(\{a\}(p), 2, 3^a) \in R$ and $(\{a\}(p), i, m) \in R$, then $(3^a, i, \langle m, p \rangle) \in \Gamma(R)$;

(b) if n is not of the form $\langle a, m \rangle$, then $(3^a, 1, n) \in \Gamma(R)$. $\square$

To establish that each Δ^1_1 relation on numbers is recursive in some D_u, we shall define primitive recursive functions F and G such that for all $a \in N^{k,0}$, $F(a) \in O$ and P_a has index $G(a)$ from $D_{F(a)}$. The result then follows from Theorem 4.12. We need two technical lemmas.

4.19 Lemma. *There exists a partial recursive function $+_O$ such that for all $u, v \in O$, $u +_O v \in O$, $|u +_O v|_O = |u|_O + |v|_O$, and if $v \neq 1$, $u <_O (u +_O v)$.*

Proof. We define $+_O$ by the Recursion Theorem to satisfy the following conditions:

(1) $u +_O 1 \simeq u$;
(2) $u +_O v^+ \simeq (u +_O v)^+$;
(3) $u +_O 3^b \simeq 3^c$, where $\{c\}(p) \simeq u +_O \{b\}(p)$.

It follows easily by induction over O that $+_O$ has the required properties. $\square$

4.20 Lemma. *There exists a primitive recursive function g such that for all $u, v \in O$, if $|u|_O \leqslant |v|_O$, then D_u is recursive in D_v with index $g(u, v)$.*

Proof. We shall define g by the Recursion Theorem simultaneously with an auxiliary function f such that for $u, v \in O$,

$$|u|_O < |v|_O \leftrightarrow \{f(v)\}(u, D_{v^+}) \simeq 0,$$

and

$$|u|_O \geqslant |v|_O \leftrightarrow \{f(v)\}(u, D_{v^+}) \simeq 1.$$

To increase legibility, we shall use the abbreviations:

$$A_{v,u} \quad \text{for} \quad \{p : \{g(u, v)\}(p, A) \simeq 0\}$$

and

$$A^{(p)} \quad \text{for} \quad \{m : \langle m, p \rangle \in A\}.$$

Thus if g is as in the statement of the lemma, we have

(*) if $u, v \in O$, $|u|_O \leqslant |v|_O$, and $A = D_v$, then $A_{v,u} = D_u$;

and without any assumption,

(**) if $3^b \in O$ and $A = D_{3^b}$, then for all p, $A^{(p)} = D_{\{b\}(p)}$.

We shall require that f and g satisfy the following conditions: for all u, v, a, b, m, and A,

(1) $\{f(1)\}(u, A) \simeq 1$; if $v \neq 1$, $\{f(v)\}(1, A) \simeq 0$;

(2) (a) $\{f(v^+)\}(u^+, A^{\omega}) \simeq \{f(v)\}(u, A)$;

(b) $\{f(v^+)\}(3^a, A^{\omega}) \simeq \begin{cases} 1, & \text{if } \exists p . \{f(v)\}(\{a\}(p), A) \simeq 1; \\ 0, & \text{otherwise}; \end{cases}$

(3) $\{f(3^b)\}(u, A^{\omega}) \simeq \begin{cases} 0, & \text{if } \exists p . \mathrm{G}(p, u, b, A) \simeq 0; \\ 1, & \text{otherwise}, \end{cases}$

where $\mathrm{G}(p, u, b, A) \simeq \{f(\{b\}(p))\}(u, (A^{(p+1)})_{\{b\}(p+1), \{b\}(p)^+})$;

(4) $\{g(1, v)\}(m, A) \simeq \mathrm{sg}(m)$;

(5) (a) $\{g(u^+, v^+)\}(m, A) \simeq \{h(g(u, v))\}(m, A)$, where h is a primitive recursive function such that

$$\{h(a)\}(m, A^{\omega}) = (\lambda n . \{a\}(n, A))^{\omega}(m)$$

whenever both sides are defined;

$$\text{(b)}\ \{g(3^a, v^+)\}(m, A^{\omega}) \simeq \{g(3^a, v)\}(m, A);$$

(6) $$\text{(a)}\ \{g(u^+, 3^b)\}(m, A) \simeq \{g(u^+, \{b\}(\bar{p}))\}(m, A^{(\bar{p})}),$$

where $\bar{p} \simeq$ "least" $p.\mathrm{G}(p, u^+, b, A) \simeq 0$ (G as in (3));

$$\text{(b)}\ \{g(3^a, 3^b)\}(\langle m, p\rangle, A) \simeq \{g(\{a\}(p), 3^b)\}(m, A).$$

To see that such f and g exist, rewrite equations (1)–(6) with the following changes:

on the left-hand sides, replace $f(v)$ by $H(e, 0, 0, v)$
and $g(u, v)$ by $H(e, 1, u, v)$;
on the right-hand sides, replace $f(v)$ by $\{e\}(0, 0, v)$
and $g(u, v)$ by $\{e\}(1, u, v)$.

Then it is not hard to verify that there is a primitive recursive function H which satisfies these rewritten equations. Note that the quantifiers on the right-hand sides of (2)(b) and (3) are accounted for in the changing of A^{ω} to A. By the Primitive Recursion Theorem, there exists an $\bar{e}$ such that $H(\bar{e}, i, u, v) \simeq \{\bar{e}\}(i, u, v)$ and we set $f(v) = \{\bar{e}\}(0, 0, v)$ and $g(u, v) = \{\bar{e}\}(1, u, v)$.

We prove by induction on $|v|$ (dropping the subscript) that f and g have the required properties. If $|v| = 0$ this is obvious, so suppose it is true with v replaced by any w such that $|w| < |v^+|$. Then

(1) $|1| < |v^+|$ is true and $\{f(v^+)\}(1, D_{v^{++}}) \simeq 0$.

(2) (a) if $u^+ \in O$, then $u \in O$, so

$$|u^+| < |v^+| \leftrightarrow |u| < |v| \leftrightarrow \{f(v)\}(u, D_{v^+}) \simeq 0$$
$$\leftrightarrow \{f(v^+)\}(u^+, D_{v^{++}}) \simeq 0.$$

(b) if $3^a \in O$, then for all p, $\{a\}(p) \in O$ and

$$|3^a| < |v^+| \leftrightarrow \forall p.|\{a\}(p)| < |v|$$
$$\leftrightarrow \forall p.\{f(v)\}(\{a\}(p), D_{v^+}) \simeq 0$$
$$\leftrightarrow \{f(v^+)\}(3^a, D_{v^{++}}) \simeq 0.$$

(4) $$\{g(1, v^+)\}(m, D_{v^+}) \simeq 0 \leftrightarrow m = 0 \leftrightarrow m \in D_1.$$

(5) (a) If $|u^+| \leqslant |v^+|$, then $|u| \leqslant |v|$, so

$$m \in D_{u^+} \leftrightarrow m \in (D_u)^{\omega} \leftrightarrow (\lambda n.\{g(u, v)\}(n, D_v))^{\omega}(m) \simeq 0$$
$$\leftrightarrow \{h(g(u, v))\}(m, (D_v)^{\omega}) \simeq 0$$
$$\leftrightarrow \{g(u^+, v^+)\}(m, D_{v^+}) \simeq 0.$$

(b) If $|3^a| \leqslant |v^+|$, then $|3^a| \leqslant |v|$, so

$$m \in D_{3^a} \leftrightarrow \{g(3^a, v)\}(m, D_v) \simeq 0$$
$$\leftrightarrow \{g(3^a, v^+)\}(m, D_{v^+}) \simeq 0.$$

Now suppose that the result holds with v replaced by any w such that $|w| < |3^b|$.

(3) We first compute that as for all p, $|\{b\}(p)| < |3^b|$, for any $u \in O$,

$$|u| < |\{b\}(p)| \leftrightarrow \{f(\{b\}(p))\}(u, D_{\{b\}(p)^+}) \simeq 0$$
$$\leftrightarrow \{f(\{b\}(p))\}(u, \lambda n.\{g(\{b\}(p)^+, \{b\}(p+1))\}(n, D_{\{b\}(p+1)})) \simeq 0$$
$$\leftrightarrow \{f(\{b\}(p))\}(u, \lambda n.\{g(\{b\}(p)^+, \{b\}(p+1))\}(n, (D_{3^b})^{(p+1)})) \simeq 0$$
$$\leftrightarrow \{f(\{b\}(p))\}(u, ((D_{3^b})^{(p+1)})_{\{b\}(p+1), \{b\}(p)^+}) \simeq 0$$
$$\leftrightarrow \mathsf{G}(p, u, b, D_{3^b}) \simeq 0.$$

The second equivalence uses the induction hypothesis, the third uses (**).

Then if $u \in O$,

$$|u| < |3^b| \leftrightarrow \exists p . |u| < |\{b\}(p)|$$
$$\leftrightarrow \exists p . \mathsf{G}(p, u, b, D_{3^b}) \simeq 0$$
$$\leftrightarrow \{f(3^b)\}(u, D_{3^b}) \simeq 0.$$

(6) We proceed by induction on $|u|$. For $u = 1$, the result is obvious by (4).

(a) If $|u^+| \leqslant |3^b|$, then $|u^+| < |3^b|$ so for some (least) $\bar{p}$, $|u^+| < |\{b\}(\bar{p})|$. Then $\mathsf{G}(\bar{p}, u^+, b, D_{3^b}) \simeq 0$ and

$$n \in D_{u^+} \leftrightarrow \{g(u^+, \{b\}(\bar{p}))\}(m, D_{\{b\}(\bar{p})}) \simeq 0$$
$$\leftrightarrow \{g(u^+, \{b\}(\bar{p}))\}(m, (D_{3^b})^{(\bar{p})}) \simeq 0 \qquad \text{(by (**))}$$
$$\leftrightarrow \{g(u^+, 3^b)\}(m, D_{3^b}) \simeq 0.$$

(b) If $|3^a| \leqslant |3^b|$, then for all p, $|\{a\}(p)| < |3^a| \leqslant |3^b|$, so

$$\langle m, p\rangle \in D_{3^a} \leftrightarrow m \in D_{\{a\}(p)}$$
$$\leftrightarrow \{g(\{a\}(p), 3^b)\}(m, D_{3^b}) \simeq 0$$
$$\leftrightarrow \{g(3^a, 3^b)\}(\langle m, p\rangle, D_{3^b}) \simeq 0. \quad \square$$

4.21 Theorem. *For all $R \subseteq {}^k\omega$, R is Δ_1^1 iff R is recursive in D_u for some $u \in O$.*

Proof. We define primitive recursive functions F and G such that for all k and all $a \in N^{k,0}$, $F(a) \in O$ and P_a is recursive in $D_{F(a)}$ with index $G(a)$. As usual, F and G are defined by effective transfinite recursion and we shall give only an informal description of the construction.

If $a \in N_0$, let $F(a) = 2$ and $G(a)$ be an index for the semi-recursive set P_a from $D_2 = \{0\}^{\omega}$. If $a \notin N_0$, set $F(a) = (3^c)^+$, where c is an index such that

$$\{c\}(0) \simeq F(\{a\}(0)) \quad \text{and} \quad \{c\}(p+1) \simeq \{c\}(p) +_O F(\{a\}(p+1)).$$

The properties of $+_O$ ensure that if for all p, $F(\{a\}(p)) \in O$, then $F(a) \in O$ and $|F(\{a\}(p))|_O < |3^c|_O < |F(a)|_O$. By 4.20 there exists a recursive function γ such that if for all p, $F(\{a\}(p)) \in O$, then $D_{F(\{a\}(p))}$ is recursive in D_{3^c} with index $\gamma(p)$. We take $G(a)$ to be an index such that

$$\{G(a)\}(\mathbf{m}, D_{F(a)}) \simeq \begin{cases} 0, & \text{if } \exists p.\{G(\{a\}(p))\}(\mathbf{m}, \lambda n.\{\gamma(p)\}(n, D_{3^c})) \simeq 1, \\ 1, & \text{otherwise.} \end{cases}$$

For the induction step in the proof that F and G are as desired, we have as induction hypothesis that for all p,

$$\{G(\{a\}(p))\}(\mathbf{m}, \lambda n.\{\gamma(p)\}(n, D_{3^c})) \simeq \{G(\{a\}(p))\}(\mathbf{m}, D_{F(\{a\}(p))}) \simeq 1$$
$$\leftrightarrow \sim \mathsf{P}_{\{a\}(p)}(\mathbf{m}).$$

Hence

$$\{G(a)\}(\mathbf{m}, D_{F(a)}) \simeq 0 \leftrightarrow \exists p. \sim \mathsf{P}_{\{a\}(p)}(\mathbf{m}) \leftrightarrow \mathsf{P}_a(\mathbf{m}). \quad \square$$

This characterization provides us with a new hierarchy of the Δ_1^1 relations on numbers known as the hyperarithmetical hierarchy. Let

$$\Sigma_\rho^{0\prime} = \{R : R \text{ is many-one reducible to some } D_u \text{ with } |u|_O \leq \rho\};$$
$$\Pi_\rho^{0\prime} = \{R : \sim R \in \Sigma_\rho^{0\prime}\},$$
$$\Delta_\rho^{0\prime} = \Sigma_\rho^{0\prime} \cap \Pi_\rho^{0\prime}.$$

If R is a relation recursive in D_u, then both R and $\sim R$ are many-one reducible to $D_u^{\mathrm{OJ}} = D_{u^+}$. Hence

$$\Delta_1^1 = \bigcup \{\Delta_\rho^{0,\prime} : \rho < \omega_1\}.$$

Furthermore, if $\rho < \sigma < \omega_1$, then $\Delta_\rho^{0,\prime} \subsetneq \Delta_\sigma^{0,\prime}$ (Exercise 4.31). By Theorem III.1.13, $\Sigma_\rho^{0,\prime} = \Sigma_\rho^0$ for $\rho < \omega$, but this is not in general true for all $\rho < \omega_1$.

4.22–4.32 Exercises

4.22. Formulate and prove a theorem which justifies recursive definitions over the sets $N^{k,l}$ of Definition 4.2.

4.23. Show that for all $\rho < \omega_1$, Σ_ρ^0 and Π_ρ^0 are closed under bounded number quantification ($\exists_<^0$ and $\forall_<^0$).

4.24. Show that for all $\rho < \omega_1$,

$$\Sigma_{\rho+1}^0 = \{\exists^0 \mathsf{Q} : \mathsf{Q} \in \Pi_\rho^0\}.$$

4.25. Give a different proof of Lemma 4.9 by defining by effective transfinite recursion a primitive recursive function f such that for all $a \in N^{k,l}$, $f(a)$ is an index for P_a as a Δ_1^1 relation — that is,

$$\mathsf{U}_1^1(((f(a))_0, \langle \mathbf{m}\rangle, \langle \boldsymbol{\alpha}\rangle) \leftrightarrow \mathsf{P}_a(\mathbf{m}, \boldsymbol{\alpha}) \leftrightarrow \sim \mathsf{U}_1^1(((f(a))_1, \langle \mathbf{m}\rangle, \langle \boldsymbol{\alpha}\rangle).$$

4.26. Show that $N^{k,l}$ (any k and l) and O are Π_1^1 complete.

4.27. Show that $\leqslant_O$ is a well-founded partial ordering.

4.28. Show that there exists a primitive recursive function h such that for any $u \in O$ and any d, if for all p, $\{d\}(p) \in O$, then $h(d) \in O$ and for all p, $|\{d\}(p)|_O < |h(d)|_O$.

4.29. Show that the ordinals $|u|_O$ for $u \in O$ are exactly the ordinals less than ω_1. (For each $u \in O$, show that the restriction of $<_O$ to $v <_O u$ is a well-ordering of type $|u|_O$. For the other direction, let $\gamma \in \mathsf{W}$ be recursive. Construct as in the proof of Theorem 4.14 a recursive $\delta \in \mathsf{W}$ such that $\|\delta\| = \omega \cdot \|\gamma\|$ and use effective transfinite recursion to define a recursive function f such that for all $p \in \mathrm{Fld}\,\delta$, $f(p) \in O$ and $|f(p)|_O \geqslant |p|_\delta$.)

4.30. Give a new proof of Corollary 2.22 by showing that for all $u \in O$, D_u is implicitly Π_2^0. (Construct by effective transfinite recursion a partial recursive function F such that for all $u \in O$ and all A

$$A = D_u \leftrightarrow \sim \mathsf{U}_2^0(F(u), \langle\ \rangle, \langle A \rangle).$$

Use an effective version of Exercise III.4.25 for defining $F(2^u)$ from $F(u)$.

4.31. Show that for all $\rho < \sigma < \omega_1$, $\Delta_\rho^{0\prime} \subsetneq \Delta_\sigma^{0\prime}$.

4.32. Recall that $U_{(\omega)}^0 = \{\langle r, a, \mathbf{m}\rangle : U_r^0(a, \langle \mathbf{m}\rangle)\}$. Show that for any R, the following are equivalent:

(i) $R \in \Sigma_\omega^0$;

(ii) $R(\mathbf{m}) \leftrightarrow \exists p . \langle f(p), a, \mathbf{m}\rangle \in U_{(\omega)}^0$ for some a and some primitive recursive f;

(iii) $R \preccurlyeq \bar{\Gamma}$ for some Σ_1^0 inductive operator Γ.

(Cf. Theorems, III.3.6–7). How does this compare with $\Sigma_\omega^{0\prime}$?

4.33 Notes. The idea of an effective version of the Borel hierarchy was developed by Addison in his thesis, Addison [1954], and announced in Addison [1955], although at that time not all of the details of the transfinite levels had been worked out. Indeed, they were never published and may not have been completely written down until the Spring of 1964 when Addison conducted a seminar on the material at Berkeley. The hyperarithmetical hierarchy based on O and the set D_u and Theorem 4.21 are due to Kleene [1955b], but Lemma 4.20 is from Spector [1955].

5. Cardinality, Measurability and Category

One of the benefits for analysts in dealing with constructively defined sets and relations is that they are more likely to be "well-behaved". We consider here some of the pleasant properties of $\mathbf{\Sigma}_1^1$ and $\mathbf{\Pi}_1^1$ relations. To simplify some of the arguments we shall deal explicitly only with subsets of ${}^\omega\omega$ (relations of rank $(0, 1)$), but all of the results hold also for relations of arbitrary rank.

5.1 Theorem. *Every $\mathbf{\Pi}_1^1$ or $\mathbf{\Sigma}_1^1$ set is both the union of an $\aleph_1$-sequence of Borel sets and the intersection of an $\aleph_1$-sequence of Borel sets.*

Proof. If suffices to prove the result for $A \in \mathbf{\Pi}_1^1$. Let P be a closed-open relation such that $\alpha \in \mathsf{A} \leftrightarrow \forall \beta\, \exists n\, \mathsf{P}(\bar{\beta}(n), \alpha)$, and $\trianglelefteq$ and $\trianglelefteq_\alpha^{\mathsf{P}}$ the relations defined in the proof of Theorem 1.1. For any t set

$$u \trianglelefteq_\alpha^{\mathsf{P}, t} v \leftrightarrow u \trianglelefteq_\alpha^{\mathsf{P}} v \wedge v \trianglelefteq t$$

and

$$\mathsf{A}_\rho^t = \{\alpha : \trianglelefteq_\alpha^{\mathsf{P}, t} \text{ has order-type} \leqslant \rho\}.$$

Thus as in the proof of Theorem 1.1,

$$\begin{aligned}\alpha \in \mathsf{A} &\leftrightarrow \trianglelefteq_\alpha^{\mathsf{P},\langle\rangle} \text{ is a well-ordering}\\ &\leftrightarrow (\exists\rho < \aleph_1)\, \alpha \in \mathsf{A}_\rho^{\langle\rangle}.\end{aligned}$$

For any $\rho < \aleph_1$, choose γ such that $\|\gamma\| = \rho$ and let F be a continuous functional such that

$$\mathsf{F}(\langle u, v\rangle, t, \alpha) = \begin{cases} 0, & \text{if } u \trianglelefteq_\alpha^{\mathsf{P},t} v;\\ 1, & \text{otherwise.}\end{cases}$$

Then

$$\alpha \in \mathsf{A}_\rho^t \leftrightarrow \mathsf{F}[t, \alpha] \leqslant_\Sigma \gamma \leftrightarrow \mathsf{F}[t, \alpha] \leqslant_\Pi \gamma$$

so that for all t, A_ρ^t is Δ_1^1, i.e., Borel. In particular, $\mathsf{A} = \bigcup\{\mathsf{A}_\rho^{\langle\rangle} : \rho < \aleph_1\}$ is a representation of A as a union of $\aleph_1$ Borel sets.

For the intersection, let

$$\mathsf{B}_\rho = \mathsf{A}_\rho^{\langle\rangle} \cup \bigcup\{\mathsf{A}_{\rho+1}^t \sim \mathsf{A}_\rho^t : t \in \omega\}.$$

Again B_ρ is clearly Borel for all countable ρ, and we claim that $\mathsf{A} = \bigcap\{\mathsf{B}_\rho : \rho < \aleph_1\}$. Suppose first that $\alpha \in \mathsf{A}$, so that $\trianglelefteq_\alpha^{\mathsf{P},\langle\rangle}$ is a well-ordering, say of order-type σ. For any $\rho < \sigma$ there exists a t such that $\trianglelefteq_\alpha^{\mathsf{P},t}$ has order-type $\rho + 1$ so that $\alpha \in \mathsf{A}_{\rho+1}^t \sim \mathsf{A}_\rho^t \subseteq \mathsf{B}_\rho$. For $\rho \geqslant \sigma$, $\alpha \in \mathsf{A}_\rho^{\langle\rangle} \subseteq \mathsf{B}_\rho$. Hence $\alpha \in \mathsf{B}_\rho$ for all $\rho < \aleph_1$.

For the converse, suppose $\alpha \in \mathsf{B}_\rho$ for all $\rho < \aleph_1$. For each t let

$$\sigma_t = \begin{cases} \text{order-type of } \trianglelefteq_\alpha^{\mathsf{P},t}, & \text{if this is an ordinal;}\\ 0, & \text{otherwise;}\end{cases}$$

and set $\bar\rho = \sup^+\{\sigma_t : t \in \omega\}$. Then $\bar\rho < \aleph_1$ and for all t, $\trianglelefteq_\alpha^{\mathsf{P},t}$ does not have order-type $\bar\rho + 1$ so $\alpha \notin \mathsf{A}_{\bar\rho+1}^t \sim \mathsf{A}_{\bar\rho}^t$. Since $\alpha \in \mathsf{B}_{\bar\rho}$, it follows that $\alpha \in \mathsf{A}_{\bar\rho}^{\langle\rangle} \subseteq \mathsf{A}$. □

5.2 Corollary. *Every $\mathbf{\Sigma}_2^1$ set is the union of an $\aleph_1$-sequence of Borel sets.*

Proof. Suppose A is $\mathbf{\Sigma}_2^1$ and $\alpha \in \mathsf{A} \leftrightarrow \exists\beta\, \mathsf{S}(\alpha, \beta)$ for some $\mathsf{S} \in \mathbf{\Pi}_1^1$. By the preceding theorem (extended to relations) there exist Borel relations S_σ such that $\mathsf{S} = \bigcup\{\mathsf{S}_\sigma : \sigma < \aleph_1\}$. Then

$$\begin{aligned}\alpha \in \mathsf{A} &\leftrightarrow \exists\beta\, (\exists\sigma < \aleph_1) \mathsf{S}_\sigma(\alpha, \beta)\\ &\leftrightarrow (\exists\sigma < \aleph_1)\, \exists\beta\, \mathsf{S}_\sigma(\alpha, \beta).\end{aligned}$$

By the theorem again there exist Borel sets $\mathsf{A}_{\sigma\tau}$ such that

$$\{\alpha : \exists\beta\, \mathsf{S}_\sigma(\alpha, \beta)\} = \bigcup\{\mathsf{A}_{\sigma\tau} : \tau < \aleph_1\}$$

and thus $\mathsf{A} = \bigcup\{\mathsf{A}_{\sigma\tau} : \sigma, \tau < \aleph_1\}$. Since $\aleph_1 \times \aleph_1$ is of power $\aleph_1$, there is a pairing function on $\aleph_1$ with (say) inverses $(\ )_0$ and $(\ )_1$. Then $\mathsf{A} = \bigcup\{\mathsf{A}_{(\sigma)_0(\sigma)_1} : \sigma < \aleph_1\}$ as required. □

The converse of this Corollary is not provable without some additional set-theoretic assumption, since if $2^{\aleph_0} = \aleph_1$, then *every* set is the union of $\aleph_1$ singletons (and singletons are Borel). (Cf. Exercise V.3.25).

We turn next to the measurability of Σ^1_1 and Π^1_1 relations. We have discussed only Lebesgue measure on ${}^\omega\omega$, but the proof will depend only on the following four properties of the measure:

(1) The class of measurable sets contains the open sets and is closed under complementation and countable union and intersection;

(2) the union of countably many sets of measure 0 has measure 0;

(3) every subset of a set of measure 0 is measurable and has measure 0;

(4) in any family of $\aleph_1$ pairwise disjoint measurable sets, at most countably many sets have positive measure.

Properties (1) and (2) are direct consequences of the countable additivity of a measure and (3) is the completeness property. For (4), if $\langle \mathsf{A}_\sigma : \sigma < \aleph_1\rangle$ is a family of pairwise disjoint measurable sets and $x_\sigma = \mathrm{mes}(\bigcup\{\mathsf{A}_\tau : \tau < \sigma\})$, then $x_0 \leqslant x_1 \leqslant \cdots \leqslant x_\sigma \leqslant x_{\sigma+1} \ldots$. The strict inequality, $x_\sigma < x_{\sigma+1}$, holds just in case $\mathrm{mes}(\mathsf{A}_\sigma) > 0$ and in this case there is a rational number y_σ such that $x_\sigma < y_\sigma < x_{\sigma+1}$. Clearly this can happen for at most countably many σ.

5.3 Theorem. *Every Σ^1_1 or Π^1_1 set is measurable with respect to any measure which satisfies* (1)–(4).

Proof. As the complement of a measurable set is measurable, it suffices to consider $\mathsf{A} \in \Pi^1_1$. Let A^t_ρ be as in the proof of Theorem 5.1 and for each t consider the family $\langle \mathsf{A}^t_{\rho+1} \sim \mathsf{A}^t_\rho : \rho < \aleph_1\rangle$. This is a family of $\aleph_1$ pairwise disjoint Borel sets (hence measurable sets by (1)) so all but countably many have measure 0. Hence there is an ordinal σ_t such that for all $\rho \geqslant \sigma_t$, $\mathrm{mes}(\mathsf{A}^t_{\rho+1} \sim \mathsf{A}^t_\rho) = 0$. Let $\bar\rho = \sup^+\{\sigma_t : t \in \omega\}$. Then $\bar\rho < \aleph_1$ and for all t, $\mathrm{mes}(\mathsf{A}^t_{\bar\rho+1} \sim \mathsf{A}^t_{\bar\rho}) = 0$. In the proof of Theorem 5.1 we showed $\mathsf{A} \subseteq \mathsf{B}_{\bar\rho}$. Hence

$$\mathsf{A} = \mathsf{A}^{\langle\,\rangle}_{\bar\rho} \cup (\mathsf{A} \cap \bigcup\{\mathsf{A}^t_{\bar\rho+1} \sim \mathsf{A}^t_{\bar\rho} : t \in \omega\}).$$

Thus A is the union of $\mathsf{A}^{\langle\,\rangle}_{\bar\rho}$, which is Borel, hence measurable, and a subset of a countable union of sets of measure 0, which is measurable by (2) and (3). □

This result cannot be extended even to Δ_2^1 relations; we shall indicate a proof of this in §V.2. Of course, countable unions and intersections of $\mathbf{\Sigma}_1^1$ and $\mathbf{\Pi}_1^1$ relations are measurable. We shall discuss further extensions of this sort in § V.4.

We turn next to the topological analogue of measurability, the Baire property.

5.4 Definition. A set A has the *Baire property* iff there exists an open set B such that both $\mathsf{A} \sim \mathsf{B}$ and $\mathsf{B} \sim \mathsf{A}$ are meager.

We aim to show that all $\mathbf{\Sigma}_1^1$ and $\mathbf{\Pi}_1^1$ sets also have the Baire property. The proof is nearly identical to that of Theorem 5.3 once we establish the following analogues of (1)–(4):

(1′) The class of sets which have the Baire property contains the open sets and is closed under complementation and countable union and intersection;

(2′) the union of countably many meager sets is meager;

(3′) every subset of a meager set is meager and has the Baire property;

(4′) in any family of pairwise disjoint sets which have the Baire property, at most countably many sets are non-meager.

We denote the topological closure of a set A by $\bar{\mathsf{A}}$. Note that $\alpha \in \bar{\mathsf{A}}$ iff, for all n, $[\bar{\alpha}(n)] \cap \mathsf{A} \neq \varnothing$. A is dense in an interval $[s]$ iff $[s] \subseteq \bar{\mathsf{A}}$. Hence A is nowhere dense iff no interval is included in $\bar{\mathsf{A}}$ iff $\sim\bar{\mathsf{A}}$ is dense iff $\overline{\sim\bar{\mathsf{A}}} = {}^{\omega}\omega$. Properties (2′) and (3′) are immediate from the definitions.

5.5 Lemma. *For any open set* B, $\bar{\mathsf{B}} \sim \mathsf{B}$ *is nowhere dense.*

Proof. If B is open, then $\bar{\mathsf{B}} \sim \mathsf{B}$ is closed so by the preceding remarks, it suffices to show that $\sim(\bar{\mathsf{B}} \sim \mathsf{B})$ is dense. This is clear from:

$$\overline{\sim(\bar{\mathsf{B}} \sim \mathsf{B})} = \overline{\sim\bar{\mathsf{B}} \cup \mathsf{B}} = \overline{\sim\bar{\mathsf{B}}} \cup \bar{\mathsf{B}} \supseteq \sim\bar{\mathsf{B}} \cup \bar{\mathsf{B}} = {}^{\omega}\omega. \quad \square$$

5.6 Lemma. *The class of sets which have the Baire property is closed under complementation and countable union and intersection.*

Proof. Suppose first that A has the Baire property and B is an open set such that $\mathsf{A} \sim \mathsf{B}$ and $\mathsf{B} \sim \mathsf{A}$ are both meager. Then $\sim\bar{\mathsf{B}}$ is open,

$$(\sim\mathsf{A}) \sim (\sim\bar{\mathsf{B}}) = \bar{\mathsf{B}} \sim \mathsf{A} \subseteq (\mathsf{B} \sim \mathsf{A}) \cup (\bar{\mathsf{B}} \sim \mathsf{B}),$$

and

$$(\sim\bar{\mathsf{B}}) \sim (\sim\mathsf{A}) = \mathsf{A} \sim \bar{\mathsf{B}} \subseteq \mathsf{A} \sim \mathsf{B}.$$

By Lemma 5.5 both of these sets are meager and thus $\sim\mathsf{A}$ also has the Baire property.

Now suppose that for all n, A_n has the Baire property and B_n is an open set such that both $A_n \sim B_n$ and $B_n \sim A_n$ are meager. Then $\bigcup\{B_n : n \in \omega\}$ is open,

$$\bigcup\{A_n : n \in \omega\} \sim \bigcup\{B_n : n \in \omega\} \subseteq \bigcup\{A_n \sim B_n : n \in \omega\},$$

and

$$\bigcup\{B_n : n \in \omega\} \sim \bigcup\{A_n : n \in \omega\} \subseteq \bigcup\{B_n \sim A_n : n \in \omega\},$$

so it follows by (2′) and (3′) that $\bigcup\{A_n : n \in \omega\}$ also has the Baire property. □

Since all open sets obviously have the Baire property, this establishes (1′). It follows immediately that all Borel sets have the Baire property.

5.7 Lemma. *For any set* A *which has the Baire property and any* s, *there is some* $t \supseteq s$ *such that one of* $A \cap [t]$ *and* $(\sim A) \cap [t]$ *is meager.*

Proof. Suppose that A has the Baire property and B is an open set such that $A \sim B$ and $B \sim A$ are meager. For any set C let

$$C^* = \{\alpha : \text{for all } n, C \cap [\bar{\alpha}(n)] \text{ is not meager}\}.$$

Clearly $C^* \subseteq \bar{C}$. Since the meagerness of $A \cap [\bar{\alpha}(n)]$ will not be affected by the addition or removal of a meager set, $A^* = B^*$ and $(\sim A)^* = (\sim B)^*$. Hence

$$A^* \cap (\sim A)^* = B^* \cap (\sim B)^* \subseteq \bar{B} \cap \overline{\sim B} \subseteq \bar{B} \sim B,$$

which is nowhere dense by Lemma 5.5. In particular, for every s there exists a function $\alpha \in [s]$ such that either $\alpha \notin A^*$ or $\alpha \notin (\sim A)^*$. If $\alpha \notin A^*$, then for some n, $A \cap [\bar{\alpha}(n)]$ is meager, and clearly n may be chosen such that $\bar{\alpha}(n) \supseteq s$. If $\alpha \notin (\sim A)^*$, then for some n, $(\sim A) \cap [\bar{\alpha}(n)]$ is meager. □

5.8 Lemma. *For any* A, *if* A *is non-meager and has the Baire property, then for some* s, $(\sim A) \cap [s]$ *is meager.*

Proof. Suppose that A has the Baire property but for all s, $(\sim A) \cap [s]$ is non-meager. Then by the preceding lemma, for every s there exists a $t \supseteq s$ such that $A \cap [t]$ is meager. Let $C = \bigcup\{[t] : A \cap [t] \text{ is meager}\}$. Clearly $A \cap C$ is meager. But $\sim C$ is nowhere dense so $A \cap (\sim C)$ is also meager and thus A is meager. □

5.9 Lemma. *In any family of pairwise disjoint sets which have the Baire property, at most countably many sets are non-meager.*

Proof. Let X be a family of pairwise disjoint sets which have the Baire property and Y the subfamily of X consisting of all non-meager sets. By the preceding lemma the function

$$\varphi(\mathsf{A}) = \text{least } s\,.\,(\sim\mathsf{A}) \cap [s] \text{ is meager}$$

is defined for all $\mathsf{A} \in Y$. It will suffice to show that φ is one-one as $\operatorname{Im}\varphi \subseteq \omega$ and hence is countable. Suppose $\varphi(\mathsf{A}) = \varphi(\mathsf{B}) = s$. Then $(\sim\mathsf{A}) \cap [s]$ and $(\sim\mathsf{B}) \cap [s]$ are both meager. Since $\mathsf{A} \cap \mathsf{B} = \varnothing$, the union of these sets is $[s]$, which is thus meager in contradiction with the Baire Category Theorem (I.2.2). □

5.10 Theorem. *Every $\boldsymbol{\Sigma}^1_1$ or $\boldsymbol{\Pi}^1_1$ set has the Baire property.*

Proof. Follow the proof of Theorem 5.3 substituting "has the Baire property" for "is measurable" and "is meager" for "has measure 0". □

Finally, we consider the question of the cardinality of $\boldsymbol{\Sigma}^1_1$ and $\boldsymbol{\Pi}^1_1$ sets. It is well known that the usual axioms of set theory, even including the Axiom of Choice, do not determine the power of the continuum. That is, it is consistent with these axioms (assuming that they alone are consistent) either that $2^{\aleph_0} = \aleph_1$ or that there are one or more distinct cardinals between $\aleph_0$ and $2^{\aleph_0}$. What we shall show is that these intermediate cardinalities cannot be realized by $\boldsymbol{\Sigma}^1_1$ sets nor, except for $\aleph_1$, by $\boldsymbol{\Sigma}^1_2$ sets.

5.11 Lemma. *For any uncountable set $\mathsf{A} \subseteq {}^{\omega}\omega$, there exist s_0 and s_1 such that $[s_0] \cap [s_1] = \varnothing$ and both $\mathsf{A} \cap [s_0]$ and $\mathsf{A} \cap [s_1]$ are uncountable.*

Proof. Let A be uncountable and set

$$B = \{s : s \in \mathrm{Sq} \wedge \mathsf{A} \cap [s] \text{ is uncountable}\}.$$

Clearly $\langle\ \rangle \in B$. Since $\mathsf{A} \cap [s] = \bigcup\{\mathsf{A} \cap [s * \langle n\rangle] : n \in \omega\}$, if $\mathsf{A} \cap [s]$ is uncountable, so is some $\mathsf{A} \cap [s * \langle n\rangle]$. Hence there there exists a unique function β such that for all k,

$$\beta(k) = \text{least } n[\bar{\beta}(k) * \langle n\rangle \in B].$$

Suppose that the conclusion of the lemma is false. Then for every s, if $\beta \notin [s]$, then $\mathsf{A} \cap [s]$ is countable. But

$$\mathsf{A} \subseteq \bigcup\{\mathsf{A} \cap [s] : \beta \notin [s]\} \cup \{\beta\}$$

and thus A is countable, contrary to assumption. □

At this point the reader should attempt to show for himself that any uncountable closed set has power $2^{\aleph_0}$. The proof of the following theorem is merely a two-dimensional version of this construction.

5.12 Theorem. *For all* $\mathsf{A} \in \Sigma_1^1$, *if* A *is uncountable, then* A *has power* $2^{\aleph_0}$.

Proof. Let A be an uncountable Σ_1^1 set. Obviously the power of A is at most $2^{\aleph_0}$ so it suffices to find a subset of A of power $2^{\aleph_0}$. Let R be a relation such that

$$\alpha \in \mathsf{A} \leftrightarrow \exists \beta\, \forall n\, R(\bar{\alpha}(n), \bar{\beta}(n)),$$

and

$$R(s,t) \wedge s' \subseteq s \wedge t' \subseteq t \rightarrow R(s', t').$$

By Lemma 5.11 choose s_0 and s_1 such that $\mathsf{A} \cap [s_0]$ and $\mathsf{A} \cap [s_1]$ are both uncountable and $[s_0] \cap [s_1] = \varnothing$. For $i = 0$ and 1 we may represent $\mathsf{A} \cap [s_i]$ in the form

$$\alpha \in \mathsf{A} \cap [s_i] \leftrightarrow \alpha \in [s_i] \wedge \exists t\, \exists \beta\, [\lg(t) = \lg(s_i) \wedge \\ \beta \in [t] \wedge \forall n\, R(\bar{\alpha}(n), \bar{\beta}(n))].$$

As there are only countably many t of length $\lg(s_i)$, there is at least one, say t_i, such that

$$\mathsf{A}_i = \{\alpha : \alpha \in [s_i] \wedge (\exists \beta \in [t_i]) \forall n\, R(\bar{\alpha}(n), \bar{\beta}(n))\}$$

is uncountable.

For the next stage, choose for $i = 0$ and 1, $s_{i,0}$ and $s_{i,1}$ such that $\mathsf{A}_i \cap [s_{i,j}]$ is uncountable, and then $t_{i,0}$ and $t_{i,1}$ extending t_i such that

$$\mathsf{A}_{i,j} = \{\alpha : \alpha \in [s_{i,j}] \wedge (\exists \beta \in [t_{i,j}]) \forall n\, R(\bar{\alpha}(n), \bar{\beta}(n))\}$$

is uncountable. Continuing in this way, we define for each (code for a) finite sequence u of 0's and 1's, sequences s_u and t_u with the following properties:

(1) $R(s_u, t_u)$;

(2) $u \subseteq v \rightarrow s_u \subseteq s_v$ and $t_u \subseteq t_v$;

(3) $s_u \subseteq s_v \rightarrow u \subseteq v$;

(4) $\lg(s_u) = \lg(t_u) \geq \lg(u)$.

For each $\gamma \in {}^\omega 2$, let α_γ be the limit of the sequences $s_{\bar{\gamma}(n)}$ and β_γ the limit of the $t_{\bar{\gamma}(n)}$ — that is, $\alpha_\gamma(n) = (s_{\bar{\gamma}(n+1)})_n$ and $\beta_\gamma(n) = (t_{\bar{\gamma}(n+1)})_n$. From (3) it follows that if $\gamma \neq \delta$, then $\alpha_\gamma \neq \alpha_\delta$, so the mapping $\gamma \mapsto \alpha_\gamma$ is 1–1. Hence $\mathsf{B} = \{\alpha_\gamma : \gamma \in {}^\omega 2\}$ has the same power as ${}^\omega 2$, that is, $2^{\aleph_0}$.

We claim that $\mathsf{B} \subseteq \mathsf{A}$. Note first that for every γ and n, $R(s_{\bar{\gamma}(n)}, t_{\bar{\gamma}(n)})$ so that also $R(\bar{\alpha}_\gamma(n), \bar{\beta}_\gamma(n))$. In particular, for every γ, $\exists\beta\,\forall n\, R(\bar{\alpha}_\gamma(n), \bar{\beta}(n))$ and thus $\alpha_\gamma \in \mathsf{A}$. □

5.13 Corollary. *For all* $\mathsf{A} \in \boldsymbol{\Sigma}^1_2$, *if* A *is uncountable, then* A *has power either* $\aleph_1$ *or* $2^{\aleph_0}$.

Proof. By Corollary 5.2, every $\boldsymbol{\Sigma}^1_2$ set A is the union of $\aleph_1$ Borel sets A_ρ $(\rho < \aleph_1)$. The sets A_ρ may clearly be chosen pairwise disjoint. By the preceding theorem, each A_ρ is either countable or of power $2^{\aleph_0}$. There are three possible cases:

(1) for some ρ, A_ρ has power $2^{\aleph_0}$: then A has power $2^{\aleph_0}$;

(2) A_ρ is countable for all ρ and A_ρ is non-empty for uncountably many ρ: then A has power $\aleph_1$;

(3) A_ρ is countable for all ρ but A_ρ is non-empty for only countably many ρ: then A is countable. □

By Theorem 3.13 all of the theorems of this section hold also for $\boldsymbol{\Sigma}^1_1$ and $\boldsymbol{\Pi}^1_1$ and $\boldsymbol{\Sigma}^1_2$ subsets of the real interval (0, 1).

5.14–5.16 Exercises

5.14. Is the following "lightface" version of Theorem 5.1 true: every Σ^1_1 or Π^1_1 set is both the union of an ω_1-sequence of Δ^1_1 sets and the intersection of an ω_1-sequence of Δ^1_1 sets? Consider separately sets of functions and sets of numbers.

5.15. Show that for any α, $\{\beta : \alpha \in \Delta^1_1[\beta]\}$ has measure either 0 or 1 (use Exercise I.2.10).

5.16 (Harrison). Show that for any Σ^1_1 set A, if $A \not\subseteq \Delta^1_1$, then A is of power $2^{\aleph_0}$. (Apply a variant of the technique used for Theorem 5.12 to the Σ^1_1 set $\mathsf{A} \sim \Delta^1_1$.) Conclude that there is no largest countable Σ^1_1 subset of ${}^\omega\omega$. Note that the relativized version of this result implies Theorem 5.12.

5.17 Notes. An excellent and thorough account of the analogies and similarities between the theories of measure and category may be found in Oxtoby [1971].

As with many of the results of this chaptr, much effort went into attempts to extend Theorems 5.3, 5.10, and 5.12 to higher levels of the analytical and projective hierarchies. The attempts were futile because the extensions are independent of ZFC. Indeed, this was forseen as early as Luzin [1930] where we read as the final paragraph:

"Ou bien, les problèmes indiqués sur les ensembles projectifs (of measure, category, and power) resterons à jamais sans solutions augmentés de quantité de problèmes nouveaux aussi naturels et aussi inabordables. Dans ce cas il est clair que le jour serait venu de réformer nos idées sur le continu arithmétique."

In §§ V.2–3 we consider the situation under the additional hypotheses of Constructibility or Projective Determinacy. The hypothesis that there exist a measurable cardinal has the somewhat surprising effect of pushing the results of this section exactly one level further: all Σ^1_2 sets are measurable, have the Baire property, and satisfy the Continuum Hypothesis (see Solovay [1969] and Shoenfield [1971b]).

6. Continuous Images

We have explored in §§1 and 2 the analogy between the classes Σ^0_1 and Π^1_1. There are, however, some ways in which Σ^0_1 resembles Σ^1_1. Consider the following two facts: for any non-empty $A \subseteq \omega$ and $\mathsf{A} \subseteq {}^\omega\omega$,

(1) $A \in \Sigma^0_1$ iff A is the image of a recursive set under a recursive function from ω into ω,

(1′) $\mathsf{A} \in \Sigma^1_1$ iff A is the image of a $\left\{\begin{matrix}\text{Borel}\\ \text{closed}\end{matrix}\right\}$ set under a continuous functional from ${}^\omega\omega$ into ${}^\omega\omega$.

(1) is just a slight variant of part of Theorem II.4.15 and (1′) follows from Lemma 6.1 below and the fact that projection is a continuous functional together with the relativized version of Lemma III.2.8. We shall investigate here corresponding analogues for the following related facts:

(2) $A \in \Sigma^0_1$ and $A \neq \varnothing \leftrightarrow A$ is the image of a total recursive function;

(3) $A \in \Sigma^0_1 \leftrightarrow A$ is the image of a recursive set under a one–one recursive function;

(4) $A \in \Sigma^0_1$ and A is infinite $\leftrightarrow A$ is the image of a total one–one recursive function (Exercise II.4.30);

(5) $A \in \Sigma^0_1 \leftrightarrow A$ is the domain of a partial function with recursive graph.

We shall use letters θ, φ, χ, and ψ to denote (partial) functionals from ${}^\omega\omega$ into ${}^\omega\omega$. Such a functional is continuous iff $\theta^{-1}([s])$ is an open set for every interval $[s]$. Note that it is *not* true that the graph of a continuous functional θ is a closed-open subset of ${}^\omega\omega \times {}^\omega\omega$ as clearly no such functional has open graph.

6.1 Lemma. *For any θ, if $\mathsf{F}(p, \alpha) \simeq \theta(\alpha)(p)$, then θ is partial continuous iff F is partial continuous.*

Proof. Suppose first that F is partial continuous. Then

$$\theta^{-1}([s]) = \{\alpha : (\forall p < \lg(s))\mathsf{F}(p, \alpha) = (s)_p\},$$

and thus $\theta^{-1}([s])$ is a finite intersection of open sets and is open. On the other hand, if θ is partial continuous,

$$\mathsf{F}^{-1}(\{n\}) = \bigcup\{\{p\} \times \theta^{-1}(\{\alpha : \alpha(p) = n\}) : p \in \omega\},$$

so $\mathsf{F}^{-1}(\{n\})$ is open and thus F is partial continuous. □

We first establish an analogue of (2). The following lemma is a refinement of Lemma 5.11.

6.2 Lemma. *For any uncountable set $\mathsf{A} \subseteq {}^{\omega}\omega$, there exists an infinite set $A \subseteq \mathrm{Sq}$ such that*

(i) *for all $s \in A$, $\mathsf{A} \cap [s]$ is uncountable;*
(ii) *for all $s, t \in A$, if $s \neq t$, then $[s] \cap [t] = \varnothing$;*
(iii) *$\mathsf{A} \sim \bigcup\{\mathsf{A} \cap [s] : s \in A\}$ is countable.*

Proof. Let B and β be defined as in the proof of Lemma 5.11. For each k set

$$A_k = \{s : s \in B \wedge \lg(s) = k+1 \wedge (\forall i < k)[(s)_i = \beta(i)] \wedge \beta(k) < (s)_k\},$$

and

$$A = \bigcup\{A_k : k \in \omega\}.$$

By definition $A \subseteq B$ and thus satisfies (i). For (ii), suppose $s, t \in A$ and $s \neq t$. If s and t belong to the same A_k, then $(s)_k \neq (t)_k$ so $[s] \cap [t] = \varnothing$. Otherwise, for some $k < l$ (say), $s \in A_k$ and $t \in A_l$, so $(t)_k = \beta(k) < (s)_k$ and again $[s] \cap [t] = \varnothing$. Hence (ii) is satisfied. For (iii), suppose $\alpha \in \mathsf{A} \sim \bigcup\{\mathsf{A} \cap [s] : s \in A\}$. Either $\alpha = \beta$ or else for some k, $\bar{\alpha}(k) = \bar{\beta}(k)$ but $\alpha(k) \neq \beta(k)$. If $\alpha(k) < \beta(k)$, then $\bar{\alpha}(k+1) \notin B$ by the definition of β. If $\beta(k) < \alpha(k)$, then again $\bar{\alpha}(k+1) \notin B$, as otherwise $\bar{\alpha}(k+1) \in A$, contrary to the assumption that α belongs to no $\mathsf{A} \cap [s]$ with $s \in A$. Hence

$$\alpha \in \bigcup\{\mathsf{A} \cap [s] : \mathsf{A} \cap [s] \text{ is countable}\} \cup \{\beta\}.$$

This set is countable so (iii) is established.

Finally, suppose that A were finite so that for some k, $A_l = \varnothing$ for all $l > k$. Then for all s such that $\lg(s) > k$, if $\mathsf{A} \cap [s]$ is uncountable, then $\beta \in [s]$. But

$$\mathsf{A} \cap [\bar{\beta}(k)] \subseteq \{\beta\} \cup \bigcup\{\mathsf{A} \cap [s] : \lg(s) > k \wedge \beta \notin [s]\}.$$

The right-hand side is countable, but $\bar{\beta}(k) \in B$ so $\mathsf{A} \cap [\bar{\beta}(k)]$ is uncountable, a contradiction. □

6.3 Lemma. *For any uncountable closed set $\mathsf{A} \subseteq {}^{\omega}\omega$, there exists a one–one total continuous functional θ such that $\mathrm{Im}\,\theta \subseteq \mathsf{A}$ and $\mathsf{A} \sim \mathrm{Im}\,\theta$ is countable.*

Proof. Let A be an uncountable closed set and set $\mathsf{A}_{\langle\,\rangle} = \mathsf{A}$. We shall define for each $u \in \mathrm{Sq}$ a set A_u such that for all u and v,

(i) A_u is uncountable;

(ii) $u \not\subseteq v \wedge v \not\subseteq u \rightarrow \mathsf{A}_u \cap \mathsf{A}_v = \emptyset$;

(iii) $\mathsf{A}_u \sim \bigcup\{\mathsf{A}_{u*\langle n\rangle} : n \in \omega\}$ is countable;

(iv) $u \subseteq v \rightarrow \mathsf{A}_v \subseteq \mathsf{A}_u$;

(v) for any β, $\bigcap\{\mathsf{A}_{\beta(k)} : k \in \omega\}$ contains a single element.

Suppose that A_u is defined and let $A_u \subseteq \mathrm{Sq}$ be as in the preceding lemma applied to the uncountable set A_u. If s_n denotes the n-th element of A_u enumerated in numerical order, set

$$\mathsf{A}_{u*\langle n\rangle} = \mathsf{A}_u \cap [s_n].$$

Properties (i)–(iii) now follow immediately from (i)–(iii) of the lemma, respectively, and (iv) is evident from the construction. (v) follows from the fact, easily proved by induction on the length of u, that A_u is closed and for all $s \in A_u$, $\lg(s) \geq \lg(u)$.

Now set $\theta(\beta) =$ the unique element of $\bigcap\{\mathsf{A}_{\bar\beta(k)} : k \in \omega\}$. For all β, $\theta(\beta) \in \mathsf{A} = \mathsf{A}_{\langle\,\rangle}$. That θ is one–one follows from (ii). If $\beta \in \theta^{-1}([s])$, then $\mathsf{A}_{\bar\beta(\lg(s))} \subseteq [s]$ so $[\bar\beta(\lg(s))] \subseteq \theta^{-1}([s])$ and thus θ is continuous. Finally, if $\alpha \in \mathsf{A} \sim \mathrm{Im}\,\theta$, then for some u, $\alpha \in \mathsf{A}_u \sim \bigcup\{\mathsf{A}_{u*\langle u\rangle} : n \in \omega\}$. Hence

$$\mathsf{A} \sim \mathrm{Im}\,\theta \subseteq \bigcup\{\mathsf{A}_u \sim \bigcup\{\mathsf{A}_{u*\langle n\rangle} : n \in \omega\} : u \in \mathrm{Sq}\}$$

which is countable by (iii). □

Note that the provision that $\mathrm{Im}\,\theta$ may differ from A by a countable set is essential as it is easy to see that $\mathrm{Im}\,\theta$ cannot contain isolated points (α is isolated in $\mathrm{Im}\,\theta$ iff for some n, α is the only member of $\mathrm{Im}\,\theta \cap [\bar\alpha(n)]$). Note that a set can have at most countably many isolated points.

6.4 Corollary. *Every non-empty closed set is the image of a total continuous functional.*

Proof. If A is countable, say $\mathsf{A} = \{\beta_n : n \in \omega\}$, we simply set $\chi(\alpha) = \beta_{\alpha(0)}$ and $\mathsf{A} = \mathrm{Im}\,\chi$. If A is uncountable, let θ be as in the preceding lemma a one–one continuous functional such that $\mathrm{Im}\,\theta \subseteq \mathsf{A}$ and $\mathsf{A} \sim \mathrm{Im}\,\theta$ is countable, say $\mathsf{A} \sim \mathrm{Im}\,\theta = \{\beta_n : n \in \omega\}$. Then if

$$\chi(\alpha) = \begin{cases} \theta(\lambda m\,.\,\alpha(m+1)), & \text{if } \alpha(0) = 0; \\ \beta_{\alpha(0)-1}, & \text{if } \alpha(0) > 0; \end{cases}$$

clearly χ is continuous and $\mathsf{A} = \mathrm{Im}\,\chi$. □

6.5 Theorem. *For any non-empty set* $A \subseteq {}^\omega\omega$, $A \in \Sigma^1_1$ *iff* A *is the image of a total continuous functional.*

Proof. First, if $A = \mathrm{Im}\,\theta$ for some total continuous θ, then

$$\alpha \in A \leftrightarrow \exists \beta\, \forall p\,.\, \theta(\beta)(p) = \alpha(p)$$

so it follows from Lemma 6.1 that $A \in \Sigma^1_1$. The converse is immediate from (1′) and Corollary 6.4. □

Consider now the following possible analogues for (3)–(5):

(3′) $A \in \Sigma^1_1 \overset{?}{\leftrightarrow} A$ is the image of a Borel set B under a continuous functional which is one–one on B;

(4′) $A \in \Sigma^1_1$ and A is uncountable $\overset{?}{\leftrightarrow}$ A is the image of a total one–one continuous functional;

(5′) $A \in \Sigma^1_1 \overset{?}{\leftrightarrow} A$ is the domain of a partial functional with Borel graph.

The implications $(\leftarrow)$ are trivially true in each case, but we shall show that all of the implications $(\rightarrow)$ are false. (This also follows from Exercise 2.27.) Of course, as we pointed out following 6.3, (4)′$(\rightarrow)$ fails for any A which has isolated points, but the following shows that it would be false even if we allowed for the exclusion of such points.

6.6 Theorem. *The image of any total one–one continuous functional is a Borel set.*

Proof. Let θ be a total one–one continuous functional. The graph of θ is Borel so there exists a Δ^1_1 relation R and a function γ such that for all α and β,

$$\theta(\alpha) = \beta \leftrightarrow R(\alpha, \beta, \gamma).$$

Then because θ is one–one,

$$\begin{aligned} \beta \in \mathrm{Im}\,\theta &\leftrightarrow \exists \alpha\, R(\alpha, \beta, \gamma) \\ &\leftrightarrow \exists!\, \alpha\, R(\alpha, \beta, \gamma) \\ &\leftrightarrow (\exists \alpha \in \Delta^1_1[\beta, \gamma])\, R(\alpha, \beta, \gamma). \end{aligned}$$

The first equivalence shows that $\mathrm{Im}\,\theta$ is Σ^1_1. The third shows that it is Π^1_1 (Theorem 2.9). □

The result of Lemma 6.3 can be extended to all Borel sets. First we give a new characterization of the class of Borel sets.

6.7 Lemma. *The class of Borel sets is the smallest class containing all closed sets and closed under countable intersection and countable disjoint union.*

Proof. Let X be the class described. Clearly $X \subseteq \mathrm{Bo}$ so it suffices to prove by induction that for all ρ, $\mathbf{\Sigma}^0_\rho \subseteq X$. For $\rho = 0$ this is true by hypothesis so we assume $\rho > 0$ and $\mathbf{\Sigma}^0_{(\rho)} \subseteq X$. As X is closed under countable intersection we have immediately that $\mathbf{\Pi}^0_\rho \subseteq X$. Let $\mathsf{R} = \bigcup\{\mathsf{P}_p : p \in \omega\}$ be any element of $\mathbf{\Sigma}^0_\rho$ with all $\mathsf{P}_p \in \mathbf{\Pi}^0_{(\rho)}$. For each p, let

$$\mathsf{Q}_p = \mathsf{P}_p \sim \bigcup\{\mathsf{P}_q : q < p\}.$$

By Lemmas 3.5 and 3.6, each $\mathsf{Q}_p \in \mathbf{\Delta}^0_\rho \subseteq \mathbf{\Pi}^0_\rho \subseteq X$. The Q_p are pairwise disjoint and $\mathsf{R} = \bigcup\{\mathsf{Q}_p : p \in \omega\}$ so $\mathsf{R} \in X$. □

6.8 Lemma. *Every Borel set is the image of a closed set* C *under a continuous functional which is one–one on* C.

Proof. Let X be the class of sets which are images of closed sets as described. Clearly all closed sets belong to X so it will suffice to show that X is closed under countable intersection and countable disjoint union.

Suppose that for all $p \in \omega$, $\mathsf{A}_p \in X$, C_p is closed, θ_p is a continuous functional which is one–one on C_p, and $\theta_p'' \mathsf{C}_p = \mathsf{A}_p$. If the sets A_p are pairwise disjoint and $\mathsf{B} = \bigcup\{\mathsf{A}_p : p \in \omega\}$, let $\mathsf{C} = \{\alpha : \lambda m \,.\, \alpha(m+1) \in \mathsf{C}_{\alpha(0)}\}$ and $\theta(\alpha) = \theta_{\alpha(0)}(\lambda m \,.\, \alpha(m+1))$. Then C is closed, θ is continuous and one–one on C, and $\theta''\mathsf{C} = \mathsf{B}$.

Now let $\mathsf{B} = \bigcap \{\mathsf{A}_p : p \in \omega\}$. Then we set

$$\mathsf{C} = \{\alpha : \forall p\, \forall q\, [(\alpha)^p \in \mathsf{C}_p \wedge \theta_p((\alpha)^p) = \theta_q((\alpha)^q)]$$
$$\wedge\, \forall t\, [t \notin \mathrm{Sq} \vee \lg(t) \neq 2 \rightarrow \alpha(t) = 0]\}$$

and $\theta(\alpha) = \theta_0((\alpha)^0)$ and claim that θ is continuous and one–one on C, C is closed, and $\theta''\mathsf{C} = \mathsf{B}$. That C is closed follows easily from Lemma 6.1. θ is clearly continuous. If $\beta \in \theta''\mathsf{C}$ then for some $\alpha \in \mathsf{C}$, $\beta = \theta_0((\alpha)^0) = \theta_p((\alpha)^p)$ for all p. Since $\theta_p((\alpha)^p) \in \theta_p''\mathsf{C}_p = \mathsf{A}_p$, $\beta \in \mathsf{B}$. On the other hand for any $\beta \in \mathsf{B}$ there exists for each p a function $\alpha_p \in \mathsf{C}_p$ such that $\theta_p(\alpha_p) = \beta$. But then there is an $\alpha \in \mathsf{C}$ such that $(\alpha)^p = \alpha_p$ for all p and thus $\beta \in \theta''\mathsf{C}$. Finally suppose for some $\alpha, \beta \in \mathsf{C}$ that $\theta(\alpha) = \theta(\beta)$. Then for all p, $\theta_p((\alpha)^p) = \theta_p((\beta)^p)$ which implies $(\alpha)^p = (\beta)^p$ since θ_p is one–one on C_p. Thus for all p and m, $\alpha(\langle p, m\rangle) = \beta(\langle p, m\rangle)$ and by the final condition on C, $\alpha = \beta$. Hence θ is one–one on C. □

We now have, in contrast with the proposed (3′) and (4′):

6.9 Theorem. *The following are equivalent for all* $\mathsf{A} \subseteq {}^\omega\omega$,

(i) A *is Borel*;

(ii) A *is the image of a Borel set* B *under a continuous functional which is one–one on* B;

(iii) A *is the image of a closed set* C *under a continuous functional which is one–one on* C;

(iv) A *is countable or for some one–one total continuous functional* θ, $\mathrm{Im}\,\theta \subseteq \mathsf{A}$ *and* $\mathsf{A} \sim \mathrm{Im}\,\theta$ *is countable.*

Proof. If A is Borel then it is the image of itself under the identity function, so (i) implies (ii). (iii) follows from (ii) by Lemma 6.8. Now suppose (iii) holds, say $\mathsf{A} = \chi''\mathsf{C}$, and by Lemma 6.3 let φ be a one–one total continuous functional such that $\mathrm{Im}\,\varphi \subseteq \mathsf{C}$ and $\mathsf{C} \sim \mathrm{Im}\,\varphi$ is countable. Then $\theta = \chi \circ \varphi$ clearly has the required properties. Finally, if A satisfies (iv), $\mathrm{Im}\,\theta$ is Borel by Theorem 6.6 and any countable set is Borel, so A is Borel. □

Finally, in contrast with the proposed (5′), we have

6.10 Corollary. *For any partial functional* θ, *if the graph of* θ *is a Borel relation, then the domain of* θ *is a Borel set.*

Proof. Let θ have Borel graph and set

$$\mathsf{B} = \{\langle \alpha, \beta \rangle : \theta(\alpha) \simeq \beta\} \quad \text{and} \quad \varphi(\gamma) = (\gamma)_0.$$

Then φ is continuous and one–one on the Borel set B and $\varphi''\mathsf{B}$ is the domain of θ. By the preceding theorem this is a Borel set. □

6.11–6.14 Exercises

6.11. Show that $\mathsf{A} \subseteq {}^{\omega}\omega$ is Σ^1_1 iff it is the image of some partial continuous functional with closed domain.

6.12. Prove that every Borel set is the domain of a functional with closed graph.

6.13. Show that if θ is continuous and 1–1 on A, then

(i) if $\mathsf{A} \in \Sigma^1_1$, then also $\theta''\mathsf{A} \in \Sigma^1_1$;

(ii) if $\mathsf{A} \notin \Delta^1_1$, then also $\theta''\mathsf{A} \notin \Delta^1_1$.

6.14. Establish the following analogue of Lemma 6.1: for any θ, if $\mathsf{F}(p, \alpha) \simeq \theta(\alpha)(p)$, then the following are equivalent:

(i) Gr_θ is Borel;

(ii) Gr_{F} is Borel;

(iii) for all s, $\theta^{-1}([s])$ is Borel.

What happens if "Borel" is replaced by 'Σ^1_1'? by 'Π^1_1'? (cf. Theorem 7.11 below).

6.15 Notes. The results of this section are all "classical" — that is, they appeared before the era of Recursion Theory. The proof of Theorem 6.6 is more modern, depending as it does on §2. There is, of course, a classical proof which goes in outline as follows: if θ is a total one–one continuous functional, use Corollary 1.12 to find for each $s \in \mathrm{Sq}$ a Borel set A_s such that

$$\theta''([s]) \subseteq \mathsf{A}_s \subseteq \sim \theta''(\sim [s])$$

and if $s \subseteq t$, then $\mathsf{A}_t \subseteq \mathsf{A}_s$. Then

$$\alpha \in \mathrm{Im}\,\theta \leftrightarrow \forall n\, \forall s\, [\lg(s) = n \wedge \alpha \in \mathsf{A}_s \cap \overline{\theta''([s])}].$$

For the flavor of this sort of Descriptive Set Theory as it was done in the good old days, peruse Ljapunov–Stschegolkov–Arsenin [1955].

7. *Uniformization*

The notion of a selection functional $\mathrm{Sel}_{\mathsf{R}}$ for a relation R played an important role in § II.4 and § 2 of this chapter. $\mathrm{Sel}_{\mathsf{R}}$ selects a number p such that $\mathsf{R}(p, \mathbf{m}, \boldsymbol{\alpha})$ holds whenever there is such a p. We consider here the analogous problem of selecting a β such that $\mathsf{R}(\mathbf{m}, \boldsymbol{\alpha}, \beta)$. The letters θ, φ, χ, and ψ denote functions from ${}^{k,l}\omega$ into ${}^{\omega}\omega$ in this section.

7.1 Definition. For any two classes X and Y of relations, X is *Y-uniformizable* (or *Y uniformizes X*) iff for every $\mathsf{R} \in X$, $\mathsf{R} \subseteq {}^{k,l+1}\omega$, there exists a partial functional $\theta : {}^{k,l}\omega \to {}^{\omega}\omega$ with $\mathrm{Gr}_\theta \in Y$ and such that for all $(\mathbf{m}, \boldsymbol{\alpha})$,

$$\exists \beta\, \mathsf{R}(\mathbf{m}, \boldsymbol{\alpha}, \beta) \leftrightarrow \mathsf{R}(\mathbf{m}, \boldsymbol{\alpha}, \theta(\mathbf{m}, \boldsymbol{\alpha})) \leftrightarrow \theta(\mathbf{m}, \boldsymbol{\alpha}) \downarrow .$$

The functional θ is said to *uniformize* R. X has the *uniformization property* iff X is X-uniformizable.

The main result of this section is that Π^1_1, $\Pi^1_1[\beta]$, and $\boldsymbol{\Pi}^1_1$ all have the uniformization property. As background we examine uniformization for simpler classes.

7.2 Theorem. *Σ^0_1 is Δ^0_2-uniformizable.*

Proof. Suppose $\mathsf{R} \in \Sigma^0_1$, say $\mathsf{R}(\mathbf{m}, \boldsymbol{\alpha}, \beta) \leftrightarrow \exists p\, \mathsf{S}(\bar{\beta}(p), \mathbf{m}, \boldsymbol{\alpha})$ with S recursive. Then the functional θ defined by:

$$\theta(\mathbf{m}, \boldsymbol{\alpha}) \simeq \beta \leftrightarrow \exists s\, \mathsf{S}(s, \mathbf{m}, \boldsymbol{\alpha}) \wedge \forall s\, [\mathsf{S}(s, \mathbf{m}, \boldsymbol{\alpha}) \wedge (\forall t < s) \sim \mathsf{S}(t, \mathbf{m}, \boldsymbol{\alpha})$$
$$\rightarrow \bar{\beta}(\lg(s)) = s \wedge (\forall m \geq \lg(s))\beta(m) = 0]$$

has Δ_2^0 graph and uniformizes R. □

Of course Σ_1^0 cannot have the uniformization property since no non-empty functional θ has open graph. The same argument shows that $\Sigma_1^0[\beta]$ is $\Delta_2^0[\beta]$-uniformizable and hence $\boldsymbol{\Sigma}_1^0$ is $\boldsymbol{\Delta}_2^0$-uniformizable.

The following simple observation allows us to apply the results of § III.4.

7.3 Lemma. *For any X and any $r > 0$, if X is Σ_r^1-uniformizable, then Δ_r^1 is a basis for X.*

Proof. Suppose X is Σ_r^1-uniformizable and A is a non-empty set in X. Let $\mathsf{R}(p, \beta) \leftrightarrow \beta \in \mathsf{A}$ and θ uniformize R with $\mathrm{Gr}_\theta \in \Sigma_r^1$. Then if γ is defined by

$$\gamma(m) = n \leftrightarrow \exists \beta\, [\theta(0) = \beta \wedge \beta(m) = n]$$
$$\leftrightarrow \forall \beta\, [\theta(0) = \beta \rightarrow \beta(m) = n],$$

$\theta(0) = \gamma \in \Delta_r^1 \cap \mathsf{A}$. □

7.4 Theorem. *Π_1^0 is not Σ_1^1-uniformizable.*

Proof. By Theorem III.4.8, Δ_1^1 is not a basis for Π_1^0. □

Again this lemma and theorem can be easily extended to show that $\Pi_1^0[\beta]$ is not $\Sigma_1^1[\beta]$-uniformizable. However $\boldsymbol{\Delta}_1^1$ is trivially a basis for $\boldsymbol{\Pi}_1^0$ (since ${}^\omega\omega \subseteq \boldsymbol{\Delta}_1^0$), so we use a different approach to prove

7.5 Theorem. *$\boldsymbol{\Pi}_1^0$ is not $\boldsymbol{\Delta}_1^1$-uniformizable.*

Proof. Let A be any set in $\boldsymbol{\Sigma}_1^1 \sim \boldsymbol{\Delta}_1^1$ and R a $\boldsymbol{\Pi}_1^0$ relation such that $\alpha \in \mathsf{A} \leftrightarrow \exists \beta\, \mathsf{R}(\alpha, \beta)$. Suppose R were uniformizable by a $\boldsymbol{\Delta}_1^1$ functional θ. Then $\mathsf{A} = \mathrm{Dm}\, \theta$ is Borel by Corollary 6.10, a contradiction. □

7.6 Lemma. *For any functional $\theta \in \Sigma_1^1$ ($\boldsymbol{\Sigma}_1^1$), there exists a functional $\psi \in \Delta_1^1$ ($\boldsymbol{\Delta}_1^1$) such that $\theta \subseteq \psi$.*

Proof. Let θ be any Σ_1^1 functional and set

$$\mathsf{R}^*(\mathbf{m}, \boldsymbol{\alpha}, \beta) \leftrightarrow \exists \gamma\, \exists p\, [\theta(\mathbf{m}, \boldsymbol{\alpha}) \simeq \gamma \wedge \bar{\gamma}(p) = \bar{\beta}(p) \wedge \gamma(p) < \beta(p)];$$
$$\mathsf{R}^{**} = \mathsf{R}^* \cup \mathrm{Gr}(\theta);$$

$$\mathsf{S}^{**}(\mathbf{m}, \alpha, \beta) \leftrightarrow \exists \gamma\, \exists p\, [\theta(\mathbf{m}, \alpha) \simeq \gamma \wedge \bar{\gamma}(p) = \bar{\beta}(p) \wedge \beta(p) < \gamma(p)];$$

$$\mathsf{S}^* = \mathsf{S}^{**} \cup \mathrm{Gr}(\theta).$$

All of these relations are clearly Σ^1_1 and they satisfy

$$\mathsf{R}^* \cap \mathsf{S}^* = \emptyset = \mathsf{R}^{**} \cap \mathsf{S}^{**}.$$

By Theorem 1.5 there exist relations P^* and $\mathsf{P}^{**} \in \Delta^1_1$ such that

$$\mathsf{R}^* \subseteq \mathsf{P}^* \subseteq \sim\mathsf{S}^* \quad \text{and} \quad \mathsf{R}^{**} \subseteq \mathsf{P}^{**} \subseteq \sim\mathsf{S}^{**}.$$

Let $\mathsf{P} = \mathsf{P}^{**} \sim \mathsf{P}^*$. Clearly $\mathrm{Gr}(\theta) \subseteq \mathsf{R}^{**} \cap \mathsf{S}^* \subseteq \mathsf{P}$. P is not in general the graph of a functional, so let

$$\mathsf{U}(\mathbf{m}, \alpha, \beta) \leftrightarrow P(\mathbf{m}, \alpha, \beta) \wedge (\exists \gamma \neq \beta) \mathsf{P}(\mathbf{m}, \alpha, \gamma).$$

$\mathsf{U} \in \Sigma^1_1$ and $\mathrm{Gr}(\theta) \cap \mathsf{U} = \emptyset$, so we again apply Theorem 1.5 to obtain a relation $\mathsf{Q} \in \Delta^1_1$ such that $\mathrm{Gr}(\theta) \subseteq \mathsf{Q} \sim \mathsf{U}$. Then $\mathsf{P} \cap \mathsf{Q}$ is the graph of a functional ψ as required. □

7.7 Corollary. *$\mathbf{\Pi}^0_1$ is not $\mathbf{\Sigma}^1_1$-uniformizable.*

Proof. By Theorem 7.5 choose R to be a $\mathbf{\Pi}^0_1$ relation which is not $\mathbf{\Delta}^1_1$-uniformizable and suppose $\theta \in \mathbf{\Sigma}^1_1$ uniformizes R. Then if ψ is any $\mathbf{\Delta}^1_1$ extension of θ, $\mathsf{R} \cap \mathrm{Gr}(\psi)$ is a $\mathbf{\Delta}^1_1$ function which uniformizes R, contrary to assumption. □

Thus the following is the best possible result.

7.8 Uniformization Theorem. *Π^1_1 and $\mathbf{\Pi}^1_1$ have the uniformization property.*

Proof. To simplify notation suppose that $\mathsf{R} \subseteq {}^{0,2}\omega$ is a Π^1_1 relation, say $\mathsf{R}(\alpha, \beta) \leftrightarrow \forall \gamma\, \exists n\, \mathsf{P}(\bar{\gamma}(n), \bar{\beta}(n), \alpha)$. Let P' be as in the proof of Theorem 1.1,

$$s \trianglelefteq^n_{\alpha,\beta} t \leftrightarrow s \trianglelefteq t \wedge t \trianglelefteq n \wedge \sim \mathsf{P}'(s, \bar{\beta}(\mathrm{lg}(s)), \alpha) \wedge \sim \mathsf{P}'(t, \bar{\beta}(\mathrm{lg}(t)), \alpha),$$

and

$$|n, \alpha, \beta| = \text{order-type of } \trianglelefteq^n_{\alpha,\beta}.$$

We define recursively relations R_n as follows: $\mathsf{R}_0 = \mathsf{R}$;

$$u_{\alpha, n} = \min\{\bar{\beta}(n) : \mathsf{R}_n(\alpha, \beta)\};$$

$$\sigma_{\alpha,n} = \min\{|n, \alpha, \beta| : \mathsf{R}_n(\alpha, \beta) \wedge \bar{\beta}(n) = u_{\alpha,n}\};$$

$$\mathsf{R}_{n+1}(\alpha, \beta) \leftrightarrow \mathsf{R}_n(\alpha, \beta) \wedge \bar{\beta}(n) = u_{\alpha,n} \wedge |n, \alpha, \beta| = \sigma_{\alpha,n}.$$

Let $\mathsf{Q}(\alpha, \beta) \leftrightarrow \forall n\, \mathsf{R}_n(\alpha, \beta)$. We shall show that $\mathsf{Q} \in \Pi^1_1$ and Q is the graph of a functional θ which uniformizes R. First note that if $\sim\exists\beta\, \mathsf{R}(\alpha, \beta)$, then also $\sim\exists\beta\, \mathsf{Q}(\alpha, \beta)$. Suppose that $\mathsf{R}(\alpha, \beta)$. Then for all n, $|n, \alpha, \beta|$ is an ordinal and there is a unique function γ_α such that for all n, $\bar{\gamma}_\alpha(n) = u_{\alpha,n}$. Clearly $\mathsf{Q}(\alpha, \beta) \rightarrow \beta = \gamma_\alpha$ so it suffices to prove that if $\exists\beta\, \mathsf{R}(\alpha, \beta)$, then $\mathsf{Q}(\alpha, \gamma_\alpha)$.

To this end we first establish the following technical lemmas:

(1) for any α such that $\exists\beta\, \mathsf{R}(\alpha, \beta)$, any p such that $\sim\mathsf{P}'(p, \bar{\gamma}_\alpha(\lg(p)), \alpha)$, and any q,

$$|p, \alpha, \gamma_\alpha| < |q, \alpha, \gamma_\alpha| \rightarrow \sigma_{\alpha,p} < \sigma_{\alpha,q};$$

(2) there exists a relation $\mathsf{S} \in \Pi^1_1$ such that for all n, α, and β,

$$\mathsf{R}_n(\alpha, \beta) \rightarrow [\mathsf{R}_{n+1}(\alpha, \beta) \leftrightarrow \mathsf{S}(n, \alpha, \beta)].$$

Suppose the hypotheses of (1) are satisfied, let $r = \max\{p, q\} + 1$, and let β be any function such that $\mathsf{R}_r(\alpha, \beta)$. Then $\bar{\beta}(\lg(p)) = \bar{\gamma}_\alpha(\lg(p))$ so also $\sim\mathsf{P}'(p, \bar{\beta}(\lg(p)), \alpha)$. Since $p \lhd q$, $|p, \alpha, \beta| < |q, \alpha, \beta|$. But since $\mathsf{R}_{p+1}(\alpha, \beta)$, $|p, \alpha, \beta| = \sigma_{\alpha,p}$ and as also $\mathsf{R}_{q+1}(\alpha, \beta)$, $|q, \alpha, \beta| = \sigma_{\alpha,q}$ and (1) is proved.

For (2) we first observe that for any n, α, and β such that $\mathsf{R}_n(\alpha, \beta)$,

$$\forall\gamma\, (\mathsf{R}_n(\alpha, \gamma) \leftrightarrow (\forall p < n)[\bar{\gamma}(p) \leqslant \bar{\beta}(p) \wedge |p, \alpha, \gamma| \leqslant |p, \alpha, \beta|])$$

and

$$\mathsf{R}_{n+1}(\alpha, \beta) \leftrightarrow \forall\gamma\, [\mathsf{R}_n(\alpha, \gamma) \rightarrow \bar{\beta}(n) \leqslant \bar{\gamma}(n) \wedge (\bar{\beta}(n) = \bar{\gamma}(n) \rightarrow |n, \alpha, \beta| \leqslant |n, \alpha, \gamma|)].$$

Using the relations $\leqslant_\Sigma$ and $\leqslant_\Pi$ of Theorem 1.4, we can define relations $\mathsf{U}_\Sigma \in \Sigma^1_1$ and $\mathsf{U}_\Pi \in \Pi^1_1$ such that if $|p, \alpha, \beta|$ and $|n, \alpha, \gamma|$ are ordinals, then

$$|p, \alpha, \gamma| \leqslant |p, \alpha, \beta| \leftrightarrow \mathsf{U}_\Sigma(p, \alpha, \beta, \gamma),$$

and

$$|n, \alpha, \beta| \leqslant |n, \alpha, \gamma| \leftrightarrow \mathsf{U}_\Pi(n, \alpha, \beta, \gamma).$$

Hence the requirements of (2) are met by

$$\mathsf{S}(n, \alpha, \beta) \leftrightarrow \forall\gamma\, [(\forall p < n)[\bar{\gamma}(p) \leqslant \bar{\beta}(p) \wedge \mathsf{U}_\Sigma(p, \alpha, \beta, \gamma)] \rightarrow \bar{\beta}(n) \leqslant \bar{\gamma}(n) \wedge (\bar{\beta}(n) = \bar{\gamma}(n) \rightarrow \mathsf{U}_\Pi(n, \alpha, \beta, \gamma))].$$

From (2) we have immediately that

$$\mathsf{Q}(\alpha,\beta)\leftrightarrow \mathsf{R}(\alpha,\beta)\wedge \forall n\, \mathsf{S}(n,\alpha,\beta)$$

and thus $\mathsf{Q}\in\Pi^1_1$. Suppose now that $\exists\beta\, \mathsf{R}(\alpha,\beta)$; we shall show $\mathsf{Q}(\alpha,\gamma_\alpha)$ and for this we first prove by induction on the ordinal $\sigma_{\alpha,n}$ that $|n,\alpha,\gamma_\alpha|\leqslant\sigma_{\alpha,n}$. Suppose this holds for all p such that $\sigma_{\alpha,p}<\sigma_{\alpha,n}$. Then by (1) and the induction hypothesis, if $|p,\alpha,\gamma_\alpha|<|n,\alpha,\gamma_\alpha|$, then $\sigma_{\alpha,p}<\sigma_{\alpha,n}$. Hence $|n,\alpha,\gamma_\alpha|\leqslant\sigma_{\alpha,n}$. Now $\mathsf{R}(\alpha,\gamma_\alpha)$ holds because $|\langle\ \rangle,\alpha,\gamma_\alpha|\leqslant\sigma_{\alpha,\langle\ \rangle}$ so $|\langle\ \rangle,\alpha,\gamma_\alpha|$ is an ordinal. Similarly, from $\mathsf{R}_n(\alpha,\gamma_\alpha)$ we conclude immediately that $\mathsf{R}_{n+1}(\alpha,\gamma_\alpha)$. Thus $\mathsf{Q}(\alpha,\gamma_\alpha)$. □

7.9 Basis Theorem. *Δ^1_2 is a basis for Σ^1_2.*

Proof. Immediate from the preceding theorem, Lemma 7.3, and Lemma III.4.7. □

We have given this result the important title, the Basis Theorem, because it will see extensive application in the latter four chapters of this book. Here we give two applications to questions arising in earlier sections. The first concerns closure ordinals for inductive operators and should be compared with Theorems 2.15 and 2.16. Here δ^1_2 is the least ordinal *not* the order-type of a Δ^1_2 wellordering of ω.

7.10 Theorem. *For any inductive operator Γ,*
(i) $\Gamma\in\Delta^1_2\rightarrow|\Gamma|<\delta^1_2$;
(ii) *Γ monotone* $\wedge\ \Gamma\in\Sigma^1_2\rightarrow|\Gamma|\leqslant\delta^1_2$.

Proof. For (i), suppose $\Gamma\in\Delta^1_2$. Using the relations defined in Theorem III.3.9 we have for all $\gamma\in\mathsf{W}$

$$|\Gamma|\leqslant\|\gamma\|\leftrightarrow\forall m\,[\mathsf{V}_\Pi(m,\gamma)\rightarrow\mathsf{V}^{()}_\Sigma(m,\gamma)].$$

Thus $\{\gamma:\gamma\in\mathsf{W}\wedge|\Gamma|\leqslant\|\gamma\|\}$ is a non-empty Σ^1_2 set and by the Basis Theorem contains a Δ^1_2 element γ_0. Thus $|\Gamma|\leqslant\|\gamma_0\|<\delta^1_2$.

For (ii), suppose Γ is monotone Σ^1_2 and $m\in\bar\Gamma$. Then using Theorem III.3.13,

$$\{\gamma:\gamma\in\mathsf{W}\wedge m\in\Gamma^{\|\gamma\|}\}=\{\gamma:\gamma\in\mathsf{W}\wedge\mathsf{V}(m,\gamma)\}$$

is a non-empty Σ^1_2 set which therefore has a Δ^1_2 element γ_0. Thus $m\in\Gamma^{\|\gamma_0\|}\subseteq\Gamma^{(\delta^1_2)}$, so $\bar\Gamma\subseteq\Gamma^{(\delta^1_2)}$. □

The second application is to continuous images. By the same resoning used to derive (1′) at the beginning of § 6, we have for any $\mathsf{A}\subseteq{}^\omega\omega$,

$$\mathsf{A} \in \boldsymbol{\Sigma}^1_2 \leftrightarrow \mathsf{A} \text{ is the image of a } \boldsymbol{\Pi}^1_1 \text{ set under a continuous functional from } {}^\omega\omega \text{ into } {}^\omega\omega.$$

In contrast with Theorem 6.9 ((i)$\leftrightarrow$(ii)), this equivalence holds also if the functional is required to be one–one on the $\boldsymbol{\Pi}^1_1$ set (cf. also Corollary 6.10).

7.11 Theorem. *For all* $\mathsf{A} \subseteq {}^\omega\omega$, A *is* $\boldsymbol{\Sigma}^1_2$ *iff* A *is the domain of a functional* $\theta : {}^\omega\omega \to {}^\omega\omega$ *which has* $\boldsymbol{\Pi}^1_1$ *graph.*

Proof. The implication ($\leftarrow$) is obvious. Conversely, if A is $\{\alpha : \exists\beta\, \mathsf{R}(\alpha, \beta)\}$ with $\mathsf{R} \in \boldsymbol{\Pi}^1_1$, let θ be a functional with $\boldsymbol{\Pi}^1_1$ graph which uniformizes R. Then $\mathsf{A} = \mathrm{Dm}\,\theta$. $\square$

7.12 Corollary. *For all* $\mathsf{A} \subseteq {}^\omega\omega$, A *is* $\boldsymbol{\Sigma}^1_2$ *iff* A *is the image of a* $\boldsymbol{\Pi}^1_1$ *set* B *under a continuous functional which is one–one on* B.

Proof. If $\mathsf{A} = \mathrm{Dm}\,\theta$ with $\theta \in \boldsymbol{\Pi}^1_1$, it suffices to take $\mathsf{B} = \{\langle\alpha, \beta\rangle : \theta(\alpha) \simeq \beta\}$ and $\varphi(\gamma) = (\gamma)_0$. $\square$

The relativized version of the Basis Theorem also gives a trivial generalization of the Spector-Gandy Theorem IV.2.9: a relation R is Σ^1_2 iff for some relation $\mathsf{P} \in \Pi^1_1$,

$$\mathsf{R}(\mathbf{m}, \boldsymbol{\alpha}) \leftrightarrow (\exists\beta \in \Delta^1_2[\boldsymbol{\alpha}])\, \mathsf{P}(\mathbf{m}, \boldsymbol{\alpha}, \beta).$$

Finally we have the following easy extension:

7.13 Theorem. Σ^1_2 *and* $\boldsymbol{\Sigma}^1_2$ *have the uniformization property.*

Proof. Let $\mathsf{R}(\mathbf{m}, \boldsymbol{\alpha}, \beta) \leftrightarrow \exists\gamma\, \mathsf{P}(\mathbf{m}, \boldsymbol{\alpha}, \beta, \gamma)$ be a Σ^1_2 relation with $\mathsf{P} \in \Pi^1_1$. Let

$$\mathsf{Q}(\mathbf{m}, \boldsymbol{\alpha}, \delta) \leftrightarrow \mathsf{P}(\mathbf{m}, \boldsymbol{\alpha}, (\delta)_0, (\delta)_1)$$

and by the Uniformization Theorem let φ be a functional with Π^1_1 graph which uniformizes Q. Then if

$$\theta(\mathbf{m}, \boldsymbol{\alpha}) \simeq \beta \leftrightarrow \exists\gamma\, [\varphi(\mathbf{m}, \boldsymbol{\alpha}) \simeq \langle\beta, \gamma\rangle],$$

θ has Σ^1_2 graph and uniformizes R. The proof for $\boldsymbol{\Sigma}^1_2$ is the same. $\square$

7.14–7.20 Exercises

7.14. Give an alternative proof for Corollary 7.7 by showing

(i) if $\mathbf{\Pi}_1^0$ were $\mathbf{\Sigma}_1^1$-uniformizable, then $\mathbf{\Sigma}_1^1$ would have the uniformization property;

(ii) if $\mathbf{\Sigma}_1^1$ had the uniformization property, then $\mathbf{\Sigma}_1^1$ would have the reduction property.

7.15. Give a simple proof that Π_1^1 uniformizes Π_1^0.

7.16. Find a Σ_1^1 subset of ${}^\omega\omega$ which has no non-empty Π_1^1 subsets.

7.17. Use the Basis Theorem to give a two-line proof of a weaker version of Exercise 5.16: every countable Σ_1^1 set of functions contains only Δ_2^1 elements.

7.18. Show that for any $\mathsf{A} \subseteq {}^\omega\omega$, if A is a model of Δ_2^1-Comprehension, then A is also a model of $\exists_2^1$-Choice. (Show that the Uniformization Theorem holds in any model of Δ_2^1-Comprehension.)

7.19. Show that Δ_2^1 is a model of the Δ_2^1-Comprehension schema (hence also of $\exists_2^1$-Choice by Exercise 7.18).

7.20. Show that Δ_2^1 has the uniformization property.

7.21 Notes. The Uniformization Theorem for $\mathbf{\Pi}_1^1$ is due to Kondô [1938]. The lightface version was announced by Addison in 1959 but never published. It has since appeared in (at least) Rogers [1967] and Shoenfield [1967].

Exercise III.3.34 suggests an alternate proof for Theorem 7.10(i).

Chapter V
Δ_2^1 and Beyond

Most of the analysis of the first level of the analytical hierarchy in Chapter IV rests on the representation of Π_1^1 sets in terms of well-orderings (Theorem IV.1.1), and for many years after these results were known there seemed to be no hope of extending any of the methods or results to higher levels. Since W is a Π_1^1 set it cannot be used directly to represent all Σ_2^1 or Π_2^1 relations, and no analogue of W at higher levels was apparent.

In § 1 we formulate the abstract pre-wellordering property and show that much of the structure of Π_1^1 and Δ_1^1 relations is due solely to the fact that Π_1^1 has this property. Furthermore, it is easily seen that Σ_2^1 also has the pre-wellordering property and this leads to the conclusion that a strong analogy exists between Π_1^1 and Σ_2^1. This correspondence will be reinforced in § VIII.3 where we discuss two generalizations of recursion theory for which Π_1^1 and Σ_2^1 are exactly the classes of "semi-recursive" relations.

The pre-wellordering property cannot be proved for other classes in the analytical hierarchy without further set-theoretical hypotheses beyond ZFC. In §§ 2 and 3 we discuss two such hypotheses — the hypothesis of constructibility (V = L) and the hypothesis of projective determinacy (PD). The principal results are (1) if V = L, then Σ_r^1 has the pre-wellordering property for all $r \geqslant 2$, whereas (2) if PD, then the classes which have the pre-wellordering property are $\Pi_1^1, \Sigma_2^1, \Pi_3^1, \Sigma_4^1, \Pi_5^1, \ldots$. These hypotheses also imply analogues of many of the results of §§ IV.5–7 for higher levels of the analytical and projective hierarchies.

We turn then to extensions of the results of §§ IV.3–4, which might be termed the study of Δ_1^1 and $\mathbf{\Delta}_1^1$ "from below". Here the results are mainly negative: no analogue of the Borel hierarchy suffices to exhaust any of the classes $\mathbf{\Delta}_r^1$ for $r \geqslant 2$, and similarly for the effective hierarchies and Δ_r^1. On the other hand, the classes of sets which comprise these analogues are themselves somewhat similar in structure to the class of (effective) Borel sets. The classical (boldface) versions lead to significant extensions of the reuslts of § IV.5, while the effective versions will be seen in §§ VI.5–6 to be closely connected with certain generalized recursion theories. Finally in § 6 we consider some facts peculiar to Δ_2^1 which lead to a hierarchy for the Δ_2^1 relations on numbers.

1. The Pre-Wellordering Property

We recall from I.1.6 that a pre-wellordering is a well-founded, transitive, reflexive, and connected relation — from being a well-ordering it lacks only antisymmetry. With any pre-wellordering $\leqslant$ is associated a norm $|\ \ |$, a function from the field of $\leqslant$ onto an ordinal such that

$$x \leqslant y \leftrightarrow |x| \leqslant |y|.$$

The image of $|\ \ |$ is called the (pre-wellorder-) type of $\leqslant$. Conversely, any function $|\ \ |$ from a set into the ordinals determines a pre-wellordering on this set by this equivalence. For example, the function $\|\ \ \|$ defined on ${}^{\omega}\omega$ by:

$$\|\gamma\| = \begin{cases} \text{the order-type of } \leqslant_\gamma, & \text{if } \gamma \in \mathsf{W}; \\ \aleph_1, & \text{otherwise}; \end{cases}$$

determines a pre-wellordering of type $\aleph_1 + 1$.

In what follows, we denote by X any one of the classes Σ_r^1 or Π_r^1 $(r \geqslant 1)$ and by V the corresponding universal relation U_r^1 or $\sim\mathsf{U}_r^1$. We sometimes think of V as the *set*, $\{\alpha : \mathsf{V}(\alpha(0), \alpha(1), \lambda p\,.\,\alpha(p+2))\}$. To improve readability, we shall write x for $(a, \langle\mathbf{m}\rangle, \langle\boldsymbol{\alpha}\rangle)$ and y for $(b, \langle\mathbf{n}\rangle, \langle\boldsymbol{\beta}\rangle)$. Then also (x, γ) denotes $(a, \langle\mathbf{m}\rangle, \langle\boldsymbol{\alpha}, \gamma\rangle)$, etc., so for example,

$$\mathsf{U}_{r+1}^1(x) \leftrightarrow \exists\gamma \sim \mathsf{U}_r^1(x, \gamma).$$

1.1 Definition. X $(= \Sigma_r^1$ or $\Pi_r^1)$ has the *pre-wellordering property* iff there exist relations $\leqslant$, $\leqslant_\Sigma$, and $\leqslant_\Pi$ such that

(i) $\leqslant$ is a pre-wellordering with field ${}^{2,1}\omega$ such that for all x and y,
 (a) $\sim\mathsf{V}(y) \rightarrow x \leqslant y$, and
 (b) $\mathsf{V}(y) \wedge x \leqslant y \rightarrow \mathsf{V}(x)$;
(ii) $\leqslant_\Sigma$ is Σ_r^1 and $\leqslant_\Pi$ is Π_r^1;
(iii) for any x and y such that either $\mathsf{V}(x)$ or $\mathsf{V}(y)$,

$$(x \leqslant_\Sigma y) \leftrightarrow (x \leqslant y) \leftrightarrow (x \leqslant_\Pi y).$$

(Cf. Exercise 1.20 for other characterizations).

Conditions (i) (a) and (b) mean that all $y \notin \mathsf{V}$ are $\leqslant$-equivalent and strictly follow all $x \in \mathsf{V}$ in the pre-wellordering. Hence the pre-wellorder type of $\leqslant$ is a successor ordinal. If X has the pre-wellordering property with notation as in the definition, then we set

$$x < y \leftrightarrow \neg(y \leqslant x), \qquad x <_\Sigma y \leftrightarrow \neg(y \leqslant_\Pi x),$$
$$\text{and} \quad x <_\Pi y \leftrightarrow \neg(y \leqslant_\Sigma x).$$

Then it is clear that $<_\Sigma$ is Σ^1_r, $<_\Pi$ is Π^1_r, and for any x and y such that either $\mathsf{V}(x)$ or $\mathsf{V}(y)$,

$$(x <_\Sigma y) \leftrightarrow (x < y) \leftrightarrow (x <_\Pi y).$$

Note also that if $\sim\mathsf{V}(y)$, then $[x < y \leftrightarrow \mathsf{V}(x)]$.

In the following, we shall write cX to denote $\{\mathsf{R}: \sim\mathsf{R} \in X\}$ and $\mathbf{X}$ to denote $\mathbf{\Sigma}^1_r$ when $X = \Sigma^1_r$, etc. In situations where X may refer to either Σ^1_r or Π^1_r we shall sometimes write $\leqslant_X$ to refer to $\leqslant_\Sigma$ in case X is Σ^1_r and to $\leqslant_\Pi$ in case X is Π^1_r. Similarly, $\leqslant_{cX}$ refers to $\leqslant_\Pi$ in case X is Σ^1_r and to $\leqslant_\Sigma$ in case X is Π^1_r.

To avoid confusion, we now write the relations of Definition IV.1.3 as $\leqslant^{\mathsf{W}}_\Sigma$, $<^{\mathsf{W}}_\Sigma$, etc.

1.2 Theorem. Π^1_1 *has the pre-wellordering property.*

Proof. Since $\sim\mathsf{U}^1_1$ is a Π^1_1 relation, there exists by Theorem IV.1.1 a recursive functional F such that for all x, $\sim\mathsf{U}^1_1(x) \leftrightarrow \mathsf{F}[x] \in \mathsf{W}$. Thus if we define

$$x \leqslant^1 y \leftrightarrow \|\mathsf{F}[x]\| \leqslant \|\mathsf{F}[y]\|;$$
$$x \leqslant^1_\Sigma y \leftrightarrow \mathsf{F}[x] \leqslant^{\mathsf{W}}_\Sigma \mathsf{F}[y];$$
$$x \leqslant^1_\Pi y \leftrightarrow \mathsf{F}[x] \leqslant^{\mathsf{W}}_\Pi \mathsf{F}[y];$$

it follows easily from Theorem IV.1.4 that the relations $\leqslant^1$, $\leqslant^1_\Sigma$, and $\leqslant^1_\Pi$ satisfy the conditions of Definition 1.1. □

1.3 Theorem. *For any* $r \geqslant 1$, *if* Π^1_r *has the pre-wellordering property, then* Σ^1_{r+1} *also has the pre-wellordering property.*

Proof. Let $\leqslant^r$, $\leqslant^1_\Sigma$, and $\leqslant^r_\Pi$ be relations which establish the pre-wellordering property for Π^1_r, $|\ \ |^r$ the norm associated with $\leqslant^r$, and $\boldsymbol{\kappa}^r + 1$ the pre-wellorder type of $\leqslant^r$. Then for any x,

$$\mathsf{U}^1_{r+1}(x) \leftrightarrow \exists\gamma \sim \mathsf{U}^1_r(x, \gamma) \leftrightarrow \exists\gamma\, (|(x, \gamma)|^r < \boldsymbol{\kappa}^r).$$

Let

$$|x|^{r+1} = \inf\{|(x, \gamma)|^r : \gamma \in {}^\omega\omega\}.$$

Then

$$(*) \qquad \mathsf{U}^1_{r+1}(x) \leftrightarrow |x|^{r+1} < \kappa^r.$$

We take $\leqslant^{r+1}$ to be the pre-wellordering determined by $|\ |^{r+1}$:

$$x \leqslant^{r+1} y \leftrightarrow |x|^{r+1} \leqslant |y|^{r+1},$$

and set

$$x \leqslant^{r+1}_{\Sigma} y \leftrightarrow \exists\gamma\, \forall\delta\, [(x, \gamma) \leqslant^r_{\Pi} (y, \delta)],$$

and

$$x \leqslant^{r+1}_{\Pi} y \leftrightarrow \forall\delta\, \exists\gamma\, [(x, \gamma) \leqslant^r_{\Sigma} (y, \delta)].$$

The provisions of clause (ii) of Definition 1.1 are clearly satisfied and those of (i) follow from (*). Towards (iii), we first observe that directly from the definition we have

$$(1) \qquad x \leqslant^{r+1} y \leftrightarrow \exists\gamma\, \forall\delta\, [(x, \gamma) \leqslant^r (y, \delta)].$$

From this, elementary logic, and the fact that $\leqslant^r$ is well founded, we conclude also

$$(2) \qquad x \leqslant^{r+1} y \leftrightarrow \forall\delta\, \exists\gamma\, [(x, \gamma) \leqslant^r (y, \delta)].$$

We claim that if either $\mathsf{U}^1_{r+1}(x)$ or $\mathsf{U}^1_{r+1}(y)$, the following four equivalences hold:

$$(3) \qquad x \leqslant^{r+1} y \leftrightarrow \exists\gamma\, \forall\delta\, [\sim\mathsf{U}^1_r(x, \gamma) \wedge (x, \gamma) \leqslant^r (y, \delta)];$$

$$(4) \qquad x \leqslant^{r+1} y \leftrightarrow \forall\delta\, \exists\gamma\, [\sim\mathsf{U}^1_r(x, \gamma) \wedge (x, \gamma) \leqslant^r (y, \delta)];$$

$$(5) \qquad x \leqslant^{r+1}_{\Sigma} y \leftrightarrow \exists\gamma\, \forall\delta\, [\sim\mathsf{U}^1_r(x, \gamma) \wedge (x, \gamma) \leqslant^r_{\Pi} (y, \delta)];$$

$$(6) \qquad x \leqslant^{r+1}_{\Pi} y \leftrightarrow \forall\delta\, \exists\gamma\, [\sim\mathsf{U}^1_r(x, \gamma) \wedge (x, \gamma) \leqslant^r_{\Sigma} (y, \delta)].$$

All of the implications ($\leftarrow$) are immediate from (1), (2), and the definitions. For (3)($\rightarrow$) assume $x \leqslant^{r+1} y$ and suppose first $\mathsf{U}^1_{r+1}(x)$. If γ_0 is such that $|(x, \gamma_0)|^r$ is as small as possible, then $\sim\mathsf{U}^1_r(x, \gamma_0)$ and for all δ, $(x, \gamma_0) \leqslant^r (y, \delta)$. If, on the other hand, $\mathsf{U}^1_{r+1}(y)$, let δ_0 be such that $\sim\mathsf{U}^1_r(y, \delta_0)$. By (1), there is some γ_0 such that for all δ, $(x, \gamma_0) \leqslant^r (y, \delta)$. In particular, $(x, \gamma_0) \leqslant^r (y, \delta_0)$ so by (i)(b) applied to $\leqslant^r$ we have $\sim\mathsf{U}^1_r(x, \gamma_0)$.

The proofs of the remaining implications ($\rightarrow$) are similar and are left to the reader. Condition (iii) of Definition 1.1 for $\leqslant^{r+1}$, $\leqslant^{r+1}_{\Sigma}$, and $\leqslant^{r+1}_{\Pi}$ is immediate from (3)–(6) and (iii) for $\leqslant^r$, $\leqslant^r_{\Sigma}$, and $\leqslant^r_{\Pi}$. □

1.4 Corollary. Σ_2^1 *has the pre-wellordering property.*

Proof. Immediate from Theorems 1.2 and 1.3. □

In the remainder of this section we shall derive a number of results which apply to any class X $(=\Sigma_r^1$ or $\Pi_r^1)$ which has the pre-wellordering property. In the main, the proofs are translations of those of the corresponding results for Π_1^1 in §§ IV.1–2 into a more general setting. As corollaries we obtain many facts about Σ_2^1 and Δ_2^1, and in §§ 2 and 3, under additional set-theoretic assumptions, also about higher levels of the analytical hierarchy.

When X is assumed to have the pre-wellordering property we shall use the notation of Definition 1.1 and in addition write $|\;\;|$ for the norm associated with $\leqslant$ and set $\kappa=\sup^+\{|x|:\mathsf{V}(x)\}$. On occasion we shall write $|a,\langle\mathbf{m}\rangle,\langle\boldsymbol{\alpha}\rangle|$ instead of $|x|$ or $|(a,\langle\mathbf{m}\rangle,\langle\boldsymbol{\alpha}\rangle)|$. We also set

$$V=\{(a,\langle\mathbf{m}\rangle):\mathsf{V}(a,\langle\mathbf{m}\rangle,\langle\;\;\rangle)\}.$$

The pre-wellordering $\leqslant$ restricted to sequences of the form $(a,\langle\mathbf{m}\rangle,\langle\;\;\rangle)$ induces a pre-wellordering of $\omega\times\omega$ in an obvious way and we write $|\;\;|_0$ for the norm associated with this pre-wellordering. Thus

$$|a,\langle\mathbf{m}\rangle|_0\leqslant|b,\langle\mathbf{n}\rangle|_0\leftrightarrow|a,\langle\mathbf{m}\rangle,\langle\;\;\rangle|\leqslant|b,\langle\mathbf{n}\rangle,\langle\;\;\rangle|,$$

and

$$|a,\langle\mathbf{m}\rangle|_0=\sup{}^+\{|b,\langle\mathbf{n}\rangle|_0:(b,\langle\mathbf{n}\rangle,\langle\;\;\rangle)<(a,\langle\mathbf{m}\rangle,\langle\;\;\rangle)\}.$$

In particular, $\kappa=\sup^+\{|a,\langle\mathbf{m}\rangle|_0: V(a,\langle\mathbf{m}\rangle)\}$ is a countable ordinal. Note that $|a,\langle\mathbf{m}\rangle|_0\leqslant|a,\langle\mathbf{m}\rangle,\langle\;\;\rangle|$ and the inequality may hold (Exercise 1.24).

1.5 Theorem. *If* X $(=\Sigma_r^1$ *or* $\Pi_r^1)$ *has the pre-wellordering property, then*

(i) X *and* $\mathbf{X}$ *have the reduction property but not the separation property;*

(ii) cX *and* $c\mathbf{X}$ *have the separation property but not the reduction property.*

Proof. By Lemmas II.4.19 and II.4.21 it suffices to show that X and $\mathbf{X}$ have the reduction property. Let R and S be any two relations in X of the same rank. Since V is universal for X there exist indices a and b such that

$$\mathsf{R}(\mathbf{m},\boldsymbol{\alpha})\leftrightarrow\mathsf{V}(a,\langle\mathbf{m}\rangle,\langle\boldsymbol{\alpha}\rangle),\quad\text{and}\quad\mathsf{S}(\mathbf{m},\boldsymbol{\alpha})\leftrightarrow\mathsf{V}(b,\langle\mathbf{m}\rangle,\langle\boldsymbol{\alpha}\rangle).$$

We set

$$\mathsf{R}^*(\mathbf{m},\boldsymbol{\alpha})\leftrightarrow\mathsf{R}(\mathbf{m},\boldsymbol{\alpha})\wedge(a,\langle\mathbf{m}\rangle,\langle\boldsymbol{\alpha}\rangle)\leqslant_X(b,\langle\mathbf{m}\rangle,\langle\boldsymbol{\alpha}\rangle);$$

$$\mathsf{S}^*(\mathbf{m},\boldsymbol{\alpha})\leftrightarrow\mathsf{S}(\mathbf{m},\boldsymbol{\alpha})\wedge(b,\langle\mathbf{m}\rangle,\langle\boldsymbol{\alpha}\rangle)<_X(a,\langle\mathbf{m}\rangle,\langle\boldsymbol{\alpha}\rangle).$$

It is straightforward to verify that $(\mathsf{R}^*, \mathsf{S}^*)$ reduces (R, S).

For R and S belonging to $\mathbf{X}$, there exist a, b, β, and γ such that

$$\mathsf{R}(\mathbf{m}, \boldsymbol{\alpha}) \leftrightarrow \mathsf{V}(a, \langle\mathbf{m}\rangle, \langle\boldsymbol{\alpha}, \beta\rangle), \quad \text{and} \quad \mathsf{S}(\mathbf{m}, \boldsymbol{\alpha}) \leftrightarrow \mathsf{V}(b, \langle\mathbf{m}\rangle, \langle\boldsymbol{\alpha}, \gamma\rangle).$$

Then we take

$$\mathsf{R}^*(\mathbf{m}, \boldsymbol{\alpha}) \leftrightarrow \mathsf{R}(\mathbf{m}, \boldsymbol{\alpha}) \wedge (a, \langle\mathbf{m}\rangle, \langle\boldsymbol{\alpha}, \beta\rangle) \leqslant_X (b, \langle\mathbf{m}\rangle, \langle\boldsymbol{\alpha}, \gamma\rangle);$$

$$\mathsf{S}^*(\mathbf{m}, \boldsymbol{\alpha}) \leftrightarrow \mathsf{S}(\mathbf{m}, \boldsymbol{\alpha}) \wedge (b, \langle\mathbf{m}\rangle, \langle\boldsymbol{\alpha}, \gamma\rangle) <_X (a, \langle\mathbf{m}\rangle, \langle\boldsymbol{\alpha}, \beta\rangle). \quad \square$$

1.6 Boundedness Theorem. *If X ($= \Sigma_r^1$ or Π_r^1) has the pre-wellordering property, then*

(i) *for any $R \in cX$, if $R \subseteq V$, then* $\sup^+\{|a, \langle\mathbf{m}\rangle|_0 : R(a, \langle\mathbf{m}\rangle)\} < \kappa$;

(ii) *if X is Π_r^1, then for any $\mathsf{R} \in c\mathbf{X}$ ($= \boldsymbol{\Sigma}_r^1$), if $\mathsf{R} \subseteq \mathsf{V}$ ($= \sim\mathsf{U}_r^1$) then* $\sup^+\{|x| : \mathsf{R}(x)\} < \boldsymbol{\kappa}$.

Proof. Suppose that for some $R \in cX$, $R \subseteq V$, the conclusion of (i) is false. Then for any b and $\mathbf{n}$,

$$V(b, \langle\mathbf{n}\rangle) \leftrightarrow \exists a\, \exists \mathbf{m} [R(a, \langle\mathbf{m}\rangle) \wedge |b, \langle\mathbf{n}\rangle|_0 \leqslant |a, \langle\mathbf{m}\rangle|_0]$$

$$\leftrightarrow \exists a\, \exists \mathbf{m} [R(a, \langle\mathbf{m}\rangle) \wedge (b, \langle\mathbf{n}\rangle, \langle\ \rangle) \leqslant_{cX} (a, \langle\mathbf{m}\rangle, \langle\ \rangle)].$$

Since cX is closed under $\exists^0$, this implies $V \in cX$, which is false as V is universal for relations on numbers in X.

If $X = \Pi_r^1$ and for some $\mathsf{R} \in c\mathbf{X}$, $\mathsf{R} \subseteq \mathsf{V}$, the conclusion of (ii) is false, then for any b, $\mathbf{n}$, and $\boldsymbol{\beta}$

$$\mathsf{V}(b, \langle\mathbf{n}\rangle, \langle\boldsymbol{\beta}\rangle) \leftrightarrow \exists a\, \exists \mathbf{m} \exists \boldsymbol{\alpha} [\mathsf{R}(a, \langle\mathbf{m}\rangle, \langle\boldsymbol{\alpha}\rangle) \wedge |b, \langle\mathbf{n}\rangle, \langle\boldsymbol{\beta}\rangle| \leqslant |a, \langle\mathbf{m}\rangle, \langle\boldsymbol{\alpha}\rangle|]$$

$$\leftrightarrow \exists a\, \exists \mathbf{m} \exists \boldsymbol{\alpha} [\mathsf{R}(a, \langle\mathbf{m}\rangle, \langle\boldsymbol{\alpha}\rangle) \wedge (b, \langle\mathbf{n}\rangle, \langle\boldsymbol{\beta}\rangle) \leqslant_{\boldsymbol{\Sigma}} (a, \langle\mathbf{m}\rangle, \langle\boldsymbol{\alpha}\rangle)].$$

Since Σ_r^1 is closed under both $\exists^0$ and $\exists^1$, this implies $\mathsf{V} \in \Sigma_r^1$, a contradiction. $\square$

Of course, the proof of (ii) does not work for $X = \Sigma_r^1$ because $\boldsymbol{\Pi}_r^1$ is not closed under $\exists^1$ — in fact, the result is false for $\boldsymbol{\Sigma}_2^1$ (cf. Exercise 1.25).

For each ordinal ρ, set

$$\mathsf{V}_\rho = \{(a, \langle\mathbf{m}\rangle, \langle\boldsymbol{\alpha}\rangle) : |a, \langle\mathbf{m}\rangle, \langle\boldsymbol{\alpha}\rangle| < \rho\}, \quad \text{and}$$

$$V_\rho = \{(a, \langle\mathbf{m}\rangle) : |a, \langle\mathbf{m}\rangle|_0 < \rho\}.$$

1.7 Theorem (Hierarchy). *If X ($= \Sigma_r^1$ or Π_r^1) has the pre-wellordering property, then*

(i) *for all relations R on numbers,*

$$R \in \Delta^1_r \leftrightarrow R \lessdot V_\rho \quad \textit{for some} \quad \rho < \kappa;$$

(ii) *if* $X = \Pi^1_r$, *then for all relations* R,

$$\mathsf{R} \in \boldsymbol{\Delta}^1_r \leftrightarrow \mathsf{R} \blacktriangleleft \mathsf{V}_\rho \quad \textit{for some} \quad \rho < \boldsymbol{\kappa}.$$

Proof. If $\rho < \kappa$, then there exists a $\langle b, \mathbf{n}\rangle \in V$ such that $\rho = |b, \langle \mathbf{n}\rangle|_0$ (we are treating V here as the *set* $\{\langle a, \mathbf{m}\rangle : V(a, \langle \mathbf{m}\rangle)\}$). Then

$$\begin{aligned} \langle a, \mathbf{m}\rangle \in V_\rho &\leftrightarrow (a, \langle \mathbf{m}\rangle, \langle\ \rangle) <_\Sigma (b, \langle \mathbf{n}\rangle, \langle\ \rangle) \\ &\leftrightarrow (a, \langle \mathbf{m}\rangle, \langle\ \rangle) <_\Pi (b, \langle \mathbf{n}\rangle, \langle\ \rangle) \end{aligned}$$

which implies that $V_\rho \in \Delta^1_r$. It follows that if $R \lessdot V_\rho$, then also $R \in \Delta^1_r$. Conversely, if $R \in \Delta^1_r$, let a be such that for all $\mathbf{m}$, $R(\mathbf{m}) \leftrightarrow \langle a, \mathbf{m}\rangle \in V$. Then $A = \{\langle a, \mathbf{m}\rangle : R(\mathbf{m})\}$ belongs to cX, so by the Boundedness Theorem $A \subseteq V_\rho$ for some $\rho < \kappa$. Thus $R \lessdot V_\rho$.

For (ii), suppose $\mathsf{R} \in \boldsymbol{\Delta}^1_r$, say $\mathsf{R} \in \Delta^1_r[\beta]$. Then for some a, $\mathsf{R}(\mathbf{m}, \boldsymbol{\alpha}) \leftrightarrow V(a, \langle \mathbf{m}\rangle, \langle \boldsymbol{\alpha}, \beta\rangle)$. Then $\mathsf{S} = \{(a, \langle \mathbf{m}\rangle, \langle \boldsymbol{\alpha}, \beta\rangle) : \mathsf{R}(\mathbf{m}, \boldsymbol{\alpha})\}$ is a $\Sigma^1_r[\beta]$ subrelation of V, so by the Boundeness Theorem, $\mathsf{S} \subseteq \mathsf{V}_\rho$ for some $\rho < \boldsymbol{\kappa}$. Then $R(\mathbf{m}, \boldsymbol{\alpha}) \leftrightarrow \mathsf{V}_\rho(a, \langle \mathbf{m}\rangle, \langle \boldsymbol{\alpha}, \beta\rangle)$, so $\mathsf{R} \blacktriangleleft \mathsf{V}_\rho$. $\square$

Note that the proof of (i) also establishes:

1.8 Corollary. *If* X $(= \Sigma^1_r$ *or* $\Pi^1_r)$ *has the pre-wellordering property, then for all relations R on numbers,* $R \in \Delta^1_r$ *iff for some a and some* $\rho < \kappa$,

$$R(\mathbf{m}) \leftrightarrow V_\rho(a, \langle \mathbf{m}\rangle). \quad \square$$

1.9 Theorem (Upper Classification). *If* X $(= \Sigma^1_r$ *or* $\Pi^1_r)$ *has the pre-wellordering property, then* $\{\alpha : \alpha \in \Delta^1_r\} \in X$.

Proof. By Corollary 1.8, for any α,

$$\begin{aligned} \alpha \in \Delta^1_r &\leftrightarrow (\exists \rho < \kappa) \exists a \forall mn\, [\alpha(m) = n \leftrightarrow V_\rho(a, \langle m, n\rangle)] \\ &\leftrightarrow (\exists u \in V) \exists a \forall mn\, ([\alpha(m) = n \rightarrow (a, \langle m, n\rangle) <_X u] \wedge \\ &\qquad \wedge [(a, \langle m, n\rangle) <_{cX} u \rightarrow \alpha(m) = n]). \quad \square \end{aligned}$$

1.10 Corollary. *If* Π^1_r *has the pre-wellordering property, then* Δ^1_r *is not a basis for* Π^1_{r-1}. *If* Σ^1_r *has the pre-wellordering property, then* Δ^1_r *is not a basis for* Π^1_r. $\square$

1.11 Selection Theorem. *If X ($=\Sigma^1_r$ or Π^1_r) has the pre-wellordering property, then for any $\mathsf{R}\in X$, there exists a partial functional* $\mathrm{Sel}_{\mathsf{R}}$ *with graph in X such that for all* $\mathbf{m}$ *and* $\boldsymbol{\alpha}$,

$$\exists p\, \mathsf{R}(p,\mathbf{m},\boldsymbol{\alpha}) \leftrightarrow \mathsf{R}(\mathrm{Sel}_{\mathsf{R}}(\mathbf{m},\boldsymbol{\alpha}),\mathbf{m},\boldsymbol{\alpha}) \leftrightarrow \mathrm{Sel}_{\mathsf{R}}(\mathbf{m},\boldsymbol{\alpha})\downarrow .$$

Proof. Suppose $\mathsf{R}\in X$ and a is an index such that $\mathsf{R}(p,\mathbf{m},\boldsymbol{\alpha})\leftrightarrow \mathsf{V}(a,\langle p,\mathbf{m}\rangle,\langle\boldsymbol{\alpha}\rangle)$. Then as in the proof of Lemma IV.2.5, it suffices to define

$$\begin{aligned}\mathrm{Sel}_{\mathsf{R}}(\mathbf{m},\boldsymbol{\alpha})\simeq p \leftrightarrow \mathsf{R}(p,\mathbf{m},\boldsymbol{\alpha}) \wedge \forall q\, [(a,\langle p,\mathbf{m}\rangle,\langle\boldsymbol{\alpha}\rangle) \leqslant_X (a,\langle q,\mathbf{m}\rangle,\langle\boldsymbol{\alpha}\rangle)]\\ \wedge (\forall q<p)[(a,\langle p,\mathbf{m}\rangle,\langle\boldsymbol{\alpha}\rangle) <_X (a,\langle q,\mathbf{m}\rangle,\langle\boldsymbol{\alpha}\rangle)]. \quad \square\end{aligned}$$

1.12 Lemma. *If X ($=\Sigma^1_r$ or Π^1_r) has the pre-wellordering property, then for every $\rho<\kappa$, there exists a set $B\in\Delta^1_r$ such that $B \lessdot V_\sigma$ for no $\sigma\leqslant\rho$.*

Proof. Similar to that of Lemma IV.2.4. $\square$

1.13 Theorem (Lower Classification). *If X ($=\Sigma^1_r$ or Π^1_r) has the pre-wellordering property, then $\{\alpha : \alpha\in\Delta^1_r\}\notin\Delta^1_r$.*

Proof. If X is Π^1_r we may proceed as in the second part of the proof of Theorem IV.2.6. Suppose now that X is Σ^1_r. Let

$$\mathsf{A}=\{\alpha : \alpha \text{ is the characteristic function of } V_\rho \text{ for some } \rho\leqslant\kappa\}.$$

First, $\mathsf{A}\in\Sigma^1_r$ since for all α,

$$\begin{aligned}\alpha\in\mathsf{A}\leftrightarrow \forall n\,[\alpha(n)\leqslant 1] \wedge \forall a\, \forall\mathbf{m}\,[\alpha(\langle a,\mathbf{m}\rangle)=0\rightarrow \langle a,\mathbf{m}\rangle\in V]\\ \wedge \forall ab\, \forall\mathbf{mn}\,[\alpha(\langle a,\mathbf{m}\rangle)=0 \wedge (b,\langle\mathbf{n}\rangle,\langle\ \rangle)\leqslant_{\Pi}(a,\langle\mathbf{m}\rangle,\langle\ \rangle)\\ \rightarrow \alpha(\langle b,\mathbf{n}\rangle)=0].\end{aligned}$$

For $\rho<\kappa$, the characteristic function of V_ρ is Δ^1_r by Theorem 1.6 whereas the characteristic function K_V of $V_\kappa=V$ is not Δ^1_r. Hence $\mathsf{A}\sim\{\alpha:\alpha\in\Delta^1_r\}=\{\mathsf{K}_V\}$. But if $\{\alpha:\alpha\in\Delta^1_r\}\in\Delta^1_r$, this set is Σ^1_r which implies, by Corollary III.2.7 (vii), that K_V is Δ^1_r, a contradiction. $\square$

It would seem at first glance that the ordinals κ and $\boldsymbol{\kappa}$ might depend on the particular pre-wellordering used to establish the pre-wellordering property. It turns out, however, that in many cases these ordinals are uniquely determined.

1.14 Definition. For any $r\geqslant 1$,

(i) $\delta^1_r=\sup^+\{\|R\| : R\in\Delta^1_r \text{ and } R \text{ is a well-ordering on } \omega\}$;
(ii) $\boldsymbol{\delta}^1_r=\sup^+\{\|\mathsf{R}\| : \mathsf{R}\in\boldsymbol{\Delta}^1_r \text{ and } \mathsf{R} \text{ is a pre-wellordering on } {}^\omega\omega\}$.

From Theorem IV.2.11 we have $\delta_1^1 = \omega_1$ and it easily follows from the techniques of that section that $\boldsymbol{\delta}_1^1 = \aleph_1$ (Exercise 1.26). Note that if R is a Δ_r^1 pre-wellordering on ω and

$$S(p,q) \leftrightarrow R(p,q) \wedge [\sim R(q,p) \vee (R(q,p) \wedge p \leqslant q)]$$

then S is a Δ_r^1 well-ordering and $\|S\| \geqslant \|R\|$. Hence δ_r^1 is also the supremum of the types of Δ_r^1 *pre*-wellorderings on ω.

1.15 Lemma. *If X $(=\Sigma_r^1$ or $\Pi_r^1)$ has the pre-wellordering property, then $\kappa \leqslant \delta_r^1$ and $\boldsymbol{\kappa} \leqslant \boldsymbol{\delta}_r^1$.*

Proof. For any ordinal $\rho < \kappa$, choose $w \in V$ such that $|w|_0 = \rho$. Then if $R_w(u,v) \leftrightarrow |u|_0 \leqslant |v|_0 < |w|_0$, R_w is a Δ_r^1 pre-wellordering of type ρ so by the preceding remark, $\rho < \delta_r^1$. Thus $\kappa \leqslant \delta_r^1$. The proof that $\boldsymbol{\kappa} \leqslant \boldsymbol{\delta}_r^1$ is similar. □

To prove the converse inequalities we shall need an effective version of the Boundeness Theorem. For the next two lemmas, let

$$R_b = \{(a, \langle \mathbf{m} \rangle) : \sim V(b, \langle a, \mathbf{m} \rangle)\}, \quad \text{and}$$

$$\mathsf{R}_b^{\beta,\delta} = \{(a, \langle \mathbf{m} \rangle, \langle \boldsymbol{\alpha} \rangle) : \sim \mathsf{V}(b, \langle a, \mathbf{m} \rangle, \langle \boldsymbol{\alpha}, \beta, \delta \rangle)\}.$$

1.16 Lemma. *If X $(=\Sigma_r^1$ or $\Pi_r^1)$ has the pre-wellordering property, then there exist primitive recursive functions f and g such that for all b, β, and δ,*

(i) *if $R_b \subseteq V$, then $V(f(b), \langle f(b) \rangle)$ and*

$$\sup{}^+\{|a, \langle \mathbf{m} \rangle|_0 : R_b(a, \langle \mathbf{m} \rangle)\} \leqslant |f(a), \langle f(a) \rangle|_0 < \kappa;$$

(ii) *if X is Π_r^1 and $\mathsf{R}_b^{\beta,\delta} \subseteq \mathsf{V}(=\sim \mathsf{U}_r^1)$, then $\mathsf{V}(g(b), \langle g(b) \rangle, \langle \beta, \delta \rangle)$ and*

$$\sup{}^+\{|a, \langle \mathbf{m} \rangle, \langle \boldsymbol{\alpha} \rangle| : \mathsf{R}_b^{\beta,\delta}(a, \langle \mathbf{m} \rangle, \langle \boldsymbol{\alpha} \rangle)\} \leqslant |g(b), \langle g(b) \rangle, \langle \beta, \delta \rangle| < \boldsymbol{\kappa}.$$

Proof. Suppose first that $R_b \subseteq V$. Since V is universal, there exists a primitive recursive function f such that for all n,

$$\sim V(f(b), \langle n \rangle) \leftrightarrow \exists a\, \exists \mathbf{m}[R_b(a, \langle \mathbf{m} \rangle) \wedge (n, \langle n \rangle, \langle\ \rangle) \leqslant_{cX} (a, \langle \mathbf{m} \rangle, \langle\ \rangle)].$$

If $\sim V(f(b), \langle f(b) \rangle)$, then for some a and $\mathbf{m}$ such that $R_b(a, \langle \mathbf{m} \rangle)$, $(f(b), \langle f(b) \rangle, \langle\ \rangle) \leqslant_{cX} (a, \langle \mathbf{m} \rangle, \langle\ \rangle)$. Since $R_b \subseteq V$, this implies $V(f(b), \langle f(b) \rangle)$, a contradiction. Hence $V(f(b), \langle f(b) \rangle)$ and thus for all a and $\mathbf{m}$ such that $R_b(a, \langle \mathbf{m} \rangle)$, $(f(b), \langle f(b) \rangle, \langle\ \rangle) \not\leqslant_{cX} (a, \langle \mathbf{m} \rangle, \langle\ \rangle)$, that is, $|a, \langle \mathbf{m} \rangle|_0 < |f(b), \langle f(b) \rangle|_0$ as required.

If $\mathsf{R}_b^{\beta,\delta} \subseteq \mathsf{V}$ and $X = \Pi^1_r$, we take g to be a primitive recursive function such that for all n,

$$\sim \mathsf{V}(g(b), \langle n\rangle, \langle \beta, \delta\rangle) \leftrightarrow \exists a\, \exists \mathbf{m}\, \exists \boldsymbol{\alpha}\, [\mathsf{R}_b^{\beta,\delta}(a, \langle \mathbf{m}\rangle, \langle \boldsymbol{\alpha}\rangle) \wedge (n, \langle n\rangle, \langle \beta, \delta\rangle) \leqslant_\Sigma (a, \langle \mathbf{m}\rangle, \langle \boldsymbol{\alpha}\rangle)]$$

and the argument proceeds similarly as above. □

1.17 Theorem. *If X ($= \Sigma^1_r$ or Π^1_r) has the pre-wellordering property, then*

(i) *for any $S \in \Delta^1_r$, if S is a well-ordering of ω, then $\|S\| < \kappa$;*

(ii) *if X is Π^1_r, then for any $\mathsf{S} \in \boldsymbol{\Delta}^1_r$, if S is a pre-wellordering of ${}^\omega\omega$, then $\|\mathsf{S}\| < \boldsymbol{\kappa}$.*

Proof. Let S be a Δ^1_r well-ordering. We shall construct a primitive recursive function h such that for all p and q,

$$(*) \qquad |p|_S < |q|_S \to |h(p), \langle h(p)\rangle|_0 < |h(q), \langle h(q)\rangle|_0.$$

It follows from (*) that for all $p \in \mathrm{Fld}(S)$, $|p|_S \leqslant |h(p), \langle h(p)\rangle|_0$, so

$$\|S\| \leqslant \sup{}^+\{|h(p), \langle h(p)\rangle|_0 : p \in \mathrm{Fld}(S)\} < \kappa$$

by the Boundedness Theorem.

Since $S \in \Delta^1_r$, there exists a primitive recursive function h_0 such that for any $q \in \mathrm{Fld}(S)$ and any e,

$$\begin{aligned} R_{h_0(e,q)} &= \{(\{e\}(p), \langle \{e\}(p)\rangle) : |p|_S < |q|_S\} \\ &= \{(b, \langle b\rangle) : \exists p\, [S(p,q) \wedge p \neq q \wedge \{e\}(p) \simeq b]\}. \end{aligned}$$

Let f be the function of the preceding lemma, by the Primitive Recursion Theorem choose $\bar{e}$ such that for all q, $\{\bar{e}\}(q) = f(h_0(\bar{e}, q))$, and set $h = \{\bar{e}\}$. We prove (*) by induction on S. Suppose that (*) holds for all p and q such that $|q|_S < |r|_S$. Then by Lemma 1.16

$$\sup{}^+\{|h(p), \langle h(p)\rangle|_0 : |p|_S < |r|_S\} \leqslant |h(r), \langle h(r)\rangle|_0$$

so that (*) holds for all p and q such that $|q|_S \leqslant |r|_S$.

For (ii), let S be a $\boldsymbol{\Delta}^1_r$ pre-wellordering — say $\mathsf{S} \in \Delta^1_r[\delta]$. We shall construct a primitive recursive functional H such that for all α and β

$$(*) \qquad |\alpha|_{\mathsf{S}} < |\beta|_{\mathsf{S}} \to |\mathsf{H}(\alpha), \langle \mathsf{H}(\alpha)\rangle, \langle \alpha, \delta\rangle| < |\mathsf{H}(\beta), \langle \mathsf{H}(\beta)\rangle, \langle \beta, \delta\rangle|.$$

Since $\mathsf{S} \in \Delta_r^1[\delta]$, there exists a primitive recursive functional H_0 such that for any $\beta \in \mathsf{Fld}(\mathsf{S})$ and any e,

$$\mathsf{R}^{\beta,\delta}_{\mathsf{H}_0(e,\beta)} = \{(\{e\}(\alpha), \langle\{e\}(\alpha)\rangle, \langle\alpha,\delta\rangle) : |\alpha|_{\mathsf{S}} < |\beta|_{\mathsf{S}}\}$$
$$= \{(b, \langle b\rangle, \gamma) : \exists\alpha\, [\mathsf{S}(\alpha,\beta) \wedge \neg\mathsf{S}(\beta,\alpha) \wedge \gamma = \langle\alpha,\delta\rangle \wedge \{e\}(\alpha) \simeq b]\}.$$

Let g be as in the preceding lemma, by the Primitive Recursion Theorem choose $\bar{e}$ such that $\{\bar{e}\}(\beta) = g(\mathsf{H}_0(\bar{e},\beta))$, and set $\mathsf{H} = \{\bar{e}\}$. The proof of (∗) is now similar to that of (∗) and the result is immediate from (∗). □

1.18 Corollary. *If X ($= \Sigma_r^1$ or Π_r^1) has the pre-wellordering property, then*

(i) $\kappa = \delta_r^1$;

(ii) *if X is Π_r^1, then $\boldsymbol{\kappa} = \boldsymbol{\delta}_r^1$.*

Proof. Immediate from 1.15 and 1.17. □

Note that it is not necessarily true that when Σ_r^1 has the pre-wellordering property, $\boldsymbol{\kappa} = \boldsymbol{\delta}_r^1$. Let κ^2 and $\boldsymbol{\kappa}^2$ be the ordinals associated with the pre-wellordering for Σ_2^1 obtained by the method of the proof of Theorem 1.3 from $\preccurlyeq^1$. Then clearly $\boldsymbol{\kappa}^2 = \boldsymbol{\delta}_1^1 = \aleph_1$, but the relation $\{(\gamma,\delta) : \gamma, \delta \in \mathsf{W} \wedge \|\gamma\| \leqslant \|\delta\|\}$ is a Π_1^1 pre-wellordering of type $\aleph_1$ so $\boldsymbol{\delta}_2^1 > \aleph_1$.

We conclude this section by listing the consequences of the pre-wellordering property for Σ_2^1.

1.19 Theorem.

(i) *Σ_2^1 and $\boldsymbol{\Sigma}_2^1$ have the reduction property but not the separation property;*

(ii) *Π_2^1 and $\boldsymbol{\Pi}_2^1$ have the separation property but not the reduction property;*

(iii) *for any $R \in \Pi_2^1$, if $R \subseteq U_2^1$, then $\sup^+\{|a, \langle\mathbf{m}\rangle|_0 : R(a, \langle\mathbf{m}\rangle)\} < \delta_2^1$;*

(iv) *for any R, $R \in \Delta_2^1 \leftrightarrow R \leqslant U_{2,\rho}^1$ for some $\rho < \delta_2^1$;*

(v) $\{\alpha : \alpha \in \Delta_2^1\} \in \Sigma_2^1 \sim \Delta_2^1$;

(vi) *Δ_2^1 is not a basis for Π_2^1;*

(vii) *for any $\mathsf{R} \in \Sigma_2^1$, there exists a partial functional $\mathrm{Sel}_{\mathsf{R}}$ with Σ_2^1 graph such that for all $\mathbf{m}$ and $\boldsymbol{\alpha}$,*

$$\exists p\, \mathsf{R}(p, \mathbf{m}, \boldsymbol{\alpha}) \leftrightarrow R(\mathrm{Sel}_{\mathsf{R}}(\mathbf{m},\boldsymbol{\alpha}), \mathbf{m}, \boldsymbol{\alpha}) \leftrightarrow \mathrm{Sel}_{\mathsf{R}}(\mathbf{m},\boldsymbol{\alpha})\downarrow;$$

(viii) *$\kappa^2 = \delta_2^1$ and $\boldsymbol{\kappa}^2 = \aleph_1 = \boldsymbol{\delta}_1^1$.* □

1.20–1.33 Exercises

1.20 (Moschovakis). Show that for $X = \Sigma_r^1$ or Π_r^1, each of the following is equivalent to the pre-wellordering property for X:

(i) for every $\mathsf{R} \in X$, there exist relation $\leqslant$, $\leqslant_\Sigma$, and $\leqslant_\Pi$ which satisfy (i)–(iii) of Definition 3.1 with 'V' replaced by 'R';

(ii) for every $\mathsf{R} \in X$, there exists an ordinal λ and a function φ mapping ${}^{k,l}\omega$ onto $\lambda+1$ such that $\mathsf{R}(x) \leftrightarrow \varphi(x) < \lambda$ and the two relations

$$x \leqslant_\varphi y \leftrightarrow \mathsf{R}(x) \wedge \varphi(x) \leqslant \varphi(y)$$

and

$$x <_\varphi y \leftrightarrow \mathsf{R}(x) \wedge \varphi(x) < \varphi(y)$$

both belong to X. (Such a φ is called an *X-norm for* R *of length* λ).

1.21. Give an alternative proof of Theorem 1.2 based on the results of Exercise III.3.33. For simplicity, consider in detail only relations on numbers.

1.22. Show that for all $r \geqslant 1$, Σ_r^0 has the pre-wellordering property (defined by making the obvious modification in Definition 1.1 (ii).)

1.23. Show that if X $(=\Sigma_r^1$ or $\Pi_r^1)$ has the pre-wellordering property, then so does $\mathbf{X}$ $(=\mathbf{\Sigma}_r^1$ or $\mathbf{\Pi}_r^1)$.

1.24 (Gandy). Show that it may happen that $|a, \langle \mathbf{m} \rangle|_0 < |a, \langle \mathbf{m} \rangle, \langle\ \rangle|$. (Take $X = \Sigma_2^1$ with the pre-wellordering defined by the proofs of Theorems 1.2 and 1.3. Apply the Basis Theorem to

$$\{\delta : \delta \in \mathsf{W} \wedge \exists \gamma\, (|a, \langle \mathbf{m} \rangle, \langle \gamma \rangle|^1 < \|\delta\| + 1) \rightarrow \exists \gamma\, (|a, \langle \mathbf{m} \rangle, \langle \gamma \rangle|^1 < \|\delta\|)\}$$

to show that not all Δ_2^1 ordinals are of the form

$$|a, \langle \mathbf{m} \rangle, \langle\ \rangle|^2 \quad \text{with} \quad \mathsf{U}_2^1(a, \langle \mathbf{m} \rangle, \langle\ \rangle).)$$

1.25. Show that the boldface boundedness property (1.6(ii)) fails for $X = \Sigma_2^1$.

1.26. Prove that $\boldsymbol{\delta}_1^1 = \aleph_1$. Show in fact that every $\mathbf{\Sigma}_1^1$ pre-wellordering of ${}^\omega\omega$ has type $< \aleph_1$. Note that there are $\mathbf{\Pi}_1^1$ pre-wellorderings of uncountable type.

1.27. (Cf. Exercise IV.2.25.) If X $(=\Sigma_r^1$ or $\Pi_r^1)$ has the pre-wellordering property, then the following two *effective choice principles* hold: for any $\mathsf{R} \in X$,

(i) if $\forall \mathbf{m} \forall \boldsymbol{\alpha} \exists p\, \mathsf{R}(p, \mathbf{m}, \boldsymbol{\alpha})$, then there exists a Δ_r^1 functional F such that $\forall \mathbf{m} \forall \boldsymbol{\alpha}\, \mathsf{R}(\mathsf{F}(\mathbf{m}, \boldsymbol{\alpha}), \mathbf{m}, \boldsymbol{\alpha})$;

(ii) if $\forall \mathbf{m} \forall \boldsymbol{\alpha}\, (\exists \beta \in \Delta_r^1[\boldsymbol{\alpha}]) \mathsf{R}(\mathbf{m}, \boldsymbol{\alpha}, \beta)$, then there exists a Δ_r^1 functional G such that $\forall \mathbf{m} \forall \boldsymbol{\alpha}\, \mathsf{R}(\mathbf{m}, \boldsymbol{\alpha}, \lambda q \,.\, \mathsf{G}(q, \mathbf{m}, \boldsymbol{\alpha}))$.

1.28. Show that if Π_r^1 has the pre-wellordering property, then the image of a $\mathbf{\Delta}_r^1$

set B under a functional $\theta : {}^\omega\omega \to {}^\omega\omega$ with $\mathbf{\Delta}_r^1$ graph which is one–one on B is $\mathbf{\Delta}_r^1$ (cf. Theorem IV.6.9).

1.29. For any X ($= \Sigma_r^1$ or Π_r^1), let $\Sigma_1^{1,X}$ be the class of relations R such that for some $\mathsf{P} \in \Delta_r^1$,

$$\mathsf{R}(\mathbf{m}, \boldsymbol{\alpha}) \leftrightarrow (\exists \beta \in \Delta_r^1[\boldsymbol{\alpha}])\, \mathsf{P}(\mathbf{m}, \boldsymbol{\alpha}, \beta).$$

Show that if X has the pre-wellordering property, then $\Sigma_1^{1,X} \subseteq X$. (Cf. Theorem 3.8 below). Show that if $X = \Sigma_r^1$ and has the uniformization property, then the converse inclusion also holds.

1.30. Show that if X ($= \Sigma_r^1$ or Π_r^1) has the pre-wellordering property, then there exists a well-ordering of ω in X of order-type δ_r^1 but every well-ordering of ω in cX has order-type less than δ_r^1.

1.31 (Martin, Solovay). A natural conjecture based on Theorem III.4.9 is: if Π_r^1 has the pre-wellordering property, then $\{\alpha : \alpha$ is recursive in some $B \in \Sigma_r^1\}$ is a basis for Σ_r^1. Show that this conjecture is false for $r > 1$ (but cf. Theorem 3.7 below). (Show that for any $B \in \Pi_r^1$ and any $\gamma \in \mathsf{W}$ such that $\|\gamma\| \geqslant \delta_r^1$, $B \in \Delta_r^1[\gamma]$. Then consider $\{\alpha : \exists \gamma\, (\gamma \in \mathsf{W} \wedge \|\gamma\| \geqslant \delta_r^1 \wedge \alpha \notin \Delta_r^1[\gamma])\}$).

1.32 (Moschovakis). Let $W^2 = \{a : \exists \beta\,.\, a \in W[\beta]\}$ and $\|a\|^2 = \inf\{\|a\|_\beta : a \in W[\beta]\}$. Show that every Σ_2^1 relation on numbers is reducible to W^2 and that if A is a Π_2^1 subset of W^2, then $\sup^+\{\|a\|^2 : a \in A\} < \delta_2^1$.

1.33. Give another proof of the existence of a Σ_2^1 relation R of order-type δ_2^1 along the following lines. Using the Uniformization Theorem IV.7.8, obtain a relation $\mathsf{R} \in \Pi_1^1$, such that

$$\exists \beta\,.\, a \in W[\beta] \leftrightarrow \exists! \beta\, \mathsf{R}(a, \beta).$$

Use this to assign to each $a \in W^2$ a function $\gamma_a \in \mathsf{W}$ with $\|\gamma_a\| \geqslant \|a\|^2$ and obtain R by "piecing together" the orderings $\leqslant \gamma_a$.

1.34 Notes. The pre-wellordering property arose from an analysis of just what is used about Π_1^1 in proving the main structure theorems. Although we have for convenience formulated it here only for the classes Σ_r^1 and Π_r^1, essentially the same definition applies to any indexable or parametrizable class of relations, (cf. Definition VI.4.7 below) and by Exercise 1.20 even this restriction is unnecessary.

The observation that Σ_2^1 has the pre-wellordering property is due to Moschovakis; an early version appears in Rogers [1967, § 16.6], but the ideas go back to Addison [1959] and by attribution there to Novikov even earlier. The property was implicit in Moschovakis [1967] and [1969] and Addison–Moschovakis [1968], but the first abstract formulation seems to be Moschovakis [1970].

2. The Hypothesis of Constructibility

The Hypothesis of Constructibility (V = L) was formulated by Gödel in 1938 to prove the relative consistency with ZF (and other axiomatizations of set theory) of the Axiom of Choice (AC) and the Generalized Continuum Hypothesis (GCH). He showed Gödel [1939]

Theorem. (i) *If* ZF *is consistent, then* ZF + (V = L) *is also consistent*;
(ii) AC *and* GCH *are theorems of* ZF + (V = L).

Since then it has been recognized that many other mathematical assertions which are not provable in ZFC — or which at least have resisted proof to the present day — *are* provable in ZF + (V = L). It follows that any such assertion is at least relatively consistent with ZFC — that is, its negation is *not* a theorem of ZFC unless ZF itself is inconsistent. We shall consider here what can be proved about the analytical and projective hierarchies in ZF + (V = L).

The general theory of constructibility is beyond the scope of this book and we shall base our discussion on results (1)–(6) below of this theory, which we state without proof. They may be found in any treatment of constructibility, for example Devlin [1973] or Mostowski [1969].

The language of set theory $\mathscr{L}_{\mathrm{ZF}}$ has the symbols $\neg$, $\wedge$, $\exists$, $\doteq$, $\dot{\in}$, and variables $x_0, x_1, \ldots$. The only terms are the variables, the atomic formulas are the expressions $x \doteq y$ and $x \dot{\in} y$ for variables x and y, and the class of formulas is the smallest class containing the atomic formulas and such that if $\mathfrak{A}$ and $\mathfrak{B}$ are formulas, so are $\neg\mathfrak{A}$, $\mathfrak{A} \wedge \mathfrak{B}$, and $\exists x_i\, \mathfrak{A}$. A structure for this language is an ordered pair $\mathfrak{M} = (M, E)$ such that M is a set and E is a binary relation on M: $E \subseteq {}^2M$. If $E = \in\restriction M = \{(u, v) : u \in v \wedge u,\ v \in M\}$, we call $\mathfrak{M}$ an $\in$*-structure* and denote it simply by $(M, \in)$ (here and for the rest of this section we suspend the convention that u, v, and w denote natural numbers). We write $\mathfrak{M} \models \mathfrak{A}[u_0, \ldots, u_{k-1}]$ to mean that the elements $u_0, \ldots, u_{k-1}$ of M satisfy the formula $\mathfrak{A}$ in $\mathfrak{M}$ when the variable x_i is interpreted as u_i. A precise definition of this relation is similar to III.5.2. A set $X \subseteq M$ is called *definable over* (M, E) iff for some formula $\mathfrak{A}$ and some $u_0, \ldots, u_{k-1} \in M$,

$$X = \{v : (M, E) \models \mathfrak{A}[v, u_0, \ldots, u_{k-1}]\}.$$

The hierarchy of constructible sets is defined by (transfinite) recursion on ordinals:

2.1 Definition. For all ordinals ρ,

(i) $L_0 = \varnothing$;

(ii) $L_{\rho+1} = \{X : X \subseteq L_\rho \text{ and } X \text{ is definable over } (L_\rho, \in)\}$;

(iii) $L_\rho = \bigcup\{L_\sigma : \sigma < \rho\}$, if ρ is a limit ordinal.

A set u is called *constructible* just in case $u \in L_\rho$ for some ρ. For constructible u, the *order* of u, Od(u), is the smallest ρ such that $u \in L_\rho$. It is easily verified from the definition that the following hold for all ordinals σ and ρ:

(1) $\sigma < \rho \rightarrow L_\sigma \cup \{L_\sigma, \sigma\} \subseteq L_\rho$;

(2) L_ρ is transitive — that is, $\forall u\, [u \in L_\rho \rightarrow u \subseteq L_\rho]$;

(3) $\mathrm{Card}(L_\rho) = \mathrm{Card}(\rho)$.

The property of being a constructible set, as with most properties of informal mathematics, may be expressed by a formula L of the language of set theory. If x denotes the unique free variable of L, then the formula $\forall x$ L, which asserts that all sets are constructible, is called the *Hypothesis of Constructibility*. In a set theory which admits proper classes as well as sets, L defines a class (also called L, the *class of constructible sets*) and the Hypothesis of Constructibility may be written 'V = L', where V is the class of all sets. In accord with common practice we shall use 'V = L' to denote this hypothesis even in ZF.

The proof that V = L implies the Axiom of Choice proceeds by showing that if V = L, then the universe may be definably well ordered. The key fact in applications to the analytical hierarchy is that the restriction of this well-ordering to ${}^\omega\omega$ is Δ^1_2. To state this precisely, we need:

(4) there exists a relation $<_L$ which well orders $\{\alpha : \alpha \text{ is constructible}\}$ with order-type $\aleph_1$ and has the property that if $\mathrm{Od}(\alpha) < \mathrm{Od}(\beta)$, then $\alpha <_L \beta$. Furthermore, there is a formula $\circledcirc$ of the language of ZF such that for any ρ and any $\alpha, \beta \in L_\rho$,

$$\alpha <_L \beta \leftrightarrow (L_\rho, \in) \models \circledcirc[\alpha, \beta].$$

2.2 Theorem. *If* V = L, *then*

(i) $<_L$ *is* Δ^1_2;

(ii) *for any* $r \geq 2$ *and any* $\mathsf{R} \in \Sigma^1_r$ (Π^1_r), *if*

$$\mathsf{P}(\mathbf{m}, \boldsymbol{\alpha}, \gamma) \leftrightarrow (\exists \beta <_L \gamma)\, \mathsf{R}(\mathbf{m}, \boldsymbol{\alpha}, \beta), \quad \textit{and}$$

$$\mathsf{Q}(\mathbf{m}, \boldsymbol{\alpha}, \gamma) \leftrightarrow (\forall \beta <_L \gamma)\, \mathsf{R}(\mathbf{m}, \boldsymbol{\alpha}, \beta),$$

then also P *and* Q *are* Σ^1_r (Π^1_r).

We shall prove the theorem after discussing some additional set-theoretic facts. The first of these is

(5) For all constructible $\alpha \in {}^{\omega}\omega$, $\mathrm{Od}(\alpha) < \aleph_1$.

From this we can already see the outline of the proof of Theorem 2.2. Under the assumption $V = L$, we have for any R,

$$(*)\quad \begin{aligned} &(\exists \beta <_L \gamma)\mathrm{R}(\mathbf{m}, \boldsymbol{\alpha}, \beta) \leftrightarrow \\ &\leftrightarrow \exists \sigma \exists \beta\, [\sigma < \aleph_1 \wedge \beta, \gamma \in L_\sigma \wedge (L_\sigma, \in) \models \circleddash[\beta, \gamma] \wedge \mathrm{R}(\mathbf{m}, \boldsymbol{\alpha}, \beta)] \\ &\leftrightarrow \forall \sigma\, [\sigma < \aleph_1 \wedge \gamma \in L_\sigma \rightarrow \exists \beta\, (\beta \in L_\sigma \\ &\qquad\qquad \wedge (L_\sigma, \in) \models \circleddash[\beta, \gamma] \wedge \mathrm{R}(\mathbf{m}, \boldsymbol{\alpha}, \beta))]. \end{aligned}$$

The reader familiar with the techniques of § III.3 (especially the discussion following III.3.8) will suggest that the first step in evaluating the complexity of these expressions is to replace quantification over countable ordinals by quantification over well-orderings. It turns out to be simpler to characterize directly the class of models L_σ ($\sigma < \aleph_1$). The other key point is that for $\sigma < \aleph_1$, L_σ is countable so that with suitable coding the quantifier '$\exists \beta \in L_\sigma$' may be replaced by a number quantifier.

A structure $\mathfrak{M}$ is called *well founded* iff the relation E is well founded — that is, there is no function $\phi : \omega \rightarrow M$ such that for all n, $E(\phi(n+1), \phi(n))$. If $\mathfrak{M}$ is well founded, then each $u \in M$ is assigned a unique ordinal number $\mathrm{hgt}(u)$ by the condition

$$\mathrm{hgt}(u) = \sup{}^+\{\mathrm{hgt}(v) : E(v, u)\}.$$

The least ordinal not assigned to any $u \in M$ is called the *height* of $\mathfrak{M}$. If $\mathfrak{M}$ is well founded and satisfies the axiom of extensionality, then there is a unique isomorphism ψ (the "collapsing map") of $\mathfrak{M}$ with an $\in$-structure $(M^\circ, \in)$. ψ is defined recursively by

$$\psi(u) = \{\psi(v) : E(v, u)\}.$$

If $\mathfrak{M}$ is a model of $V = L$ together with a certain finite collection of axioms of ZF, then in fact M° must be exactly L_ρ for ρ the height of $\mathfrak{M}$:

(6) there exists a theorem $\mathfrak{C}$ of $\mathrm{ZF} + (V = L)$ such that for any ρ, any well-founded model of $\mathfrak{C}$ of height ρ is isomorphic to $(L_\rho, \in)$. Furthermore, for any $\sigma < \aleph_1$, there exists a ρ such that $\sigma < \rho < \aleph_1$ and $(L_\rho, \in) \models \mathfrak{C}$.

The effect of (6) is to allow us in (*) to replace quantification over the sets L_ρ ($\rho < \aleph_1$) by quantification over well-founded models of $\mathfrak{C}$.

For any $\varepsilon \in {}^{\omega}\omega$, let $\mathfrak{M}_\varepsilon = (\omega, \{(m, n) : \varepsilon(\langle m, n\rangle) = 0\})$. Any countable structure for the language of ZF is isomorphic to some $\mathfrak{M}_\varepsilon$, so we may restrict attention to these. Let $\ulcorner\ \urcorner$ denote a fixed Gödel numbering of this language.

2.3 Lemma. *There exist relations* $\mathsf{Wf} \in \Pi^1_1$ *and* $\mathsf{Mod} \in \Delta^1_1$ *such that for all* $\mathbf{m}$, ε, *and* $\mathfrak{A}$,

(i) $\mathsf{Wf}(\varepsilon) \leftrightarrow \mathfrak{M}_\varepsilon$ *is well founded*;

(ii) $\mathsf{Mod}(\ulcorner\mathfrak{A}\urcorner, \langle\mathbf{m}\rangle, \varepsilon) \leftrightarrow \mathfrak{M}_\varepsilon \models \mathfrak{A}[\mathbf{m}]$.

Sketch of proof. *For* (i) *we have*

$$\mathfrak{M}_\varepsilon \text{ is well founded} \leftrightarrow \neg\exists\alpha\,\forall m\,[\varepsilon(\langle\alpha(m+1), \alpha(m)\rangle) = 0].$$

(ii) may be proved by constructing a Σ^0_1 family of inductive operators Γ_ε such that for all $\mathfrak{A}$, $\mathbf{m}$ and ε,

$$\mathfrak{M}_\varepsilon \models \mathfrak{A}[\mathbf{m}] \leftrightarrow \exists t.\langle t, \ulcorner\mathfrak{A}\urcorner, \langle\mathbf{m}\rangle\rangle \in \bar{\Gamma}_\varepsilon,$$

and applying Theorem III.3.17 (cf. Theorem III.3.6 and Exercise III.5.19). □

Suppose now that $\mathfrak{M}_\varepsilon$ is a well-founded model of $\mathfrak{C}$ and define $\theta_\varepsilon : \omega \to \omega$ as follows:

$$\theta_\varepsilon(0) = \text{the unique } u \in \omega \text{ such that } \forall n.\,\varepsilon(\langle n, u\rangle) \neq 0;$$

$$\theta_\varepsilon(m+1) = \text{the unique } u \in \omega \text{ such that}$$
$$\forall n\,[\varepsilon(\langle n, u\rangle) = 0 \leftrightarrow \varepsilon(\langle n, \theta_\varepsilon(m)\rangle) = 0 \text{ or } n = \theta_\varepsilon(m)].$$

Then $\theta_\varepsilon(m)$ is the element of the model $\mathfrak{M}_\varepsilon$ which plays the role of the natural number m. In particular, if ψ_ε is the unique isomorphism of $\mathfrak{M}_\varepsilon$ with some $(\mathrm{L}_\rho, \in)$, then $\psi_\varepsilon(\theta_\varepsilon(m)) = m$. Similarly, we extend θ_ε by setting

$$\theta_\varepsilon(\beta) \simeq u \leftrightarrow \forall mn\,(\beta(m) = n \leftrightarrow \mathfrak{M}_\varepsilon \models ((x_1, x_2) \in x_0)[u, \theta_\varepsilon(m), \theta_\varepsilon(n)]).$$

If $\theta_\varepsilon(\beta) \simeq u$, then u plays the role of β in $\mathfrak{M}_\varepsilon$ and $\psi_\varepsilon(\theta_\varepsilon(\beta)) = \beta$. It follows from Lemma 2.3 that the relation '$\theta_\varepsilon(\beta) \simeq u$' is Δ^1_1.

We can now conclude the proof of Theorem 2.2. By $(*)$ together with (6) we have

$$(\exists\beta <_{\mathrm{L}} \gamma)\,\mathsf{R}(\mathbf{m}, \boldsymbol{\alpha}, \beta) \leftrightarrow$$
$$\leftrightarrow \exists\varepsilon\,\exists\beta\,\exists uv\,[\mathfrak{M}_\varepsilon \models \mathfrak{C} \wedge \mathfrak{M}_\varepsilon \text{ is well founded} \wedge \theta_\varepsilon(\beta) \simeq u \wedge \theta_\varepsilon(\gamma) \simeq v \wedge$$
$$\mathfrak{M}_\varepsilon \models \mathfrak{S}[u, v] \wedge \mathsf{R}(\mathbf{m}, \boldsymbol{\alpha}, \beta)]$$
$$\leftrightarrow \forall\varepsilon\,\forall v\,[\mathfrak{M}_\varepsilon \models \mathfrak{C} \wedge \mathfrak{M}_\varepsilon \text{ is well founded} \wedge \theta_\varepsilon(\gamma) \simeq v \to$$
$$\exists u\,(\mathfrak{M}_\varepsilon \models \mathfrak{S}[u, v] \wedge \exists\beta\,(\theta_\varepsilon(\beta) \simeq u) \wedge \forall\beta\,[\theta_\varepsilon(\beta) \simeq u \to \mathsf{R}(\mathbf{m}, \boldsymbol{\alpha}, \beta)])].$$

If $\mathsf{R} \in \Sigma^1_r$ $(r \geq 2)$, then the first equivalence shows that also $\mathsf{P} \in \Sigma^1_r$. Similarly, if $\mathsf{R} \in \Pi^1_r$, the second equivalence shows $\mathsf{P} \in \Pi^1_r$. The results for Q are immediate by complementation and (i) follows from the equivalence: $\alpha <_{\mathrm{L}} \gamma \leftrightarrow (\exists \beta <_{\mathrm{L}} \gamma)[\alpha = \beta]$. □

2.4 Theorem. *If* $\mathrm{V} = \mathrm{L}$, *then for all* $r \geq 2$, Σ^1_r *has the pre-wellordering property.*

Proof. Under the assumption $\mathrm{V} = \mathrm{L}$, $<_{\mathrm{L}}$ well orders ${}^{\omega}\omega$ in type $\aleph_1$ and therefore assigns to each γ a countable ordinal $|\gamma|_{\mathrm{L}}$. For each $x = (a, \langle \mathbf{m} \rangle, \langle \boldsymbol{\alpha} \rangle)$ we define

$$|x|^r_{\mathrm{L}} = \begin{cases} \inf\{|\gamma|_{\mathrm{L}} : \sim \mathsf{U}^1_{r-1}(x, \gamma)\}, & \text{if } \mathsf{U}^1_r(x); \\ \aleph_1, & \text{otherwise.} \end{cases}$$

Then the relation $\leq^r_{\mathrm{L}}$ defined by

$$x \leq^r_{\mathrm{L}} y \leftrightarrow |x|^r_{\mathrm{L}} \leq |y|^r_{\mathrm{L}}$$

is a pre-wellordering on ${}^{2,1}\omega$. It is easy to check that the conditions of Definition 1.1 are satisfied if we set

$$x \leq^r_{\mathrm{L},\Sigma} y \leftrightarrow \exists \gamma\, [\sim \mathsf{U}^1_{r-1}(x, \gamma) \wedge (\forall \delta <_{\mathrm{L}} \gamma) \mathsf{U}^1_{r-1}(y, \delta)],$$

and

$$x \leq^r_{\mathrm{L},\Pi} y \leftrightarrow \forall \delta\, [\sim \mathsf{U}^1_{r-1}(y, \delta) \rightarrow (\exists \gamma \leq_{\mathrm{L}} \delta) \sim \mathsf{U}^1_{r-1}(x, \gamma)]. \quad \square$$

In accord with the notation of the preceding section, let $|\ \ |^r_{\mathrm{L},0}$ be the norm associated with the pre-wellordering induced on U^1_r by $\leq^r_{\mathrm{L}}$,

$$\kappa^r_{\mathrm{L}} = \sup{}^+\{|x|^r_{\mathrm{L}} : \mathsf{U}^1_r(x)\},$$

and

$$\kappa^r_{\mathrm{L}} = \sup{}^+\{|a, \langle \mathbf{m} \rangle|^r_{\mathrm{L},0} : U^1_r(a, \langle \mathbf{m} \rangle)\}.$$

2.5 Theorem. *If* $\mathrm{V} = \mathrm{L}$, *then for all* $r \geq 2$,

(i) Σ^1_r *and* $\boldsymbol{\Sigma}^1_r$ *have the reduction property but not the separation property*;

(ii) Π^1_r *and* $\boldsymbol{\Pi}^1_r$ *have the separation property but not the reduction property*;

(iii) *for any* $R \in \Pi^1_r$, *if* $R \subseteq U^1_r$, *then*

$$\sup{}^+\{|a, \langle \mathbf{m} \rangle|^r_{\mathrm{L},0} : R(a, \langle \mathbf{m} \rangle)\} < \delta^1_r;$$

(iv) *for any* R, $R \in \Delta^1_r \leftrightarrow R \leq U^1_{r,\rho}$ *for some* $\rho < \delta^1_r$;

(v) $\{\alpha : \alpha \in \Delta_r^1\} \in \Sigma_r^1 \sim \Delta_r^1$;
(vi) Δ_r^1 *is not a basis for* Π_r^1;
(vii) *for any* $\mathsf{R} \in \Sigma_r^1$, *there exists a partial functional* $\mathrm{Sel}_{\mathsf{R}}$ *with* Σ_r^1 *graph such that for all* $\mathbf{m}$ *and* $\boldsymbol{\alpha}$,

$$\exists p\, \mathsf{R}(p, \mathbf{m}, \boldsymbol{\alpha}) \leftrightarrow \mathsf{R}(\mathrm{Sel}_{\mathsf{R}}(\mathbf{m}, \boldsymbol{\alpha}), \mathbf{m}, \boldsymbol{\alpha}) \leftrightarrow \mathrm{Sel}_{\mathsf{R}}(\mathbf{m}, \boldsymbol{\alpha}) \downarrow;$$

(viii) $\kappa_{\mathrm{L}}^r = \delta_r^1$ *and* $\boldsymbol{\kappa}_{\mathrm{L}}^r = \aleph_1 = \boldsymbol{\delta}_1^1$.

Proof. Immediate from Theorem 2.4 and the results of § 1. □

The Hypothesis of Constructibility also has consequences concerning the properties of Σ_1^1 and Π_1^1 sets considered in § IV.5–7. As regards uniformization, we have

2.6 Theorem. *If* $\mathrm{V} = \mathrm{L}$, *then for all* $r \geq 2$, Σ_r^1 *and* $\boldsymbol{\Sigma}_r^1$ *have the uniformization property.*

Proof. Suppose $\mathrm{V} = \mathrm{L}$ and

$$\mathsf{R}(\mathbf{m}, \boldsymbol{\alpha}, \beta) \leftrightarrow \exists \gamma\, \mathsf{P}(\mathbf{m}, \boldsymbol{\alpha}, \beta, \gamma)$$

with $\mathsf{P} \in \Pi_{r-1}^1$. Let

$$\mathsf{Q}(\mathbf{m}, \boldsymbol{\alpha}, \beta, \gamma) \leftrightarrow \mathsf{P}(\mathbf{m}, \boldsymbol{\alpha}, \beta, \gamma) \wedge (\forall \delta <_{\mathrm{L}} \langle \beta, \gamma \rangle) \sim \mathsf{P}(\mathbf{m}, \boldsymbol{\alpha}, (\delta)_0, (\delta)_1)$$

and

$$\theta(\mathbf{m}, \boldsymbol{\alpha}) \simeq \beta \leftrightarrow \exists \gamma\, \mathsf{Q}(\mathbf{m}, \boldsymbol{\alpha}, \beta, \gamma).$$

Then θ uniformizes R and $\mathsf{Q} \in \Delta_r^1$ so $\mathrm{Gr}_\theta \in \Sigma_r^1$. □

2.7 Corollary. *If* $\mathrm{V} = \mathrm{L}$, *then for all* $r \geq 2$, Δ_r^1 *is a basis for* Σ_r^1. □

The main conclusion we draw concerning the results of § IV.5 is that they cannot be extended to higher levels of the projective hierarchy without assumptions which contradict $\mathrm{V} = \mathrm{L}$. Consider first the results on cardinality. Of course, if $\mathrm{V} = \mathrm{L}$, then the Continuum Hypothesis holds and *every* uncountable subset of ${}^\omega\omega$ has power $2^{\aleph_0}$. We are interested, however, in the method used to prove Theorem IV.5.12. The construction shows that any uncountable $\boldsymbol{\Sigma}_1^1$ set has a perfect subset (which therefore has power $2^{\aleph_0}$).

2.8 Theorem. *If* $\mathrm{V} = \mathrm{L}$, *then there exists an uncountable* Π_1^1 *subset of* ${}^\omega\omega$ *which has no perfect subset.*

Proof. Let $W^* = \{\gamma : \gamma \in W \wedge \neg(\exists\delta <_L \gamma)(|\gamma| = |\delta|)\}$. For each countable ordinal ρ, W^* contains exactly one function γ such that $\|\gamma\| = \rho$ and thus W^* is uncountable. It follows from Theorem 2.2 that $W^* \in \Sigma^1_2$, so let R be a Π^1_1 relation such that $\gamma \in W^* \leftrightarrow \exists\beta\, R(\gamma, \beta)$. By the Uniformization Theorem there exists a Π^1_1 relation $S \subseteq R$ such that $\gamma \in W^* \leftrightarrow \exists\beta\, S(\gamma, \beta) \leftrightarrow \exists!\beta\, S(\gamma, \beta)$. Let $B = \{\langle\gamma, \beta\rangle : S(\gamma, \beta)\}$. B is Π^1_1, and as the projection function $\langle\gamma, \beta\rangle \mapsto \gamma$ is one–one from B onto W^*, B is uncountable. Suppose B had a perfect subset P, and let $C \subseteq W^*$ be the projection of P. On the one hand, C is the one–one image of the uncountable set P so is uncountable. But P is closed so by (1) of § IV.6, $C \in \mathbf{\Sigma}^1_1$ (indeed by Theorem IV.6.9, C is Borel). Hence, by the Boundedness Theorem, $C \subseteq W_\rho$ for some $\rho < \aleph_1$. Since $W^* \cap W_\rho$ is countable, this is a contradiction. □

Concerning measurability and the Baire property, we need the following two standard results. The proof of Fubini's Theorem may be found in almost any text on Measure Theory, while that of the Kuratowski–Ulam Theorem is in Oxtoby [1971] and Kuratowski [1966]. For any $R \subseteq {}^{0,2}\omega$, let

$$R^\alpha = \{\beta : R(\alpha, \beta)\} \quad \text{and} \quad R_\beta = \{\alpha : R(\alpha, \beta)\}.$$

We denote the usual Lebesgue measure in the plane also by mes. The phrase "for almost all α (measure) (category)" means "for all α except those in some set (of measure 0) (which is meager)".

Fubini's Theorem. *For any measurable relation $R \subseteq {}^{0,2}\omega$, the following are equivalent:*

(i) $\mathrm{mes}(R) = 0$;
(ii) $\mathrm{mes}(R^\alpha) = 0$ *for almost all* α *(measure)*;
(iii) $\mathrm{mes}(R_\beta) = 0$ *for almost all* β *(measure)*.

Kuratowski–Ulam Theorem. *For any relation $R \subseteq {}^{0,2}\omega$ which has the Baire Property, the following are equivalent:*

(i) R *is meager*;
(ii) R^α *is meager for almost all* α *(category)*;
(iii) R_β *is meager for almost all* β *(category)*.

2.9 Theorem. *If* $V = L$, *then there exists a* Δ^1_2 *relation which is neither measurable nor has the Baire Property.*

Proof. The relation is $<_L$. Since the order-type of $<_L$ is $\aleph_1$, $<_{L,\beta}$ is countable for each β and thus is meager and of measure 0. Similarly, for each α, $<^\alpha_L$ is the complement of a countable set, hence is comeager and of measure 1, hence is not meager (by the Baire Category Theorem) and is not of measure 0 (by additivity). These facts contradict the preceding theorems if $<_L$ is either measurable or has the Baire Property.

2.10–2.13 Exercises

2.10. Without the hypothesis V = L, $\{\alpha : \alpha$ is constructible$\}$ may be a proper subset of ${}^{\omega}\omega$. Show that it is Σ^1_2.

2.11. Show that if V = L, then for all $r \geqslant 2$ and all β,

$$\beta \in \Delta^1_r \leftrightarrow \beta \in \mathrm{L}_{\delta^1_r} \leftrightarrow |\beta|_{\mathrm{L}} < \delta^1_r.$$

(Use the Basis Theorem (2.7) and the fact that the functions belonging to any L_σ form an initial segment in the $<_{\mathrm{L}}$ ordering).

2.12. Show that if V = L, then for all r, Δ^1_r is a model of the Δ^1_r-Comprehension schema.

2.13 (Spector [1958]–Addison [1959a]). Show that there exist α and β such that neither $\alpha \in \Delta^1_1[\beta]$ nor $\beta \in \Delta^1_1[\alpha]$ (α and β have incomparable hyperdegrees), but that if V = L, then any two functions are Δ^1_2-comparable. (For the first part use Fubini's Theorem).

2.14 Notes. The history of the consequences of V = L for the analytical and projective hierarchies is rather complex. A good summary of it appears in Addison [1959a]. That Theorem 2.2 leads to the pre-wellordering property for all Σ^1_r ($r \geqslant 2$) was obvious as soon as this property was formulated. Much of the material of this section may be found also in Devlin [1973] and Mostowski [1969].

In most of the literature the assertion V = L is called the *Axiom* of Constructibility. We have used the term "hypothesis" to reflect more accurately the light in which this assertion is regarded by most logicians.

3. The Hypothesis of Projective Determinacy

Although the Hypothesis of Constructibility leads to a reasonably pleasant and elegant world of sets, it has not been accepted by many as a true statement about the intuitive world of sets. The case for the intuitive truth of any assertion is supported by the "correctness" of its consequences, but there seems to be little support to be gained from the structure of the analytical hierarchy described in the preceding section. In this section we discuss an alternative hypothesis (PD) which leads to a quite different picture of the analytical hierarchy. The reader may judge for himself whether or not these results are arguments in favor of the intuitive truth of PD.

Determinacy is an assertion concerning the existence of strategies for a certain class of infinite two-person games. With each set $A \subseteq {}^\omega\omega$ we associate a game as follows. Players I and II choose alternately the values $\varepsilon(0), \varepsilon(1), \varepsilon(2), \ldots$ of a function ε. If, after an ω-sequence of moves, the completed function ε belongs to A, then I is the winner, if $\varepsilon \notin A$, then II wins.

Let $\varepsilon_{\mathrm{I}}(m) = \varepsilon(2m)$ and $\varepsilon_{\mathrm{II}}(m) = \varepsilon(2m+1)$. Player I plays the number $\varepsilon_{\mathrm{I}}(m)$ at his m-th turn and player II plays the number $\varepsilon_{\mathrm{II}}(m)$ at his. We say that I plays according to the *strategy* γ iff for all m, $\varepsilon_{\mathrm{I}}(m) = \gamma(\bar{\varepsilon}(2m))$. Similarly, II plays according to the strategy δ iff for all m, $\varepsilon_{\mathrm{II}}(m) = \delta(\bar{\varepsilon}(2m+1))$. If each player plays according to his respective strategy γ or δ, the unique function generated is denoted by $\gamma \# \delta$ — that is, for all m,

$$(\gamma \# \delta)(2m) = \gamma(\overline{(\gamma \# \delta)}(2m)) \quad \text{and} \quad (\gamma \# \delta)(2m+1) = \delta(\overline{(\gamma \# \delta)}(2m+1)).$$

A strategy γ is called *winning for* I iff $\forall\delta\,[\gamma \# \delta \in A]$. δ is *winning for* II iff $\forall\gamma\,[\gamma \# \delta \notin A]$. A is *determined* iff either I or II has a winning strategy. Det(X) means that all $A \in X$ are determined. The *Hypothesis of Projective Determinacy* (PD) is the assertion $\mathrm{Det}(\boldsymbol{\Delta}^1_{(\omega)})$.

It would be inappropriate here to enter into a full-scale study of all the consequences of PD or other forms of determinacy. Our main aim here is to show that under the assumption PD, the classes $\Pi^1_3, \Sigma^1_4, \Pi^1_5, \ldots$ have the pre-wellordering property. However, to orient the reader who is completely unfamiliar with determinacy, we shall mention a few of the simpler general facts about it. Others are treated in the exercises.

Using the Axiom of Choice, one can easily construct a non-determined set — given a function from a cardinal λ onto ${}^\omega\omega$, $\langle \gamma_\sigma : \sigma < \lambda \rangle$, one constructs A in λ stages to ensure at stage σ that γ_σ is not a winning strategy for either I or II (Exercise 3.13). Thus $\neg\mathrm{Det}(\mathbf{P}({}^\omega\omega))$ is a theorem of ZFC. Without the Axiom of Choice, however, there seems to be no way to construct a non-determined set, and it may well be that $\mathrm{Det}(\mathbf{P}({}^\omega\omega))$ is relatively consistent with ZF. It is known, however, that even the consistency of $\mathrm{ZF} + \mathrm{Det}(\Delta^1_2)$ cannot be proved in the theory ZF + (ZF is consistent) (Friedman [1971]).

In another direction, $\mathrm{Det}(\boldsymbol{\Delta}^1_1)$ is a theorem of ZFC (Martin [1975]). $\mathrm{Det}(\boldsymbol{\Sigma}^1_1 \cup \boldsymbol{\Pi}^1_1)$ is provable from the existence of a measurable cardinal (Martin [1970]), but $\neg\mathrm{Det}(\Sigma^1_1 \cup \Pi^1_1)$ holds in ZF + (V = L) and is therefore relatively consistent with ZFC (Corollary 3.11).

We shall continue to work in ZFC even when we assume PD. The preceding remarks do not imply that the theory ZFC + PD is inconsistent because the non-determined set constructed above is not projective (or definable in any way). It is worth noting, however, that the results of this section depend only on two special forms of the Axiom of Choice. The first is used in the general development of the analytical and projective hierarchies to prove that Σ^1_r and Π^1_r are closed under number quantification and in the last part of the proof of Theorem 3.1:

$$\forall p\, \exists \beta\, \mathsf{R}(p, \mathbf{m}, \boldsymbol{\alpha}, \beta) \leftrightarrow \exists \beta\, \forall p\, \mathsf{R}(p, \mathbf{m}, \boldsymbol{\alpha}, (\beta)^p).$$

This equivalence for projective R in fact follows from PD (Exercise 3.14). The other use of the Axiom of Choice is in the proof of Theorem 3.1 below when we use the alternative condition (4′) of I.6 for well-foundedness. To prove the equivalence of (4′) with (4) requires not the full Axiom of Choice, but only the weaker axiom of Dependent Choice (DC). Thus the results we obtain are all theorems of ZF + DC + PD.

3.1 Theorem. *If* PD, *then for any* $r > 0$, *if* Σ^1_r *has the pre-wellordering property, then* Π^1_{r+1} *also has the pre-wellordering property.*

Proof. As in the preceding sections we shall write x for $(a, \langle\mathbf{m}\rangle, \langle\boldsymbol{\alpha}\rangle)$, (x, γ) for $(a, \langle\mathbf{m}\rangle, \langle\boldsymbol{\alpha}, \gamma\rangle)$, etc. Let $\leqslant^r$, $\leqslant^r_\Sigma$ and $\leqslant^r_\Pi$ be relations which establish the pre-wellordering property for Σ^1_r and $|\ \ |^r$ the norm associated with $\leqslant^r$. We aim to define relations $\leqslant^{r+1}$, $\leqslant^{r+1}_\Sigma$, and $\leqslant^{r+1}_\Pi$ which establish the pre-wellordering property for Π^1_{r+1}.

To motivate the definitions, we first consider a construction which does *not* work. Since $\sim\mathsf{U}^1_{r+1}(x) \leftrightarrow \forall\gamma\, \mathsf{U}^1_r(x, \gamma)$, the method used in the proof of Theorem 1.3 suggests that we define

$$|x\dot{|}^{r+1} = \sup{}^+\{|x, \gamma|^r : \gamma \in {}^\omega\omega\}$$

and define $\dot{\leqslant}^{r+1}$ so that $|\ \ \dot{|}^{r+1}$ is the associated norm (the dot signifies that these are *not* our eventual definitions). Corresponding to formula (2) in that proof we have

$$(\dot{2}) \qquad x \,\dot{\leqslant}^{r+1} y \leftrightarrow \forall\gamma\, \exists\delta\, [(x, \gamma) \leqslant^r (y, \delta)]$$

so that if we set

$$x \,\dot{\leqslant}^{r+1}_\Pi\, y \leftrightarrow \forall\gamma\, \exists\delta\, [(x, \gamma) \leqslant^r_\Sigma (y, \delta)]$$

we can prove as before that $\dot{\leqslant}^{r+1}$ and $\dot{\leqslant}^{r+1}_\Pi$ coincide when one of the arguments lies in $\sim\mathsf{U}^1_{r+1}$. However, the equivalence corresponding to (1) is false:

$$(\dot{1}) \qquad x \,\dot{\leqslant}^{r+1} y \not\leftrightarrow \exists\delta\, \forall\gamma\, [(x, \gamma) \leqslant^r (y, \delta)].$$

The implication ($\rightarrow$) fails because it is not the case that every non-empty set of ordinals has a largest element. Thus there is no good candidate for $\dot{\leqslant}^{r+1}_\Sigma$.

The contribution of determinacy is essentially to provide a new sort of "quantifier" which avoids this problem. We set

$$x \leqslant^{r+1} y \leftrightarrow \forall\gamma\, \exists\delta\, [(x, (\gamma \# \delta)_{\mathrm{I}}) \leqslant^{r} (y, (\gamma \# \delta)_{\mathrm{II}})];$$
$$x \leqslant^{r+1}_{\Sigma} y \leftrightarrow \exists\delta\, \forall\gamma\, [(x, (\gamma \# \delta)_{\mathrm{I}}) \leqslant^{r}_{\Pi} (y, (\gamma \# \delta)_{\mathrm{II}})];$$
$$x \leqslant^{r+1}_{\Pi} y \leftrightarrow \forall\gamma\, \exists\delta\, [(x, (\gamma \# \delta)_{\mathrm{I}}) \leqslant^{r}_{\Sigma} (y, (\gamma \# \delta)_{\mathrm{II}})].$$

It is clear that $\leqslant^{r+1}_{\Sigma}$ is Σ^1_{r+1} and $\leqslant^{r+1}_{\Pi}$ is Π^1_{r+1}. For any x and y, let

$$\mathsf{A}_{xy} = \{\varepsilon : \neg[(x, \varepsilon_{\mathrm{I}}) \leqslant^{r} (y, \varepsilon_{\mathrm{II}})]\}.$$

Thus

(1) $\qquad x \leqslant^{r+1} y \leftrightarrow$ Player I does not have a winning strategy in A_{xy}.

Suppose that one of $\sim\mathsf{U}^1_{r+1}(x)$ or $\sim\mathsf{U}^1_{r+1}(y)$. Then for all ε, $\mathsf{U}^1_r(x, \varepsilon_{\mathrm{I}})$ or $\mathsf{U}^1_r(y, \varepsilon_{\mathrm{II}})$, so that for all ε

(2) $\qquad (x, \varepsilon_{\mathrm{I}}) \leqslant^{r}_{\Sigma} (y, \varepsilon_{\mathrm{II}}) \leftrightarrow (x, \varepsilon_{\mathrm{I}}) \leqslant^{r} (y, \varepsilon_{\mathrm{II}}) \leftrightarrow (x, \varepsilon_{\mathrm{I}}) \leqslant^{r}_{\Pi} (y, \varepsilon_{\mathrm{II}})$

by clause (iii) of Definition 1.1. It follows that A_{xy} is $\mathbf{\Delta}^1_r$ and thus, by the assumption PD, is determined. Hence, whenever one of $\sim\mathsf{U}^1_{r+1}(x)$ or $\sim\mathsf{U}^1_{r+1}(y)$,

(3) $\qquad x \leqslant^{r+1} y \leftrightarrow$ player II has a winning strategy in A_{xy}

$$\leftrightarrow \exists\delta\, \forall\gamma\, [(x, (\gamma \# \delta)_{\mathrm{I}}) \leqslant^{r} (y, (\gamma \# \delta)_{\mathrm{II}})].$$

It follows immediately from (2), (3) and the definitions that whenever one of $\sim\mathsf{U}^1_{r+1}(x)$ or $\sim\mathsf{U}^1_{r+1}(y)$,

(4) $\qquad x \leqslant^{r+1}_{\Sigma} y \leftrightarrow x \leqslant^{r+1} y \leftrightarrow x \leqslant^{r+1}_{\Pi} y,$

that is, condition (iii) of Definition 1.1 is satisfied.

It remains to check condition (i). Since V here is $\sim\mathsf{U}^1_{r+1}$, (i)(a) and (b) become
(a) $\mathsf{U}^1_{r+1}(y) \to x \leqslant^{r+1} y$;
(b) $\sim\mathsf{U}^1_{r+1}(y) \wedge x \leqslant^{r+1} y \to \sim\mathsf{U}^1_{r+1}(x)$.
For (a), suppose $\mathsf{U}^1_{r+1}(y)$, so for some δ, $\sim\mathsf{U}^1_r(y, \delta)$. If player II plays in the game A_{xy} so that $\varepsilon_{\mathrm{II}} = \delta$, then by the corresponding property for $\leqslant^r$, $(x, \varepsilon_{\mathrm{I}}) \leqslant^r (y, \varepsilon_{\mathrm{II}})$ and thus II wins. Hence II has a winning strategy in A_{xy}, hence I does not so $x \leqslant^{r+1} y$.

For (b), suppose $\sim\mathsf{U}^1_{r+1}(y)$, so for all δ, $\sim\mathsf{U}^1_r(y, \delta)$, and $x \leqslant^{r+1} y$. If ε is any play of A_{xy} in which II follows his winning strategy, then $(x, \varepsilon_{\mathrm{I}}) \leqslant^r (y, \varepsilon_{\mathrm{II}})$. Since $\mathsf{U}^1_r(y, \varepsilon_{\mathrm{II}})$, the corresponding property for $\leqslant^r$ implies also $\mathsf{U}^1_r(x, \varepsilon_{\mathrm{I}})$. Clearly player I may realize any function as ε_{I} and thus for all γ, $\mathsf{U}^1_r(x, \gamma)$—that is, $\sim\mathsf{U}^1_{r+1}(x)$.

Finally, we prove that $\leqslant^{r+1}$ is a pre-wellordering. First let δ be any function such that for all s and n, $\delta(s * \langle n \rangle) = n$. Then for any γ, if $\varepsilon = \gamma \# \delta$, then

$$\varepsilon_{\text{II}}(m) = \delta(\bar{\varepsilon}(2m+1)) = \varepsilon(2m) = \varepsilon_{\text{I}}(m),$$

and thus by the reflexivity of $\preccurlyeq^r$, $(x, \varepsilon_{\text{I}}) \preccurlyeq^r (x, \varepsilon_{\text{II}})$. In other words, δ is a winning strategy for II in the game A_{xx}, so I has no winning strategy and thus $x \preccurlyeq^{r+1} x$ — that is, $\preccurlyeq^{r+1}$ is reflexive.

We next establish that $\preccurlyeq^{r+1}$ is connected. Suppose $y \not\preccurlyeq^{r+1} x$, so I has a winning strategy, say γ^0, for A_{yx}. Hence

$$\forall \delta\, [(y, (\gamma^0 \# \delta)_{\text{I}}) \not\preccurlyeq^r (x, (\gamma^0 \# \delta)_{\text{II}})].$$

Then because $\preccurlyeq^r$ is connected,

$$\forall \delta\, [(x, (\gamma^0 \# \delta)_{\text{II}}) \preccurlyeq^r (x, (\gamma^0 \# \delta)_{\text{I}})].$$

We aim to show that player I does not have a winning strategy in A_{xy} and thus that $x \preccurlyeq^{r+1} y$. For this it will suffice to show

(5) there exists δ such that for any γ there exists δ_0 such that if $\varepsilon = (\gamma \# \delta)$ and $\varepsilon^0 = (\gamma^0 \# \delta^0)$, then $\varepsilon_{\text{I}} = \varepsilon^0_{\text{II}}$ and $\varepsilon^0_{\text{I}} = \varepsilon_{\text{II}}$.

Given (5), we have for any γ,

$$(x, \varepsilon_{\text{I}}) = (x, \varepsilon^0_{\text{II}}) \preccurlyeq^r (y, \varepsilon^0_{\text{I}}) = (y, \varepsilon_{\text{II}})$$

so that δ is a winning strategy for II.

To prove (5) we define δ and δ^0 by the recursive conditions:

$$\begin{aligned} &\delta^0(\bar{\varepsilon}^0(2m+1)) = \gamma(\bar{\varepsilon}(2m)); \quad \text{and} \\ &\delta(\bar{\varepsilon}(2m+1)) = \gamma^0(\bar{\varepsilon}^0(2m)). \end{aligned} \tag{6}$$

The solution to these equations may best be visualized by considering the following diagram:

A_{xy}	I	$m_0 = \gamma(\langle\ \rangle)$		$m_2 = \gamma(\langle m_0, n_0\rangle)$	$\ldots$	ε_{I}
	II		n_0	n_2	$\ldots$	ε_{II}
A_{yx}	I	$n_0 = \gamma^0(\langle\ \rangle)$		$n_2 = \gamma^0(\langle n_0, m_0\rangle)$	$\ldots$	ε^0_{I}
	II		m_0	m_2	$\ldots$	$\varepsilon^0_{\text{II}}$.

As player I plays in A_{xy} according to his strategy γ, player II uses these moves to construct δ^0 which he then plays against γ^0 in A_{yx} to determine his moves in A_{xy}. δ describes this "strategy".

We show next that $\leqslant^{r+1}$ is transitive. Suppose $x \leqslant^{r+1} y$ and $y \leqslant^{r+1} z$. If $U^1_{r+1}(z)$, then by (a) above, $x \leqslant^{r+1} z$ as desired, so we assume $\sim U^1_{r+1}(z)$. By (b) above, we have then $\sim U^1_{r+1}(y)$ and $\sim U^1_{r+1}(x)$, so condition (3) holds for both pairs (x, y) and (y, z). Let δ^0 and δ^1 be winning strategies for II in A_{xy} and A_{yz}, respectively — that is,

$$\begin{aligned} &\forall \gamma\, [(x, (\gamma \# \delta^0)_{\mathrm{I}}) \leqslant^r (y, (\gamma \# \delta^0)_{\mathrm{II}})], \quad \text{and} \\ (7) \qquad &\forall \gamma\, [(y, (\gamma \# \delta^1)_{\mathrm{I}}) \leqslant^r (z, (\gamma \# \delta^1)_{\mathrm{II}})]. \end{aligned}$$

We aim to show that player I does not have a winning strategy in A_{xz}. For this it will suffice to show:

$$\begin{aligned} &\text{there exists } \delta \text{ such that for any } \gamma \text{ there exist } \gamma^0 \text{ and } \gamma^1 \text{ such that if} \\ (8) \qquad &\varepsilon = (\gamma \# \delta) \quad \text{and} \quad \varepsilon^i = (\gamma^i \# \delta^i)\ (i = 0, 1), \quad \text{then} \\ &\varepsilon_{\mathrm{I}} = \varepsilon^0_{\mathrm{I}}, \qquad \varepsilon^0_{\mathrm{II}} = \varepsilon^1_{\mathrm{I}}, \quad \text{and} \quad \varepsilon^1_{\mathrm{II}} = \varepsilon_{\mathrm{II}}, \end{aligned}$$

as if (8) holds, then for any γ,

$$(x, \varepsilon_{\mathrm{I}}) = (x, \varepsilon^0_{\mathrm{I}}) \leqslant^r (y, \varepsilon^0_{\mathrm{II}}) = (y, \varepsilon^1_{\mathrm{I}}) \leqslant^r (z, \varepsilon^1_{\mathrm{II}}) = (z, \varepsilon_{\mathrm{II}})$$

so that by the transitivity of $\leqslant^r$, $(x, \varepsilon_{\mathrm{I}}) \leqslant^r (z, \varepsilon_{\mathrm{II}})$ and thus δ is a winning strategy for II. Again, the solution to (8) may best by visualized by means of a diagram:

A_{xz}	I	$m_0 = \gamma(\langle\ \rangle)$		$m_2 = \gamma(\langle m_0, p_0\rangle)$		$\ldots\ \varepsilon_{\mathrm{I}}$
	II		p_0		p_2	$\ldots\ \varepsilon_{\mathrm{II}}$
A_{xy}	I	m_0		m_2		$\ldots\ \varepsilon^0_{\mathrm{I}}$
	II	$n_0 = \delta^0(\langle m_0\rangle)$		$n_2 = \delta^0(\langle m_0, n_0, m_2\rangle)$		$\ldots\ \varepsilon^0_{\mathrm{II}}$
A_{yz}	I	n_0		n_2		$\ldots\ \varepsilon^1_{\mathrm{I}}$
	II	$p_0 = \delta^1(\langle n_0\rangle)$		$p_2 = \delta^1(\langle n_0, p_0, n_2\rangle)$		$\ldots\ \varepsilon^1_{\mathrm{II}}$.

As player I plays according to γ in A_{xz}, player II constructs γ^0 and γ^1 as indicated, plays them against δ^0 and δ^1 in A_{xy} and A_{yz}, respectively, and uses the resulting moves in A_{yz} as his moves in A_{xz}.

Finally, suppose that $\leqslant^{r+1}$ is not well founded so there exist x_i such that for all $i \in \omega$, $x_{i+1} \preccurlyeq^{r+1} x_i$ but $x_i \not\preccurlyeq^{r+1} x_{i+1}$. Then in each game $\mathsf{A}_{x_i, x_{i+1}}$, player I has a winning strategy, say γ^i. We aim to show:

$$\text{(9)} \qquad \begin{array}{l} \text{there exist strategies } \delta^i \ (i \in \omega) \text{ such that for all } i, \\ \text{if } \varepsilon^i = (\gamma^i \# \delta^i), \text{ then } \varepsilon_{\mathrm{I}}^{i+1} = \varepsilon_{\mathrm{II}}^{i}. \end{array}$$

From (9), it follows that for all i,

$$(x_i, \varepsilon_{\mathrm{I}}^i) \not\preccurlyeq^r (x_{i+1}, \varepsilon_{\mathrm{II}}^i) = (x_{i+1}, \varepsilon_{\mathrm{I}}^{i+1})$$

which contradicts the well-foundedness of $\preccurlyeq^r$. Strategies δ^i satisfying (9) may be constructed as in the following diagram:

$$\begin{array}{llllllll}
 & \mathrm{I} & m_0 = \gamma^0(\langle\ \rangle) & & m_2 = \gamma^0(\langle m_0, n_0\rangle) & & \dots & \varepsilon_{\mathrm{I}}^0 \\
\mathsf{A}_{x_0 x_1} & & & & & & & \\
 & \mathrm{II} & & n_0 & & n_2 & \dots & \varepsilon_{\mathrm{II}}^0 \\
 & \mathrm{I} & n_0 = \gamma^1(\langle\ \rangle) & & n_2 = \gamma^1(\langle n_0, p_0\rangle) & & \dots & \varepsilon_{\mathrm{I}}^1 \\
\mathsf{A}_{x_1 x_2} & & & & & & & \\
 & \mathrm{II} & & p_0 & & p_2 & \dots & \varepsilon_{\mathrm{II}}^1 \\
 & \mathrm{I} & p_0 = \gamma^2(\langle\ \rangle) & & p_2 = \gamma^2(\langle p_0, q_0\rangle) & & \dots & \varepsilon_{\mathrm{I}}^2 \\
\mathsf{A}_{x_2 x_3} & & & & & & & \\
 & \mathrm{II} & & q_0 & & q_2 & \dots & \varepsilon_{\mathrm{II}}^2 . \\
\cdot & & & & & & & \\
\cdot & & & & & & & \\
\cdot & & & & & & &
\end{array}$$

□

3.2 Theorem. *If* PD, *then for all odd* r, Π_r^1 *and* Σ_{r+1}^1 *have the pre-wellordering property.*

Proof. By induction using Theorems 1.2, 1.3, and 3.1. □

To express concisely the properties of the analytical and projective hierarchies which now follow from the general results of § 1, let $\preccurlyeq_{\mathrm{D}}^1$ be $\preccurlyeq^1$ as defined in the proof of Theorem 1.2 and for all $r > 1$, let $\preccurlyeq_{\mathrm{D}}^r$ be the pre-wellorderings which arise by application of the proofs of Theorems 1.3 and 3.1. Let $|\ \ |_{\mathrm{D}}^r$ be the norm associated with $\preccurlyeq_{\mathrm{D}}^r$, and $\boldsymbol{\kappa}_{\mathrm{D}}^r = \sup^+\{|x|_{\mathrm{D}}^r : \mathsf{V}_r(x)\}$, where V_r is $\sim\mathsf{U}_r^1$, if r is odd, and U_r^1, if r is even. Similarly, $|\ \ |_{\mathrm{D},0}^r$ is the norm associated with the restriction of $\preccurlyeq_{\mathrm{D}}^r$ to sequences of the form $(a, \langle \mathbf{m}\rangle, \langle\ \rangle)$ and $\kappa_{\mathrm{D}}^r = \sup^+\{|a, \langle \mathbf{m}\rangle|_{\mathrm{D},0}^r : V_r(a, \langle \mathbf{m}\rangle)\}$, where V_r is $\sim U_r^1$, if r is odd, and U_r^1, if r is even. Note that there is no reason to believe that V_r is an initial segment of V_r with respect to $\leqslant^r_{\mathrm{D}}$ so that we may have $|a,\langle\mathbf{m}\rangle|^r_{\mathrm{D},0}<|a,\langle\mathbf{m}\rangle,\langle\ \ \rangle|^r_{\mathrm{D}}$.

3.3 Theorem. *If* PD, *then for all odd r,*

(i) Π^1_r *and* $\boldsymbol{\Pi}^1_r$ *have the reduction property but not the separation property;*

(ii) Σ^1_r *and* $\boldsymbol{\Sigma}^1_r$ *have the separation property but not the reduction property;*

(iii) (a) *for any* $R\in\Sigma^1_r$, *if* $R\subseteq\sim U^1_r$, *then*

$$\sup{}^+\{|a,\langle\mathbf{m}\rangle|^r_{\mathrm{D},0}:R(a,\langle\mathbf{m}\rangle)\}<\delta^1_r,$$

(b) *for any* $\mathsf{R}\in\boldsymbol{\Sigma}^1_r$, *if* $\mathsf{R}\subseteq\sim\mathsf{U}^1_r$, *then*

$$\sup{}^+\{|x|^r_{\mathrm{D}}:\mathsf{R}(x)\}<\boldsymbol{\delta}^1_r;$$

(iv) (a) *for all R,* $R\in\Delta^1_r\leftrightarrow R\lessdot\sim U^1_{r,\rho}$ *for some* $\rho<\delta^1_r$;

(b) *for all* R, $\mathsf{R}\in\boldsymbol{\Delta}^1_r\leftrightarrow\mathsf{R}\lessdot\sim\mathsf{U}^1_{r,\rho}$ *for some* $\rho<\boldsymbol{\delta}^1_r$;

(v) $\{\alpha:\alpha\in\Delta^1_r\}\in\Pi^1_r\sim\Delta^1_r$;

(vi) Δ^1_r *is not a basis for* Π^1_{r-1};

(vii) *for any* $\mathsf{R}\in\Pi^1_r$, *there exists a partial functional* $\mathrm{Sel}_{\mathsf{R}}$ *with* Π^1_r *graph such that for all* $\mathbf{m}$ *and* $\boldsymbol{\alpha}$,

$$\exists p\,\mathsf{R}(p,\mathbf{m},\boldsymbol{\alpha})\leftrightarrow\mathsf{R}(\mathrm{Sel}_{\mathsf{R}}(\mathbf{m},\boldsymbol{\alpha}),\mathbf{m},\boldsymbol{\alpha})\leftrightarrow\mathrm{Sel}_{\mathsf{R}}(\mathbf{m},\boldsymbol{\alpha})\downarrow;$$

(viii) $\kappa^r_{\mathrm{D}}=\delta^1_r$ *and* $\boldsymbol{\kappa}^r_{\mathrm{D}}=\boldsymbol{\delta}^1_r$. □

3.4 Theorem. *If* PD, *then for all even* $r\geqslant 2$,

(i) Σ^1_r *and* $\boldsymbol{\Sigma}^1_r$ *have the reduction property but not the separation property;*

(ii) Π^1_r *and* $\boldsymbol{\Pi}^1_r$ *have the separation property but not the reduction property;*

(iii) *for any* $R\in\Pi^1_r$, *if* $R\subseteq U^1_r$, *then*

$$\sup{}^+\{|a,\langle\mathbf{m}\rangle|^r_{\mathrm{D},0}:R(a,\langle\mathbf{m}\rangle)\}<\delta^1_r;$$

(iv) *for all R,* $R\in\Delta^1_r\leftrightarrow R\lessdot U^1_{r,\rho}$ *for some* $\rho<\delta^1_r$;

(v) $\{\alpha:\alpha\in\Delta^1_r\}\in\Sigma^1_r\sim\Delta^1_r$;

(vi) Δ^1_r *is not a basis for* Π^1_r;

(vii) *for any* $\mathsf{R}\in\Sigma^1_r$, *there exists a partial functional* $\mathrm{Sel}_{\mathsf{R}}$ *wtih* Σ^1_r *graph such that for all* $\mathbf{m}$ *and* $\boldsymbol{\alpha}$,

$$\exists p\,\mathsf{R}(p,\mathbf{m},\boldsymbol{\alpha})\leftrightarrow\mathsf{R}(\mathrm{Sel}_{\mathsf{R}}(\mathbf{m},\boldsymbol{\alpha}),\mathbf{m},\boldsymbol{\alpha})\leftrightarrow\mathrm{Sel}_{\mathsf{R}}(\mathbf{m},\boldsymbol{\alpha})\downarrow;$$

(viii) $\kappa^r_{\mathrm{D}}=\delta^1_r$ *and* $\boldsymbol{\kappa}^r_{\mathrm{D}}=\boldsymbol{\kappa}^{r-1}_{\mathrm{D}}=\boldsymbol{\delta}^1_{r-1}$. □

In the remainder of this section we shall give a brief survey of some of the other consequences of PD and other forms of determinacy for the analytical and projective hierarchies. The theory is very rich; in fact, nearly all of the theory of Π_1^1 and Σ_2^1 applies to Π_r^1 and Σ_{r+1}^1 for all odd r under the hypothesis PD. We put in parentheses following each result the number of the result(s) which it generalizes.

3.5 Theorem (Moschovakis [1971a]). *If* PD, *then for all odd r*, Π_r^1, $\mathbf{\Pi}_r^1$, Σ_{r+1}^1 *and* $\mathbf{\Sigma}_{r+1}^1$ *have the uniformization property* (IV.7.8 *and* IV.7.13). □

3.6 Corollary. *If* PD, *then for all even* $r \geqslant 2$, Δ_r^1 *is a basis for* Σ_r^1 (IV.7.9). □

3.7 Theorem. *If* PD, *then for all odd r, and any* $\beta \notin \Delta_r^1$ *which is implicitly* Π_r^1, $\{\alpha : \alpha$ *is* Δ_r^1 *in* $\beta\}$ *is a basis for* Σ_r^1 (III.4.7 — *cf. Exercise* 1.31). □

3.8 Theorem (Moschovakis). *If* PD, *then for all odd r and all* $\mathsf{R} \subseteq {}^{k,l}\omega$, $\mathsf{R} \in \Pi_r^1$ *iff for some* Π_{r-1}^1 *relation* P,

$$\mathsf{R}(\mathbf{m}, \boldsymbol{\alpha}) \leftrightarrow (\exists \beta \in \Delta_r^1[\boldsymbol{\alpha}])\mathsf{P}(\mathbf{m}, \boldsymbol{\alpha}, \beta) \qquad \text{(IV.2.9).}\ \square$$

3.9 Theorem (Davis [1964], Mycielski–Swierczkowski [1964]). *If* PD, *then every uncountable projective set has a perfect subset* (IV.5.12) *and every projective set is Lebesgue measurable* (IV.5.3) *and has the Baire property* (IV.5.10). □

3.10 Theorem (Moschovakis). *If* PD, *then for all odd r and all* $\mathsf{A} \subseteq {}^{\omega}\omega$, *the following are equivalent*:

(i) $\mathsf{A} \in \mathbf{\Delta}_r^1$;

(ii) A *is the image of a* $\mathbf{\Delta}_r^1$ *set* B *under a continuous functional which is one–one on* B;

(iii) A *is the image of a* $\mathbf{\Pi}_{r-1}^1$ *set* C *under a continuous functional which is one–one on* C. (IV.6.9 — *cf. Exercise* 1.28). □

In each of the preceding theorems where the hypothesis PD is assumed, the proof establishes a somewhat sharper theorem. In the proof of Theorem 3.1, to define the pre-wellordering $\leqslant^{r+1}$ from $\leqslant^{r}$ we needed only the determinacy of certain $\mathbf{\Delta}_r^1$ sets. Hence, for example, that Π_3^1 has the pre-wellordering property requires only Det($\mathbf{\Delta}_2^1$). Similarly, Theorem 3.9 can be refined to show that if every $\mathbf{\Pi}_r^1$ set is determined, then every uncountable $\mathbf{\Pi}_r^1$ set has a perfect subset. Thus with Theorem 2.8 we have a clear measure of the incompatibility of V = L and PD:

3.11 Corollary. *If* V = L, *then* $\neg$Det($\mathbf{\Pi}_1^1$). □

A parallel result we simply mention is

3.12 Theorem (Solovay). *If* $\mathrm{Det}(\boldsymbol{\Delta}_2^1)$, *then there exists a* Δ_3^1 *non-constructible subset of* ω (*Friedman* [1971]). □

The conclusion of this Theorem is optimal in that all Σ_2^1 and Π_2^1 subsets of ω are provably constructible (Shoenfield [1962] and Theorem VIII.3.7).

The ordinals δ_r^1 are all countable, but the ordinals $\boldsymbol{\delta}_r^1$ may be quite large. It turns out that their size relative to the sequence $\aleph_0, \aleph_1, \ldots$ depends rather delicately on just what set-theoretical assumptions are used. Of course, if the Axiom of Choice is assumed, then $2^{\aleph_0} = \aleph_\sigma$ for some σ and any pre-wellordering of ${}^\omega\omega$ has length less than $\aleph_{\sigma+1}$. In any case $\boldsymbol{\delta}_1^1 = \aleph_1$ and $\boldsymbol{\delta}_2^1 \leqslant \aleph_2$ Martin [1977]. Under the assumption of PD together with the Axiom of Choice, one has also $\boldsymbol{\delta}_3^1 \leqslant \aleph_3$ and $\boldsymbol{\delta}_4^1 \leqslant \aleph_4$, for all r, $\boldsymbol{\delta}_r^1 < \boldsymbol{\delta}_{r+1}^1$, and if r is odd, $\boldsymbol{\delta}_{r+1}^1$ is not larger than the next cardinal greater than $\boldsymbol{\delta}_r^1$. It is conjectured that in this theory $\boldsymbol{\delta}_r^1 \leqslant \aleph_r$ for all r.

If one assumes $\mathrm{Det}(\mathbf{P}({}^\omega\omega))$, the determinacy of all subsets of ${}^\omega\omega$, then it is no longer consistent to use the Axiom of Choice. It seems, however, to be consistent to assume the Axiom of Dependent Choice. In this theory it can be proved that all $\boldsymbol{\delta}_r^1$ are regular cardinals and for odd r, $\boldsymbol{\delta}_{r+1}^1$ is the least cardinal greater than $\boldsymbol{\delta}_r^1$. Thus $\boldsymbol{\delta}_2^1 = \aleph_2$ here. Curiously, however, all $\aleph_n$ with $2 < n \leqslant \omega$ are singular in this theory and it turns out that $\boldsymbol{\delta}_3^1 = \aleph_{\omega+1}$ and $\boldsymbol{\delta}_4^1 = \aleph_{\omega+2}$. In fact, for all odd r, $\boldsymbol{\delta}_r^1$ is the least cardinal greater than some cardinal with cofinality ω. It would then be expected that $\boldsymbol{\delta}_5^1 = \aleph_{\omega\cdot 2+1}$, but it is known that $\boldsymbol{\delta}_5^1$ is still larger.

We mentioned earlier that it is not possible to prove the consistency of ZFC + PD without some strong hypotheses. Hence the results of this section do not provide us with any proofs of consistency or independence. Harrington [1978] has shown that it is consistent with ZFC that neither of the classes Σ_r^1 or Π_r^1 $(r \geqslant 3)$ has either the reduction or separation properties. It follows that reduction for Σ_r^1 $(r \geqslant 3)$ and separation for Π_r^1 $(r \geqslant 3)$ are independent of ZFC. For each α and n, let $\mathfrak{A}_{r,n}$ be an assertion in the language of set theory as follows:

$\mathfrak{A}_{r,0}$: Σ_r^1 has the reduction property;

$\mathfrak{A}_{r,1}$: Π_r^1 has the reduction property;

$\mathfrak{A}_{r,n+2}$: neither Σ_r^1 nor Π_r^1 has the reduction property.

It is an open question in general for which α, the assertion $\forall r\, \mathfrak{A}_{r,\alpha(r)}$ is consistent with ZFC. The same question with "reduction" replaced by "separation" or "pre-wellordering" is also open.

3.13–3.32 Exercises

3.13. Complete the proof sketched above that if the Axiom of Choice holds, then there is a non-determined set. (*Hint*: construct two sequences α_σ and β_σ

and take $\mathsf{A} = \{\alpha_\sigma : \sigma < \lambda\}$. The role of α_σ (β_σ) is to ensure that γ_σ is not a winning strategy for player II (I).)

3.14 (Mycielski [1964]). Let A_n ($n \in \omega$) be a countable family of subsets of ${}^\omega\omega$ such that the relation "$\beta \in \mathsf{A}_n$" is projective. Show that if PD, then there exists a choice function θ such that $\forall n . \theta(n) \in \mathsf{A}_n$. Derive from this fact that Σ^1_r and Π^1_r are closed under number quantification.

3.15 (Gale–Stewart [1953]). Prove that every open game is determined. (If I has no winning strategy, then player II may win by ensuring that at each stage of the game he still has a chance.)

3.16 (Martin [1968]). The following is an outline of an alternative proof for Theorem 3.1. We write "$\gamma \leqslant \delta$" for "γ is recursive in δ" and "$\gamma \equiv \delta$" for "$\gamma \leqslant \delta$ and $\delta \leqslant \gamma$".

(i) Let A be any determined subset of ${}^\omega\omega$ which is closed under $\equiv$ (A may be thought of as a set of degrees.) Show that there is a function γ such that either

$$(\forall \delta \geqslant \gamma)\ \delta \in \mathsf{A} \quad \text{or} \quad (\forall \delta \geqslant \gamma)\ \delta \notin \mathsf{A}.$$

(ii) We say $\overline{\text{mes}}(\mathsf{A}) = 1$ if the first alternative holds, $\overline{\text{mes}}(\mathsf{A}) = 0$ if the second holds. Show that if PD, then $\overline{\text{mes}}$ is a countably additive measure on the projective sets of degrees.

(iii) Let $\leqslant^r$, $\leqslant^r_\Sigma$, and $\leqslant^r_\Pi$ be relations which establish the pre-wellordering property for Σ^1_r and $|\ \ |^r$ the norm associated with $\leqslant^r$. For any ε, let

$$|x|^r = \sup{}^+\{|x, \gamma|^r : \gamma \leqslant \varepsilon\}$$

and set

$$x \leqslant^{r+1} y \leftrightarrow \overline{\text{mes}}(\{\varepsilon : |x|^r_\varepsilon \leqslant |y|^r_\varepsilon\}) = 1.$$

Show that if PD then there exist relations $\leqslant^{r+1}_\Sigma$ and $\leqslant^{r+1}_\Pi$ which together with $\leqslant^{r+1}$ establish the pre-wellordering property for Π^1_{r+1}.

3.17 (Blackwell [1967]). Complete the following game-theoretic proof that Π^1_1 has the reduction property. Suppose

$$\mathsf{R}(\mathbf{m}, \boldsymbol{\alpha}) \leftrightarrow \forall \beta\, \exists p\, \mathsf{P}(\bar{\beta}(p), \mathbf{m}, \boldsymbol{\alpha})$$

and

$$\mathsf{S}(\mathbf{m}, \boldsymbol{\alpha}) \leftrightarrow \forall \gamma\, \exists q\, \mathsf{Q}(\bar{\gamma}(q), \mathbf{m}, \boldsymbol{\alpha})$$

with P and Q closed-open. For each $(\mathbf{m}, \boldsymbol{\alpha})$, set

$$\mathsf{A}_{\mathbf{m},\boldsymbol{\alpha}} = \{\varepsilon : \exists p\, [\mathsf{P}(\bar{\varepsilon}_{\mathrm{II}}(p), \mathbf{m}, \boldsymbol{\alpha}) \wedge (\forall q \leqslant p) \neg \mathsf{Q}(\bar{\varepsilon}_{\mathrm{I}}(q), \mathbf{m}, \boldsymbol{\alpha})]\}.$$

Let

$\mathsf{R}^*(\mathbf{m}, \boldsymbol{\alpha}) \leftrightarrow \mathsf{R}(\mathbf{m}, \boldsymbol{\alpha})$ and player I has a winning strategy for $\mathsf{A}_{\mathbf{m},\boldsymbol{\alpha}}$;

$\mathsf{S}^*(\mathbf{m}, \boldsymbol{\alpha}) \leftrightarrow \mathsf{S}(\mathbf{m}, \boldsymbol{\alpha})$ and player II has a winning strategy for $\mathsf{A}_{\mathbf{m},\boldsymbol{\alpha}}$.

Show that $(\mathsf{R}^*, \mathsf{S}^*)$ reduces (R, S). (*Hint*: $\mathsf{A}_{\mathbf{m},\boldsymbol{\alpha}}$ is an open, hence determined game by Exercise 3.15.)

The following series of exercises leads to a proof of Theorem 3.5. We say that X ($= \Sigma^1_r$ or Π^1_r) has the *scale property* iff for all $n \in \omega$ there exists relations $\leqslant_n$, $\leqslant_{\Sigma,n}$ and $\leqslant_{\Pi,n}$ which satisfy $(\mathrm{i})_n$ and $(\mathrm{iii})_n$ as in Definition 1.1,

(ii)′ the relations $\mathsf{R}_\Sigma(n, x, y) \leftrightarrow x \leqslant_{\Sigma,n} y$ and $\mathsf{R}_\Pi(n, x, y) \leftrightarrow x \leqslant_{\Pi,n} y$ are Σ^1_r and Π^1_r, respectively, and if $|\ \ |_n$ is the norm associated with $\leqslant_n$,

(iv) for any x and any ordinals λ_n, if $\langle x_i : i \in \omega\rangle$ is a sequence of elements of V which converges to x and $\forall n\, \exists i_n\, (\forall i \geqslant i_n)\, |x_i|_n = \lambda_n$, then $\mathsf{V}(x)$ and $\forall n . |x|_n \leqslant \lambda_n$.

We say X has the *nice scale property* iff there exist $\leqslant_n$, etc. which in addition satisfy the following (when $x = \langle a, \langle \mathbf{m}\rangle, \langle \boldsymbol{\alpha}\rangle\rangle$, we write $x(n)$ for $\langle a, \mathbf{m}, \alpha_0(n), \ldots, \alpha_{l-1}(n)\rangle$):

(v) for any x and y, if $|x|_n \leqslant |y|_n$, then also $|x|_m \leqslant |y|_m$ for all $m < n$; if $|x|_n = |y|_n$, then $x(m) = y(m)$ for all $m \leqslant n$.

Note that (iv) and (v) together imply

(vi) if $\langle x_i : i \in \omega\rangle$ is a sequence such that $\forall i\, \mathsf{V}(x_i)$ and there exist $n_0 < n_1 < \cdots$ and ordinals λ_{n_k} such that $\forall k\, \exists j_k\, (\forall i \geqslant j_k) . |x_i|_{n_k} = \lambda_{n_k}$, then $\langle x_i : i \in \omega\rangle$ converges to some $x \in \mathsf{V}$ and for *all* n there are ordinals λ_n such that $\exists i_n\, (\forall i \geqslant i_n) . |x_i|_n = \lambda_n$, and $|x|_n \leqslant \lambda_n$.

3.18. Show that if X has the scale property, then X has the nice scale property. (If $\leqslant_n$, etc., establish the scale property, set

$$x^{(n)} = (|x|_0, x(0), |x|_1, x(1), \ldots, |x|_n, x(n))$$

and

$$x \leqslant^*_n y \leftrightarrow x^{(n)} \text{ precedes } y^{(n)} \text{ lexicographically or } x, y \notin \mathsf{V}.)$$

3.19. Show that if Π^1_r has the nice scale property, then Π^1_r has the uniformization property. (Suppose $\mathsf{R}(x, \gamma) \leftrightarrow \mathsf{V}(a, x, \gamma)$ is a Π^1_r relation. Let $\lambda_{x,n} = \inf\{|a, x, \gamma|_n : \mathsf{R}(x, \gamma)\}$ and for each n choose γ_n such that $|a, x, \gamma_n|_n = \lambda_{x,n}$.

Show that the (a, x, γ_n) converge to some (a, x, γ_x) such that $\mathsf{R}(x, \gamma_x)$ and that the function θ such that $\theta(x) = \gamma_x$ has Π^1_r graph.)

3.20. Show that Π^1_1 has the scale property. (To each x assign a linear ordering as usual such that $\sim\mathsf{U}^1_1(x)$ iff $\leqslant_x$ is a well-ordering. Set $p \leqslant_{x,n} q$ iff $p \leqslant_x q \wedge q <_x n$, and take

$$\begin{aligned} x \leqslant_n y \quad \text{iff} \quad & \|\leqslant_x\| < \|\leqslant_y\| \\ \text{or} \quad & (\|\leqslant_x\| = \|\leqslant_y\| < \aleph_1 \wedge \|\leqslant_{x,n}\| \leqslant \|\leqslant_{y,n}\|) \\ \text{or} \quad & (\|\leqslant_x\| = \|\leqslant_y\| = \aleph_1)). \end{aligned}$$

Towards (iv), show that if $\langle x_i : i \in \omega \rangle$ converges to x and for $i \geqslant i_n$, $\|\leqslant_{x_i}\| = \bar{\lambda}$ and $\|\leqslant_{x_i,n}\| = \bar{\lambda}_n$, then the function $n \mapsto \bar{\lambda}_n$ is order-preserving on the field of $\leqslant_x$.)

3.21. Show that if Π^1_r has the scale property, then also Σ^1_{r+1} has the scale property. (Let $\leqslant^r_n$, etc., be a nice scale for Π^1_r and let θ be constructed as in Exercise 3.19 to uniformize $\sim\mathsf{U}^1_r$. Thus θ has Π^1_r graph and $\mathsf{U}^1_{r+1} = \mathrm{Dm}\,\theta$. Set

$$|x|^{r+1}_n = \begin{cases} |x, \theta(x)|^r_n, & \text{if} \quad \mathsf{U}^1_{r+1}(x); \\ \kappa^r_n, & \text{otherwise}, \end{cases}$$

let $\leqslant^{r+1}_n$ be the associated pre-wellordering,

$$x \leqslant^{r+1}_{\Sigma,n} y \leftrightarrow \exists\gamma\, \exists\delta\, [\theta(x) \simeq \gamma \wedge \theta(y) \simeq \delta \wedge (x, \gamma) \leqslant^r_{\Pi,n} (y, \delta)],$$

and

$$x \leqslant^{r+1}_{\Pi,n} y \leftrightarrow \forall\delta\, \exists\gamma\, [\theta(y) \simeq \delta \rightarrow (x, \gamma) \leqslant^r_{\Sigma,n} (y, \delta)].)$$

3.22. Show that if PD, then for any $r \geqslant 1$, if Σ^1_r has the scale property, then also Π^1_{r+1} has the scale property. (Let $\leqslant^r_n$, etc. be a nice scale for Σ^1_r and let $s_0, s_1, \ldots$ be a recursive one–one enumeration of codes for finite sequences of natural numbers such that $s_0 = \langle\ \rangle$ and if s_i is a proper initial segment of s_j, then $i < j$. Set

$$\begin{aligned} &\mathsf{A}^n_{x,y} = \{\varepsilon : \neg[(x, s_n * \varepsilon_{\mathrm{I}}) \leqslant^r_n (y, s_n * \varepsilon_{\mathrm{II}})]\}; \\ &x \lessapprox^{r+1}_n y \leftrightarrow \text{player II has a winning strategy for } \mathsf{A}^n_{xy}; \\ &x \leqslant^{r+1}_n y \leftrightarrow x \lessapprox^{r+1}_0 y \vee (\sim\mathsf{U}^1_{r+1}(x) \wedge x \lessapprox^{r+1}_0 y \wedge \\ &\qquad\qquad y \lessapprox^{r+1}_0 x \wedge x \lessapprox^{r+1}_n y) \vee \mathsf{U}^1_{r+1}(y). \end{aligned}$$

Define $\leqslant^{r+1}_{\Sigma,n}$ and $\leqslant^{r+1}_{\Pi,n}$ and prove (i), (ii)′, and (iii) analogously to the proof of Theorem 3.1. For (iv), suppose $\langle x_i : i \in \omega \rangle$ converges to x and

$\forall n\,(\forall i \geq n).\,|x_i|_n^{r+1} = \lambda_n$. For any fixed β, let $s_{n_i} = \bar{\beta}(i)$, and choose δ^i to be winning strategies for player II in $A^{n_i}_{x_{n_i+1}, x_{n_i}}$. Construct strategies γ^i such that if $\varepsilon^i = \gamma^i \# \delta^i$, then $\varepsilon^i_{\mathrm{I}} = \langle\beta(i)\rangle * \varepsilon^{i+1}_{\mathrm{II}}$ (hence $s_{n_i} * \varepsilon^i_{\mathrm{I}} = s_{n_{i+1}} * \varepsilon^{i+1}_{\mathrm{II}}$). For each m, the ordinals $|(x_{n_i+1}, s_{n_i} * \varepsilon^i_{\mathrm{I}})|^r_{n_j}$ decrease with i, for $i \geq j$, and are thus eventually constant. Then $(x_{n_i+1}, s_{n_i} * \varepsilon^i_{\mathrm{I}})$ converges to (x, β) and $\mathsf{U}^1_r(x, \beta)$. To see that $|x|_n^{r+1} \leq \lambda_n$, it suffices to show that player II has a winning strategy in A^n_{x, x_n}. Fix n and a strategy γ for player I. For each k and all $m \geq k$, player II wins $\mathsf{A}^k_{x_m, x_k}$, say by strategy $\delta^{m,k}$. Show that there exist $m_0 = n < m_1 < m_2 < \ldots$ and strategies γ_i and δ such that if $\varepsilon = (\gamma \# \delta)$ and for all i, $\varepsilon^i = (\gamma^i \# \delta^{m_{i+1}, m_i})$, then $s_{m_i} = s_n * \bar{\varepsilon}_{\mathrm{I}}(i)$, $\varepsilon^i_{\mathrm{I}} = \langle\varepsilon_{\mathrm{I}}(i)\rangle * \varepsilon^{i+1}_{\mathrm{II}}$ (hence $s_{m_i} * \varepsilon^i_{\mathrm{I}} = s_{m_{i+1}} * \varepsilon^{i+1}_{\mathrm{II}}$), and $\varepsilon_{\mathrm{II}} = \varepsilon^0_{\mathrm{II}}$. It follows that the ordinals $|(x_{m_{i+1}}, s_{m_i} * \varepsilon^i_{\mathrm{I}})|^r_{m_j}$ decrease with i, for $i \geq j$, hence are eventually constant, say at λ'_j. Then $(x_{m_{i+1}}, s_{m_i} * \varepsilon^i_{\mathrm{I}})$ converges to $(x, s_n * \varepsilon_{\mathrm{I}})$ and

$$|(x, s_n * \varepsilon_{\mathrm{I}})|^r_n \leq \lambda'_0 \leq |(x_{m0}, s_{m0} * \varepsilon^0_{\mathrm{II}})|^r_{m0} = |(x_n, s_n * \varepsilon_{\mathrm{II}})|^r_n.)$$

3.23 (Davis [1964]). Fill in the following sketch of a proof of the first clause of Theorem 3.9. Given $\mathsf{A} \subseteq {}^{\omega}2$, consider the game played as before except that player I at each of his turns may play any finite sequence of natural numbers. Show first that if player I has a winning strategy for this game, then A has a perfect subset. If player II has a winning strategy δ, for each (code for a) finite sequence s of 0's and 1's, let β_s be the unique function in ${}^{\omega}2$ such that

$$\text{for all } i < \lg(s),\ \beta_s(i) = (s)_i, \quad \text{and}$$
$$\text{for all } i \geq \lg(s),\ \beta_s(i) \neq \delta(\bar{\beta}_s(i)).$$

Show that in this case $\mathsf{A} \subseteq \{\beta_s : s \in \omega\}$ and thus A is countable.

There now remain two steps in the proof:

(i) if every uncountable projective subset of ${}^{\omega}2$ has a perfect subset, then the same is true for every uncountable projective subset of ${}^{\omega}\omega$;

(ii) if PD, then for every projective set $\mathsf{A} \subseteq {}^{\omega}2$, the game described above is determined.

Towards (ii), to each $\varepsilon \in {}^{\omega}2$ assign a function $\varepsilon^* \in {}^{\omega}2$ by interpreting the even values of ε as codes for finite segments of ε^* as follows: if $\lg(\varepsilon(0)) = n_0$, then $\bar{\varepsilon}^*(n_0) = \varepsilon(0)$ and $\varepsilon^*(n_0) = \varepsilon(1)$; if $\lg(\varepsilon(2)) = n_1$, then $\bar{\varepsilon}^*(n_0 + 1 + n_1) = \varepsilon(0) * \langle\varepsilon(1)\rangle * \varepsilon(1)$ and $\varepsilon^*(n_0 + 1 + n_1) = \varepsilon(3)$; etc. Given a projective set A, let $\mathsf{B} = \{\varepsilon : \varepsilon^* \in \mathsf{A}\}$. Then B is also projective and a winning strategy for either player in the ordinary game associated with B can be converted into one for the same player in the new game associated with A. Finally, verify that this proof gives also the refined version of Theorem 3.9 needed for Corollary 3.11.

3.24 (Solovay). Show that if all sets are determined, then the following choice principle holds: for any function $\varphi : \aleph_1 \to P({}^{\omega}\omega)$ such that if $\sigma < \tau < \aleph_1$, then

$\emptyset \neq \varphi(\tau) \subseteq \varphi(\sigma)$, there exists a continuous functional $\theta : {}^\omega\omega \to {}^\omega\omega$ such that for all $\gamma \in W$, $\theta(\gamma) \in \varphi(\|\gamma\|)$.

3.25. Use the preceding exercise to show that if all sets are determined, then every union of an $\aleph_1$-sequence of Borel sets is Σ^1_2 (cf. Corollary IV.5.2).

3.26 (Mycielski [1964]). (For readers versed in set theory). Show that if all sets are determined, then $\aleph_1$ is a strongly inaccessible cardinal in the constructible universe L. Conclude from this that the relative consistency of the statement "all sets are determined" with ZF cannot be proved in ZF. (It is obvious that $\aleph_1$ is still regular in L. If $\aleph_1$ were a successor cardinal $\aleph^L_{\sigma+1}$ in L, then show that there exists a set of relations $R_\rho \subseteq \aleph^L_\sigma \times \aleph^L_\sigma$ of power $\aleph_1$. Since $\aleph^L_\sigma$ is countable, this leads to a subset of ${}^\omega\omega$ of power $\aleph_1$. Since under the hypothesis ${}^\omega\omega$ is not well-orderable, this yields $\aleph_1 < 2^{\aleph_0}$, which contradicts an extension of Exercise 3.23. The proof is completed by use of Gödel's second incompleteness theorem.)

3.27. Show that if PD, then for all r there exist α and β which are Δ^1_r-incomparable. (Cf. Exercise 2.13.)

3.28. Show that if PD, then Δ^1_3 is a model of the Δ^1_3-Comprehension schema.

3.29. Strengthen Corollary 3.11 to: if Det $(\mathbf{\Pi}^1_1)$, then $\{\alpha : \alpha$ is constructible$\}$ is countable. (If Det $(\mathbf{\Pi}^1_1)$ but $\{\alpha : \alpha$ is constructible$\}$ is uncountable, then using Exercises 2.10 and 3.23 it contains a perfect subset. Use Lemma IV.6.3 and a modification of Theorem 2.2 to construct a $\mathbf{\Delta}^1_2$ well-ordering of ${}^\omega\omega$ and reach a contradiction via Theorem 2.8.)

3.30 (Wadge, Martin). For sets $A, B \subseteq {}^\omega\omega$, let

$$[A, B] = \{\varepsilon : \varepsilon_I \in A \leftrightarrow \varepsilon_{II} \notin B\}.$$

(These are known as *Wadge Games.*) Set

$A \boxed{\leq} B$ iff player II has a winning strategy in either

$$[A, B] \quad \text{or} \quad [A, \sim B]$$

and

$$A \boxed{<} B \quad \text{iff} \quad A \boxed{\leq} B \quad \text{but } not \quad B \boxed{\leq} A.$$

Show that if PD, then $\boxed{\leq}$ restricted to projective sets is a pre-wellordering. (For well-foundedness, suppose $\forall i. A_{i+1} \boxed{<} A_i$. Then I has a winning strategy in both

$[A_i, A_{i+1}]$ and $[A_i, \sim A_{i+1}]$, say (by Exercise 3.14) γ_i^0 and γ_i^1, respectively. For any $\delta \in {}^\omega 2$, imitate the proof of well-foundedness in Theorem 3.1 using the strategies $\gamma_i^{\delta(i)}$ to construct functions $\varepsilon^{i,\delta}$. Show that for any finite sequence s of 0's and 1's,

$$\text{mes}(\{\delta : \varepsilon_I^{i,\delta} \in A_i\} \cap [s]) = \tfrac{1}{2} \cdot \text{mes}([s]).$$

Derive a contradiction from the 0–1 Law, Exercise I.2.10.)

3.31. Show that the closed-open subsets of ${}^\omega\omega$ are pre-wellordered by $\boxed{\leq}$ in type $\aleph_1$.

3.32 (Martin). Show that if *every* Wadge game is determined, then for any $X \subseteq \aleph_1$, X is $\mathbf{\Pi}_1^1$ *in the codes* — that is, Code $X = \{\gamma : \|\gamma\| \in X\}$ is $\mathbf{\Pi}_1^1$. (Consider the game [Code X, W].)

3.33 Notes. The question of determinacy of infinite games was first discussed in Gale–Stewart [1953], but the first suggestion of its relevance for set theory came in Mycielski–Steinhaus [1962]. Since closed-open games are essentially finite, the Gale–Stewart proof that open games are determinate was an extension of Von Neumann's proof of determinacy for finite two-person games. Determinacy for $\mathbf{\Sigma}_2^0$ was proved by P. Wolfe in 1956, for $\mathbf{\Sigma}_3^0$ by Morton Davis [1964], and for $\mathbf{\Sigma}_4^0$ by J. Paris [1972]. Friedman [1971] shows that $\mathbf{\Sigma}_5^0$ determinacy is not provable in Zermelo set theory (ZF without the replacement schema).

The inspiration for applications of determinacy to the projective hierarchy was Blackwell's [1967] game-theoretic proof of the reduction property for $\mathbf{\Pi}_1^1$ (Exercise 3.17). Theorem 3.1 appeared almost simultaneously in Martin [1968] and Addison–Moschovakis [1968]. There followed an avalanche of results which is still rolling. As a guide for further reading, we suggest beginning with the survey articles, Fenstad [1971a] and Moschovakis [1973]. A comprehensive treatment will appear in Moschovakis [1979?] and Martin [1979?].

4. *Classical Hierarchies in* $\mathbf{\Delta}_r^1$

We turn now to analogues of the Borel and effective Borel hierarchies. The main results are negative: for $r \geq 2$ there is no way of building up $\mathbf{\Delta}_r^1$ or Δ_r^1 from below as the Borel and effective Borel hierarchies do for $\mathbf{\Delta}_1^1$ and Δ_1^1, respectively (Theorems IV.3.3 and IV.4.12).

The operations $\bigcup$ and $\bigcap$ which generate the class of Borel relations may be thought of as applied either to a countable *set* of relations or to a countable *family* — that is, a function $\langle P_p : p \in \omega \rangle$:

$$\bigcup\langle \mathsf{P}_p : p \in \omega\rangle = \{(\mathbf{m}, \boldsymbol{\alpha}) : \exists p \,.\, \mathsf{P}_p(\mathbf{m}, \boldsymbol{\alpha})\}.$$

We shall consider classes of relations which are constructed by use of other operations on families. One such operation is $\mathscr{A}$ discussed in Exercise III.2.19:

$$\mathscr{A}\langle \mathsf{P}_p : p \in \omega\rangle = \{(\mathbf{m}, \boldsymbol{\alpha}) : \exists \beta \, \forall p \,.\, \mathsf{P}_{\bar{\beta}(p)}(\mathbf{m}, \boldsymbol{\alpha})\}.$$

4.1 Definition. An *operation* is a function Φ which for any k and l, to any family $\langle \mathsf{P}_p : p \in \omega\rangle$ of relations of rank (k, l) assigns a relation $\Phi\langle \mathsf{P}_p : p \in \omega\rangle$ of rank (k, l). An operation Φ is *positive analytic* iff

(i) for any constant family $\langle \mathsf{P} : p \in \omega\rangle$, $\Phi\langle \mathsf{P} : p \in \omega\rangle = \mathsf{P}$;

(ii) for any families $\langle \mathsf{P}_p : p \in \omega\rangle$ and $\langle \mathsf{Q}_p : p \in \omega\rangle$, if for all p, $\mathsf{P}_p \subseteq \mathsf{Q}_p$, then $\Phi\langle \mathsf{P}_p : p \in \omega\rangle \subseteq \Phi\langle \mathsf{Q}_p : p \in \omega\rangle$;

(iii) for any family $\langle \mathsf{P}_p : p \in \omega\rangle$ of relations of rank (k, l) and any $\psi : {}^{k',l'}\omega \to {}^{k,l}\omega$,

$$\psi^{-1}(\Phi\langle \mathsf{P}_p : p \in \omega\rangle) = \Phi\langle \psi^{-1}(\mathsf{P}_p) : p \in \omega\rangle.$$

It is obvious that $\bigcup$, $\bigcap$, and $\mathscr{A}$ are positive analytic operations. For any $\mathsf{B} \subseteq \mathbf{P}(\omega)$, let Θ_{B} be the operation defined by

$$\Theta_{\mathsf{B}}\langle \mathsf{P}_p : p \in \omega\rangle = \{(\mathbf{m}, \boldsymbol{\alpha}) : (\exists A \in \mathsf{B})(\forall p \in A)\, \mathsf{P}_p(\mathbf{m}, \boldsymbol{\alpha})\}.$$

B is called a *base* of the operation Θ_{B}. It is similarly easy to check that for any B, if $\mathsf{B} \neq \varnothing$ and $\varnothing \notin \mathsf{B}$, then Θ_{B} is a positive analytic operation. Conversely, if Φ is positive analytic, let

$$\mathsf{B}(\Phi) = \Phi\langle \{A : p \in A\} : p \in \omega\rangle.$$

4.2 Theorem. *For every positive analytic operation Φ, $\mathsf{B}(\Phi)$ is a base of Φ — that is, $\Phi = \Theta_{\mathsf{B}(\Phi)}$.*

Proof. Let $\langle \mathsf{P}_p : p \in \omega\rangle$ be a fixed family of relations of rank (k, l). For each $(\mathbf{m}, \boldsymbol{\alpha}) \in {}^{k,l}\omega$, let $\psi_{\mathbf{m},\boldsymbol{\alpha}}$ be the constant function with domain $\mathbf{P}(\omega)$ and value $(\mathbf{m}, \boldsymbol{\alpha})$. Note that for any R,

(1) $$\mathsf{R}(\mathbf{m}, \boldsymbol{\alpha}) \leftrightarrow \psi^{-1}_{\mathbf{m},\boldsymbol{\alpha}}(\mathsf{R}) \neq \varnothing \leftrightarrow \psi^{-1}_{\mathbf{m},\boldsymbol{\alpha}}(\mathsf{R}) = \mathbf{P}(\omega).$$

Similarly, if for any $B \subseteq \omega$, ψ_B is the constant function with domain $\mathbf{P}(\omega)$ and value B, we have for all p

(2) $$p \in B \leftrightarrow \psi_B^{-1}(\{A : p \in A\}) \neq \varnothing \leftrightarrow \psi_B^{-1}(\{A : p \in A\}) = \mathbf{P}(\omega).$$

For a fixed $(\mathbf{m}, \boldsymbol{\alpha})$, let $C = \{p : \mathsf{P}_p(\mathbf{m}, \boldsymbol{\alpha})\}$. Then by (1) and (2), for all p

(3) $\psi^{-1}_{\mathbf{m},\boldsymbol{\alpha}}(\mathsf{P}_p) = \psi^{-1}_C(\{A : p \in A\}).$

Hence we have

$$\begin{aligned}(\mathbf{m}, \boldsymbol{\alpha}) \in \Phi\langle \mathsf{P}_p\rangle &\leftrightarrow \psi^{-1}_{\mathbf{m},\boldsymbol{\alpha}}(\Phi\langle \mathsf{P}_p\rangle) \neq \varnothing && \text{by (1)}\\ &\leftrightarrow \Phi\langle \psi^{-1}_{\mathbf{m},\boldsymbol{\alpha}}(\mathsf{P}_p)\rangle \neq \varnothing && \text{by (iii) of 4.1}\\ &\leftrightarrow \Phi\langle \psi^{-1}_C(\{A : p \in A\})\rangle \neq \varnothing && \text{by (3) and (ii) of 4.1}\\ &\leftrightarrow \psi^{-1}_C(\mathsf{B}(\Phi)) \neq \varnothing && \text{by (iii) of 4.1}\\ &\leftrightarrow C \in \mathsf{B}(\Phi)\\ &\rightarrow (\mathbf{m}, \boldsymbol{\alpha}) \in \Theta_{\mathsf{B}(\Phi)}\langle \mathsf{P}_p\rangle.\end{aligned}$$

On the other hand, if $(\mathbf{m}, \boldsymbol{\alpha}) \in \Theta_{\mathsf{B}(\Phi)}\langle \mathsf{P}_p\rangle$ and C is any set in $\mathsf{B}(\Phi)$ such that $(\forall p \in C)\mathsf{P}_p(\mathbf{m}, \boldsymbol{\alpha})$, then $\psi^{-1}_C(\{A : p \in A\}) \subseteq \psi^{-1}_{\mathbf{m},\boldsymbol{\alpha}}(\mathsf{P}_p)$ and the implications $(\leftarrow)$ above all hold, so we conclude $(\mathbf{m}, \boldsymbol{\alpha}) \in \Phi\langle \mathsf{P}_p\rangle$. Hence $\Phi\langle \mathsf{P}_p\rangle = \Theta_{\mathsf{B}(\Phi)}\langle \mathsf{P}_p\rangle$ as required. □

In particular, the class of positive analytic operations coincides with the class of operations Θ_{B} with $\mathsf{B} \neq \varnothing$ and $\varnothing \notin \mathsf{B}$. $\mathsf{B}(\Phi)$ is called the *canonical base* for Φ and is easily seen to be the unique base B for Φ which satisfies the condition: for all A and B, if $A \in \mathsf{B}$ and $A \subseteq B$, then $B \in \mathsf{B}$.

Positive analytic operations may also be thought of as quantifiers or operators on relations:

(4)
$$\begin{aligned}(\Phi \mathsf{R})(\mathbf{m}, \boldsymbol{\alpha}) &\leftrightarrow (\exists A \in \mathsf{B}(\Phi))(\forall p \in A)\mathsf{R}(p, \mathbf{m}, \boldsymbol{\alpha})\\ &\leftrightarrow \{p : \mathsf{R}(p, \mathbf{m}, \boldsymbol{\alpha})\} \in \mathsf{B}(\Phi).\end{aligned}$$

In this notation, $\bigcup$ coincides with $\exists^0$ and $\bigcap$ with $\forall^0$.

4.3 Definition. For any operation Φ, the *dual* Φ° of Φ is defined by:

$$\Phi^\circ\langle \mathsf{P}_p : p \in \omega\rangle = {\sim}\Phi\langle {\sim}\mathsf{P}_p : p \in \omega\rangle.$$

It is trivial that if Φ is positive analytic, so is Φ° — in fact, by a direct computation we see that

(5) $\mathsf{B}(\Phi^\circ) = \{B : (\forall A \in \mathsf{B}(\Phi))A \cap B \neq \varnothing\}.$

As an immediate consequence of (5) we have for any family $\langle \mathsf{P}_p : p \in \omega\rangle$:

$$(6)\quad \begin{aligned}&(\exists B \in \mathsf{B}(\Phi^\circ))(\forall p \in B)\mathsf{P}_p(\mathbf{m}, \boldsymbol{\alpha}) \leftrightarrow (\forall A \in \mathsf{B}(\Phi))(\exists p \in A)\mathsf{P}_p(\mathbf{m}, \boldsymbol{\alpha});\\ &(\forall B \in \mathsf{B}(\Phi^\circ))(\exists p \in B)\mathsf{P}_p(\mathbf{m}, \boldsymbol{\alpha}) \leftrightarrow (\exists A \in \mathsf{B}(\Phi))(\forall p \in A)\mathsf{P}_p(\mathbf{m}, \boldsymbol{\alpha}).\end{aligned}$$

Another useful characterization is given by:

4.4 Lemma. *For any positive analytic operation Φ,*

$$\mathsf{B}(\Phi^\circ) = \{B : {\sim}B \notin \mathsf{B}(\Phi)\}.$$

Proof. If $B \in \mathsf{B}(\Phi^\circ)$, then $B \cap A \neq \varnothing$ for all $A \in \mathsf{B}(\Phi)$ so clearly ${\sim}B$ cannot be among such A. Conversely, if ${\sim}B \notin \mathsf{B}(\Phi)$, since $\mathsf{B}(\Phi)$ is closed under superset, for no $A \in \mathsf{B}(\Phi)$ is $A \subseteq {\sim}B$. Hence for every $A \in \mathsf{B}(\Phi)$, $A \cap B \neq \varnothing$ — that is, $B \in \mathsf{B}(\Phi^\circ)$. □

4.5 Definition. For any operation Φ, $\nabla(\Phi)$ is the smallest class of relations containing the closed-open relations and closed under Φ and Φ°.

Thus $\nabla(\bigcup) = \nabla(\bigcap) =$ the class of Borel relations and Theorem IV.3.3 asserts that $\nabla(\bigcup) = \mathbf{\Delta}_1^1$. The main result of this section is that if $r \geq 2$, then $\mathbf{\Delta}_r^1 \neq \nabla(\Phi)$ for *any* positive analytic operation Φ. We call Φ a $\mathbf{\Delta}_r^1$ *operation* iff $\mathbf{\Delta}_r^1$ is closed under Φ. Then we have:

4.6 Theorem. *For any positive analytic operation Φ, and any $r \geq 1$, Φ is a $\mathbf{\Delta}_r^1$ operation iff $\mathsf{B}(\Phi) \in \mathbf{\Delta}_r^1$.*

Proof. Since $\mathsf{B}(\Phi)$ results from applying Φ to a family of open relations, if Φ is a $\mathbf{\Delta}_r^1$ operation, then $\mathsf{B}(\Phi) \in \mathbf{\Delta}_r^1$. Suppose now that $\mathsf{B}(\Phi) \in \mathbf{\Delta}_r^1$ and let $\langle \mathsf{P}_p : p \in \omega\rangle$ be any family of $\mathbf{\Delta}_r^1$ relations. Let $\mathsf{P}(p, \mathbf{m}, \boldsymbol{\alpha}) \leftrightarrow \mathsf{P}_p(\mathbf{m}, \boldsymbol{\alpha})$. We essentially showed in the proof of Theorem III.1.16 that also $\mathsf{P} \in \mathbf{\Delta}_r^1$. Then by Theorem 4.2 we have

$$(\mathbf{m}, \boldsymbol{\alpha}) \in \Phi\langle \mathsf{P}_p : p \in \omega\rangle \leftrightarrow (\exists A \in \mathsf{B}(\Phi))(\forall p \in A)\mathsf{P}(p, \mathbf{m}, \boldsymbol{\alpha})$$

which yields immediately that $\Phi\langle \mathsf{P}_p : p \in \omega\rangle \in \mathbf{\Sigma}_r^1$. On the other hand, it follows from Lemma 4.4 that also $\mathsf{B}(\Phi^\circ) \in \mathbf{\Delta}_r^1$ and by (6) we have

$$(\mathbf{m}, \boldsymbol{\alpha}) \in \Phi\langle \mathsf{P}_p : p \in \omega\rangle \leftrightarrow (\forall B \in \mathsf{B}(\Phi^\circ))(\exists p \in B)\mathsf{P}(p, \mathbf{m}, \boldsymbol{\alpha})$$

which implies that $\Phi\langle \mathsf{P}_p : p \in \omega\rangle \in \mathbf{\Pi}_r^1$. □

We aim next to show that for any $\mathbf{\Delta}_r^1$ operation Φ, $\nabla(\Phi) \subsetneq \mathbf{\Delta}_r^1$ for all $r \geq 2$. To this end we introduce the operator $*$ on operations and prove that if Φ is $\mathbf{\Delta}_r^1$, so is Φ^* and that $\nabla(\Phi) \subsetneq \nabla(\Phi^*) \subseteq \mathbf{\Delta}_r^1$. The operator $*$ turns out to be closely related to certain sorts of inductive definability.

The idea behind $*$ is an attempt to generalize the relationship between $\bigcup$ and $\mathscr{A}$. For any operation Φ, let $\mathbf{\Sigma}_1^{\Phi}$ be the class of relations of the form $\Phi\langle \mathsf{P}_p : p \in \omega\rangle$ with all P_p closed-open, $\mathbf{\Pi}_1^{\Phi}$ the class of complements of such relations, and $\mathbf{\Delta}_1^{\Phi} = \mathbf{\Sigma}_1^{\Phi} \cap \mathbf{\Pi}_1^{\Phi}$. Then $\mathbf{\Sigma}_1^{\cup} = \mathbf{\Sigma}_1^0$ and $\mathbf{\Sigma}_1^{\mathscr{A}} = \mathbf{\Sigma}_1^1$ and Theorem IV.3.3 may be stated in the form: $\mathbf{\nabla}(\bigcup) = \mathbf{\Delta}_1^{\mathscr{A}}$. We shall define $*$ in such a way that $\mathbf{\Delta}_1^{\cup^*} = \mathbf{\Delta}_1^{\mathscr{A}}$ (Exercise 4.15) and for all positive analytic Φ, $\mathbf{\nabla}(\Phi) \subseteq \mathbf{\Delta}_1^{\Phi^*}$, where in general the inclusion may be proper.

To motivate the definition of $*$, consider the operation $\mathscr{A}$ in the form

$$\mathscr{A}\langle \mathsf{P}_p : p \in \omega\rangle(\mathbf{m}, \boldsymbol{\alpha}) \leftrightarrow \exists p_0 \exists p_1 \cdots \forall n\, \mathsf{P}_{\langle p_0, \ldots, p_{n-1}\rangle}(\mathbf{m}, \boldsymbol{\alpha}).$$

In this form it is obvious that $\Sigma_1^{\mathscr{A}}$ is closed under $\exists^0$ and $\bigcup$, but less obvious that it is closed under $\forall^0$ and $\bigcap$. In Exercise III.3.22 we showed that $\Sigma_1^{\mathscr{A}}$ also coincides with the class of relations R expressible in the form

$$\mathsf{R}(\mathbf{m}, \boldsymbol{\alpha}) \leftrightarrow \exists p_0 \forall p_1 \exists p_2 \forall p_3 \cdots \forall n\, \mathsf{P}_{\langle p_0, \ldots, p_{n-1}\rangle}(\mathbf{m}, \boldsymbol{\alpha})$$

with all P_s closed-open. This expression is interpreted to mean that player I has a winning strategy in the game determined by $\{\varepsilon : \forall n\, \mathsf{P}_{\bar{\varepsilon}(n)}(\mathbf{m}, \boldsymbol{\alpha})\}$. This leads to:

4.7 Definition. For any positive analytic operation Φ, Φ^* is the operation such that for any family $\langle \mathsf{P}_s : s \in \omega\rangle$,

$$\begin{aligned}\Phi^*\langle \mathsf{P}_s : s \in \omega\rangle(\mathbf{m}, \boldsymbol{\alpha}) \leftrightarrow (\exists A_0 \in \mathsf{B}(\Phi))(\forall p_0 \in A_0)(\forall A_1 \in \mathsf{B}(\Phi))\\ (\exists p_1 \in A_1)(\exists A_2 \in \mathsf{B}(\Phi)) \cdots \forall n\, \mathsf{P}_{\langle p_0, \ldots, p_{n-1}\rangle}(\mathbf{m}, \boldsymbol{\alpha}),\end{aligned}$$

where the right-hand expression is true just in case player I has a winning strategy in the game $\mathscr{G}_{\mathbf{m},\boldsymbol{\alpha}}$ played as follows: I chooses $A_0 \in \mathsf{B}(\Phi)$, then II chooses both $p_0 \in A_0$ and $A_1 \in \mathsf{B}(\Phi)$, then I chooses both $p_1 \in A_1$ and $A_2 \in \mathsf{B}(\Phi)$, etc; I wins iff $\forall n\, \mathsf{P}_{\langle p_0, \ldots, p_{n-1}\rangle}(\mathbf{m}, \boldsymbol{\alpha})$.

There is no difficulty in verifying that if Φ is positive analytic, so is Φ^*. The canonical base $\mathsf{B}(\Phi^*)$ may be described as follows. Call a set C a *Φ-fan* iff C is a set of sequence numbers closed under subsequence, $\langle\ \rangle \in C$, and for every $s \in C$,

$$\text{(7)}\qquad \begin{aligned}&\text{if } \lg(s) \text{ is even, then } (\exists A \in \mathsf{B}(\Phi))(\forall p \in A)[s * \langle p\rangle \in C];\\ &\text{if } \lg(s) \text{ is odd, then } (\forall A \in \mathsf{B}(\Phi))(\exists p \in A)[s * \langle p\rangle \in C].\end{aligned}$$

Then

$$\mathsf{B}(\Phi^*) = \{B : \exists C\,[C \subseteq B \wedge C \text{ is a } \Phi\text{-fan}]\}.$$

Indeed, a Φ-fan C such that $(\forall s \in C)\mathsf{P}_s(\mathbf{m}, \boldsymbol{\alpha})$ is a natural way of encoding a winning strategy for player I in the game $\mathscr{G}_{\mathbf{m}, \boldsymbol{\alpha}}$.

4.8 Theorem. *For any positive analytic operation Φ, $\mathbf{\Delta}_1^{\Phi^*}$ is closed under both Φ and Φ°. Hence $\nabla(\Phi) \subseteq \mathbf{\Delta}_1^{\Phi^*}$.*

Proof. It will suffice to show that $\mathbf{\Sigma}_1^{\Phi^*}$ is closed under both Φ and Φ°, as then by complementation so is $\mathbf{\Pi}_1^{\Phi^*}$. By condition (i) of Definition 4.1, $\mathbf{\Delta}_1^{\Phi^*}$ contains all closed-open relations, so the second assertion follows immediately from the first.

Let $\langle \mathsf{P}_p : p \in \omega \rangle$ be a family of $\mathbf{\Sigma}_1^{\Phi^*}$ relations and $\langle \mathsf{P}_{p,s} : s \in \omega \rangle$ corresponding families of closed-open relations such that $\mathsf{P}_p = \Phi^*\langle \mathsf{P}_{p,s} : s \in \omega \rangle$. Let $\langle \mathsf{Q}_t : t \in \omega \rangle$ be a family of closed-open relations such that for all p, q, and s,

$$\mathsf{Q}_{\langle p \rangle} = {}^{k,l}\omega \quad \text{and} \quad \mathsf{Q}_{\langle p, q \rangle * s} = \mathsf{P}_{p,s}.$$

Then

$$
\begin{aligned}
(\mathbf{m}, \boldsymbol{\alpha}) \in \Phi\langle \mathsf{P}_p : p \in \omega \rangle &\leftrightarrow (\exists A \in \mathsf{B}(\Phi))(\forall p \in A)[(\mathbf{m}, \boldsymbol{\alpha}) \in \Phi^*\langle \mathsf{P}_{p,s} : s \in \omega \rangle] \\
&\leftrightarrow (\exists A \in \mathsf{B}(\Phi))(\forall p \in A)(\forall B \in \mathsf{B}(\Phi))(\exists q \in B)\,[(\mathbf{m}, \boldsymbol{\alpha}) \in \Phi^*\langle \mathsf{Q}_{\langle p, q \rangle * s} \rangle] \\
&\leftrightarrow (\mathbf{m}, \boldsymbol{\alpha}) \in \Phi^*\langle \mathsf{Q}_t : t \in \omega \rangle.
\end{aligned}
$$

The last equivalence may be seen by reflecting on the notion of strategy. The proof that $\mathbf{\Sigma}_1^{\Phi^*}$ is closed under Φ° is similar except that we take

$$\mathsf{Q}_{\langle q \rangle} = {}^{k,l}\omega \quad \text{and} \quad \mathsf{Q}_{\langle q, p \rangle * s} = \mathsf{P}_{p,s}. \quad \square$$

The dual operation $\Phi^{*\circ}$ may also be expressed in terms of the existence of a winning strategy for a game. Observe first that the game $\mathscr{G}_{\mathbf{m}, \boldsymbol{\alpha}}$ is open from the point of view of player II: to win he must force the choice of a finite sequnce $\langle p_0, \ldots, p_{n-1} \rangle$ such that $\sim\mathsf{P}_{\langle p_0, \ldots, p_{n-1} \rangle}(\mathbf{m}, \boldsymbol{\alpha})$ and from that point on the moves are irrelevant. The argument sketched for Exercise 3.15 shows that $\mathscr{G}_{\mathbf{m}, \boldsymbol{\alpha}}$ is determined. Let $\langle \mathsf{P}_s : s \in \omega \rangle$ be a fixed family, $\mathscr{G}_{\mathbf{m}, \boldsymbol{\alpha}}$ the associated games which define Φ^*, and $\tilde{\mathscr{G}}_{\mathbf{m}, \boldsymbol{\alpha}}$ the games associated with the family $\langle \sim \mathsf{P}_s : s \in \omega \rangle$. Then

$$
\begin{aligned}
&(\mathbf{m}, \boldsymbol{\alpha}) \in \Phi^{*\circ}\langle \mathsf{P}_s : s \in \omega \rangle \\
&\quad \leftrightarrow \text{I does not have a winning strategy in } \tilde{\mathscr{G}}_{\mathbf{m}, \boldsymbol{\alpha}} \\
(8) \qquad &\quad \leftrightarrow \text{II does have a winning strategy in } \tilde{\mathscr{G}}_{\mathbf{m}, \boldsymbol{\alpha}} \\
&\quad \leftrightarrow (\forall A_0 \in \mathsf{B}(\Phi))(\exists p_0 \in A_0)(\exists A_1 \in \mathsf{B}(\Phi)) \\
&\qquad\qquad (\forall p_1 \in A_1) \cdots \exists n\, \mathsf{P}_{\langle p_0, \ldots, p_{n-1} \rangle}(\mathbf{m}, \boldsymbol{\alpha}).
\end{aligned}
$$

Towards showing that if Φ is $\mathbf{\Delta}^1_r$, then so is Φ^*, we first prove an analogue of Theorem III.3.16.

4.9 Definition. For any decomposable inductive operator Γ over ${}^{k,l}\omega$, Γ is *Φ-positive* iff P_Γ belongs to the smallest set X of relations such that:

(i) for any closed-open relation R, $\{(\mathbf{m}, \boldsymbol{\alpha}, \delta) : \mathsf{R}(\mathbf{m}, \boldsymbol{\alpha})\} \in X$;

(ii) for any continuous functional F, $\{(\mathbf{m}, \boldsymbol{\alpha}, \delta) : \delta(\mathsf{F}(\mathbf{m}, \boldsymbol{\alpha})) = 0\} \in X$;

(iii) X is closed under (countable) $\bigcup$, $\bigcap$, Φ, and Φ°.

4.10 Theorem. *For any positive analytic operation Φ and any $\mathsf{R} \in \mathbf{\Pi}^{\Phi^*}_1$, there exists a decomposable Φ-positive inductive operator Γ such that for all $\mathbf{m}$ and $\boldsymbol{\alpha}$,*

$$\mathsf{R}(\mathbf{m}, \boldsymbol{\alpha}) \leftrightarrow \langle \mathbf{m}, \langle\ \rangle\rangle \in \bar{\Gamma}_{\boldsymbol{\alpha}}.$$

Proof. Let R be any relation in $\mathbf{\Pi}^{\Phi^*}_1$ and $\langle \mathsf{P}_s : s \in \omega\rangle$ a family of closed-open relations such that

$$\mathsf{R}(\mathbf{m}, \boldsymbol{\alpha}) \leftrightarrow (\mathbf{m}, \boldsymbol{\alpha}) \in \Phi^{*\circ}\langle \mathsf{P}_s : s \in \omega\rangle.$$

We may clearly assume that if $s \subseteq t$, then $\mathsf{P}_s \subseteq \mathsf{P}_t$. Let Γ be the Φ-positive operator defined by:

$$\langle \mathbf{m}, s\rangle \in \Gamma_{\boldsymbol{\alpha}}(D) \leftrightarrow \mathsf{P}_s(\mathbf{m}, \boldsymbol{\alpha}) \vee (\forall A \in \mathsf{B}(\Phi))(\exists p \in A)(\exists B \in \mathsf{B}(\Phi))$$
$$(\forall q \in B)[\langle \mathbf{m}, s * \langle p, q\rangle\rangle \in D].$$

We claim that for all s of even length and all $\mathbf{m}$ and $\boldsymbol{\alpha}$,

$$(9) \qquad \langle \mathbf{m}, s\rangle \in \bar{\Gamma}_{\boldsymbol{\alpha}} \leftrightarrow (\mathbf{m}, \boldsymbol{\alpha}) \in \Phi^{*\circ}\langle \mathsf{P}_{s*t} : t \in \omega\rangle$$

which for $s = \langle\ \rangle$ is the desired result. To establish the claim, let, for each α,

$$C_{\boldsymbol{\alpha}} = \{\langle \mathbf{m}, s\rangle : \lg(s) \text{ is even} \wedge (\mathbf{m}, \boldsymbol{\alpha}) \in \Phi^{*\circ}\langle \mathsf{P}_{s*t} : t \in \omega\rangle\}.$$

We follow the proof of Theorem III.3.2 and first show $\Gamma_{\boldsymbol{\alpha}}(C_{\boldsymbol{\alpha}}) \subseteq C_{\boldsymbol{\alpha}}$, which implies $\bar{\Gamma}_{\boldsymbol{\alpha}} \subseteq C_{\boldsymbol{\alpha}}$ and thus the implication $(\rightarrow)$ of (9). Suppose $\langle \mathbf{m}, s\rangle \in \Gamma_{\boldsymbol{\alpha}}(C_{\boldsymbol{\alpha}})$. If $\mathsf{P}_s(\mathbf{m}, \boldsymbol{\alpha})$, then II may play any strategy and win the game $\mathscr{G}^{\sim}_{\mathbf{m}, \boldsymbol{\alpha}, s}$ associated with the family $\langle \sim\mathsf{P}_{s*t} : t \in \omega\rangle$, so $\langle \mathbf{m}, s\rangle \in C_{\boldsymbol{\alpha}}$. Otherwise we have

$$(\forall A \in \mathsf{B}(\Phi))(\exists p \in A)(\exists B \in \mathsf{B}(\Phi))(\forall q \in B)[\langle \mathbf{m}, s * \langle p, q\rangle\rangle \in C_{\boldsymbol{\alpha}}].$$

By use of (8) this easily implies $\langle \mathbf{m}, s\rangle \in C_{\boldsymbol{\alpha}}$.

To show $C_{\boldsymbol{\alpha}} \subseteq \bar{\Gamma}_{\boldsymbol{\alpha}}$, we suppose $\langle \mathbf{m}, s\rangle \notin \bar{\Gamma}_{\boldsymbol{\alpha}}$ and show that in this case I has a winning strategy in $\mathscr{G}^{\sim}_{\mathbf{m},\alpha,s}$. Let $D_\alpha = \{t : \langle \mathbf{m}, s * t\rangle \notin \bar{\Gamma}_\alpha\}$. By assumption $\langle\ \rangle \in D_\alpha$. Since $\Gamma_\alpha(\bar{\Gamma}_\alpha) = \bar{\Gamma}_\alpha$, we have for any t,

$$(10) \quad t \in D_\alpha \to (\exists A \in \mathsf{B}(\Phi))(\forall p \in A)(\forall B \in \mathsf{B}(\Phi))(\exists q \in B)[t * \langle p, q\rangle \in D_\alpha].$$

Now I may win $\mathscr{G}^{\sim}_{\mathbf{m},\alpha,s}$ by the following strategy. Applying (10) to $t = \langle\ \rangle$ she picks $A_0 \in \mathsf{B}(\Phi)$ such that for any choice of $p_0 \in A_0$ and $A_1 \in \mathsf{B}(\Phi)$ there is a $p_1 \in A_1$ which she may choose such that $\langle p_0, p_1\rangle \in D_\alpha$. Applying (10) now to $t = \langle p_0, p_1\rangle$, there is a proper choice of $A_2 \in \mathsf{B}(\Phi)$ such that for any $p_2 \in A_2 \cdots \langle p_0, p_1, p_2, p_3\rangle \in D_\alpha$. Thus I has a strategy which ensures that for all even n, $\langle p_0, \ldots, p_{n-1}\rangle \in D_\alpha$ and thus in particular that $\sim\mathsf{P}_{\langle p_0,\ldots,p_{n-1}\rangle}(\mathbf{m}, \alpha)$. By the initial assumption on the P_r, this implies that for *all* n, $\sim\mathsf{P}_{\langle p_0,\ldots,p_{n-1}\rangle}(\mathbf{m}, \alpha)$ and thus I wins $\mathscr{G}^{\sim}_{\mathbf{m},\alpha,s}$. □

4.11 Corollary. *For all $r \geq 2$ and any $\mathbf{\Delta}_r^1$ positive analytic operation Φ,*
(i) *Φ^* is also $\mathbf{\Delta}_r^1$;*
(ii) $\nabla(\Phi) \subsetneq \mathbf{\Delta}_r^1$.

Proof. Suppose $r \geq 2$ and Φ is a $\mathbf{\Delta}_r^1$ positive analytic operation. It follows directly from the definitions (and Theorem III.2.5) that any Φ-positive inductive operation Γ is $\mathbf{\Delta}_r^1$ and hence from the boldface version of Theorem III.3.18(i) that $\bar{\Gamma} \in \mathbf{\Delta}_r^1$. Hence by the preceding theorem, $\mathbf{\Pi}_1^{\Phi^*} \subseteq \mathbf{\Delta}_r^1$ and thus also $\mathbf{\Sigma}_1^{\Phi^*} \subseteq \mathbf{\Delta}_r^1$. In particular, $\mathsf{B}(\Phi^*) \in \mathbf{\Sigma}_1^{\Phi^*}$ so $\mathsf{B}(\Phi^*) \in \mathbf{\Delta}_r^1$ and thus Φ^* is a $\mathbf{\Delta}_r^1$ operation. That $\nabla(\Phi) \subseteq \mathbf{\Delta}_r^1$ is now immediate from Theorem 4.8. That this inclusion is proper follows by a standard diagonal argument: if

$$\mathsf{V}(\langle \mathbf{m}\rangle, \langle \alpha\rangle, \beta) \leftrightarrow (\exists A \in \mathsf{B}(\Phi^*))(\forall p \in A)\, \mathsf{U}_1^0(\langle \mathbf{m}\rangle, \langle \alpha\rangle, (\beta)^p).$$

Then V is a $\mathbf{\Delta}_r^1$ relation universal for $\mathbf{\Sigma}_1^{\Phi^*}$. Hence $\mathsf{V} \in \mathbf{\Delta}_r^1 \sim \mathbf{\Delta}_1^{\Phi^*}$, so also $\mathsf{V} \in \mathbf{\Delta}_r^1 \sim \nabla(\Phi)$. □

The class $\nabla(\Phi)$ may be decomposed into a hierarchy just as were the Borel relations. We set $\mathbf{\Sigma}_0^{\Phi} = \mathbf{\Pi}_0^{\Phi}$ = the class of closed-open relations and take

$$\mathbf{\Sigma}_\rho^{\Phi} = \{\Phi\langle \mathsf{P}_p : p \in \omega\rangle\text{: all } \mathsf{P}_p \text{ have the same rank and belong to } \mathbf{\Pi}_{(\rho)}^{\Phi}\}.$$

The classes $\mathbf{\Pi}_\rho^{\Phi}$, $\mathbf{\Delta}_\rho^{\Phi}$, etc. are defined analogously as in Definition IV.3.4. It is immediate that all of these classes are included in $\nabla(\Phi)$ and indeed that $\nabla(\Phi) = \mathbf{\Delta}_{(\aleph_1)}^{\Phi}$. The analogue of Lemma IV.3.5 holds with the same proof and the same is true for the parts of Lemma IV.3.6 which concern expansion, and composition and substitution of continuous functionals. It is not, however, true for all positive analytic Φ that $\mathbf{\Sigma}_\rho^{\Phi}$ is closed under $\bigcup$ and $\exists^0$ (for example, if Φ is $\bigcap$ or $\Phi\langle \mathsf{P}_p : p \in \omega\rangle = \mathsf{P}_0$). The hierarchy theorem (corresponding to IV.3.11) is clearly false for the second of these examples. These results hold, however, for a restricted class of operations:

4.12 Definition. An operation Φ is *normal* iff there exist primitive recursive functions f and g such that for any family $\langle \mathsf{P}_p : p \in \omega \rangle$,

(i) $\Phi\langle \Phi\langle \mathsf{P}_{\langle p,q\rangle} : q \in \omega\rangle : p \in \omega\rangle = \Phi\langle \mathsf{P}_{f(r)} : r \in \omega\rangle$;

(ii) $\bigcup\langle \mathsf{P}_p : p \in \omega\rangle = \Phi\langle \mathsf{P}_{g(p)} : p \in \omega\rangle$.

For normal operations Φ there is no difficulty in imitating the proofs of IV.3.10–11 to show that for all ρ such that $0<\rho<\aleph_1$, $\Sigma^\Phi_\rho \not\subseteq \Delta^\Phi_\rho$ and $\Delta^\Phi_{\rho+1} \not\subseteq \Sigma^\Phi_\rho \cup \Pi^\Phi_\rho$ (cf. Exercise 4.16). Of course $\bigcup$, $\bigcap$, $\mathscr{A}$, and $\mathscr{A}^\circ$ are normal and it can be verified by an elementary but tedious computation that if Φ is normal, so is Φ^* (cf. Hinman [1969]).

The characterization of Theorem 4.10 also yields some extensions of the results of § IV.5. Let Φ be a normal operation. If Γ is any Φ-positive inductive operator, it is easy to prove by induction on ρ that for all $\rho < \aleph_1$, $\Gamma^\rho \in \nabla(\Phi)$. Suppose $\mathsf{A} \in \Pi^{\Phi^*}_1$ and Γ is a Φ-positive inductive operator such that for all α,

$$\alpha \in \mathsf{A} \leftrightarrow \langle\ \rangle \in \bar{\Gamma}_\alpha.$$

Set

$$\mathsf{A}^t_\rho = \{\alpha : t \in \Gamma^{(\rho)}_\alpha\}.$$

Then just as in Theorem IV.5.1, $\mathsf{A} = \bigcup\{\mathsf{A}^{\langle\rangle}_\rho : \rho < \aleph_1\}$ and thus A is the union of an $\aleph_1$-sequence of sets belonging to $\nabla(\Phi)$. Similarly, if

$$\mathsf{B}_\rho = \mathsf{A}^{\langle\rangle}_\rho \cup \bigcup\{\mathsf{A}^t_{\rho+1} \sim \mathsf{A}^t_\rho : t \in \omega\}.$$

Each B_ρ $(\rho < \aleph_1)$ belongs to $\nabla(\Phi)$ and $\mathsf{A} = \bigcap\{\mathsf{B}_\rho : \rho < \aleph_1\}$.

Let us say that an operation Φ *preserves measurability* (*preserves the Baire property*) iff whenever all P_p $(p \in \omega)$ are measurable (have the Baire property) so is (does) $\Phi\langle \mathsf{P}_p : p \in \omega\rangle$. Obviously, if Φ preserves measurability (the Baire property) then all members of $\nabla(\Phi)$ are measurable (have the Baire property). Of course $\bigcup$ preserves both properties and Theorems IV.5.3 and IV.5.10 show essentially that $\mathscr{A}$ also preserves both. To formulate this more precisely, for any class Y of relations, let $\Sigma^\Phi_1(Y)$ be the class of relations of the form $\Phi\langle \mathsf{P}_p : p \in \omega\rangle$ with $\mathsf{P}_p \in Y$, and define the class of $(\Phi; Y)$-positive inductive operators by introducing Y as an initial class. Then the proof of Theorem 4.10 is easily modified to yield that for any $\mathsf{R} \in \Pi^{\Phi^*}_1(Y)$ there is a decomposable $(\Phi; Y)$-positive inductive operator Γ such that for all $\mathbf{m}$ and $\boldsymbol{\alpha}$,

$$\mathsf{R}(\mathbf{m}, \boldsymbol{\alpha}) \leftrightarrow \langle \mathbf{m}, \langle\ \rangle\rangle \in \bar{\Gamma}_{\boldsymbol{\alpha}}.$$

4.13 Theorem. *For any positive analytic operation Φ, if Φ preserves measurability (the Baire property), then so does Φ^*.*

Proof. Suppose that Φ preserves measurability and let Y be the class of measurable sets. It will suffice to show that every $\mathsf{A} \in \Pi_1^{\Phi^*}(Y)$ is measurable. Let Γ be a $(\Phi; Y)$-positive inductive operator as above and set

$$\mathsf{A}_\rho^t = \{\alpha : t \in \Gamma_\alpha^{(\rho)}\}.$$

Since Φ, Φ°, $\bigcup$, and $\bigcap$ all preserve measurability, it is easy to prove by induction on ρ that all A_ρ^t $(\rho < \aleph_1)$ are measurable. The proof concludes exactly as for Theorem IV.5.3. The proof for the Baire property is similar. □

Let $\Psi_0 = \bigcup$ and $\Psi_{r+1} = \Psi_r^*$. Then it follows that for all r, Ψ_r preserves both measurability and the Baire property and thus all sets in $\nabla(\Psi_r)$ are measurable and have the Baire property. By appropriately "joining" at limit ordinals, one can construct a sequence Ψ_ρ $(\rho < \aleph_1)$ of positive analytic operations such that all members of $\bigcup\{\nabla(\Psi_\rho) : \rho < \aleph_1\}$ are measurable and have the Baire property. These are known classically as the R-sets and form a proper subclass of Δ_2^1.

4.14–4.17 Exercises

4.14. Compute $\mathsf{B}(\bigcup)$, $\mathsf{B}(\bigcap)$, $\mathsf{B}(\mathscr{A})$ and $\mathsf{B}(\mathscr{A}^\circ)$.

4.15 Verify that $\Delta_1^{\cup^*} = \Delta_1^{\mathscr{A}}$ (cf. Exercise III.3.22).

4.16. Prove the hierarchy theorem for normal operations (discussion following Definition 4.12).

4.17. For any positive analytic operation Φ and any class Y of relations, $\nabla(\Phi; Y)$ is the smallest class of relations including Y and closed under Φ and Φ°. Show that

$$\nabla(\bigcup; \Sigma_1^1 \cup \Pi_1^1) \subsetneq \Delta_1^{\mathscr{A}}(\Sigma_1^1 \cup \Pi_1^1).$$

(Show that the right side contains a relation universal for the left).

4.18 Notes. The operation $\mathscr{A}$ was defined by Suslin in 1917. Its original importance was in affording the first "constructive" method of obtaining a non-Borel set, but it also led directly to the definition and study of the analytical operations in the 1920's. Kantorovitch–Livenson [1932] and [1933] is a good survey of this work, which includes many of the results of this section. The operator * is ascribed there to Kolmogorov, but he seems not to have published any account of it. The main facts were established in Ljapunov [1953]. The presentation in terms of games is new here and was also discovered independently by Aczel [1975].

The class of R-sets is not the largest class all of whose members are measurable and have the Baire property. A theorem due to Solovay and published in Fenstad–Normann [1974] asserts that all provably $\mathbf{\Delta}_2^1$ relations are measurable and have the Baire property. R-sets and some extensions discussed in Ljapunov [1953] are all provably $\mathbf{\Delta}_2^1$.

5. *Effective Hierarchies in* Δ_r^1

The effective Borel hierarchy of § IV.4 is derived from the classical Borel hierarchy by restricting the generating operation of countable union to families of relations which are recursively enumerable (relative to a given indexing). No ingenuity is needed to apply the same techniques to the hierarchies of the preceding section and obtain a class $\Delta_{(\omega_1)}^{\Phi}$ of relations "effectively generated" by any positive analytic operation Φ. Unfortunately, a more accurate analogy is obtained by a more complex procedure.

The problem is that the ordinal ω_1, the number of levels in the effective Borel hierarchy, is not only the least non-recursive ordinal, but also the least non-effective-Borel ordinal — that is, the least ordinal which is not the order type of an effective Borel ($= \Delta_1^1$) well-ordering of ω. If Φ is a more powerful operation ($\mathscr{A}$, for example), ω_1 is represented already by a Π_1^{Φ} well-ordering of ω (Theorem IV.2.11 — cf. Definition 5.2 below), and it seems natural that Σ_ρ^{Φ} should be defined for all ρ for which a well-ordering of type ρ occurs in the hierarchy. The apparent circularity in this idea is avoided by a "boot-strap" procedure: at each level of the hierarchy, Φ is applied to families which are recursively enumerable *in* a relation which occurs at some previous level. This will require that the indices and relations be generated simultaneously. It is by no means obvious that the resulting construction has the desired property and indeed this is proved most naturally by the methods of § VI.5, where we shall also give further evidence for the "naturalness" of the construction.

In the second part of the section we shall consider briefly a similar generalization of the second hierarchy of § IV.4 obtained by iterating a jump operator J over a set of notations for ordinals.

It would be entirely understandable if the reader were to blanch slightly at the prospect of heaping new complexities on the already complicated and somewhat tedious proofs of § IV.4. It is a sad fact that relatively clear intuitions often require masses of unpleasant calculation for their justification. We shall attempt in this section to give proofs in sufficient detail such that the average reader will be able to get an intuitive grasp of the ideas without too much pain, and the dedicated reader will be able to reconstruct complete proofs.

As a start in this direction, we consider here only relations on numbers.

5.1 Definition. For any positive analytic Φ and each k, $N^{\Phi,k}$ is the smallest subset of ω such that for all $a \in N^{\Phi,k}$, there exist relations $P_a^{\Phi} \subseteq {}^k\omega$ which satisfy the following conditions: for all a, b, and c,

(i) if $(c)_1 = k$ and $(c)_2 = 0$, then $\langle 7, c\rangle \in N^{\Phi,k}$ and $P^{\Phi}_{\langle 7,c\rangle} = \mathrm{Dm}\{c\}$;

(ii) if $b \in N^{\Phi,k}$ and for all p, $\{a\}(p, P_b^{\Phi}) \in N^{\Phi,k}$, then $\langle a, b\rangle \in N^{\Phi,k}$ and $P^{\Phi}_{\langle a,b\rangle} = \Phi\langle \sim P^{\Phi}_{\{a\}(p,P_b^{\Phi})} : p \in \omega\rangle$.

It is immediate that for $a \in N^{\Phi,k}$, P_a^{Φ} denotes a unique k-ary relation. The definition may be viewed as a single inductive definition of the relation

$$V^{\Phi,k} = \{(a, i, \mathbf{m}) : a \in N^{\Phi,k} \wedge ([i = 0 \wedge P_a^{\Phi}(\mathbf{m})] \vee [i = 1 \wedge \sim P_a^{\Phi}(\mathbf{m})])\}.$$

We denote by $\omega_1[\Phi]$ the closure ordinal of this inductive definition. The sets $N^{\Phi,k}_{(\rho)}$, N^{Φ}_{ρ}, N^{Φ}, etc. are defined as in § IV.4.

5.2 Definition. For all $\rho < \omega_1[\Phi]$,

(i) $\Sigma_\rho^{\Phi} = \{P_a^{\Phi} : a \in N_\rho^{\Phi}\}$;

(ii) $\Pi_\rho^{\Phi} = \{\sim P_a : a \in N_\rho^{\Phi}\}$;

(iii) $\Delta_\rho^{\Phi} = \Sigma_\rho^{\Phi} \cap \Pi_\rho^{\Phi}$;

(iv) $\nabla(\Phi) = \{P_a^{\Phi} : a \in N^{\Phi}\}$.

The reader may find it curious that we have arranged things so that Σ_1^{Φ} consists of relations obtained by applying Φ to families of co-semi-recursive relations rather than to families of recursive relations. The reasons for this are purely technical and are of no concern except in the proof of the hierarchy theorem where it is essential that the sets of indices corresponding to the various levels are relatively simple — see Theorem IV.4.14.

It is not in general true that $\Sigma_\rho^{\cup}$ coincides with Σ_ρ^0 as defined in § IV.4. It is relatively easy to prove that for all ρ, $\Sigma_\rho^0 \subseteq \Sigma_\rho^{\cup}$ and it follows from Theorem 5.3 below that $\nabla(\cup) \subseteq \Delta_1^1$, so the classes $\Sigma_\rho^{\cup}$ form an alternative hierarchy on Δ_1^1.

We shall call Φ a Δ_r^1 *operation* iff (equivalently) $\mathsf{B}(\Phi) \in \Delta_r^1$ or for any $\mathsf{R} \in \Delta_r^1$, also $\Phi\mathsf{R} \in \Delta_r^1$ (cf. (4) of § 4 and Exercise 5.9).

5.3 Theorem. *For all $r \geq 1$ and any positive analytic Δ_r^1 operation Φ, $\nabla(\Phi) \subseteq \{R : R \in \Delta_r^1\}$, and if $r \geq 2$ the inclusion is proper.*

Proof. It will suffice to show that if Φ is Δ_r^1, then the relations $V^{\Phi,k}$ defined above belong to Π_1^1, if $r = 1$, and Δ_r^1, if $r \geq 2$. This establishes the inclusion and a standard diagonal argument shows $V^{\Phi,k} \notin \nabla(\Phi)$. By Theorems III.3.1 and 10, it furthermore suffices to construct a monotone operator $\Gamma \in \Delta_r^1$ such that $\bar{\Gamma} = V^{\Phi,k}$. For any R, let

$$N_R = \{b : \forall \mathbf{m}\,(\exists i \leq 1)\, R(b, i, \mathbf{m})\};$$

$$R_b = \{\mathbf{m} : R(b, 0, \mathbf{m})\};$$

$$Ix_R(a, b) \leftrightarrow b \in N_R \wedge \forall p . \{a\}(p, R_b) \in N_R.$$

If $R \subseteq V^{\Phi,k}$, then $N_R \subseteq N^{\Phi,k}$, and for any $b \in N_R$, $R_b = P_b^{\Phi}$. We now define Γ similarly as in the proof of Lemma IV.4.9: for any $R \subseteq {}^{k+2}\omega$, all i, and all $\mathbf{m} \in {}^k\omega$,

(i) if $(c)_1 = k$ and $(c)_2 = 0$, then
  (1) if $\{c\}(\mathbf{m})\downarrow$, then $(\langle 7, c\rangle, 0, \mathbf{m}) \in \Gamma(R)$;
  (2) if $\{c\}(\mathbf{m})\uparrow$, then $(\langle 7, c\rangle, 1, \mathbf{m}) \in \Gamma(R)$;

(ii) if $Ix_R(a, b)$, then
  (1) if $(\exists A \in \mathrm{B}(\Phi))(\forall p \in A)[(\{a\}(p, R_b), 1, \mathbf{m}) \in R]$, then $(\langle a, b\rangle, 0, \mathbf{m}) \in \Gamma(R)$;
  (2) if $(\forall A \in \mathrm{B}(\Phi))(\exists p \in A)[(\{a\}(p, R_b), 0, \mathbf{m}) \in R]$, then $(\langle a, b\rangle, 1, \mathbf{m}) \in \Gamma(R)$.

We leave it to the reader to check that this suffices. □

The properties of the hierarchy corresponding to IV.4.5–7 require in general that Φ be normal, but under this hypothesis the proofs are similar to the earlier ones. (For closure under finite intersection use Exercise 5.7.) The hierarchy theorem (corresponding to Theorem IV.4.15) also holds for all normal Φ, although its proof is substantially more complicated.

The definition of the class of *effective Φ-positive* inductive operators Γ is obtained from Definition 4.9 of the preceding section by replacing "closed-open" and "continuous" by "semi-recursive" and "recursive". Then the proof of Theorem 4.10 is easily adapted to show that for any $R \in \Pi_1^{\Phi^*}$, there is an effective Φ-positive inductive operator Γ such that for all m,

$$R(\mathbf{m}) \leftrightarrow \langle \mathbf{m}, \langle\ \rangle\rangle \in \bar{\Gamma}.$$

If Φ is normal, it is also true that for any inductive operator Γ which is effective Φ-positive, $\bar{\Gamma} \in \Pi_1^{\Phi^*}$ so we have that for all R, $R \in \Pi_1^{\Phi^*}$ iff $R \ll \bar{\Gamma}$ for some effective Φ-positive Γ. In particular, the operator Γ defined in the proof of Theorem 5.3 is effective Φ-positive, so that if Φ is normal, $V^{\Phi,k} \in \Pi_1^{\Phi^*}$. From this it follows that $\nabla(\Phi) \subseteq \Delta_1^{\Phi^*}$, the effective analogue of Theorem 4.8 (see Theorems VI.6.14–19 below).

In § VI.5 we shall introduce a notion of recursion relative to a positive analytic operation Φ, It turns out that for normal Φ, $\nabla(\Phi)$ consists exactly of the relations recursive in Φ and $\omega_1[\Phi]$ is the least ordinal not the order-type of a well-ordering of ω recursive in Φ, hence the least non-$\nabla(\Phi)$ ordinal as discussed in the introduction to this section. In § VI.6 we consider another notion of recursion relative to Φ (recursion in $\Phi^{\#}$). The set of relations recursive in Φ in this second sense is exactly $\Delta_1^{\Phi^*}$. Thus for $\Phi = \bigcup$ the two senses coincide but for more powerful Φ($\mathscr{A}$, for example), $\nabla(\Phi)$ is a proper subclass of $\Delta_1^{\Phi^*}$.

We now turn to the generalization of the second hierarchy of § IV.4. We started there with a set O of notations for the recursive ordinals and constructed sets D_u by applying the jump operator iteratively. Our generalization will consist in replacing the ordinary jump operator by a general jump operator J and extending the set of notations by allowing for recursions relative to previously generated sets much as we extended the set of indices N in Definition 5.1.

5.4 Definition. A *jump operator* is a function $\mathsf{J}: {}^\omega\omega \to {}^\omega 2$ such that there exists an index d and a primitive recursive function h such that for all a, α, and β,

(i) α^{oJ} is recursive in $\mathsf{J}(\alpha)$ with index d;

(ii) if α is recursive in β with index a, then $\mathsf{J}(\alpha)$ is recursive in $\mathsf{J}(\beta)$ with index $h(a)$.

Of course, oJ is a jump operator. For other examples, consider for any $i \leqslant 1$ and $r > 0$:

$$\mathsf{J}_r^i(\alpha)(\langle a, \mathbf{m}\rangle) = \begin{cases} 0, & \text{if } \mathsf{U}_r^i(a, \langle \mathbf{m}\rangle, \langle \boldsymbol{\alpha}\rangle); \\ 1, & \text{otherwise.} \end{cases}$$

J_1^0 is the ordinary jump, oJ, and J_1^1 is the *hyperjump*, hJ, (Exercise IV.2.32). Any jump operator has a natural extension to sets defined by $\mathsf{J}(A) = \{m : \mathsf{J}(\mathsf{K}_A)(m) = 0\}$. Propertities (i) and (ii) of the definition hold also with α and β replaced by A and B. To avoid confusion we shall now write $\mathsf{oJ}(A)$ instead of A^{oJ}.

5.5 Definition. For any jump operator J, $<^{\mathsf{J}}$ is the smallest subset of $\omega \times \omega$ such that for all w in the field of $<^{\mathsf{J}}$ there exist sets D_w^{J} which satisfy the following conditions:

(i) $1 <^{\mathsf{J}} 2$, $D_1^{\mathsf{J}} = \{0\}$, and $D_2^{\mathsf{J}} = \mathsf{J}(D_1^{\mathsf{J}})$;

(ii) if $u <^{\mathsf{J}} v$, then $v <^{\mathsf{J}} 2^v$ and $D_{2^v}^{\mathsf{J}} = \mathsf{J}(D_v^{\mathsf{J}})$;

(iii) if $u \in \mathrm{Fld}(<^{\mathsf{J}})$, $\{a\}(0, D_u^{\mathsf{J}}) \simeq u$, and for all p, $\{a\}(p, D_u^{\mathsf{J}}) <^{\mathsf{J}} \{a\}(p+1, D_u^{\mathsf{J}})$, then for all p, $\{a\}(p, D_u^{\mathsf{J}}) <^{\mathsf{J}} 3^a 5^u$ and

$$D_{3^a 5^u}^{\mathsf{J}} = \{\langle m, p\rangle : m \in D_{\{a\}(p, D_u^{\mathsf{J}})}^{\mathsf{J}}\};$$

(iv) if $u <^{\mathsf{J}} v$ and $v <^{\mathsf{J}} w$, then $u <^{\mathsf{J}} w$.

We write O^{J} for the field of $<^{\mathsf{J}}$ and assign ordinals $|u|^{\mathsf{J}}$ to $u \in O^{\mathsf{J}}$ as in § IV.4. We set

$$\nabla(\mathsf{J}) = \{R : R \text{ is recursive in } D_u^{\mathsf{J}}, \text{ for some } u \in O^{\mathsf{J}}\}.$$

Again $O^{o\mathsf{J}}$ and $D_u^{o\mathsf{J}}$ are not identical with O and D_u as defined in § IV.4, but are in a sense equivalent to them. First, it is not (too) hard to show by effective transfinite recursion that there are primitive recursive functions f and g such that for all $u \in O$, D_u is recursive in $D_{f(u)}^{o\mathsf{J}}$ with index $g(u)$. Hence, by Theorem IV.4.21, $\Delta_1^1 \subseteq \nabla(o\mathsf{J})$. The converse inclusion follows from the next theorem.

We call J a Δ_r^1 *jump operator* iff the relation $\mathsf{P}_\mathsf{J}(m, A) \leftrightarrow m \in \mathsf{J}(A)$ is a Δ_r^1 relation.

5.6 Theorem. *For all $r \geq 1$ and any Δ_r^1 jump operator* J, $\nabla(\mathsf{J}) \subseteq \{R : R \in \Delta_r^1\}$ *and if $r \geq 2$, the inclusion is proper.*

Proof. Suppose J is a Δ_r^1 jump operator. As in the proof of Lemma IV.4.18, modified similarly as for Theorem 5.3, we can define a Δ_r^1 monotone operator Γ such that

$$(u, i, m) \in \bar{\Gamma} \leftrightarrow u \in O^{\mathsf{J}} \wedge ([i = 0 \wedge m \in D_{u]}^{\mathsf{J}}] \vee [i = 1 \wedge m \notin D_{u]}^{\mathsf{J}}] \vee [i = 2 \wedge u <^{\mathsf{J}} m]).$$

If $r = 1$, then $\bar{\Gamma} \in \Pi_1^1$ so all $D_u^{\mathsf{J}} \in \Delta_1^1$; if $r \geq 2$, then $\bar{\Gamma} \in \Delta_r^1$ and all $D_u^{\mathsf{J}} \in \Delta_r^1$, but a diagonal argument shows that $\bar{\Gamma} \notin \nabla(\mathsf{J})$. □

In § VI.5 we shall also introduce a notion of recursion relative to a jump operator and show that $\nabla(\mathsf{J})$ consists exactly of the relations recursive in J. If Φ is a positive analytic operation such that P_J is Φ-positive, then J will be recursive in Φ and thus $\nabla(\mathsf{J}) = \{R : R \text{ recursive in } \mathsf{J}\} \subseteq \{R : R \text{ recursive in } \Phi\} = \nabla(\Phi)$.

5.7–5.9 Exercises

5.7. Show that for any positive analytic operator Φ and any families $\langle \mathsf{P}_p \rangle$ and $\langle \mathsf{Q}_q \rangle$, $\Phi\langle \mathsf{P}_p \rangle \cap \Phi\langle \mathsf{Q}_q \rangle = \Phi\langle \Phi\langle \mathsf{P}_p \cap \mathsf{Q}_q \rangle\rangle$.

5.8. Show that for any positive analytic Φ, $\nabla(\Phi)$ is closed under the quantifier Φ.

5.9. Adapt the proof of Theorem 4.6 to show that for $r \geq 1$, $\mathsf{B}(\Phi) \in \Delta_r^1$ iff Δ_r^1 is closed under the quantifier Φ.

5.10 Notes. The idea of generating sets and indices simultaneously originates with Kleene [1963]. The method was exploited in Clarke [1964] and Enderton [1964]. The constructions discussed here are essentially those of Hinman [1966] and [1969] and Shoenfield [1968].

It is not only for reasons of simplicity that we have restricted attention here to hierarchies of sets of numbers. If we attempt to generalize Definition 5.1 to be parallel to Definition IV.4.1–2, we come to expressions $\{a\}(p, \mathsf{P}_b^{\Phi})$ which at the present stage of the theory are not even defined. Although such recursions relative to functionals are defined in Chapter VI, the resulting construction is not a natural one. The reason is that among the relations P_b^{Φ} for b in some early stage of $N^{\Phi,k,l}$ there is one that is recursively equivalent to Φ (as a functional). Thus all enumerating functions recursive in Φ are already available at this stage and the "boot-strap" nature of the construction is lost. This objection can be overcome by restricting the enumerating functions to be recursive in the relations on numbers P_b^{Φ} for $b \in N^{\Phi,k,0}$. In either case, however, the hierarchy does not in general exhaust the class of relations recursive in Φ. This is discussed further in Hinman [1969] and proved in Hinman [1966].

6. *A Hierarchy for* Δ_2^1

In the preceding section we have seen that most of the characterizations of Δ_1^1 in terms of simpler relations do not have any natural extensions to characterizations of Δ_r^1 for $r \geqslant 2$. There is, however, one characterization which does extend, that of Corollary IV.2.22:

$$R \in \Delta_1^1 \leftrightarrow R \in \Delta_1^0[\gamma], \text{ for some implicitly } \Pi_1^0 \text{ function } \gamma.$$

The extension leads to some interesting results on implicitly Π_1^1 functions as well as a pleasant hierarchy of the Δ_2^1 relations on numbers.

6.1 Theorem. *For all* $R \subseteq {}^k\omega$,

$$R \in \Delta_2^1 \leftrightarrow R \in \Delta_1^1[\gamma], \textit{ for some implicitly } \Pi_1^1 \textit{ function } \gamma,$$

Proof. If R satisfies the right side, then $R \in \Delta_2^1$ by Examples III.2.3 and Corollary III.2.13. For the converse implication, suppose $R \in \Delta_2^1$ and let $\mathsf{K}_R'(m) = \mathsf{K}_R((m)_0, \ldots, (m)_{k-1})$. Then $\mathsf{K}_R' \in \Delta_2^1$ so there exists a Π_1^1 relation R such that

$$\mathsf{K}_R'(m) = n \leftrightarrow \exists \beta\, \mathsf{R}(m, n, \beta).$$

Hence for any α,

$$\alpha = \mathsf{K}_R' \leftrightarrow \exists \beta\, \forall m\, \forall n\, [\alpha(m) = n \rightarrow \mathsf{R}(m, n, (\beta)^{\langle m, n \rangle})].$$

If we denote the right side of this equivalence by $\exists\beta\, \mathsf{S}(\alpha, \beta)$, then $\mathsf{S} \in \Pi_1^1$ so we may apply the Uniformization Theorem to obtain a relation $\mathsf{S}' \in \Pi_1^1$ such that

$$\alpha = \mathsf{K}'_R \leftrightarrow \exists!\, \beta\, \mathsf{S}'(\alpha, \beta) \leftrightarrow \exists\beta\, \mathsf{S}'(\alpha, \beta).$$

Then S' holds of a unique pair of functions (K'_R, β) and thus the function $\gamma = \langle \mathsf{K}'_R, \beta\rangle$ is implicitly Π_1^1. Clearly R is $\Delta_1^1[\gamma]$. □

Note that we actually proved that every Δ_2^1 relation is recursive in some implicitly Π_1^1 function. Before constructing the hierarchy for Δ_2^1, we establish some facts about the implicitly Π_1^1 functions which are interesting in their own right. First, it follows immediately from the theorem that $\{\alpha : \alpha$ is recursive in some $B \in \Sigma_1^1\}$ is not a basis for even the class of Π_1^1 singletons (cf. Theorem III.4.7).

If γ is implicitly Π_1^1, then there are (infinitely many) recursive relations P such that for any α,

$$\alpha = \gamma \leftrightarrow \forall\beta\, \exists n\, \mathsf{P}(\bar{\beta}(n), \alpha) \leftrightarrow \mathsf{F}_{\mathsf{P}}[\alpha] \in \mathsf{W},$$

where F_{P} is the functional F constructed in the proof of Theorem IV.1.1. We temporarily call such a P a *matrix* for γ. Note that for each matrix P for γ, $\mathsf{F}_{\mathsf{P}}[\gamma] \in \mathsf{W}$.

6.2 Definition. For any implicitly Π_1^1 function γ,

$$\chi(\gamma) = \inf\{\|\mathsf{F}_{\mathsf{P}}[\gamma]\| : \mathsf{P} \text{ is a matrix for } \gamma\}.$$

It is clear that $\chi(\gamma)$ is always a countable ordinal — in fact:

6.3 Lemma. *For any implicitly Π_1^1 function γ,*

$$\chi(\gamma) < \omega_1[\gamma] < \delta_2^1.$$

Proof. For any matrix P for γ, $\mathsf{F}_{\mathsf{P}}[\gamma]$ is clearly the order-type of a well-ordering recursive in γ so that $\|\mathsf{F}_{\mathsf{P}}[\gamma]\| < \omega_1[\gamma]$. Since $\gamma \in \Delta_2^1$, $\omega_1[\gamma] < \delta_2^1$ by Exercise IV.2.31. □

6.4 Theorem. *For any β and any implicitly Π_1^1 function γ,*

$$\gamma \in \Delta_1^1[\beta] \leftrightarrow \chi(\gamma) < \omega_1[\beta].$$

Proof. The implication $(\rightarrow)$ is immediate from the preceding Lemma and a relativized version of Theorem IV.2.11. For $(\leftarrow)$, suppose that $\chi(\gamma) < \omega_1[\beta]$ and let P be a matrix for γ such that $F_P[\gamma] = \chi(\gamma)$. Choose $\delta \in W$, δ recursive in β, such that $\|\delta\| = \chi(\gamma)$. Then for any α,

$$\alpha = \gamma \leftrightarrow F_P[\alpha] \in W \leftrightarrow F_P[\alpha] \leqslant_\Sigma^W \delta \leftrightarrow F_P[\alpha] \leqslant_\Pi^W \delta.$$

Thus $\{\gamma\} \in \Delta_1^1[\beta]$, so also $\gamma \in \Delta_1^1[\beta]$ by the relativized version of Corollary III.2.7(vii). □

6.5 Corollary. *For any implicitly Π_1^1 functions γ and δ,*
(i) $\chi(\gamma) \leqslant \chi(\delta) \rightarrow \gamma \in \Delta_1^1[\delta]$;
(ii) $\gamma \in \Delta_1^1[\delta]$ *or* $\delta \in \Delta_1^1[\gamma]$.

Proof. (i) is immediate from 6.3 and 6.4, and (ii) is immediate from (i). □

To interpret these results most succinctly, we return to the notion of hyperdegree introduced in § IV.2 following Theorem IV.2.12, and the ordering $\leqslant_1^1$ on them. We call a hyperdegree x *implicitly* Π_1^1 iff some implicitly Π_1^1 function belongs to x. We may extend the function χ to implicitly Π_1^1 hyperdegrees x by setting

$$\chi(x) = \inf\{\chi(\gamma) : \gamma \in x \text{ and } \gamma \text{ is implicitly } \Pi_1^1\}.$$

6.6 Corollary. *The relation $\leqslant_1^1$ restricted to implicitly Π_1^1 hyperdegrees is a well-ordering.*

Proof. That the ordering is linear is exactly 6.5(ii). From the contrapositive of 6.5(i) we have for any implicitly Π_1^1 functions γ and δ,

$$\delta <_1^1 \gamma \rightarrow \chi(\delta) < \chi(\gamma)$$

from which it follows that for any implicitly Π_1^1 hyperdegrees x and y,

$$y <_1^1 x \rightarrow \chi(y) < \chi(x).$$

Since $\chi(x)$ and $\chi(y)$ are ordinals, this implies that the ordering is well-founded. □

We return now to the construction of hierarchy for the Δ_2^1 relations on numbers. The most obvious choice for the levels of the hierarchy is the sequence of sets

$$X(\sigma) = \{R : R \in \Delta_1^1[\gamma], \text{ for some implicitly } \Pi_1^1 \text{ function } \gamma \text{ such that } \chi(\gamma) < \sigma\}.$$

However, it is immediate from Theorem 6.4 that for some $\sigma < \omega_1$, $X(\sigma) = X(\tau) = X(\omega_1) = \Delta_1^1$ for all τ, $\sigma \leqslant \tau \leqslant \omega_1$. To obtain a properly increasing sequence of sets we select only certain of the $X(\sigma)$.

For any hyperdegree x and any $\alpha, \beta \in x$, $\Delta_1^1[\alpha] = \Delta_1^1[\beta]$ and then from Theorem IV.2.14, $\omega_1[\alpha] = \omega_1[\beta]$. We denote these common values by $\Delta_1^1[x]$ and $\omega_1[x]$.

6.7 Definition. For all $\sigma > 0$,

(i) $x_0 = \{\alpha : \alpha \in \Delta_1^1\}$ (the zero hyperdegree);

(ii) if there exists an implicitly Π_1^1 hyperdegree x such that for all $\tau < \sigma$, $x_\tau <_1^1 x$, then x_σ is the $\leqslant_1^1$-least such x; otherwise $x_\sigma = x_0$;

(iii) $Z_\sigma = X(\omega_1[x_\sigma])$;

(iv) $\kappa = (\text{least } \sigma > 0)\ [x_\sigma = x_0]$.

Note that by Theorem 6.4, for $\sigma < \kappa$, $Z_\sigma = \{R : R \in \Delta_1^1[x_\sigma]\}$.

6.8 Theorem. (i) *For all* $\sigma < \tau < \kappa$, $Z_\sigma \subsetneq Z_\tau \subseteq \Delta_2^1$;

(ii) *for any* R, $R \in \Delta_2^1$ *iff* $R \in Z_\sigma$ *for some* $\sigma < \kappa$.

Proof. (i) is immediate from the remark preceding the theorem. For (ii), suppose $R \in \Delta_2^1$. By Theorem 6.1 choose an implicitly Π_1^1 function γ such that $R \in \Delta_1^1[\gamma]$. Then there exists a $\sigma < \kappa$ such that $\text{hydg}(\gamma) \leqslant_1^1 x_\sigma$ and thus $R \in \Delta_1^1[x_\sigma]$ so $R \in Z_\sigma$. □

It remains only to evaluate the length κ of this hierarchy. Since any γ which is implicitly Π_1^1 is Δ_2^1, $\sigma < \omega_1[x_\sigma] < \delta_2^1$ for all $\sigma < \kappa$, and thus $\kappa \leqslant \delta_2^1$.

6.9 Theorem. $\kappa = \delta_2^1$.

Proof. Suppose to the contrary that $\kappa < \delta_2^1$. If κ were a successor ordinal $\lambda + 1$, then x_λ would be $<_1^1$-greatest among all implicitly Π_1^1 hyperdegrees. By Theorem 6.4 this implies that $\chi(\gamma) < \omega_1[x_\lambda]$ for all implicitly Π_1^1 functions γ and this leads easily to the conclusion that the class of Δ_2^1 functions is Δ_2^1, contrary to Theorem 1.19(v). Hence κ is a limit ordinal.

We shall derive a contradiction by constructing an implicitly Π_1^1 function γ such that $x_\sigma <_1^1 \text{hydg}(\gamma)$ for all $\sigma < \kappa$. By the Uniformization Theorem, there exists a Π_1^1 relation V such that for all a and γ,

$$\mathsf{V}(a, \gamma) \rightarrow \sim \mathsf{U}_1^1(a, \langle\ \rangle, \langle\gamma\rangle), \quad \text{and}$$

$$\exists! \gamma\, \mathsf{V}(a, \gamma) \leftrightarrow \exists \gamma \sim \mathsf{U}_1^1(a, \langle\ \rangle, \langle\gamma\rangle).$$

Thus γ is implicitly Π_1^1 iff for some a, γ is the unique function satisfying $\mathsf{V}(a, \gamma)$. Let ε be a Δ_2^1 well-ordering such that $\|\varepsilon\| = \kappa$ and consider the following set of functions:

$$A = \{\alpha : \forall p \in \mathrm{Fld}(\varepsilon) \exists \gamma\, (V(\alpha(p), \gamma) \wedge (\forall q <_\varepsilon p) \exists \beta\, [V(\alpha(q), \beta) \wedge \beta <^1_1 \gamma])\}.$$

A is Σ^1_2 and if $\alpha \in A$, then for all $p \in \mathrm{Fld}(\varepsilon)$, if $\sigma = |p|_\varepsilon$, then the unique γ such that $V(\alpha(p), \gamma)$ satisfies $x_\sigma \leq^1_1 \mathrm{hydg}(\gamma)$. Furthermore, since each $|p|_\varepsilon$ is less than κ, there always is such a γ and thus $A \neq \varnothing$. By the Basis Theorem choose a fixed $\alpha \in A \cap \Delta^1_2$.

Let δ be an implicitly Π^1_1 function such that both α and ε are recursive in δ, say with indices a and e, respectively. Let

$$\begin{aligned} B = \{\gamma : (\gamma)^0 = \delta \wedge \\ &\forall p\, [p \in \mathrm{Fld}(\lambda m . \{e\}(m, (\gamma)^0)) \rightarrow V(\{a\}(p, (\gamma)^0), (\gamma)^{p+1})] \wedge \\ &\forall r\, [(r \notin \mathrm{Sq} \vee \mathrm{lg}(r) \neq 2 \vee ((r)_0 \dot{-} 1) \notin \mathrm{Fld}(\lambda m . \{e\}(m, (\gamma)^0))) \rightarrow \gamma(r) = 0]\}. \end{aligned}$$

Clearly B is a Π^1_1 set with a unique member γ, which is thus implicitly Π^1_1. For any $\tau < \kappa$, there exists a $p \in \mathrm{Fld}(\varepsilon)$ such that $|p|_\varepsilon = \tau + 1$ and thus

$$x_\tau <^1_1 x_{\tau+1} \leq^1_1 \mathrm{hydg}((\gamma)^{p+1}) \leq^1_1 \mathrm{hydg}(\gamma).$$

It follows that $x_\kappa \neq x_0$, a contradiction. □

6.10–6.17 Exercises

6.10. Show that not all Δ^1_2 functions are implicitly Π^1_1 (use the existence of incomparable hyperdegrees from Exercise 2.13).

6.11. Prove

(i) if β is implicitly Π^1_1 and β and γ are each Δ^1_1 in the other, then also γ is implicitly Π^1_1;

(ii) if β is implicitly Π^1_1, then so is the hyperjump of β;

(iii) in Definition 6.7, $x_{\sigma+1}$ is the hyperjump of x_σ (use Theorems 6.4 and IV.2.14).

6.12. Show that for any implicitly Π^1_1 functions γ and δ,

$$\gamma <^1_1 \delta \leftrightarrow \omega_1[\gamma] < \omega_1[\delta].$$

6.13. Show that for any implicitly Π^1_1 function γ, γ is implicitly Π^0_1 iff $\chi(\gamma) \leq 1$.

6.14. Show that

$$\sup{}^+\{\|\gamma\| : \gamma \in W \wedge \gamma \text{ is implicitly } \Pi^1_1\} = \delta^1_2.$$

6.15. Use the relation V defined in the proof of Theorem 6.9 together with Theorem 6.1 to give a new proof that $\{\alpha : \alpha \in \Delta^1_2\} \in \Sigma^1_2$.

6.16. Combine the results of the preceding two exercises to contruct a Σ^1_2 well-ordering of length δ^1_2.

6.17. What parts of this section may be generalized to Δ^1_r under either of the hypotheses $V = L$ or PD?

6.18 Notes. The results of this section are due to Suzuki [1964].

Part C

Generalized Recursion Theories

Each of the measures of complexity we have discussed in Parts A and B can be seen as an analysis of the means required to define a class of relations into two components — one constructive or effective and the other in some way non-constructive. Recursive relations are from our point of view entirely constructive. The arithmetical and analytical relations are generated from them by the non-constructive operation of quantification over an infinite set. For the relativized and boldface hierarchies, the parameters from ${}^{\omega}\omega$ are another non-constructive component. In the case of inductive definability, the operators are defined non-constructively, but we have implicitly extended our concept of constructivity to include the process of iteration over the ordinals.

The generalized recursion theories we study in this part are an elaboration of this dichotomy. In each case the "recursive" functionals are those which are computable by essentially the same constructive means as in Chapter II together with some non-constructive one. In Chapters VI and VII the non-constructive component consists in the introduction of various fixed functionals; in Chapter VIII it is the extension of the fundamental domain from ω to various ordinals κ or to the class of all ordinals.

As might be expected, there are many connections with the earlier theory. Let

$$\mathsf{E}(\alpha) = \begin{cases} 0, & \text{if } \exists m \,.\, \alpha(m) = 0; \\ 1, & \text{otherwise.} \end{cases}$$

Then the relations (semi-) recursive in E are exactly the (Π^1_1) Δ^1_1 relations (Theorem VI.1.7). The relations recursive in a normal operation Φ or a jump operator J turn out to be exactly those encompassed by the hierarchies of § V.5. (Π^1_1) Δ^1_1 is also exactly the class of relations which are (semi-) recursive when the fundamental domain is enlarged from ω to ω_1 (Theorem VIII.3.4). If computations are carried out over the ordinal δ^1_2 (or over $\aleph_1$), the (semi-) recursive relations on numbers are exactly the (Σ^1_2) Δ^1_2 relations (Theorem VIII.3.7 and Corollary VIII.5.10).

Chapter VI
Recursion in a Type-2 Functional

In the notes to § II.5 we described an alternative approach to ordinary recursion theory in which the notion "F is partial recursive in β" is defined first for functions $F: {}^k\omega \to \omega$ and a fixed β and used to derive the notion of a recursive functional. For the development of the theory of recursion with type-2 arguments, we have the same choice "one type up". We may take as our primary notion either partial recursiveness of functionals $\mathbb{F}: {}^{k,l,l'}\omega \to \omega$ (with arguments of type $(\mathbf{m}, \boldsymbol{\alpha}, \mathsf{I})$) or the notion "$\mathsf{F}$ is partial recursive in I" for functionals $\mathsf{F}: {}^{k,l}\omega \to \omega$ and a fixed I. We choose the second course for two reasons. First, the technical details of the theory involving only one fixed type-2 parameter are considerably simpler. Second, the primary objects under study remain functionals and relations on ${}^{k,l}\omega$, and thus the theory forms a natural extension of that of earlier chapters. Partial recursiveness for functions $\mathbb{F}: {}^{k,l,l'}\omega \to \omega$ will be discussed in § 7.

In § 1 we establish the simplest properties of recursion relative to a functional and discuss the connections with the arithmetical and analytical hierarchies. The most important technical feature which distinguishes recursion relative to a functional from ordinary recursion is the absence of normal form theorems as in § II.3. This lack necessitates substantially longer and more involved proofs of many of the results corresponding to those in §§ II.3–4 and these occupy §§ 2–4 of this chapter. The fact that Δ_1^1 is the class of relations recursive in E suggests that the class of relations recursive in other functionals may share some of the properties of Δ_1^1. This is already evident in §§ 3 and 4 and is carried further by the discussion of hierarchies in § 5. In § 6 we discuss a flaw in this analogy and an alternative analogy related to the operator $*$ of § V.4.

1. Basic Properties

To avoid unecessary complications we shall restrict ourselves to the study of recursion relative to a single total functional $\mathsf{I}: {}^{\omega}\omega \to \omega$. Other cases may be reduced to this one as indicated at the end of this section. Let us consider first the intuitive notion of a functional F being mechanically calculable relative to I. As for recursion relative to a type-1 function, we imagine an idealized computer prepared to accept inputs of the form $(\mathbf{m}, \boldsymbol{\alpha})$ and connected to a memory device M which contains the graph of I. Of course, the graph of I is a set of power $2^{\aleph_0}$ so this M must be larger than the memory units we have previously considered. Somewhat more troublesome is the fact that in order to obtain from M a desired value $\mathsf{I}(\beta)$, the computer must in some way present M with the argument β. As β is an infinite object, it is not obvious how we should imagine this to be accomplished. We have, however, a precedent in the mechanism for presenting the computer with (infinite) inputs $(\mathbf{m}, \boldsymbol{\alpha})$. Accordingly, we imagine a subsidiary infinite memory unit M' large enough to store a single function β. Then to obtain a value $\mathsf{I}(\beta)$ during the course of a computation, the computer "loads" M' with the graph of β, whereupon M responds with the desired value $\mathsf{I}(\beta)$.

This model leads immediately to two observations. First, computations relative to I must in general be infinite, as the computer cannot be expected to load M' with the graph of β in a finite time, and it must in general be decided during the course of the computation for which β the value $\mathsf{I}(\beta)$ is needed, so that M' cannot be loaded before the start of the computation. Second, during any given computation the only values $\mathsf{I}(\beta)$ which may be obtained are those for which β itself is computable — in fact, the calculation of the values of β is a part of the computation. Thus if F is partial computable in I, the value of $\mathsf{F}(\mathbf{m}, \boldsymbol{\alpha})$ depends on the values $\mathsf{I}(\beta)$ at most for β which are computable in I and $\boldsymbol{\alpha}$ (cf. Exercise 1.13).

The formal definition is very similar to that of § II.2 with an additional clause to introduce values of I.

1.1 Definition. For any total functional $\mathsf{I}: {}^{\omega}\omega \to \omega$, the set $\Omega[\mathsf{I}]$ is the smallest set such that for all k, l, n, p, q, r, and s, all $i < k$ and $j < l$, and all $(\mathbf{m}, \boldsymbol{\alpha}) \in {}^{k,l}\omega$,

(0), (1), (2) — identical to the corresponding clauses of Definition II.2.1;

(3) for any b and any β, if for all p, $(b, p, \mathbf{m}, \boldsymbol{\alpha}, \beta(p)) \in \Omega[\mathsf{I}]$, then $(\langle 3, k, l, b\rangle, \mathbf{m}, \boldsymbol{\alpha}, \mathsf{I}(\beta)) \in \Omega[\mathsf{I}]$.

In contrast with Ω, $\Omega[\mathsf{I}]$ is not a closure under finitary functions. It is, however, a closure under functions of rank ω, as clause (3) is equivalent to the requirement that $\Omega[\mathsf{I}]$ be closed under all the functions φ_β, where

$$\varphi_\beta(\{(b, p, \mathbf{m}, \boldsymbol{\alpha}, \beta(p)) : p \in \omega\}) = (\langle 3, k, l, b\rangle, \mathbf{m}, \boldsymbol{\alpha}, \mathsf{I}(\beta)).$$

Hence the inductive operator is $\aleph_1$-compact and, by the intended result of

Exercise I.3.10, this implies that the inductive definition has closure ordinal at most $\aleph_1$. If we associate the ordinal of the level at which a given sequence $(a, \mathbf{m}, \boldsymbol{\alpha}, n)$ occurs with the "length" of the corresponding computation $\{a\}^{\mathsf{I}}(\mathbf{m}, \boldsymbol{\alpha}) \simeq n$, we see that although computations relative to I may be infinite, they are all countable.

The first part of the theory of recursion relative to a functional I now proceeds almost exactly as in § II.2. The proof that for all a, $\mathbf{m}$, and $\boldsymbol{\alpha}$ there is at most one n such that $(a, \mathbf{m}, \boldsymbol{\alpha}, n) \in \Omega[\mathsf{I}]$ differs from that of Lemma II.2.2 only in that the induction is now on all countable ordinals. We write

$$\{a\}^{\mathsf{I}}(\mathbf{m}, \boldsymbol{\alpha}) \simeq n \leftrightarrow (a, \mathbf{m}, \boldsymbol{\alpha}, n) \in \Omega[\mathsf{I}].$$

A functional F is *partial recursive in* I iff for some a, $\mathsf{F} = \{a\}^{\mathsf{I}}$; a is called an *index of* F *from* I. F is *recursive in* I iff F is partial recursive in I and total, and R is *recursive in* I iff $\mathsf{K_R}$ is recursive in I. R is *semi-recursive in* I iff R is the domain of some functional partial recursive in I, and *co-semi-recursive in* I iff $\sim$R is semi-recursive in I.

As in Remark II.2.4, it is easily verified that each clause in the Definition has its intended meaning. Thus

$$\{\langle 3, k, l, b\rangle\}^{\mathsf{I}}(\mathbf{m}, \boldsymbol{\alpha}) \simeq \mathsf{I}(\lambda p . \{b\}^{\mathsf{I}}(p, \mathbf{m}, \boldsymbol{\alpha})),$$

and the class of functionals partial recursive in I is closed under functional composition. It is clear from the definition that $\Omega \subseteq \Omega[\mathsf{I}]$. Hence if $\{a\}(\mathbf{m}, \boldsymbol{\alpha}) \simeq n$, then also $\{a\}^{\mathsf{I}}(\mathbf{m}, \boldsymbol{\alpha}) \simeq n$. In particular, if $\{a\}$ is total, $\{a\} = \{a\}^{\mathsf{I}}$, so every recursive functional is recursive in I.

The functions Sb_i of Lemma II.2.5 also have the property that

$$\{\mathrm{Sb}_i(a, m_0, \ldots, m_i)\}^{\mathsf{I}}(m_{i+1}, \ldots, m_{k-1}, \boldsymbol{\alpha}) \simeq \{a\}^{\mathsf{I}}(\mathbf{m}, \boldsymbol{\alpha}),$$

and exactly the same proof as before establishes the I-*Recursion Theorem*: *for any functional* F *partial recursive in* I, *there exists an* $\bar{e} \in \omega$ *such that for all* $(\mathbf{m}, \boldsymbol{\alpha})$,

$$\{\bar{e}\}^{\mathsf{I}}(\mathbf{m}, \boldsymbol{\alpha}) \simeq \mathsf{F}(\bar{e}, \mathbf{m}, \boldsymbol{\alpha}).$$

From this follows exactly as in § II.2 that the class of functionals partial recursive in I is closed under primitive recursion, course-of-values recursion, and unbounded search. Hence by Corollary II.3.3, every partial recursive functional is also partial recursive in I. Definition by cases with relations recursive in I and functionals partial recursive in I is established exactly as for Theorem II.2.9. Similarly, the class of relations recursive in I is a Boolean algebra closed under composition with functionals recursive in I.

Computations relative to I may also be thought of as being arranged in labeled trees. Nodes corresponding to clauses (0)–(2) are as before, while for clause (3) we have nodes of the form:

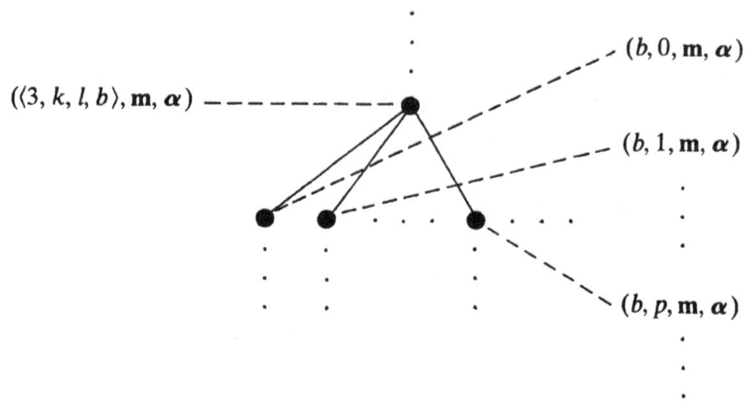

Of course, branching is no longer finite, so the tree may be infinite without having an infinite branch.

Exactly as in Lemma II.4.2 we may prove that every relation recursive in I is also semi-recursive in I, but not conversely. Similarly, the class of relations semi-recursive in I is closed under finite intersection and bounded universal quantification — in fact,

1.2 Lemma. *For any* I, *the class of relations semi-recursive in* I *is closed under universal number quantification* ($\forall^0$).

Proof. Suppose $\mathsf{R}(p, \mathbf{m}, \boldsymbol{\alpha}) \leftrightarrow \{a\}^{\mathsf{I}}(p, \mathbf{m}, \boldsymbol{\alpha})\downarrow$. Then

$$\forall p\, \mathsf{R}(p, \mathbf{m}, \boldsymbol{\alpha}) \leftrightarrow \mathsf{I}(\lambda p\,.\,\{a\}^{\mathsf{I}}(p, \mathbf{m}, \boldsymbol{\alpha}))\downarrow$$

so $\forall^0 \mathsf{R} = \mathrm{Dm}\{\langle 3, k, l, a\rangle\}^{\mathsf{I}}$ and is thus semi-recursive in I. □

The remaining part of the theory of semi-recursive relations depends on the normal form theorems of § II.3. As we have no analogue for these for recursion relative to I, the corresponding results will be proved in a different way in §§ 2–4.

To illustrate the notion of recursion relative to a functional, we now consider some examples. Consider the functionals defined as follows:

$$\mathsf{E}(\alpha) = \begin{cases} 0, & \text{if } \exists m\,.\,\alpha(m) = 0; \\ 1, & \text{otherwise;} \end{cases}$$

$$\mathsf{E}_1(\alpha) = \begin{cases} 0, & \text{if} \quad \exists\beta\,\forall n\,.\,\alpha(\bar{\beta}(n)) = 0; \\ 1, & \text{otherwise}; \end{cases}$$

$$\mathsf{oJ}((\langle a, \mathbf{m}\rangle) * \alpha) = \begin{cases} 0, & \text{if} \quad \{a\}(\mathbf{m}, \alpha)\downarrow; \\ 1, & \text{otherwise}; \end{cases}$$

where (as defined in § I.1) $((\langle a, \mathbf{m}\rangle) * \alpha)(0) = \langle a, \mathbf{m}\rangle$ and $((\langle a, \mathbf{m}\rangle) * \alpha)(p+1) = \alpha(p)$. The functional oJ is of course just a natural encoding of the ordinary jump operator of the same name:

$$\mathsf{oJ}((\langle a, \mathbf{m}\rangle) * \alpha) = \alpha^{\mathsf{oJ}}(\langle a, \mathbf{m}\rangle).$$

1.3 Lemma. *For all* R,
(i) *if* R *is arithmetical, then* R *is recursive in* E;
(ii) *if* $\mathsf{R} \in \Sigma^1_1 \cup \Pi^1_1$, *then* R *is recursive in* E_1.

Proof. Consider the set X of all relations recursive in E. By the preceding remarks, X contains all recursive relations, so for (i) it suffices to show that X is closed under $\exists^0$ (as it is then closed under $\forall^0$ by complementation). Suppose R is recursive in E. Then

$$\begin{aligned} \mathsf{K}_{\exists^0\mathsf{R}}(\mathbf{m}, \boldsymbol{\alpha}) &= \begin{cases} 0, & \text{if} \quad \exists p\,.\,\mathsf{K}_\mathsf{R}(p, \mathbf{m}, \boldsymbol{\alpha}) = 0; \\ 1, & \text{otherwise}; \end{cases} \\ &= \mathsf{E}(\lambda p\,.\,\mathsf{K}_\mathsf{R}(p, \mathbf{m}, \boldsymbol{\alpha})). \end{aligned}$$

Hence if $\mathsf{K}_\mathsf{R} = \{b\}^\mathsf{E}$, then $\mathsf{K}_{\exists^0\mathsf{R}} = \{\langle 3, k, l, b\rangle\}^\mathsf{E}$, so $\exists^0\mathsf{R}$ is also recursive in E.

For (ii), if $\mathsf{R} \in \Sigma^1_1$, then for some recursive S,

$$\mathsf{R}(\mathbf{m}, \boldsymbol{\alpha}) \leftrightarrow \exists\beta\,\forall n\,\mathsf{S}(\bar{\beta}(n), \mathbf{m}, \boldsymbol{\alpha}).$$

Hence

$$\mathsf{K}_\mathsf{R}(\mathbf{m}, \boldsymbol{\alpha}) = \mathsf{E}_1(\lambda p\,.\,\mathsf{K}_\mathsf{S}(p, \mathbf{m}, \boldsymbol{\alpha})),$$

so K_R is recursive in E_1. The result for Π^1_1 again follows by closure under complementation. □

Of course, it follows that E is recursive in E_1.

1.4 Lemma. *Each of* E *and* oJ *is recursive in the other.*

Proof. Let a be an index such that $\{a\}(\alpha) \simeq$ least $m . \alpha(m) = 0$. Then clearly $\mathsf{E}(\alpha) = {}^{\mathrm{o}}\mathsf{J}((\langle a \rangle) * \alpha)$, so if b is an index such that $\{b\}^{{}^{\mathrm{o}}\mathsf{J}}(0, \alpha) = a$ and $\{b\}^{{}^{\mathrm{o}}\mathsf{J}}(p+1, \alpha) = \alpha(p)$, then $\mathsf{E} = \{\langle 3, 0, 1, b \rangle\}^{{}^{\mathrm{o}}\mathsf{J}}$; so E is recursive in ${}^{\mathrm{o}}\mathsf{J}$. On the other hand from Theorem II.3.1 we have

$$ {}^{\mathrm{o}}\mathsf{J}((\langle a, \mathbf{m} \rangle) * \alpha) = \mathsf{E}(\lambda u . \mathsf{K}_{\mathsf{T}}(a, \langle \mathbf{m} \rangle, u, \langle \alpha \rangle)) $$

from which we similarly conclude that ${}^{\mathrm{o}}\mathsf{J}$ is recursive in E. □

We shall prove below that the relations recursive in E are exactly the Δ^1_1 relations, but we can already guess that more than the arithmetical relations are recursive in E. Recall the sets D_r of Theorem III.1.13:

$$ D_0 = \{0\} \quad \text{and} \quad D_{r+1} = (D_r)^{{}^{\mathrm{o}}\mathsf{J}}. $$

It is not hard to see that there is a primitive recursive function f such that for all r, $f(r)$ is an index of D_r from E. Then the set

$$ \{\langle r, m \rangle : m \in D_r\} = \{\langle r, m \rangle : \{f(r)\}^{\mathsf{E}}(m) = 0\} $$

is also recursive in E but, by III.1.13 and the Arithmetical Hierarchy Theorem, is not arithmetical. Similarly, Lemma 1.3(ii) is far from optimal. We shall return in §§ 5 and 6 to the class of relations recursive in E_1.

Upper bounds on the class of relations recursive in a given functional I again follow from the fundamental results on inductive definability of § III.3. Let

$$ \mathsf{V}^{\mathsf{I}}(a, \langle \mathbf{m} \rangle, n, \langle \boldsymbol{\alpha} \rangle) \leftrightarrow (a, \mathbf{m}, \boldsymbol{\alpha}, n) \in \Omega[\mathsf{I}]. $$

1.5 Theorem. *For any* I,

(i) *if* $\mathsf{I} \in \Delta^1_1$, *then* $\mathsf{V}^{\mathsf{I}} \in \Pi^1_1$;

(ii) *for all* $r \geq 2$, *if* $\mathsf{I} \in \Delta^1_r$, *then* $\mathsf{V}^{\mathsf{I}} \in \Delta^1_r$.

Proof. For each l and each $\boldsymbol{\alpha} \in {}^{l}({}^{\omega}\omega)$, we define operators $\Gamma_{\boldsymbol{\alpha},0}, \ldots, \Gamma_{\boldsymbol{\alpha},3}$ and $\Lambda_{\boldsymbol{\alpha},3}$ as follows. $\Gamma_{\boldsymbol{\alpha},q}$ corresponds to clause (q) of Definition 1.1. For any $A \subseteq \omega$,

$$ \begin{aligned} \Gamma_{\boldsymbol{\alpha},0}(A) = \{&\langle\langle 0, k, l, 0, n \rangle, \langle \mathbf{m} \rangle, n \rangle : k, n \in \omega \wedge \mathbf{m} \in {}^{k}\omega\} \\ &\cup \cdots \cup \{\langle\langle 0, k, l, 3, i, j \rangle, \langle \mathbf{m} \rangle, \alpha_j(m_i) \rangle : k \in \omega \wedge \mathbf{m} \in {}^{k}\omega \wedge \\ &\qquad\qquad\qquad i < k \wedge j < l\} \\ &\cup \cdots \cup \{\langle\langle 0, k+2, l, 5 \rangle, \langle p, q, \mathbf{m} \rangle, \mathrm{Sb}_0(p, q) \rangle : k, p, q \in \omega, \\ &\qquad\qquad\qquad \mathbf{m} \in {}^{k}\omega\}; \end{aligned} $$

$$ \begin{aligned} \Gamma_{\boldsymbol{\alpha},1}(A) = \{&\langle\langle 1, k, l, b, c_0, \ldots, c_{k'-1} \rangle, \langle \mathbf{m} \rangle, n \rangle : \text{for some } k', q_0, \ldots, \\ &q_{k'-1}, (\forall i < k')[\langle c_i, \langle \mathbf{m} \rangle, q_i \rangle \in A \wedge \langle b, \langle \mathbf{q} \rangle, n \rangle \in A]\}; \end{aligned} $$

$$\Gamma_{\alpha,2}(A)=\{\langle\langle 2,k+1,l\rangle,\langle b,\mathbf{m}\rangle,n\rangle:\langle b,\langle\mathbf{m}\rangle,n\rangle\in A\};$$

$$\Gamma_{\alpha,3}(A)=\{\langle\langle 3,k,l,b\rangle,\langle\mathbf{m}\rangle,n\rangle:\exists\beta\,(\forall p\,[\langle b,\langle p,\mathbf{m}\rangle,\beta(p)\rangle\in A]\wedge \mathsf{I}(\beta)=n)\};$$

$$\Lambda_{\alpha,3}(A)=\{\langle\langle 3,k,l,b\rangle,\langle\mathbf{m}\rangle,n\rangle:\forall p\,\exists q\,[\langle b,\langle p,\mathbf{m}\rangle,q\rangle\in A]\wedge \wedge\forall\beta\,(\forall p\,\forall q\,[\langle b,\langle p,\mathbf{m}\rangle,q\rangle\in A\rightarrow\beta(p)=q]\rightarrow\mathsf{I}(\beta)=n)\}.$$

First, note that all of these operators are monotone. Hence, if we set

$$\Gamma_{\langle\alpha\rangle}(A)=\Gamma_{\alpha,0}(A)\cup\Gamma_{\alpha,1}(A)\cup\Gamma_{\alpha,2}(A)\cup\Gamma_{\alpha,3}(A),$$

and

$$\Lambda_{\langle\alpha\rangle}(A)=\Gamma_{\alpha,0}(A)\cup\Gamma_{\alpha,1}(A)\cup\Gamma_{\alpha,2}(A)\cup\Lambda_{\alpha,3}(A),$$

then $\Gamma_{\langle\alpha\rangle}$ and $\Lambda_{\langle\alpha\rangle}$ are also monotone. Furthermore it is clear from the definition of $\Omega[\mathsf{I}]$ that

$$(a,\mathbf{m},\alpha,n)\in\Omega[\mathsf{I}]\leftrightarrow\langle a,\langle\mathbf{m}\rangle,n\rangle\in\bar{\Gamma}_{\langle\alpha\rangle}$$

that is, $\mathsf{V}^{\mathsf{I}}=\bar{\Gamma}$, where Γ is the decomposable monotone operator defined by the family $\{\Gamma_{\langle\alpha\rangle}:l\in\omega,\ \alpha\in{}^{l}({}^{\omega}\omega)\}$.

Although it is not in general true that $\Gamma_{\alpha,3}(A)=\Lambda_{\alpha,3}(A)$, this is easily seen to be true whenever A has the property

$$(*)\qquad \langle a,\langle\mathbf{m}\rangle,n\rangle\in A\wedge\langle a,\langle\mathbf{m}\rangle,n'\rangle\in A\rightarrow n=n'.$$

An easy inductive argument shows that for all ordinals σ, $\Lambda^{\sigma}_{\langle\alpha\rangle}$ satisfies (*) and hence coincides with $\Gamma^{\sigma}_{\langle\alpha\rangle}$. Thus $\mathsf{V}^{\mathsf{I}}=\bar{\Lambda}$ as well.

We complete the proof by evaluating the complexity of Γ and Λ. If $\mathsf{I}\in\Delta^1_r$, then $\Gamma\in\Sigma^1_r$ and $\Lambda\in\Pi^1_r$. For all $r\geqslant 1$ this implies $\mathsf{V}^{\mathsf{I}}\in\Pi^1_r$ and for all $r\geqslant 2$, $\mathsf{V}^{\mathsf{I}}\in\Sigma^1_r$. □

1.6 Corollary. *For any* I,

(i) *if* $\mathsf{I}\in\Delta^1_1$, *then for all* R,

R *semi-recursive in* $\mathsf{I}\rightarrow\mathsf{R}\in\Pi^1_1$;

R *recursive in* $\mathsf{I}\rightarrow\mathsf{R}\in\Delta^1_1$;

(ii) *for any* $r\geqslant 2$, *if* $\mathsf{I}\in\Delta^1_r$, *then for all* R,

R *semi-recursive in* $\mathsf{I}\rightarrow\mathsf{R}\in\Delta^1_r$;

in particular, $\{\mathsf{R}:\mathsf{R}$ *recursive in* $\mathsf{I}\}$ *is a proper subset of* Δ^1_r.

Proof. If R is semi-recursive in I, then for some a,

$$\mathsf{R}(\mathbf{m}, \boldsymbol{\alpha}) \leftrightarrow \exists n\, \mathsf{V}^{\mathsf{I}}(a, \langle \mathbf{m} \rangle, n, \langle \boldsymbol{\alpha} \rangle),$$

and the implications are immediate from the theorem. □

In particular, we have

$$\{\mathsf{R} : \mathsf{R} \text{ is recursive in } \mathsf{E}\} \subseteq \Delta_1^1 \quad \text{and} \quad \{\mathsf{R} : \mathsf{R} \text{ is recursive in } \mathsf{E}_1\} \subsetneq \Delta_2^1.$$

To prove the converse of the first of these, we must "borrow" two results from later sections:

(†) (Corollary 2.11) for any I, the class of functionals partial recursive in I is closed under functional substitution — that is, for any functionals G and H partial recursive in I, if

$$\mathsf{F}(\mathbf{m}, \boldsymbol{\alpha}) \simeq \mathsf{G}(\mathbf{m}, \boldsymbol{\alpha}, \lambda p\,.\,\mathsf{H}(p, \mathbf{m}, \boldsymbol{\alpha})),$$

then F is also partial recursive in I;

(††) (Corollary 4.3) the class of relations semi-recursive in E is closed under existential number quantification ($\exists^0$).

The basic tool is the following *Ordinal Comparison Lemma*:

1.7 Lemma. *There exists a functional* H *partial recursive in* E *such that for all* γ *and* δ,

(i) $\gamma \in \mathsf{W}$ *and* $\|\gamma\| \leqslant \|\delta\| \rightarrow \mathsf{H}(\gamma, \delta) \simeq 0$;

(ii) $\delta \in \mathsf{W}$ *and* $\|\delta\| < \|\gamma\| \rightarrow \mathsf{H}(\gamma, \delta) \simeq 1$.

Proof. Let the dual functional E° be defined by:

$$\mathsf{E}^\circ(\alpha) = 1 \dot{-} \mathsf{E}(\lambda p\, [1 \dot{-} \alpha(p)])$$

so that

$$\mathsf{E}^\circ(\alpha) = \begin{cases} 0, & \text{if } \forall p\,.\,\alpha(p) = 0; \\ 1, & \text{if } \exists p\,.\,\alpha(p) \neq 0. \end{cases}$$

Let

$$\mathsf{F}(e, \gamma, \delta) \simeq \begin{cases} 0, & \text{if } \|\gamma\| = 0; \\ 1, & \text{if } \|\gamma\| \neq 0 \wedge \|\delta\| = 0; \\ \mathsf{E}^\circ(\lambda p\,.\,\mathsf{E}(\lambda q\,.\,\{e\}^{\mathsf{E}}(\gamma \restriction p, \delta \restriction q))), & \text{otherwise.} \end{cases}$$

The relation $\|\gamma\|=0$ is arithmetical, so by Lemma 1.3 and (†), F is partial recursive in E. Hence we may apply the E-Recursion Theorem to obtain an index $\bar{e}$ such that

$$\mathsf{F}(\bar{e}, \gamma, \delta) \simeq \{\bar{e}\}^{\mathsf{E}}(\gamma, \delta).$$

If we set $\mathsf{H}=\{\bar{e}\}^{\mathsf{E}}$, then H is partial recursive in E and

$$\mathsf{H}(\gamma, \delta) \simeq \begin{cases} 0, & \text{if } \|\gamma\|=0; \\ 1, & \text{if } \|\gamma\| \neq 0 \wedge \|\delta\|=0; \\ \mathsf{E}^{\circ}(\lambda p\,.\,\mathsf{E}(\lambda q\,.\,\mathsf{H}(\gamma\restriction p, \delta\restriction q))), & \text{otherwise.} \end{cases}$$

In other words, if both $\|\gamma\|$ and $\|\delta\|$ are non-zero, then $\mathsf{H}(\gamma, \delta)$ is defined just in case $\mathsf{H}(\gamma\restriction p, \delta\restriction q)$ is defined for all p and q, and if so, then

(1) $\quad \mathsf{H}(\gamma, \delta) \simeq 0 \leftrightarrow \forall p\, \exists q\,.\,\mathsf{H}(\gamma\restriction p, \delta\restriction q) \simeq 0;$

(2) $\quad \mathsf{H}(\gamma, \delta) \simeq 1 \leftrightarrow \exists p\, \forall q\,.\,\mathsf{H}(\gamma\restriction p, \delta\restriction q) \simeq 1.$

Clauses (i) and (ii) are vacuous unless one of γ and δ belongs to W. Hence we may prove (i) and (ii) by induction on $\sigma=\min\{\|\gamma\|, \|\delta\|\}$ for $\sigma<\aleph_1$. If $\sigma=0$, then either $\|\gamma\|=0$, so $\mathsf{H}(\gamma, \delta) \simeq 1$ in accord with (i), or $\|\delta\|=0<\|\gamma\|$, so $\mathsf{H}(\gamma, \delta) \simeq 1$ in accord with (ii). We assume now that $\sigma>0$ and as induction hypothesis that (i) and (ii) hold for all γ_0 and δ_0 such that $\min\{\|\gamma_0\|, \|\delta_0\|\}<\sigma$.

Suppose first that $\gamma \in \mathsf{W}$ and $\|\gamma\| \leq \|\delta\|$. Then by (8) of § I.1, for all p, $\gamma\restriction p \in \mathsf{W}$ and $\|\gamma\restriction p\|<\|\gamma\|=\sigma$. Hence for all p and q, $\min\{\|\gamma\restriction p\|, \|\delta\restriction q\|\}<\sigma$ so from the induction hypothesis we have

(3) $\quad \|\gamma\restriction p\| \leq \|\delta\restriction q\| \rightarrow \mathsf{H}(\gamma\restriction p, \delta\restriction q) \simeq 0;$

(4) $\quad \|\delta\restriction q\|<\|\gamma\restriction p\| \rightarrow \mathsf{H}(\gamma\restriction p, \delta\restriction q) \simeq 1.$

As one of these holds for each p and q, we have $\mathsf{H}(\gamma\restriction p, \delta\restriction q)$ defined for all p, q. Since $\|\gamma\| \leq \|\delta\|$,

$$\forall p\, \exists q\,.\,\|\gamma\restriction p\| \leq \|\delta\restriction q\|,$$

so by (1) and (3), $\mathsf{H}(\gamma, \delta) \simeq 0$.

If $\delta \in \mathsf{W}$ and $\|\delta\|<\|\gamma\|$, then for all q, $\|\delta\restriction q\|<\|\delta\|=\sigma$, so again (3) and (4) hold for all p, q. Furthermore, there exists a $\bar{p}$ such that $\|\delta\| \leq \|\gamma\restriction \bar{p}\|$ and hence $\|\delta\restriction q\|<\|\gamma\restriction \bar{p}\|$ for all q. By (2) and (4), $\mathsf{H}(\gamma, \delta) \simeq 1$. □

1.8 Theorem. $\Delta_1^1=\{\mathsf{R} : \mathsf{R}$ *is recursive in* $\mathsf{E}\}$.

Proof. The inclusion ($\supseteq$) is immediate from Corollary 1.6. Suppose $\mathsf{R} \in \Delta_1^1$, so both R and $\sim\mathsf{R}$ are Π^1_1. By Theorem IV.1.1 there exist recursive functionals F and G such that for all $(\mathbf{m}, \boldsymbol{\alpha})$,

$$\mathsf{R}(\mathbf{m}, \boldsymbol{\alpha}) \leftrightarrow \mathsf{F}[\mathbf{m}, \boldsymbol{\alpha}] \in \mathsf{W} \quad \text{and} \quad \sim\mathsf{R}(\mathbf{m}, \boldsymbol{\alpha}) \leftrightarrow \mathsf{G}[\mathbf{m}, \boldsymbol{\alpha}] \in \mathsf{W}.$$

Then it is routine to check that for all $(\mathbf{m}, \boldsymbol{\alpha})$,

$$\mathsf{K}_{\mathsf{R}}(\mathbf{m}, \boldsymbol{\alpha}) = \mathsf{H}(\mathsf{F}[\mathbf{m}, \boldsymbol{\alpha}], \mathsf{G}[\mathbf{m}, \boldsymbol{\alpha}]),$$

which by (†) implies that K_{R} is recursive in E. □

1.9 Corollary. *For any* $R \subseteq {}^k\omega$,

$$R \in \Pi^1_1 \leftrightarrow R \textit{ is semi-recursive in } \mathsf{E}.$$

Proof. The implication $(\leftarrow)$ is part of Corollary 1.6. Let $P(a) \leftrightarrow \{a\}^{\mathsf{E}}(a)\downarrow$. Clearly P is semi-recursive in E and thus is Π^1_1. However a diagonal argument shows P is not recursive in E and hence by the preceding theorem not Δ^1_1. By Theorem IV.1.8 there exists a recursive function f such that

$$P(a) \leftrightarrow f(a) \in W.$$

However, as $P \notin \Delta^1_1$, it follows from Theorem IV.2.2 that

$$\sup{}^+\{\|f(a)\| : P(a)\} = \omega_1.$$

Thus we have for all c,

$$\begin{aligned} c \in W &\leftrightarrow \exists a\,(P(a) \wedge \|c\| \leqslant \|f(a)\|) \\ &\leftrightarrow \{c\} \text{ is a total unary function and} \\ &\qquad\qquad \exists a\,(P(a) \wedge \mathsf{H}(\{c\}, \{f(a)\}) \simeq 0). \end{aligned}$$

The relation defined by the expression in parentheses is semi-recursive in E, so it follows from (††) that W is semi-recursive in E. Every $R \in \Pi^1_1$ is many-one reducible to W and thus is also semi-recursive in E. □

Exercise 1.16 provides an outline for extending Corollary 1.9 to all $\mathsf{R} \subseteq {}^{k,l}\omega$:

1.10 Corollary. $\Pi^1_1 = \{\mathsf{R} : \mathsf{R} \textit{ is semi-recursive in } \mathsf{E}\}$. □

It is worth noting that Lemma 1.7 (together with Corollary 1.6) provides a new proof of the pre-wellordering property for Π^1_1. Indeed, we have

$$\gamma \leqslant_{\Pi}^{\mathsf{W}} \delta \leftrightarrow \gamma \in \mathsf{W} \wedge \mathsf{H}(\gamma, \delta) \simeq 0, \quad \text{and}$$
$$\gamma \leqslant_{\Sigma}^{\mathsf{W}} \delta \leftrightarrow \delta \notin \mathsf{W} \vee \mathsf{H}(\gamma, \delta) \not\simeq 1.$$

We shall use a similar technique in § 3 to show that whenever E is recursive in I, then the class of relations semi-recursive in I has the pre-wellordering property.

Recursion relative to several functionals is defined by the usual coding. We call F *partial recursive in* $\mathbf{I} = (\mathsf{I}_0, \ldots, \mathsf{I}_n)$ iff F is partial recursive in the coded sequence $\langle \mathbf{I} \rangle$ where

$$\langle \mathbf{I} \rangle(\alpha) = \langle \mathsf{I}_0(\alpha), \ldots, \mathsf{I}_n(\alpha) \rangle.$$

Similarly, a functional F is *partial recursive in* **I** *and* β iff for some G partial recursive in **I**, $\mathsf{F}(\mathbf{m}, \boldsymbol{\alpha}) \simeq \mathsf{G}(\mathbf{m}, \boldsymbol{\alpha}, \beta)$.

Corresponding to the ordinary jump oJ on functions, we have for functionals the superjump sJ:

$$\mathsf{I}^{\mathsf{sJ}}((\langle a, \mathbf{m} \rangle) * \alpha) = \begin{cases} 0, & \text{if } \{a\}^{\mathsf{I}}(\mathbf{m}, \alpha) \downarrow; \\ 1, & \text{otherwise.} \end{cases}$$

The following is proved just as is Theorem II.5.7:

1.11 Theorem. *For any* I *and* $\boldsymbol{\beta}$ *and all* $\mathsf{R} \subseteq {}^{k,l}\omega$, *if* R *is semi-recursive in* I *(and* $\boldsymbol{\beta}$*), then* R *is recursive in* I^{sJ} *(and* $\boldsymbol{\beta}$*).* □

It follows that I^{sJ} is not recursive in I. Note that by Theorem 1.5, if $r \geqslant 2$ and $\mathsf{I} \in \Delta_r^1$, then also $\mathsf{I}^{\mathsf{sJ}} \in \Delta_r^1$.

1.12–1.20 Exercises

1.12. Show that the monotone operator which inductively defines $\Omega[\mathsf{I}]$ has closure ordinal exactly $\aleph_1$.

1.13 (Tugué [1960]). Show that if H and I are two functionals such that $\mathsf{H}(\beta) = \mathsf{I}(\beta)$ for all β recursive in I, then for all relations R on numbers, R is (semi-) recursive in I iff R is (semi-) recursive in H.

1.14. The class of functionals *primitive recursive in* I is the smallest class of total functionals which contains the initial functionals (of Definition II.1.1) and is closed under functional composition, primitive recursion, and I-Application, (I-$\mathrm{Ap}^{k,l}$), where for any functional G, I-$\mathrm{Ap}^{k,l}$(G) is the functional F of rank (k, l) such that

(a) if G is of rank $(k+1, l)$, then

$$\mathsf{F}(\mathbf{m}, \boldsymbol{\alpha}) \simeq \mathsf{I}(\lambda p \,.\, \mathsf{G}(p, \mathbf{m}, \boldsymbol{\alpha}));$$

(b) otherwise, $\mathsf{F}(\mathbf{m}, \boldsymbol{\alpha}) \simeq 0$.

Characterize the class of relations primitive recursive in E.

1.15. Give an alternative proof of Theorem 1.8 by constructing a primitive recursive function f such that for all $a \in N$ (Definition IV.4.1), $f(a)$ is an index of P_a from E, and applying Theorem IV.4.12.

1.16. Using the fact that the relation $\{a\}^{\mathsf{E}}(a, \boldsymbol{\alpha})\downarrow$ is $\Pi^1_1 \sim \Delta^1_1$, show that the relation $c \in W[\boldsymbol{\alpha}]$ is semi-recursive in E. Then apply the result of Exercise IV.1.24 to prove Corollary 1.10.

1.17. Suppose that F is partial recursive in E and $F: W \to W$. Show that there exists a function $G: W \to W$ partial recursive in E such that for all $c \in W$,

$$\|G(c)\| = \sup{}^{+}\{\|F(d)\| : \|d\| < \|c\|\}.$$

1.18. Show that for any relation R on numbers recursive in E_1,

$$\exists \beta\, \forall p\, R(\bar{\beta}(p), m) \leftrightarrow (\exists \beta \text{ recursive in } \mathsf{E}_1)\forall p\, R(\bar{\beta}(p), m).$$

Does this also hold with E in place of E_1? Does it hold with I in place of E_1 for any functional I such that E_1 is recursive in I?

1.19 (Hinman [1969]). For any I, let Iri (the set of I-recursive indices) be the smallest set such that for $u \in \mathsf{Iri}$ there exist functions $[u]$ which satisfy the following conditions:

(i) $0 \in \mathsf{Iri}$ and $[0](m) = 0$;

(ii) if $v \in \mathsf{Iri}$ and a is an index of a partial recursive functional $\{a\}$ such that for all m, $\{a\}(m, [v])\downarrow$, then $\langle 1, v, a\rangle \in \mathsf{Iri}$ and $[\langle 1, v, a\rangle](m) = \{a\}(m, [v])$;

(iii) if $v \in \mathsf{Iri}$ and for all m, $[v](m) \in \mathsf{Iri}$, then $\langle 2, v\rangle \in \mathsf{Iri}$ and $[\langle 2, v\rangle](m) = [[v](m)](m)$;

(iv) if $v \in \mathsf{Iri}$, then $\langle 3, v\rangle \in \mathsf{Iri}$ and $[\langle 3, v\rangle](m) = \mathsf{I}([v])$. Show that for any α, α is recursive in I iff for some $u \in \mathsf{Iri}$, $\alpha = [u]$.

1.20 (Gandy). Prove that each of $\mathsf{E}^{\#\mathsf{J}}$ and E_1 is recursive in the other.

1.21 Notes. The definition of recursive functionals with objects of types ≥ 2 as arguments is due to Kleene [1959]. The formulation here is somewhat different from Kleene's. Kleene also proved Theorem 1.8 (by a method similar to that of Exercise 1.15). The only other basic treatment of recursion in higher types in print is Gandy [1967]. The superjump was invented and studied in Gandy [1967a]. Kleene also developed analogues of several of the other basic characterizations of ordinary recursion theory for higher types and showed them all equivalent (see Kleene [1963] for references).

2. Substitution Theorems

The first goal of this section is to establish that the class of functionals partial recursive in I is closed under functional substitution ((†) of the preceding section). Note that this is more than was true for ordinary recursion theory (cf. Exercise II.4.27 and Theorem II.3.9). As for ordinary recursion theory (cf. II.5.2), (†) leads directly to:

(I) If β is recursive in I (and γ) and F is partial recursive in I and β, then F is partial recursive in I (and γ).

We are then led naturally to the following question — is it true also that

(II) if H is recursive in I and F is partial recursive in H, then F is partial recursive in I?

We shall see that although (II) is true, it becomes false if reference to I is omitted. Here we have only the weaker:

(III) if H is recursive and F is recursive in H, then F is recursive.

The proof of (†) is substantially more complicated than the corresponding proof for ordinary recursion theory. This is due to the lack of simple normal forms for recursion relative to a functional. We must, rather, rely directly on the definition. A related complication is that we cannot prove the results as stated but must first prove effective versions. For example, for (†) we show that there exists a primitive recursive function f such that if a and d are indices of G and H from I, respectively, then $f(a, d)$ is an index of F from I. The definitions are all by effective transfinite recursion.

First we see why (II) fails if the phrase "in I" is omitted:

2.1 Theorem. *There exists a recursive functional* H *and a functional* F *partial recursive in* H *such that* F *is not partial recursive.*

Proof. Let H be *any* (total) recursive functional, and R any Π_1^0 relation which is not recursive. By Lemma 1.2, R is semi-recursive in H, so there exists a functional F partial recursive in H such that R is the domain of F. Then Dm F is not semi-recursive so F is not partial recursive. □

We shall approach (†) by way of a series of approximations. We first treat one part of the special case in which H depends only on the function arguments $\boldsymbol{\alpha}$:

2.2 Lemma. *There exists a primitve recursive function f_0 such that for all* I, a, d, $\mathbf{m}$, $\boldsymbol{\alpha}$, *and* n,

$$\{a\}^{\mathsf{I}}(\mathbf{m}, \boldsymbol{\alpha}, \lambda p.\{d\}^{\mathsf{I}}(p, \boldsymbol{\alpha})) \simeq n \to \{f_0(a, d)\}^{\mathsf{I}}(\mathbf{m}, \boldsymbol{\alpha}) \simeq n.$$

Proof. We shall define first an auxiliary primitive recursive function h and then apply the Primitive Recursion Theorem (II.2.6) to obtain an index $\bar{e}$ such that for all a and d,

$$\{\bar{e}\}(a, d) \simeq h(\bar{e}, a, d).$$

$\{\bar{e}\}$ is the desired f_0.

We define h by course-of-values recursion on a and by cases according to the clauses of Definition 1.1 as follows: for any a, d, e, k, and l,

(0) if $a = \langle 0, k, l+1, i, \ldots\rangle$ for $i = 0, 1, 2, 4$ or 5, then $h(e, a, d) = \langle 0, k, l, i, \ldots\rangle$;

if $a = \langle 0, k, l+1, 3, i, j\rangle$ with $j < l$, then $h(e, a, d) = \langle 0, k, l, 3, i, j\rangle$;

if $a = \langle 0, k, l+1, 3, i, l\rangle$, then $h(e, a, d)$ is an index such that $\{h(e, a, d)\}^{\mathsf{I}}(\mathbf{m}, \boldsymbol{\alpha}) \simeq \{d\}^{\mathsf{I}}(m_i, \boldsymbol{\alpha})$;

(1) if for some $k', b, c_0, \ldots, c_{k'-1}$, $a = \langle 1, k, l+1, b, c_0, \ldots, c_{k'-1}\rangle$, then $h(e, a, d) = \langle 1, k, l, h(e, b, d), h(e, c_0, d), \ldots, h(e, c_{k'-1}, d)\rangle$;

(2) if $a = \langle 2, k+1, l+1\rangle$, let c be the natural index such that

$$\{c\}^{\mathsf{I}}(e, a, d, b, \mathbf{m}, \boldsymbol{\alpha}) \simeq \{\{e\}(b, d)\}^{\mathsf{I}}(\mathbf{m}, \boldsymbol{\alpha})$$

and set $h(e, a, d) = \mathrm{Sb}_2(c, e, a, d)$;
(by a *natural index* for a given computation we mean the index which codes instructions to perform the computation in the manner indicated — in this case to compute first $\{e\}(b, d)$ and then use the result as an index applied to $(\mathbf{m}, \boldsymbol{\alpha})$. For the record we want here

$$c = \langle 1, k+4, l, \langle 2, k+1, l\rangle, e', \langle 0, k+4, l, 1, 4\rangle, \ldots, \langle 0, k+4, l, 1, 3+k\rangle\rangle$$

where

$$e' = \langle 1, k+4, l, e, \langle 0, k+4, l, 1, 3\rangle, \langle 0, k+4, l, 1, 2\rangle\rangle).$$

(3) if for some b, $a = \langle 3, k, l+1, b\rangle$, then $h(e, a, d) = \langle 3, k, l, h(e, b, d)\rangle$;

(4) if a is none of these forms, $h(e, a, d) = 0$.

If $\lambda p.\{d\}^{\mathsf{I}}(p, \boldsymbol{\alpha})$ is not a total function, the implication of the theorem is trivial, so let d and $\boldsymbol{\alpha}$ be fixed such that $\lambda p.\{d\}^{\mathsf{I}}(p, \boldsymbol{\alpha})$ is a total function β. We prove by induction over $\Omega[\mathsf{I}]$ that for all a, $\mathbf{m}$, and n,

$$\{a\}^{\mathsf{I}}(\mathbf{m}, \boldsymbol{\alpha}, \beta) \simeq n \to \{f_0(a, d)\}^{\mathsf{I}}(\mathbf{m}, \boldsymbol{\alpha}) \simeq n.$$

Let X denote the class of sequences $(a, \mathbf{m}, \boldsymbol{\alpha}, \beta, n)$ in $\Omega[\mathsf{I}]$ such that $\{f_0(a, d)\}^{\mathsf{I}}(\mathbf{m}, \boldsymbol{\alpha}) \simeq n$. We must show that X satisfies the closure conditions which define $\Omega[\mathsf{I}]$. If a is of the form $\langle 0, k, l+1, \ldots\rangle$ and $\{a\}^{\mathsf{I}}(\mathbf{m}, \boldsymbol{\alpha}, \beta) \simeq n$, then it is obvious by inspection that for any e, $\{h(e, a, d)\}^{\mathsf{I}}(\mathbf{m}, \boldsymbol{\alpha}) \simeq n$ so in particular, $\{f_0(a, d)\}^{\mathsf{I}}(\mathbf{m}, \boldsymbol{\alpha}) \simeq n$. Suppose next that $a = \langle 1, k, l+1, b, c_0, \ldots, c_{k'-1}\rangle$ and for each $i < k'$ there exists a q_i such that

$$(c_i, \mathbf{m}, \boldsymbol{\alpha}, \beta, q_i) \in X \quad \text{and} \quad (b, \mathbf{q}, \boldsymbol{\alpha}, \beta, n) \in X.$$

Because $X \subseteq \Omega[\mathsf{I}]$ by definition, we have then

$$\{c_i\}^{\mathsf{I}}(\mathbf{m}, \boldsymbol{\alpha}, \beta) \simeq q_i \quad \text{and} \quad \{b\}^{\mathsf{I}}(\mathbf{q}, \boldsymbol{\alpha}, \beta) \simeq n,$$

so

$$\{a\}^{\mathsf{I}}(\mathbf{m}, \boldsymbol{\alpha}) \simeq n.$$

Furthermore, by the definition of X,

$$\{f_0(c_i, d)\}^{\mathsf{I}}(\mathbf{m}, \boldsymbol{\alpha}) \simeq q_i \quad \text{and} \quad \{f_0(b, d)\}^{\mathsf{I}}(\mathbf{q}, \boldsymbol{\alpha}) \simeq n.$$

Hence, since

$$f_0(a, d) = \langle 1, k, l, f_0(b, d), f_0(c_0, d), \ldots, f_0(c_{k'-1}, d)\rangle,$$

also

$$\{f_0(a, d)\}^{\mathsf{I}}(\mathbf{m}, \boldsymbol{\alpha}) \simeq n$$

so $(a, \mathbf{m}, \boldsymbol{\alpha}, \beta, n) \in X$.

If $a = \langle 2, k+1, l+1\rangle$ and $(b, \mathbf{m}, \boldsymbol{\alpha}, \beta, n) \in X$, then

$$\{a\}^{\mathsf{I}}(b, \mathbf{m}, \boldsymbol{\alpha}, \beta) \simeq \{b\}^{\mathsf{I}}(\mathbf{m}, \boldsymbol{\alpha}, \beta) \simeq n$$

and

$$\{f_0(b, d)\}^{\mathsf{I}}(\mathbf{m}, \boldsymbol{\alpha}) \simeq n.$$

In this case, $f_0(a, d) = \mathrm{Sb}_2(c, \bar{e}, a, d)$ with c chosen exactly to make the following true:

$$\begin{aligned}\{f_0(a, d)\}^{\mathsf{I}}(b, \mathbf{m}, \boldsymbol{\alpha}) &\simeq \{c\}^{\mathsf{I}}(\bar{e}, a, d, b, \mathbf{m}, \boldsymbol{\alpha})\\ &\simeq \{\{\bar{e}\}(b, d)\}^{\mathsf{I}}(\mathbf{m}, \boldsymbol{\alpha})\\ &\simeq \{f_0(b, d)\}^{\mathsf{I}}(\mathbf{m}, \boldsymbol{\alpha}) \simeq n.\end{aligned}$$

Hence $(a, b, \mathbf{m}, \boldsymbol{\alpha}, \beta, n) \in X$. (In an attempt to rescue the reader from total confusion here we make two observations. First, the naturalness of c is not used here and will be needed only in the proof of Theorem 2.8. Second, the reader may be wondering why we did not use the simpler definition $h(e, a, d) = h(e, b, d)$ in this case. This is not possible because b comes from the argument list and not from the index, as in case (3), so it is not necessarily true that $b < a$ as would be required for the course-of-values recursion.)

Finally, suppose $a = \langle 3, k, l+1, b\rangle$ and γ is a function such that for all p, $(b, p, \mathbf{m}, \boldsymbol{\alpha}, \beta, \gamma(p)) \in X$. Then for all p,

$$\{b\}^{\mathsf{I}}(p, \mathbf{m}, \boldsymbol{\alpha}, \beta) = \gamma(p) \quad \text{and} \quad \{a\}^{\mathsf{I}}(\mathbf{m}, \boldsymbol{\alpha}, \beta) = \mathsf{I}(\gamma).$$

Furthermore, for all p,

$$\{f_0(b, d)\}^{\mathsf{I}}(p, \mathbf{m}, \boldsymbol{\alpha}) = \gamma(p).$$

Hence, as $f_0(a, d) = \langle 3, k, l, f_0(b, d)\rangle$,

$$\begin{aligned}\{f_0(a, d)\}^{\mathsf{I}}(\mathbf{m}, \boldsymbol{\alpha}) &= \mathsf{I}(\lambda p \,.\, \{f_0(b, d)\}^{\mathsf{I}}(p, \mathbf{m}, \boldsymbol{\alpha}))\\ &= \mathsf{I}(\gamma) = n,\end{aligned}$$

so again $(a, \mathbf{m}, \boldsymbol{\alpha}, \beta, n) \in X$. □

2.3 Corollary. *There exists a primitive recursive function* f_1 *such that for all* I, a, d, $\mathbf{m}$, $\boldsymbol{\alpha}$, *and* n,

$$\{a\}^{\mathsf{I}}(\mathbf{m}, \boldsymbol{\alpha}, \lambda p \,.\, \{d\}^{\mathsf{I}}(p, \mathbf{m}, \boldsymbol{\alpha})) \simeq n \rightarrow \{f_1(a, d)\}^{\mathsf{I}}(\mathbf{m}, \boldsymbol{\alpha}) \simeq n.$$

Proof. Let h be a primitive recursive function such that for all d, p, $\mathbf{m}$, and $\boldsymbol{\alpha}$,

$$\{d\}^{\mathsf{I}}(p, \mathbf{m}, \boldsymbol{\alpha}) \simeq \{h(d, \langle \mathbf{m}\rangle)\}^{\mathsf{I}}(p, \boldsymbol{\alpha}).$$

Then we take $f_1(a, d)$ to be an index such that

$$\{f_1(a, d)\}^{\mathsf{I}}(\mathbf{m}, \boldsymbol{\alpha}) \simeq \{f_0(a, h(d, \langle \mathbf{m}\rangle))\}^{\mathsf{I}}(\mathbf{m}, \boldsymbol{\alpha}).$$

The result now follows from Lemma 2.2 by a direct computation. □

With this corollary, we have in a sense completed half of the proof of (†): if G and H are partial recursive in I, say with indices a and d from I, and

$$\mathsf{F}(\mathbf{m}, \boldsymbol{\alpha}) \simeq \mathsf{G}(\mathbf{m}, \boldsymbol{\alpha}, \lambda p \,.\, \mathsf{H}(p, \mathbf{m}, \boldsymbol{\alpha})),$$

then we have shown the existence of a functional F' partial recursive in I (with index $f_1(a, d)$ from I) such that $\mathsf{F} \subseteq \mathsf{F}'$. In fact F' may be properly larger than F. We shall complete the proof by showing that for $\mathbf{m}$ and $\boldsymbol{\alpha}$ such that $\lambda p . \mathsf{H}(p, \mathbf{m}, \boldsymbol{\alpha})$ is total, $\mathsf{F}(\mathbf{m}, \boldsymbol{\alpha}) \simeq \mathsf{F}'(\mathbf{m}, \boldsymbol{\alpha})$, and then observing that

$$\mathsf{F}(\mathbf{m}, \boldsymbol{\alpha}) \simeq n \leftrightarrow \mathsf{I}(\lambda p . \mathsf{H}(p, \mathbf{m}, \boldsymbol{\alpha}))\downarrow \wedge \mathsf{F}'(\mathbf{m}, \boldsymbol{\alpha}) \simeq n$$

so that F itself is partial recursive.

2.4 Definition. For each I and each $x \in \Omega[\mathsf{I}]$, the set $\mathrm{Sbc}(x)$ of *subcomputations of* x is defined recursively as follows: for any k, l, $\mathbf{m}$, $\boldsymbol{\alpha}$, and n,

(0) if $x = (\langle 0, k, l, \ldots\rangle, \mathbf{m}, \boldsymbol{\alpha}, n)$, then $\mathrm{Sbc}(x) = \varnothing$;

(1) if for some $k', b, c_0, \ldots, c_{k'-1}$, $x = (\langle 1, k, l, b, c_0, \ldots, c_{k'-1}\rangle, \mathbf{m}, \boldsymbol{\alpha}, n)$, and $q_0, \ldots, q_{k'-1}$ are the unique numbers such that for all $i < k'$ $y_i = (c_i, \mathbf{m}, \boldsymbol{\alpha}, q_i) \in \Omega[\mathsf{I}]$ and $z = (b, \mathbf{q}, \boldsymbol{\alpha}, n) \in \Omega[\mathsf{I}]$, then

$$\mathrm{Sbc}(x) = \bigcup\{\mathrm{Sbc}(y_i) : i < k'\} \cup \mathrm{Sbc}(z) \cup \{y_i : i < k'\} \cup \{z\};$$

(2) if for some b, $x = (\langle 2, k+1, l\rangle, b, \mathbf{m}, \boldsymbol{\alpha}, n)$, then

$$y = (b, \mathbf{m}, \boldsymbol{\alpha}, n) \in \Omega[\mathsf{I}] \quad \text{and} \quad \mathrm{Sbc}(x) = \mathrm{Sbc}(y) \cup \{y\};$$

(3) if for some b and β, $x = (\langle 3, k, l, b\rangle, \mathbf{m}, \boldsymbol{\alpha}, \mathsf{I}(\beta))$ and for all p, $y_p = (b, p, \mathbf{m}, \boldsymbol{\alpha}, \beta(p)) \in \Omega[\mathsf{I}]$, then

$$\mathrm{Sbc}(x) = \bigcup\{\mathrm{Sbc}(y_p) : p \in \omega\} \cup \{y_p : p \in \omega\}.$$

This recursive definition is justified by the monomorphic character of the definition of $\Omega[\mathsf{I}]$ as in Theorem I.3.5 (cf. Exercise I.3.13).

We use the notion of subcomputation to derive a new principle of proof by induction over $\Omega[\mathsf{I}]$ which is related to ordinary proofs by induction over $\Omega[\mathsf{I}]$ as course-of-values induction over ω is related to ordinary induction over ω (cf. Exercise I.3.12).

2.5 Theorem. *For any* $X \subseteq \Omega[\mathsf{I}]$, *if for all* $x \in \Omega[\mathsf{I}]$,

$$\mathrm{Sbc}(x) \subseteq X \rightarrow x \in X,$$

then $X = \Omega[\mathsf{I}]$.

Proof. Suppose X satisfies the hypothesis of the theorem and set

$$Y = \{x : x \in X \wedge \mathrm{Sbc}(x) \subseteq X\}.$$

We show $Y = \Omega[\mathsf{I}]$ by ordinary induction over $\Omega[\mathsf{I}]$. If $x \in \Omega[\mathsf{I}]$ is of the form $x = (\langle 0, k, l, \ldots\rangle, \mathbf{m}, \boldsymbol{\alpha}, n)$, then $\mathrm{Sbc}(x) = \emptyset \subseteq X$ so $x \in X$ and thus also $x \in Y$.

The three inductive clauses are treated similarly and we do only (3). Suppose $x = (\langle 3, k, l, b\rangle, \mathbf{m}, \boldsymbol{\alpha}, \mathsf{I}(\beta))$ and for all p, $y_p = (b, p, \mathbf{m}, \boldsymbol{\alpha}, \beta(p)) \in Y$. By the definition of Y, $\mathrm{Sbc}(y_p) \subseteq X$ for all p. Hence as $y_p \in X$, $\mathrm{Sbc}(x) \subseteq X$ and thus $x \in X$ and $x \in Y$. □

For a given $x = (a, \mathbf{m}, \boldsymbol{\alpha}, n) \in \Omega[\mathsf{I}]$, the set $\{x\} \cup \mathrm{Sbc}(x)$ consists of exactly those sequences $(b, \mathbf{p}, \boldsymbol{\beta}, q)$ such that $\{b\}^{\mathsf{I}}(\mathbf{p}, \boldsymbol{\beta}) \simeq q$ and a node labeled $(b, \mathbf{p}, \boldsymbol{\beta})$ occurs in the tree for the computation $\{a\}^{\mathsf{I}}(\mathbf{m}, \boldsymbol{\alpha}) \simeq n$. If we denote by $\mathrm{ISbc}(x)$ the *immediate* subcomputations of x — that is, those which appear in the tree immediately below x, then ordinary induction over $\Omega[\mathsf{I}]$ may be seen as deriving the conclusion $X = \Omega[\mathsf{I}]$ from the hypothesis $\mathrm{ISbc}(x) \subseteq X \rightarrow x \in X$.

Another way to view Theorem 2.5 is in terms of the levels of the inductive definition of $\Omega[\mathsf{I}]$. If Γ_{I} is the monotone operator implicit in Definition 1.1 such that $\Omega[\mathsf{I}] = \bar{\Gamma}_{\mathsf{I}}$, then for any σ and x,

$$x \in \Gamma_{\mathsf{I}}^{\sigma} \rightarrow \mathrm{Sbc}(x) \subseteq \Gamma_{\mathsf{I}}^{(\sigma)}.$$

Furthermore, if $X \subseteq \Omega[\mathsf{I}]$ satisfies $\Gamma_{\mathsf{I}}^{(\sigma)} \subseteq X \rightarrow \Gamma_{\mathsf{I}}^{\sigma} \subseteq X$, then clearly $X = \Omega[\mathsf{I}]$. The hypothesis of 2.5 is a refinement or "localization" of this condition in that membership of a given $x \in \Gamma_{\mathsf{I}}^{\sigma}$ in X requires only that the "relevant part" of $\Gamma_{\mathsf{I}}^{(\sigma)}$, $\mathrm{Sbc}(x)$, be included in X.

Before establishing the rest of (†), we need two technical lemmas whose proofs we leave to the reader (Exercise 2.20).

2.6 Lemma. *For all x and $y \in \Omega[\mathsf{I}]$,*

$$y \in \mathrm{Sbc}(x) \rightarrow \mathrm{Sbc}(y) \subseteq \mathrm{Sbc}(x). \quad \square$$

2.7 Lemma. *For any k, l, $i < k$, a, n, and $(\mathbf{m}, \boldsymbol{\alpha}) \in {}^{k,l}\omega$, if $\{a\}^{\mathsf{I}}(\mathbf{m}, \boldsymbol{\alpha}) \simeq n$, then*

$$(a, \mathbf{m}, \boldsymbol{\alpha}, n) \in \mathrm{Sbc}(\mathrm{Sb}_i(a, m_0, \ldots, m_i), m_{i+1}, \ldots, m_{k-1}, \boldsymbol{\alpha}, n). \quad \square$$

2.8 Theorem. *For all I, a, d, $\mathbf{m}$, $\boldsymbol{\alpha}$, and n, if $\lambda p.\{d\}^{\mathsf{I}}(p, \boldsymbol{\alpha})$ is a total function, then*

$$\{f_0(a, d)\}^{\mathsf{I}}(\mathbf{m}, \boldsymbol{\alpha}) \simeq n \rightarrow \{a\}^{\mathsf{I}}(\mathbf{m}, \boldsymbol{\alpha}, \lambda p.\{d\}^{\mathsf{I}}(p, \boldsymbol{\alpha})) \simeq n.$$

Proof. We shall apply Theorem 2.5 to the set

$$X = \{x : x \in \Omega[\mathsf{I}], \text{ and for all } a, d, \mathbf{m}, \boldsymbol{\alpha}, \text{ and } n, \text{ if } \lambda p.\{d\}^{\mathsf{I}}(p, \boldsymbol{\alpha}) \text{ is total and } x = (f_0(a, d), \mathbf{m}, \boldsymbol{\alpha}, n), \text{ then } \{a\}^{\mathsf{I}}(\mathbf{m}, \boldsymbol{\alpha}, \lambda p.\{d\}^{\mathsf{I}}(p, \boldsymbol{\alpha})) \simeq n\}.$$

We need thus to prove that for all $x \in \Omega[\mathrm{l}]$, if $\mathrm{Sbc}(x) \subseteq X$, then $x \in X$. If x is not of the form $x = (f_0(a, d), \mathbf{m}, \boldsymbol{\alpha}, n)$ with $\lambda p.\{d\}^{\mathrm{l}}(p, \boldsymbol{\alpha})$ total, there is nothing to prove, so we assume that it is of this form, that $\mathrm{Sbc}(x) \subseteq X$, and let β be the total function $\lambda p.\{d\}^{\mathrm{l}}(p, \boldsymbol{\alpha})$. We consider the cases (0)–(4) under which a may fall.

(0) If $a = \langle 0, k, l+1, \ldots \rangle$, the result is clear.

(1) If for some $k', b, c_0, \ldots, c_{k'-1}$, $a = \langle 1, k, l+1, b, c_0, \ldots, c_{k'-1} \rangle$, then $f_0(a, d) = \langle 1, k, l, f_0(b, d), f_0(c_0, d), \ldots, f_0(c_{k'-1}, d) \rangle$.
Since by assumption $\{f_0(a, d)\}^{\mathrm{l}}(\mathbf{m}, \boldsymbol{\alpha}) \simeq n$, there exist $q_0, \ldots, q_{k'-1}$ such that for all $i < k'$,

$$\{f_0(c_i, d)\}^{\mathrm{l}}(\mathbf{m}, \boldsymbol{\alpha}) \simeq q_i \quad \text{and} \quad \{f_0(b, d)\}^{\mathrm{l}}(\mathbf{q}, \boldsymbol{\alpha}) \simeq n.$$

Furthermore, for all $i < k'$,

$$(f_0(c_i, d), \mathbf{m}, \boldsymbol{\alpha}, q_i) \in \mathrm{Sbc}(x) \subseteq X,$$

and

$$(f_0(b, d), \mathbf{q}, \boldsymbol{\alpha}, n) \in \mathrm{Sbc}(x) \subseteq X.$$

Hence by the definition of X we have

$$\{c_i\}^{\mathrm{l}}(\mathbf{m}, \boldsymbol{\alpha}, \beta) \simeq q_i \quad \text{and} \quad \{b\}^{\mathrm{l}}(\mathbf{q}, \boldsymbol{\alpha}, \beta) \simeq n$$

and hence $\{a\}^{\mathrm{l}}(\mathbf{m}, \boldsymbol{\alpha}, \beta) \simeq n$, so $x \in X$.

(2) Suppose now that $a = \langle 2, k+1, l+1 \rangle$,

$$x = (f_0(a, d), b, \mathbf{m}, \boldsymbol{\alpha}, n) \in \Omega[\mathrm{l}], \quad \text{and} \quad \mathrm{Sbc}(x) \subseteq X.$$

Let c be as in the definition of f_0 so that

$$f_0(a, d) = \mathrm{Sb}_2(c, \bar{e}, a, d).$$

Thus

$$\begin{aligned} n \simeq \{f_0(a, d)\}^{\mathrm{l}}(b, \mathbf{m}, \boldsymbol{\alpha}) &\simeq \{c\}^{\mathrm{l}}(\bar{e}, a, d, b, \mathbf{m}, \boldsymbol{\alpha}) \\ &\simeq \{\{\bar{e}\}(b, d)\}^{\mathrm{l}}(\mathbf{m}, \boldsymbol{\alpha}) \\ &\simeq \{f_0(b, d)\}^{\mathrm{l}}(\mathbf{m}, \boldsymbol{\alpha}). \end{aligned}$$

It follows from Lemma 2.7 that

$$(c, \bar{e}, a, d, b, \mathbf{m}, \boldsymbol{\alpha}, n) \in \mathrm{Sbc}(x).$$

The choice of c as the natural index for the indicated computation ensures that

$$(f_0(b, d), \mathbf{m}, \boldsymbol{\alpha}, n) \in \mathrm{Sbc}((c, \bar{e}, a, d, b, \mathbf{m}, \boldsymbol{\alpha}, n)).$$

Thus by Lemma 2.6, we have

$$(f_0(b, d), \mathbf{m}, \boldsymbol{\alpha}, n) \in \mathrm{Sbc}(x) \subseteq X.$$

Hence, by the definition of X, $\{b\}^{\mathsf{I}}(\mathbf{m}, \boldsymbol{\alpha}, \beta) \simeq n$ and thus also $\{a\}^{\mathsf{I}}(b, \mathbf{m}, \boldsymbol{\alpha}, \beta) \simeq n$, so $x \in X$.

(3) If $a = \langle 3, k, l+1, b\rangle$ for some b, then $f_0(a, d) = \langle 3, k, l, f_0(b, d)\rangle$, and

$$n \simeq \{f_0(a, d)\}^{\mathsf{I}}(\mathbf{m}, \boldsymbol{\alpha}) \simeq \mathsf{I}(\lambda p \,.\, \{f_0(b, d)\}^{\mathsf{I}}(p, \mathbf{m}, \boldsymbol{\alpha})).$$

Thus $\lambda p \,.\, \{f_0(b, d)\}^{\mathsf{I}}(p, \mathbf{m}, \boldsymbol{\alpha})$ is a total function γ (because I of it is defined) and for each p,

$$(f_0(b, d), p, \mathbf{m}, \boldsymbol{\alpha}, \gamma(p)) \in \mathrm{Sbc}(x) \subseteq X,$$

so for each p, $\{b\}^{\mathsf{I}}(p, \mathbf{m}, \boldsymbol{\alpha}, \beta) \simeq \gamma(p)$. Thus

$$\{a\}^{\mathsf{I}}(\mathbf{m}, \boldsymbol{\alpha}, \beta) \simeq \mathsf{I}(\lambda p \,.\, \{b\}^{\mathsf{I}}(p, \mathbf{m}, \boldsymbol{\alpha}, \beta)) \simeq \mathsf{I}(\gamma) \simeq n.$$

(4) If a is none of these forms, $f_0(a, d) = 0$, so $x \notin \Omega[\mathsf{I}]$. □

2.9 Corollary. *For all* I, a, d, $\mathbf{m}$, $\boldsymbol{\alpha}$, *and* n, *if* $\lambda p \,.\, \{d\}^{\mathsf{I}}(p, \mathbf{m}, \boldsymbol{\alpha})$ *is total, then*

$$\{f_1(a, d)\}^{\mathsf{I}}(\mathbf{m}, \boldsymbol{\alpha}) \simeq n \to \{a\}^{\mathsf{I}}(\mathbf{m}, \boldsymbol{\alpha}, \lambda p \,.\, \{d\}^{\mathsf{I}'}(p, \mathbf{m}, \boldsymbol{\alpha})) \simeq n.$$

Proof. Suppose $\lambda p \,.\, \{d\}^{\mathsf{I}}(p, \mathbf{m}, \boldsymbol{\alpha})$ is a total function β and $\{f_1(a, d)\}^{\mathsf{I}}(\mathbf{m}, \boldsymbol{\alpha}) \simeq n$. Then with notation as in Corollary 2.3,

$$\{f_0(a, h(d, \langle \mathbf{m}\rangle))\}^{\mathsf{I}}(\mathbf{m}, \boldsymbol{\alpha}) \simeq n$$

and by the preceding theorem,

$$\{a\}^{\mathsf{I}}(\mathbf{m}, \boldsymbol{\alpha}, \lambda p \,.\, \{h(d, \langle \mathbf{m}\rangle)\}^{\mathsf{I}}(p, \boldsymbol{\alpha})) \simeq n.$$

But h was chosen just so that for all p,

$$\{h(d, \langle \mathbf{m}\rangle)\}^{\mathsf{I}}(p, \boldsymbol{\alpha}) \simeq \{d\}^{\mathsf{I}}(p, \mathbf{m}, \boldsymbol{\alpha}) \simeq \beta(p)$$

so that $\{a\}^{\mathsf{I}}(\mathbf{m}, \boldsymbol{\alpha}, \beta) \simeq n$. □

Now to establish (†) it remains only to remove the hypothesis that $\lambda p.\{d\}^{\mathsf{I}}(p, \mathbf{m}, \boldsymbol{\alpha})$ be total. This we achieve by a simple trick.

2.10 Theorem. *There exists a primitive recursive function f such that for all* I, *a, d,* $\mathbf{m}$ *and* $\boldsymbol{\alpha}$,

$$\{f(a, d)\}^{\mathsf{I}}(\mathbf{m}, \boldsymbol{\alpha}) \simeq \{a\}^{\mathsf{I}}(\mathbf{m}, \boldsymbol{\alpha}, \lambda p.\{d\}^{\mathsf{I}}(p, \mathbf{m}, \boldsymbol{\alpha})).$$

Proof. We want $f(a, d)$ to be an index such that

$$\{f(a, d)\}^{\mathsf{I}}(\mathbf{m}, \boldsymbol{\alpha}) \simeq 0 \cdot \mathsf{I}(\lambda p.\{d\}^{\mathsf{I}}(p, \mathbf{m}, \boldsymbol{\alpha})) + \{f_1(a, d)\}^{\mathsf{I}}(\mathbf{m}, \boldsymbol{\alpha}).$$

Clearly such an f can be defined explicitly from f_1 and indices for $+$ and $\cdot$ from I. Then

$$\begin{aligned}\{f(a, d)\}^{\mathsf{I}}(\mathbf{m}, \boldsymbol{\alpha}) \simeq n &\leftrightarrow \mathsf{I}(\lambda p.\{d\}^{\mathsf{I}}(p, \mathbf{m}, \boldsymbol{\alpha}))\downarrow \wedge \{f_1(a, d)\}^{\mathsf{I}}(\mathbf{m}, \boldsymbol{\alpha}) \simeq n\\ &\leftrightarrow \lambda p.\{d\}^{\mathsf{I}}(p, \mathbf{m}, \boldsymbol{\alpha}) \text{ is total} \wedge \{f_1(a, d)\}^{\mathsf{I}}(\mathbf{m}, \boldsymbol{\alpha}) \simeq n\\ &\leftrightarrow \{a\}^{\mathsf{I}}(\mathbf{m}, \boldsymbol{\alpha}, \lambda p.\{d\}^{\mathsf{I}}(p, \mathbf{m}, \boldsymbol{\alpha})) \simeq n.\end{aligned}$$

The last equivalence uses Corollary 2.9 and the fact that computations are defined only for total arguments. □

2.11 Corollary (†). *For every* I, *the class of functionals partial recursive in* I *is closed under functional substitution.* □

The reader should ontrast this with the result of Exercise II.4.27 to see that the use of I in the proof of Theorem 2.10 is really essential.

To derive (I) from this we need that the class of functionals partial recursive in I is closed under expansion. This could have been proved earlier, but the method of proof is similar to that of Lemma 2.2 and at this point becomes an exercise.

2.12 Lemma. *There exists a primitive recursive function* f_2 *such that for all* I, *d, k, l,* $(\mathbf{m}, \boldsymbol{\alpha}) \in {}^{k,l}\omega$, $\mathbf{p}$, *and* $\boldsymbol{\beta}$,

$$\{f_2(d, k, l)\}^{\mathsf{I}}(\mathbf{p}, \mathbf{m}, \boldsymbol{\alpha}, \boldsymbol{\beta}) \simeq \{d\}^{\mathsf{I}}(\mathbf{p}, \boldsymbol{\beta}).$$

Proof. Exercise 2.21. □

2.13 Corollary (I). *For any* I, F, β, *and* γ, *if* β *is recursive in* I *(and* γ*) and* F *is partial recursive in* I *and* β, *then* F *is partial recursive in* I *(and* γ*).*

Proof. We prove the version without $\boldsymbol{\gamma}$. Suppose $\mathsf{F}(\mathbf{m}, \boldsymbol{\alpha}) \simeq \mathsf{G}(\mathbf{m}, \boldsymbol{\alpha}, \beta)$ and a and d are indices of G and β, respectively, from I. Then

$$\begin{aligned}\mathsf{F}(\mathbf{m}, \boldsymbol{\alpha}) &\simeq \{a\}^{\mathsf{I}}(\mathbf{m}, \boldsymbol{\alpha}, \lambda p.\{d\}^{\mathsf{I}}(p)) \\ &\simeq \{a\}^{\mathsf{I}}(\mathbf{m}, \boldsymbol{\alpha}, \lambda p.\{f_2(d, k, l)\}^{\mathsf{I}}(p, \mathbf{m}, \boldsymbol{\alpha})) \\ &\simeq \{f(a, f_2(d, k, l))\}^{\mathsf{I}}(\mathbf{m}, \boldsymbol{\alpha}). \quad \square\end{aligned}$$

We are now in a position to prove (the effective version of) (II). The technique is an extension of that used above.

2.14 Theorem. *There exists a primitive recursive function g such that for all* I, a, d, $\mathbf{m}$, *and* $\boldsymbol{\alpha}$, *if* $\lambda\beta.\{d\}^{\mathsf{I}}(\beta)$ *is a total functional* H, *then*

$$\{g(a, d)\}^{\mathsf{I}}(\mathbf{m}, \boldsymbol{\alpha}) \simeq \{a\}^{\mathsf{H}}(\mathbf{m}, \boldsymbol{\alpha}).$$

Proof. We define g as in the proof of Lemma 2.2 via the Recursion Theorem and an auxiliary primitive recursive function h, which is defined by cases as follows: for any a, d, e, k, and l,

(0) if $a = \langle 0, k, l, \ldots\rangle$, then $h(e, a, d) = a$;

(1) if for some k', b, $c_0, \ldots, c_{k'-1}$, $a = \langle 1, k, l, b, c_0, \ldots, c_{k'-1}\rangle$, then $h(e, a, d) = \langle 1, k, l, h(e, b, d), h(e, c_0, d), \ldots, h(e, c_{k'-1}, d)\rangle$;

(2) if $a = \langle 2, k+1, l\rangle$, then h is defined exactly as in the corresponding case in the proof of Lemma 2.2;

(3) if for some b, $a = \langle 3, k, l, b\rangle$, then $h(e, a, d)$ is the natural index such that

$$\begin{aligned}\{h(e, a, d)\}^{\mathsf{I}}(\mathbf{m}, \boldsymbol{\alpha}) \simeq 0 \cdot \mathsf{I}(\lambda p.\{h(e, b, d)\}^{\mathsf{I}}(p, \mathbf{m}, \boldsymbol{\alpha})) \\ + \{f(f_2(d, k, l), h(e, b, d))\}^{\mathsf{I}}(\mathbf{m}, \boldsymbol{\alpha}),\end{aligned}$$

where f and f_2 are from 2.10 and 2.12, respectively;

(4) if a is of none of these forms, then $h(e, a, d) = 0$.

It is clear that h is primitve recursive, so let $\bar{e}$ be an index such that $\{\bar{e}\}(a, d) = h(\bar{e}, a, d)$ and take $g = \{\bar{e}\}$. We need to prove that for all a, $\mathbf{m}$, $\boldsymbol{\alpha}$, and n,

$$\{g(a, d)\}^{\mathsf{I}}(\mathbf{m}, \boldsymbol{\alpha}) \simeq n \leftrightarrow \{a\}^{\mathsf{H}}(\mathbf{m}, \boldsymbol{\alpha}) \simeq n$$

under the assumption that $\mathsf{H} = \lambda\beta.\{d\}^{\mathsf{I}}(\beta)$ is total.

The implication $(\leftarrow)$ is proved by induction over $\Omega[\mathsf{H}]$ much as in Lemma 2.2. Let X denote the set of sequences $(a, \mathbf{m}, \boldsymbol{\alpha}, n) \in \Omega[\mathsf{H}]$ such that $\{g(a, d)\}^{\mathsf{I}}(\mathbf{m}, \boldsymbol{\alpha}) \simeq n$. We must show that X satisfies the closure conditions which define $\Omega[\mathsf{H}]$.

(0) if $a = \langle 0, k, l, \ldots\rangle$ and $\{a\}^{\mathsf{H}}(\mathbf{m}, \boldsymbol{\alpha}) \simeq n$, then $g(a, d) = a$, so $\{g(a, d)\}^{\mathsf{I}}(\mathbf{m}, \boldsymbol{\alpha}) \simeq \{a\}^{\mathsf{I}}(\mathbf{m}, \boldsymbol{\alpha}) \simeq \{a\}^{\mathsf{H}}(\mathbf{m}, \boldsymbol{\alpha}) \simeq n$. Thus $(a, \mathbf{m}, \boldsymbol{\alpha}, n) \in X$.

(1)
(2) } These are treated exactly as in the proof of Lemma 2.2.

(3) If for some b, $a = \langle 3, k, l, b\rangle$, then for some β, $(b, p, \mathbf{m}, \boldsymbol{\alpha}, \beta(p)) \in X$ for all p. Then for all p,

$$\{g(b, d)\}^{\mathsf{I}}(p, \mathbf{m}, \boldsymbol{\alpha}) \simeq \beta(p)$$

so

$$\begin{aligned}\{g(a, d)\}^{\mathsf{I}}(\mathbf{m}, \boldsymbol{\alpha}) &\simeq 0 \cdot \mathsf{I}(\beta) + \{f(f_2(d, k, l), g(b, d))\}^{\mathsf{I}}(\mathbf{m}, \boldsymbol{\alpha}) \\ &\simeq \{f_2(d, k, l)\}^{\mathsf{I}}(\mathbf{m}, \boldsymbol{\alpha}, \beta) \\ &\simeq \{d\}^{\mathsf{I}}(\beta) \\ &\simeq \mathsf{H}(\beta) \\ &\simeq \mathsf{H}(\lambda p . \{b\}^{\mathsf{H}}(p, \mathbf{m}, \boldsymbol{\alpha})) \simeq n.\end{aligned}$$

Thus $(a, \mathbf{m}, \boldsymbol{\alpha}, n) \in X$.

For the implication $(\rightarrow)$ we apply Theorem 2.5 to the set $X = \{x : x \in \Omega[\mathsf{I}]$, and for all a, d, $\mathbf{m}$, $\boldsymbol{\alpha}$, and n, if

$$x = (g(a, d), \mathbf{m}, \boldsymbol{\alpha}, n) \quad \text{then} \quad \{a\}^{\mathsf{H}}(\mathbf{m}, \boldsymbol{\alpha}) \simeq n\}.$$

We need to prove that for all $x \in \Omega[\mathsf{I}]$, if $\mathrm{Sbc}(x) \subseteq X$, then $x \in X$. If x is not of the form $x = (g(a, d), \mathbf{m}, \boldsymbol{\alpha}, n) \in \Omega[\mathsf{I}]$, then there is nothing to prove, so we assume that it is of this form and that $\mathrm{Sbc}(x) \subseteq X$. We consider the case (0)–(4) under which a may fall.

(0)
(1) } These cases are treated as in the proof of Theorem 2.8.
(2)

(3) If for some b, $a = \langle 3, k, l, b\rangle$, then since $\{g(a, d)\}^{\mathsf{I}}(\mathbf{m}, \boldsymbol{\alpha}) \downarrow$, it follows from the definition of g in this case that $\mathsf{I}(\lambda p . \{g(b, d)\}^{\mathsf{I}}(p, \mathbf{m}, \boldsymbol{\alpha})) \downarrow$ and thus that for some β, $\{g(b, d)\}^{\mathsf{I}}(p, \mathbf{m}, \boldsymbol{\alpha}) \simeq \beta(p)$ for all p. Furthermore, for all p,

$$(g(b, d), p, \mathbf{m}, \boldsymbol{\alpha}, \beta(p)) \in \mathrm{Sbc}(x) \subseteq X$$

(this is why $h(e, a, d)$ must be a natural index in this case). Thus, for all p, $\{b\}^{\mathsf{H}}(p, \mathbf{m}, \boldsymbol{\alpha}) \simeq \beta(p)$ and

$$\begin{aligned}\{a\}^{\mathsf{H}}(\mathbf{m}, \boldsymbol{\alpha}) &\simeq \mathsf{H}(\lambda p . \{b\}^{\mathsf{H}}(p, \mathbf{m}, \boldsymbol{\alpha})) \\ &\simeq \mathsf{H}(\beta) \\ &\simeq \{g(a, d)\}^{\mathsf{I}}(\mathbf{m}, \boldsymbol{\alpha}) \simeq n,\end{aligned}$$

by the computation in the first half of the proof.

(4) If a is of none of these forms, $g(a, d) = 0$ so $x \notin \Omega[\mathsf{I}]$. □

2.15 Corollary (II). *For any* F, H, *and* I, *if* H *is recursive in* I *and* F *is partial recursive in* H, *then* F *is partial recursive in* I. □

Of course, we used the same trick in the induction step (3) of the definition of g as we did in Theorem 2.11 and it is clear that the proof will not work without I.

2.16 Corollary. *The relation "recursive in" is transitive among total functionals.* □

2.17 Corollary. *For any* H *and* I, *if* H *is recursive in* I, *then* H^{sJ} *is recursive in* I^{sJ}.

Proof. Suppose H is recursive in I with index d. Then with g as in Theorem 2.14

$$\begin{aligned}\mathsf{H}^{sJ}((\langle a, \mathbf{m}\rangle) * \alpha) = 0 &\leftrightarrow \{a\}^{\mathsf{H}}(\mathbf{m}, \alpha) \text{ is defined}\\ &\leftrightarrow \{g(a, d)\}^{\mathsf{I}}(\mathbf{m}, \alpha) \text{ is defined}\\ &\leftrightarrow \mathsf{I}^{sJ}((\langle g(a, d)\rangle, \langle \mathbf{m}\rangle) * \alpha) \text{ is defined.} \quad \Box\end{aligned}$$

2.18 Theorem. *There exists a primitive recursive function* g' *such that for all* a, d, $\mathbf{m}$, α, *and* n, *if* $\lambda\beta.\{d\}(\beta)$ *is a total recursive functional* H, *then*

$$\{a\}^{\mathsf{H}}(\mathbf{m}, \alpha) \simeq n \rightarrow \{g'(a, d)\}(\mathbf{m}, \alpha) \simeq n.$$

Proof. Exercise 2.22. □

2.19 Corollary (III). *If* H *is recursive and* F *is recursive in* H, *then* F *is recursive.*

Proof. If F has index a from H and H is recursive with index d, then $\mathsf{F} \subseteq \{g'(a, d)\}$ by the preceding theorem. Since F is total, these functionals are equal. □

2.20–2.29 Exercises

2.20. Prove Lemmas 2.6 and 2.7. Along the same lines, show that if a is any I-index for a functional F partial recursive in I and $\bar{e}$ is the natural index given by the proof of the I-Recursion Theorem such that

$$\{\bar{e}\}(\mathbf{m}, \alpha) \simeq \mathsf{F}(\bar{e}, \mathbf{m}, \alpha)$$

then

$$(a, \bar{e}, \mathbf{m}, \boldsymbol{\alpha}, \mathsf{F}(\bar{e}, \mathbf{m}, \boldsymbol{a})) \in \mathrm{Sbc}(\bar{e}, \mathbf{m}, \boldsymbol{\alpha}, \mathsf{F}(\bar{e}, \mathbf{m}, \boldsymbol{\alpha})).$$

2.21. Prove Lemma 2.12.

2.22. Prove Theorem 2.18.

2.23. Let I be any functional such that E_1 is recursive in I and denote by $\omega_1[\mathsf{I}]$ the least ordinal not the order-type of a well-ordering of ω recursive in I. Show that

(i) for any β recursive in I, $\omega_1[\beta] < \omega_1[\mathsf{I}]$;

(ii) for any $\sigma < \omega_1[\mathsf{I}]$, there exists a β recursive in I such that $\sigma < \omega_1[\beta]$.

2.24. Show that for any I such that E is recursive in I and any relation P semi-recursive in I, if

$$\mathsf{R}(\mathbf{m}, \boldsymbol{\alpha}) \leftrightarrow \forall \beta\, \exists p\, \mathsf{P}(\bar{\beta}(p), \mathbf{m}, \boldsymbol{\alpha})$$

then also R is semi-recursive in I.

2.25. (Cf. Exercise II.4.31). For any I and any partial function f of rank 1, let $\Omega[\mathsf{I}, f]$ be defined as is $\Omega[\mathsf{I}]$ with the following additional clause:

$$\text{if} \quad f(p) \simeq n, \quad \text{then} \quad (\langle 0, k+1, l, 6\rangle, p, \mathbf{m}, \boldsymbol{\alpha}, n) \in \Omega[\mathsf{I}, f].$$

We write

$$\{a\}^{\mathsf{I}}(\mathbf{m}, \boldsymbol{\alpha}, f) \simeq n \leftrightarrow (a, \mathbf{m}, \boldsymbol{\alpha}, n) \in \Omega[\mathsf{I}, f].$$

We have thus defined the class of functionals partial recursive in I with arguments of type $(\mathbf{m}, \boldsymbol{\alpha}, f)$. Show that for any such functional F partial recursive in I,

(i) there exists an ordinary functional G partial recursive in I such that

$$\mathsf{G}(e, \mathbf{m}, \boldsymbol{\alpha}) \simeq \mathsf{F}(\mathbf{m}, \boldsymbol{\alpha}, \{e\}^{\mathsf{I}});$$

(ii) (*First* I-*Recursion Theorem*) there exists a function $\bar{f}$ partial recursive in I such that for all p, $\mathsf{F}(p, \bar{f}) \simeq \bar{f}(p)$ and for any h, if also for all p, $\mathsf{F}(p, h) \simeq h(p)$, then $\bar{f} \subseteq h$.

(iii) Formulate and prove a version of (ii) which allows for the presence of parameters.

Hint. (i) is immediate from a minor modification of Lemma 2.9. For (ii), show that if G is chosen correctly in (i) with, say, index b from I, then if $\mathsf{G}(e, p) \simeq n$ and g is a partial function defined by:

$$g(q) \simeq r \leftrightarrow (e, q, r) \in \mathrm{Sbc}((b, e, p, n)),$$

then $\mathsf{F}(p, g) \simeq n$ (i.e. all of the values of $\{e\}$ which are "necessary" in order that $\mathsf{F}(p, \{e\}) \simeq n$ are computed as subcomputations of the computation $\mathsf{G}(e, p) \simeq n$). Then show that if $\bar{e}$ is defined as in the proof of the I-Recursion Theorem such that $\{\bar{e}\}^{\mathsf{I}}(m) \simeq \mathsf{G}(\bar{e}, m)$, then $\bar{f} = \{\bar{e}\}^{\mathsf{I}}$ is the required solution.

2.26. For any I and any inductive operator Γ on ω, Γ is *positive semi-recursive* in I iff there exists a functional F partial recursive in I in the sense of the preceding Exercise such that for any m and f

$$\mathsf{F}(m, f) \simeq 0 \leftrightarrow m \in \Gamma(\{p : f(p) \simeq 0\}).$$

(i) Use the First I-Recursion Theorem to show that for any Γ which is positive semi-recursive in I, $\bar{\Gamma}$ is semi-recursive in I.

(ii) Prove that every Π^1_1 set of numbers is semi-recursive in I (for arbitrary I!).

(iii) Discuss the relationship of this to Theorem 1.8 and Corollary 1.9. Does it follow that every Δ^1_1 set is recursive in an arbitrary I?

2.27. Characterize the classes of relations recursive and semi-recursive in the functional $\lambda\alpha.0$.

2.28. A functional I is called *effectively discontinuous* iff there exists a function F recursive in I such that the sequence of functions $F_p = \lambda m.F(p, m)$ converges to a function G, but the sequence of values $\mathsf{I}(F_p)$ does not converge to $\mathsf{I}(G)$. Show that I is effectively discontinuous iff E is recursive in I.

2.29. Show that for any I, E is recursive in I iff $\{\alpha : \alpha$ is recursive in $\mathsf{I}\}$ is closed under the ordinary jump oJ. (Suppose E is not recursive in I so by the preceding exercise I is not effectively discontinuous. For each s let $\beta_s = ((s)_0, \ldots, (s)_{\lg(s)-1}) * (\lambda m.0)$ and $\gamma(s) = \mathsf{I}(\beta_s)$. Show that every α recursive in I is recursive in γ^{oJ}.)

2.30 Notes. (I), (II) for total F, and (III) and their effective versions are due to Kleene [1963]. The improved versions 2.10 and (II) (2.14) first appear in Hinman [1966]. Kleene [1963] gave the counterexample 2.1. Exercises 2.28 and 2.29 are due to Grilliot [1971].

3. Ordinal Comparison

This section is devoted to the proof of a technical result, the Ordinal Comparison Theorem, which is the key to the theory of relations semi-recursive in a type-2 functional. This theorem is closely related, in both statement and proof, to

Lemma 1.7. There we were able, using E, to compare the ordinals assigned to members of W, a set to which all relations semi-recursive in E can be reduced. For an arbitrary functional I such that E is recursive in I, we shall compare the ordinals assigned to members of U^{I}, a relation universal for the class of relations semi-recursive in I.

3.1 Definition. For any I and $\boldsymbol{\alpha}$,

(i) $\mathsf{U}^{\mathsf{I}}(a, \langle \mathbf{m} \rangle, \langle \boldsymbol{\alpha} \rangle) \leftrightarrow \{a\}^{\mathsf{I}}(\mathbf{m}, \boldsymbol{\alpha})\downarrow$;

(ii) $U^{\mathsf{I}}_{\boldsymbol{\alpha}} = \{\langle a, \mathbf{m} \rangle : \mathsf{U}^{\mathsf{I}}(a, \langle \mathbf{m} \rangle, \langle \boldsymbol{\alpha} \rangle)\}$;

(iii) $U^{\mathsf{I}} = U^{\mathsf{I}}_{\varnothing}$.

To each element of U^{I} we assign an ordinal which intuitively measures how "long" the corresponding computation is:

$$|(a, \langle \mathbf{m} \rangle, \langle \boldsymbol{\alpha} \rangle)|^{\mathsf{I}} = \text{least } \sigma.\, (a, \mathbf{m}, \boldsymbol{\alpha}, \{a\}^{\mathsf{I}}(\mathbf{m}, \boldsymbol{\alpha})) \in \Omega[\mathsf{I}]^{\sigma},$$

where, as usual, $\Omega[\mathsf{I}]^{\sigma}$ is the σ-th stage of the inductive definition of $\Omega[\mathsf{I}]$. If $\sim\mathsf{U}^{\mathsf{I}}(a, \langle \mathbf{m} \rangle, \langle \boldsymbol{\alpha} \rangle)$, we set $|(a, \langle \mathbf{m} \rangle, \langle \boldsymbol{\alpha} \rangle)|^{\mathsf{I}} = \aleph_1$. Similarly, for any $\boldsymbol{\alpha}$,

$$|\langle a, \mathbf{m} \rangle|^{\mathsf{I}}_{\boldsymbol{\alpha}} = |(a, \langle \mathbf{m} \rangle, \langle \boldsymbol{\alpha} \rangle)|^{\mathsf{I}}.$$

We often write $|a, \mathbf{m}, \boldsymbol{\alpha}|^{\mathsf{I}}$ and $|a, \mathbf{m}|^{\mathsf{I}}_{\boldsymbol{\alpha}}$ and omit the sub- and superscripts when they are clear from the context.

3.2 Lemma. *For any* I, a, k, l, *and* $(\mathbf{m}, \boldsymbol{\alpha}) \in {}^{k,l}\omega$,

(0) *if* $a = \langle 0, k, l, \ldots \rangle$ *and* $\langle a, \mathbf{m} \rangle \in U^{\mathsf{I}}_{\boldsymbol{\alpha}}$, *then* $|a, \mathbf{m}|^{\mathsf{I}}_{\boldsymbol{\alpha}} = 0$;

(1) *if for some* $b, c_0, \ldots, c_{k'-1}$, $a = \langle 1, k, l, b, c_0, \ldots, c_{k'-1} \rangle$ *and* $\langle a, \mathbf{m} \rangle \in U^{\mathsf{I}}_{\boldsymbol{\alpha}}$, *then*

$$|a, \mathbf{m}|^{\mathsf{I}}_{\boldsymbol{\alpha}} = \max\{|c_i, \mathbf{m}|^{\mathsf{I}}_{\boldsymbol{\alpha}} + 1 : i < k'\} \cup \{|b, \mathbf{q}|^{\mathsf{I}}_{\boldsymbol{\alpha}} + 1\},$$

where for all $i < k'$, $\{c_i\}^{\mathsf{I}}(\mathbf{m}, \boldsymbol{\alpha}) \simeq q_i$;

(2) *if* $a = \langle 2, k+1, l \rangle$, *and* $\langle a, b, \mathbf{m} \rangle \in U^{\mathsf{I}}_{\boldsymbol{\alpha}}$, *then*

$$|a, b, \mathbf{m}|^{\mathsf{I}}_{\boldsymbol{\alpha}} = |b, \mathbf{m}|^{\mathsf{I}}_{\boldsymbol{\alpha}} + 1;$$

(3) *if for some* b, $a = \langle 3, k, l, b \rangle$ *and* $\langle a, \mathbf{m} \rangle \in U^{\mathsf{I}}_{\boldsymbol{\alpha}}$, *then*

$$|a, \mathbf{m}|^{\mathsf{I}}_{\boldsymbol{\alpha}} = \sup{}^{+}\{|b, p, \mathbf{m}|^{\mathsf{I}}_{\boldsymbol{\alpha}} : p \in \omega\}.$$

Proof. Immediate from the definitions. □

3.3 Ordinal Comparison Theorem. *For any* I *such that* E *is recursive in* I, *there exists a functional* H *partial recursive in* I *such that for all* u^0, u^1, *and* $\boldsymbol{\alpha}$,

(i) *if* $u^0 \in U'_\alpha$ *and* $|u^0|'_\alpha \leq |u^1|'_\alpha$, *then* $\mathsf{H}(u^0, u^1, \langle\alpha\rangle) \simeq 0$;
(ii) *if* $u^1 \in U'_\alpha$ *and* $|u^1|'_\alpha < |u^0|'_\alpha$, *then* $\mathsf{H}(u^0, u^1, \langle\alpha\rangle) \simeq 1$.

Proof. To simplify things slightly, we shall give the proof for the case $\alpha = \varnothing$; the general case merely requires including $\langle\alpha\rangle$ as a parameter throughout. The basic intuition behind the proof is the same as for Lemma 1.7: to compare the ordinals $|u^0|$ and $|u^1|$, H compares the ordinals assigned to the "immediate subcomputations" of u^0 and u^1. If at least one of u^0 and u^1 belongs to U', some of these will be smaller than $\min\{|u^0|, |u^1|\}$ and a recursion will be established. Of course, the partial recursiveness of H is ensured by means of the I-Recursion Theorem: we define first a functional F partial recursive in I, choose $\bar{e}$ such that $\{\bar{e}\}'(u^0, u^1) \simeq \mathsf{F}(\bar{e}, u^0, u^1)$, and set $\mathsf{H} = \{\bar{e}\}'$.

The definition of F is divided into 16 cases labeled (r, s), with $0 \leq r, s \leq 3$ plus two "otherwise" cases. In case (r, s) we define $\mathsf{F}(e, u^0, u^1)$ for all u^0 and u^1 such that $u^i = \langle a^i, \mathbf{m}^i\rangle$, a^0 is an index appropriate for the arguments $\mathbf{m}^0$ under clause (r) of the definition of $\Omega[\mathsf{I}]$, and a^1 is an index appropriate for the arguments $\mathbf{m}^1$ under clause (s). "Appropriate" means merely that $(a^0)_0 = r$, $(a^0)_1 = \lg(\mathbf{m}^0)$, and $(a^0)_2 = 0$, etc., not that the corresponding computation is defined. If u^i is not of the form $\langle a^i, \mathbf{m}^i\rangle$ with a^i an appropriate index, we say that u^i is not of the *proper form*.

First, if u^1 is not of the proper form, we set $\mathsf{F}(e, u^0, u^1) \simeq 0$, while if u^1 is of the proper form but u^0 is not $\mathsf{F}(e, u^0, u^1) \simeq 1$.

Cases $(0, s)$ *with* $0 \leq s \leq 3$. $\mathsf{F}(e, u^0, u^1) \simeq 0$.

Cases $(r, 0)$ *with* $1 \leq r \leq 3$. $\mathsf{F}(e, u^0, u^1) \simeq 1$.

Case (1,2). The assumption of proper form for this case means that for some b^0, c, and b^1, $k^0 = \lg(\mathbf{m}^0)$, and $k^1 = \lg(\mathbf{m}^1)$, $u^0 = \langle\langle 1, k^0, 0, b^0, c\rangle, \mathbf{m}^0\rangle$ and $u^1 = \langle\langle 2, k^1 + 1, 0\rangle, b^1, \mathbf{m}^1\rangle$ (for simplicity we have taken $k^{0\prime} = 1$). Our aim is that the computation of H should proceed according to the following "flow diagram":

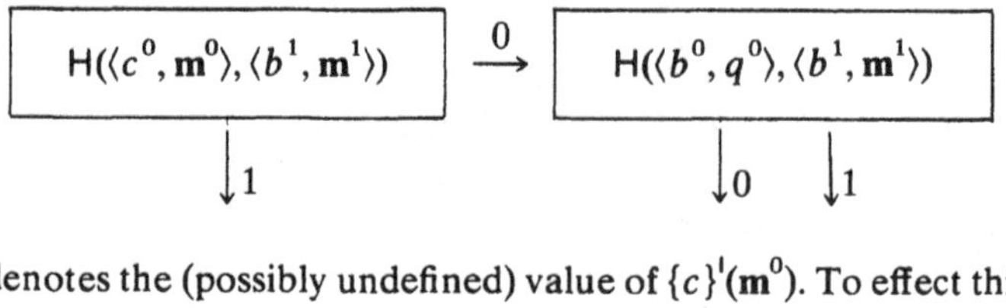

where q^0 denotes the (possibly undefined) value of $\{c\}'(\mathbf{m}^0)$. To effect this, we set

$$\mathsf{F}(e, u^0, u^1) \simeq \mathsf{G}(\{e\}'(\langle c^0, \mathbf{m}^0\rangle, \langle b^1, \mathbf{m}^1\rangle), e, u^0, u^1)$$

where

$$G(n, e, u^0, u^1) \simeq \begin{cases} \{e\}'(\langle b^0, q^0\rangle, \langle b^1, \mathbf{m}^1\rangle), & \text{if} \quad n = 0; \\ 1, \quad \text{if} \quad n = 1. \end{cases}$$

Case (1,1). In this and succeeding cases we shall give only the recursive conditions we want H to satisfy — in some cases by means of a flow diagram as above — and leave it to the reader to verify that F may be defined so that the resulting H does indeed satisfy these conditions. Here again we assume for simplicity that $k^{0'} = k^{1'} = 1$. The computation of $\mathsf{H}\ (u^0, u^1)$ is to proceed according to the diagram:

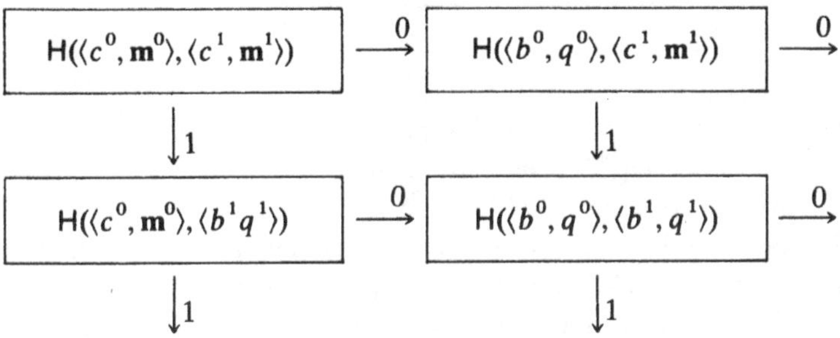

where q^i denotes the (possibly undefined) value of $\{c^i\}^{\mathsf{I}}(\mathbf{m}^i)$.

Case (2,1).

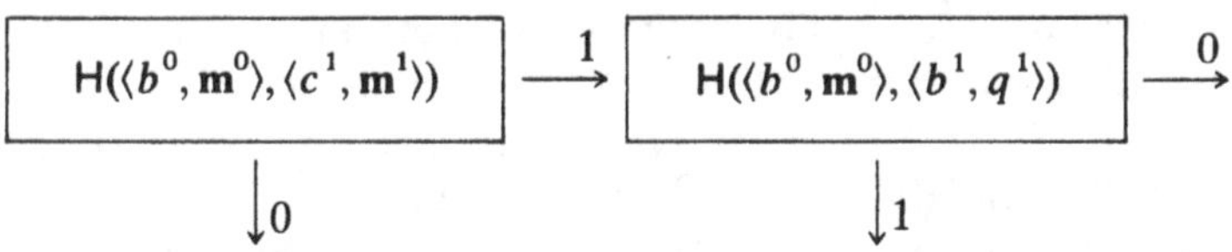

Case (2,2). $\mathsf{H}(\langle b^0, \mathbf{m}^0\rangle, \langle b^1, \mathbf{m}^1\rangle)$.

Case (1,3).

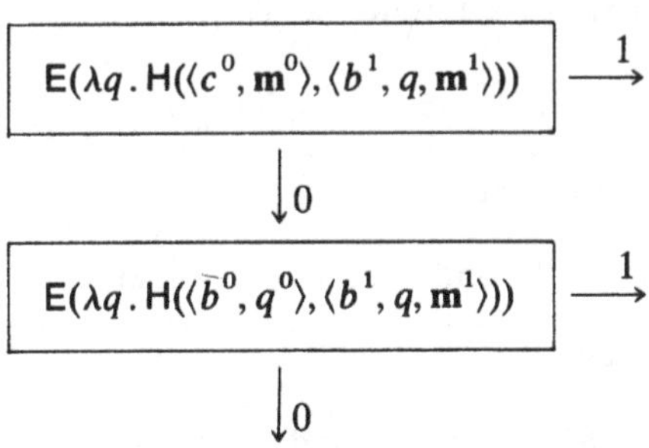

Case (3,1).

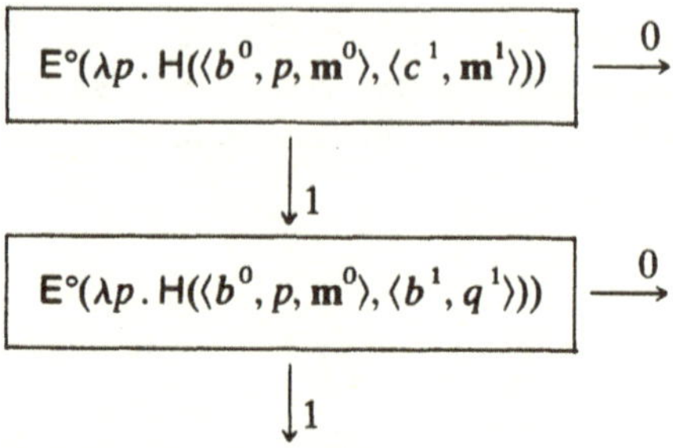

Case (2,3). $\mathsf{E}(\lambda q\,.\,\mathsf{H}(\langle b^0, \mathbf{m}^0\rangle, \langle b^1, q, \mathbf{m}^1\rangle))$.

Case (3,2). $\mathsf{E}^\circ(\lambda p\,.\,\mathsf{H}(\langle b^0, p, \mathbf{m}^0\rangle, \langle b^1, \mathbf{m}^1\rangle))$.

Case (3,3). $\mathsf{E}^\circ(\lambda p\,.\,\mathsf{E}(\lambda q\,.\,H(\langle b^0, p, \mathbf{m}^0\rangle, \langle b^1, q, \mathbf{m}^1\rangle)))$.

We first observe that F is indeed partial recursive in I as all case distinctions are recursive and the indicated computations require I, E, or E° and are thus recursive in I by the assumption that E is recursive in I. Thus the I-Recursion Theorem applies and there exists an H which satisfies the stated conditions.

Clauses (i) and (ii) of the theorem are vacuous unless either u^0 or u^1 belongs to U^{I}. Hence, it suffices to prove (i) and (ii) by induction on $\sigma = \min\{|u^0|, |u^1|\}$ for $\sigma < \aleph_1$. If u^1 is not of the proper form, then $|u^0| \leq |u^1| = \aleph_1$ and $\mathsf{H}(u^0, u^1) \simeq 0$ in accord with (i). If u^1 is of the proper form but u^0 is not, then $\mathsf{H}(u^0, u^1) \simeq 1$ in accord with the fact that at most clause (ii) can apply. Hence, for the induction we may assume that (u^0, u^1) falls under one of the cases (r, s) with $0 \leq r, s \leq 3$.

If $\sigma = 0$, then either $|u^0| = 0$ or $|u^1| = 0$ or both. If $|u^0| = 0$ (cases $(0, s)$, $0 \leq s \leq 3$), then $\mathsf{H}(u^0, u^1) \simeq 0$ in accord with (i). If $|u^1| = 0$ while $|u^0| \neq 0$ (cases $(r, 0)$, $1 \leq r \leq 3$), then $\mathsf{H}(u^0, u^1) \simeq 1$ in accord with (ii).

Suppose now that $0 < \sigma = \min\{|u^0|, |u^1|\} < \aleph_1$ and that for any v^0 and v^1, clauses (i) and (ii) hold for v^0 and v^1 provided that $\min\{|v^0|, |v^1|\} < \sigma$. We treat three of the nine remaining cases in some detail and leave it to the reader to check the others.

Case (1,2). Let u^0 and u^1 be as in the definition of F for this case. Suppose first that $u^0 \in U^{\mathsf{I}}$ and $|u^0| \leq |u^1|$. Then for some q^0, $\{c^0\}^{\mathsf{I}}(\mathbf{m}^0) \simeq q^0$ and $\{b^0\}^{\mathsf{I}}(q^0)\downarrow$. Furthermore, by Lemma 3.2 (1),

(1) $$|c^0, \mathbf{m}^0| < |u^0| = \sigma \quad \text{and} \quad |b^0, q^0| < |u^0| = \sigma.$$

From (1) and Lemma 3.2(2) follows

(2) $$|c^0, \mathbf{m}^0| \leq |b^1, \mathbf{m}^1| \quad \text{and} \quad |b^0, q^0| \leq |b^1, \mathbf{m}^1|,$$

and from (1) it is obvious that

(3) $$\min\{|c^0, \mathbf{m}^0|, |b^1, \mathbf{m}^1|\} < \sigma \quad \text{and} \quad \min\{|b^0, q^0|, |b^1, \mathbf{m}^1|\} < \sigma.$$

Then from (2), (3), and the induction hypothesis we have

$$\text{(4)} \qquad \mathsf{H}(\langle c^0, \mathbf{m}^0\rangle, \langle b^1, \mathbf{m}^1\rangle) \simeq 0 \quad \text{and} \quad \mathsf{H}(\langle b^0, q^0\rangle, \langle b^1, \mathbf{m}^1\rangle) \simeq 0.$$

From (4) and the flow diagram for H in this case it is evident that $\mathsf{H}(u^0, u^1) \simeq 0$ in accord with (i).

Suppose now that $u^1 \in U^1$ and $|u^1| < |u^0|$ so that $|b^1, \mathbf{m}^1| < |u^1| = \sigma$. If $|b^1, \mathbf{m}^1| < |c^0, \mathbf{m}^0|$, then we see as above that

$$\mathsf{H}(u^0, u^1) \simeq \mathsf{H}(\langle c^0, \mathbf{m}^0\rangle, \langle b^1, \mathbf{m}^1\rangle) \simeq 1,$$

in accord with (ii). Otherwise $|c^0, \mathbf{m}^0| \leqslant |b^1, \mathbf{m}^1|$. Since $|b^1, \mathbf{m}^1| < \aleph_1$, also $|c^0, \mathbf{m}^0| < \aleph_1$, so $\langle c^0, \mathbf{m}^0\rangle \in U^1$ and there exists a number $q^0 \simeq \{c^0\}^1(\mathbf{m}^0)$. Note that in this case there is no necessity that $\{b^0\}^1(q^0)$ be defined. If it were the case that $|b^0, q^0| \leqslant |b^1, \mathbf{m}^1|$, we would have by Lemma 3.2(1,2),

$$|u^0| = \max\{|c^0, \mathbf{m}^0| + 1, |b^0, q^0| + 1\} \leqslant |b^1, \mathbf{m}^1| + 1 = u^1,$$

contrary to the present assumption. Hence $|b^1, \mathbf{m}^1| < |b^0, q^0|$, and by the induction hypothesis we have

$$\mathsf{H}(\langle c^0, \mathbf{m}^0\rangle, \langle b^1, \mathbf{m}^1\rangle) \simeq 0 \quad \text{and} \quad \mathsf{H}(\langle b^0, q^0\rangle, \langle b^1, \mathbf{m}^1\rangle) \simeq 1.$$

Referring again to the flow diagram, we see that $\mathsf{H}(u^0, u^1) \simeq 1$ in accord with (ii).

Case (1,1). Let $u^i = \langle\langle 1, k^i, 0, b^i, c^i\rangle, \langle \mathbf{m}^i\rangle\rangle$ and suppose first that $u^0 \in U^1$ and $|u^0| \leqslant |u^1|$. Then there exists a $q^0 \simeq \{c^0\}^1(\mathbf{m}^0)$, (1) holds, and by the induction hypothesis, for any v,

$$\text{(5)} \qquad |c^0, \mathbf{m}^0| \leqslant |v| \rightarrow \mathsf{H}(\langle c^0, \mathbf{m}^0\rangle, v) \simeq 0;$$

$$\text{(6)} \qquad |v| < |c^0, \mathbf{m}^0| \rightarrow \mathsf{H}(\langle c^0, \mathbf{m}^0\rangle, v) \simeq 1;$$

$$\text{(7)} \qquad |b^0, q^0| \leqslant |v| \rightarrow \mathsf{H}(\langle b^0, q^0\rangle, v) \simeq 0;$$

$$\text{(8)} \qquad |v| < |b^0, q^0| \rightarrow \mathsf{H}(\langle b^0, q^0\rangle, v) \simeq 1.$$

Suppose first that $|c^0, \mathbf{m}^0| \leqslant |c^1, \mathbf{m}^1|$ so that by (5), $\mathsf{H}(\langle c^0, \mathbf{m}^0\rangle, \langle c^1, \mathbf{m}^1\rangle) \simeq 0$. If also $|b^0, q^0| \leqslant |c^1, \mathbf{m}^1|$, then by (7) and the flow diagram,

$$\mathsf{H}(u^0, u^1) \simeq \mathsf{H}(\langle b^0, q^0\rangle, \langle c^1, \mathbf{m}^1\rangle) \simeq 0$$

in accord with (i). If, on the other hand, $|c^1, \mathbf{m}^1| < |b^0, q^0|$, then $\langle c^1, \mathbf{m}^1\rangle \in U^1$ so there exists a $q^1 \simeq \{c^1\}^1(\mathbf{m}^1)$. If it were the case that $|b^1, q^1| < |b^0, q^0|$, then

$$|u^1| = \max\{|c^1, \mathbf{m}^1| + 1, |b^1, q^1| + 1\} < |b^0, q^0| + 1 \leqslant |u^0|$$

contrary to assumption. Hence $|b^0, q^0| \leq |b^1, q^1|$, and by (8) and (7) we have

$$\mathsf{H}(\langle b^0, q^0\rangle, \langle c^1, \mathbf{m}^1\rangle) \simeq 1 \quad \text{and} \quad \mathsf{H}(\langle b^0, q^0\rangle, \langle b^1, q^1\rangle) \simeq 0,$$

so again $\mathsf{H}(u^0, u^1) \simeq 0$ in accord with (i).

The other possibility under (i) is that $|c^1, \mathbf{m}^1| < |c^0, \mathbf{m}^0|$, so that by (6), $\mathsf{H}(\langle c^0, \mathbf{m}^0\rangle, \langle c^1, \mathbf{m}^1\rangle) \simeq 1$ and there exists a $q^1 \simeq \{c^1\}^{\mathsf{I}}(\mathbf{m}^1)$. In this case it is impossible that $|b^1, q^1| < |c^0, \mathbf{m}^0|$ or $|b^1, q^1| < |b^0, q^0|$, as either leads to the conclusion $|u^1| < |u^0|$, contrary to assumption. Hence by (5) and (7) we have

$$\mathsf{H}(\langle c^0, \mathbf{m}^0\rangle, \langle b^1, q^1\rangle) \simeq 0 \quad \text{and} \quad \mathsf{H}(\langle b^0, q^0\rangle, \langle b^1, q^1\rangle) \simeq 0,$$

whence $\mathsf{H}(u^0, u^1) \simeq 0$ as required.

The argument for clause (ii) is similar and is omitted.

Case (3,3). This case is similar to the main case encountered in the proof of Lemma 1.7. By the case hypothesis we have $u^i = \langle\langle 3, k^i, 0, b^i\rangle, \mathbf{m}^i\rangle$. The definition of H in this case is such that $\mathsf{H}(u^0, u^1)$ is defined just in case for all p and q, $\mathsf{H}(\langle b^0, p, \mathbf{m}^0\rangle, \langle b^1, q, \mathbf{m}^1\rangle)$ is defined, and if so, then for all p and q,

(9) $\quad \mathsf{H}(u^0, u^1) \simeq 0 \leftrightarrow \forall p\, \exists q\,.\, \mathsf{H}(\langle b^0, p, \mathbf{m}^0\rangle, \langle b^1, q, \mathbf{m}^1\rangle) \simeq 0;$

(10) $\quad \mathsf{H}(u^0, u^1) \simeq 1 \leftrightarrow \exists p\, \forall q\,.\, \mathsf{H}(\langle b^0, p, \mathbf{m}^0\rangle, \langle b^1, q, \mathbf{m}^1\rangle) \simeq 1.$

Suppose first that $u^0 \in U^{\mathsf{I}}$ and $|u^0| \leq |u^1|$. Then

$$\{\langle 3, k^0, 0, b^0\rangle\}^{\mathsf{I}}(\mathbf{m}^0) \simeq \mathsf{I}(\lambda p\,.\,\{b^0\}^{\mathsf{I}}(p, \mathbf{m}^0))$$

is defined, so for all p, $\langle b^0, p, \mathbf{m}^0\rangle \in U^{\mathsf{I}}$. Furthermore, by Lemma 3.2(3), for all p, $|b^0, p, \mathbf{m}^0| < |u^0| = \sigma$. Hence, for all p and q,

$$\min\{|b^0, p, \mathbf{m}^0|, |b^1, q, \mathbf{m}^1|\} < \sigma,$$

so from the induction hypothesis we have for all p and q,

(11) $\quad |b^0, p, \mathbf{m}^0| \leq |b^1, q, \mathbf{m}^1| \to \mathsf{H}(\langle b^0, p, \mathbf{m}^0\rangle, \langle b^1, q, \mathbf{m}^1\rangle) \simeq 0;$

(12) $\quad |b^1, q, \mathbf{m}^1| < |b^0, p, \mathbf{m}^0| \to \mathsf{H}(\langle b^0, p, \mathbf{m}^0\rangle, \langle b^1, q, \mathbf{m}^1\rangle) \simeq 1.$

As one of these holds for each p and q, $\mathsf{H}(\langle b^0, p, \mathbf{m}^0\rangle, \langle b^1, q, \mathbf{m}^1\rangle)$ is defined for all p and q and thus $\mathsf{H}(u^0, u^1)$ is defined. If $u^1 \notin U^{\mathsf{I}}$, then for some q, $\{b^1\}^{\mathsf{I}}(q, \mathbf{m}^1)$ is undefined and thus $|b^1, q, \mathbf{m}^1| = \aleph_1$. In this case, $\mathsf{H}(u^0, u^1) \simeq 0$ by (9) and (11). If $u^1 \in U^{\mathsf{I}}$, then by the assumption and Lemma 3.2(3), for each p,

$$|b^0, p, \mathbf{m}^0| < |u^0| \leq |u^1| = \sup{}^{+}\{|b^1, q, \mathbf{m}^1| : q \in \omega\}.$$

Hence for each p, there is a q such that $|b^0, p, \mathbf{m}^0| \leqslant |b^1, q, \mathbf{m}^1|$ and thus again $\mathsf{H}(u^0, u^1) \simeq 0$ by (9) and (11).

The subcase (ii) is handled similarly by use of (10) and (12). □

At this point it may come as a shock to the reader to learn that the preceding theorem is in fact a simplified version of the full Ordinal Comparison Theorem. Fortunately, the proof of the latter is an obvious translation of the preceding proof. To state it, we return to the notation of § V.1 and set $x^i = (a^i, \langle \mathbf{m}^i \rangle, \langle \boldsymbol{\alpha}^i \rangle)$. Similarly, we write

$$\mathsf{H}(x^0, x^1) \quad \text{for} \quad \mathsf{H}(\langle a^0, \mathbf{m}^0 \rangle, \langle a^1, \mathbf{m}^1 \rangle, \langle \boldsymbol{\alpha}^0 \rangle, \langle \boldsymbol{\alpha}^1 \rangle).$$

3.4 Theorem. *For any* I *such that* E *is recursive in* I, *there exists a functional* H *partial recursive in* I *such that for all* x^0 *and* x^1,

(i) *if* $\mathsf{U}^{\mathsf{I}}(x^0)$ *and* $|x^0|^{\mathsf{I}} \leqslant |x^1|^{\mathsf{I}}$, *then* $\mathsf{H}(x^0, x^1) \simeq 0$;

(ii) *if* $\mathsf{U}^{\mathsf{I}}(x^1)$ *and* $|x^1|^{\mathsf{I}} < |x^0|^{\mathsf{I}}$, *then* $\mathsf{H}(x^0, x^1) \simeq 1$.

Proof. Follows the proof of Theorem 3.3 including everywhere $\boldsymbol{\alpha}^0$ and $\boldsymbol{\alpha}^1$ as parameters. (Cf. Exercise 3.7). □

3.5–3.7 Exercises

3.5. Write out explicitly the definition of F for cases (1,1) and (1,3) in the proof of Theorem 3.3.

3.6. Work out the flow diagram for arbitrary $k^{0'}$ and $k^{1'}$ for cases (1,2) and (1,1) in the proof of Theorem 3.3.

3.7. Sketch the proof of Theorem 3.4 by indicating what modifications are necessary to include the parameters $\boldsymbol{\alpha}$.

3.8 Notes. The method of ordinal comparison and its application to the Selection Theorem (4.1 below) were announced in Gandy [1967a]. Moschovakis [1967] contains the first published proof.

4. *Relations Semi-Recursive in a Type-2 Functional*

The theory of relations semi-recursive in a type-2 functional I (such that E is recursive in I) is a blend of the theory of (absolutely) semi-recursive relations and the theory of classes of relations which have the pre-wellordering property. We develop first that part of the theory which parallels the theory of semi-recursive relations of § II.4 and then study the effect of the pre-wellordering property. Both parts depend heavily on the Ordinal Comparison Theorem of the preceding section and we must therefore assume throughout that I is a fixed functional such that E is recursive in I.

4.1 Selection Theorem (Gandy). *There exists a functional* $\mathrm{Sel}^{\mathsf{I}}$ *partial recursive in* I *such that for all* $a, \mathbf{m}$, *and* $\boldsymbol{\alpha}$, *the following are equivalent*:

(i) $\exists p.\{a\}^{\mathsf{I}}(p, \mathbf{m}, \boldsymbol{\alpha})\downarrow$;

(ii) $\{a\}^{\mathsf{I}}(\mathrm{Sel}^{\mathsf{I}}(a, \langle\mathbf{m}\rangle, \langle\boldsymbol{\alpha}\rangle), \mathbf{m}, \boldsymbol{\alpha})\downarrow$.

Proof. For any a, let a^+ denote an index such that

$$\{a^+\}^{\mathsf{I}}(p, \mathbf{m}, \boldsymbol{\alpha}) \simeq \{a\}^{\mathsf{I}}(p+1, \mathbf{m}, \boldsymbol{\alpha}).$$

There is a primitive recursive function $a \mapsto a^+$. Let H be the functional defined in the Ordinal Comparison Theorem (3.3) and F the functional computed according to the following flow diagram:

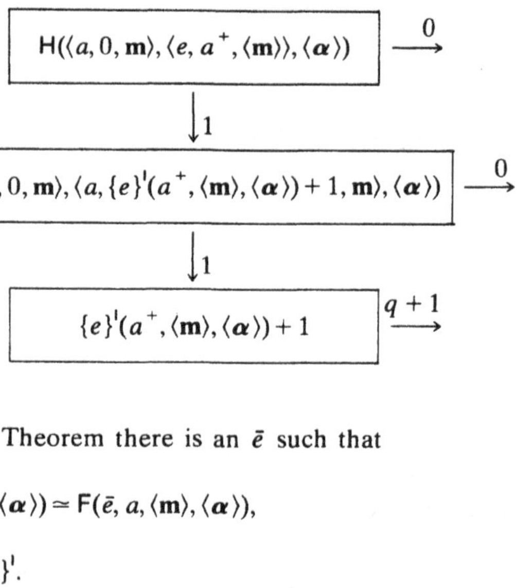

By the I-Recursion Theorem there is an $\bar{e}$ such that

$$\{\bar{e}\}^{\mathsf{I}}(a, \langle\mathbf{m}\rangle, \langle\boldsymbol{\alpha}\rangle) \simeq \mathsf{F}(\bar{e}, a, \langle\mathbf{m}\rangle, \langle\boldsymbol{\alpha}\rangle),$$

and we set $\mathrm{Sel}^{\mathsf{I}} = \{\bar{e}\}^{\mathsf{I}}$.

Let $p_{a,\mathbf{m},\boldsymbol{\alpha}}$ denote the least p such that $\{a\}^{\mathsf{I}}(p, \mathbf{m}, \boldsymbol{\alpha})$ is defined if there is such a p. We prove the implication (i)$\rightarrow$(ii) by induction on $p_{a,\mathbf{m},\boldsymbol{\alpha}}$; that (ii)$\rightarrow$(i) is evident.

Suppose first that $p_{a,\mathbf{m},\boldsymbol{\alpha}} = 0$ — that is, $\{a\}^{\mathsf{I}}(0, \mathbf{m}, \boldsymbol{\alpha})$ is defined. Then $\langle a, 0, \mathbf{m}\rangle \in U^{\mathsf{I}}_{\boldsymbol{\alpha}}$, so for $n = 0$ or 1,

$$\mathsf{H}(\langle a, 0, \mathbf{m}\rangle, \langle\bar{e}, a^+, \langle\mathbf{m}\rangle\rangle, \langle\boldsymbol{\alpha}\rangle) \simeq n.$$

If $n = 0$, then

$$\mathrm{Sel}^{\mathrm{I}}(a, \langle \mathbf{m} \rangle, \langle \boldsymbol{\alpha} \rangle) \simeq 0$$

as is appropriate. If $n = 1$, then $|\bar{e}, a^+, \langle \mathbf{m} \rangle| < |a, 0, \mathbf{m}|$ and thus for some q, $\{\bar{e}\}^{\mathrm{I}}(a^+, \langle \mathbf{m} \rangle, \langle \boldsymbol{\alpha} \rangle) \simeq q$. Again because $\langle a, 0, \mathbf{m} \rangle \in U^{\mathrm{I}}_{\boldsymbol{\alpha}}$, we have that for $r = 0$ or 1,

$$\mathsf{H}(\langle a, 0, \mathbf{m} \rangle, \langle a, q+1, \mathbf{m} \rangle, \langle \boldsymbol{\alpha} \rangle) \simeq r.$$

If $r = 0$, we have again $\mathrm{Sel}^{\mathrm{I}}(a, \langle \mathbf{m} \rangle, \langle \boldsymbol{\alpha} \rangle) \simeq 0$ as is appropriate. If $r = 1$, it follows that $|a, q+1, \mathbf{m}| < |a, 0, \mathbf{m}|$ and thus that $\{a\}^{\mathrm{I}}(q+1, \mathbf{m}, \boldsymbol{\alpha})$ is defined. Furthermore, in this case

$$\begin{aligned} \mathrm{Sel}^{\mathrm{I}}(a, \langle \mathbf{m} \rangle, \langle \boldsymbol{\alpha} \rangle) &\simeq \{\bar{e}\}^{\mathrm{I}}(a^+, \langle \mathbf{m} \rangle, \langle \boldsymbol{\alpha} \rangle) + 1 \\ &\simeq q + 1 \end{aligned}$$

which is thus a suitable value. Note that in this case it is *not* the least p which is selected.

Now suppose that $p_{a,\mathbf{m},\boldsymbol{\alpha}} > 0$. By definition, $p_{a,\mathbf{m},\boldsymbol{\alpha}} = p_{a^+,\mathbf{m},\boldsymbol{\alpha}} + 1$, so by the induction hypothesis there exists a q such that

$$\{\bar{e}\}^{\mathrm{I}}(a^+, \langle \mathbf{m} \rangle, \langle \boldsymbol{\alpha} \rangle) \simeq \mathrm{Sel}^{\mathrm{I}}(a^+, \langle \mathbf{m} \rangle, \langle \boldsymbol{\alpha} \rangle) \simeq q$$

and

$$\{a\}^{\mathrm{I}}(q+1, \mathbf{m}, \boldsymbol{\alpha}) \simeq \{a^+\}^{\mathrm{I}}(q, \mathbf{m}, \boldsymbol{\alpha}),$$

which is defined. In particular, $\langle \bar{e}, a^+, \langle \mathbf{m} \rangle \rangle \in U^{\mathrm{I}}_{\boldsymbol{\alpha}}$. By the assumption, $\langle a, 0, \mathbf{m} \rangle \notin U^{\mathrm{I}}_{\boldsymbol{\alpha}}$, so $|\bar{e}, a^+, \langle \mathbf{m} \rangle| < |a, 0, \mathbf{m}|$ and thus

$$\mathsf{H}(\langle a, 0, \mathbf{m} \rangle, \langle \bar{e}, a^+, \langle \mathbf{m} \rangle \rangle, \langle \boldsymbol{\alpha} \rangle) \simeq 1.$$

Similarly, $|a, q+1, \mathbf{m}| < |a, 0, \mathbf{m}|$, so

$$\mathsf{H}(\langle a, 0, \mathbf{m} \rangle, \langle a, q+1, \mathbf{m} \rangle, \langle \boldsymbol{\alpha} \rangle) \simeq 1.$$

Then

$$\begin{aligned} \mathrm{Sel}^{\mathrm{I}}(a, \langle \mathbf{m} \rangle, \langle \boldsymbol{\alpha} \rangle) &\simeq \{\bar{e}\}^{\mathrm{I}}(a^+, \langle \mathbf{m} \rangle, \langle \boldsymbol{\alpha} \rangle) + 1 \\ &\simeq q + 1, \end{aligned}$$

which is a suitable value. □

The proofs of the following corollaries are exactly the same as the proofs of the corresponding results in § II.4.

4.2 Corollary. *For any relation* R *semi-recursive in* I, *there exists a functional* $\mathrm{Sel}_{\mathsf{R}}$ *partial recursive in* I *such that for all* $\mathbf{m}$ *and* $\boldsymbol{\alpha}$,

$$\exists p \,.\, \mathsf{R}(p, \mathbf{m}, \boldsymbol{\alpha}) \leftrightarrow \mathsf{R}(\mathrm{Sel}_{\mathsf{R}}(\mathbf{m}, \boldsymbol{\alpha}), \mathbf{m}, \boldsymbol{\alpha}). \quad \square$$

4.3 Corollary. *The class of relations semi-recursive in* I *is closed under finite union and existential number quantification* ($\exists^0$), *hence also under bounded existential quantification* ($\exists^0_<$). $\square$

4.4 Corollary. *The class of functionals partial recursive in* I *and relations semi-recursive in* I *is closed under definition by positive cases.* $\square$

4.5 Corollary. *A relation is recursive in* I *iff it is both semi-recursive in* I *and co-semi-recursive in* I. $\square$

4.6 Corollary. *For any partial functional* F, F *is partial recursive in* I *iff* Gr_{F} *is semi-recursive in* I *and* F *is recursive in* I *iff* F *is total and* Gr_{F} *is recursive in* I. $\square$

In § V.1 we defined the pre-wellordering property only for the classes Σ^1_r and Π^1_r, but essentially the same definition applies to any indexable class of relations:

4.7 Definition. For any class X of relations, X has the *pre-wellordering property* iff there exists a relation V universal for X and relations $\preccurlyeq$, $\preccurlyeq_+$, and $\preccurlyeq_-$ such that:

(i) $\preccurlyeq$ is a pre-wellordering with field ${}^{2,1}\omega$ such that for all x and y,
  (a) $\sim\mathsf{V}(y) \rightarrow x \preccurlyeq y$, and
  (b) $\mathsf{V}(y) \wedge x \preccurlyeq y \rightarrow \mathsf{V}(x)$;
(ii) $\preccurlyeq_+$ belongs to X and $\preccurlyeq_-$ is the complement of a member of X;
(iii) for any x and y such that either $\mathsf{V}(x)$ or $\mathsf{V}(y)$,

$$(x \preccurlyeq_+ y) \leftrightarrow (x \preccurlyeq y) \leftrightarrow (x \preccurlyeq_- y).$$

4.8 Theorem. *The class of relations semi-recursive in* I *has the pre-wellordering property.*

Proof. With H as in Theorem 3.4, we set $\mathsf{V} = \mathsf{U}^{\mathsf{I}}$;

$$x \preccurlyeq y \leftrightarrow |x|^{\mathsf{I}} \leqslant |y|^{\mathsf{I}};$$

$$x \preccurlyeq_+ y \leftrightarrow \mathsf{U}^{\mathsf{I}}(x) \wedge \mathsf{H}(x, y) \simeq 0;$$

$$x \preccurlyeq_- y \leftrightarrow \sim\mathsf{U}^{\mathsf{I}}(y) \vee \mathsf{H}(x, y) \not\simeq 1.$$

Conditions (i) and (ii) of Definition 4.7 are immediate, and condition (iii) is easy to verify from the properties of H expressed in Theorem 3.4. □

Although we cannot now directly apply the results of § V.1, as they are stated only for Σ_r^1 and Π_r^1, in many cases the proofs depend only on properties shared by the class of relations semi-recursive in I. In particular, the classes of relations semi-recursive in I and recursive in I are closed under both kinds of number quantification ($\exists^0$ and $\forall^0$) (we continue to assume throughout that E is recursive in I). Of course, for E itself, the results are just those of §§ IV.1–2.

4.9 Theorem. (i) *The class of relations semi-recursive in* I *has the reduction property but not the separation property*;

(ii) *the class of relations co-semi-recursive in* I *has the separation property but not the reduction property.*

Proof. Exactly as for Theorem V.1.4. This could also be proved directly from the Selection Theorem as in Exercise II.4.33. □

As in § V.1, let $|\ \ |_0^{\mathsf{I}}$ be the norm induced on $\omega \times \omega$ by the restriction of the pre-wellordering $\preccurlyeq$ to sequences of type $(a, \langle \mathbf{m}\rangle, \langle\ \ \rangle)$, so that

$$|a, \langle \mathbf{m}\rangle|_0^{\mathsf{I}} = \sup{}^+\{|b, \langle \mathbf{n}\rangle|_0^{\mathsf{I}} : (b, \langle \mathbf{n}\rangle, \langle\ \ \rangle) \prec (a, \langle \mathbf{m}\rangle, \langle\ \ \rangle)\},$$

and set

$$\kappa^{\mathsf{I}} = \sup{}^+\{|a, \langle \mathbf{m}\rangle|_0^{\mathsf{I}} : U^{\mathsf{I}}(a, \langle \mathbf{m}\rangle)\}.$$

(In fact, $|a, \langle \mathbf{m}\rangle|_0^{\mathsf{I}} = |a, \langle \mathbf{m}\rangle, \langle\ \ \rangle|^{\mathsf{I}}$ (Exercise 4.27)).

4.10 Boundedness Theorem. *For any set A co-semi-recursive in* I, *if* $A \subseteq U^{\mathsf{I}}$, *then* $\sup^+\{|u|^{\mathsf{I}} : u \in A\} < \kappa^{\mathsf{I}}$.

Proof. Similar to that of Theorem V.1.5. □

For any ρ, let $U_\rho^{\mathsf{I}} = \{u : u \in U^{\mathsf{I}} \wedge |u|^{\mathsf{I}} < \rho\}$.

4.11 Hierarchy Theorem. *For all relations R on numbers,*

$$R \text{ is recursive in } \mathsf{I} \leftrightarrow R \preccurlyeq U_\rho^{\mathsf{I}} \text{ for some } \rho < \kappa^{\mathsf{I}}$$
$$\leftrightarrow R = \{\mathbf{m} : \langle a, \mathbf{m}\rangle \in U_\rho^{\mathsf{I}}\} \text{ for some } \rho < \kappa^{\mathsf{I}} \text{ and } a \in \omega.$$

Proof. As for Theorem V.1.6. □

4.12. Upper Classification Theorem. *$\{\alpha : \alpha$ is recursive in $\mathsf{I}\}$ is semi-recursive in I.*

Proof. As for Theorem V.1.8 using the second part of Corollary 4.6. A different proof is obtained from the equivalence

$$\alpha \text{ is recursive in } \mathsf{I} \leftrightarrow \exists a \, \forall mn \, [\alpha(m) = n \rightarrow \{a\}^{\mathsf{I}}(m) \simeq n]. \quad \square$$

4.13. Corollary. *The set of functions recursive in I is not a basis for the class of sets co-semi-recursive in I.* $\square$

There are, of course, some properties of the class of semi-recursive relations which are not shared by the class of relations semi-recursive in I. For example it is not true that every relation semi-recursive in I is of the form $\exists^0\mathsf{P}$ with P recursive in I, as every such relation is itself recursive in I (remember, we are assuming that E is recursive in I). All of the equivalences of Theorem II.4.15 fail if we replace (semi-, partial) recursive by (semi-, partial) recursive in I. If a function $f : {}^k\omega \rightarrow \omega$ is recursive in I, then so is $\mathrm{Im}(f)$. On the other hand, from the assumption that a functional F is recursive in I we can conclude only that $\mathrm{Im}(\mathsf{F})$ is a non-empty set of the form $\exists^1\mathsf{P}$ for P a relation recursive in I:

$$n \in \mathrm{Im}(\mathsf{F}) \leftrightarrow \exists \boldsymbol{\alpha} \, [\exists \mathbf{m} \, . \, \mathsf{F}(\mathbf{m}, \boldsymbol{\alpha}) \simeq n].$$

This estimate cannot be improved, for if A is any non-empty set of the form $\exists^1\mathsf{P}$ with P recursive in I, then $A = \mathrm{Im}(\mathsf{F})$, where

$$\mathsf{F}(m, \alpha) = \begin{cases} m, & \text{if } \mathsf{P}(m, \alpha); \\ \bar{m}, & \text{otherwise}; \end{cases}$$

where $\bar{m}$ is some fixed element of A. Thus, for example, we have by Theorem 1.8, for any $A \neq \varnothing$,

$$A \in \Sigma^1_1 \leftrightarrow A = \mathrm{Im}(\mathsf{F}) \text{ for some functional } \mathsf{F} \text{ recursive in } \mathsf{E}.$$

By a similar argument we see that the images of functionals partial recursive in I are exactly those of the form $\exists^1\mathsf{P}$ with P semi-recursive in I. Hence by Corollary 1.10,

$$A \in \Sigma^1_2 \leftrightarrow A = \mathrm{Im}(\mathsf{F}) \text{ for some function } \mathsf{F} \text{ partial recursive in } \mathsf{E}.$$

One result of this sort remains:

4.14 Theorem. *For any $A \subseteq \omega$, A is semi-recursive in I iff $A = \mathrm{Im}(f)$ for some function f partial recursive in I.*

Proof. If A is semi-recursive in I and

$$f(m) \simeq \begin{cases} m, & \text{if} \quad m \in A; \\ \text{undefined}, & \text{otherwise}; \end{cases}$$

then f is partial recursive in I by Corollary 4.4 and $\mathrm{Im}(f) = A$. Conversely, if f is partial recursive in I, then

$$n \in \mathrm{Im}(f) \leftrightarrow \exists m \,.\, f(m) \simeq n,$$

so $\mathrm{Im}(f)$ is semi-recursive in I by Corollaries 4.3 and 4.6. □

The astute reader will also have noticed that there is no result here corresponding to the Lower Classification Theorem V.1.13. Instead, we have

4.15 Theorem. *There exists a functional* I *such that* E *is recursive in* I *and* $\{\alpha : \alpha$ *is recursive in* $\mathsf{I}\}$ *is recursive in* I.

Proof. Let I be the functional defined by:

$$\mathsf{I}(\alpha) = \begin{cases} \mathsf{E}(\alpha), & \text{if} \quad \alpha \in \Delta_1^1; \\ \mathsf{E}(\alpha) + 2, & \text{otherwise}. \end{cases}$$

That E is recursive in I is evident from the fact that

$$\mathsf{E}(\alpha) = 0 \quad \text{iff} \quad (\mathsf{I}(\alpha) = 0 \text{ or } \mathsf{I}(\alpha) = 2).$$

For all α recursive in E (i.e. in Δ_1^1), $\mathsf{E}(\alpha) = \mathsf{I}(\alpha)$, so by the result of Exercise 1.13

$$\{\alpha : \alpha \text{ is recursive in } \mathsf{I}\} = \{\alpha : \alpha \text{ is recursive in } \mathsf{E}\}.$$

Since $\alpha \in \Delta_1^1$ iff $\mathsf{I}(\alpha) \leq 1$, this set is recursive in I. □

Corresponding to the result of Exercises IV.2.25 and V.1.27, we have the following *effective choice principle*:

4.16 Theorem. *For any relation* R *semi-recursive in* I,

(i) *if* $\forall \mathbf{m} \forall \boldsymbol{\alpha} \exists p\, \mathsf{R}(p, \mathbf{m}, \boldsymbol{\alpha})$, *then there is an* F *recursive in* I *such that* $\forall \mathbf{m} \forall \boldsymbol{\alpha}\, \mathsf{R}(\mathsf{F}(\mathbf{m}, \boldsymbol{\alpha}), \mathbf{m}, \boldsymbol{\alpha})$;

(ii) *if* $\forall \mathbf{m} \forall \boldsymbol{\alpha}\, (\exists \beta$ recursive in I and $\boldsymbol{\alpha})\, \mathsf{R}(\mathbf{m}, \boldsymbol{\alpha}, \beta)$, *then there is a* G *recursive in* I *such that* $\forall \mathbf{m} \forall \boldsymbol{\alpha}\, \mathsf{R}(\mathbf{m}, \boldsymbol{\alpha}, \lambda q \,.\, \mathsf{G}(q, \mathbf{m}, \boldsymbol{\alpha}))$.

Proof. For (i) we take simply $\mathsf{F} = \mathrm{Sel}_{\mathsf{R}}$. F is partial recursive in I and is total by the hypothesis, so F is recursive in I. For (ii), let

$$\mathsf{S}(a, \mathbf{m}, \boldsymbol{\alpha}) \leftrightarrow \mathsf{R}(\mathbf{m}, \boldsymbol{\alpha}, \lambda q . \{a\}^{\mathsf{I}}(q, \mathbf{m}, \boldsymbol{\alpha})).$$

S is semi-recursive in I by Theorem 2.10 and we may set

$$\mathsf{G}(q, \mathbf{m}, \boldsymbol{\alpha}) \simeq \{\mathrm{Sel}_{\mathsf{S}}(\mathbf{m}, \boldsymbol{\alpha})\}^{\mathsf{I}}(q, \mathbf{m}, \boldsymbol{\alpha}).$$

We turn next to characterizing the ordinal κ^{I} (analogously to Corollary V.1.17). Let $\omega_1[\mathsf{I}]$ denote the least ordinal not the order-type of a well-ordering of ω recursive in I.

4.17 Theorem. $\kappa^{\mathsf{I}} = \omega_1[\mathsf{I}]$.

Proof. To establish $\kappa^{\mathsf{I}} \leq \omega_1[\mathsf{I}]$, we show that to every $u \in U^{\mathsf{I}}$ there corresponds a well-ordering of type $|u|^{\mathsf{I}}$ recursive in I. First note that for each $u \in U^{\mathsf{I}}$, the sets

$$\{v : v \in U^{\mathsf{I}} \wedge |v|^{\mathsf{I}} < |u|^{\mathsf{I}}\} \quad \text{and} \quad \{v : v \in U^{\mathsf{I}} \wedge |v|^{\mathsf{I}} = |u|^{\mathsf{I}}\}$$

are recursive in I (use the pre-wellordering property). For each $u \in U^{\mathsf{I}}$ the first of these sets is pre-wellordered in type $|u|^{\mathsf{I}}$ by a relation which is recursive in I. We obtain a *well*-ordering of the same type by a refinement. Let

$$A_u = \{v : v \in U^{\mathsf{I}} \wedge |v|^{\mathsf{I}} < |u|^{\mathsf{I}} \wedge \forall w\, (|w|^{\mathsf{I}} = |v|^{\mathsf{I}} \rightarrow v \leq w)\}.$$

A_u is recursive in I and contains a unique notation for each ordinal less than $|u|^{\mathsf{I}}$. Hence if

$$R_u(v, w) \leftrightarrow v,\ w \in A_u \wedge v \preccurlyeq_+ w,$$

then R_u is the required well-ordering.

For the converse, let γ be any element of W recursive in I. We recall that $|p|_\gamma$ denotes the ordinal $\|\gamma \upharpoonright p\|$. Let $\bar{p}$ be such that $|\bar{p}|_\gamma = 0$. We shall find an index $\bar{e}$ such that for all q, $\langle \bar{e}, q\rangle \in U^{\mathsf{I}}$ and $|\bar{e}, q|^{\mathsf{I}} \geq |q|_\gamma$. From this it follows that $\kappa^{\mathsf{I}} \geq \|\gamma\|$. As $\|\gamma\|$ is an arbitrary ordinal less than $\omega_1[\mathsf{I}]$, the result follows.

Let f and g be functions partial recursive in I defined as follows:

$$f(p, q) = \begin{cases} p, & \text{if } p <_\gamma q; \\ \bar{p}, & \text{otherwise}; \end{cases}$$

$$g(e, q) = \begin{cases} 0, & \text{if } q = \bar{p}; \\ \mathsf{I}(\lambda p . \{e\}^{\mathsf{I}}(f(p, q))), & \text{otherwise}. \end{cases}$$

By the I-Recursion Theorem there exists an index $\bar{e}$ such that $g(\bar{e},q)\simeq\{\bar{e}\}^{\mathsf{I}}(q)$. We prove by induction on $|q|_\gamma$ that $\langle\bar{e},q\rangle\in U^{\mathsf{I}}$ and $|\bar{e},q|^{\mathsf{I}}\geq|q|_\gamma$.

If $|q|_\gamma=0$, then $q=\bar{p}$ so

$$\{\bar{e}\}(q)\simeq g(\bar{e},\bar{p})\simeq 0$$

and thus $\langle\bar{e},q\rangle\in U^{\mathsf{I}}$ and $|\bar{e},q|\geq 0=|q|_\gamma$.

Now suppose $|q|_\gamma>0$. Note that for all p, $|f(p,q)|_\gamma<|q|_\gamma$ so by the induction hypothesis, $\langle\bar{e},f(p,q)\rangle\in U^{\mathsf{I}}$. But then $\lambda p.\{\bar{e}\}^{\mathsf{I}}(f(p,q))$ is total so $g(\bar{e},q)$ is defined and thus $\langle\bar{e},q\rangle\in U^{\mathsf{I}}$. Let b be a natural index for g from I — that is, one such that

$$(\bar{e},f(p,q),\{\bar{e}\}^{\mathsf{I}}(f(p,q)))\in \mathrm{Sbc}(b,\bar{e},q,g(\bar{e},q))$$

and thus

$$|\bar{e},f(p,q)|^{\mathsf{I}}<|b,\bar{e},q|^{\mathsf{I}}.$$

From the proof of the Recursion Theorem it is easy to verify (Exercise 2.20) that we may assume that $\bar{e}$ is chosen in such a way that

$$(b,\bar{e},q,g(\bar{e},q))\in \mathrm{Sbc}(\bar{e},q,g(\bar{e},q))$$

and thus

$$|b,\bar{e},q|^{\mathsf{I}}<|\bar{e},q|^{\mathsf{I}}.$$

By the induction hypothesis we have for all $p<_\gamma q$,

$$|p|_\gamma\leq|\bar{e},p|^{\mathsf{I}}=|\bar{e},f(p,q)|^{\mathsf{I}}<|\bar{e},q|^{\mathsf{I}},$$

from which it follows that $|q|_\gamma\leq|\bar{e},q|^{\mathsf{I}}$. □

Of course, this, as well as most of the other results of this section, may be relativized to yield theorems about the class of relations semi-recursive in I and $\boldsymbol{\alpha}$. In particular, we obtain from the preceding theorem that $\kappa^{\mathsf{I}}_{\boldsymbol{\alpha}}=\omega_1[\mathsf{I},\boldsymbol{\alpha}]$ — that is, $\sup^+\{|u|^{\mathsf{I}}:u\in U^{\mathsf{I}}_{\boldsymbol{\alpha}}\}$ is the least ordinal not recursive in I and $\boldsymbol{\alpha}$. From this it follows that

$$\sup^+\{|x|^{\mathsf{I}}:x\in \mathsf{U}^{\mathsf{I}}\}=\aleph_1.$$

In other words, the inductive definition of $\Omega[\mathsf{I}]$ has exactly $\aleph_1$ stages.

We conclude this section by establishing the analogue for recursion in I of the

Spector–Gandy Theorem (IV.2.9). We denote by $\Sigma_1^{1,\mathsf{I}}$ the class of relations R such that for some relation P recursive in I,

$$\mathsf{R}(\mathbf{m}, \boldsymbol{\alpha}) \leftrightarrow (\exists \beta \text{ recursive in } \mathsf{I}, \boldsymbol{\alpha})\, \mathsf{P}(\mathbf{m}, \boldsymbol{\alpha}, \beta).$$

4.18 Theorem. $\Sigma_1^{1,\mathsf{I}} = \{\mathsf{R} : \mathsf{R} \text{ is semi-recursive in } \mathsf{I}\}$.

Proof. If $\mathsf{R} \in \Sigma_1^{1,\mathsf{I}}$ and satisfies the preceding equivalence for some relation P recursive in I, then

$$\mathsf{R}(\mathbf{m}, \boldsymbol{\alpha}) \leftrightarrow \exists a\, \mathsf{P}(\mathbf{m}, \boldsymbol{\alpha}, \lambda p . \{a\}^{\mathsf{I}}(p, \boldsymbol{\alpha}))$$

and R is semi-recursive in I by Corollaries 2.11 and 4.3.

We shall prove the converse inclusion only for relations on numbers. For this it will clearly suffice to show $U^{\mathsf{I}} \in \Sigma_1^{1,\mathsf{I}}$. For any v, let

$$\beta_v(u^0, u^1) = \begin{cases} 0, & \text{if } |u^0| \leqslant |u^1| \text{ and } |u^0| < |v|; \\ 1, & \text{if } |u^1| < |u^0| \text{ and } |u^1| < |v|; \\ 2, & \text{otherwise}; \end{cases}$$

$$\gamma_v(\langle a, \mathbf{m} \rangle) = \begin{cases} \{a\}^{\mathsf{I}}(\mathbf{m}), & \text{if } |a, \mathbf{m}| < |v|; \\ 0, & \text{otherwise}. \end{cases}$$

We aim to define a relation P recursive in I such that for any v,

(i) if $v \in U^{\mathsf{I}}$, then $\mathsf{P}(v, \beta_v, \gamma_v)$ holds, and

(ii) if $v \notin U^{\mathsf{I}}$, then for any β and γ such that $\mathsf{P}(v, \beta, \gamma)$, and any $\langle a, \mathbf{m} \rangle \in U^{\mathsf{I}}$, $\{a\}^{\mathsf{I}}(\mathbf{m}) \simeq \gamma(\langle a, \mathbf{m} \rangle)$.

From (i) and (ii) we can conclude that for all v,

(iii) $v \in U^{\mathsf{I}} \leftrightarrow (\exists \beta, \gamma \text{ recursive in } \mathsf{I}) \mathsf{P}(v, \beta, \gamma)$

as follows. For the implication ($\rightarrow$) it suffices to show that for all $v \in U^{\mathsf{I}}$, β_v and γ_v are recursive in I. Using the ordinal comparison functional H of Theorem 3.3, $\beta_v(u^0, u^1)$ may be computed according to the following flow diagram:

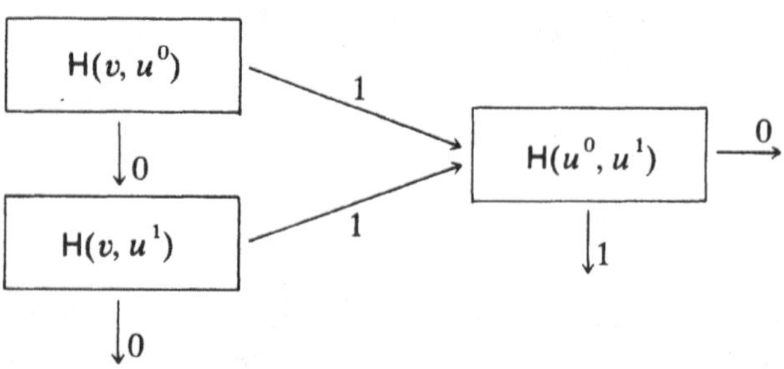

Since $v \in U^{\mathsf{I}}$, the values in the left-hand boxes are defined, and if the right-hand box is reached, also one of u^0 and u^1 belongs to U^{I}. Hence β_v is recursive in I. Then

$$\gamma_v(\langle a, \mathbf{m}\rangle) = \begin{cases} \{a\}^{\mathsf{I}}(\mathbf{m}), & \text{if} \quad \beta_v(\langle a, \mathbf{m}\rangle, \langle a, \mathbf{m}\rangle) = 0; \\ 0, & \text{otherwise}; \end{cases}$$

so γ_v is also recursive in I.

For the implication ($\leftarrow$) of (iii), suppose $v \notin U^{\mathsf{I}}$ but that β and γ are functions recursive in I such that $\mathsf{P}(v, \beta, \gamma)$ holds. Then there exists an index a such that for all m, $\{a\}^{\mathsf{I}}(m) = \gamma(\langle m, m\rangle) + 1$. In particular, $\langle a, a\rangle \in U^{\mathsf{I}}$ so by (ii),

$$\gamma(\langle a, a\rangle) = \{a\}^{\mathsf{I}}(a) = \gamma(\langle a, a\rangle) + 1,$$

a contradiction.

To define P and establish (i) and (ii) we return to the proof of Theorem 3.3. For $\mathsf{P}(v, \beta, \gamma)$ to hold will require, roughly speaking, that β satisfy the recursive conditions imposed on H in that proof and that γ satisfy the recursive conditions inherent in the definition of $\Omega[\mathsf{I}]$, in each case only for arguments which precede v in the pre-wellordering of U^{I}. In order that P be recursive in I and not merely semi-recursive in I, we replace occurrences of $\{c\}^{\mathsf{I}}(\mathbf{m})$ in the recursive conditions on H by $\gamma(\langle c, \mathbf{m}\rangle)$.

First, for $0 \leq r, s \leq 4$ we define relations $\mathsf{Q}_{r,s}$ as follows. $\mathsf{Q}_{r,s}(u^0, u^1, \beta, \gamma)$ holds only if (u^0, u^1) falls under case (r, s), with "4" signifying "not of the proper form". Then $\mathsf{Q}_{r,s}(u^0, u^1, \beta, \gamma)$ is to hold when, with β replacing H, the answer 0 is obtained from the flow diagram for the corresponding case of Theorem 3.3. Thus, we set:

$$\mathsf{Q}_{r,4}(u^0, u^1, \beta, \gamma) \leftrightarrow 0 = 0 \qquad (0 \leq r \leq 4);$$

$$\mathsf{Q}_{4,s}(u^0, u^1, \beta, \gamma) \leftrightarrow 0 = 1 \qquad (0 \leq s \leq 3);$$

$$\mathsf{Q}_{0,s}(u^0, u^1, \beta, \gamma) \leftrightarrow 0 = 0 \qquad (0 \leq s \leq 3);$$

$$\mathsf{Q}_{r,0}(u^0, u^1, \beta, \gamma) \leftrightarrow 0 = 1 \qquad (1 \leq r \leq 3);$$

$$\mathsf{Q}_{1,2}(u^0, u^1, \beta, \gamma) \leftrightarrow \beta(\langle c^0, \mathbf{m}^0\rangle, \langle b^1, \mathbf{m}^1\rangle) = 0 \quad \text{and} \quad \beta(\langle b^0, q^0\rangle, \langle b^1, \mathbf{m}^1\rangle) = 0,$$

where $q^0 = \gamma(\langle c^0, \mathbf{m}^0\rangle)$;

$$\mathsf{Q}_{2,1}(u^0, u^1, \beta, \gamma) \leftrightarrow \beta(\langle b^0, \mathbf{m}^0\rangle, \langle c^1, \mathbf{m}^1\rangle) = 0 \quad \text{or} \quad [\beta(\langle b^0, \mathbf{m}^0\rangle, \langle c^1, \mathbf{m}^1\rangle) = 1 \text{ and } \beta(\langle b^0, \mathbf{m}^0\rangle, \langle b^1, q^1\rangle) = 0],$$

where $q^1 = \gamma(\langle c^1, \mathbf{m}^1\rangle)$;

$$Q_{2,3}(u^0, u^1, \beta, \gamma) \leftrightarrow \exists q\,[\beta(\langle b^0, \mathbf{m}^0\rangle, \langle b^1, q, \mathbf{m}^1\rangle) = 0];$$
$$Q_{3,3}(u^0, u^1, \beta, \gamma) \leftrightarrow \forall p\,\exists q\,[\beta(\langle b^0, p, \mathbf{m}^0\rangle, \langle b^1, q, \mathbf{m}^1\rangle) = 0].$$

The relations $Q_{r,s}$ for the remaining cases are obtained similarly from the corresponding cases in the proof of Theorem 3.3. Let

$$Q(u^0, u^1, \beta, \gamma) \leftrightarrow (\exists r \leq 4)(\exists s \leq 4) Q_{r,s}(u^0, u^1, \beta, \gamma).$$

The conditions on γ are expressed by a relation S which may be thought of as saying "γ is *locally correct for* β":

$$S(\beta, \gamma) \leftrightarrow \text{for all } u, k, \quad \text{and} \quad m \in {}^k\omega, \quad \text{if} \quad \beta(u, u) = 0,$$

then the following hold:

(0) if $u = \langle\langle 0, k, 0, \ldots\rangle, \mathbf{m}\rangle$ and $\langle 0, k, 0, \ldots\rangle$ is an index of the proper form for $\mathbf{m}$, then $\gamma(u) = \{\langle 0, k, 0, \ldots\rangle\}^{\mathsf{I}}(\mathbf{m})$;

(1) if for some $b, c_0, \ldots, c_{k'-1}$, $u = \langle\langle 1, k, 0, b, c_0, \ldots, c_{k'-1}\rangle, \mathbf{m}\rangle$, then $\gamma(u) = \gamma(\langle b, \gamma(\langle c_0, \mathbf{m}\rangle), \ldots, \gamma(\langle c_{k'-1}, \mathbf{m}\rangle)\rangle)$;

(2) if for some b, $u = \langle\langle 2, k+1, 0\rangle, b, \mathbf{m}\rangle$, then $\gamma(u) = \gamma(\langle b, \mathbf{m}\rangle)$;

(3) if for some b, $u = \langle\langle 3, k, 0, b\rangle, \mathbf{m}\rangle$, then $\gamma(u) = \mathsf{I}(\lambda p\,.\,\gamma(\langle b, p, \mathbf{m}\rangle))$.

Finally, we set

$P(v, \beta, \gamma) \leftrightarrow S(\beta, \gamma)$ and for all u^0 and u^1,

$$[\beta(u^0, u^1) = 0 \leftrightarrow Q(u^0, u^1, \beta, \gamma) \wedge \sim Q(v, u^0, \beta, \gamma)]$$

and

$$[\beta(u^0, u^1) = 1 \leftrightarrow \sim Q(u^0, u^1, \beta, \gamma) \wedge \sim Q(v, u^1, \beta, \gamma)].$$

Since Q and S are each recursive in I, so is P.

Towards (i), suppose $v \in U^{\mathsf{I}}$. Then it is obvious from the definitions that $S(\beta_v, \gamma_v)$ holds. We need thus only show that for all u^0 and u^1,

$$\text{(iv)} \qquad \beta_v(u^0, u^1) = 0 \leftrightarrow Q(u^0, u^1, \beta_v, \gamma_v) \wedge \sim Q(v, u^0, \beta_v, \gamma_v),$$

and

$$\text{(v)} \qquad \beta_v(u^0, u^1) = 1 \leftrightarrow \sim Q(u^0, u^1, \beta_v, \gamma_v) \wedge \sim Q(v, u^1, \beta_v, \gamma_v).$$

These will, in turn, be derived from:

(vi) $\min\{|u^0|,|u^1|\}\leqslant|v|\to[|u^0|\leqslant|u^1|\leftrightarrow \mathsf{Q}(u^0,u^1,\beta_v,\gamma_v)]$.

The proof of (vi) is by induction on $\sigma=\min\{|u^0|,|u^1|\}$ and is closely parallel to the corresponding part of the proof of Theorem 3.3. We examine only one case to indicate the relationship between the proofs. Suppose $u^0=\langle\langle 1,k^0,0,b^0,c^0\rangle,\mathbf{m}^0\rangle$ and $u^1=\langle\langle 2,k'+1,0\rangle,b^1,\mathbf{m}^1\rangle$, $\min\{|u^0|,|u^1|\}\leqslant|v|$, and first that $|u^0|\leqslant|u^1|$. Then $|u^0|\leqslant|v|$, for some q^0, $\{c^0\}^{\mathsf{l}}(\mathbf{m}^0)\simeq q^0$, and

$$|c^0,\mathbf{m}^0|<|u^0|\leqslant|v| \quad\text{and}\quad |b^0,q^0|<|u^0|\leqslant|v|.$$

By Lemma 3.2,

$$|c^0,\mathbf{m}^0|\leqslant|b^1,\mathbf{m}^1| \quad\text{and}\quad |b^0,q^0|\leqslant|b^1,\mathbf{m}^1|,$$

and thus

$$\beta_v(\langle c^0,\mathbf{m}^0\rangle,\langle b^1,\mathbf{m}^1\rangle)=0=\beta_v(\langle b^0,q^0\rangle,\langle b^1,\mathbf{m}^1\rangle).$$

Furthermore, since $|c^0,\mathbf{m}^0|<|v|$, $\gamma_v(\langle c^0,\mathbf{m}^0\rangle)=q^0$ and we have $\mathsf{Q}_{1,2}(u^0,u^1,\beta_v,\gamma_v)$ and hence $\mathsf{Q}(u^0,u^1,\beta_v,\gamma_v)$. If, on the other hand, $|u^1|<|u^0|$, then $|b^1,\mathbf{m}^1|<|u^1|\leqslant|v|$. If $|b^1,\mathbf{m}^1|<|c^0,\mathbf{m}^0|$, then $\beta_v(\langle c^0,\mathbf{m}^0\rangle,\langle b^1,\mathbf{m}^1\rangle)=1$ which implies $\sim\mathsf{Q}(u^0,u^1,\beta_v,\gamma_v)$. Otherwise, $|c^0,\mathbf{m}^0|\leqslant|b^1,\mathbf{m}^1|$, and thus $\beta_v(\langle c^0,\mathbf{m}^0\rangle,\langle b^1,\mathbf{m}^1\rangle)=0$. But then $|c^0,\mathbf{m}^0|<|v|$, so $\{c^0\}^{\mathsf{l}}(\mathbf{m}^0)\simeq\gamma_v(\langle c^0,\mathbf{m}^0\rangle)\simeq q^0$ (say) and necessarily $|b^1,\mathbf{m}^1|<|b^0,q^0|$ (by the assumption $|u^1|<|u^0|$), so $\beta_v(\langle b^0,q^0\rangle,\langle b^1,\mathbf{m}^1\rangle)=1$ and again we have $\sim\mathsf{Q}(u^0,u^1,\beta_v,\gamma_v)$.

Now to derive (iv) from (vi), suppose first that $\beta_v(u^0,u^1)=0$ so that $|u^0|\leqslant|u^1|$ and $|u^0|<|v|$. Since $\min\{|u^0|,|u^1|\}=\min\{|u^0|,|v|\}=|u^0|<|v|$, (vi) yields immediately the right-hand side of (iv). Conversely, since $\min\{|v|,|u^0|\}\leqslant|v|$, if the right-hand side of (iv) holds, then from (vi) and $\sim\mathsf{Q}(v,u^0,\beta_v,\gamma_v)$ we conclude that $|u^0|<|v|$. Then also $\min\{|u^0|,|u^1|\}\leqslant|u^0|<|v|$, so from (vi) and $\mathsf{Q}(u^0,u^1,\beta_v,\gamma_v)$ we conclude that $|u^0|\leqslant|u^1|$ and thus that $\beta_v(u^0,u^1)=0$. The proof of (v) from (vi) is similar. This concludes the proof of (i).

For (ii), suppose $v\notin U^{\mathsf{l}}$ and that β and γ are any functions such that $\mathsf{P}(v,\beta,\gamma)$ holds. We propose to prove that for all u^0, u^1, a, and $\mathbf{m}$,

(vii) $u^0\in U^{\mathsf{l}}\wedge|u^0|\leqslant|u^1|\to\beta(u^0,u^1)=0$;

(viii) $u^1\in U^{\mathsf{l}}\wedge|u^1|<|u^0|\to\beta(u^0,u^1)=1$;

(ix) $\langle a,\mathbf{m}\rangle\in U^{\mathsf{l}}\to\gamma(\langle a,\mathbf{m}\rangle)=\{a\}^{\mathsf{l}}(\mathbf{m})$.

As in the proof of Theorem 3.3, the proof of (vii)–(ix) is again by induction on $\min\{|u^0|,|u^1|\}=|\langle a,\mathbf{m}\rangle|=\sigma<\aleph_1$. If $|u^0|=0$, then $\mathsf{Q}_{0,s}(u^0,u^1,\beta,\gamma)$ holds for

u^1 falling under any case (s) $(0 \leqslant s \leqslant 4)$ and $\sim \mathsf{Q}_{r,0}(v, u^0, \beta, \gamma)$ holds for r such that v falls under case (r) $(1 \leqslant r \leqslant 4$, because $v \notin U^1$ and thus cannot fall under case (0)). Hence by the definition of P, $\beta(u^0, u^1) = 0$ in accord with (vii). If, on the other hand, $|u^1| = 0 < |u^0|$, then $\sim \mathsf{Q}_{r,0}(u^0, u^1, \beta, \gamma)$ and $\sim \mathsf{Q}_{r',0}(v, u^1, \beta, \gamma)$ for some r and r' $(1 \leqslant r, r' \leqslant 4)$ representing the cases applicable to u^0 and v, respectively. Hence $\beta(u^0, u^1) = 1$ in accord with (viii). Finally, if $|\langle a, \mathbf{m}\rangle| = 0$, then it follows from (vii) that $\beta(\langle a, \mathbf{m}\rangle, \langle a, \mathbf{m}\rangle) = 0$ and thus from $\mathsf{S}(v, \beta, \gamma)$ we have $\gamma(\langle a, \mathbf{m}\rangle) = \{a\}^1(\mathbf{m})$.

Now suppose $\sigma > 0$ and as induction hypothesis that (vii)–(ix) hold for all u^0, u^1, a, and $\mathbf{m}$ with $\min\{|u^0|, |u^1|\} < \sigma$ and $|\langle a, \mathbf{m}\rangle| < \sigma$. We first claim that for all u^0 and u^1,

$$\text{(x)} \qquad \min\{|u^0|, |u^1|\} \leqslant \sigma \rightarrow [|u^0| \leqslant |u^1| \leftrightarrow \mathsf{Q}(u^0, u^1, \beta, \gamma)].$$

As in the case of (vi), the proof of (x) by cases is virtually a copy of the corresponding argument for Theorem 3.3. We omit details and only mention that at those points in the argument where for (vi) we appealed to the fact that some ordinal is less than $|v|$ to justify an assertion concerning β_v or γ_v, we here appeal to the fact that the corresponding ordinal is less than σ and apply the relevant clause (vii)–(ix) of the induction hypothesis.

To complete the induction step in the proof of (vii)–(ix), let $\min\{|u^0|, |u^1|\} = \sigma$ and suppose first that $|u^0| \leqslant |u^1|$. Then by (x), $\mathsf{Q}(u^0, u^1, \beta, \gamma)$ holds. Since $v \notin U^1$, $\min\{|v|, |u^0|\} = |u^0| = \sigma$ and thus as $|v| \nleqslant |u^0|$, by (x) again $\sim \mathsf{Q}(v, u^0, \beta, \gamma)$. Hence $\beta(u^0, u^1) \simeq 0$ as required by (vii). If $|u^1| < |u^0|$, then

$$\min\{|v|, |u^1|\} = |u^1| = \sigma < |v|,$$

and two more applications of (x) yield $\sim \mathsf{Q}(u^0, u^1, \beta, \gamma)$ and $\sim \mathsf{Q}(v, u^1, \beta, \gamma)$, so $\beta(u^0, u^1) = 1$ as required by (viii). Finally, suppose $|a, \mathbf{m}| = \sigma$. If, for example, $a = \langle 1, k, 0, b, c\rangle$, then $|c, \mathbf{m}| < \sigma$ so by (ix) of the induction hypothesis $\gamma(\langle c, \mathbf{m}\rangle) = \{c\}^1(\mathbf{m})$. Also $|b, \{c\}^1(\mathbf{m})| < \sigma$ so $\gamma(\langle b, \{c\}^1(\mathbf{m})\rangle) = \{b\}^1(\{c\}^1(\mathbf{m}))$. By (vii), $\beta(\langle a, \mathbf{m}\rangle, \langle a, \mathbf{m}\rangle) = 0$ so since $\mathsf{S}(\beta, \gamma)$ holds, we have

$$\gamma(\langle a, \mathbf{m}\rangle) = \gamma(b, \gamma(\langle a, \mathbf{m}\rangle)) = \{b\}^1(\{c\}^1(\mathbf{m})) = \{a\}^1(\mathbf{m}).$$

The other cases are similar. □

4.19–4.31 Exercises

4.19. Can Sel^1 be defined to have the additional property present in other cases that

$$\exists p \,.\, \{a\}^1(p, \mathbf{m}, \boldsymbol{\alpha}) \downarrow \leftrightarrow \mathrm{Sel}^1(\mathbf{m}, \boldsymbol{\alpha}) \downarrow ?$$

4.20. Modify the proof of the Selection Theorem 4.1 to obtain a functional $\mathrm{Sel}_0^{\mathsf{I}}$ with the additional property that for all p,

$$|a, \mathrm{Sel}_0^{\mathsf{I}}(a, \langle \mathbf{m} \rangle, \langle \boldsymbol{\alpha} \rangle), \mathbf{m}|_{\boldsymbol{\alpha}}^{\mathsf{I}} \leq |a, p, \mathbf{m}|_{\boldsymbol{\alpha}}^{\mathsf{I}}.$$

4.21. Show that if we do not assume that E is recursive in I, the results of §§ 3 and 4 may fail. In particular, show that it may happen that both R and ~R are semi-recursive in I but R is not recursive in I.

4.22. Show that for any I such that E is recursive in I, there exist functions f and g recursive in I such that for all u and m,

(i) if $u \in U^{\mathsf{I}}$ and $|u|^{\mathsf{I}} < \omega$, then $f(u) = |u|^{\mathsf{I}}$

(ii) $g(m) \in U^{\mathsf{I}}$ and $|g(m)|^{\mathsf{I}} = m$.

4.23. Suppose E is recursive in I and F is a function partial recursive in I such that $F : U^{\mathsf{I}} \to U^{\mathsf{I}}$. Show that there exists a function $G : U^{\mathsf{I}} \to U^{\mathsf{I}}$ partial recursive in I such that for all $u \in U^{\mathsf{I}}$

$$|G(u)|^{\mathsf{I}} = \sup{}^{+}\{|F(v)|^{\mathsf{I}} : |v|^{\mathsf{I}} < |u|^{\mathsf{I}}\}.$$

4.24. Prove the following *Effective Boundedness Principle*: there exists a primitive recursive function h such that for any d, if $\forall p . \{d\}^{\mathsf{I}}(p) \in U^{\mathsf{I}}$, then $h(d) \in U^{\mathsf{I}}$ and $\forall p . |\{d\}^{\mathsf{I}}(p)|^{\mathsf{I}} < |h(d)|^{\mathsf{I}}$.

4.25. For any I such that E is recursive in I, there exists a well-ordering of ω of order-type $\omega_1[\mathsf{I}]$ which is semi-recursive in I.

4.26. Show that if E is recursive in I, then $\mathsf{W}_{\omega_1[\mathsf{I}]}$ is semi-recursive in I and if E_1 is recursive in I, then $\mathsf{W}_{\omega_1[\mathsf{I}]}$ is recursive in I. (Define a functional G partial recursive in I such that for all $\gamma \in \mathsf{W}$ and all u,

$$\mathsf{G}(u, \gamma) \simeq \begin{cases} 0, & \text{if } |u|^{\mathsf{I}} < \|\gamma\|; \\ 1, & \text{otherwise.)} \end{cases}$$

4.27. Show that $|a, \langle \mathbf{m} \rangle|_0^{\mathsf{I}} = |a, \langle \mathbf{m} \rangle, \langle\ \ \rangle|^{\mathsf{I}}$.

4.28. For any I such that E is recursive in I, if A and B are sets of numbers both semi-recursive in I but not recursive in I, then A is recursive in I and B and B is recursive in I and A. (Cf. Corollary IV.2.13).

4.29 (cf. Exercise IV.2.24). The hierarchy of Theorem 4.11 is deficient in that it may happen that for some $\rho < \sigma < \kappa^{\mathsf{I}}$ no new relations are reducible to some U_{τ}^{I},

$(\tau < \sigma)$ that are not already reducible to some U^{I}_{τ} $(\tau < \rho)$. This can be remedied by omitting superfluous levels in the hierarchy. Let

$$X = \{\sigma : \sigma < \kappa^{\mathsf{I}} \wedge U^{\mathsf{I}}_{\sigma} \not\leqslant U^{\mathsf{I}}_{\rho} \text{ for any } \rho < \sigma\}$$

and

$$\bar{\kappa}^{\mathsf{I}} = \text{order-type of } X.$$

Show that if E is recursive in I, then $\bar{\kappa}^{\mathsf{I}} = \kappa^{\mathsf{I}}$.

Hint. Let

$$\bar{U}^{\mathsf{I}} = \{u : u \in U^{\mathsf{I}} \wedge |u|^{\mathsf{I}} \in X \wedge \forall v\,(|v|^{\mathsf{I}} = |u|^{\mathsf{I}} \rightarrow u \leqslant v)\}$$

and suppose that for some $t \in U^{\mathsf{I}}$, $\bar{\kappa}^{\mathsf{I}} = |t|^{\mathsf{I}} < \kappa^{\mathsf{I}}$. Set

$$\begin{aligned} P(u, v) \leftrightarrow (|u|^{\mathsf{I}} \not< |t|^{\mathsf{I}} \wedge v = 0) \vee (|u|^{\mathsf{I}} < |t|^{\mathsf{I}} \wedge v \in U^{\mathsf{I}} \wedge \\ \forall \alpha\,[\text{if } \alpha \text{ is an ordinal-preserving map of} \\ \{r : |r|^{\mathsf{I}} \leqslant |u|^{\mathsf{I}}\} \text{ into } \{s : s \in \bar{U}^{\mathsf{I}} \wedge |s|^{\mathsf{I}} \leqslant |v|^{\mathsf{I}}\}, \\ \text{then } \alpha(u) = v]). \end{aligned}$$

Show that P is semi-recursive in I, there is a function β recursive in I such that $\forall u\, P(u, \beta(u))$ and

$$v \in \bar{U}^{\mathsf{I}} \leftrightarrow \exists u\,(|u|^{\mathsf{I}} < |t|^{\mathsf{I}} \wedge \beta(u) = v).$$

4.30. Show that for any $v \notin U^{\mathsf{I}}$ and any β and γ, if $\mathsf{P}(v, \beta, \gamma)$ holds (P from the proof of Theorem 4.18), then β is not recursive in I.

4.31 (Gandy). Show that $\{\alpha : \alpha$ is recursive in $\mathsf{E}_1\}$ is a model of the Δ^1_2 Comprehension schema.

4.32 Notes. To anyone who has reached this note legitimately — that is, by following the proof of Theorem 4.18 — we offer our congratulations and suggest that some strong refreshment is in order. Try combining some hard-frozen strawberries, raspberries, or peaches in a blender with enough dark rum so that the result is a stiff mush (add powdered sugar if the fruit was not sweetened). Pour into a stemmed cocktail glass and relax! For an alternative, see the Notes to Barwise [1975, § II.6].

A simpler proof of a weaker version of Theorem 4.18 is sketched in Exercise 6.26.

5. Hierarchies of Relations Recursive in a Type-2 Functional

In § V.5 we constructed hierarchies of relations obtained by repeated application of positive analytic operations Φ and jump operators J. Here we shall study the relationship of these hierarchies with recursion in a functional. With any jump operator J we associate the functional I_J defined by

$$\mathsf{I}_\mathsf{J}((p)*\alpha) = \mathsf{J}(\alpha)(p)$$

and say that a relation or functional is *recursive in* J iff it is recursive in I_J. We write $\{a\}^\mathsf{J}$ instead of $\{a\}^{\mathsf{I}_\mathsf{J}}$. Recall that $\nabla(\mathsf{J})$ denotes the class of relations on numbers which are recursive in some set D^J_u $(u \in O^\mathsf{J})$ of Definition V.5.5. The main result of this section is that for any jump operator J,

$$\nabla(\mathsf{J}) = \{R : R \text{ is recursive in } \mathsf{J}\}.$$

Note that for oJ this equation follows from Theorem IV.4.21, Lemma 1.4 and Theorem 1.8; both sides are Δ^1_1.

With any positive analytic operation Φ we may associate the functional I_Φ defined by

$$\mathsf{I}_\Phi(\alpha) = \begin{cases} 0, & \text{if } (\exists A \in \mathsf{B}(\Phi))(\forall p \in A).\,\alpha(p) = 0; \\ 1, & \text{otherwise}; \end{cases}$$

and say that a relation or functional is *recursive in* Φ iff it is recursive in I_Φ. Then, for example, $\mathsf{I}_\cup = \mathsf{E}$ and $\mathsf{I}_\mathscr{A} = \mathsf{E}_1$. Similar methods lead to the result that for normal Φ,

$$\nabla(\Phi) = \{R : R \text{ is recursive in } \Phi\},$$

but we shall not carry out the details. A crucial point is that if Φ is normal, then E is recursive in Φ (Exercise 5.10).

5.1 Lemma. *For any jump operator* J, $\nabla(\mathsf{J}) \subseteq \{R : R$ is recursive in J$\}$.

Proof. We shall define a primitive recursive function f such that for all $u \in O^\mathsf{J}$, $\{f(u)\}^\mathsf{J}$ is the characteristic function of D^J_u. *Let* h be a primitive recursive function which satisfies the following conditions for any J:

$$\{h(e,1)\}^{\mathsf{J}} = \mathrm{sg}^{+};$$
$$\{h(e,2^{u})\}^{\mathsf{J}} = \mathsf{J}(\{\{e\}(u)\}^{\mathsf{J}});$$
$$\{h(e,3^{a}5^{u})\}^{\mathsf{J}}(\langle m,p\rangle) = \{\{e\}(\{a\}(p,\{\{e\}(u)\}^{\mathsf{J}}))\}^{\mathsf{J}}(m).$$

By the Primitive Recursion Theorem choose $\bar{e}$ such that for all u, $\{\bar{e}\}(u) = h(\bar{e},u)$ and set $f = \{\bar{e}\}$. It is straightforward to prove by induction on the well-founded relation $<^{\mathsf{J}}$ that f is as required. □

The proof of the converse will follow the pattern of IV.4.19–21.

5.2 Lemma. *For any jump operator* J, *there exists a primitive recursive function* g *such that for all* $u, v \in O^{\mathsf{J}}$, $|u|^{\mathsf{J}} \leqslant |v|^{\mathsf{J}}$, *then* D^{J}_{u} *is recursive in* D^{J}_{v} *with index* $g(u,v)$.

Proof. We shall define g by the Recursion Theorem simultaneously with an auxiliary function f such that for $u, v \in O^{\mathsf{J}}$,

$$|u|^{\mathsf{J}} < |v|^{\mathsf{J}} \leftrightarrow \{f(v)\}(u, D^{\mathsf{J}}_{v^{+}}) \simeq 0$$

and

$$|u|^{\mathsf{J}} \geqslant |v|^{\mathsf{J}} \leftrightarrow \{f(v)\}(u, D^{\mathsf{J}}_{v^{+}}) \simeq 1.$$

In the proof we shall use the following abbreviations:

$$A_{v,u} \quad \text{for} \quad \{p : \{g(u,v)\}(p,A) \simeq 0\},$$
$$A^{(p)} \quad \text{for} \quad \{m : \langle m,p\rangle \in A\},$$
$$\gamma_a(p) \quad \text{for} \quad \{a\}(p, A_{v,u}),$$

and

$$\delta_b(p) \quad \text{for} \quad \{b\}(p, A^{(0)}).$$

Thus if g satisfies the conclusion of the Lemma, we have:

(*) if $u, v \in O^{\mathsf{J}}$, $|u|^{\mathsf{J}} \leqslant |v|^{\mathsf{J}}$, and $A = D^{\mathsf{J}}_{v}$, then $A_{v,u} = D^{\mathsf{J}}_{u}$ and for all p, $\gamma_a(p) = \{a\}(p, D^{\mathsf{J}}_{u})$;

and without any assumption,

(**) if $3^{b}5^{v} \in O^{\mathsf{J}}$ and $A = D^{\mathsf{J}}_{3^{b}5^{v}}$, then $A^{(0)} = D^{\mathsf{J}}_{v}$, and for all p, $\delta_b(p) = \{b\}(p, D^{\mathsf{J}}_{v})$ and $A^{(p)} = D^{\mathsf{J}}_{\delta_b(p)}$.

We require that f and g satisfy the following conditions: for all u, v, a, b, m, and A,

(1) $\{f(1)\}(u, A) \simeq 1$; if $v \neq 1$, $\{f(v)\}(1, A) \simeq 0$;

(2) (a) $\{f(v^+)\}(u^+, \mathsf{J}(A)) \simeq \{f(v)\}(u, A)$;

$$\text{(b) } \{f(v^+)\}(3^a 5^u, \mathsf{J}(\mathsf{J}(A))) = \begin{cases} \mathsf{F}(u, v, a, \mathsf{J}(\mathsf{J}(A))), & \text{if } \{f(v)\}(u, \mathsf{J}(A)) \simeq 0; \\ 1, & \text{otherwise;} \end{cases}$$

where

$$\mathsf{F}(u, v, a, \mathsf{J}(\mathsf{J}(A))) \simeq \begin{cases} 1, & \text{if } \exists p . \{f(v)\}(\gamma_a(p), \mathsf{J}(A)) \simeq 1; \\ 0, & \text{otherwise.} \end{cases}$$

(3) $$\{f(3^b 5^v)\}(u, \mathsf{J}(A)) \simeq \begin{cases} 0, & \text{if } \exists p . \mathsf{G}(p, u, b, A) \simeq 0; \\ 1, & \text{otherwise;} \end{cases}$$

where

$$\mathsf{G}(p, u, b, A) \simeq \{f(\delta_b(p))\}(u, (A^{(p+1)})_{\delta_b(p+1), \delta_b(p)^+});$$

(4) $\{g(1, v)\}(m, A) \simeq \mathrm{sg}^+(m)$;

(5) (a) $\{g(u^+, v^+)\}(m, A) \simeq \{h(g(u, v))\}(m, A)$,

where h is the function of Definition V.5.4;

(b) $\{g(3^a 5^u, v^+)\}(m, \mathsf{J}(A)) \simeq \{g(3^a 5^u, v)\}(m, A)$;

(6) (a) $\{g(u^+, 3^b 5^v)\}(m, A) \simeq \{g(u^+, \delta_b(\bar{p}))\}(m, A^{(\bar{p})})$,

where $\bar{p} \simeq$ "least" $p . \mathsf{G}(p, u^+, b, A) \simeq 0$ (G as in (3));

(b) $\{g(3^a 5^u, 3^b 5^v)\}(\langle m, p \rangle, A) \simeq \{g(\alpha(p), 3^b 5^v)\}(m, A)$,

where $\alpha(p) \simeq \{a\}(p, A_{3^b 5^v, u})$.

The proof that there exist primitive recursive f and g which satisfy these conditions is essentially as for Lemma IV.4.20. The fact that $\mathsf{oJ}(A)$ is recursive in $\mathsf{J}(A)$ is needed to account for the quantifiers. We prove by induction on $|v|$ $(= |v|^{\mathsf{J}})$ that f and g have the required properties. If $|v| = 0$ this is obvious, so suppose it is true with v replaced by any w such that $|w| < |v^+|$. Then $(*)$ holds for this v and any u and

(1) $|1| < |v^+|$ is true and $\{f(v^+)\}(1, D^{\mathsf{J}}_{v^{++}}) \simeq 0$.

(2) (a) If $u^+ \in O^J$, then $u \in O^J$, so

$$|u^+| < |v^+| \leftrightarrow |u| < |v| \leftrightarrow \{f(v)\}(u, D^J_{v^+}) \simeq 0$$
$$\leftrightarrow \{f(v^+)\}(u^+, D^J_{v^{++}}) \simeq 0.$$

(b) If $3^a 5^u \in O^J$, then $u \in O^J$ and for all p, $\{a\}(p, D^J_u) \in O^J$. Thus

$$|3^a 5^u| < |v^+| \leftrightarrow |u| < |v| \quad \text{and} \quad \forall p . |\{a\}(p, D^J_u)| < |v|$$
$$\leftrightarrow \{f(v)\}(u, D^J_{v^+}) \simeq 0 \quad \text{and} \quad \forall p . \{f(v)\}(\{a\}(p, D^J_u), D^J_{v^+}) \simeq 0$$
$$\leftrightarrow \{f(v^+)\}(3^a 5^u, D^J_{v^{++}}) \simeq 0.$$

The last equivalence follows from (*).

(5) (a) If $|u^+| \leqslant |v^+|$, then $|u| \leqslant |v|$, so

$$m \in D^J_{u^+} \leftrightarrow m \in \mathsf{J}(D^J_u)$$
$$\leftrightarrow \mathsf{J}(\lambda n . \{g(u, v)\}(n, D^J_v))(m) \simeq 0$$
$$\leftrightarrow \{h(g(u, v))\}(m, D^J_{v^+}) \simeq 0$$
$$\leftrightarrow \{g(u^+, v^+)\}(m, D^J_{v^+}) \simeq 0.$$

(b) If $|3^a 5^u| \leqslant |v^+|$, then $|3^a 5^u| \leqslant |v|$, so

$$m \in D^J_{3^a 5^u} \leftrightarrow \{g(3^a 5^u, v)\}(m, D^J_v) \simeq 0$$
$$\leftrightarrow \{g(3^a 5^u, v^+)\}(m, D^J_{v^+}) \simeq 0.$$

Now suppose that the result holds with v replaced by any w such that $|w| < |3^b 5^v|$.

(3) First note that by (**) and the induction hypothesis,

$$((D^J_{3^b 5^v})^{(p+1)})_{\delta_b(p+1), \delta_b(p)^+} = D^J_{\delta_b(p)^+}.$$

Then as in the corresponding part of the proof of IV.4.20, for any $u \in O^J$ and any p,

$$|u| < |\delta_b(p)| \leftrightarrow \mathsf{G}(p, u, b, D^J_{3^b 5^v}) \simeq 0.$$

Hence

$$|u| < |3^b 5^v| \leftrightarrow \exists p . |u| < |\delta_b(p)|$$
$$\leftrightarrow \exists p . \mathsf{G}(p, u, b, D^J_{3^b 5^v}) \simeq 0$$
$$\leftrightarrow \{f(3^b 5^v)\}(u, D^J_{(3^b 5^v)^+}) \simeq 0.$$

(6) We proceed by induction on $|u|$.

(a) If $|u^+| \leq |3^b 5^v|$, then $|u^+| < |3^b 5^v|$ so (by (**)) for some (least) $\bar{p}$, $|u^+| < |\delta_b(\bar{p})|$. Then $\mathsf{G}(\bar{p}, u^+, b, D^{\mathsf{J}}_{3^b 5^v}) \simeq 0$ and

$$\begin{aligned} m \in D^{\mathsf{J}}_{u^+} &\leftrightarrow \{g(u^+, \delta_b(\bar{p}))\}(m, D^{\mathsf{J}}_{\delta_b(\bar{p})}) \simeq 0 \\ &\leftrightarrow \{g(u^+, \delta_b(\bar{p}))\}(m, (D^{\mathsf{J}}_{3^b 5^v})^{(\bar{p})}) \simeq 0 \\ &\leftrightarrow \{g(u^+, 3^b 5^v)\}(m, D^{\mathsf{J}}_{3^b 5^v}) \simeq 0. \end{aligned}$$

(b) If $|3^a 5^u| \leq |3^b 5^v|$, then $|u| < |3^b 5^v|$ and for all p, $|\{a\}(p, D^{\mathsf{J}}_u)| < |3^b 5^v|$. By the induction hypothesis on u,

$$(D^{\mathsf{J}}_{3^b 5^v})_{3^b 5^v, u} = D^{\mathsf{J}}_u$$

so that $\alpha(p) = \{a\}(p, D^{\mathsf{J}}_u)$. Then

$$\begin{aligned} \langle m, p\rangle \in D^{\mathsf{J}}_{3^a 5^u} \leftrightarrow m \in D^{\mathsf{J}}_{\alpha(p)} & \\ &\leftrightarrow \{g(\alpha(p), 3^b 5^v)\}(m, D^{\mathsf{J}}_{3^b 5^v}) \simeq 0 \\ &\leftrightarrow \{g(3^a 5^u, 3^b 5^v)\}(\langle m, p\rangle, D^{\mathsf{J}}_{3^b 5^v}) \simeq 0. \quad \square \end{aligned}$$

5.3 Lemma. *For any jump operator* J, *there exists a partial recursive function* $+^{\mathsf{J}}$ *such that for all* $u, v \in O^{\mathsf{J}}$, $u +^{\mathsf{J}} v \in O^{\mathsf{J}}$, $|u +^{\mathsf{J}} v|^{\mathsf{J}} = |u|^{\mathsf{J}} + |v|^{\mathsf{J}}$, *and if* $v \neq 1$, *then* $u <^{\mathsf{J}} (u +^{\mathsf{J}} v)$.

Proof. We define $+^{\mathsf{J}}$ by the Recursion Theorem to satisfy the following conditions:

(1) $u +^{\mathsf{J}} 1 \simeq u$;

(2) $u +^{\mathsf{J}} v^+ \simeq (u +^{\mathsf{J}} v)^+$;

(3) $u +^{\mathsf{J}} 3^b 5^v \simeq 3^c 5^{u +^{\mathsf{J}} v}$, where c is an index such that

$$\{c\}(p, D^{\mathsf{J}}_{u +^{\mathsf{J}} v}) \simeq u +^{\mathsf{J}} \{b\}(p, D^{\mathsf{J}}_v).$$

That is, with g as in the preceding lemma,

$$\{c\}(p, A^{\cdot}) \simeq u +^{\mathsf{J}} \{b\}(p, \lambda n.\{g(v, u +^{\mathsf{J}} v)\}(n, A)).$$

If follows easily by induction over O^{J} that $+^{\mathsf{J}}$ has the required properties. $\square$

5.4 Corollary (Effective Boundedness). *For any jump operator* J, *there exists a primitive recursive function* h *such that for any* $u \in O^{\mathsf{J}}$ *and any* d, *if for all* p,

$\{d\}(p, D^{\mathrm{J}}_u) \in O^{\mathrm{J}}$, *then* $h(d, u) \in O^{\mathrm{J}}$, $|u|^{\mathrm{J}} < |h(d, u)|^{\mathrm{J}}$, *and for all* p, $|\{d\}(p, D^{\mathrm{J}}_u)|^{\mathrm{J}} < |h(d, u)|^{\mathrm{J}}$.

Proof. We set $h(d, u) = 3^c 5^u$, where c is an index such that

$$\{c\}(0, A) \simeq u, \quad \text{and}$$

$$\{c\}(p + 1, A) \simeq \{c\}(p, A) +^{\mathrm{J}} \{d\}(p, A)^+.$$

The required properties of h follow easily from the properties of $+^{\mathrm{J}}$. □

5.5 Theorem. *For any jump operator* J, *there exist primitive recursive functions* f *and* g *such that for any* a *and* $\mathbf{m}$, *if* $\{a\}^{\mathrm{J}}(\mathbf{m})$ *is defined, then* $f(a, \langle\mathbf{m}\rangle) \in O^{\mathrm{J}}$ *and*

$$\{a\}^{\mathrm{J}}(\mathbf{m}) \simeq \{g(a)\}(\mathbf{m}, D^{\mathrm{J}}_{f(a, \langle\mathbf{m}\rangle)}).$$

Proof. We shall use the notation $A_{v,u}$ as in the proof of Lemma 5.2 and the function h of the preceding Corollary. In particular if $u \in O^{\mathrm{J}}$ and for all p, $\{d\}(p, D^{\mathrm{J}}_u) \in O^{\mathrm{J}}$,

$$(D^{\mathrm{J}}_{h(d,u)})_{h(d,u), \{d\}(p, D^{\mathrm{J}}_u)} = D^{\mathrm{J}}_{\{d\}(p, D^{\mathrm{J}}_u)}.$$

We require that f and g satisfy the following conditions for all a and $\mathbf{m}$,
(0) If $a = \langle 0, k, 0, \ldots\rangle$, then $f(a, \langle\mathbf{m}\rangle) = 1$ and

$$\{g(a)\}(\mathbf{m}, D^{\mathrm{J}}_1) \simeq \{a\}^{\mathrm{J}}(\mathbf{m});$$

(1) if $a = \langle 1, k, 0, b, c_1, c_2\rangle$ (say), then $f(a, \langle\mathbf{m}\rangle) = h(d_1, h(d_0, 1))$, where

$$\{d_0\}(p, D^{\mathrm{J}}_1) \simeq \begin{cases} f(c_1, \langle\mathbf{m}\rangle), & \text{if} \quad p = 0; \\ f(c_2, \langle\mathbf{m}\rangle), & \text{if} \quad p > 0; \end{cases}$$

$$\{d_1\}(p, A) \simeq f(b, \langle\{g(c_1)\}(\mathbf{m}, A_{h(d_0,1), f(c_1, \langle\mathbf{m}\rangle)}), \{g(c_2)\}(\mathbf{m}, A_{h(d_0,1), f(c_2, \langle\mathbf{m}\rangle)})\rangle);$$

$$\{g(a)\}(\mathbf{m}, A) \simeq \{g(b)\}(q_1, q_2, A_{f(a, \langle\mathbf{m}\rangle), f(b, \langle q_1, q_2\rangle)}),$$

where

$$q_i \simeq \{g(c_i)\}(\mathbf{m}, A_{f(a, \langle\mathbf{m}\rangle), f(c_i, \langle\mathbf{m}\rangle)});$$

(2) if $a = \langle 2, k + 1, 0\rangle$, then $f(a, \langle b, \mathbf{m}\rangle) = h(d_2, 1)$, where

$$\{d_2\}(p, D^{\mathrm{J}}_1) \simeq f(b, \langle\mathbf{m}\rangle);$$

$$\{g(a)\}(b, \mathbf{m}, A) \simeq \{g(b)\}(\mathbf{m}, A_{f(a, \langle b, \mathbf{m}\rangle), f(b, \langle\mathbf{m}\rangle)});$$

(3) if $a = \langle 3, k, l, b\rangle$, then $f(a, \langle \mathbf{m}\rangle) = h(d_3, 1)^+$, where $\{d_3\}(p, D^{\mathsf{J}}_1) \simeq f(b, \langle p, \mathbf{m}\rangle)$; $g(a)$ is an index such that

$$\{g(a)\}(\mathbf{m}, \mathsf{J}(A)) \simeq \mathsf{J}(\lambda p\,.\{d_4\}(p+1, \mathbf{m}, A))(\{d_4\}(0, A)),$$

where

$$\{d_4\}(p, \mathbf{m}, A) \simeq \{g(b)\}(p, \mathbf{m}, A_{h(d_3,1), f(b, \langle p, \mathbf{m}\rangle)}).$$

The proof that there exist such f and g is again similar to the corresponding proof in Lemma IV.4.20. We prove that f and g have the required properties by induction on $\Omega[\mathsf{J}]$ $(= \Omega[\mathsf{I}_{\mathsf{J}}])$.

(0) If $a = \langle 0, k, 0, \ldots\rangle$, then the result is clear.

(1) If $a = \langle 1, k, 0, b, c_1, c_2\rangle$, $\{c_i\}^{\mathsf{J}}(\mathbf{m}) \simeq q_i$ $(i = 1, 2)$, and $\{a\}^{\mathsf{J}}(\mathbf{m}) \simeq \{b\}^{\mathsf{J}}(q_1, q_2) \simeq n$, then by the induction hypothesis, $f(c_i, \langle \mathbf{m}\rangle) \in O^{\mathsf{J}}$. Hence for all p, $\{d_0\}(p, D^{\mathsf{J}}_1) \in O^{\mathsf{J}}$, $h(d_0, 1) \in O^{\mathsf{J}}$, $|f(c_i, \langle \mathbf{m}\rangle)|^{\mathsf{J}} < |h(d_0, 1)|^{\mathsf{J}}$ $(i = 1, 2)$, and

$$(D^{\mathsf{J}}_{h(d_0,1)})_{h(d_0,1), f(c_i, \langle \mathbf{m}\rangle)} = D^{\mathsf{J}}_{f(c_i, \langle \mathbf{m}\rangle)}.$$

By the induction hypothesis for g, $\{g(c_i)\}(\mathbf{m}, D^{\mathsf{J}}_{f(c_i, \langle \mathbf{m}\rangle)}) \simeq q_i$, so for all p, $\{d_1\}(p, D^{\mathsf{J}}_{h(d_0,1)}) \simeq f(b, \langle q_1, q_2\rangle) \in O^{\mathsf{J}}$, and thus $f(a, \langle \mathbf{m}\rangle) = h(d_1, h(d_0, 1)) \in O^{\mathsf{J}}$. Furthermore, by the induction hypothesis,

$$\{g(a)\}(\mathbf{m}, D^{\mathsf{J}}_{f(a, \langle \mathbf{m}\rangle)}) \simeq \{g(b)\}(q_1, q_2, D^{\mathsf{J}}_{f(b, \langle q_1, q_2\rangle)}) \simeq \{b\}^{\mathsf{J}}(q_1, q_2) \simeq n.$$

(2) If $a = \langle 2, k+1, 0\rangle$ and $\{a\}^{\mathsf{J}}(b, \mathbf{m}) \simeq \{b\}^{\mathsf{J}}(\mathbf{m}) \simeq n$, then for all p, $\{d_2\}(p, D^{\mathsf{J}}_1) \simeq f(b, \langle \mathbf{m}\rangle) \in O^{\mathsf{J}}$, so $f(a, \langle b, \mathbf{m}\rangle) \simeq h(d_2, 1) \in O^{\mathsf{J}}$. Furthermore,

$$\{g(a)\}(b, \mathbf{m}, D^{\mathsf{J}}_{f(a, \langle b, \mathbf{m}\rangle)}) \simeq \{g(b)\}(\mathbf{m}, D^{\mathsf{J}}_{f(b, \langle \mathbf{m}\rangle)}) \simeq \{b\}^{\mathsf{J}}(\mathbf{m}) \simeq n.$$

(3) If $a = \langle 3, k, 0, b\rangle$ and $\{a\}^{\mathsf{J}}(\mathbf{m}) \simeq \mathsf{I}_{\mathsf{J}}(\lambda p\,.\{b\}^{\mathsf{J}}(p, \mathbf{m})) \simeq n$, then for all p, $\{d_2\}(p, D^{\mathsf{J}}_1) \simeq f(b, \langle p, \mathbf{m}\rangle) \in O^{\mathsf{J}}$, so $f(a, \langle \mathbf{m}\rangle) \simeq h(d_3, 1)^+ \in O^{\mathsf{J}}$. Then also

$$\begin{aligned}
\{g(a)\}(\mathbf{m}, D^{\mathsf{J}}_{f(a, \langle \mathbf{m}\rangle)}) &\simeq \mathsf{J}(\lambda p\,.\{d_4\}(p+1, \mathbf{m}, D^{\mathsf{J}}_{h(d_3,1)}))(\{d_4\}(0, D^{\mathsf{J}}_{h(d_3,1)}))\\
&\simeq \mathsf{J}(\lambda p\,.\{g(b)\}(p+1, \mathbf{m}, D^{\mathsf{J}}_{f(b, \langle p+1, \mathbf{m}\rangle)}))(\{g(b)\}(0, \mathbf{m}, D^{\mathsf{J}}_{f(b, \langle 0, \mathbf{m}\rangle)}))\\
&\simeq \mathsf{J}(\lambda p\,.\{b\}^{\mathsf{J}}(p+1, \mathbf{m}))(\{b\}^{\mathsf{J}}(0, \mathbf{m}))\\
&\simeq \mathsf{I}_{\mathsf{J}}(\lambda p\,.\{b\}^{\mathsf{J}}(p, \mathbf{m})) \simeq n. \quad \square
\end{aligned}$$

5.6 Corollary. *For any jump operator* J *and any* $R \subseteq {}^k\omega$, *R is recursive in* J *iff for some* $u \in O^{\mathsf{J}}$, *R is recursive in* D^{J}_u.

Proof. Immediate from Lemma 5.1, Corollary 5.4, and Theorem 5.5. $\square$

Since for any jump operator J and any A, oJ(A) is recursive in J(A), it follows that J(A) is not recursive in A. In particular, for all $u \in O^{\mathsf{J}}, D^{\mathsf{J}}_{u^+}$ is not recursive in D^{J}_u, whence by Lemma 5.2, if $|u|^{\mathsf{J}} < |v|^{\mathsf{J}}$, then D^{J}_v is not recursive in D^{J}_u. Thus if we set

$$\Delta^{\mathsf{J}}_\sigma = \{R : R \text{ is recursive in some } D^{\mathsf{J}}_u \text{ such that } |u|^{\mathsf{J}} \leqslant \sigma\},$$

the classes $\Delta^{\mathsf{J}}_\sigma$ form a properly increasing hierarchy of length

$$\lambda^{\mathsf{J}} = \sup{}^+\{|u|^{\mathsf{J}} : u \in O^{\mathsf{J}}\},$$

which includes exactly the relations on numbers which are recursive in J. The evaluation of λ^{J} provides no surprise:

5.7 Theorem. *For any jump operator* J, $\lambda^{\mathsf{J}} = \omega_1[\mathsf{J}]$.

Sketch of proof. An easy induction over $\Omega[\mathsf{J}]$ shows that for all a and $\mathbf{m}$, if $\{a\}^{\mathsf{J}}(\mathbf{m})$ is defined, then $|\langle a, \mathbf{m}\rangle|^{\mathsf{J}} \leqslant |f(a, \langle \mathbf{m}\rangle)|^{\mathsf{J}}$, where the first ordinal is that assigned in § 3, the second is the ordinal assigned to members of O^{J}, and f is the function of Theorem 5.5. It then follows from Theorem 4.16 that $\lambda^{\mathsf{J}} \geqslant \omega_1[\mathsf{J}]$. Conversely, it is possible to define a primitive recursive function h such that for all $u \in O^{\mathsf{J}}$, $h(u)$ is an index from D^{J}_u of $\{(v, w) : v <^{\mathsf{J}} w <^{\mathsf{J}} u\}$ which is a well-ordering of type $|u|^{\mathsf{J}}$ and is recursive in J. It follows that $\omega_1[\mathsf{J}] \geqslant \lambda^{\mathsf{J}}$ (Exercise 5.11). □

These results provide a hierarchy for the relations on numbers recursive in any jump operator. Happily, for any functional I in which E is recursive there is a jump operator J_I of the same degree (each of I and J_I is recursive in the other) so that we obtain a hierarchy for the relations on numbers recursive in any such I.

Let

$$\mathsf{J}_\mathsf{I}(\alpha)(\langle a, n\rangle) = \begin{cases} 0, & \text{if } \mathsf{I}(\lambda p.\{a\}(p, \alpha)) \simeq n; \\ 1, & \text{otherwise.} \end{cases}$$

5.8 Lemma. *For any* I, J_I *is a jump operator.*

Proof. Let $n_0 = \mathsf{I}(\lambda p.0)$ and f be a primitive recursive function such that for all a, p, $\mathbf{m}$, and α,

$$\{f(a, \langle \mathbf{m}\rangle)\}(p, \alpha) \simeq 0 \cdot \{a\}(\mathbf{m}, \alpha).$$

Then

$$\begin{aligned}\alpha^{\mathrm{oJ}}((a, \mathbf{m})) = 0 &\leftrightarrow \{a\}(\mathbf{m}, \alpha) \text{ is defined}\\ &\leftrightarrow \forall p . \{f(a, \langle \mathbf{m} \rangle)\}(p, \alpha) \simeq 0\\ &\leftrightarrow \mathsf{J}_{\mathsf{I}}(\alpha)((f(a, \langle \mathbf{m} \rangle), n_0)) = 0\\ &\leftrightarrow \{d\}((a, \mathbf{m}), \mathsf{J}_{\mathsf{I}}(\alpha)) = 0\end{aligned}$$

for an appropriate index d. Condition (ii) of Definition V.5.4 follows easily from Theorem 2.10. □

5.9 Theorem. *For any* I, *if* E *is recursive in* I, *then* I *and* J_{I} *are each recursive in the other.*

Proof. Suppose that E is recursive in I. Then one can decide recursively in I whether or not $\lambda p . \{a\}(p, \alpha)$ is total and, if it is, whether or not $\mathsf{I}(\lambda p . \{a\}(p, \alpha)) = n$. Hence J_{I} is recursive in I. For the converse, let a_0 be an index such that $\{a_0\}(m, \alpha) = \alpha(m)$. Then

$$\mathsf{I}(\alpha) = n \leftrightarrow \mathsf{J}_{\mathsf{I}}(\alpha)((a_0, n)) = 0,$$

so Gr_{I} is recursive in J_{I}. Since oJ is recursive in J_{I}, so is E by Lemma 1.4. Hence by Corollary 4.6, I is recursive in J_{I}. □

5.10–5.11 Exercises

5.10. Show that if Φ is a normal positive analytic operation, then E is recursive in Φ.

5.11. Complete the proof of Theorem 5.7. Give an alternative proof of the inequality $\lambda^{\mathsf{J}} \geqslant \omega_1[\mathsf{J}]$ by showing that for any $\gamma \in \mathsf{W}$ recursive in J, there exists a function f recursive in J such that for all $p \in \mathrm{Fld}(\gamma)$, $|p|_\gamma \leqslant |f(p)|^{\mathsf{J}}$.

5.12 Notes. The construction of a hierarchy for the relations on numbers recursive in a type-2 functional is due independently to Hinman [1966] and Shoenfield [1968]. The method here is Shoenfield's. A proof of the result mentioned just before Lemma 5.1 may be found in Hinman [1969].

6. *Extended Functionals*

The notion of extended functionals arises from the following question. Recall that a positive analytic operation Φ has a natural interpretation as a quantifier:

$$(\Phi \mathsf{R})(\mathbf{m}, \boldsymbol{\alpha}) \leftrightarrow (\exists A \in \mathsf{B}(\Phi))(\forall p \in A)\, \mathsf{R}(p, \mathbf{m}, \boldsymbol{\alpha}).$$

These would be natural objects of study even in the absence of their classical origins. In § 5 we defined the notion of recursion relative to Φ. It is immediate from the definition that for any positive analytic operation Φ,

(1) $\{\mathsf{R} : \mathsf{R}$ is recursive in $\Phi\}$ is closed under the quantifier Φ.

Since "(semi-) recursive in $\cup$" coincides with "(semi-) recursive in E", it follows from either Corollary 1.10 or Corollary 4.3 that

(2) $\{\mathsf{R} : \mathsf{R}$ is semi-recursive in $\cup\}$ $(=\Pi^1_1)$ is closed under the quantifier $\cup\,(=\exists^0)$.

We ask, therefore, if also (2) holds with "$\cup$" replaced by an arbitrary "Φ". The answer is no (Corollary 6.3). To restore the analogy we define a "functional" $\Phi^{\#}$ with the property that $\{\mathsf{R} : \mathsf{R}$ is semi-recursive in $\Phi^{\#}\}$ is closed under the quantifier Φ. $\Phi^{\#}$ is not a functional in the sense that we have used the term as its domain includes some partial unary functions from ω into ω as well as all total unary functions. Although the previous definitions and results do not apply directly, it turns out that the theory of recursion relative to $\Phi^{\#}$ is very much like the theory of recursion relative to an ordinary functional. In the latter part of the section we investigate the (close) connections among $\Phi^{\#}$, Φ^*, and Φ-positive inductive definitions.

6.1 Lemma. *For any functional* I, *there exists a relation* P^{I} *semi-recursive in* I *such that for all* u, v, *and* $\boldsymbol{\alpha}$,

(i) $u \in U^{\mathsf{I}}_{\boldsymbol{\alpha}} \wedge \mathsf{P}^{\mathsf{I}}(u, v, \langle\boldsymbol{\alpha}\rangle) \rightarrow |v|^{\mathsf{I}}_{\boldsymbol{\alpha}} < |u|^{\mathsf{I}}_{\boldsymbol{\alpha}}$;

(ii) $u \notin U^{\mathsf{I}}_{\boldsymbol{\alpha}} \rightarrow \exists v\, [v \notin U^{\mathsf{I}}_{\boldsymbol{\alpha}} \wedge \mathsf{P}^{\mathsf{I}}(u, v, \langle\boldsymbol{\alpha}\rangle)]$.

Proof. Intuitively, $\mathsf{P}^{\mathsf{I}}(u, v, \langle\boldsymbol{\alpha}\rangle)$ means that v represents a computation relative to $\boldsymbol{\alpha}$ which is an immediate subcomputation of that represented by u. We define $\mathsf{P}^{\mathsf{I}}(u, v, \langle\boldsymbol{\alpha}\rangle)$ to hold just in case one of the following holds for some $k, \mathbf{m} \in {}^k\omega$, and $l = \lg(\boldsymbol{\alpha})$:

(1) for some $b, c_0, \ldots, c_{k'-1}$, $u = \langle\langle 1, k, l, b, c_0, \ldots, c_{k'-1}\rangle, \mathbf{m}\rangle$, and either $v = \langle c_i, \mathbf{m}\rangle$ for some $i < k'$ or there exist $q_0, \ldots, q_{k'-1}$ such that for all $i < k'$, $\{c_i\}^{\mathsf{I}}(\mathbf{m}, \boldsymbol{\alpha}) \simeq q_i$ and $v = \langle b, \mathbf{q}\rangle$;

(2) for some b, $u = \langle\langle 2, k+1, l\rangle, b, \mathbf{m}\rangle$ and $v = \langle b, \mathbf{m}\rangle$;

(3) for some b, $u = \langle\langle 3, k, l, b\rangle, \mathbf{m}\rangle$ and for some p,

$$v = \langle b, p, \mathbf{m}\rangle;$$

(4) u is of none of these forms, $|u|^{\mathsf{I}}_{\boldsymbol{\alpha}} \neq 0$, and $v = 0$.

Suppose first that $u \in U^{\mathsf{I}}_{\boldsymbol{\alpha}}$ and $\mathsf{P}^{\mathsf{I}}(u, v, \langle\boldsymbol{\alpha}\rangle)$ holds. Then u falls under one of clauses (1)–(3) and it is clear from Lemma 3.2 that in each case $|v|^{\mathsf{I}}_{\boldsymbol{\alpha}} < |u|^{\mathsf{I}}_{\boldsymbol{\alpha}}$. For (ii), suppose that $u \notin U^{\mathsf{I}}_{\boldsymbol{\alpha}}$. If u is not of the proper form for $\boldsymbol{\alpha}$, then by clause (4), $\mathsf{P}^{\mathsf{I}}(u, 0, \langle\boldsymbol{\alpha}\rangle)$ holds and (ii) is satisfied becasuse $0 \notin U^{\mathsf{I}}_{\boldsymbol{\alpha}}$. Otherwise u satisfies the hypotheses of one of (1)–(3). Suppose $u = \langle\langle 1, k, l, b, c_0, \ldots, c_{k'-1}\rangle, \mathbf{m}\rangle$. If for some $i < k'$, $\langle c_i, \mathbf{m}\rangle \notin U^{\mathsf{I}}_{\boldsymbol{\alpha}}$, let v be such a $\langle c_i, \mathbf{m}\rangle$. Otherwise, there exist $q_0, \ldots, q_{k'-1}$ such that $\{c_i\}^{\mathsf{I}}(\mathbf{m}, \boldsymbol{\alpha}) \simeq q_i$ and $\langle b, \mathbf{q}\rangle \notin U^{\mathsf{I}}_{\boldsymbol{\alpha}}$, as otherwise we would have $u \in U^{\mathsf{I}}_{\boldsymbol{\alpha}}$, contrary to assumption. Thus $v = \langle b, \mathbf{q}\rangle$ satisfies the conclusion of (ii). The other cases are similar. □

6.2 Theorem. *For any functional* I *and all* a, $\mathbf{m}$, *and* $\boldsymbol{\alpha}$,

$$\{a\}^{\mathsf{I}}(\mathbf{m}, \boldsymbol{\alpha}) \textit{ is defined} \leftrightarrow \neg \exists\beta\, \forall p\, [\mathsf{P}^{\mathsf{I}}(\beta(p), \beta(p+1), \langle\boldsymbol{\alpha}\rangle) \wedge \beta(0) = \langle a, \mathbf{m}\rangle].$$

Proof. Suppose first that $\{a\}^{\mathsf{I}}(\mathbf{m}, \boldsymbol{\alpha})$ is defined but for some β, $\beta(0) = \langle a, \mathbf{m}\rangle$ and for all p, $\mathsf{P}^{\mathsf{I}}(\beta(p), \beta(p+1), \langle\boldsymbol{\alpha}\rangle)$. Since $\langle a, \mathbf{m}\rangle \in U^{\mathsf{I}}_{\boldsymbol{\alpha}}$ by assumption, it follows by induction from Lemma 6.1(i) that for all p, $\beta(p) \in U^{\mathsf{I}}_{\boldsymbol{\alpha}}$ and thus for all p, $|\beta(p+1)|^{\mathsf{I}}_{\boldsymbol{\alpha}} < |\beta(p)|^{\mathsf{I}}_{\boldsymbol{\alpha}}$, a contradiction.

If, on the other hand, $\{a\}^{\mathsf{I}}(\mathbf{m}, \boldsymbol{\alpha})$ is not defined, clause (ii) of the preceding lemma guarantees that there is a unique function β such that $\beta(0) = \langle a, \mathbf{m}\rangle$ and for all p,

$$\beta(p+1) = \text{least } v[v \notin U^{\mathsf{I}}_{\boldsymbol{\alpha}} \text{ and } \mathsf{P}^{\mathsf{I}}(\beta(p), v, \langle\boldsymbol{\alpha}\rangle)]. \quad \square$$

6.3 Corollary. *For any functional* I *such that* E *is recursive in* I, *there exists a relation* R *semi-recursive in* I *such that* $\mathscr{A}\mathsf{R}(= \{(\mathbf{m}, \boldsymbol{\alpha}) : \exists\beta\, \forall p\, \mathsf{R}(\bar{\beta}(p), \mathbf{m}, \boldsymbol{\alpha})\})$ *is not semi-recursive in* I. *In particular, the class of relations semi-recursive in* $\mathscr{A}$ ($=$ *semi-recursive in* E_1) *is not closed under* $\mathscr{A}$.

Proof. Let

$$\mathsf{R}(s, \langle\mathbf{m}\rangle, \langle\boldsymbol{\alpha}\rangle) \leftrightarrow [(\forall p < \lg(s) \dot{-} 1)\mathsf{P}^{\mathsf{I}}((s)_p, (s)_{p+1}, \langle\boldsymbol{\alpha}\rangle) \wedge (s)_0 = \langle a, \mathbf{m}\rangle].$$

By the preceding theorem $\mathscr{A}\mathsf{R} = {\sim}\mathsf{U}^{\mathsf{I}}$. If $\mathscr{A}\mathsf{R}$ were semi-recursive in I, then by Corollary 4.5, U^{I} would be recursive in I, a contradiction. □

For any positive analytic operation Φ, we define the *extended functional* $\Phi^{\#}$ as follows: for any *partial* function f from ω into ω,

$$\Phi^{\#}(f) \simeq \begin{cases} 0, & \text{if } (\exists A \in \mathsf{B}(\Phi))(\forall p \in A).\, f(p) \simeq 0; \\ 1, & \text{if } (\forall A \in \mathsf{B}(\Phi))(\exists p \in A)(\exists n > 0).\, f(p) \simeq n; \\ \text{undefined}, & \text{otherwise.} \end{cases}$$

Note that $\Phi^{\#}$ is an extension of the functional I_{Φ} defined in § 5. In fact, for any partial function f,

$$\Phi^{\#}(f) \simeq n \leftrightarrow \forall \alpha\, [f \subseteq \alpha \rightarrow \mathsf{I}_{\Phi}(\alpha) = n].$$

Note also that $\Phi^{\#}$ is *consistent* — that is, if $\Phi^{\#}(f) \simeq n$ and $f \subseteq g$, then also $\Phi^{\#}(g) \simeq n$. We shall write $\mathsf{E}^{\#}$ for $\cup^{\#}$ and $\mathsf{E}_1^{\#}$ for $\mathscr{A}^{\#}$. Thus, for any partial f

$$\mathsf{E}_1^{\#}(f) \simeq \begin{cases} 0, & \text{if } \exists \beta\, \forall p\,.\, f(\bar{\beta}(p)) \simeq 0; \\ 1, & \text{if } \forall \beta\, \exists p\, (\exists n > 0)\,.\, f(\bar{\beta}(p)) \simeq n; \\ \text{undefined,} & \text{otherwise.} \end{cases}$$

Recursion relative to extended functionals may be defined in nearly the same way as for ordinary functionals:

6.4 Definition. For any extended functional $\Phi^{\#}$, $\Omega[\Phi^{\#}]$ is the smallest set such that for all k, l, n, p, q, r, and s, all $i < k$, and $j < l$, and all $(\mathbf{m}, \boldsymbol{\alpha}) \in {}^{k,l}\omega$,

(0)
(1) } identical to the corresponding clauses of Definition II.2.1;
(2)

(3)$^{\#}$ for any b and any f such that $\Phi^{\#}(f) \simeq n$, if for all p and q,

$$f(p) \simeq q \rightarrow (b, p, \mathbf{m}, \boldsymbol{\alpha}, q) \in \Omega[\Phi^{\#}],$$

then $(\langle 3, k, l, b \rangle, \mathbf{m}, \boldsymbol{\alpha}, n) \in \Omega[\Phi^{\#}]$.

The proof that for all a, $\mathbf{m}$, and $\boldsymbol{\alpha}$ there is at most one n such that $(a, \mathbf{m}, \boldsymbol{\alpha}, n) \in \Omega[\Phi^{\#}]$ depends on the consistency of $\Phi^{\#}$ but otherwise proceeds as in the previous cases and we set

$$\{a\}^{\Phi^{\#}}(\mathbf{m}, \boldsymbol{\alpha}) \simeq n \leftrightarrow (a, \mathbf{m}, \boldsymbol{\alpha}, n) \in \Omega[\Phi^{\#}]$$

and define the other relevant notions as in § 1. The remainder of the theory of §§ 1 and 2 may now be carried over to recursion in an extended functional with exactly the same proofs.

We show first that replacing Φ by $\Phi^{\#}$ has the intended effect.

6.5 Theorem. *For any extended functional $\Phi^{\#}$, $\{\mathsf{R} : \mathsf{R}$ is semi-recursive in $\Phi^{\#}\}$ is closed under the quantifier Φ.*

Proof. Let R be semi-recursive in $\Phi^{\#}$, say with semi-index a from $\Phi^{\#}$, and suppose $\mathsf{S} = \Phi\mathsf{R}$. Then $\mathsf{S}(\mathbf{m}, \boldsymbol{\alpha}) \leftrightarrow \Phi^{\#}(\lambda p\,.\,0 \cdot \{a\}^{\Phi^{\#}}(p, \mathbf{m}, \boldsymbol{\alpha}))$ is defined, so S is also semi-recursive in $\Phi^{\#}$. □

Since $\Phi^{\#}$ is an extension of Φ, it is to be expected that functionals (partial) recursive in Φ are also (partial) recursive in $\Phi^{\#}$. For total functionals this is evident as clause $(3)^{\#}$ implies clause (3) for I_Φ and thus $\Omega[\Phi] \subseteq \Omega[\Phi^{\#}]$. Hence if $\{a\}^{\Phi}$ is total, $\{a\}^{\Phi} = \{a\}^{\Phi^{\#}}$. Clearly, however, more indices define computations relative to $\Phi^{\#}$ than to Φ.

Rather than investigate further the subject of extended functionals in full generality we shall concentrate on the particular example $\mathsf{E}_1^{\#}$. The properties of recursion relative to $\mathsf{E}_1^{\#}$ are illustrative of those of all extended functionals and the proofs which establish these properties will benefit greatly from the removal of one layer of technical complexity (see the remarks following Corollary 6.16). $\mathsf{E}_1^{\#}$ is also of particular interest because of its close connection with Σ_1^1 inductive definitions (Theorem 6.14) and the superjump (Theorem 6.11 and Theorem VII.1.8).

6.6 Theorem. *There exists a primitive recursive function g such that for all a,* **m**, *and* $\boldsymbol{\alpha}$,

$$\{a\}^{\mathsf{E}_1}(\mathbf{m}, \boldsymbol{\alpha}) \simeq \{g(a)\}^{\mathsf{E}_1^{\#}}(\mathbf{m}, \boldsymbol{\alpha}).$$

Proof. The proof is similar to that of Theorem 2.14. We define a primitive recursive function h just as there except that in case (3) we take $h(e, a)$ to be the natural index such that

$$\{h(e, a)\}^{\mathsf{E}_1^{\#}}(\mathbf{m}, \boldsymbol{\alpha}) \simeq \mathsf{E}_1^{\#}(\lambda s\,[0 \cdot \{h(e, b)\}^{\mathsf{E}_1^{\#}}(\mathrm{lg}(s), \mathbf{m}, \boldsymbol{\alpha})]) + \mathsf{E}_1^{\#}(\lambda p\,.\{h(e, b)\}^{\mathsf{E}_1^{\#}}(p, \mathbf{m}, \boldsymbol{\alpha})).$$

Applying the Primitive Recursion Theorem we obtain an index $\bar{e}$ such that $h(\bar{e}, a) = \{\bar{e}\}(a)$ and set $g = \{\bar{e}\}$.

We need to prove that for all a, **m**, $\boldsymbol{\alpha}$, and n,

$$\{a\}^{\mathsf{E}_1}(\mathbf{m}, \boldsymbol{\alpha}) \simeq n \leftrightarrow \{g(a)\}^{\mathsf{E}_1^{\#}}(\mathbf{m}, \boldsymbol{\alpha}) \simeq n.$$

The implication $(\rightarrow)$ is proved by induction over $\Omega[\mathsf{E}_1]$. Cases (0)–(2) are treated exactly as in Theorem 2.14. Suppose that $a = \langle 3, k, l, b\rangle$ and $\{a\}^{\mathsf{E}_1}(\mathbf{m}, \boldsymbol{\alpha}) \simeq n$. Then $\lambda p\,.\{b\}^{\mathsf{E}_1}(p, \mathbf{m}, \boldsymbol{\alpha})$ is a total function β and $\mathsf{E}_1(\beta) = n$. By the induction hypothesis, for all p, $\{g(b)\}^{\mathsf{E}_1^{\#}}(p, \mathbf{m}, \boldsymbol{\alpha}) \simeq \beta(p)$. Hence

$$\mathsf{E}_1^{\#}(\lambda p\,.\{g(b)\}^{\mathsf{E}_1^{\#}}(p, \mathbf{m}, \boldsymbol{\alpha})) \simeq \mathsf{E}_1(\beta) = n.$$

Furthermore, $\mathrm{lg}(\bar{\beta}(p)) = p$, and

$$\forall p\,[0 \cdot \{g(b)\}^{\mathsf{E}_1^{\#}}(p, \mathbf{m}, \boldsymbol{\alpha}) \simeq 0],$$

so

$$\exists\gamma\,\forall p\,[0\cdot\{g(b)\}^{\mathsf{E}_1^{\#}}(\lg(\bar{\gamma}(p)),\mathbf{m},\boldsymbol{\alpha})\simeq 0],$$

and thus $\{g(a)\}^{\mathsf{E}_1^{\#}}(\mathbf{m},\boldsymbol{\alpha})\simeq n$.

The converse implication is established by induction on subcomputations for $\mathsf{E}_1^{\#}$. We leave the details to the reader. □

It now follows immediately that every functional partial recursive in E_1 is also partial recursive in $\mathsf{E}_1^{\#}$ (in particular, E is recursive in $\mathsf{E}_1^{\#}$) and every relation semi-recursive in E_1 is also semi-recursive in $\mathsf{E}_1^{\#}$. In particular, $\mathsf{U}^{\mathsf{E}_1}$ is semi-recursive in $\mathsf{E}_1^{\#}$ and from Theorem 6.5 and the proof of Corollary 6.3 we see that also $\sim\mathsf{U}^{\mathsf{E}_1}$ is semi-recursive in $\mathsf{E}_1^{\#}$. We should like to conclude from this that $\mathsf{U}^{\mathsf{E}_1}$ is in fact *recursive* in $\mathsf{E}_1^{\#}$. For this we need the analogue of Corollary 4.5 and thus, in turn, an ordinal comparison theorem.

We could proceed much as in § 3 to compare directly the ordinals of computations relative to $\mathsf{E}_1^{\#}$. The main additional complication is that in contrast with Lemma 3.2(3) we have if $a=\langle 3,k,l,b\rangle$ and $\{a\}^{\mathsf{E}_1^{\#}}(\mathbf{m},\boldsymbol{\alpha})\simeq 0$, then

$$|a,\mathbf{m}|_{\boldsymbol{\alpha}}^{\mathsf{E}_1^{\#}}=\inf_\beta \sup_p^{+} |b,\bar{\beta}(p),\mathbf{m}|_{\boldsymbol{\alpha}}^{\mathsf{E}_1^{\#}}$$

whereas if $\{a\}^{\mathsf{E}_1^{\#}}(\mathbf{m},\boldsymbol{\alpha})\simeq 1$, then

$$|a,\mathbf{m}|_{\boldsymbol{\alpha}}^{\mathsf{E}_1^{\#}}=\sup_\beta^{+} \inf_p |b,\bar{\beta}(p),\mathbf{m}|_{\boldsymbol{\alpha}}^{\mathsf{E}_1^{\#}}.$$

Fortunately we have available here a somewhat simpler method akin to that used in Lemma 1.7. We define inductively a relation M to which all relations semi-recursive in $\mathsf{E}_1^{\#}$ can be reduced. It then suffices to compare the ordinals associated with elements of M.

We define M to be the smallest relation such that for all $d\in\mathrm{Pri}$ and all $\boldsymbol{\alpha}$,

(i) $(0,\langle\boldsymbol{\alpha}\rangle)\in\mathsf{M}$;
(ii) if $\exists\beta\,\forall p.([d](\bar{\beta}(p),\boldsymbol{\alpha}),\langle\boldsymbol{\alpha}\rangle)\in\mathsf{M}$, then $(\langle 0,d\rangle,\langle\boldsymbol{\alpha}\rangle)\in\mathsf{M}$;
(iii) if $\forall\beta\,\exists p.([d](\bar{\beta}(p),\boldsymbol{\alpha}),\langle\boldsymbol{\alpha}\rangle)\in\mathsf{M}$, then $(\langle 1,d\rangle,\langle\boldsymbol{\alpha}\rangle)\in\mathsf{M}$.

Then for each $\boldsymbol{\alpha}$ we set

$$M_{\boldsymbol{\alpha}}=\{u:(u,\langle\boldsymbol{\alpha}\rangle)\in\mathsf{M}\}\quad\text{and}\quad M=M_{\varnothing}.$$

Of course, if $\mathsf{M}=\bar{\Gamma}$, then $M_{\boldsymbol{\alpha}}=\bar{\Gamma}_{\langle\boldsymbol{\alpha}\rangle}$, where Γ_β are the components of the decomposable operator Γ. In view of the (relative) simplicity of the definition of M, the following result is somewhat surprising:

6.7 Theorem. *There exists a primitive recursive function g such that for all a,* $\mathbf{m}$*,* $\boldsymbol{\alpha}$*, and n,*

$$\{a\}^{\mathsf{E}_1^\#}(\mathbf{m}, \boldsymbol{\alpha}) \simeq n \leftrightarrow g(a, \langle \mathbf{m} \rangle, n) \in M_{\boldsymbol{\alpha}}.$$

Proof. We shall prove the version without parameters ($\boldsymbol{\alpha} = \varnothing$). Note first that there exist primitive recursive functions $h_\forall$ and $h_\exists$ such that for any $d \in \text{Pri}$,

$$h_\forall(d, \langle \mathbf{m} \rangle) \in M \leftrightarrow \forall p . [d](p, \mathbf{m}) \in M,$$

and

$$h_\exists(d, \langle \mathbf{m} \rangle) \in M \leftrightarrow \exists p . [d](p, \mathbf{m}) \in M.$$

Namely, $h_\forall(d, \langle \mathbf{m} \rangle) = \langle 0, f(d, \langle \mathbf{m} \rangle) \rangle$ and $h_\exists(d, \langle \mathbf{m} \rangle) = \langle 1, f(d, \langle \mathbf{m} \rangle) \rangle$, where f is a primitive recursive function such that

$$[f(d, \langle \mathbf{m} \rangle)](s) = [d](\text{lg}(s), \mathbf{m}).$$

Now we shall specify the recursion conditions that g should satisfy and leave it to the reader to provide the details of application of the Recursion Theorem. The conditions are as usual by cases corresponding to those in the definition of $\Omega[\mathsf{E}_1^\#]$.

(0) If $a = \langle 0, k, 0, \dots \rangle$, then $g(a, \langle \mathbf{m} \rangle, n) = h_\exists(d_0, \langle a, \mathbf{m}, n \rangle)$, where

$$[d_0](u, a, \mathbf{m}, n) = \begin{cases} 0, & \text{if } \ \mathsf{T}(a, \langle \mathbf{m} \rangle; u, \langle \ \ \rangle) \wedge (u)_0 = n; \\ 1, & \text{otherwise.} \end{cases}$$

(1) If $a = \langle 1, k, 0, b, c_0, c_1 \rangle$, then $g(a, \langle \mathbf{m} \rangle, n) = h_\exists(d_1, \langle a, \mathbf{m}, n \rangle)$, where

$$[d_1](q_1, a, \mathbf{m}, n) = h_\exists(d_2, \langle q_1, a, \mathbf{m}, n \rangle),$$

$$[d_2](q_0, q_1, a, \mathbf{m}, n) = h_\forall(d_3, \langle q_0, q_1, a, \mathbf{m}, n \rangle), \quad \text{and}$$

$$[d_3](p, q_0, q_1, a, \mathbf{m}, n) = \begin{cases} g(b, \langle q_0, q_1 \rangle, n), & \text{if } \ p = 0; \\ g(c_0, \langle \mathbf{m} \rangle, q_0), & \text{if } \ p = 1; \\ g(c_1, \langle \mathbf{m} \rangle, q_1), & \text{if } \ p \geqslant 2. \end{cases}$$

(2) If $a = \langle 2, k+1, 0 \rangle$, then $g(a, \langle b, \mathbf{m} \rangle, n) = h_\exists(d_4, \langle b, \mathbf{m}, n \rangle)$, where

$$[d_4](p, b, \mathbf{m}, n) = g(b, \langle \mathbf{m} \rangle, n).$$

(3)$^\#$ If $a = \langle 3, k, 0, b \rangle$, then $g(a, \langle \mathbf{m} \rangle, 0) = \langle 0, f_1(a, \langle \mathbf{m} \rangle, n) \rangle$, and $g(a, \langle \mathbf{m} \rangle, 1) = \langle 1, h_\exists(d_5, \langle a, \mathbf{m}, n \rangle) \rangle$, where

$$[f_1(a, \langle \mathbf{m} \rangle, n)](s) = g(b, \langle s, \mathbf{m} \rangle, 0),$$

and

$$[d_5](q, a, \mathbf{m}, n) = \begin{cases} 1, & \text{if} \quad q = 0; \\ g(b, \langle s, \mathbf{m}\rangle, q), & \text{if} \quad q > 0. \end{cases}$$

The proof that if $\{a\}^{\mathsf{E}_1^\#}(\mathbf{m}) \simeq n$, then $g(a, \langle\mathbf{m}\rangle, n) \in M$ is by induction on $\Omega[\mathsf{E}_1^\#]$ with cases as follows.

(0) If $\{a\}^{\mathsf{E}_1^\#}(\mathbf{m}) \simeq n$ and $a = \langle 0, k, 0, \dots\rangle$, then also $\{a\}(\mathbf{m}) \simeq n$ and for some u, $\mathsf{T}(a, \langle\mathbf{m}\rangle, u, \langle\ \rangle)$ and $(u)_0 = n$. Hence for some u, $[d_0](u, a, \mathbf{m}, n) = 0 \in M$ and thus $g(a, \langle\mathbf{m}\rangle, n) \in M$.

(1) If $\{a\}^{\mathsf{E}_1^\#}(\mathbf{m}) \simeq \{b\}^{\mathsf{E}_1^\#}(q_0, q_1) \simeq n$, where $\{c_i\}^{\mathsf{E}_1^\#}(\mathbf{m}) \simeq q_i$, then successively

$$\exists q_0 \exists q_1 \forall p . [d_3](p, q_0, q_1, a, \mathbf{m}, n) \in M,$$

$$\exists q_0 \exists q_1 . [d_2](q_0, q_1, a, \mathbf{m}, n) \in M,$$

$$\exists q_1 . [d_1](q_1, a, \mathbf{m}, n) \in M, \quad \text{and}$$

$$g(a, \langle\mathbf{m}\rangle, n) \in M.$$

The cases (2) and $(3)^\#$ are similar. The proof of the converse implication is by induction over M and is left to the reader. □

6.8 Corollary. *There exists a primitive recursive function f such that for all a, $\mathbf{m}$, $\boldsymbol{\alpha}$, and n,*

$$\{a\}^{\mathsf{E}_1^\#}(\mathbf{m}, \boldsymbol{\alpha}) \text{ is defined} \leftrightarrow f(a, \langle\mathbf{m}\rangle) \in M_{\boldsymbol{\alpha}}.$$

Proof. Let $f(a, \langle\mathbf{m}\rangle) = h_\exists(d, \langle a, \langle\mathbf{m}\rangle\rangle)$, where d is an index such that

$$[d](n, a, \langle\mathbf{m}\rangle) = g(a, \langle\mathbf{m}\rangle, n),$$

with g and $h_\exists$ as in the preceding theorem. □

For $u \in M_{\boldsymbol{\alpha}}$, $|u|^\#_{\boldsymbol{\alpha}}$ denotes the least σ such that $u \in M^\sigma_{\boldsymbol{\alpha}}$; otherwise, $|u|^\#_{\boldsymbol{\alpha}} = \aleph_1$. Then it is immediate that

$$|0|^\#_{\boldsymbol{\alpha}} = 0, \quad \text{and if} \quad d \in \mathrm{Pri}^{1,l}, \quad \text{then}$$

$$|\langle 0, d\rangle|^\#_{\boldsymbol{\alpha}} = \inf\{\sup{}^+\{|[d](\bar{\beta}(p), \boldsymbol{\alpha})|^\#_{\boldsymbol{\alpha}} : p \in \omega\} : \beta \in {}^\omega\omega\};$$

$$|\langle 1, d\rangle|^\#_{\boldsymbol{\alpha}} = \sup{}^+\{\inf\{|[d](\bar{\beta}(p), \boldsymbol{\alpha})|^\#_{\boldsymbol{\alpha}} : p \in \omega\} : \beta \in {}^\omega\omega\}.$$

In what follows we shall abbreviate these expressions by

$$|\langle 0, d\rangle|^{\#}_{\alpha} = \inf_{\beta} \sup^{+}_{p} |d, \beta, p|^{\#}_{\alpha},$$

etc.

6.9 Ordinal Comparison Theorem. *There exists a functional* H *partial recursive in* $\mathsf{E}_1^{\#}$ *such that for all* u, v, *and* $\boldsymbol{\alpha}$,

(i) *if* $u \in M_{\alpha}$ *and* $|u|^{\#}_{\alpha} \leqslant |v|^{\#}_{\alpha}$, *then* $\mathsf{H}(u, v, \langle\boldsymbol{\alpha}\rangle) \simeq 0$;
(ii) *if* $v \in M_{\alpha}$ *and* $|v|^{\#}_{\alpha} < |u|^{\#}_{\alpha}$, *then* $\mathsf{H}(u, v, \langle\boldsymbol{\alpha}\rangle) \simeq 1$.

Proof. We shall again prove the version without parameters. For this proof only we set

$$\mathsf{A}_1^{\#}(f) \simeq 1 \dot{-} \mathsf{E}_1^{\#}(\lambda m \,.\, 1 \dot{-} f(m))$$

so that

$$\mathsf{A}_1^{\#}(f) \simeq \begin{cases} 0, & \text{if} \quad \forall\beta\,\exists p \,.\, f(\bar\beta(p)) \simeq 0; \\ 1, & \text{if} \quad \exists\beta\,\forall p\,(\exists n > 0)\,.\, f(\bar\beta(p)) \simeq n; \\ \text{undefined}, & \text{otherwise.} \end{cases}$$

By the $\mathsf{E}_1^{\#}$-Recursion Theorem there exists a functional H partial recursive in $\mathsf{E}_1^{\#}$ which satisfies the following conditions for all u and v, and all d and $e \in \mathrm{Pri}^{1,0}$:

(1) $\mathsf{H}(0, v) \simeq 0$;
(2) $\mathsf{H}(u+1, 0) \simeq 1$;
(3) $\mathsf{H}(u+1, v+1) \simeq 0$, if $v+1$ is not of the form $\langle 0, c\rangle$ or $\langle 1, c\rangle$ with $c \in \mathrm{Pri}^{1,0}$;
(4) $\mathsf{H}(u+1, \langle 0, e\rangle) \simeq \mathsf{H}(u+1, \langle 1, e\rangle) \simeq 1$, if $u+1$ is not of the form $\langle 0, c\rangle$ or $\langle 1, c\rangle$ with $c \in \mathrm{Pri}^{1,0}$;
(5) $\mathsf{H}(\langle 0, d\rangle, \langle 0, e\rangle) \simeq \mathsf{E}_1^{\#}(\lambda s \,.\, \mathsf{A}_1^{\#}(\lambda t \,.\, \mathsf{H}([d](s), [e](t))))$;
(6) $\mathsf{H}(\langle 1, d\rangle, \langle 0, e\rangle) \simeq \mathsf{A}_1^{\#}(\lambda s \,.\, \mathsf{A}_1^{\#}(\lambda t \,.\, \mathsf{H}([d](s), [e](t))))$;
(7) $\mathsf{H}(\langle 0, d\rangle, \langle 1, e\rangle) \simeq \mathsf{E}_1^{\#}(\lambda t \,.\, \mathsf{E}_1^{\#}(\lambda s \,.\, \mathsf{H}([d](s), [e](t))))$;
(8) $\mathsf{H}(\langle 1, d\rangle, \langle 1, e\rangle) \simeq \mathsf{A}_1^{\#}(\lambda s \,.\, \mathsf{E}_1^{\#}(\lambda t \,.\, \mathsf{H}([d](s), [e](t))))$.

The proof that H satisfies (i) and (ii) is by induction on $\sigma = \min\{|u|^{\#}, |v|^{\#}\}$. Clauses (1) and (2) of the definition of H correspond to the case $\sigma = 0$; (3) and (4) correspond to cases in which either u or v does not belong to M by virtue of being of the wrong form. Both of these are easily seen to be in accord with (i) and (ii), so we consider the four cases $u = \langle i, d\rangle$ and $v = \langle j, e\rangle$, with $d, e \in \mathrm{Pri}^{1,0}$ and at least one of u and v a member of M. For the case $i = j = 0$, we have

$$\begin{aligned} |\langle 0, d\rangle|^{\#} \leqslant |\langle 0, e\rangle|^{\#} &\leftrightarrow \inf_{\beta} \sup^{+}_{p} |d, \beta, p|^{\#} \leqslant \inf_{\gamma} \sup^{+}_{q} |e, \gamma, q|^{\#} \\ &\leftrightarrow \exists\beta\,\forall\gamma \,.\, \sup^{+}_{p} |d, \beta, p|^{\#} \leqslant \sup^{+}_{q} |e, \gamma, q|^{\#} \\ &\leftrightarrow \exists\beta\,\forall\gamma\,\forall p\,\exists q \,.\, |d, \beta, p|^{\#} \leqslant |e, \gamma, q|^{\#} \\ &\leftrightarrow \exists\beta\,\forall p\,\forall\gamma\,\exists q \,.\, |d, \beta, p|^{\#} \leqslant |e, \gamma, q|^{\#}. \end{aligned}$$

If these hold, then $\exists\beta\,\forall p\,.\,|d,\beta,p|^{*}<\sigma$, so by the induction hypothesis,

$$\exists\beta\,\forall p\,\forall\gamma\,\exists q\,.\,\mathsf{H}([d](\bar\beta(p)),[e](\bar\gamma(q)))\simeq 0$$

and thus by clause (5), $\mathsf{H}(\langle 0,d\rangle,\langle 0,e\rangle)\simeq 0$ as required by (i). On the other hand, if these are false, then

$$\forall\beta\,\exists p\,\exists\gamma\,\forall q\,.\,\mathsf{H}([d](\bar\beta(p)),[e](\bar\gamma(p)))\simeq 1$$

so $\mathsf{H}(\langle 0,d\rangle,\langle 0,e\rangle)\simeq 1$.

Consider next the case $i=0$ and $j=1$. We have

$$\begin{aligned}|\langle 0,d\rangle|^{*}\leqslant|\langle 1,e\rangle|^{*}&\leftrightarrow \inf_\beta \sup^{+}_p|d,\beta,p|^{*}\leqslant \sup^{+}_\gamma \inf_q|e,\gamma,q|^{*}\\&\leftrightarrow\exists\beta\,\exists\gamma\,.\,\sup^{+}_p|d,\beta,q|^{*}\leqslant\inf_q|e,\gamma,q|^{*}+1\\&\leftrightarrow\exists\beta\,\exists\gamma\,\forall p\,\forall q\,.\,|d,\beta,p|^{*}\leqslant|e,\gamma,q|^{*}\\&\leftrightarrow\exists\gamma\,\forall q\,\exists\beta\,\forall p\,.\,|d,\beta,p|^{*}\leqslant|e,\gamma,q|^{*}.\end{aligned}$$

The last implication ($\rightarrow$) is by first-order logic. For ($\leftarrow$), for any γ_0 choose q_0 to minimize $|e,\gamma_0,q_0|^{*}$ and β_0 such that for all p, $|d,\beta_0,p|^{*}\leqslant|e,\gamma_0,q_0|^{*}$. Then for all p and q, $|d,\beta_0,p|^{*}\leqslant|e,\gamma_0,q|^{*}$. (Note that this sort of argument would not suffice to obtain the prefix $\exists\beta\,\forall p\,\exists\gamma\,\forall q$, which accounts for the interchange of s and t in clause (7) of the definition of H.) Now just as above we have, under the assumption that one of $\langle 0,d\rangle$ and $\langle 1,e\rangle$ belongs to M,

$$\begin{aligned}|\langle 0,d\rangle|^{*}\leqslant|\langle 1,e\rangle|^{*}&\rightarrow\exists\gamma\,\forall q\,\exists\beta\,\forall p\,.\,\mathsf{H}([d](\bar\beta(p)),[e](\bar\gamma(q)))\simeq 0\\&\rightarrow\mathsf{H}(\langle 0,d\rangle,\langle 1,e\rangle)\simeq 0,\end{aligned}$$

and

$$\begin{aligned}|\langle 1,e\rangle|^{*}<|\langle 0,d\rangle|^{*}&\rightarrow\forall\gamma\,\exists q\,\forall\beta\,\exists p\,.\,\mathsf{H}([d](\bar\beta(p)),[e](\bar\gamma(q)))\simeq 1\\&\rightarrow\mathsf{H}(\langle 0,d\rangle,\langle 1,e\rangle)\simeq 1.\end{aligned}$$

The other two cases are based similarly on the equivalences:

$$\begin{aligned}|\langle 1,d\rangle|^{*}\leqslant|\langle 0,e\rangle|^{*}&\leftrightarrow \sup^{+}_\beta \inf_p|d,\beta,p|^{*}\leqslant \inf_\gamma \sup^{+}_q|e,\gamma,q|^{*}\\&\leftrightarrow\forall\beta\,\forall\gamma\,.\,\inf_p|d,\beta,p|^{*}<\sup^{+}_q|e,\gamma,q|^{*}\\&\leftrightarrow\forall\beta\,\forall\gamma\,\exists p\,\exists q\,.\,|d,\beta,p|^{*}\leqslant|e,\gamma,q|^{*}\\&\leftrightarrow\forall\beta\,\exists p\,\forall\gamma\,\exists q\,.\,|d,\beta,p|^{*}\leqslant|e,\gamma,q|^{*};\end{aligned}$$

and

$$|\langle 1, d\rangle|^{\#} \leq |\langle 1, e\rangle|^{\#} \leftrightarrow \sup_{\beta}^{+} \inf_p |d, \beta, p|^{\#} \leq \sup_{\gamma}^{+} \inf_q |e, \gamma, q|^{\#}$$
$$\leftrightarrow \forall\beta\, \exists\gamma \,.\, \inf_p |d, \beta, p|^{\#} \leq \inf_q |e, \gamma, q|^{\#}$$
$$\leftrightarrow \forall\beta\, \exists\gamma\, \exists p\, \forall q \,.\, |d, \beta, p|^{\#} \leq |e, \gamma, q|^{\#}$$
$$\leftrightarrow \forall\beta\, \exists p\, \exists\gamma\, \forall q \,.\, |d, \beta, p|^{\#} \leq |e, \gamma, q|^{\#}. \quad \square$$

From Theorems 6.7 and 6.9 we can derive results corresponding to theose of §4 for the class of relations semi-recursive in $\mathsf{E}_1^{\#}$:

6.10 Corollary.

(i) *There exists a functional* $\mathrm{Sel}^{\mathsf{E}_1^{\#}}$ *partial recursive in* $\mathsf{E}_1^{\#}$ *such that for all* a, $\mathbf{m}$, *and* $\boldsymbol{\alpha}$,

$$\exists p \,.\, \{a\}^{\mathsf{E}_1^{\#}}(p, \mathbf{m}, \boldsymbol{\alpha}) \text{ is defined} \leftrightarrow \{a\}^{\mathsf{E}_1^{\#}}(\mathrm{Sel}^{\mathsf{E}_1^{\#}}(a, \langle\mathbf{m}\rangle, \langle\boldsymbol{\alpha}\rangle), \mathbf{m}, \boldsymbol{\alpha}) \text{ is defined};$$

(ii) *for any relation* R *semi-recursive in* $\mathsf{E}_1^{\#}$, *there exists a functional* $\mathrm{Sel}_{\mathsf{R}}$ *partial recursive in* $\mathsf{E}_1^{\#}$ *such that for all* $\mathbf{m}$ *and* $\boldsymbol{\alpha}$,

$$\exists p\, \mathsf{R}(p, \mathbf{m}, \boldsymbol{\alpha}) \leftrightarrow \mathsf{R}(\mathrm{Sel}_{\mathsf{R}}(\mathbf{m}, \boldsymbol{\alpha}), \mathbf{m}, \boldsymbol{\alpha});$$

(iii) *the class of relations semi-recursive in* $\mathsf{E}_1^{\#}$ *is closed under finite union and bounded and unbounded existential number quantification*;

(iv) *the class of functionals partial recursive in* $\mathsf{E}_1^{\#}$ *and relations semi-recursive in* $\mathsf{E}_1^{\#}$ *is closed under definition by positive cases*;

(v) *a relation is recursive in* $\mathsf{E}_1^{\#}$ *iff it is both semi-recursive in* $\mathsf{E}_1^{\#}$ *and co-semi-recursive in* $\mathsf{E}_1^{\#}$;

(vi) *for any partial functional* F, F *is partial recursive in* $\mathsf{E}_1^{\#}$ *iff* Gr_{F} *is semi-recursive in* $\mathsf{E}_1^{\#}$ *and* F *is recursive in* $\mathsf{E}_1^{\#}$ *iff* F *is total and* Gr_{F} *is recursive in* $\mathsf{E}_1^{\#}$.

Proof. For (i) we proceed as for Theorem 4.1 except that in place of $\mathsf{H}(\langle a, \mathbf{m}\rangle, \langle b, \mathbf{n}\rangle, \langle\boldsymbol{\alpha}\rangle)$, we write $\mathsf{H}(f(a, \langle\mathbf{m}\rangle), f(b, \langle\mathbf{n}\rangle), \langle\boldsymbol{\alpha}\rangle)$, with f the function of Corollary 6.8. (ii)–(vi) then follow as in §4. $\square$

In particular, it follows from (v) and the discussion following Theorem 6.6 that $\mathsf{U}^{\mathsf{E}_1}$ is recursive in $\mathsf{E}_1^{\#}$. In fact,

6.11 Theorem. *For any functional* I, *if* I *is recursive in* $\mathsf{E}_1^{\#}$, *then also* I^{sJ} *is recursive in* $\mathsf{E}_1^{\#}$.

Proof. Suppose that I is recursive in $\mathsf{E}_1^{\#}$. An obvious modification of the proof of

Theorem 6.6 establishes that there exists a primitive recursive function g such that for all a, $\mathbf{m}$, and $\boldsymbol{\alpha}$,

$$\{a\}^{\mathsf{I}}(\mathbf{m}, \boldsymbol{\alpha}) \simeq \{g(a)\}^{\mathsf{E}_1^{\#}}(\mathbf{m}, \boldsymbol{\alpha}).$$

Hence U^{I} and $\sim\mathsf{U}^{\mathsf{I}}$ are both semi-recursive in $\mathsf{E}_1^{\#}$, so by 6.10(v), U^{I} is recursive in $\mathsf{E}_1^{\#}$. Since

$$\mathsf{I}^{\mathrm{sJ}}((\langle a, \mathbf{m}\rangle) * \alpha) = \begin{cases} 0, & \text{if } \mathsf{U}^{\mathsf{I}}(a, \langle \mathbf{m}\rangle, \langle \alpha\rangle); \\ 1, & \text{otherwise,} \end{cases}$$

also I^{sJ} is recursive in $\mathsf{E}_1^{\#}$. □

Thus if we define a sequence of functionals E_r by:

$$\mathsf{E}_0 = \mathsf{E} \quad \text{and} \quad \mathsf{E}_{r+1} = (\mathsf{E}_r)^{\mathrm{sJ}},$$

then E_r is recursive in E_s iff $r \leqslant s$ and all E_r are recursive in $\mathsf{E}_1^{\#}$ (cf. Exercise 1.20). The functionals E_r will play a role in § VIII.4, and in § VII.1. Theorem 6.11 will be extended to show that the type-3 functional corresponding to the superjump **sJ** is recursive in $\mathsf{E}_1^{\#}$ (Theorem VII.1.8).

We turn now to the relationship of $\mathsf{E}_1^{\#}$ to inductive definability and the $*$-operation.

6.12 Theorem. *For any Σ_1^1 monotone operator Γ over ω, $\bar{\Gamma}$ is semi-recursive in $\mathsf{E}_1^{\#}$.*

Proof. Suppose that Γ is Σ_1^1 and monotone, say

$$\mathsf{P}_\Gamma(m, \alpha) \leftrightarrow \exists \gamma\, \forall p\, R(\bar{\alpha}(p), \bar{\gamma}(p), m)$$

with R recursive. Since Γ is monotone,

$$m \in \Gamma(A) \leftrightarrow \exists B\, [B \subseteq A \wedge m \in \Gamma(B)].$$

Thus,

$$\begin{aligned} \mathsf{P}_\Gamma(m, \alpha) &\leftrightarrow \exists \beta\, \exists \gamma\, [\forall n\, (\beta(n) = 0 \rightarrow \alpha(n) = 0) \wedge \forall p\, R(\bar{\beta}(p), \bar{\gamma}(p), m)] \\ &\leftrightarrow \exists \beta\, \forall n\, [((\beta)_0(n) = 0 \rightarrow \alpha(n) = 0) \wedge R(\overline{(\beta)_0}(n), \overline{(\beta)_1}(n), m)]. \end{aligned}$$

Let G be the function partial recursive in $\mathsf{E}_1^{\#}$ which computes according to the following flow diagram:

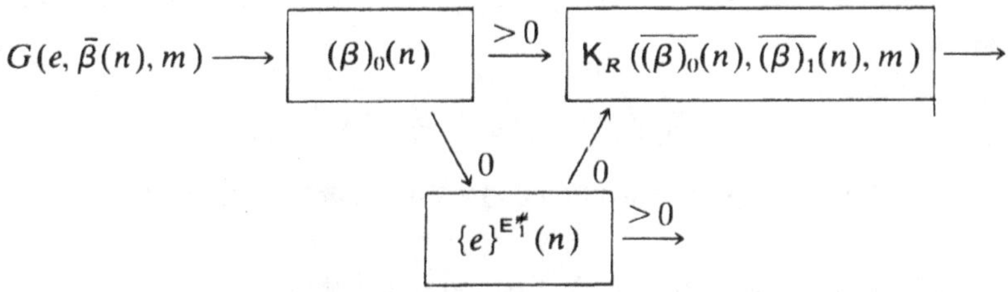

Set $H(e, m) \simeq \mathsf{E}_1^\#(\lambda s.\, G(e, s, m))$ and let $\bar{e}$ be the natural index (provided by the proof of the $\mathsf{E}_1^\#$-Recursion Theorem) such that $H(\bar{e}, m) \simeq \{\bar{e}\}^{\mathsf{E}_1^\#}(m)$. We claim that

$$m \in \bar{\Gamma} \leftrightarrow \{\bar{e}\}^{\mathsf{E}_1^\#}(m) \simeq 0.$$

The implication ($\rightarrow$) is proved by a straightforward induction over $\bar{\Gamma}$. The implication ($\leftarrow$) is proved by an induction on subcomputations using the fact that if $\{\bar{e}\}^{\mathsf{E}_1^\#}(m) \simeq 0$, then for some β and all n, the computation of $G(\bar{e}, \bar{\beta}(n), m)$ is a subcomputation and thus so is that of $\{\bar{e}\}^{\mathsf{E}_1^\#}(n) \simeq 0$ for all n such that $(\beta)_0(n) = 0$. Hence all such n belong to $\bar{\Gamma}$ by the induction hypothesis. □

The inductive definition of M (and of M) has both Σ_1^1 and Π_1^1 clauses — that is, there are monotone operators $\Gamma_0 \in \Sigma_1^1$ and $\Gamma_1 \in \Pi_1^1$ such that if for all A, $\Gamma(A) = \Gamma_0(A) \cup \Gamma_1(A)$, then $\bar{\Gamma} = M$. Because of the special form of Γ_1 and the fact that Π_1^1 relations are all Π_1^0 inductively definable, we can in fact replace Γ by a Σ_1^1 operator:

6.13 Lemma. *Let Γ_0 and Γ_1 be monotone operators defined by*

$$m \in \Gamma_0(A) \leftrightarrow \exists \gamma\, \forall p\, \mathsf{R}_0(\bar{\gamma}(p), m, A),$$

and

$$m \in \Gamma_1(A) \leftrightarrow \forall \delta\, \exists q\, \mathsf{R}_1(\bar{\delta}(q), m, A),$$

where R_0 and R_1 are recursive relations and R_1 satisfies:

$$(*) \qquad \mathsf{R}_1(s, m, A) \wedge A \subseteq B \rightarrow \mathsf{R}_1(s, m, B),$$

and set $\Gamma(A) = \Gamma_0(A) \cup \Gamma_1(A)$. Then there exists a Σ_1^1 monotone operator Λ such that for all m,

$$m \in \bar{\Gamma} \leftrightarrow \langle 0, m \rangle \in \bar{\Lambda}.$$

Proof. For any set A, let $A_{(0)} = \{m : \langle 0, m\rangle \in A\}$ and let Λ be the operator defined by:

(1) $\langle 0, m\rangle \in \Lambda(A) \leftrightarrow m \in \Gamma_0(A_{(0)}) \vee \langle 1, m, \langle\ \rangle\rangle \in A;$

(2) $\langle 1, m, s\rangle \in \Lambda(A) \leftrightarrow \mathsf{R}_1(s, m, A_{(0)}) \vee \forall n . \langle 1, m, s * \langle n\rangle\rangle \in A.$

Clearly Λ is Σ_1^1 and monotone. We first establish that

(3) $\forall\delta\, \exists q\, \mathsf{R}_1(s * \bar{\delta}(q), m, \bar{\Lambda}_{(0)}) \rightarrow \langle 1, m, s\rangle \in \bar{\Lambda};$

(4) $\langle 1, m, s\rangle \in \Lambda^{\sigma} \rightarrow \forall\delta\, \exists q\, \mathsf{R}_1(s * \bar{\delta}(q), m, (\Lambda^{\sigma})_{(0)}).$

For (3), suppose that $\langle 1, m, s\rangle \notin \bar{\Lambda}$. Then as in the latter part of the proof of Theorem III.3.2 there exists a unique function δ such that

$$\delta(q) = \text{least } n . \langle 1, m, s * \bar{\delta}(q) * \langle n\rangle\rangle \notin \bar{\Lambda}.$$

In particular, $\forall q \sim \mathsf{R}_1(s * \bar{\delta}(q), m, \bar{\Lambda}_{(0)})$.

For (4), we proceed by induction on σ and assume as induction hypothesis that (4) holds for all $\tau < \sigma$ in place of σ. Suppose that $\langle 1, m, s\rangle \in \Lambda^{\sigma}$. Then either $\mathsf{R}_1(s, m, (\Lambda^{(\sigma)})_{(0)})$ or $\forall n . \langle 1, m, s * \langle n\rangle\rangle \in \Lambda^{(\sigma)}$. In the first case, by the property $(*)$ of R_1, $\mathsf{R}_1(s, m, (\Lambda^{\sigma})_{(0)})$ and thus $\forall\delta\, \mathsf{R}_1(s * \bar{\delta}(0), m, (\Lambda^{\sigma})_{(0)})$. In the second case, for each n there exists a $\tau_n < \sigma$ such that $\langle 1, m, s * \langle n\rangle\rangle \in \Lambda^{\tau_n}$ and hence, by the induction hypothesis, $\forall\delta\, \exists q\, \mathsf{R}_1(s * \langle n\rangle * \bar{\delta}(q), m, (\Lambda^{\tau_n})_{(0)})$. Then, again by $(*)$, $\forall n\, \forall\delta\, \exists q\, \mathsf{R}_1(s * \langle n\rangle * \bar{\delta}(q), m, (\Lambda^{\sigma})_{(0)})$ and thus $\forall\delta\, \exists q\, \mathsf{R}_1(s * \bar{\delta}(q), m, (\Lambda^{\sigma})_{(0)})$.

To prove that $\bar{\Gamma} \subseteq \bar{\Lambda}_{(0)}$ we show that $\Gamma(\bar{\Lambda}_{(0)}) \subseteq \bar{\Lambda}_{(0)}$. First, if $m \in \Gamma_0(\bar{\Lambda}_{(0)})$, then by (1), $\langle 0, m\rangle \in \Lambda(\bar{\Lambda}) = \bar{\Lambda}$, so $m \in \bar{\Lambda}_{(0)}$. Suppose that $m \in \Gamma_1(\bar{\Lambda}_{(0)})$, so $\forall\delta\, \exists q\, \mathsf{R}_1(\bar{\delta}(q), m, \bar{\Lambda}_{(0)})$. Then by (3), $\langle 1, m, \langle\ \rangle\rangle \in \bar{\Lambda}$ and thus by (1), $\langle 0, m\rangle \in \bar{\Lambda}$ so $m \in \bar{\Lambda}_{(0)}$.

For the converse, we prove by induction on σ that $(\Lambda^{\sigma})_{(0)} \subseteq \bar{\Gamma}$. As induction hypothesis we assume that $(\Lambda^{(\sigma)})_{(0)} \subseteq \bar{\Gamma}$. Then if $m \in (\Lambda^{\sigma})_{(0)}$, either $m \in \Gamma_0((\Lambda^{(\sigma)})_{(0)})$ or $\langle 1, m, \langle\ \rangle\rangle \in \Lambda^{(\sigma)}$. In the first case we have by the induction hypothesis and the monotonicity of Γ_0 that $m \in \Gamma_0(\bar{\Gamma}) \subseteq \bar{\Gamma}$. In the second case, there exists an ordinal $\tau < \sigma$ such that $\langle 1, m, \langle\ \rangle\rangle \in \Lambda^{\tau}$. By (4), $\forall\delta\, \exists q\, \mathsf{R}_1(\bar{\delta}(q), m, (\Lambda^{\tau})_{(0)})$, that is $m \in \Gamma_1((\Lambda^{\tau})_{(0)})$. By the induction hypothesis and the monotonicity of Γ_1, $m \in \Gamma_1(\bar{\Gamma}) \subseteq \bar{\Gamma}$. □

6.14 Theorem. *For all $R \subseteq {}^k\omega$, the following are equivalent:*

(i) *R is semi-recursive in $\mathsf{E}_1^{\#}$;*

(ii) *R is reducible to the closure of a monotone Σ_1^1 operator;*

(iii) $R \in \Pi_1^{\mathscr{A}*}$.

Proof. (i)$\to$(ii): The Π^1_1 clause in the inductive defintion of M clearly satisfies the condition ($*$) of the preceding lemma. Hence M is reducible to the closure of a monotone Σ^1_1 operator, and by Corollary 6.8, so is every R semi-recursive in $\mathsf{E}^{\#}_1$.

(ii)$\to$(i): Immediate from Theorem 6.12.

(iii)$\to$(ii): From Definition V.5.2 we see easily that a relation R is $\Pi^{\mathscr{A}^*}_1$ just in case there is a recursive relation P such that

$$(**) \qquad R(\mathbf{m}) \leftrightarrow (\forall A \in \mathsf{B}(\mathscr{A}^*))(\exists t \in A)\exists q\, P(t, q, \mathbf{m}).$$

Let Γ be the monotone operator defined by:

$$\langle \mathbf{m}, s\rangle \in \Gamma(D) \leftrightarrow \exists q\, P(s, q, \mathbf{m}) \vee \forall \beta\, \exists q\, \exists \gamma\, \forall q . \langle \mathbf{m}, s * \langle \bar{\beta}(p), \bar{\gamma}(q)\rangle\rangle \in D.$$

The proof of Theorem V.4.10 (with obvious simple modifications) shows that

$$R(\mathbf{m}) \leftrightarrow \langle \mathbf{m}, \langle\ \rangle\rangle \in \bar{\Gamma}.$$

As it stands, Γ is nowhere near being a Σ^1_1 operator. However, it is not hard to see that this equivalence holds also for the operator Γ defined by:

$$\begin{aligned}\langle \mathbf{m}, s\rangle \in \Gamma(D) \leftrightarrow\ & \exists q\, P(s, q, \mathbf{m}) \vee \\ & \vee [\mathrm{lg}(s) \text{ is even} \wedge \forall \beta\, \exists p . \langle \mathbf{m}, s * \langle \bar{\beta}(p)\rangle\rangle \in D] \\ & \vee [\mathrm{lg}(s) \text{ is odd} \wedge \exists \gamma\, \forall q . \langle \mathbf{m}, s * \langle \bar{\gamma}(q)\rangle\rangle \in D].\end{aligned}$$

This Γ is in the form specified in the hypothesis of Lemma 6.13 and thus may be in turn replaced by a Σ^1_1 monotone operator.

(ii)$\to$(iii): Let Γ be a monotone Σ^1_1 operator, say

$$m \in \Gamma(A) \leftrightarrow \exists \gamma\, \forall q\, \mathsf{R}(\bar{\gamma}(q), m, A)$$

for a recursive relation R. Because Γ is monotone, we have

$$\begin{aligned} m \in \Gamma(A) & \leftrightarrow \exists B\, [B \subseteq A \wedge m \in \Gamma(B)] \\ & \leftrightarrow \exists B\, \exists \gamma\, \forall q\, \forall r\, [(r \notin B \vee r \in A) \wedge \mathsf{R}(\bar{\gamma}(q), m, B)].\end{aligned}$$

For any $A \neq \omega$ we have then

$$m \in \Gamma(A) \leftrightarrow \exists B\, \exists \gamma\, \forall q\, \forall r\, [(r \notin B \wedge \mathsf{R}(\bar{\gamma}(q), m, B)) \vee r \in A].$$

Hence for a suitable recursive relation S,

$$(\dagger) \qquad m \in \Gamma(A) \leftrightarrow \exists \beta\, \forall p\, [S(\bar{\beta}(p), m) \vee (p)_0 \in A].$$

Let P and Q be recursive relations such that

$$Q(\langle s_0,\ldots,s_n\rangle,m)\leftrightarrow S(s_0,m)\vee(\exists i<n)S(s_{i+1},(\lg(s_i))_0),$$

$P(\langle s_0\rangle,m)$ is always false, and for all n, $s_0,\ldots,s_{2n+2}$,

$$P(\langle s_0,\ldots,s_{2n+2}\rangle,m)\leftrightarrow P(\langle s_0,\ldots,s_{2n+1}\rangle,m)\leftrightarrow Q(\langle s_1,s_3,s_5,\ldots,s_{2n+1}\rangle,m).$$

We shall show that

$$m\in\bar{\Gamma}\leftrightarrow(\forall A\in \mathrm{B}(\mathscr{A}^*))(\exists t\in A)P(t,m)$$

which by (**) above implies that $\bar{\Gamma}\in\Pi_1^{\mathscr{A}^*}$.

First observe that by formula (8) of § V.4, the right side of this equivalence is equivalent to

$$(\forall\beta_0\exists p_0)(\exists\beta_1\forall p_1)(\forall\beta_2\exists p_2)\cdots\exists n P(\langle\bar{\beta}_0(p_0),\ldots,\bar{\beta}_n(p_n)\rangle,m)$$

and hence to

$$(\exists\beta_0\forall p_0)(\exists\beta_1\forall p_1)(\exists\beta_2\forall p_2)\cdots\exists n Q(\langle\bar{\beta}_0(p_0),\ldots,\bar{\beta}_n(p_n)\rangle,m),$$

which we abbreviate by: $m\in Q^*$. We recall from § V.4 that $m\in Q^*$ just in case player I has a winning strategy in the game $\mathscr{G}_m$ played as follows: the players play alternately, at his n-th turn player I chooses β_n and player II chooses p_n, and player I wins just in case $\exists n Q(\langle\bar{\beta}_0(p_0),\ldots,\bar{\beta}_n(p_n)\rangle,m)$. Thus we need to prove:

$$m\in\bar{\Gamma}\leftrightarrow\text{player I has a winning strategy for } \mathscr{G}_m.$$

Suppose first that $m\notin\bar{\Gamma}$; we shall describe a winning strategy for player II in $\mathscr{G}_m$. Since $m\notin\Gamma(\bar{\Gamma})$, we have by (†),

$$\forall\beta_0\exists p_0[\sim S(\bar{\beta}_0(p_0),m)\wedge(p_0)_0\notin\bar{\Gamma}]$$

so let player II at his first turn respond to I's choice of β_0 with such a p_0. Then, since $(p_0)_0\notin\Gamma(\bar{\Gamma})$, again by (†),

$$\forall\beta_1\exists p_1[\sim S(\bar{\beta}_1(p_1),(p_0)_0)\wedge(p_1)_0\notin\bar{\Gamma}],$$

and player II at his second turn responds to player I's choice of β_1 with such a p_1. It is clear that by following this strategy, player II ensures that for all n,

$$\sim S(\bar{\beta}_0(p_0),m)\wedge(\forall i<n)\sim S(\bar{\beta}_{i+1}(p_{i+1}),(p_i)_0)$$

and thus that for all n, $\sim Q(\langle\bar{\beta}_0(p_0), \ldots, \bar{\beta}_n(p_n)\rangle, m)$. Hence this is a winning strategy for player II.

For the converse implication we prove that $\Gamma(Q^*) \subseteq Q^*$. Suppose that $m \in \Gamma(Q^*)$ so by (†),

$$\exists\gamma\, \forall q\, [S(\bar{\gamma}(q), m) \vee (q)_0 \in Q^*].$$

Then by the definition of Q^*,

$$\exists\gamma\, \forall q\, [S(\bar{\gamma}(q), m) \vee (\exists\beta_0 \forall p_0) \cdots \exists n\, Q(\langle\bar{\beta}_0(p_0), \ldots, \bar{\beta}_n(p_n)\rangle, (q)_0)]$$

or equivalently,

$$\exists\gamma\, \forall q\, [S(\bar{\gamma}(q), m) \vee (\exists\beta_0 \forall p_0) \cdots \exists n \\ (S(\bar{\beta}_0(p_0), (q)_0) \vee (\exists i < n)\, S(\bar{\beta}_{i+1}(p_{i+1}), (p_i)_0))].$$

By a rule of first-order logic, which is easily seen to apply also to the current situation involving infinite strings of quantifiers,

$$\exists\gamma\, \forall q\, \exists\beta_0 \forall p_0 \cdots \exists n\, [S(\bar{\gamma}(q), m) \\ \vee S(\bar{\beta}_0(p_0), (q_0)) \vee (\exists i < n)\, S(\bar{\beta}_{i+1}(p_{i+1}), (p_i)_0)].$$

From this follows immediately that $m \in Q^*$, as required. □

6.15 Corollary. *For all $R \subseteq {}^k\omega$, the following are equivalent:*

(i) *R is recursive in $\mathsf{E}_1^\#$;*

(ii) *R and $\sim R$ are each reducible to the closure of a Σ_1^1 monotone operator;*

(iii) *$R \in \Delta_1^{\mathscr{A}^*}$.*

Proof. Immediate from Theorem 6.14 and Corollary 6.10(v). □

6.16 Corollary. *$\nabla(\mathscr{A})$ is a proper subclass of $\Delta_1^{\mathscr{A}^*}$.*

Proof. U^{E_1} is recursive in $\mathsf{E}_1^\#$, hence belongs to $\Delta_1^{\mathscr{A}^*}$, but U^{E_1} is not recursive in E_1 and thus by the results stated at the beginning of § 5, U^{E_1} does not belong to $\nabla(\mathscr{A})$. □

We note that the proof of Theorem 6.14 shows that a relation R is reducible to the closure of a Σ_1^1 monotone operator iff it is in the form

$$\exists\beta_0 \forall p_0 \exists\beta_1 \forall p_1 \cdots \exists n\, P(\langle\bar{\beta}_0(p_0), \ldots, \bar{\beta}_n(p_n)\rangle, \mathbf{m}),$$

with P recursive. The reader should contrast this with Exercise III.3.23.

We conclude this section with a brief discussion of what is needed to extend 6.6–6.16 to extended functionals $\Phi^{\#}$ other than $\mathsf{E}_1^{\#}$. For 6.6 it will suffice that Φ be a normal operation — the crucial point is to have the power of $\mathsf{E}^{\#}$ available during a computation to check that certain functions are total, and if Φ is normal, then $\mathsf{E}^{\#}$ is recursive in $\Phi^{\#}$. The same is true for 6.7–6.10. Theorem 6.11 depends on the special property of the quantifier $\mathscr{A}$ expressed in Corollary 6.3. It holds for any extended functional $\Phi^{\#}$ which arises from an operation Φ which is strongly normal: Φ is *strongly normal* iff Φ is normal and there exists a primitive recursive function h such that for any family $\langle \mathsf{P}_p : p \in \omega\rangle$ of relations,

$$\mathscr{A}\langle \mathsf{P}_p : p \in \omega\rangle = \Phi\langle \mathsf{P}_{h(p)} : p \in \omega\rangle.$$

When Φ is strongly normal, $\mathsf{E}_1^{\#}$ is computable in terms of $\Phi^{\#}$ and thus if I is recursive in $\Phi^{\#}$, so is I^{sJ}. The remaining results are summarized in

6.17 Theorem. *For any normal positive analytic operation Φ and all $R \subseteq {}^k\omega$, the following are equivalent*:

(i) *R is semi-recursive in $\Phi^{\#}$*;

(ii) *R is reducible to the closure of an effective Φ-positive inductive operator*;

(iii) $R \in \Pi_1^{\Phi^*}$.

Sketch of proof. That (i)→(ii) is immediate from the observation that the inductive operator which defines $M[\Phi^{\#}]$ is Φ-positive. The proof that (ii)→(i) is nearly the same as that of Theorem 6.12, where the main trick was to observe that every Σ_1^1 monotone operator is in fact a Σ_1^1 *positive* operator. The implication (iii)→(ii) is just the effective version of Theorem V.4.10. For the implication (ii)→(iii), the main difficulty lies in showing that every Φ-positive operator has a normal form similar to (†) in the proof of Theorem 6.14. This can be done and the proof completed much as before. □

6.18 Corollary. *For any normal positive analytic operation Φ and all $R \subseteq {}^k\omega$, the following are equivalent*:

(i) *R is recursive in $\Phi^{\#}$*;

(ii) *R and $\sim R$ are each reducible to the closure of an effective Φ-positive inductive operator*;

(iii) $R \in \Delta_1^{\Phi^*}$. □

6.19 Corollary. *For any strongly normal positive analytic operation Φ, $\nabla(\Phi)$ is a proper subclass of $\Delta_1^{\Phi^*}$.* □

6.20–6.26 Exercises

6.20. Show that the set of relations semi-recursive in $\mathsf{E}_1^{\#}$ is a proper subset of Δ_2^1.

6.21. Show that for any relation R on numbers semi-recursive in $\mathsf{E}_1^{\#}$, $\exists\beta\,\forall p\,R(\bar{\beta}(p), m) \leftrightarrow (\exists\beta$ recursive in $\mathsf{E}_1^{\#})\forall p\,R(\bar{\beta}(p), m)$. (cf. Exercise 1.18)

6.22. Show that M is semi-recursive in $\mathsf{E}_1^{\#}$ without using 6.12–6.14. (Show that if A is semi-recursive in $\mathsf{E}_1^{\#}$ but not recursive in $\mathsf{E}_1^{\#}$, and $A \subseteq M$, then $\{|u|^{\#} : u \in A\}$ is unbounded in $\{|u|^{\#} : u \in M\}$. Then

$$v \in M \leftrightarrow \exists u \in A\,.\,|v|^{\#} \leq |u|^{\#}.)$$

6.23. Give another proof that for any Σ_1^1 monotone operator Γ, $\bar{\Gamma}$ is semi-recursive in $\mathsf{E}_1^{\#}$ (6.12) by the method of Exercises 2.25–26.

6.24. Show that in Lemma 6.13 the hypothesis $(*)$ is superfluous; every monotone operator $\Gamma_1 \in \Pi_1^1$ may be defined in the given form by a recursive relation R_1 which satisfies $(*)$.

6.25. Use the method of Exercises VI.2.26 to show that in contrast with Corollary 6.3, if E is recursive in I, R is semi-recursive in I, and

$$\mathsf{S}(\mathbf{m}, \boldsymbol{\alpha}) \leftrightarrow \forall\beta\,\exists p\,\mathsf{R}(\bar{\beta}(p), \mathbf{m}, \boldsymbol{\alpha}),$$

then also S is semi-recursive in I.

6.26. Fill in the following sketch of an alternative proof for Theorem 4.18 for any functional I such that E_1 is recursive in I. Let P be a relation such that $\mathsf{P}(u, v, \langle\boldsymbol{\alpha}\rangle, \gamma)$ holds under exactly the same conditions as does $\mathsf{P}'(u, v, \langle\boldsymbol{\alpha}\rangle)$ in the proof of Lemma 6.1 except that in clause (1) we replace the condition

$$\{c_i\}^{\mathsf{I}}(\mathbf{m}, \boldsymbol{\alpha}) \simeq q_i \quad \text{by} \quad \gamma(\langle c_i, \mathbf{m}\rangle) = q_i.$$

Clearly P is recursive. Let $<_{\boldsymbol{\alpha},\gamma}$ denote the transitive closure of the relation $<'_{\boldsymbol{\alpha},\gamma}$ defined by:

$$v <'_{\boldsymbol{\alpha},\gamma} u \leftrightarrow \mathsf{P}(u, v, \langle\boldsymbol{\alpha}\rangle, \gamma).$$

Say that a function δ is *closed for* $\boldsymbol{\alpha}, \gamma$ iff for all u and v,

$$\delta(u) = 0 \wedge v <_{\boldsymbol{\alpha},\gamma} u \rightarrow \delta(v) = 0.$$

δ is *well-founded for* $\boldsymbol{\alpha}, \gamma$ iff

$$\neg\exists\beta\,\forall p\,[\delta(\beta(p)) = 0 \wedge \beta(p+1) <_{\boldsymbol{\alpha},\gamma} \beta(p)].$$

Finally, say that γ is *locally correct for* $\boldsymbol{\alpha}, \delta$ iff for all k, $\mathbf{m} \in {}^k\omega$, and $l = \lg(\boldsymbol{\alpha})$, and all u such that $\delta(u) = 0$:

(0) if $u = \langle\langle 0, k, l, \ldots\rangle, \mathbf{m}\rangle$ and $\langle 0, k, l, \ldots\rangle$ is an index of the proper form for $(\mathbf{m}, \boldsymbol{\alpha})$, then $\gamma(u) = \{\langle 0, k, l, \ldots\rangle\}^{\mathsf{I}}(\mathbf{m}, \boldsymbol{\alpha})$;

(1) if for some $b, c_0, \ldots, c_{k'-1}$, $u = \langle\langle 1, k, l, b, c_0, \ldots, c_{k'-1}\rangle, \mathbf{m}\rangle$ then $\gamma(n) = \gamma(\langle b, \gamma(\langle c_0, \mathbf{m}\rangle), \ldots, \gamma(\langle c_{k'-1}, \mathbf{m}\rangle)\rangle)$;

(2) if for some b, $u = \langle\langle 2, k+1, l\rangle, b, \mathbf{m}\rangle$, then $\gamma(n) = \gamma(\langle b, \mathbf{m}\rangle)$;

(3) if for some b, $u = \langle\langle 3, k, l, b\rangle, \mathbf{m}\rangle$, then $\gamma(n) = \mathsf{I}(\lambda p \,.\, \gamma(\langle b, p, \mathbf{m}\rangle))$.

Now it suffices to show that for all u and $\boldsymbol{\alpha}$,

$$u \in U^{\mathsf{I}}_{\boldsymbol{\alpha}} \leftrightarrow (\exists \gamma, \delta \text{ recursive in } \mathsf{I}, \boldsymbol{\alpha})\,[\gamma \text{ is locally correct for } \boldsymbol{\alpha}, \delta \wedge \delta(u) = 0 \wedge \delta \text{ is closed for } \boldsymbol{\alpha}, \gamma \wedge \delta \text{ is well-founded for } \boldsymbol{\alpha}, \gamma].$$

For the implication $(\rightarrow)$, consider the functions

$$\gamma_u(\langle a, \mathbf{m}\rangle) = \begin{cases} \{a\}^{\mathsf{I}}(\mathbf{m}, \boldsymbol{\alpha}), & \text{if } \ |a, \mathbf{m}|^{\mathsf{I}}_{\boldsymbol{\alpha}} \leqslant |u|^{\mathsf{I}}_{\boldsymbol{\alpha}}; \\ 0, & \text{otherwise;} \end{cases}$$

and

$$\delta_u(v) = \begin{cases} 0, & \text{if } \ |v|^{\mathsf{I}}_{\boldsymbol{\alpha}} \leqslant |u|^{\mathsf{I}}_{\boldsymbol{\alpha}}; \\ 1, & \text{otherwise.} \end{cases}$$

For the implication $(\leftarrow)$, suppose that γ and δ are recursive in $\mathsf{I}, \boldsymbol{\alpha}$ and satisfy the condition in brackets. Prove by induction on $<_{\boldsymbol{\alpha},\gamma}$ that for all a and $\mathbf{m}$, if $\delta(\langle a, \mathbf{m}\rangle) = 0$ and $\langle a, \mathbf{m}\rangle \leqslant_{\boldsymbol{\alpha},\gamma} u$, then $\langle a, \mathbf{m}\rangle \in U^{\mathsf{I}}_{\boldsymbol{\alpha}}$ and

$$\gamma(\langle a, \mathbf{m}\rangle) = \{a\}^{\mathsf{I}}(\mathbf{m}, \boldsymbol{\alpha}).$$

6.27 Notes. The functionals $\Phi^{\#}$ were introduced in Hinman [1969] to prove Corollary 6.19. Many of the results of this section appear there. The key Theorem 6.2 is inspired by a similar result of Moschovakis [1967] (Theorem VII.2.10 below). The connections between $\mathsf{E}^{\#}_1$ and Σ^1_1 inductive definability are due to Aczel [1970]. The implication (ii)→(iii) of Theorem 6.14 was originally (or Moschovakis [1974, Chapter 4]). Exercise 6.26 is due to Gandy and Moschovakis.

7. Recursive Type-3 Functionals and Relations

Recall that a functional $\mathbb{F}$ is of *type* 3 iff it has arguments $(\mathbf{m}, \boldsymbol{\alpha}, \mathbf{I})$ of types 0, 1, and 2 — that is $\mathbb{F}$ is a function from ${}^{k,l,l'}\omega = ({}^k\omega)\times{}^l({}^\omega\omega)\times{}^{l'}({}^{({}^\omega\omega)}\omega)$ into ω. Similarly a relation $\mathbb{R}$ is of type 3 iff it is a subset of some ${}^{k,l,l'}\omega$. In this section we study the notion of such functionals and relations being recursive or definable from recursive relations by quantification over ω, ${}^\omega\omega$, and ${}^{({}^\omega\omega)}\omega$. In Chapter VII we shall study the properties of recursion relative to a fixed type-3 functional.

The definition of recursiveness for type-3 functionals is essentially the same as that of recursiveness relative to a fixed functional I. The main difference is that we now explicitly allow for a finite sequence of type-2 arguments (rather than one) and think of them as genuine arguments (rather than parameters). In keeping with the discussions in the Notes to § II.5 and the Introduction to this chapter, it will be obvious that for a fixed I, a functional F is partial recursive in I iff for some partial recursive type-3 functional $\mathbb{G}$,

$$\mathsf{F}(\mathbf{m}, \boldsymbol{\alpha}) \simeq \mathbb{G}(\mathbf{m}, \boldsymbol{\alpha}, \mathsf{I}).$$

7.1 Definition. Ω^3 is the smallest set such that for all k, l, l', p, q, r, and s, all $i<k$, $j<l$, and $j'<l'$, and all $(\mathbf{m}, \boldsymbol{\alpha}, \mathbf{I}) \in {}^{k,l,l'}\omega$.

(0) $(\langle 0, k, l, l', 0, n\rangle, \mathbf{m}, \boldsymbol{\alpha}, \mathbf{I}, n) \in \Omega^3$; the remaining parts of (0) and clauses (1) and (2) are similar modifications of the corresponding parts of Definition II.2.1 to accomodate the sequence $\mathbf{I}$ of l' type-2 arguments;

(3) for any b and any β, if for all p, $(b, p, \mathbf{m}, \boldsymbol{\alpha}, \mathbf{I}, \beta(p)) \in \Omega^3$, then $(\langle 3, k, l, l', j', b\rangle, \mathbf{m}, \boldsymbol{\alpha}, \mathbf{I}, \mathsf{I}_{j'}(\beta)) \in \Omega^3$.

Just as in previous cases we can prove easily that for all a, $\mathbf{m}$, $\boldsymbol{\alpha}$, and $\mathbf{I}$, there is at most one n such that $(a, \mathbf{m}, \boldsymbol{\alpha}, \mathbf{I}, n) \in \Omega^3$ and set

$$\{a\}^3(\mathbf{m}, \boldsymbol{\alpha}, \mathbf{I}) \simeq n \quad \text{iff} \quad (a, \mathbf{m}, \boldsymbol{\alpha}, \mathbf{I}, n) \in \Omega^3.$$

In particular, it follows from clause (3) that

$$\{\langle 3, k, l, l', j', b\rangle\}^3(\mathbf{m}, \boldsymbol{\alpha}, \mathbf{I}) \simeq \mathsf{I}_{j'}(\lambda p\,.\,\{b\}^3(p, \mathbf{m}, \boldsymbol{\alpha}, \mathbf{I})).$$

A type-3 functional $\mathbb{F}$ is *partial recursive* iff $\mathbb{F} = \{a\}^3$ for some a and *recursive* iff it is partial recursive and total. A type-3 relation $\mathbb{R}$ is *recursive* iff its characteristic functional $\mathbb{K}_{\mathbb{R}}$ is recursive and *semi-recursive* iff $\mathbb{R} = \operatorname{Dm}\mathbb{G}$ for some partial recursive $\mathbb{G}$.

The Recursion Theorem and the consequent closure properties may be established just as in § 1. In particular,

7.2 Lemma. *For any partial recursive functional* F, *if*

$$\mathbb{F}(\mathbf{m}, \boldsymbol{\alpha}, \mathbf{I}) \simeq \mathsf{F}(\mathbf{m}, \boldsymbol{\alpha}),$$

then $\mathbb{F}$ *is also partial recursive. Conversely, if* $\mathbb{F}$ *is partial recursive of rank* $(k, l, 0)$ *then* $\mathbb{F}$ *is partial recursive in the previous sense.* □

The techniques of § 2 yield the following substitution results:

7.3 Theorem. *For any partial recursive functionals* $\mathbb{G}$ *and* $\mathbb{H}$,

(i) *if* $\mathbb{F}(\mathbf{m}, \boldsymbol{\alpha}, \mathbf{I}) \simeq \mathbb{G}(\mathbf{m}, \boldsymbol{\alpha}, \lambda p . \mathbb{H}(p, \mathbf{m}, \boldsymbol{\alpha}, \mathbf{I}), \mathbf{I})$ *and* $\lg(\mathbf{I}) > 0$, *then* $\mathbb{F}$ *is also partial recursive*;

(ii) *there exists a partial recursive functional* $\mathbb{F}$ *such that for all* $\mathbf{m}$, n, $\boldsymbol{\alpha}$, *and* $\mathbf{I}$,

$$\mathbb{G}(\mathbf{m}, \boldsymbol{\alpha}, \mathbf{I}, \lambda \beta . \mathbb{H}(\mathbf{m}, \boldsymbol{\alpha}, \beta, \mathbf{I})) \simeq n \rightarrow \mathbb{F}(\mathbf{m}, \boldsymbol{\alpha}, \mathbf{I}) \simeq n,$$

and if $\lg(\mathbf{I}) > 0$ *and* $\lambda \beta . \mathbb{H}(\mathbf{m}, \boldsymbol{\alpha}, \beta, \mathbf{I})$ *is total, then also*

$$\mathbb{F}(\mathbf{m}, \boldsymbol{\alpha}, \mathbf{I}) \simeq n \rightarrow \mathbb{G}(\mathbf{m}, \boldsymbol{\alpha}, \mathbf{I}, \lambda \beta . \mathbb{H}(\mathbf{m}, \boldsymbol{\alpha}, \beta, \mathbf{I})) \simeq n.$$

Proof. The proof of (i) may be obtained by a suitable adaptation of that of Theorem 2.10. The hypothesis $\lg(\mathbf{I}) > 0$ is necessary because of Theorem 2.1. (ii) is proved by a similar modification of the proof of Theorem 2.14. (Cf. VII.1.6(ii) below.) □

7.4 Corollary. *For any recursive functionals* $\mathbb{G}$ *and* $\mathbb{H}$, *if*

$$\mathbb{F}(\mathbf{m}, \boldsymbol{\alpha}, \mathbf{I}) \simeq \mathbb{G}(\mathbf{m}, \boldsymbol{\alpha}, \mathbf{I}, \lambda \beta . \mathbb{H}(\mathbf{m}, \boldsymbol{\alpha}, \beta, \mathbf{I})),$$

then $\mathbb{F}$ *is also recursive.* □

The results of §§ 3 and 4 cannot be generalized to recursion over ${}^{k,l,l'}\omega$ because of the hypothesis in these theorems that E be recursive in I. In fact, we have

7.5 Theorem. *There exists a relation* $\mathbb{R} \subseteq {}^{1,0,1}\omega$ *such that both* $\mathbb{R}$ *and* $\sim\mathbb{R}$ *are semi-recursive, but* $\mathbb{R}$ *is not recursive.*

Proof. Let G be a partial recursive function such that

$$\{m : \lambda p . G(p, m) \text{ is total}\} \text{ is } \Pi^0_1 \text{ but not recursive}$$

and set

$$\mathbb{R}(m, \mathsf{I}) \leftrightarrow \mathsf{I}(\lambda p . G(p, m)) \text{ is defined.}$$

Clearly $\mathbb{R}$ is semi-recursive. Furthermore, by the assumption and Lemma 7.2, since

$$\sim\mathbb{R}(m, \mathsf{I}) \leftrightarrow \lambda p\,.\,G(p, m) \text{ is not total,}$$

$\sim\mathbb{R}$ is also semi-recursive. Suppose, however, that $\mathbb{R}$ is recursive and set $F(m) = \mathbb{K}_{\mathbb{R}}(m, \lambda\beta\,.\,0)$. It follows from 7.2 and 7.4 that F is a recursive function, but

$$F(m) = 0 \leftrightarrow \mathbb{R}(m, \lambda\beta\,.\,0) \leftrightarrow \lambda p\,.\,G(p, m) \text{ is total,}$$

which contradicts the assumption on G. □

We consider next the relations over ${}^{k,l,l'}\omega$ obtained from the recursive ones by quantification. The classes of arithmetical and analytical relations and the subclasses Σ^0_r, Σ^1_r, etc. may be defined exactly as in §§ III.1–2. We shall write Σ^i_r etc. ambiguously to refer also to these classes of relations over ${}^{k,l,l'}\omega$. Let $\exists^2\mathbb{P}$ denote the relation $\mathbb{R}$ defined by

$$\mathbb{R}(\mathbf{m}, \boldsymbol{\alpha}, \mathsf{I}) \leftrightarrow \exists \mathsf{H}\,.\,\mathbb{P}(\mathbf{m}, \boldsymbol{\alpha}, \mathsf{I}, \mathsf{H})$$

and similarly for $\forall^2\mathbb{R}$. Then we set, for all r,

$$\Sigma^2_0 = \Pi^2_0 = \text{the class of analytical relations over } {}^{k,l,l'}\omega;$$
$$\Sigma^2_{r+1} = \{\exists^2\mathbb{P} : \mathbb{P} \in \Pi^2_r\};$$
$$\Pi^2_{r+1} = \{\forall^2\mathbb{P} : \mathbb{P} \in \Sigma^2_r\};$$
$$\Delta^2_r = \Sigma^2_r \cap \Pi^2_r.$$

The properties of the arithmetical hierarchy expressed by III.1.4–7 hold here with appropriate modifications by the same proofs. Similarly, the properties of the analytical hierarchy expressed by III.2.4–7 are easily extended to the analytical hierarchy over ${}^{k,l,l'}\omega$. The situation concerning universal relations and hierarchy theorems is, however, somewhat more complicated. On the one hand, the arithmetical and analytical hierarchy theorems follow immediately from those of §§ III.1–2, since a relation R which is $\Sigma^i_r \sim \Delta^i_r$ in the previous sense is still $\Sigma^i_r \sim \Delta^i_r$ as a relation over ${}^{k,l,0}\omega$. This approach does not, however, suggest any way to prove that Σ^2_1 is properly larger than Δ^2_1. The proofs of these results in §§ III.1–2 all spring from the fact that there is a universal Σ^0_1 relation U^0_1. Not too surprisingly, this is no longer true in the present context:

7.6 Lemma. *There is no relation universal for the class of Σ^0_1 relations over ${}^{k,l,l'}\omega$.*

Proof. Suppose $\mathbb{U}$ were such a universal relation and let U be defined by

$$\mathsf{U}(a, \langle \mathbf{m} \rangle, \langle \boldsymbol{\alpha} \rangle) \leftrightarrow \mathsf{U}(a, \langle \mathbf{m} \rangle, \langle \boldsymbol{\alpha} \rangle, \langle \mathsf{E} \rangle).$$

Then clearly U would be universal for relations Σ_1^0 in E — that is, of the form $\exists^0\mathsf{R}$ with R recursive in E. But if R is recursive in E, so is $\exists^0\mathsf{R}$, so in fact U would be universal for the class of relations recursive in E. A standard diagonal argument shows that this is impossible. □

In find a universal relation, we must go to Π_1^1.

7.7 Theorem. *There exists a relation $\mathbb{U}_1^1 \subseteq {}^{2,1,1}\omega$ which is universal for the class of Π_1^1 relations over ${}^{k,l,l'}\omega$.*

Proof. Let

$$\mathbb{U}_1^1(a, \langle \mathbf{m} \rangle, \langle \boldsymbol{\alpha} \rangle, \langle \mathbf{I} \rangle) \leftrightarrow \forall \beta\, \exists p\, [\{a\}^3(p, \mathbf{m}, \boldsymbol{\alpha}, \beta, \mathbf{I}) \text{ is defined}].$$

It is clear that for any Π_1^1 relation $\mathbb{R}$ there is a number a such that

$$\mathbb{R}(\mathbf{m}, \boldsymbol{\alpha}, \mathbf{I}) \leftrightarrow \mathbb{U}_1^1(a, \langle \mathbf{m} \rangle, \langle \boldsymbol{\alpha} \rangle, \langle \mathbf{I} \rangle)$$

and it remains to show that $\mathbb{U}_1^1$ is Π_1^1. This is accomplished by defining a Π_1^1 family of monotone operators $\Lambda_{\langle \boldsymbol{\alpha} \rangle, \langle \mathbf{I} \rangle}$ over ω such that

$$\{a\}^3(\mathbf{m}, \boldsymbol{\alpha}, \mathbf{I}) \simeq n \leftrightarrow \langle a, \langle \mathbf{m} \rangle, n \rangle \in \bar{\Lambda}_{\langle \boldsymbol{\alpha} \rangle, \langle \mathbf{I} \rangle}.$$

Λ is defined much as in the proof of Theorem 1.5:

$$\Lambda_{\langle \boldsymbol{\alpha} \rangle, \langle \mathbf{I} \rangle}(A) = \Gamma_{\boldsymbol{\alpha}, \mathbf{I}, 0}(A) \cup \cdots \cup \Gamma_{\boldsymbol{\alpha}, \mathbf{I}, 2}(A) \cup \Lambda_{\boldsymbol{\alpha}, \mathbf{I}, 3}(A)$$

where, for example,

$$\begin{aligned} \Lambda_{\boldsymbol{\alpha}, \mathbf{I}, 3}(A) = \{\langle\langle 3, k, l, l', j', b \rangle, \langle \mathbf{m} \rangle, n \rangle :\ & \forall p\, \exists q\, [\langle b, \langle p, \mathbf{m} \rangle, q \rangle \in A] \wedge \\ & \forall \beta\, (\forall p\, \forall q\, [\langle b, \langle p, \mathbf{m} \rangle, q \rangle \in A \rightarrow \beta(p) = q] \rightarrow \mathsf{I}_{j'}(\beta) = n)\}. \end{aligned}$$

If follows easily that Λ is as required and that $\bar{\Lambda} \in \Pi_1^1$. Thus also $\mathbb{U}_1^1$ is Π_1^1. □

We may now define, for all $r \geqslant 1$, relations $\mathbb{U}_r^1$ such that $\mathbb{U}_r^1$ is universal for the class of Π_r^1 relations over ${}^{k,l,l'}\omega$:

$$\mathbb{U}_{r+1}^1(a, \langle \mathbf{m} \rangle, \langle \boldsymbol{\alpha} \rangle, \langle \mathbf{I} \rangle) \leftrightarrow \exists \beta \sim \mathbb{U}_r^1(a, \langle \mathbf{m} \rangle, \langle \boldsymbol{\alpha}, \beta \rangle, \langle \mathbf{I} \rangle).$$

The following analogue of Lemma III.2.9 will allow us to extend this to the classes Σ_r^2:

7.8 Lemma. $\Sigma_1^2 = (\exists^2\mathbb{P} : \mathbb{P} \in \Pi_1^1\}$.

Proof. The inclusion ($\supseteq$) is immediate from the definitions. For the converse we show that the class of relations of the form $\exists^2\mathbb{P}$ with $\mathbb{P} \in \Pi_1^1$ is closed under $\exists^1$, $\forall^1$, and $\exists^2$. This follows easily from the equivalences:

$$\exists\beta\, \exists \mathsf{H}\, \mathbb{P}(\mathbf{m}, \boldsymbol{\alpha}, \beta, \mathbf{I}, \mathsf{H}) \leftrightarrow \exists \mathsf{H}\, \mathbb{P}(\mathbf{m}, \boldsymbol{\alpha}, \lambda p\,.\,(\mathsf{H})^{p+1}(\lambda q\,.\,0), \mathbf{I}, (\mathsf{H})^0);$$

$$\exists \mathsf{G}\, \exists \mathsf{H}\, \mathbb{P}(\mathbf{m}, \boldsymbol{\alpha}, \mathbf{I}, \mathsf{G}, \mathsf{H}) \leftrightarrow \exists \mathsf{H}\, \mathbb{P}(\mathbf{m}, \boldsymbol{\alpha}, \mathbf{I}, (\mathsf{H})^0, (\mathsf{H})^1);$$

$$\forall\beta\, \exists \mathsf{H}\, \mathbb{P}(\mathbf{m}, \boldsymbol{\alpha}, \beta, \mathbf{I}, \mathsf{H}) \leftrightarrow \exists \mathsf{H}\, \forall\beta\, \mathbb{P}(\mathbf{m}, \boldsymbol{\alpha}, \beta, \mathbf{I}, (\mathsf{H})^\beta);$$

where $(\mathsf{H})^p(\alpha) = \mathsf{H}((p) * \alpha)$ and $(\mathsf{H})^\beta(\alpha) = \mathsf{H}(\langle\beta, \alpha\rangle)$. □

Thus if we set, for $r \geqslant 1$,

$$\mathbb{U}_1^2(a, \langle\mathbf{m}\rangle, \langle\boldsymbol{\alpha}\rangle, \langle\mathbf{I}\rangle) \leftrightarrow \exists \mathsf{H}\,.\, \mathbb{U}_1^1(a, \langle\mathbf{m}\rangle, \langle\boldsymbol{\alpha}\rangle, \langle\mathbf{I}, \mathsf{H}\rangle)$$

and

$$\mathbb{U}_{r+1}^2(a, \langle\mathbf{m}\rangle, \langle\boldsymbol{\alpha}\rangle, \langle\mathbf{I}\rangle) \leftrightarrow \exists \mathsf{H} \sim \mathbb{U}_r^2(a, \langle\mathbf{m}\rangle, \langle\boldsymbol{\alpha}\rangle, \langle\mathbf{I}, \mathsf{H}\rangle)$$

and

$$\mathbb{U}_{r+1}^2(a, \langle\mathbf{m}\rangle, \langle\boldsymbol{\alpha}\rangle, \langle\mathbf{I}\rangle) \leftrightarrow \exists \mathsf{H} \sim \mathbb{U}_r^2(a, \langle\mathbf{m}\rangle, \langle\boldsymbol{\alpha}\rangle, \langle\mathbf{I}, \mathsf{H}\rangle)$$

we have relations $\mathbb{U}_r^2$ universal for Σ_r^2 and $\sim\mathbb{U}_r^2$ universal for Π_r^2. Then just as in §§ III.1–2,

7.9 Theorem (The Functional Quantifier Hierarchy). *For all* $r \geqslant 1$,
(i) $\Sigma_r^2 \not\subseteq \Delta_r^2$ *and* $\Pi_r^2 \not\subseteq \Delta_r^2$;
(ii) $\Delta_{r+1}^2 \not\subseteq \Sigma_r^2 \cup \Pi_r^2$. □

We turn now to the consideration of inductive definitions of subsets of ${}^{k,l}\omega$ which we were forced to abandon in § III.3. For simplicity we restrict attention to inductively defined subsets of ${}^\omega\omega$; extension to the general case is obtained by the usual codings. If Γ is an operator over ${}^\omega\omega$, we set

$$\mathbb{P}_\Gamma(\alpha, \mathsf{I}) \leftrightarrow \alpha \in \Gamma(\mathsf{Z}_\mathsf{I}),$$

where $\mathsf{Z}_\mathsf{I} = \{\beta : \mathsf{I}(\beta) = 0\}$, and classify Γ as Σ_r^i, etc. according to the classification of $\mathbb{P}_\Gamma$. Parallel to Theorem III.3.1, we have:

7.10 Theorem. *For any* $r > 0$ *and any monotone operator* Γ *over* ${}^\omega\omega$,

$$\Gamma \in \Pi^2_r \to \bar{\Gamma} \in \Pi^2_r.$$

Proof. We have

$$\alpha \in \bar{\Gamma} \leftrightarrow \forall \mathsf{A}\,(\Gamma(\mathsf{A}) \subseteq \mathsf{A} \to \alpha \in \mathsf{A})$$
$$\leftrightarrow \forall \mathsf{I}\,(\forall \beta\,[\mathbb{P}_\Gamma(\beta, \mathsf{I}) \to \mathsf{I}(\beta) = 0] \to \mathsf{I}(\alpha) = 0). \quad \square$$

To obtain further results of this sort, we shall need to examine the closure ordinals of operators over ${}^\omega\omega$. By Lemma I.3.2, $|\Gamma|$ is less than the least cardinal greater than $2^{\aleph_0}$. Hence the ordinals $\leq |\Gamma|$ can be coded as well-orderings of subsets of ${}^\omega\omega$. For any I, let

$$\leqslant_{\mathsf{I}} = \{(\gamma, \delta) : \mathsf{I}(\langle \gamma, \delta \rangle) = 0\},$$

and let

$$\mathsf{W} = \{\mathsf{I} : \leqslant_{\mathsf{I}} \text{ is a well-ordering}\}.$$

In contrast to the fact that $\mathsf{W} \in \Pi^1_1 \sim \Delta^1_1$ (Corollary IV.1.2), we have here

7.11 Lemma. $\mathsf{W} \in \Delta^2_1$; *in fact*, $\mathsf{W} \in \Delta^1_{(\omega)}$.

Proof. To express that $\leqslant_{\mathsf{I}}$ is a linear ordering clearly requires only function quantifiers. That $\leqslant_{\mathsf{I}}$ is well founded is expressed by: $\neg \exists \alpha\, \forall p\,.\,(\alpha)^{p+1} <_{\mathsf{I}} (\alpha)^p$, which is also analytical. $\square$

For $\mathsf{I} \in \mathsf{W}$, we denote by $\|\mathsf{I}\|$ the order-type of $\leqslant_{\mathsf{I}}$ and define $\mathsf{I} \restriction \gamma$ and $|\gamma|_{\mathsf{I}}$ in the natural way so that the analogues of (8)–(11) of § I.1 hold. Then for any Γ over ${}^\omega\omega$,

$$\alpha \in \bar{\Gamma} \leftrightarrow \exists \mathsf{I}\,(\mathsf{I} \in \mathsf{W} \wedge \alpha \in \Gamma^{\|\mathsf{I}\|})$$
$$\leftrightarrow \forall \mathsf{I}\,(\mathsf{I} \in \mathsf{W} \wedge \Gamma^{\|\mathsf{I}\|} \subseteq \Gamma^{(\|\mathsf{I}\|)} \to \alpha \in \Gamma^{\|\mathsf{I}\|})$$

and to classify $\bar{\Gamma}$ it remains to evaluate the complexity of $\Gamma^{\|\mathsf{I}\|}$ and $\Gamma^{(\|\mathsf{I}\|)}$.

7.12 Theorem. *For any $r > 0$ and any inductive operator $\Gamma \in \Delta^2_r$ over ${}^\omega\omega$, there exist relations $\mathsf{V}^{()}_\Sigma$ and $\mathsf{V}_\Sigma \in \Sigma^2_r$ and $\mathsf{V}^{()}_\Pi$ and $\mathsf{V}_\Pi \in \Pi^2_r$ such that for all $\mathsf{I} \in \mathsf{W}$ and all α,*

(i) $\alpha \in \Gamma^{(\|\mathsf{I}\|)} \leftrightarrow \mathsf{V}^{()}_\Sigma(\alpha, \mathsf{I}) \leftrightarrow \mathsf{V}^{()}_\Pi(\alpha, \mathsf{I})$;

(ii) $\alpha \in \Gamma^{\|\mathsf{I}\|} \leftrightarrow \mathsf{V}_\Sigma(\alpha, \mathsf{I}) \leftrightarrow \mathsf{V}_\Pi(\alpha, \mathsf{I})$.

Proof. Similar to that of Theorem III.3.9. $\square$

7.13 Corollary. *For any $r>0$ and any inductive operator Γ over ${}^{\omega}\omega$,*

$$\Gamma \in \Delta_r^2 \rightarrow \bar{\Gamma} \in \Delta_r^2. \quad \square$$

Parallel to Theorem III.3.13 we have

7.14 Theorem. *For any $r>0$ and any monotone operator $\Gamma \in \Sigma_r^2$ (Π_r^2), there exist relations $\mathbb{V}^{(\ )}$ and $\mathbb{V} \in \Sigma_r^2$ (Π_r^2) such that for all $\mathsf{I} \in \mathbb{W}$ and all α,*

(i) $\alpha \in \Gamma^{(\|\mathsf{I}\|)} \leftrightarrow \mathbb{V}^{(\ )}(\alpha, \mathsf{I})$;

(ii) $\alpha \in \Gamma^{\|\mathsf{I}\|} \leftrightarrow \mathbb{V}(\alpha, \mathsf{I})$. $\square$

7.15 Corollary. *For any $r>0$ and any monotone operator Γ over ${}^{\omega}\omega$,*

$$\Gamma \in \Sigma_r^2 \rightarrow \bar{\Gamma} \in \Sigma_r^2. \quad \square$$

Very little more is known about the classes Σ_r^2 and Π_r^2. It follows from the Hypothesis of Constructibility (V = L) that there is a Δ_1^2 well-ordering of ${}^{({}^{\omega}\omega)}\omega$ and thus that for all $r \geq 1$, Σ_r^2 has the pre-wellordering property. Determinacy of all subsets of ${}^{\omega}\omega$ also implies that Σ_1^2 has the pre-wellordering property and the proposition that Π_1^2 has the pre-wellordering property is known to be consistent with ZFC.

For investigations under the Hypothesis of Determinacy or some other hypothesis which contradicts the Axiom of Choice, it is useful to note that the theory of the functional quantifier hierarchy can be developed without this axiom (cf. discussion preceding Theorem V.3.1). One use of choice is in the last of the three equivalences used in the proof of Lemma 7.8. Unfortunately, choice seems essential for the lemma and to avoid it we must find a new way to define the Σ_r^2 universal relations. The idea is provided by Theorems III.3.6–7. Following the pattern of these theorems we may define a Σ_1^1 inductive operator Γ such that $|\Gamma| = \omega$ and $\bar{\Gamma}$ is a relation $\mathbb{U}_{(\omega)}^1$ which is universal for $\Delta_{(\omega)}^1$. Although $\mathbb{U}_{(\omega)}^1$ cannot be $\Delta_{(\omega)}^1$, it is easily seen to be Δ_1^2. Hence we may take

$$\mathbb{U}_1^2(a, \langle \mathbf{m} \rangle, \langle \boldsymbol{\alpha} \rangle, \langle \mathbf{I} \rangle) \leftrightarrow \exists \mathsf{H} \,.\, \mathbb{U}_{(\omega)}^1(a, \langle \mathbf{m} \rangle, \langle \boldsymbol{\alpha} \rangle, \langle \mathbf{I}, \mathsf{H} \rangle)$$

and $\mathbb{U}_{r+1}^2$ as before.

The same equivalence is required to show that Σ_r^2 and Π_r^2 are closed under function quantification ($\exists^1$ and $\forall^1$). Here we cannot avoid choice altogether, but can replace it with the weaker *Collection Principle*:

$$\forall \beta\, \exists \mathsf{H} \,.\, \mathbb{R}(\mathbf{m}, \boldsymbol{\alpha}, \beta, \mathbf{I}, \mathsf{H}) \rightarrow \exists \mathsf{H}\, \forall \beta\, \exists \gamma \,.\, \mathbb{R}(\mathbf{m}, \boldsymbol{\alpha}, \beta, \mathbf{I}, (\mathsf{H})^{\gamma}).$$

This principle holds in $L[{}^{\omega}\omega]$, the class of sets constructible from ${}^{\omega}\omega$ (cf. Moschovakis [1970]). Many people have conjectured that some strong form of determinacy also holds in this model.

The final use of the Axiom of Choice occurs in the discussion of the closure ordinals of inductive definitions over ${}^{\omega}\omega$. If ${}^{\omega}\omega$ is not well-orderable, then "the least cardinal greater than $2^{\aleph_0}$" is meaningless and we need a new bound for these closure ordinals. Fortunately, this is easy to compute. Any inductive operator Γ over any set X induces a pre-wellordering $\leqslant_\Gamma$ of X of type $|\Gamma|+1$:

$$x \leqslant_\Gamma y \leftrightarrow \forall \sigma\,[y \in \Gamma^\sigma \rightarrow x \in \Gamma^\sigma].$$

Hence if we set

$$o(X) = \sup{}^{+}\{\sigma : \text{there exists a pre-wellordering of } X \text{ of type } \sigma\},$$

then $|\Gamma| < o(X)$. Furthermore, it is no harder to code ordinals with pre-wellorderings. Thus 7.11–12 and 7.14 may be proved with W replaced by $\mathbf{p}\mathsf{W} = \{\mathsf{I} : \leqslant_{\mathsf{I}} \text{ is a pre-wellordering}\}$, so 7.13 and 7.15 hold also in a set theory without the Axiom of Choice.

7.16–7.22 Exercises

7.16. Sketch a proof of Theorem 7.3.

7.17. Does there exist a subset of ${}^{({}^{\omega}\omega)}\omega$ which is Σ^0_1 but not recursive?

7.18 (Tugué [1960]). Show that the class of Σ^1_1 relations over ${}^{k,l,l'}\omega$ does not have the separation property.

7.19. Formulate and prove substitution theorems for recursive type-3 functionals analogous to those of § 2.

7.20. Formulate and prove substitution theorems for the functional quantifier hierarchy analogous to those for the arithmetical and analytical hierarchies (III.1.11 and III.2.11). Consider substitution for both type-1 and type-2 arguments.

7.21. Suppose that F is defined from G and H by the following recursion:

$$\mathsf{F}(0, \mathbf{m}, \boldsymbol{\alpha}, \boldsymbol{\beta}) \simeq \mathsf{G}(\mathbf{m}, \boldsymbol{\alpha}, \boldsymbol{\beta}),$$

$$\mathsf{F}(p+1, \mathbf{m}, \boldsymbol{\alpha}, \boldsymbol{\beta}) \simeq \mathsf{H}(p, \mathbf{m}, \boldsymbol{\alpha}, \boldsymbol{\beta}, \lambda\gamma\,.\,\mathsf{F}(p, \mathbf{m}, \boldsymbol{\alpha}, \gamma)).$$

If G and H are recursive, does it follow that also F is recursive?

7.22. Prove the fact mentioned following Corollary 7.15 that if $V = L$, there is a Δ^2_1 well-ordering of ${}^{({}^{\omega}\omega)}\omega$ and for all r, Σ^2_r has the pre-wellordering property.

7.23 Notes. The results of this section are largely from the folk literature.

Chapter VII
Recursion in a Type-3 Functional

We first hasten to assure the reader that this chapter is not the second in an infinite sequence. Although there are several important differences between the theories of recursion relative to functionals of types 2 and 3, most of the theory of recursion relative to functionals of types greater than 3 can be obtained from type-3 theory with essentially only notational changes. This is discussed in § 4.

In § 1 we consider the basic definitions and facts about recursion in a type-3 functional and examples which illustrate the differences between types 2 and 3. For example, although E and oJ are each recursive in the other, the same is not true of their type-3 analogues $\mathbb{E}$ and $\mathbf{sJ}$. In § 2 we see that although the basic structure of the class of relations semi-recursive in a type-3 functional is superficially similar to the corresponding structure of type-2, the differences begin to be more important. Finally in § 3 we see that with respect to hierarchies the situation for recursion in a type-3 functional is very different from that for type-2.

1. Basic Properties

We shall consider in detail only the notion of recursion relative to a single fixed total type-3 functional $\mathbb{I}$: ${}^{({}^{\omega}\omega)}\omega \to \omega$. From this may easily be derived, by the usual sorts of coding, notions of recursion relative to several type-3 functionals and, as in § VI.7, the notion of a recursive type-4 functional. Although we are still primarily interested in the properties of relations and functionals over ${}^{k,l}\omega$, we shall also need the notion of relative recursiveness among type-3 functionals. For this reason we state the basic definitions in terms of functionals and relations over ${}^{k,l,l'}\omega$.

Consider first the intuitive notion of a functional F being calculable relative to $\mathbb{I}$. We must stretch our imagination one step further to conceive of an idealized computer prepared to receive inputs of the form $(\mathbf{m}, \boldsymbol{\alpha}, \mathbf{I})$ and connected to a memory device M which contains the graph of $\mathbb{I}$. The inputs are considered to be stored before the beginning of the computation in infinite memory devices, which may be consulted at any time during the computation. Arguments β are presented to one of the $\mathsf{I}_{j'}$ just as before via a subsidiary infinite memory unit M' which is "loaded" with the graph of β. To apply $\mathbb{I}$, the computer requires another memory unit M'' which can be "loaded" with the graph of a functional F in order to obtain the value of $\mathbb{I}(\mathsf{F})$. As before, the "loading" of M' and M'' is considered to occur during the computation, which is in general therefore infinite in length. In fact, since it takes at least $2^{\aleph_0}$ "steps" to load M'' with the graph of a functional F, computations relative to $\mathbb{I}$ are generally uncountable. Clearly, the value of a computation depends on values $\mathbb{I}(\mathsf{F})$ only for F which are themselves computable from $\mathbb{I}$, $\mathbf{I}$, and $\boldsymbol{\alpha}$.

The precise definition is obtained from Definition VI.7.1 by adding a clause which insures that

$$\{\langle 4, k, l, l'\ b\rangle\}^{\mathbb{I}}(\mathbf{m}, \boldsymbol{\alpha}, \mathbf{I}) \simeq \mathbb{I}(\lambda\beta.\{b\}^{\mathbb{I}}(\mathbf{m}, \boldsymbol{\alpha}, \beta, \mathbf{I})).$$

1.1 Definition. For any total functional $\mathbb{I}: {}^{({}^{\omega}\omega)}\omega \to \omega$, $\Omega[\mathbb{I}]$ is the smallest set such that for all k, l, l', n, p, q, r, and s, *all* $i < k$, $j < k$, and $j' < l'$, and all $(\mathbf{m}, \boldsymbol{\alpha}, \mathbf{I}) \in {}^{k,l,l'}\omega$,

(0)
(1)
(2)
(3) } identical to the corresponding clauses of Definition VI. 7.1;

(4) for any b and any F, if for all β, $(b, \mathbf{m}, \boldsymbol{\alpha}, \beta, \mathbf{I}, \mathsf{F}(\beta)) \in \Omega[\mathbb{I}]$, then $(\langle 4, k, l, l', b\rangle, \mathbf{m}, \boldsymbol{\alpha}, \mathbf{I}, \mathbb{I}(\mathsf{F})) \in \Omega[\mathbb{I}]$.

The definition of $\Omega[\mathbb{I}]$ is easily interpreted as a closure under functions of rank $2^{\aleph_0}$ so that the closure ordinal is at most the least cardinal greater than $2^{\aleph_0}$. Without the Axiom of Choice we can conclude that the closure ordinal is at most $o({}^{({}^{\omega}\omega)}\omega)$. The reader should be now find it obvious that for all a, $\mathbf{m}$, $\boldsymbol{\alpha}$, and $\mathbf{I}$, there is at most one n such that $(a, \mathbf{m}, \boldsymbol{\alpha}, \mathbf{I}, n) \in \Omega[\mathbb{I}]$. We set

$$\{a\}^{\mathbb{I}}(\mathbf{m}, \boldsymbol{\alpha}, \mathbf{I}) \simeq n \quad \text{iff} \quad (a, \mathbf{m}, \boldsymbol{\alpha}, \mathbf{I}, n) \in \Omega[\mathbb{I}].$$

A functional $\mathbb{F}$ is *partial recursive in* $\mathbb{I}$ iff $\mathbb{F} = \{a\}^{\mathbb{I}}$ for some $a \in \omega$, etc. Since $\Omega^3 \subseteq \Omega[\mathbb{I}]$, every recursive functional $\mathbb{F}$ is recursive in $\mathbb{I}$. The $\mathbb{I}$-Recursion Theorem is proved exactly as in all other cases and it follows that the class of functionals partial recursive in $\mathbb{I}$ is closed under primitive and course-of-values recursion and unbounded search. It follows from Corollary II.3.3 that every partial recursive type-2 functional F is also partial recursive in $\mathbb{I}$ and by a slightly more complicated proof that the same is true for type-3 functionals $\mathbb{F}$. Definition by cases with relations recursive in $\mathbb{I}$ and functionals partial recursive in $\mathbb{I}$ is established as is Theorem II.2.12. The class of relations recursive in $\mathbb{I}$ is a Boolean algebra closed under composition with functionals recursive in $\mathbb{I}$.

Computations relative to $\mathbb{I}$ may be thought of as arranged in labeled trees. Nodes corresponding to clauses (0)–(3) are as before, while for clause (4) we have nodes of the form:

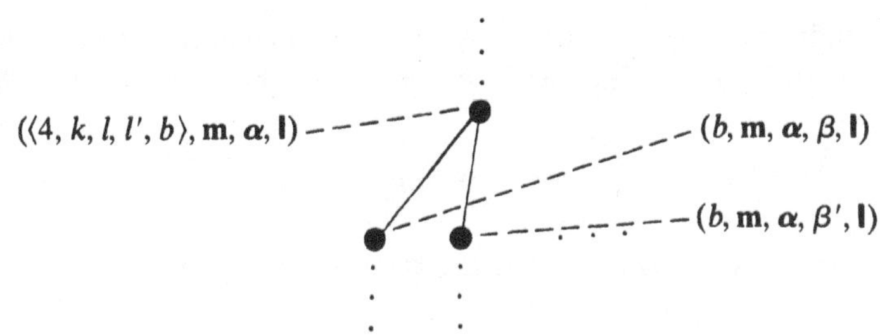

where $\beta, \beta', \ldots$ includes all members of ${}^{\omega}\omega$. Of course, it is not necessary that the successor nodes be arranged in a well-ordered sequence.

Every relation recursive in $\mathbb{I}$ is also semi-recursive in $\mathbb{I}$ but not conversely. The class of relations semi-recursive in $\mathbb{I}$ is closed under finite intersection and bounded universal number quantification. Analogously to Lemma VI.1.2, the class of relations semi-recursive in $\mathbb{I}$ is seen to be closed under universal function quantification ($\forall^1$) and hence also under universal number quantification ($\forall^0$).

To illustrate the theory, we consider the type-3 analogues of the functionals E and oJ:

$$\mathbb{E}(\mathsf{I}) = \begin{cases} 0, & \text{if } \exists\alpha\,.\,\mathsf{I}(\alpha) = 0; \\ 1, & \text{otherwise;} \end{cases}$$

$$\mathsf{sJ}((a, \mathbf{m}, \alpha) * \mathsf{I}) = \begin{cases} 0, & \text{if } \{a\}^{\mathsf{I}}(\mathbf{m}, \alpha) \text{ is defined;} \\ 1, & \text{otherwise;} \end{cases}$$

where $(a, \mathbf{m}, \alpha) * \mathsf{I}$ is the functional F such that for any γ, $\mathsf{F}((0) * \gamma) = \mathsf{I}(\gamma)$, $\mathsf{F}((1) * \gamma) = \langle a, \mathbf{m}\rangle$, and for all p, $\mathsf{F}((p+2) * \gamma) = \alpha(p)$. Of course, sJ is just the natural encoding of the superjump:

$$\mathsf{sJ}((a, \mathbf{m}, \alpha) * \mathsf{I}) = \mathsf{I}^{\mathsf{sJ}}((\langle a, \mathbf{m}\rangle) * \alpha).$$

1.2 Lemma. *For all $\mathbb{R}$, if $\mathbb{R}$ is analytical, then $\mathbb{R}$ is recursive in $\mathbb{E}$.*

Proof. For any $\mathbb{R}$,

$$\mathbb{K}_{\exists^1\mathbb{R}}(\mathbf{m}, \boldsymbol{\alpha}, \mathbf{I}) = \mathbb{E}(\lambda\beta\,.\,\mathbb{K}_{\mathbb{R}}(\mathbf{m}, \boldsymbol{\alpha}, \beta, \mathbf{I}))$$

so that the class of relations recursive in $\mathbb{E}$ is closed under $\exists^1$. It follows by complementation that it is closed under $\forall^1$, hence also under $\exists^0$ and $\forall^0$. Since this class contans all recursive relations, it contains all arithmetical relations and hence all analytical relations. □

1.3 Corollary. *sJ is recursive in $\mathbb{E}$.*

Proof. It suffices to show that (the graph of) sJ is analytical. For each I, let $\Gamma_{\alpha,\mathsf{I}}$ be a monotone operator defined as in the proof of Theorem VI.1.5 such that

$$\{a\}^{\mathsf{I}}(\mathbf{m}, \alpha) \simeq n \leftrightarrow \langle a, \langle \mathbf{m} \rangle, n \rangle \in \bar{\Gamma}_{\alpha,\mathsf{I}}.$$

From the definition it is clear that the relation

$$\mathbb{R}(p, \alpha, A, \mathsf{I}) \leftrightarrow p \in \Gamma_{\alpha,\mathsf{I}}(A)$$

is analytical. Since

$$p \in \bar{\Gamma}_{\alpha,\mathsf{I}} \leftrightarrow \forall A [\Gamma_{\alpha,\mathsf{I}}(A) \subseteq A \rightarrow p \in A],$$

also the relation

$$\mathbb{S}(p, \alpha, \mathsf{I}) \leftrightarrow p \in \bar{\Gamma}_{\alpha,\mathsf{I}}$$

is analytical. Finally,

$$\begin{aligned} \mathsf{sJ}((a, \mathbf{m}, \alpha) * \mathsf{I}) = r \leftrightarrow & [r = 0 \wedge \exists n . \langle a, \langle \mathbf{m} \rangle, n \rangle \in \bar{\Gamma}_{\alpha,\mathsf{I}}] \vee \\ & [r = 1 \wedge \forall n . \langle a, \langle \mathbf{m} \rangle, n \rangle \notin \bar{\Gamma}_{\alpha,\mathsf{I}}] \end{aligned}$$

and thus sJ is analytical. □

The analogy with the results of § VI.1 suggests that we should expect also that $\mathbb{E}$ is recursive in sJ and that the class of relations recursive in either is Δ_1^2. This is, however, far from the case, and we show this next. For any $\mathbb{I}$, let

$$\mathbb{V}^{\mathbb{I}}(a, \langle \mathbf{m} \rangle, n, \langle \boldsymbol{\alpha} \rangle, \langle \mathbf{I} \rangle) \leftrightarrow (a, \mathbf{m}, \boldsymbol{\alpha}, \mathbf{I}, n) \in \Omega[\mathbb{I}].$$

1.4 Theorem. *For any $\mathbb{I}$ and any $r \geq 1$, if $\mathbb{I} \in \Delta_r^2$, then $\mathbb{V}^{\mathbb{I}} \in \Delta_r^2$.*

Proof. For each $\mathbf{I}$ we define operators $\Gamma_{\mathbf{I},0}, \ldots, \Gamma_{\mathbf{I},4}$ and $\Lambda_{\mathbf{I},4}$ much as in the proof of Theorem VI.1.5 but "one type up". For any $\mathsf{R} \subseteq \omega \times {}^{\omega}\omega$,

$$\Gamma_{\mathsf{I},0}(\mathsf{R}) = \{(\langle\langle 0, k, l, l', 0, n\rangle, \langle\mathbf{m}\rangle, n\rangle, \langle\boldsymbol{\alpha}\rangle) : (\mathbf{m}, \boldsymbol{\alpha}) \in {}^{k,l}\omega\}$$
$$\cup \cdots \cup \{(\langle\langle 0, k+2, l, l', 5\rangle, \langle p, q, \mathbf{m}\rangle, \mathrm{Sb}_0(p, q)\rangle, \langle\boldsymbol{\alpha}\rangle) : k, p, q \in \omega \wedge (\mathbf{m}, \boldsymbol{\alpha}) \in {}^{k,l}\omega\}$$

$\left.\begin{matrix}\Gamma_{\mathsf{I},1} \\ \Gamma_{\mathsf{I},2}\end{matrix}\right\}$ defined by similar modifications of the corresponding clauses of Theorem VI.1.5;

$$\Gamma_{\mathsf{I},3}(\mathsf{R}) = \{(\langle\langle 3, k, l, l', j', b\rangle, \langle\mathbf{m}\rangle, n\rangle, \langle\boldsymbol{\alpha}\rangle) : j' < l' \wedge \exists\beta\,[\forall p\,\mathsf{R}(\langle b, \langle p, \mathbf{m}\rangle, \beta(p)\rangle, \langle\boldsymbol{\alpha}\rangle) \wedge \mathsf{I}_{j'}(\beta) = n]\};$$

$$\Gamma_{\mathsf{I},4}(\mathsf{R}) = \{(\langle\langle 4, k, l, l', b\rangle, \langle\mathbf{m}\rangle, n\rangle, \langle\boldsymbol{\alpha}\rangle) : \exists\mathsf{F}\,[\forall\beta\,\mathsf{R}(\langle b, \langle\mathbf{m}\rangle, \mathsf{F}(\beta)\rangle, \langle\boldsymbol{\alpha}, \beta\rangle) \wedge \mathbb{I}(\mathsf{F}) = n]\};$$

$$\Lambda_{\mathsf{I},4}(\mathsf{R}) = \{(\langle\langle 4, k, l, l', b\rangle, \langle\mathbf{m}\rangle, n\rangle, \langle\boldsymbol{\alpha}\rangle) : \forall\beta\,\exists q\,\mathsf{R}(\langle b, \langle\mathbf{m}\rangle, q\rangle, \langle\boldsymbol{\alpha}, \beta\rangle) \wedge \forall\mathsf{F}\,[\forall\beta\,\forall q\,(\mathsf{R}(\langle b, \langle\mathbf{m}\rangle, q\rangle, \langle\boldsymbol{\alpha}, \beta\rangle) \rightarrow \mathsf{F}(\beta) = q) \rightarrow \mathbb{I}(\mathsf{F}) = n]\}.$$

Defining $\Gamma_{\langle\mathsf{I}\rangle}$ and $\Lambda_{\langle\mathsf{I}\rangle}$ as in VI.1.5, one can prove as before that

$$(a, \mathbf{m}, \boldsymbol{\alpha}, \mathsf{I}, n) \in \Omega[\mathbb{I}] \leftrightarrow (\langle a, \langle\mathbf{m}\rangle, n\rangle, \langle\boldsymbol{\alpha}\rangle) \in \bar{\Gamma}_{\langle\mathsf{I}\rangle} \leftrightarrow (\langle a, \langle\mathbf{m}\rangle, n\rangle, \langle\boldsymbol{\alpha}\rangle) \in \bar{\Lambda}_{\langle\mathsf{I}\rangle}.$$

If $\mathbb{I} \in \Delta^2_r$, then clearly the decomposable operator Γ is Σ^2_r and Λ is Π^2_r. It then follows from VI.7.10 and VI.7.15 that $\mathbb{V}^{\mathbb{I}} \in \Delta^2_r$. □

This proof depends essentially on the fact that the subcomputations of a computation $\{a\}^{\mathbb{I}}(\mathbf{m}, \boldsymbol{\alpha}, \mathsf{I})$ all have the same list I of type-2 arguments, so that I may be treated as a parameter. The type-1 arguments may not be so treated, as subcomputations of $\{a\}^{\mathbb{I}}(\mathbf{m}, \boldsymbol{\alpha}, \mathsf{I})$ will in general involve longer lists $\boldsymbol{\alpha}, \boldsymbol{\beta}$ of type-1 arguments. Of course, Theorem VI.1.5 works exactly because for computations relative to a type-2 functional I, the type-1 arguments do behave as parameters.

1.5 Corollary. *For any $r \geq 1$ and any $\mathbb{I} \in \Delta^2_r$,*
(i) *if $\mathbb{R}$ is semi-recursive in $\mathbb{I}$, then $\mathbb{R} \in \Delta^2_r$;*
(ii) $\{\mathbb{R} : \mathbb{R}$ is recursive in $\mathbb{I}\}$ *is a proper subset of* Δ^2_r. □

In particular, the class of relations recursive in $\mathbb{E}$ is a proper subset of Δ^2_1.

Before we consider further the relationship between $\mathbb{E}$ and $\mathbf{s}\mathbb{J}$, we discuss the substitution properties of recursion in a type-3 functional. The basic results are the following:

1.6 Theorem. *There exist primitive recursive functions f, g, and h such that for all $\mathbb{I}$, a, d, $\mathbf{m}$, $\boldsymbol{\alpha}$, and I,*

(i) $\{f(a,d)\}^{\mathsf{I}}(\mathbf{m},\boldsymbol{\alpha},\mathsf{I}) \simeq \{a\}^{\mathsf{I}}(\mathbf{m},\boldsymbol{\alpha},\lambda p.\{d\}^{\mathsf{I}}(p,\mathbf{m},\boldsymbol{\alpha},\mathsf{I}),\mathsf{I})$;
(ii) $\{g(a,d)\}^{\mathsf{I}}(\mathbf{m},\boldsymbol{\alpha},\mathsf{I}) \simeq \{a\}^{\mathsf{I}}(\mathbf{m},\boldsymbol{\alpha},\mathsf{I},\lambda\beta.\{d\}^{\mathsf{I}}(\mathbf{m},\boldsymbol{\alpha},\beta,\mathsf{I}))$;
(iii) *if* $\lambda \mathsf{G}.\{d\}^{\mathsf{I}}(\mathsf{G})$ *is a total type-3 functional* $\mathbb{H}$, *then*

$$\{h(a,d)\}^{\mathsf{I}}(\mathbf{m},\boldsymbol{\alpha},\mathsf{I}) \simeq \{a\}^{\mathbb{H}}(\mathbf{m},\boldsymbol{\alpha},\mathsf{I}).$$

Proof. The proof of (i) is very similar to that of Theorem VI.2.10. (ii) should be seen as a uniform version of Theorem VI.2.14; we first define a function g_1 which works if $\lambda\beta.\{d\}^{\mathsf{I}}(\mathbf{m},\boldsymbol{\alpha},\beta,\mathsf{I})$ is total and then choose g such that

$$\{g(a,d)\}^{\mathsf{I}}(\mathbf{m},\boldsymbol{\alpha},\mathsf{I}) \simeq 0\cdot \mathbb{I}(\lambda\beta.\{d\}^{\mathsf{I}}(\mathbf{m},\boldsymbol{\alpha},\beta,\mathsf{I})) + \{g_1(a,d)\}^{\mathsf{I}}(\mathbf{m},\boldsymbol{\alpha},\mathsf{I}).$$

(iii) is related to (ii) as is VI.2.14 to VI.2.10. □

1.7. Corollary. *For any* $\mathbb{F}$, $\mathbb{H}$, $\mathbb{I}$, β, *and* I, *if* $\mathbb{F}$ *is partial recursive in* β, I, *and* $\mathbb{H}$, *and* β, I, *and* $\mathbb{H}$ *are recursive in* $\mathbb{I}$, *then* $\mathbb{F}$ *is partial recursive in* $\mathbb{I}$. *In particular, the relation "recursive in" is transitive among objects of type-3.* □

Our next aim is to show that sJ is recursive in $\mathsf{E}_1^{\#}$. We have not in fact defined the notion of a type-3 functional being recursive in an extended type-2 functional. What is needed, of course, is a set $\Omega^3[\mathsf{E}_1^{\#}]$ defined exactly as in VI.7.1 but with an extra clause to handle application of $\mathsf{E}_1^{\#}$. We leave it to the reader to formulate explicitly these definitions and convince him- or herself that this notion has all of the usual properties. In particular, we shall need that if a functional $\mathbb{F}$ is recursive in $\mathbb{H}$ and $\mathbb{H}$ is recursive in $\mathsf{E}_1^{\#}$, then $\mathbb{F}$ is recursive in $\mathsf{E}_1^{\#}$.

1.8 Theorem. sJ *is recursive in* $\mathsf{E}_1^{\#}$.

Proof. The proof is essentially a uniform version of that of Theorem VI.6.11. First, the theory of recursion in $\mathsf{E}_1^{\#}$ must be extended to allow for type-2 arguments. In particular, there are sets $M_{\boldsymbol{\alpha},\mathsf{I}}$ and a primitive recursive function f similar to that of Corollary VI.6.8 such that for all a, $\mathbf{m}$, $\boldsymbol{\alpha}$, I and n,

$$\{a\}^{\mathsf{E}_1^{\#}}(\mathbf{m},\boldsymbol{\alpha},\mathsf{I}) \text{ is defined} \leftrightarrow f(a,\langle\mathbf{m}\rangle)\in M_{\boldsymbol{\alpha},\mathsf{I}}.$$

Corresponding to the ordinal comparison functional H of Theorem VI.6.9, there is a functional $\mathbb{H}$ partial recursive in $\mathsf{E}_1^{\#}$ such that for all u, v, $\boldsymbol{\alpha}$, and I,
(i) if $u\in M_{\boldsymbol{\alpha},\mathsf{I}}$ and $|u|^{\#}_{\boldsymbol{\alpha},\mathsf{I}} \leqslant |v|^{\#}_{\boldsymbol{\alpha},\mathsf{I}}$, then $\mathbb{H}(u,v,\langle\boldsymbol{\alpha}\rangle,\mathsf{I})\simeq 0$;
(ii) if $v\in M_{\boldsymbol{\alpha},\mathsf{I}}$ and $|v|^{\#}_{\boldsymbol{\alpha},\mathsf{I}} < |u|^{\#}_{\boldsymbol{\alpha},\mathsf{I}}$, then $\mathbb{H}(u,v,\langle\boldsymbol{\alpha}\rangle,\mathsf{I})\simeq 1$.

Now, by the methods of Theorem VI.6.11, construct primitive recursive functions g and h such that

$$\{a\}^{\mathsf{I}}(\mathbf{m},\boldsymbol{\alpha}) \text{ is defined} \leftrightarrow \{g(a)\}^{\mathsf{E}_1^{\#}}(\mathbf{m},\boldsymbol{\alpha},\mathsf{I}) \text{ is defined},$$

and

$$\{a\}^{\mathsf{I}}(\mathbf{m}, \boldsymbol{\alpha}) \text{ is undefined} \leftrightarrow \{h(a)\}^{\mathsf{E}_1^\#}(\mathbf{m}, \boldsymbol{\alpha}, \mathsf{I}) \text{ is defined.}$$

Then we claim that

$$\mathsf{sJ}((a, \mathbf{m}, \alpha) * \mathsf{I}) = \mathbb{H}(f(g(a), \langle \mathbf{m} \rangle), f(h(a), \langle \mathbf{m} \rangle), \langle \alpha \rangle, \mathsf{I}).$$

If $\{a\}^{\mathsf{I}}(\mathbf{m}, \alpha)$ is defined, then $f(g(a), \langle \mathbf{m} \rangle) \in M_{\alpha, \mathsf{I}}$ and $f(h(a), \langle \mathbf{m} \rangle) \notin M_{\alpha, \mathsf{I}}$, so

$$|f(g(a), \langle \mathbf{m} \rangle)|^\#_{\alpha, \mathsf{I}} < |f(h(a), \langle \mathbf{m} \rangle)|^\#_{\alpha, \mathsf{I}}$$

and thus $\mathbb{H}(f(g(a), \langle \mathbf{m} \rangle), f(h(a), \langle \mathbf{m} \rangle), \langle \alpha \rangle, \mathsf{I}) = 0$. If $\{a\}^{\mathsf{I}}(\mathbf{m}, \alpha)$ is undefined, then the inequality is reversed and the value is 1, as required. □

1.9 Corollary.

(i) *The class of relations recursive in* sJ *is a proper subset of* Δ^1_2;

(ii) $\mathbb{E}$ *is not recursive in* sJ.

Proof. Since the graph of $\mathsf{E}_1^\#$ is Δ^1_2, there is a Δ^1_2 decomposable operator Γ such that for all a, $\mathbf{m}$, $\boldsymbol{\alpha}$, and n

$$\langle a, \mathbf{m}, n \rangle \in \bar{\Gamma}_{\boldsymbol{\alpha}} \leftrightarrow \{a\}^{\mathsf{E}_1^\#}(\mathbf{m}, \boldsymbol{\alpha}) \simeq n$$

(cf. Exercise VI.6.20). In particular every relation semi-recursive in $\mathsf{E}_1^\#$ is Δ^1_2 and the class of relations recursive in $\mathsf{E}_1^\#$ is a proper subclass of Δ^1_2. (i) follows by Theorem 1.8 and the remarks preceding it. Then (ii) is immediate from Lemma 1.2 and Corollary 1.7. □

Note that it is *not* true that every relation semi-recursive in sJ is semi-recursive in $\mathsf{E}_1^\#$. E_1 is equivalent to E^{sJ} (Exercise VI.1.20), hence E_1 is recursive in sJ and thus every Σ^1_1 relation is recursive in sJ. Since the class of relations semi-recursive in sJ is closed under $\forall^1$, every Π^1_2 relation is semi-recursive in sJ. From this we can also conclude that there are relations R such that both R and $\sim R$ are semi-recursive in sJ, but R is not recursive in sJ. We shall see in the next section that a functional in which $\mathbb{E}$ is recursive does not exhibit this pathological behavior.

1.10–1.13 Exercises

1.10. Show that any partial recursive type-3 functional $\mathbb{F}$ is partial recursive in any type-3 $\mathbb{I}$.

1.11. Show in set theory without the Axiom of Choice that the closure ordinal of the monotone operator which defines $\Omega[\mathsf{I}]$ is at most $o({}^{\omega}\omega)$, the least ordinal not the type of a pre-wellordering of ${}^{\omega}\omega$.

1.12. Fill in some of the details in the proof of Theorem 1.6.

1.13. Define $\mathbf{b}\mathbb{E}$, the *bounded type-3 quantifier functional* by:

$$\mathbf{b}\mathbb{E}(\mathsf{I}) = \begin{cases} 0, & \text{if } \exists\alpha\,[\alpha \text{ recursive in } \mathsf{E}, \mathsf{I} \wedge \mathsf{I}(\alpha) = 0]; \\ 1, & \text{otherwise.} \end{cases}$$

Show that $\mathbf{b}\mathbb{E}$ and $\mathbf{s}\mathsf{J}$ are each recursive in the other. (That $\mathbf{b}\mathbb{E}$ is recursive in $\mathbf{s}\mathsf{J}$ is easy. For the converse let P be defined as in the proof of Lemma VI.6.1 and $<_{\boldsymbol{\alpha}}$ the transitive closure of the relation $<'_{\boldsymbol{\alpha}}$ defined by:

$$v <'_{\boldsymbol{\alpha}} u \leftrightarrow \mathsf{P}(u, v, \langle\boldsymbol{\alpha}\rangle).$$

(cf. Exercise VI.6.26). For any $\langle a, \mathbf{m}\rangle \in U^{\mathsf{I}}_{\boldsymbol{\alpha}}$, let $\delta_{a,\mathbf{m},\boldsymbol{\alpha}}$ be the characteristic function of $<_{\boldsymbol{\alpha}}$ restricted to $u \leqslant_{\boldsymbol{\alpha}} \langle a, \mathbf{m}\rangle$; otherwise, $\delta_{a,\mathbf{m},\boldsymbol{\alpha}} = \lambda p.0$. Show that $\mathsf{C}^{\mathsf{I}}_{\boldsymbol{\alpha}} = \{\delta\colon \text{for some } \langle a, \mathbf{m}\rangle \in U^{\mathsf{I}}_{\boldsymbol{\alpha}},\ \delta = \delta_{a,\mathbf{m},\boldsymbol{\alpha}}\}$ is recursive in E_1 and I, uniformly in $\boldsymbol{\alpha}$. Let

$$\mathsf{I}_{a,\mathbf{m},\boldsymbol{\alpha}}(\delta) = \begin{cases} 0, & \text{if } \delta \in \mathsf{C}^{\mathsf{I}}_{\boldsymbol{\alpha}} \text{ and } \delta = \delta_{a,\mathbf{m},\boldsymbol{\alpha}}; \\ 1, & \text{if } \delta \in \mathsf{C}^{\mathsf{I}}_{\boldsymbol{\alpha}} \text{ and } \delta \neq \delta_{a,\mathbf{m},\boldsymbol{\alpha}}; \\ 2, & \text{otherwise.} \end{cases}$$

Show that every function recursive in E and I is recursive in $\mathsf{I}_{a,\mathbf{m},\boldsymbol{\alpha}}$ and that

$$\mathbf{s}\mathsf{J}((a, \mathbf{m}, \boldsymbol{\alpha}) * \mathsf{I}) = \mathbf{b}\mathbb{E}(\mathsf{I}_{a,\mathbf{m},\boldsymbol{\alpha}}).\ )$$

1.14. Notes. The basic definition here is again due to Kleene [1959]. Theorem 1.4 for $\mathbb{E}$ was announced in Hinman [1964] and proved in Hinman [1966] but never published. The substitution theorems (1.6) are again due to Kleene [1963] with improvements from Hinman [1966]. Theorem 1.8 appears in Aczel [1970] where it is attributed to Gandy. Corollary 1.9 and Exercise 1.13 are due to Gandy [1967a]. Some variations on Exercise 1.13 appear in Aczel–Hinman [1974].

2. *Relations Semi-Recursive in a Type-3 Functional*

When $\mathbb{E}$ is recursive in I, the class of functionals semi-recursive in I is similar in structure to the class of relations semi-recursive in a type-2 functional. In most instances, $\mathbb{E}$ plays a role analogous to that of E. This analogy fails at one point —

it is not in general true that the class of relations semi-recursive in $\mathbb{I}$ is closed under $\exists^1$. This fact is closely related to the considerations of § VI.6.

For any $\mathbb{I}$, we set

$$\mathbb{U}^{\mathbb{I}}(a, \langle \mathbf{m} \rangle, \langle \boldsymbol{\alpha} \rangle, \langle \mathbf{I} \rangle) \leftrightarrow \{a\}^{\mathbb{I}}(\mathbf{m}, \boldsymbol{\alpha}, \mathbf{I}) \text{ is defined;}$$

$$\mathbb{U}^{\mathbb{I}}_{\mathbf{I}}(a, \langle \mathbf{m} \rangle, \langle \boldsymbol{\alpha} \rangle) \leftrightarrow \mathbb{U}^{\mathbb{I}}(a, \langle \mathbf{m} \rangle, \langle \boldsymbol{\alpha} \rangle, \langle \mathbf{I} \rangle);$$

$$\mathbb{U}^{\mathbb{I}} = \mathbb{U}^{\mathbb{I}}_{\varnothing}.$$

Of course, $\mathbb{U}^{\mathbb{I}}$ is universal for the class of relations semi-recursive in $\mathbb{I}$. To each element of $\mathbb{U}^{\mathbb{I}}$ we assign an ordinal which measures the "length" of the corresponding computation:

$$|a, \mathbf{m}, \boldsymbol{\alpha}, \mathbf{I}|^{\mathbb{I}} = \text{least } \sigma . (a, \mathbf{m}, \boldsymbol{\alpha}, \mathbf{I}, \{a\}^{\mathbb{I}}(\mathbf{m}, \boldsymbol{\alpha}, \mathbf{I})) \in \Omega[\mathbb{I}]^{\sigma}.$$

If $\sim\mathbb{U}^{\mathbb{I}}(a, \langle \mathbf{m} \rangle, \langle \boldsymbol{\alpha} \rangle, \langle \mathbf{I} \rangle)$, we set $|a, \mathbf{m}, \boldsymbol{\alpha}, \mathbf{I}|^{\mathbb{I}} = o({}^{\omega}\omega)$. Then an appropriate modification of Lemma VI.3.2 holds with the additional clause:

(4) if for some b, $a = \langle 4, k, l, l', b \rangle$ and $\{a\}^{\mathbb{I}}(\mathbf{m}, \boldsymbol{\alpha}, \mathbf{I})$ is defined, then $|a, \mathbf{m}, \boldsymbol{\alpha}, \mathbf{I}|^{\mathbb{I}} = \sup^{+}\{|b, \mathbf{m}, \boldsymbol{\alpha}, \beta, \mathbf{I}|^{\mathbb{I}} : \beta \in {}^{\omega}\omega\}$.

In the following we shall write (as in several earlier sections) x^i for $(a^i, \langle \mathbf{m}^i \rangle, \langle \boldsymbol{\alpha}^i \rangle)$ and $\mathbb{H}(x^0, x^1, \langle \mathbf{I} \rangle)$ for $\mathbb{H}(a^0, a^1, \langle \mathbf{m}^0 \rangle, \langle \mathbf{m}^1 \rangle, \langle \boldsymbol{\alpha}^0 \rangle, \langle \boldsymbol{\alpha}^1 \rangle, \langle \mathbf{I} \rangle)$. We take $\mathbb{I}$ to be a fixed type-3 functional such that $\mathbb{E}$ is recursive in $\mathbb{I}$, and write $|x|^{\mathbb{I}}_{\mathbf{I}}$ for $|x, \mathbf{I}|^{\mathbb{I}}$.

2.1 Ordinal Comparison Theorem. *There exists a functional $\mathbb{H}$ partial recursive in $\mathbb{I}$ such that for all x^0, x^1, and $\mathbf{I}$,*

(i) *if $x^0 \in \mathbb{U}^{\mathbb{I}}_{\mathbf{I}}$ and $|x^0|^{\mathbb{I}}_{\mathbf{I}} \leqslant |x^1|^{\mathbb{I}}_{\mathbf{I}}$, then $\mathbb{H}(x^0, x^1, \langle \mathbf{I} \rangle) \simeq 0$;*

(ii) *if $x^1 \in \mathbb{U}^{\mathbb{I}}_{\mathbf{I}}$ and $|x^1|^{\mathbb{I}}_{\mathbf{I}} < |x^0|^{\mathbb{I}}_{\mathbf{I}}$, then $\mathbb{H}(x^0, x^1, \langle \mathbf{I} \rangle) \simeq 1$.*

Proof. We shall not give many details here as the reader may easily supply most of them by translating the proof of Theorem VI.3.3, the Ordinal Comparison Theorem for recursion relative to a type-2 functional. For simplicity let $\mathbf{I} = \varnothing$ and write $\mathbb{H}(x^0, x^1)$ for $\mathbb{H}(x^0, x^1, \langle\ \rangle)$. $\mathbb{H}$ is defined by effective transfinite recursion via the $\mathbb{I}$-Recursion Theorem and a functional $\mathbb{F}$ partial recursive in $\mathbb{I}$. $\mathbb{F}$ is defined in 25 cases labeled (r, s) for $0 \leqslant r, s \leqslant 4$ plus two "otherwise" cases. In all cases except those in which $r = 4$ or $s = 4$, the definition is almost exactly as before. We shall consider cases (2,4) and (3,4) and leave the remainder to the reader.

Case (2,4). Suppose $x^0 = (\langle 2, k^0 + 1, l^0, 0 \rangle, \langle b^0, \mathbf{m}^0 \rangle, \langle \boldsymbol{\alpha}^0 \rangle)$ and $x^1 = (\langle 4, k^1, l^1, 0, b \rangle, \langle \mathbf{m}^1 \rangle, \langle \boldsymbol{\alpha}^1 \rangle)$. We define $\mathbb{F}$ in such a way as to ensure that

$$\mathbb{H}(x^0, x^1) \simeq \mathbb{E}(\lambda\beta . \mathbb{H}((b^0, \langle \mathbf{m}^0 \rangle, \langle \boldsymbol{\alpha}^0 \rangle), (b^1, \langle \mathbf{m}^1 \rangle, \langle \boldsymbol{\alpha}^1, \beta \rangle))).$$

If $x^0 \in \mathsf{U}^{\mathbb{I}}$ and $|x^0|^{\mathbb{I}} \leq |x^1|^{\mathbb{I}}$, then for some β,

$$|b^0, \mathbf{m}^0, \boldsymbol{\alpha}^0|^{\mathbb{I}} \leq |b^1, \mathbf{m}^1, \boldsymbol{\alpha}^1, \beta|^{\mathbb{I}}$$

and thus $\mathbb{H}(x^0, x^1) \simeq 0$. If $x^1 \in \mathsf{U}^{\mathbb{I}}$ and $|x^1|^{\mathbb{I}} < |x^0|^{\mathbb{I}}$, then for all β,

$$|b^1, \mathbf{m}^1, \boldsymbol{\alpha}^1, \beta|^{\mathbb{I}} < |b^0, \mathbf{m}^0, \boldsymbol{\alpha}^0|^{\mathbb{I}}$$

and $\mathbb{H}(x^0, x^1) \simeq 1$.

Case (3,4). Suppose $x^0 = (\langle 3, k^0, l^0, 0, b^0\rangle, \langle \mathbf{m}^0\rangle, \langle \boldsymbol{\alpha}^0\rangle)$ and $x^1 = (\langle 4, k^1, l^1, 0, b^1\rangle, \langle \mathbf{m}^1\rangle, \langle \boldsymbol{\alpha}^1\rangle)$. We define $\mathbb{F}$ in such a way that

$$\mathbb{H}(x^0, x^1) \simeq \mathsf{E}^{\circ}(\lambda p\,.\,\mathbb{E}(\lambda\beta\,.\,\mathbb{H}((b^0, \langle p, \mathbf{m}^0\rangle, \langle \boldsymbol{\alpha}^0\rangle), (b^1, \langle \mathbf{m}^1\rangle, \langle \boldsymbol{\alpha}^1, \beta\rangle)))))$$

where E° is the "dual" of E, which is, of course, recursive in $\mathbb{E}$ and hence in $\mathbb{I}$. If $x^0 \in \mathsf{U}^{\mathbb{I}}$ and $|x^0|^{\mathbb{I}} \leq |x^1|^{\mathbb{I}}$, then for all p there exists a β such that

$$|b^0, p, \mathbf{m}^0, \boldsymbol{\alpha}^0|^{\mathbb{I}} \leq |b^1, \mathbf{m}^1, \boldsymbol{\alpha}^1, \beta|^{\mathbb{I}}$$

and thus $\mathbb{H}(x^0, x^1) \simeq 0$. If $x^1 \in \mathsf{U}^{\mathbb{I}}$ and $|x^1|^{\mathbb{I}} < |x^0|^{\mathbb{I}}$, then for some p and all β,

$$|b^1, \mathbf{m}^1, \boldsymbol{\alpha}^1, \beta|^{\mathbb{I}} < |b^0, p, \mathbf{m}^0, \boldsymbol{\alpha}^0|^{\mathbb{I}}$$

and $\mathbb{H}(x^0, x^1) \simeq 1$. □

2.2 Selection Theorem. *There exists a function* $\mathrm{Sel}^{\mathbb{I}}$ *partial recursive in* $\mathbb{I}$ *such that for all* a, $\mathbf{m}$, $\boldsymbol{\alpha}$, *and* $\mathbf{I}$, *the following are equivalent*:

(i) $\exists p\,.\,\{a\}^{\mathbb{I}}(p, \mathbf{m}, \boldsymbol{\alpha}, \mathbf{I})$ *is defined*;

(ii) $\{a\}^{\mathbb{I}}(\mathrm{Sel}^{\mathbb{I}}(a, \langle\mathbf{m}\rangle, \langle\boldsymbol{\alpha}\rangle, \langle\mathbf{I}\rangle), \mathbf{m}, \boldsymbol{\alpha}, \mathbf{I})$ *is defined.*

Proof. Exactly as for Theorem VI.4.1. □

2.3 Corollary. *For any relation* $\mathbb{R}$ *semi-recursive in* $\mathbb{I}$, *there exists a functional* $\mathrm{Sel}_{\mathbb{R}}$ *partial recursive in* $\mathbb{I}$ *such that for all* $\mathbf{m}$, $\boldsymbol{\alpha}$, *and* $\mathbf{I}$,

$$\exists p\, R(p, \mathbf{m}, \boldsymbol{\alpha}, \mathbf{I}) \leftrightarrow \mathbb{R}(\mathrm{Sel}_{\mathbb{R}}(\mathbf{m}, \boldsymbol{\alpha}, \mathbf{I}), \mathbf{m}, \boldsymbol{\alpha}, \mathbf{I}). \quad \square$$

We may now derive corollaries corresponding to VI.4.3–6, 8–9 with very similar proofs, which we leave to the reader. The boundedness and hierarchy results take a little more care. Let

$$\boldsymbol{\kappa}^{\mathbb{I}} = \sup{}^{+}\{|x|^{\mathbb{I}} : x \in \mathsf{U}^{\mathbb{I}}\}.$$

Since $\mathsf{U}^{\mathbb{I}}$ is of power $2^{\aleph_0}$ there is no reason to expect that $\boldsymbol{\kappa}^{\mathbb{I}}$ is a countable ordinal (cf. Exercise 2.15). As in § V.1 it will be useful to consider the ordinals associated with the "countable part" of $\mathsf{U}^{\mathbb{I}}$. Let

$$U^{\mathbb{I}} = \{\langle a, \mathbf{m}\rangle : \mathsf{U}^{\mathbb{I}}(a, \langle \mathbf{m}\rangle, \langle\ \ \rangle)\}.$$

The pre-wellordering on $\mathsf{U}^{\mathbb{I}}$ associated with the norm $|\ \ |^{\mathbb{I}}$ induces a pre-wellordering on $U^{\mathbb{I}}$ in an obvious way. We denote by $|\ \ |_0^{\mathbb{I}}$ the norm associated with this pre-wellordering and set

$$\kappa^{\mathbb{I}} = \sup{}^{+}\{|a, \mathbf{m}|_0^{\mathbb{I}} : U^{\mathbb{I}}(\langle a, \mathbf{m}\rangle)\}.$$

Note that $U^{\mathbb{I}}$ is *not* an initial segment of $\mathsf{U}^{\mathbb{I}}$ (Exercise 2.18) so $|\ \ |_0^{\mathbb{I}}$ is not simply the restriction of $|\ \ |^{\mathbb{I}}$ to $U^{\mathbb{I}}$. We use the notations $\mathsf{U}_\rho^{\mathbb{I}}$ and $U_\rho^{\mathbb{I}}$ as in § V.1.

2.4 Boundedness Theorem. (i) *for any A co-semi-recursive in $\mathbb{I}$, if $A \subsetneq U^{\mathbb{I}}$, then* $\sup^{+}\{|u|_0^{\mathbb{I}} : u \in A\} < \kappa^{\mathbb{I}}$;

(ii) *for any β and any R co-semi-recursive in $\mathbb{I}$ and β, if $\mathsf{R} \subsetneq \mathsf{U}^{\mathbb{I}}$, then* $\sup^{+}\{|x|^{\mathbb{I}} : x \in \mathsf{R}\} < \boldsymbol{\kappa}^{\mathbb{I}}$.

Proof. The proof of Theorem V.1.5 applies because the class of relations co-semi-recursive in $\mathbb{I}$ is closed under both $\exists^0$ and $\exists^1$. □

2.5 Hierarchy Theorem. (i) *for all relations R on numbers, R is recursive in $\mathbb{I}$ iff $R \ll U_\rho^{\mathbb{I}}$ for some $\rho < \kappa^{\mathbb{I}}$*;

(ii) *for all relations R, R is recursive in $\mathbb{I}$ and some β iff $\mathsf{R} \ll \mathsf{U}_\rho^{\mathbb{I}}$ for some $\rho < \boldsymbol{\kappa}^{\mathbb{I}}$.* □

Part (ii) of this theorem of course also provides a hierarchy for the relations R which are recursive in $\mathbb{I}$ alone, but it is somewhat unnatural in that although there are only countably many such R, the hierarchy is indexed by uncountably many ordinals, so that "most" of the levels are empty. The same phenomenon occurs at the odd levels beyond the first of the analytical hierarchy under the hypothesis PD (cf. Theorems V.1.6 and V.3.3 (iv)(b)). At the first level, of course, there is a natural hierarchy of length ω_1 for the class of Δ_1^1 relations of types 1 and 2 (Theorem IV.2.2); the situation is similar here. For each $\rho < \kappa^{\mathbb{I}}$, let $\bar{\rho}$ be the unique ordinal such that for some $u \in U^{\mathbb{I}}$, $|u|_0^{\mathbb{I}} = \rho$ and $|u|^{\mathbb{I}} = \bar{\rho}$.

2.6 Theorem. *For all relations R, R is recursive in $\mathbb{I}$ iff $\mathsf{R} \ll \mathsf{U}_{\bar{\rho}}^{\mathbb{I}}$ for some $\rho < \kappa^{\mathbb{I}}$.*

Proof. For any $\rho < \kappa^{\mathbb{I}}$ and any u such that $|u|_0^{\mathbb{I}} = \rho$,

$$x \in \mathsf{U}_{\bar{\rho}}^{\mathbb{I}} \leftrightarrow |x|^{\mathbb{I}} < \bar{\rho} = |u|^{\mathbb{I}} \leftrightarrow \mathbb{H}(x, (u, \langle\ \ \rangle)) = 0,$$

where $\mathbb{H}$ is the ordinal comparison functional of Theorem 2.1. Hence $\mathsf{U}^{\mathbb{I}}_{\bar\rho}$ is recursive in $\mathbb{I}$ and thus so is any R reducible to it.

For the implication ($\rightarrow$), let R be recursive in $\mathbb{I}$, say with index a from $\mathbb{I}$, and let b and c be indices such that

$$\{b\}^{\mathbb{I}}(\mathbf{m}) \simeq \mathbb{I}(\lambda\alpha_0 \,.\, \mathbb{I}(\lambda\alpha_1 \cdots \mathbb{I}(\lambda\alpha_{l-1} \,.\, \{a\}^{\mathbb{I}}(\mathbf{m}, \boldsymbol{\alpha})) \cdots));$$

$$\{c\}^{\mathbb{I}}(\ \ ) \simeq \mathbb{I}(\lambda(\beta_0 \,.\, \mathbb{I}(\lambda\beta_1 \cdots \mathbb{I}(\lambda\beta_{k-1} \,.\, \{b\}^{\mathbb{I}}(\beta_0(0), \ldots, \beta_{k-1}(0))) \cdots)).$$

Clearly for each $\mathbf{m}$ and $\boldsymbol{\alpha}$,

$$|a, \mathbf{m}, \boldsymbol{\alpha}|^{\mathbb{I}} \leqslant |b, \mathbf{m}, \varnothing|^{\mathbb{I}} \leqslant |c, \varnothing, \varnothing|^{\mathbb{I}}.$$

Let d be an index such that

$$\{d\}^{\mathbb{I}}(\mathbf{m}, \boldsymbol{\alpha}) \simeq \begin{cases} 0, & \text{if } \{a\}^{\mathbb{I}}(\mathbf{m}, \boldsymbol{\alpha}) = 0; \\ \text{undefined}, & \text{otherwise}; \end{cases}$$

and there exists a number q such that if $\{a\}^{\mathbb{I}}(\mathbf{m}, \boldsymbol{\alpha}) = 0$, then $|d, \mathbf{m}, \boldsymbol{\alpha}|^{\mathbb{I}} < |a, \mathbf{m}, \boldsymbol{\alpha}|^{\mathbb{I}} + q$ (i.e., d is a natural index). Then if $\rho = |c, \varnothing|^{\mathbb{I}}_0 + q$, $\rho < \kappa^{\mathbb{I}}$ and

$$\begin{aligned} \mathsf{R}(\mathbf{m}, \boldsymbol{\alpha}) &\leftrightarrow (d, \mathbf{m}, \boldsymbol{\alpha}) \in \mathsf{U}^{\mathbb{I}} \\ &\leftrightarrow |d, \mathbf{m}, \boldsymbol{\alpha}|^{\mathbb{I}} < |x, \varnothing, \varnothing|^{\mathbb{I}} + q \leqslant \bar\rho \\ &\leftrightarrow (d, \mathbf{m}, \boldsymbol{\alpha}) \in \mathsf{U}^{\mathbb{I}}_{\bar\rho}. \quad \square \end{aligned}$$

Of course, it is not true in general that if $\rho < \sigma$ then strictly more relations are reducible to $\mathsf{U}^{\mathbb{I}}_{\bar\tau}$ for $\tau \leqslant \sigma$ than for $\tau \leqslant \rho$ (cf. Exercise 2.16).

2.7 Upper Classification Theorem. $\{\alpha : \alpha \text{ is recursive in } \mathbb{I}\}$ *is semi-recursive in* $\mathbb{I}$.

Proof. As for Theorem VI.4.12. $\square$

We leave to the reader the formulation and proof of choice principles parallel to Theorem VI.4.15 (cf. Exercise 2.17).

The ordinals $\kappa^{\mathbb{I}}$ and $\boldsymbol{\kappa}^{\mathbb{I}}$ may be evaluated much as in previous similar cases. Let $\omega_1[\mathbb{I}]$ be the least ordinal not recursive in $\mathbb{I}$ and $\boldsymbol{\omega}_1[\mathbb{I}]$ the least ordinal not the order-type of a pre-wellordering of ${}^{\omega}\omega$ recursive in $\mathbb{I}$ and some β.

2.8 Theorem. $\kappa^{\mathbb{I}} = \omega_1[\mathbb{I}]$ *and* $\boldsymbol{\kappa}^{\mathbb{I}} = \boldsymbol{\omega}_1[\mathbb{I}]$.

Proof. Similar to that of Theorem VI.4.17 (cf. Exercise 2.19). $\square$

To this point there has been a close analogy between the roles played by E and 𝔼 in recursion relative to functionals of types 2 and 3, respectively. The remaining results of this section indicate some differences. One that we have already seen is that in contrast with the fact that the relations recursive in E are exactly the Δ_1^1 relations, those recursive in 𝔼 form a proper subclass of Δ_1^2. The basis for this discrepancy is that the property of well-foundedness for type-2 relations is Δ_1^2 (in fact $\Delta_{(\omega)}^1$, Lemma VI.7.11), whereas well-foundedness for type-1 relations is Π_1^1 but not Δ_1^1. We now exploit the same fact in a different way to show that the class of relations semi-recursive in a type-3 functional 𝕀 is not closed under existential function quantification ($\exists^1$) (cf. discussion at the beginning of § VI.6).

2.9 Lemma. *There exists a relation* $\mathsf{P}^{\mathbb{I}}$ *semi-recursive in* 𝕀 *such that for all* u, v, $\boldsymbol{\alpha}$, *and* $\boldsymbol{\beta}$,

(i) $(u,\langle\boldsymbol{\alpha}\rangle)\in \mathsf{U}^{\mathbb{I}} \wedge \mathsf{P}^{\mathbb{I}}(u,v,\langle\boldsymbol{\alpha}\rangle,\langle\boldsymbol{\beta}\rangle)\rightarrow |v,\boldsymbol{\beta}|^{\mathbb{I}}<|u,\boldsymbol{\alpha}|^{\mathbb{I}}$;

(ii) $(u,\langle\boldsymbol{\alpha}\rangle)\notin \mathsf{U}^{\mathbb{I}}\rightarrow \exists v\,\exists\boldsymbol{\beta}\,[(v,\langle\boldsymbol{\beta}\rangle)\notin \mathsf{U}^{\mathbb{I}}\wedge \mathsf{P}^{\mathbb{I}}(u,v,\langle\boldsymbol{\alpha}\rangle,\langle\boldsymbol{\beta}\rangle)]$.

Proof. Much as in the proof of Lemma VI.6.1, we define $\mathsf{P}^{\mathbb{I}}(u,v,\langle\boldsymbol{\alpha}\rangle,\langle\boldsymbol{\beta}\rangle)$ to hold just in case one of the following holds for some k, $\mathbf{m}\in {}^k\omega$, and $l=\lg(\boldsymbol{\alpha})$:

(1) for some $b, c_0,\ldots,c_{k'-1}$, $u=\langle\langle 1,k,l,0,b,c_0,\ldots,c_{k'-1}\rangle,\mathbf{m}\rangle$, $\boldsymbol{\beta}=\boldsymbol{\alpha}$, and either $v=\langle c_i,\mathbf{m}\rangle$ for some $i<k'$ or there exist $q_0,\ldots,q_{k'-1}$ such that for all $i<k'$, $\{c_i\}^{\mathbb{I}}(\mathbf{m},\boldsymbol{\alpha})\simeq q_i$ and $v=\langle b,\mathbf{q}\rangle$;

(2) for some b, $u=\langle\langle 2,k+1,l,0\rangle,b,\mathbf{m}\rangle$, $\boldsymbol{\beta}=\boldsymbol{\alpha}$, and $v=\langle b,\mathbf{m}\rangle$;

(4) for some b, $u=\langle\langle 4,k,l,0,b\rangle,\mathbf{m}\rangle$, $v=\langle b,\mathbf{m}\rangle$, and for some γ, $\boldsymbol{\beta}=\boldsymbol{\alpha}*(\gamma)$;

(5) u is of none of these forms, $|u,\boldsymbol{\alpha}|^{\mathbb{I}}\neq 0$, $\boldsymbol{\beta}=\boldsymbol{\alpha}$, and $v=0$.

Because we are (for simplicity) omitting type-2 parameters ($l'=0$), there is no need for a clause (3).

The proof that $\mathsf{P}^{\mathbb{I}}$ satisfies (i) and (ii) is essentially as before. □

2.10 Theorem *For all* a, $\mathbf{m}$, *and* $\boldsymbol{\alpha}$,

$$\{a\}^{\mathbb{I}}(\mathbf{m},\boldsymbol{\alpha}) \text{ is defined} \leftrightarrow \neg\exists\delta\,\exists\varepsilon\,\forall p\,[\mathsf{P}^{\mathbb{I}}(\delta(p),\delta(p+1),(\varepsilon)^p,(\varepsilon)^{p+1})\wedge \delta(0)=\langle a,\mathbf{m}\rangle\wedge(\varepsilon)^0=\langle\boldsymbol{\alpha}\rangle].$$

Proof. This follows from Lemma 2.9 in very much the same way that Theorem VI.6.2 follows from Lemma VI.6.1. The difference is that the descending sequence of undefined computations can no longer be described by a Suslin quantifier ($\mathscr{A}$) as the type-1 function arguments may vary. □

2.11 Corollary. *The class of relations semi-recursive in* 𝕀 *is not closed under existential function quantification* ($\exists^1$). □

As in § VI.6, this part of the analogy between E and 𝔼 can be repaired by replacing $\mathbb{E}$ by an extended type-3 functional $\mathbb{E}^{\#}$ defined by: for any partial functional $\mathsf{F} : {}^{\omega}\omega \to \omega$,

$$\mathbb{E}^{\#}(\mathsf{F}) \simeq \begin{cases} 0, & \text{if } \exists\alpha . \mathsf{F}(\alpha) \simeq 0; \\ 1, & \text{if } \forall\alpha\,(\exists n > 0) . \mathsf{F}(\alpha) \simeq n; \\ \text{undefined,} & \text{otherwise.} \end{cases}$$

The theory of recursion relative to $\mathbb{E}^{\#}$ may be developed in a way closely parallel with the theory of recursion relative to $\mathsf{E}_1^{\#}$ in § VI.6. It is clear that the class of relations semi-recursive in $\mathbb{E}^{\#}$ is closed under $\exists^1$. That every relation semi-recursive in $\mathbb{E}$ is also semi-recursive in $\mathbb{E}^{\#}$ may be proved as in VI.6.6 and thus both $\mathsf{U}^{\mathbb{E}}$ and $\sim\mathsf{U}^{\mathbb{E}}$ are semi-recursive in $\mathbb{E}^{\#}$. To establish an ordinal comparison theorem for $\mathbb{E}^{\#}$ we define inductively a relation M as follows: M is the smallest relation such that for all $d \in \mathrm{Pri}$ and all $\boldsymbol{\alpha}$,

(i) $(0, \langle\boldsymbol{\alpha}\rangle) \in \mathsf{M}$;

(ii) if $\exists\beta\,([d](\boldsymbol{\alpha}, \beta), \langle\boldsymbol{\alpha}, \beta\rangle) \in \mathsf{M}$, then $(\langle 0, d\rangle, \langle\boldsymbol{\alpha}\rangle) \in \mathsf{M}$;

(iii) if $\forall\beta\,([d](\boldsymbol{\alpha}, \beta), \langle\boldsymbol{\alpha}, \beta\rangle) \in \mathsf{M}$, then $(\langle 1, d\rangle, \langle\boldsymbol{\alpha}\rangle) \in \mathsf{M}$.

The analogues of Theorems VI.6.7–9 all hold with very similar proofs, and we conclude in particular that $\mathsf{U}^{\mathbb{E}}$ is recursive in $\mathbb{E}^{\#}$.

If we include type-2 parameters everywhere, we can similarly prove that $\mathbb{U}^{\mathbb{E}}$ is recursive in $\mathbb{E}^{\#}$. Let $s\mathcal{J}$ be the type-4 jump operator defined by:

$$s\mathcal{J}(\mathbb{I})((a, \mathbf{m}, \alpha) * \mathsf{I}) = \begin{cases} 0, & \text{if } \{a\}^{\mathbb{I}}(\mathbf{m}, \alpha, \mathsf{I}) \text{ is defined;} \\ 1, & \text{otherwise.} \end{cases}$$

Then the same argument shows that if $\mathbb{I}$ is recursive in $\mathbb{E}^{\#}$, so is $s\mathcal{J}(\mathbb{I})$. Thus if we set

$$\mathbb{E}_0 = \mathbb{E} \quad \text{and} \quad \mathbb{E}_{r+1} = s\mathcal{J}(\mathbb{E}_r),$$

then $\mathbb{E}_r$ is recursive in $\mathbb{E}_s$ iff $r \leqslant s$ and all $\mathbb{E}_r$ are recursive in $\mathbb{E}^{\#}$. The relations recursive in $\mathbb{E}$ are called *hyperanalytical* while those recursive in $\mathbb{E}^{\#}$ are called *hyperprojective*. Note that it follows from Corollary VI.7.13 that the relations semi-recursive in $\mathbb{E}^{\#}$ (*semi-hyperprojective*) still form a subset of Δ_1^2. The subset is proper because it is not closed under complementation.

We conclude this section by establishing an analogue of the Spector–Gandy Theorem (VI.4.18) for recursion in $\mathbb{I}$. The proof is a modification of that sketched in Exercise VI.6.26 and is substantially simpler than that of VI.4.18. We denote by $\Sigma_1^{2,\mathbb{I}}$ the class of relations $\mathbb{R}$ such that for some $\mathbb{S}$ recursive in $\mathbb{I}$,

$$\mathbb{R}(\mathbf{m}, \boldsymbol{\alpha}, \mathbf{I}) \leftrightarrow (\exists \mathsf{F} \text{ recursive in } \boldsymbol{\alpha}, \mathbf{I}, \text{ and } \mathbb{I})\, \mathbb{S}(\mathbf{m}, \boldsymbol{\alpha}, \mathbf{I}, \mathsf{F}).$$

2.12 Theorem. *For all $\mathbb{R}$, $\mathbb{R} \in \Sigma_1^{2,\mathbb{I}}$ iff $\mathbb{R}$ is semi-recursive in $\mathbb{I}$.*

Proof. The implication $(\rightarrow)$ is an immediate consequence of the Substitution Theorem (1.6): if $\mathbb{R}$ and $\mathbb{S}$ are as above, then

$$\mathbb{R}(\mathbf{m}, \boldsymbol{\alpha}, \mathbf{I}) \leftrightarrow \exists a \,.\, \mathbb{S}(\mathbf{m}, \boldsymbol{\alpha}, \mathbf{I}, \lambda\beta \,.\, \{a\}^{\mathbb{I}}(\boldsymbol{\alpha}, \beta, \mathbf{I})).$$

For simplicity, we shall prove the implication $(\leftarrow)$ only for relations with arguments of types 0 and 1; the general case follows by an obvious modification. It will suffice to show that $\mathsf{U}^{\mathbb{I}} \in \Sigma_1^{2,\mathbb{I}}$. The method is analogous to that of Exercise VI.6.26. Let $\mathbb{P}$ be a relation such that $\mathbb{P}(u, v, \langle\boldsymbol{\alpha}\rangle, \langle\boldsymbol{\beta}\rangle, \mathsf{G})$ holds under exactly the same conditions as does $\mathsf{P}^{\mathbb{I}}(u, v, \langle\boldsymbol{\alpha}\rangle, \langle\boldsymbol{\beta}\rangle)$ in Lemma 2.9 except that in clause (1) we replace the condition

$$\{c_i\}^{\mathbb{I}}(\mathbf{m}, \boldsymbol{\alpha}) \simeq q_i \quad \text{by} \quad \mathsf{G}(\langle c_i, \mathbf{m}\rangle, \langle\boldsymbol{\alpha}\rangle) = q_i.$$

Clearly $\mathbb{P}$ is recursive in $\mathbb{I}$. Let $<_{\mathsf{G}}$ denote the transitive closure of the relation $<'_{\mathsf{G}}$ defined by

$$(v, \langle\boldsymbol{\beta}\rangle) <'_{\mathsf{G}} (u, \langle\boldsymbol{\alpha}\rangle) \leftrightarrow \mathbb{P}(u, v, \langle\boldsymbol{\alpha}\rangle, \langle\boldsymbol{\beta}\rangle, \mathsf{G}).$$

We say that a functional H is *closed for* G iff for all u, v, $\boldsymbol{\alpha}$, and $\boldsymbol{\beta}$,

$$\mathsf{H}(u, \langle\boldsymbol{\alpha}\rangle) = 0 \wedge (v, \langle\boldsymbol{\beta}\rangle) <_{\mathsf{G}} (u, \langle\boldsymbol{\alpha}\rangle) \rightarrow \mathsf{H}(v, \langle\boldsymbol{\beta}\rangle) = 0.$$

H is *well-founded for* G iff

$$\neg \exists\beta\, \exists\gamma\, \forall p\, [\mathsf{H}(\beta(p), (\gamma)^p) = 0 \wedge (\beta(p+1), (\gamma)^{p+1}) <_{\mathsf{G}} (\beta(p), (\gamma)^p)].$$

Finally, we shall say that G is *locally correct for* H iff for all k, l, and $(\mathbf{m}, \boldsymbol{\alpha}) \in {}^{k,l}\omega$ and all u such that $\mathsf{H}(u, \langle\boldsymbol{\alpha}\rangle) = 0$:

(0) if $u = \langle\langle 0, k, l, 0, \ldots\rangle, \mathbf{m}\rangle$ and $\langle 0, k, l, 0, \ldots\rangle$ is an index of the proper form for $(\mathbf{m}, \boldsymbol{\alpha})$, then $\mathsf{G}(u, \langle\boldsymbol{\alpha}\rangle) = \{\langle 0, k, l, 0, \ldots\rangle\}^{\mathbb{I}}(\mathbf{m}, \boldsymbol{\alpha})$;

(1) if for some $b, c_0, \ldots, c_{k'-1}$, $u = \langle\langle 1, k, l, 0, b, c_0, \ldots, c_{k'-1}\rangle, \mathbf{m}\rangle$, then

$$\mathsf{G}(u, \langle\boldsymbol{\alpha}\rangle) = \mathsf{G}(\langle b, \mathsf{G}(\langle c_0, \mathbf{m}\rangle, \langle\boldsymbol{\alpha}\rangle), \ldots, \mathsf{G}(\langle c_{k'-1}, \mathbf{m}\rangle, \langle\boldsymbol{\alpha}\rangle)\rangle, \langle\boldsymbol{\alpha}\rangle);$$

(2) if for some b, $u = \langle\langle 2, k+1, l, 0\rangle, b, \mathbf{m}\rangle$, then

$$\mathsf{G}(u, \langle\boldsymbol{\alpha}\rangle) = \mathsf{G}(\langle b, \mathbf{m}\rangle, \langle\boldsymbol{\alpha}\rangle);$$

(4) if for some b, $u = \langle\langle 4, k, l, 0, b\rangle, \mathbf{m}\rangle$, then

$$\mathsf{G}(u, \langle\boldsymbol{\alpha}\rangle) = \mathbb{I}(\lambda\delta \,.\, \mathsf{G}(\langle b, \mathbf{m}\rangle, \langle\boldsymbol{\alpha}, \delta\rangle)).$$

We now claim that for all d, $\mathbf{n}$, and $\boldsymbol{\beta}$,

$$\mathsf{U}^{\mathbb{I}}(d, \langle \mathbf{n} \rangle, \langle \boldsymbol{\beta} \rangle) \leftrightarrow (\exists \mathsf{G}, \mathsf{H} \text{ recursive in } \mathbb{I}, \boldsymbol{\beta})[\mathsf{G} \text{ is locally correct for } \mathsf{H} \wedge \\ \mathsf{H}(\langle d, \mathbf{n} \rangle, \langle \boldsymbol{\beta} \rangle) = 0 \wedge \mathsf{H} \text{ is closed for } \mathsf{G} \wedge \mathsf{H} \text{ is well-founded for } \mathsf{G}].$$

For the implication ($\rightarrow$), suppose that $\mathsf{U}^{\mathbb{I}}(d, \langle \mathbf{n} \rangle, \langle \boldsymbol{\beta} \rangle)$ holds and let

$$\mathsf{G}(\langle a, \mathbf{m} \rangle, \langle \boldsymbol{\alpha} \rangle) = \begin{cases} \{a\}^{\mathbb{I}}(\mathbf{m}, \boldsymbol{\alpha}), & \text{if } \ |a, \mathbf{m}, \boldsymbol{\alpha}|^{\mathbb{I}} \leqslant |d, \mathbf{n}, \boldsymbol{\beta}|^{\mathbb{I}}; \\ 0, & \text{otherwise}; \end{cases}$$

$$\mathsf{H}(\langle a, \mathbf{m} \rangle, \langle \boldsymbol{\alpha} \rangle) = \begin{cases} 0, & \text{if } \ |a, \mathbf{m}, \boldsymbol{\alpha}|^{\mathbb{I}} \leqslant |d, \mathbf{n}, \boldsymbol{\beta}|^{\mathbb{I}}; \\ 1, & \text{otherwise}. \end{cases}$$

It follows from the Ordinal Comparison Theorem (2.1) that G and H are recursive in $\mathbb{I}$, $\boldsymbol{\beta}$, and it is straightforward to check that they satisfy the other conditions.

Suppose now that for some fixed d, $\mathbf{n}$, and $\boldsymbol{\beta}$, G and H are functionals recursive in $\mathbb{I}$, $\boldsymbol{\beta}$ which satisfy the condition in brackets. Let $<$ denote the restriction of $<_{\mathsf{G}}$ to $\{(u, \langle \boldsymbol{\alpha} \rangle) : \mathsf{H}(u, \langle \boldsymbol{\alpha} \rangle) = 0\}$. Then $<$ is a well-founded partial order whose field contains $(\langle d, \mathbf{n} \rangle, \langle \boldsymbol{\beta} \rangle)$ and it will suffice to prove by induction on $<$ that for all pairs $(\langle a, \mathbf{m} \rangle, \langle \boldsymbol{\alpha} \rangle)$ in its field,

$$(*) \qquad \mathsf{G}(\langle a, \mathbf{m} \rangle, \langle \boldsymbol{\alpha} \rangle) \simeq \{a\}^{\mathbb{I}}(\mathbf{m}, \boldsymbol{\alpha}).$$

By the definition of $\mathbb{P}$, the proof breaks into four cases where $\mathbf{m} \in {}^{k}\omega$ and $l = \lg(\boldsymbol{\alpha})$.

(1) if for some $b, c_0, \ldots, c_{k'-1}$, $a = \langle 1, k, l, 0, b, c_0, \ldots, c_{k'-1} \rangle$, then because H is closed for G, for all $i < k'$

$$(\langle c_i, \mathbf{m} \rangle, \langle \boldsymbol{\alpha} \rangle) < (\langle a, \mathbf{m} \rangle, \langle \boldsymbol{\alpha} \rangle) \quad \text{and} \quad (\langle b, \mathbf{q} \rangle, \langle \boldsymbol{\alpha} \rangle) < (\langle a, \mathbf{m} \rangle, \langle \boldsymbol{\alpha} \rangle)$$

where $q_i = \mathsf{G}(\langle c_i, \mathbf{m} \rangle, \langle \boldsymbol{\alpha} \rangle)$. Then by the induction hypothesis, for all $i < k'$, $\{c_i\}^{\mathbb{I}}(\mathbf{m}, \boldsymbol{\alpha}) \simeq q_i$ and thus by the induction hypothesis again and the local correctness of G,

$$\mathsf{G}(\langle a, \mathbf{m} \rangle, \langle \boldsymbol{\alpha} \rangle) = \mathsf{G}(\langle b, \mathbf{q} \rangle, \langle \boldsymbol{\alpha} \rangle) \simeq \{b\}^{\mathbb{I}}(\mathbf{q}, \boldsymbol{\alpha}) \simeq \{a\}^{\mathbb{I}}(\mathbf{m}, \boldsymbol{\alpha}).$$

(2)–(4) Similarly.

(5) If $\langle a, \mathbf{m} \rangle$ is of none of the preceding forms, then necessarily

$|\langle a, \mathbf{m}\rangle, \langle \boldsymbol{\alpha}\rangle|^{\mathsf{I}} = 0$ — that is, $a = \langle 0, k, l, 0, \ldots \rangle$ and is an index. Otherwise, $(\langle a, \mathbf{m}\rangle, \langle \boldsymbol{\alpha}\rangle), (0, \langle \boldsymbol{\alpha}\rangle), (0, \langle \boldsymbol{\alpha}\rangle) \ldots$ would be a descending $<$-chain. Hence (*) is satisfied by clause (0) of the definition of local correctness. □

2.13–2.21 Exercises

2.13. Show that if $\mathbb{H}$ and $\mathbb{I}$ are two type-3 functionals such that $\mathbb{H}(\mathsf{G}) = \mathbb{I}(\mathsf{G})$ for all G recursive in $\mathbb{I}$ and some $\beta \in {}^{\omega}\omega$, then for all R, R is (semi-) recursive in $\mathbb{I}$ iff R is (semi-) recursive in $\mathbb{H}$. (cf. Exercise VI.1.13).

2.14. Show that $s\mathscr{J}(\mathsf{sJ})$ and $\mathbb{E}$ are each recursive in the other.

2.15. For any $\mathbb{I}$ such that $\mathbb{E}$ is recursive in $\mathbb{I}$, $\kappa^{\mathsf{I}} > \aleph_1$.

2.16. Formulate and prove the analogue of Exercise VI.4.29 for the hierarchies of Theorems 2.5 and 2.6.

2.17. Formulate and prove choice principles for recursion in a type-3 functional $\mathbb{I}$ parallel to those of Theorem VI.4.15.

2.18. Show that U^{I} is not an initial segment of U^{I} — that is, there exists $u \in U^{\mathsf{I}}$ and $\sigma < |u|^{\mathsf{I}}$ such that $\sigma \neq |v|^{\mathsf{I}}$ for any $v \in U^{\mathsf{I}}$.

2.19. Prove Theorem 2.8.

2.20. Show that there exists a type-3 functional $\mathbb{I}$ such that $\mathbb{E}$ is recursive in $\mathbb{I}$ and $\{\mathsf{I} : \mathsf{I}$ is recursive in $\mathbb{I}$ and some $\beta \in {}^{\omega}\omega\}$ is recursive in $\mathbb{I}$ (cf. Theorem VI.4.15). Is this true without the reference to β?

2.21. Show that for all R, the following are equivalent:

(i) R is semi-recursive in $\mathbb{E}^{\#}$;

(ii) R is reducible to the closure $\bar{\Gamma}$ of a Δ^1_2 monotone operator Γ over ${}^{\omega}\omega$;

(iii) for some recursive relation P,

$$\mathsf{R}(\mathbf{m}, \boldsymbol{\alpha}) \leftrightarrow \forall \beta_0 \exists \beta_1 \forall \beta_2 \exists \beta_3 \cdots \exists n\, \mathsf{P}(\mathbf{m}, \boldsymbol{\alpha}, \langle \beta_0, \ldots, \beta_{n-1} \rangle).$$

2.22 Notes. The basic theory here was worked out by Moschovakis [1967]. The functional $\mathbb{E}^{\#}$ and the (semi-) hyperprojective relations are studied in Hinman–Moschovakis [1971].

3. Hierarchies of Relations Recursive in a Type-3 Functional

In Theorem 2.6 of the preceding section we showed that when $\mathbb{E}$ is recursive in a type-3 functional $\mathbb{I}$, there is a natural hierarchy of (countable) length $\kappa^{\mathbb{I}}$ on the type-2 relations recursive in $\mathbb{I}$. It results from the previous construction of a hierarchy of uncountable length and the observation that there is a natural way of picking out a countable "cofinal" subhierarchy. Since there are only countably many relations recursive in $\mathbb{I}$, it seems somewhat unnatural that such a roundabout procedure should be necessary. In the case of recursion relative to a type-2 functional I there was an alternative method of generating inductively the type-1 relations recursive in I, that of iterating the jump operator J_{I} over a set of notations for ordinals less than $\omega_1[\mathsf{I}]$ (cf. Corollary VI.5.6). The aim of this section is to show that if $\mathbf{s}\mathbb{J}$ is recursive in $\mathbb{I}$, then a similar construction for a type-3 $\mathbb{I}$ breaks down before $\omega_1[\mathbb{I}]$ and classifies only a (small!) subclass of the relations recursive in $\mathbb{I}$.

For any type-3 functional $\mathbb{I}$, let $\mathbb{J}_{\mathbb{I}}: {}^{({}^{\omega}\omega)}\omega \to {}^{({}^{\omega}\omega)}\omega$ be defined by:

$$\mathbb{J}_{\mathbb{I}}(\mathsf{F})((\langle a, n\rangle) * \alpha) = \begin{cases} 0, & \text{if } \mathbb{I}(\lambda\beta.\{a\}^{\mathsf{F}}(\alpha, \beta)) \simeq n; \\ 1, & \text{otherwise.} \end{cases}$$

3.1 Lemma. *There exists an index d and a primitive recursive function h such that for all c,* F, G, *and* $\mathbb{I}$,

(i) $\mathbf{s}\mathbb{J}(\mathsf{F})$ *is recursive in* $\mathbb{J}_{\mathbb{I}}(\mathsf{F})$ *with index* d;

(ii) *if* F *is recursive in* G *with index c, then* $\mathbb{J}_{\mathbb{I}}(\mathsf{F})$ *is recursive in* $\mathbb{J}_{\mathbb{I}}(\mathsf{G})$ *with index* $h(c)$.

(*cf. Definition* V.5.4).

Proof. Let $n_0 = \mathbb{I}(\lambda\beta.0)$ and let f be a primitive recursive function such that for all a, $\mathbf{m}$, α, and β,

$$\{f(a, \langle\mathbf{m}\rangle)\}^{\mathsf{F}}(\alpha, \beta) \simeq 0 \cdot \{a\}^{\mathsf{F}}(\mathbf{m}, \alpha).$$

Then

$$\begin{aligned} \mathbf{s}\mathbb{J}(\mathsf{F})((\langle a, \mathbf{m}\rangle) * \alpha) = 0 &\leftrightarrow \{a\}^{\mathsf{F}}(\mathbf{m}, \alpha) \text{ is defined} \\ &\leftrightarrow \forall\beta.\{f(a, \langle\mathbf{m}\rangle)\}^{\mathsf{F}}(\alpha, \beta) \simeq 0 \\ &\leftrightarrow \mathbb{J}_{\mathbb{I}}(\mathsf{F})((\langle f(a, \langle\mathbf{m}\rangle), n_0\rangle) * \alpha) = 0 \\ &\leftrightarrow \{d\}((\langle a, \mathbf{m}\rangle) * \alpha, \mathbb{J}_{\mathbb{I}}(\mathsf{F})) = 0 \end{aligned}$$

for an appropriate d.

For (ii), suppose that for all γ, $\mathsf{F}(\gamma)=\{c\}^{\mathsf{G}}(\gamma)$ and let g be the primitive recursive function of Theorem VI.2.14 such that for any a, α, and β,

$$\{g(a,c)\}^{\mathsf{G}}(\alpha,\beta)\simeq\{a\}^{\mathsf{F}}(\alpha,\beta).$$

Then

$$\begin{aligned}\mathsf{J}_{\mathbb{I}}(\mathsf{F})(((a,n))*\alpha)=0 &\leftrightarrow \mathbb{I}(\lambda\beta.\{a\}^{\mathsf{F}}(\alpha,\beta))=n\\ &\leftrightarrow \mathbb{I}(\lambda\beta.\{g(a,c)\}^{\mathsf{G}}(\alpha,\beta))=n\\ &\leftrightarrow \mathsf{J}_{\mathbb{I}}(\mathsf{G})(((g(a,c),n))*\alpha)=0\end{aligned}$$

and from this it is easy to define an appropriate h. □

For the remainder of this section, let $\mathbb{I}$ be a fixed type-3 functional such that $s\mathsf{J}$ is recursive in $\mathbb{I}$. We write simply J for $\mathsf{J}_{\mathbb{I}}$ and construct a hierarchy in a way entirely analgous to that of § VI.5. For a minor convenience we use functionals instead of sets as the "backbone" of the hierarchy.

3.2 Definition. $<^{\mathsf{J}}$ is the smallest subset of $\omega\times\omega$ such that for all w in the field of $<^{\mathsf{J}}$ there exist functionals $\mathsf{H}^{\mathsf{J}}_w$ which satisfy the following conditions:

(i) $1<^{\mathsf{J}}2$, $\mathsf{H}^{\mathsf{J}}_1=\lambda\beta.0$, and $\mathsf{H}^{\mathsf{J}}_2=\mathsf{J}(\mathsf{H}^{\mathsf{J}}_1)$;

(ii) if $u<^{\mathsf{J}}v$, then $v<^{\mathsf{J}}2^v$ and $\mathsf{H}^{\mathsf{J}}_{2^v}=\mathsf{J}(\mathsf{H}^{\mathsf{J}}_v)$;

(iii) if $u\in\mathrm{Fld}(<^{\mathsf{J}})$, for all p $\{a\}(p,\mathsf{H}^{\mathsf{J}}_u)<^{\mathsf{J}}\{a\}(p+1,\mathsf{H}^{\mathsf{J}}_u)$, and $\{a\}(0,\mathsf{H}^{\mathsf{J}}_u)\simeq u$, then for all p,

$$\{a\}(p,\mathsf{H}^{\mathsf{J}}_u)<^{\mathsf{J}}3^a5^u \quad\text{and}\quad \mathsf{H}^{\mathsf{J}}_{3^a5^u}((p)*\alpha)=\mathsf{H}^{\mathsf{J}}_{\{a\}(p,\mathsf{H}^{\mathsf{J}}_u)}(\alpha);$$

(iv) if $u<^{\mathsf{J}}v$ and $v<^{\mathsf{J}}w$, then $u<^{\mathsf{J}}w$.

We write O^{J} for the field of $<^{\mathsf{J}}$, assign ordinals $|u|^{\mathsf{J}}$ to $u\in O^{\mathsf{J}}$ as in § IV.4, and set

$$\nabla(\mathsf{J})=\{\mathsf{R}:\mathsf{R}\text{ is recursive in }\mathsf{H}^{\mathsf{J}}_u\text{ for some }u\in O^{\mathsf{J}}\}.$$

We shall show that $\nabla(\mathsf{J})$ is a proper subset of the set of relations recursive in $\mathbb{I}$ by defining a functional K which is recursive in $\mathbb{I}$ but has the property that for all $u\in O^{\mathsf{J}}$, $\mathsf{H}^{\mathsf{J}}_u$ is recursive in K.

For $\gamma\in\mathsf{W}$, we define K^{γ} by recursion on $\|\gamma\|$ by:

$$\mathsf{K}^{\gamma}((p)*\alpha)=\begin{cases}\mathrm{o}\mathsf{J}(\alpha)(p), & \text{if } \|\gamma\|=0;\\ \mathsf{J}(\mathsf{K}^{(\gamma\restriction p)})(\alpha), & \text{otherwise.}\end{cases}$$

Roughly speaking, K^{γ} is the result of $\|\gamma\|$ applications of J. Set

$$\mathsf{K}(\langle\alpha,\gamma\rangle) = \begin{cases} \mathsf{K}^{\gamma}(\alpha), & \text{if} \quad \gamma \in \mathsf{W}; \\ 0, & \text{otherwise.} \end{cases}$$

3.3 Lemma. *K is recursive in $\mathbb{I}$.*

Proof. Since $\mathbf{s}\mathsf{J}$ is recursive in $\mathbb{I}$, so is E_1 and hence so is W. Let

$$\mathsf{G}(e,(p)*\alpha,\gamma) \simeq \begin{cases} \mathrm{o}\mathsf{J}(\alpha)(p), & \text{if} \quad \gamma \in \mathsf{W} \quad \text{and} \quad \|\gamma\| = 0; \\ \mathsf{J}(\lambda\beta.\{e\}^{\mathbb{I}}(\beta,\gamma \restriction p))(\alpha), & \text{if} \quad \gamma \in \mathsf{W} \quad \text{and} \quad \|\gamma\| > 0; \\ 0, & \text{otherwise.} \end{cases}$$

Because $\mathbf{s}\mathsf{J}$ is recursive in $\mathbb{I}$, J is calculable from $\mathbb{I}$ and thus G is partial recursive in $\mathbb{I}$. By the $\mathbb{I}$-Recursion Theorem there exists an index $\bar{e}$ such that

$$\{\bar{e}\}^{\mathbb{I}}((p)*\alpha,\gamma) \simeq \mathsf{G}(\bar{e},(p)*\alpha,\gamma).$$

A straightforward induction on $\|\gamma\|$ shows that for all γ,

$$\{\bar{e}\}^{\mathbb{I}}(\alpha,\gamma) = \mathsf{K}(\langle\alpha,\gamma\rangle)$$

and thus K is recursive in $\mathbb{I}$. $\square$

For each $u \in O^{\mathsf{J}}$, let δ_u be defined by:

$$\delta_u(\langle v,w\rangle) = \begin{cases} 0, & \text{if} \quad v \leqslant^{\mathsf{J}} w \leqslant^{\mathsf{J}} u; \\ 1, & \text{otherwise.} \end{cases}$$

An easy induction on $|u|^{\mathsf{J}}$ shows that for all $u \in O^{\mathsf{J}}$, $\delta_u \in \mathsf{W}$ and $\|\delta_u\| = |u|^{\mathsf{J}}$. Note that if $v <^{\mathsf{J}} u$, then $\delta_u \restriction v = \delta_v$.

3.4 Lemma. *There exist primitive recursive functions f and g such that for all $u \in O^{\mathsf{J}}$,*

(i) *δ_u is recursive in $\mathsf{H}_u^{\mathsf{J}}$ with index $f(u)$;*
(ii) *$\mathsf{H}_u^{\mathsf{J}}$ is recursive in K^{δ_u} with index $g(u)$.*

Proof. The existence of f may be established much as in Lemma VI.5.2 and we leave the details to Exercise 3.7. For (ii), we define g by effective transfinite recursion to satisfy the following conditions. If $u = 1$, $g(u)$ is any index of the recursive functional $\mathsf{H}_1^{\mathsf{J}}$ from K^{δ_1}. If $u = 2^v$, then $g(u)$ is an index for the following computation: from K^{δ_u} compute $\mathsf{J}(\mathsf{K}^{\delta_v}) = \mathsf{J}(\mathsf{K}^{\delta_u \restriction v})$ and from this (by 3.1(ii)) compute $\mathsf{J}(\mathsf{F})$, where F is the functional recursive in K^{δ_v} with index $g(v)$.

Since F is $\mathsf{H}_v^{\mathsf{J}}$, $g(u)$ is indeed an index for $\mathsf{H}_u^{\mathsf{J}}$ from K^{δ_u}. If $u = 3^a 5^v$, let $\varepsilon(p) = \{a\}(p, \mathsf{H}_v^{\mathsf{J}})$ and let $g(u)$ be an index from K^{δ_u} for the functional G described as follows. Let I be the functional recursive in $\mathsf{K}^{\delta_u \restriction v}$ with index $g(v)$ and set $\varepsilon'(p) = \{a\}(p, \mathsf{I})$. For each p, let G_p be the functional recursive in $\mathsf{K}^{\delta_u \restriction \varepsilon'(p)}$ with index $g(\varepsilon'(p))$. Finally, $\mathsf{G}((p) * \alpha) = \mathsf{G}_p(\alpha)$. It is now straightforward to prove by induction over O^{J} that (ii) is satisfied. □

3.5 Theorem. *There exists a primitive recursive function h such that for all $u \in O^{\mathsf{J}}$, $\mathsf{H}_u^{\mathsf{J}}$ is recursive in K with index $h(u)$.*

Proof. We again define h by effective transfinite recursion. $h(1)$ is any index for the recursive functional $\mathsf{H}_1^{\mathsf{J}}$ from K. If $u = 2^v$ and $\mathsf{H}_v^{\mathsf{J}}$ is recursive in K with index $h(v)$, then from K we can, by Lemma 3.4(i), compute an index for δ_v, and thence one for δ_u. By 3.4(ii), $\mathsf{H}_u^{\mathsf{J}}$ is recursive in K^{δ_u}, hence in K alone. We take $h(u)$ to be any index which describes this computation.

If $u = 3^a 5^v$ and $\varepsilon(p) = \{a\}(p, \mathsf{H}_v^{\mathsf{J}})$, then using $h(v)$ we can compute an index of ε from K. Then the indices $h(\varepsilon(p))$ uniformly compute $\mathsf{H}_{\varepsilon(p)}^{\mathsf{J}}$ from K and thus lead to an index of $\mathsf{H}_u^{\mathsf{J}}$ from K. □

3.6 Corollary. *There exists a set $A \subseteq \omega$ which is recursive in $\mathbb{I}$ but not recursive in any $\mathsf{H}_u^{\mathsf{J}}$ $(u \in O^{\mathsf{J}})$.*

Proof. Let $A = \{a : \{a\}^{\mathsf{K}}(a) \text{ is defined}\}$. A is recursive in $\mathsf{sJ}(\mathsf{K})$, hence in $\mathbb{I}$. A standard diagonal argument shows that A is not recursive in K and thus by Theorem 3.5 A is not recursive in any $\mathsf{H}_u^{\mathsf{J}}$ $(u \in O^{\mathsf{J}})$. □

It can also be shown that the type-1 functions and relations recursive in K are exactly those recursive in some $\mathsf{H}_u^{\mathsf{J}}$ $(u \in O^{\mathsf{J}})$ (Aczel–Hinman [1974, Theorem 1.10]). In the particular case $\mathbb{I} = \mathsf{sJ}$ there is a direct way to construct a countable hierarchy for the type-1 relations recursive in $\mathbb{I}$ (Aczel–Hinman [1974, Corollary 4.20]).

3.7 Exercise. Prove part (i) of Lemma 3.4.

3.8. Notes. The failure of hierarchies of the type discussed here to exhaust the class of objects recursive in $\mathbb{I}$ was shown in Moschovakis [1967] for $\mathbb{I}$ such that $\mathbb{E}$ is recursive in $\mathbb{I}$ and extended in Aczel–Hinman [1974] to all $\mathbb{I}$ such that sJ is recursive in $\mathbb{I}$.

4. Higher Types

For the most part, the definitions and results for recursion relative to functionals of types 4 and higher are straightforward generalizations of those for type 3. Rather than discuss these for arbitrary types, we sketch them for type 4 and leave it to the reader to formulate the general results. There is one new phenomenon which appears at type 4, a selection theorem which implies that when $\mathscr{E}$ (the type-4 quantifier functional) is recursive in a type-4 functional $\mathscr{I}$, then the class of relations semi-recursive in $\mathscr{I}$ is closed under existential function quantification ($\exists^1$).

We shall use script letters to denote functions and relations of type 4 — $\mathscr{E}$, $\mathscr{F}$, $\mathscr{G}$, $\mathscr{H}$, $\mathscr{I}$, and $\mathscr{J}$ for functions from ${}^{k,l,l',l''}\omega$ into ω and $\mathscr{R}$, $\mathscr{S}$, $\mathscr{T}$. $\mathscr{U}$, and $\mathscr{V}$ for subsets of ${}^{k,l,l',l''}\omega$ (a typical element of ${}^{k,l,l',l''}\omega$ is of the form $(\mathbf{m}, \boldsymbol{\alpha}, \mathsf{I}, \vec{\mathbb{I}})$). As with types 2 and 3, we define recursion relative to a single type-4 functional $\mathscr{I}$ which takes a single argument $\mathbb{F}$. Examples of such functionals are:

$$\mathscr{E}(\mathbb{F}) = \begin{cases} 0, & \text{if } \exists \mathsf{G}\,.\,\mathbb{F}(\mathsf{G}) = 0; \\ 1, & \text{otherwise;} \end{cases}$$

$$s\mathscr{J}((a, \mathbf{m}, \alpha, \mathsf{I}) * \mathbb{F}) = \begin{cases} 0, & \text{if } \{a\}^{\mathbb{F}}(\mathbf{m}, \alpha, \mathsf{I}) \text{ is defined;} \\ 1, & \text{otherwise.} \end{cases}$$

Recursion relative to other sorts of type-4 functionals or relative to several type-4 functionals is then defined in the usual way by coding.

4.1 Definition. For any $\mathscr{I}$, $\Omega[\mathscr{I}]$ is the smallest set such that for all k, l, l', l'', n, p, q, r, and s, all $i < k$, $j < l$, $j' < l'$, and $j'' < l''$, and all $(\mathbf{m}, \boldsymbol{\alpha}, \mathsf{I}, \vec{\mathbb{I}}) \in {}^{k,l,l',l''}\omega$,

(0)–(4) identical to the corresponding clauses of Definition 1.1 with the addition of the parameter l'' and the provision in clause (4) for indices $\langle 4, k, l, l', l'', j'', b\rangle$ for the application of $\mathbb{I}_{j''}$;

(5) for any b and any $\mathbb{F}$,

$$\text{if for all } \mathsf{H},\ (b, \mathbf{m}, \boldsymbol{\alpha}, \mathsf{I}, \mathsf{H}, \vec{\mathbb{I}}, \mathbb{F}(\mathsf{H})) \in \Omega[\mathscr{I}],$$
$$\text{then } (\langle 5, k, l, l', l'', b\rangle, \mathbf{m}, \boldsymbol{\alpha}, \mathsf{I}, \vec{\mathbb{I}}, \mathscr{I}(\mathbb{F})) \in \Omega[\mathscr{I}].$$

As always, we set

$$\{a\}^{\mathscr{I}}(\mathbf{m}, \boldsymbol{\alpha}, \mathsf{I}, \vec{\mathbb{I}}) \simeq n \quad \text{iff} \quad (a, \mathbf{m}, \boldsymbol{\alpha}, \mathsf{I}, \vec{\mathbb{I}}, n) \in \Omega[\mathscr{I}].$$

and say that $\mathscr{F}$ is *partial recursive in* $\mathscr{I}$ iff $\mathscr{F} = \{a\}^{\mathscr{I}}$ for some a, etc. The import of clause (5) is that

$$\{\langle 5, k, l, l', l'', b\rangle\}^{\mathscr{I}}(\mathbf{m}, \boldsymbol{\alpha}, \mathsf{I}, \vec{\mathbb{I}}) \simeq \mathscr{I}(\lambda \mathsf{H}\,.\,\{b\}^{\mathscr{I}}(\mathbf{m}, \boldsymbol{\alpha}, \mathsf{I}, \mathsf{H}, \vec{\mathbb{I}})).$$

The simplest properties of recursion in a type-4 functional are established exactly as in previous cases and we leave their formulation and proof to the reader. Note that the class of relations semi-recursive in $\mathcal{I}$ is closed under universal (type-2) functional quantification ($\forall^2$). The substitution theorems are related to those for type 3 (Theorem 1.6) as those are related to the results for type 2 (Theorems VI.2.10, 14). The new cases here involve primitive recursive functions h and h' such that for all $\mathcal{I}$, a, d, $\mathbf{m}$, $\boldsymbol{\alpha}$, I and $\vec{\mathbb{I}}$,

(i) $\{h(a,d)\}^{\mathcal{I}}(\mathbf{m},\boldsymbol{\alpha},\mathsf{I},\vec{\mathbb{I}}) \simeq \{a\}^{\mathcal{I}}(\mathbf{m},\boldsymbol{\alpha},\mathsf{I},\vec{\mathbb{I}},\lambda\mathsf{H}.\{d\}^{\mathcal{I}}(\mathbf{m},\boldsymbol{\alpha},\mathsf{I},\mathsf{H},\vec{\mathbb{I}}))$;

(ii) if $\lambda\mathbb{G}.\{d\}^{\mathcal{I}}(\mathbb{G})$ is a total type-4 functional $\mathcal{H}$, then

$$\{h'(a,d)\}^{\mathcal{I}}(\mathbf{m},\boldsymbol{\alpha},\mathsf{I},\vec{\mathbb{I}}) \simeq \{a\}^{\mathcal{H}}(\mathbf{m},\boldsymbol{\alpha},\mathsf{I},\vec{\mathbb{I}}).$$

The ideas involved in the discussion of $s\mathcal{I}$ and $\mathbb{E}^{\#}$ following Corollary 2.11 can be extended to show that $s\mathcal{I}$ is recursive in $\mathbb{E}^{\#}$ and thus that the class of relations recursive in $s\mathcal{I}$ is a proper subset of Δ^2_1. On the other hand, all Π^2_1 relations are semi-recursive in $s\mathcal{I}$ and all $\Delta^2_{(\omega)}$ relations are recursive in $\mathscr{E}$. It follows from the proof of Theorem 1.4 that $s\mathcal{I}$ is recursive in $\mathscr{E}$, but the preceding remarks imply that $\mathscr{E}$ is not recursive in $s\mathcal{I}$. The classes Σ^3_r, Π^3_r, and Δ^3_r may be defined in the natural way and one may prove similarly as for Corollary VI.7.12 that for any inductive operator Γ over ${}^{({}^\omega\omega)}\omega$ and any $r \geq 1$, if $\Gamma \in \Delta^3_r$, then also $\bar{\Gamma} \in \Delta^3_r$. In particular we obtain by the method of Theorem 1.4 that for any $r \geq 1$, if $\mathcal{I} \in \Delta^3_r$, then the class of relations semi-recursive in $\mathcal{I}$ forms a proper subset of Δ^3_r. Of course, the example $\mathscr{E}$ shows that no Δ^2_r has this closure property.

For the rest of this section we shall assume that $\mathcal{I}$ is a fixed type-4 functional such that $\mathscr{E}$ is recursive in $\mathcal{I}$. For simplicity, we shall omit mention of type-3 parameters; type-2 arguments can no longer be regarded as parameters. We use x and y to denote typical elements $(\mathbf{m},\boldsymbol{\alpha},\mathsf{I})$ of ${}^{k,l,l'}\omega$, $\langle\mathsf{x}\rangle$ to denote the corresponding triple $(\langle\mathbf{m}\rangle,\langle\boldsymbol{\alpha}\rangle,\langle\mathsf{I}\rangle)$, $\langle\mathsf{x},\beta\rangle$ for $(\langle\mathbf{m}\rangle,\langle\boldsymbol{\alpha},\beta\rangle,\langle\mathsf{I}\rangle)$, etc.

Set

$$\mathbb{U}^{\mathcal{I}}(a,\langle\mathsf{x}\rangle) \leftrightarrow \{a\}^{\mathcal{I}}(\mathsf{x}) \text{ is defined}$$

and

$$|a,\mathsf{x}|^{\mathcal{I}} = \text{least } \sigma.(a,\mathsf{x},\{a\}^{\mathcal{I}}(\mathsf{x})) \in \Omega[\mathcal{I}]^{\sigma}$$

if such a σ exists; otherwise, $o({}^{({}^\omega\omega)}\omega)$ (the least ordinal not the type of a pre-wellordering of ${}^{({}^\omega\omega)}\omega$). The ordinals of computations introduced by clause (5) are given by: if $a = \langle 5,k,l,l',0,b\rangle$ and $\{a\}^{\mathcal{I}}(\mathsf{x})$ is defined, then

$$|a,\mathsf{x}|^{\mathcal{I}} = \sup{}^{+}\{|b,\mathsf{x},\mathsf{H}|^{\mathcal{I}} : \mathsf{H} \in {}^{({}^\omega\omega)}\omega\}.$$

4.2 Ordinal Comparison Theorem. *There exists a functional $\mathbb{H}$ partial recursive in $\mathcal{I}$ such that for all a^0, a^1, x^0, and x^1,*

(i) *if $(a^0,\langle\mathsf{x}^0\rangle) \in \mathbb{U}^{\mathcal{I}}$ and $|a^0,\mathsf{x}^0|^{\mathcal{I}} \leq |a^1,\mathsf{x}^1|^{\mathcal{I}}$, then*

$$\mathbb{H}(a^0, a^1, \langle \mathsf{x}^0\rangle, \langle \mathsf{x}^1\rangle) \simeq 0;$$

(ii) *if* $(a^1, \langle \mathsf{x}^1\rangle) \in \mathsf{U}^{\mathcal{I}}$ *and* $|a^1, \mathsf{x}^1|^{\mathcal{I}} < |a^0, \mathsf{x}^0|^{\mathcal{I}}$, *then*

$$\mathbb{H}(a^0, a^1, \langle \mathsf{x}^0\rangle, \langle \mathsf{x}^1\rangle) \simeq 1.$$

Proof. The definition of $\mathbb{H}$ is entirely analogous to that in Theorem 2.1 and thus (in more detail) to that in Theorem VI.3.3. □

From this we obtain exactly as in § VI.4 and § 2 of this chapter the Selection Theorem and all the structural consequences analogous to 2.3–2.8. The proof of Lemma 2.9 can similarly be generalized and we obtain a relation $\mathbb{P}^{\mathcal{I}}$ semi-recursive in $\mathcal{I}$ such that for all a and $\mathsf{x} = (\mathbf{m}, \boldsymbol{\alpha}, \mathbf{I})$,

$$\{a\}^{\mathcal{I}}(\mathsf{x}) \text{ is defined} \leftrightarrow \neg \exists \mathsf{F}\, \exists \delta\, \exists \varepsilon\, \forall p\, [\mathbb{P}^{\mathcal{I}}(\delta(p), \delta(p+1), (\varepsilon)^p, (\varepsilon)^{p+1}, (\mathsf{F})^p, (\mathsf{F})^{p+1}) \\ \wedge\, \delta(0) = \langle a, \mathbf{m}\rangle \wedge (\varepsilon)^0 = \langle \boldsymbol{\alpha}\rangle \wedge (\mathsf{F})^0 = \langle \mathbf{I}\rangle].$$

Intuitively, $\mathbb{P}^{\mathcal{I}}(\langle a, \mathbf{m}\rangle, \langle b, \mathbf{n}\rangle, \langle \boldsymbol{\alpha}\rangle, \langle \boldsymbol{\beta}\rangle, \langle \mathbf{I}\rangle, \langle \mathbf{H}\rangle)$ means that the computation $(b, \mathbf{n}, \boldsymbol{\beta}, \mathbf{H})$ is an immediate predecessor of $(a, \mathbf{m}, \boldsymbol{\alpha}, \mathbf{I})$. In particular, the class of relations semi-recursive in $\mathcal{I}$ is not closed under existential functional quantification ($\exists^2$). We leave to the reader the definition and study of the functional $\mathscr{E}^{\#}$ and the formulation and proof of the analogue for $\mathcal{I}$ of Theorem 2.12 (Spector–Gandy Theorem).

We come now to showing that the class of relations semi-recursive in $\mathcal{I}$ *is* closed under existential function quantification ($\exists^1$).

4.3 Theorem. *There exists a functional* $\mathbb{F}$ *partial recursive in* $\mathcal{I}$ *such that for any* a *and* x, *the following are equivalent*:

(i) $\exists \beta . \{a\}^{\mathcal{I}}(\mathsf{x}, \beta)$ *is defined*;

(ii) $\forall \beta . \mathbb{F}(a, \langle \mathsf{x}, \beta\rangle)$ *is defined and* $\exists \beta . \mathbb{F}(a, \langle \mathsf{x}, \beta\rangle) \simeq 0$.

From this follows immediately that if $\mathbb{R}$ is semi-recursive in $\mathcal{I}$, say with semi-index a from $\mathcal{I}$, then

$$\exists \beta\, \mathbb{R}(\mathsf{x}, \beta) \leftrightarrow \mathbb{E}(\lambda \beta . \mathbb{F}(a, \langle \mathsf{x}, \beta\rangle)) \simeq 0$$

so that $\exists^1 \mathbb{R}$ is also semi-recursive in $\mathcal{I}$.

The proof of Theorem 4.3 is rather involved if all details are included, but the basic idea is relatively simple and we shall confine our discussion to this.

Proof of Theorem 4.3. The point of the functional $\mathbb{F}$ is to select from $\mathsf{B}_{a,\mathsf{x}} = \{\beta : \{a\}^{\mathcal{I}}(\mathsf{x}, \beta) \text{ is defined}\}$, whenever it is non-empty, a non-empty subset $\mathsf{A}_{a,\mathsf{x}} = \{\beta : \mathbb{F}(a, \langle \mathsf{x}, \beta\rangle) \simeq 0\}$ which is recursive in x and $\mathcal{I}$, uniformly in a and x — that is, with index from x and $\mathcal{I}$ computable from a independently of x. $\mathsf{A}_{a,\mathsf{x}}$ will be recursive (not merely semi-recursive) in x and $\mathcal{J}$ because when $\mathsf{B}_{a,\mathsf{x}} \neq \emptyset$, $\lambda\beta\, \mathbb{F}(a, \langle \mathsf{x}, \beta\rangle)$ is total.

It turns out to be more convenient to prove a slightly more general result. Let φ, ψ, and θ denote functions from ${}^{\omega}\omega$ into ${}^{2,1,1}\omega$, so $\varphi(\beta) = (a, m, \alpha, \mathsf{I})$. We may think of φ as a quadruple $(\mathsf{F}, \mathsf{G}, \mathsf{H}, \mathsf{I})$ of type-2 functionals, where

$$\varphi(\beta) = (\mathsf{F}(\beta), \mathsf{G}(\beta), \lambda p\,.\,\mathsf{H}(p, \beta), \lambda\gamma\,.\,\mathsf{I}(\beta, \gamma))$$

and thus as a single type-2 functional. We shall define a function $\mathbb{G}$ partial recursive in $\mathcal{J}$ which selects a set $\mathsf{A}_\varphi = \{\beta : \mathbb{G}(\beta, \varphi) \simeq 0\}$ such that whenever $\mathsf{B}_\varphi = \{\beta : \varphi(\beta) \in \mathbb{U}^{\mathcal{J}}\}$ is non-empty, then $\mathsf{A}_\varphi \subseteq \mathsf{B}_\varphi$ and A_φ is recursive in φ and $\mathcal{J}$, uniformly in φ. Then $\mathbb{F}$ is obtained by applying this result to the function $\varphi_{a,\mathsf{x}}$ defined by: $\varphi_{a,\mathsf{x}}(\beta) = (a, \langle \mathsf{x}, \beta\rangle)$.

For any φ, let

$$\min \varphi = \inf\{|\varphi(\beta)|^{\mathcal{J}} : \beta \in {}^{\omega}\omega\}$$

and

$$\boldsymbol{\kappa}^{\mathcal{J}} = \sup{}^{+}\{|a, \mathsf{x}|^{\mathcal{J}} : \mathbb{U}^{\mathcal{J}}(a, \langle \mathsf{x}\rangle)\}.$$

Clearly, $\exists\beta\,.\,\varphi(\beta) \in \mathbb{U}^{\mathcal{J}} \leftrightarrow \min \varphi < \boldsymbol{\kappa}^{\mathcal{J}}$.

We shall determine $\mathbb{G}$ so that for all β and φ,

(a) if $\min \varphi < \boldsymbol{\kappa}^{\mathcal{J}}$, then

$$\mathbb{G}(\beta, \varphi) = \begin{cases} 0, & \text{if } \ |\varphi(\beta)|^{\mathcal{J}} = \min \varphi; \\ 1, & \text{otherwise}; \end{cases}$$

(b) $\mathbb{G}(\beta, \varphi) \simeq 0 \rightarrow \varphi(\beta) \in \mathbb{U}^{\mathcal{J}}$.

In fact, we need only be concerned about condition (a), for if $\mathbb{G}$ satisfies (a), and $\varphi(\beta) = (a, \langle \mathsf{x}\rangle)$, then the functional computed according to the following flow diagram satisfies both (a) and (b):

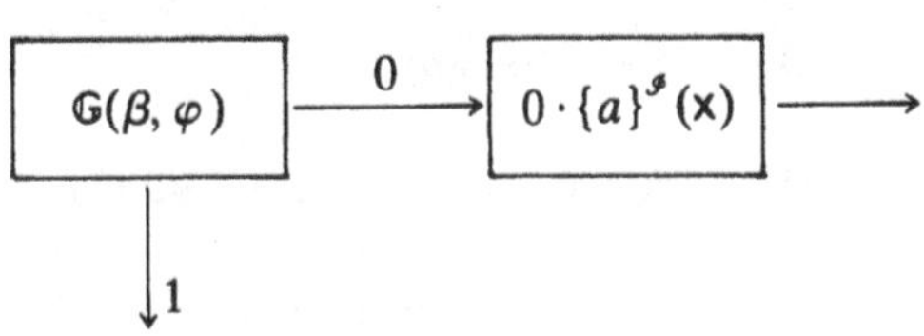

We may thus define $\mathbb{G}$ by effective transfinite recursion on $\min \varphi$ for $\min \varphi < \boldsymbol{\kappa}^{\mathcal{J}}$. Suppose that for all ψ such that $\min \psi < \min \varphi$ and all β, $\mathbb{G}(\beta, \psi)$ has been defined to satisfy (a); we shall describe how to compute $\mathbb{G}(\beta, \varphi)$. If $\min \varphi = 0$, this is clear and we assume $\min \varphi > 0$.

Let $\circledS$ denote the "augmented" immediate predecessor relation for computations relative to $\mathcal{I}$:

$$(a',\mathbf{x}')\circledS(a,\mathbf{x}) \leftrightarrow \mathbb{P}^{\mathcal{I}}(a,a',\mathbf{x},\mathbf{x}') \quad \text{or} \quad a = \langle 0,k,l,l',0,\ldots\rangle$$

$$\text{is an index appropriate for } \mathbf{x} \text{ and}$$

$$(a',\mathbf{x}') = (a,\mathbf{x}).$$

The relation $\circledS_0$ is the restriction of $\circledS$ to those cases where the fact that $(a',\mathbf{x}')$ is an immediate predecessor of $(a,\mathbf{x})$ can be determined without performing any computations. Precisely, let $\mathbb{P}_0^{\mathcal{I}}$ be the relation defined by replacing clause (1) of the definition of $\mathbb{P}^{\mathcal{I}}(u,v,\langle\boldsymbol{\alpha}\rangle,\langle\boldsymbol{\beta}\rangle,\langle\mathbf{H}\rangle,\langle\mathbf{I}\rangle)$ by

(1′) for some $b, c_0,\ldots,c_{k'-1}$, $u = \langle\langle 1,k,l,l',0,b,c_0,\ldots,c_{k'-1}\rangle,\mathbf{m}\rangle$, $\boldsymbol{\beta}=\boldsymbol{\alpha}$, $\mathbf{H} = \mathbf{I}$, and $v = \langle c_i,\mathbf{m}\rangle$ for some $i<k'$;

Then the definition of $\circledS_0$ is obtained from that of $\circledS$ by replacing $\mathbb{P}^{\mathcal{I}}$ by $\mathbb{P}_0^{\mathcal{I}}$. Thus $\circledS_0$ is a subrelation of $\circledS$ which is recursive in $\mathcal{I}$. Although in general $\circledS$ is only semi-recursive in $\mathcal{I}$, it is clear that for any $(a,\mathbf{x})$,

(c) if for all $(a',\mathbf{x}')\circledS_0(a,\mathbf{x})$, $\{a'\}^{\mathcal{I}}(\mathbf{x}')$ is defined, then $\{(a',\mathbf{x}'):(a',\mathbf{x}')\circledS(a,\mathbf{x})\}$ is recursive in $\mathbf{x}$ and $\mathcal{I}$, uniformly in $\mathbf{x}$.

Let

$$\mathsf{B} = \{\beta: \text{for all } (a,\mathbf{x})\circledS_0\,\varphi(\beta),\ |a,\mathbf{x}|^{\mathcal{I}} < \min\varphi\};$$

$$\mathsf{C} = \{\beta: \text{for all } (a,\mathbf{x})\circledS\,\varphi(\beta),\ |a,\mathbf{x}|^{\mathcal{I}} < \min\varphi\}.$$

Clearly $\mathsf{C}\subseteq\mathsf{B}$ and $|\varphi(\beta)|^{\mathcal{I}} = \min\varphi$ iff $\beta\in\mathsf{C}$, so it will suffice to show that C is recursive in $\mathcal{I}$ and φ, uniformly in φ, as we may then set $\mathbb{G}(\beta,\varphi) = \mathsf{K}_{\mathsf{C}}(\beta)$. Our approach to this is to show that B is recursive in $\mathcal{I}$ and φ and to define a set $\mathbb{A}$ also recursive in $\mathcal{I}$ and φ such that

(d) for all $\theta\in\mathbb{A}$, $\min\theta<\min\varphi$, but

(e) $\min\varphi = \sup^{+}\{\min\theta:\theta\in\mathbb{A}\}$.

Given such $\mathbb{A}$ and B, we have for any β,

$$\beta\in\mathsf{C}\leftrightarrow\beta\in\mathsf{B} \text{ and for all } (a,\mathbf{x})\circledS\,\varphi(\beta),\ (\exists\theta\in\mathbb{A}).\,|a,\mathbf{x}|^{\mathcal{I}}\leqslant\min\theta.$$

For $\beta\in\mathsf{B}$, the condition $(a,\mathbf{x})\circledS\,\varphi(\beta)$ is recursive in $\mathcal{I}$ and φ by (c). The quantifier $(\exists\theta\in\mathbb{A})$ is computable from $\mathscr{E}$ and $\mathbb{A}$, hence from $\mathcal{I}$ and φ. Finally, for $\theta\in\mathbb{A}$, it follows from (d) and the induction hypothesis that

(f) $$|a,\mathbf{x}|^{\mathcal{I}}\leqslant\min\theta\leftrightarrow\exists\gamma\,(\mathbb{G}(\gamma,\theta)=0\wedge|a,\mathbf{x}|^{\mathcal{I}}\leqslant|\theta(\gamma)|^{\mathcal{I}}).$$

That the condition on the right side is recursive in $\mathscr{I}$ and φ is immediate from the Ordinal Comparison Theorem and the fact that $\mathbb{E}$ is recursive in $\mathscr{I}$ and we have thus verified that C is recursive in $\mathscr{I}$ and φ.

Let λ denote the least cardinal greater than $2^{\aleph_0}$. For $\sigma < \lambda$ we define $\mathbb{A}_\sigma$ and B_σ by recursion as follows. Let $\mathsf{B}_{(\sigma)} = \bigcup\{\mathsf{B}_\tau : \tau < \sigma\}$. Then

$$\mathbb{A}_\sigma = \{\theta\colon \forall\beta\,[\beta \in \mathsf{B}_{(\sigma)} \wedge \theta(\beta) \circledcirc \varphi(\beta)] \vee [\beta \notin \mathsf{B}_{(\sigma)} \wedge \theta(\beta) \circledcirc_0 \varphi(\beta)]\};$$

$$\mathsf{B}_\sigma = \{\beta\colon \text{for all } (a, \mathsf{x}) \circledcirc_0 \varphi(\beta)\ (\exists\theta \in \mathbb{A}_\sigma).\, |a, \mathsf{x}|^{\mathscr{I}} \leq \min\theta\}.$$

It is straightforward to prove by induction that for $\sigma \leq \rho$, $\mathbb{A}_\sigma \subseteq \mathbb{A}_\rho$ and $\mathsf{B}_\sigma \subseteq \mathsf{B}_\rho \subseteq \mathsf{B}$. Suppose that for all $\sigma < \lambda$, $\mathsf{B}_{(\sigma)} \subsetneq \mathsf{B}_\sigma$. Then by associating with each $\beta \in \mathsf{B}_{(\lambda)}$ the least σ such that $\beta \in \mathsf{B}_\sigma$, we obtain a pre-wellordering of ${}^\omega\omega$ of length λ, a contradiction. Hence, for some $\bar{\sigma} < \lambda$, $\mathsf{B}_{(\bar{\sigma})} = \mathsf{B}_{\bar{\sigma}}$. It follows that for $\rho \geq \bar{\sigma}$, $\mathbb{A}_{\bar{\sigma}} = \mathbb{A}_\rho$ and $\mathsf{B}_{\bar{\sigma}} = \mathsf{B}_\rho$. We set $\mathbb{A} = \mathbb{A}_{\bar{\sigma}}$ and show below that $\mathsf{B} = \mathsf{B}_{\bar{\sigma}}$.

It is immediate that (d) and the inequality ($\geq$) of (e) are satisfied. Suppose that $\min\varphi > \sup^+\{\min\theta : \theta \in \mathbb{A}\}$. We shall construct a function ψ with the following two mutually contradictory properties: $\psi \in \mathbb{A}$ and $(\forall\theta \in \mathbb{A})\forall\beta\,(|\psi(\beta)|^{\mathscr{I}} > \min\theta)$. For $\beta \in \mathsf{B}_{\bar{\sigma}}$,

$$\sup{}^+\{\min\theta : \theta \in \mathbb{A}\} < \min\varphi \leq \varphi(\beta) = \sup{}^+\{|a, \mathsf{x}|^{\mathscr{I}} : (a, \mathsf{x}) \circledcirc \varphi(\beta)\}.$$

Hence there exists a pair $(a, \mathsf{x}) \circledcirc \varphi(\beta)$ such that for all $\theta \in \mathbb{A}$, $|a, \mathsf{x}|^{\mathscr{I}} > \min\theta$ and we set $\psi(\beta)$ equal to some such pair. For $\beta \notin \mathsf{B}_{\bar{\sigma}}$, there exists by the definition of $\mathsf{B}_{\bar{\sigma}}$ a pair $(a, \mathsf{x}) \circledcirc_0 \varphi(\beta)$ such that for all $\theta \in \mathbb{A}_{\bar{\sigma}} = \mathbb{A}$, $|a, \mathsf{x}|^{\mathscr{I}} > \min\theta$ and we set $\psi(\beta)$ equal to some such pair. That ψ has the announced properties follows from the fact that $\mathsf{B}_{(\bar{\sigma})} = \mathsf{B}_{\bar{\sigma}}$.

It remains to show that $\mathbb{A}$ and B are recursive in $\mathscr{I}$ and φ. Note first that $\mathsf{B} = \mathsf{B}_{\bar{\sigma}}$, since if $\beta \in \mathsf{B}$, then by (e), for all $(a, \mathsf{x}) \circledcirc_0 \varphi(\beta)$, $|a, \mathsf{x}|^{\mathscr{I}} < \sup^+\{\min\theta : \theta \in \mathbb{A}\}$, so $\beta \in \mathsf{B}_\sigma$. Recall that $\mathbb{W}$ denotes the set of type-2 functionals I which code well-orderings of ${}^\omega\omega$. $\mathbb{W}$ is recursive in $\mathscr{I}$ and contains codes for all ordinals less than λ. We claim that there exist functionals $\mathbb{H}$ and $\mathbb{I}$ recursive in $\mathscr{I}$ such that for all $\mathsf{I} \in \mathbb{W}$, β, and φ,

$$\mathbb{H}(\theta, \varphi, \mathsf{I}) = 0 \leftrightarrow \theta \in \mathbb{A}_{\|\mathsf{I}\|} \quad\text{and}\quad \mathbb{I}(\beta, \varphi, \mathsf{I}) = 0 \leftrightarrow \beta \in \mathsf{B}_{\|\mathsf{I}\|}.$$

From this it follows that

$$\theta \in \mathbb{A} \leftrightarrow (\exists \mathsf{I} \in \mathbb{W}).\, \mathbb{H}(\theta, \varphi, \mathsf{I}) = 0$$

and

$$\beta \in \mathsf{B} \leftrightarrow (\exists \mathsf{I} \in \mathbb{W}).\, \mathbb{I}(\beta, \varphi, \mathsf{I}) = 0$$

so both are recursive in $\mathcal{J}$. $\mathbb{H}$ and $\mathbb{I}$ are defined by effective transfinite recursion parallel to the recursive definitions of $\mathbb{A}_\sigma$ and B_σ. The equivalence (f) is used again to justify the recursiveness of B_σ and in the definition of $\mathbb{H}$ we use the equivalence, for $\mathsf{I} \in \mathbb{W}$,

$$\begin{aligned} \beta \in \mathsf{B}_{(\|\mathsf{I}\|)} &\leftrightarrow \exists \mathsf{J}\,(\|\mathsf{J}\| < \|\mathsf{I}\| \wedge \beta \in \mathsf{B}_{\|\mathsf{J}\|}) \\ &\leftrightarrow \exists \mathsf{J}\,(\mathsf{J} < \mathsf{I} \wedge \beta \in \mathsf{B}_{\|\mathsf{J}\|}) \end{aligned}$$

where

$$\mathsf{J} < \mathsf{I} \leftrightarrow \exists \mathsf{F}\,(\mathsf{F} \text{ is an isomorphism of } \leqslant_{\mathsf{J}} \text{ into a proper initial segment of } \leqslant_{\mathsf{I}}). \quad \square$$

4.4–4.5 Exercises

4.4. Prove that $s\mathcal{J}$ is recursive in $\mathbb{E}^{\#}$.

4.5. Sketch a proof that for all $r \geqslant 1$ and any inductive operator Γ over ${}^{({}^{\omega}\omega)}\omega$, if $\Gamma \in \Delta^3_r$, then also $\bar{\Gamma} \in \Delta^3_r$.

4.6 Notes. Theorem 4.3 was announced in Grilliot [1969] and is known as the Grilliot Selection Theorem. The proof there was incorrect and our proof is from Harrington–MacQueen [1976].

Chapter VIII
Recursion on Ordinals

In the preceding two chapters we have put our attention on the generalized recursion theories which arise from ordinary recursion theory by the introduction of functionals of types 2 and higher. As we saw in § VI.7, these theories can equally well be viewed as theories of effective computability over domains which include such functionals. In this chapter, we shall study theories of effective computability over (initial segments of) the class of ordinal numbers.

The intuitive notion of effective computability for functions from ordinals to ordinals is again based on an idealized computing machine M which is equipped to store and manipulate ordinals according to a given program and produce an ordinal as answer. Many features of the "language" in which the programs for M are to be written are unchanged from earlier notions of computability, but there are two new basic instructions. These ensure that if H is a partial computable functions, then so are F and G defined by:

$$F(\rho, \boldsymbol{\mu}) \simeq \sup^{+}_{\pi<\rho} H(\pi, \boldsymbol{\mu}),$$

and

$$G(\boldsymbol{\mu}) \simeq \text{"least" } \pi \,.\, H(\pi, \boldsymbol{\mu}) \simeq 0.$$

The justification for including these schemes lies in a generalization of the notion of finiteness, which we call *metafiniteness.*

In ordinary recursion theory an object is finite iff it is in a one-to-one correspondence with a natural number — that is, with an element of the fundamental domain. If this were the only property of finite objects considered, we might say that an object is metafinite iff it is in a one-to-one correspondence with an ordinal. Of course, given the Axiom of Choice, this would make every set metafinite! Instead, we observe that in addition, every finite set of natural numbers is computable and indeed is in a *computable* one-to-one correspondence with a natural number. Hence we shall call an object *metafinite* iff it is in a computable one-to-one relationship with an ordinal.

Our basic principle of intuitive calculability over the ordinals is that a metafinite sequence of computations may be regarded as a completed totality and an answer drawn from the sequence of results of these computations. Thus, from the metafinite sequence of values $\langle H(\pi, \boldsymbol{\mu}) : \pi < \rho \rangle$, we can compute the least ordinal greater than all members of the sequence. Similarly, if for some ν, $H(\nu, \boldsymbol{\mu}) \simeq 0$ but for all $\pi < \nu$, $H(\pi, \boldsymbol{\mu})$ is defined with a non-zero value, then from the metafinite sequence $\langle H(\pi, \boldsymbol{\mu}) : \pi \leq \nu \rangle$ we can compute that $G(\boldsymbol{\mu}) \simeq \nu$.

In fact, our main concern will be with various refinements of this notion of computability. The most important of these consists simply in restricting attention to ordinals less than some fixed κ. For $\kappa = \omega$ we get a new characterization of ordinary recursion theory (on numbers) — in this case, the two new schemes are superfluous. In general, we shall study the class of κ-partial recursive functions. Many of the simplest properties of ordinary recursion theory are shared by κ-recursion theory for all κ, but it soon becomes evident that not all κ are equally well suited to support a recursion theory. Those that are suited are called *recursively regular* or *admissible* and studied in § 2.

In § 3 we show, among other things, that the ω_1-(semi-) recursive relations over ω are exactly the (Π^1_1) Δ^1_1 relations, and the $\aleph_1$-(semi-) recursive relations over ω are exactly the (Σ^1_2) Δ^1_2 relations. From one point of view these results reinforce the analogies and similarities of structure we have already observed; from another, they support the naturalness of the definition of κ-recursiveness. Both of these conclusions are further confirmed in § 4, where we show that for any type-2 functional I such that E is recursive in I, every $\omega_1[\mathsf{I}]$-(semi-) recursive relation on numbers is (semi-) recursive in I and, under certain conditions, which are satisfied by $\mathsf{E}, \mathsf{E}_1, \mathsf{E}_2, \ldots$, the converse holds as well.

In the next two sections, we investigate the general structure of the ordinals using measures of complexity defined in terms of ordinal recursion. Some of these are closely analogous to so-called "large cardinals" of set theory. Here again there are interesting connections with the analytical hierarchy and type-2 recursion. For example, the least stable ordinal is δ^1_2, the least ordinal not the order type of a Δ^1_2 well-ordering of ω, and the least recursively inaccessible ordinal is $\omega_1[\mathsf{E}_1]$.

In § 7 we explore the close connection between ordinal recursion and the hierarchy of constructible sets. Most of the recursion-theoretic properties of an ordinal κ correspond to model-theoretic properties of L_κ, the class of sets constructible in fewer than κ steps.

Some of our previous notational conventions are changed for this chapter. The letters f, g, h and F, G, H, I now denote k-ary partial functions from Or into Or; $R, S, T, \ldots$ are k-ary relations over Or.

1. Recursive Ordinal Functions

When we expand our fundamental domain from ω to a larger ordinal κ or to all of Or, some notions of ordinary recursion theory have obvious counterparts whereas others may have either more or fewer than one natural version. Such things as the successor and projection functions need no explanation and we assume that the reader is familiar with the operations of ordinal arithmetic: for all ordinals μ and ρ,

$$\mu+0=\mu, \qquad \mu+(\rho+1)=(\mu+\rho)+1,$$
$$\text{and if } \rho \text{ is a limit ordinal, } \mu+\rho=\sup^{+}_{\pi<\rho}(\mu+\pi);$$
$$\mu\cdot 0=0, \qquad \mu\cdot(\rho+1)=(\mu\cdot\rho)+\mu,$$
$$\text{and if } \rho \text{ is a limit ordinal, } \mu\cdot\rho=\sup^{+}_{\pi<\rho}(\mu\cdot\pi);$$
$$\mu^0=1, \qquad \mu^{\rho+1}=\mu^{\rho}\cdot\mu,$$
$$\text{and if } \rho \text{ is a limit ordinal, } \mu^{\rho}=\sup^{+}_{\pi<\rho}(\mu^{\pi});$$
$$\mu \dot{-} \rho = \begin{cases} \text{least } \nu(\mu=\rho+\nu), & \text{if } \rho\leqslant\mu; \\ 0, & \text{otherwise.} \end{cases}$$

We shall say that F arises from G, H, and I by *Primitive Recursion* iff for all $\boldsymbol{\mu}$ and ρ,

$$F(0,\boldsymbol{\mu})\simeq G(\boldsymbol{\mu});$$
$$F(\rho+1,\boldsymbol{\mu})\simeq H(F(\rho,\boldsymbol{\mu}),\rho,\boldsymbol{\mu});$$

and if ρ is a limit ordinal,

$$F(\rho,\boldsymbol{\mu})\simeq I(\sup^{+}_{\pi<\rho}F(\pi,\boldsymbol{\mu}),\rho,\boldsymbol{\mu}).$$

Ordinal addition, multiplication, and exponentiation are examples as is the *predecessor function*:

$$\text{Pd}(0)=0, \qquad \text{Pd}(\rho+1)=\rho, \quad \text{and if } \rho \text{ is a limit ordinal,} \quad \text{Pd}(\rho)=\rho.$$

The notions of expansion, bounded search, definition by cases, relational and functional composition, and bounded quantification all have natural extensions to functions and relations on ordinals; we encourage the reader to verify this for himself. For example, F arises from G by *bounded search* iff

$$F(\rho,\boldsymbol{\mu})\simeq \text{“least”}\ \pi<\rho\,.\,G(\pi,\boldsymbol{\mu})\simeq 0$$
$$\simeq \begin{cases} \pi, \text{ if } G(\pi,\boldsymbol{\mu})\simeq 0 \text{ and } (\forall\sigma<\pi)(\exists\nu>0)\,.\,G(\sigma,\boldsymbol{\mu})\simeq\nu; \\ \rho, \text{ if } (\forall\pi<\rho)(\exists\nu>0)\,.\,G(\pi,\boldsymbol{\mu})\simeq\nu; \\ \text{undefined, otherwise.} \end{cases}$$

In ordinary recursion theory, by unbounded search we mean a search through all ordinals less than ω. In the present context it will be useful to provide for varying "lengths" of searches. We say that F arises from G by *λ-search* and write

$$F(\boldsymbol{\mu}) \simeq \lambda\text{-"least" } \pi . G(\pi, \boldsymbol{\mu}) \simeq 0$$

iff

$$F(\boldsymbol{\mu}) \simeq \begin{cases} \pi, \text{ if } \pi < \lambda,\ G(\pi, \boldsymbol{\mu}) \simeq 0, \quad \text{and} \quad (\forall \sigma < \pi)(\exists \nu > 0)\ G(\sigma, \boldsymbol{\mu}) \simeq \nu; \\ \text{undefined}, \quad \text{if there is no such } \pi < \lambda. \end{cases}$$

Similarly, F arises from G by *unbounded search*,

$$F(\boldsymbol{\mu}) \simeq \text{"least" } \pi . G(\pi, \boldsymbol{\mu}) \simeq 0$$

when the same condition holds without the restriction to $\pi < \lambda$.

We shall need simple "coding" functions for finite sequences of ordinals. Since ordinals do not have unique prime power decompositions ($2^\omega = 3^\omega = \omega$), we must use a different technique. Many approaches are possible here, but the following is one of the simplest. We first define a coding function $\langle\langle\ \ \rangle\rangle$ for pairs of ordinals and then treat k-tuples by iterated pairing. Let

$$\langle\langle \mu, \nu \rangle\rangle = 3^{\mu+\nu} + 3^{\mu}.$$

Since for any μ and ν, $3^{\mu+\nu} < 3^{\mu+\nu} + 3^{\mu} < 3^{\mu+\nu+1}$, if we set

$$f(\sigma) \simeq \text{least } \pi < \sigma . \sigma < 3^{\pi+1},$$

then $f(\langle\langle \mu, \nu \rangle\rangle) = \mu + \nu$. Let

$$(\sigma)^0 = f(\sigma \dot{-} 3^{f(\sigma)}) \quad \text{and} \quad (\sigma)^1 = f(\sigma) \dot{-} (\sigma)^0.$$

Then

$$(\langle\langle \mu, \nu \rangle\rangle)^0 = f(3^{\mu+\nu} + 3^{\mu}) \dot{-} 3^{\mu+\nu} = f(3^{\mu}) = \mu$$

and

$$(\langle\langle \mu, \nu \rangle\rangle)^1 = (\mu + \nu) \dot{-} \mu = \nu.$$

It follows that $\langle\langle\ \ \rangle\rangle$ is one-one. Now set

$$\langle\ \ \rangle^0 = 1, \quad \text{and for all } k \text{ and all } \boldsymbol{\mu} = (\mu_0, \ldots, \mu_k),$$
$$\langle \boldsymbol{\mu} \rangle^{k+1} = \langle\langle \mu_0, \langle \mu_1, \ldots, \mu_k \rangle^k \rangle\rangle.$$

Clearly each $\langle\ \rangle^k$ is one-one. To see that the different $\langle\ \rangle^k$ have disjoint images, let g be the function defined for $\rho < \omega$ by:

$$g(0, \sigma) = \sigma \quad \text{and} \quad g(\rho + 1, \sigma) = (g(\rho, \sigma))^1.$$

Then if we set

$$\lg(\sigma) = \text{least } \rho < \sigma \,.\, g(\rho, \sigma) = 1,$$

then $\lg(\langle \mu_0, \ldots, \mu_{k-1}\rangle^k) = k$. We may thus omit the superscript k without ambiguity. In terms of g we have also decoding functions:

$$(\sigma)_i = (g(i, \sigma))^0.$$

To define a concatenation function $*$, let, for $\rho < \omega$,

$$h(0, \sigma, \tau) = \tau, \quad \text{and}$$
$$h(\rho + 1, \sigma, \tau) = \langle\langle(\sigma)_{\lg(\sigma) \dot{-} (\rho+1)}, h(\rho, \sigma, \tau)\rangle\rangle.$$

Then if

$$\sigma * \tau = h(\lg(\sigma), \sigma, \tau),$$

we have easily

$$\langle \boldsymbol{\mu} \rangle * \langle \boldsymbol{\nu} \rangle = \langle \boldsymbol{\mu}, \boldsymbol{\nu} \rangle.$$

Finally, if we set

$$\mathrm{Sq}(0, \sigma) \leftrightarrow \sigma = 1;$$
$$\mathrm{Sq}(\rho + 1, \sigma) \leftrightarrow (\exists \pi < \sigma)(\exists \tau < \sigma)[\mathrm{Sq}(\rho, \tau) \wedge \sigma = \langle\langle \pi, \tau \rangle\rangle];$$
$$\mathrm{Sq}(\rho, \sigma) \leftrightarrow 0 = 1, \quad \text{if} \quad \rho \text{ is a limit ordinal};$$

and

$$\mathrm{Sq}(\sigma) \leftrightarrow (\exists \rho < \sigma) \mathrm{Sq}(\rho, \sigma);$$

then $\mathrm{Sq}(\sigma)$ iff for some $\boldsymbol{\mu}$, $\sigma = \langle \boldsymbol{\mu} \rangle$.

The formal definition of the classes of recursive ordinal functions that we shall study proceeds similarly as in previous cases. For each κ and λ we define a set $\Omega_{\kappa\lambda}$ of sequences of the form $(a, \boldsymbol{\mu}, \nu)$, where a is a natural number and $\mu_0, \ldots, \mu_{k-1}, \nu$ are ordinals less than κ (the role of λ is explained in the next paragraph). Clauses (0)–(2) of the definition are the natural extensions of the corresponding clauses of Definition II.2.1 to ordinal arguments and values with two exceptions. First, there is no clause for applying function arguments, as we have none. Second, only the constant functions with natural number values are included. This restriction is unavoidable if we want to preserve the property that indices (programs) be finite objects.

Clauses (3) and (4) will introduce the operations of $\sup^+$ and λ-search and we shall have:

$$\{\langle 3, k+1, b\rangle\}_{\kappa\lambda}(\rho, \boldsymbol{\mu}) \simeq \sup^+_{\pi<\rho}\{b\}_{\kappa\lambda}(\pi, \boldsymbol{\mu}),$$

and

$$\{\langle 4, k, b\rangle\}_{\kappa\lambda}(\boldsymbol{\mu}) \simeq \lambda\text{-“least”}\ \pi.\{b\}_{\kappa\lambda}(\pi, \boldsymbol{\mu}) \simeq 0.$$

To avoid uninteresting pathological cases, we shall always assume that κ is a limit ordinal.

1.1 Definition. For any κ and $\lambda, \Omega_{\kappa\lambda}$ is the smallest set such that for all $k, n \in \omega$, all $i < k$, all $\boldsymbol{\mu} \in {}^k\kappa$, and all ν, π, ρ, σ, and $\tau < \kappa$,

(0) $(\langle 0, k, 0, n\rangle, \boldsymbol{\mu}, n) \in \Omega_{\kappa\lambda}$;
$(\langle 0, k, 1, i\rangle, \boldsymbol{\mu}, \mu_i) \in \Omega_{\kappa\lambda}$;
$(\langle 0, k, 2, i\rangle, \boldsymbol{\mu}, \mu_i + 1) \in \Omega_{\kappa\lambda}$;
$(\langle 0, k+4, 4\rangle, \pi, \rho, \sigma, \tau, \boldsymbol{\mu}, \pi) \in \Omega_{\kappa\lambda}$, if $\sigma = \tau$;
$(\langle 0, k+4, 4\rangle, \pi, \rho, \sigma, \tau, \boldsymbol{\mu}, \rho) \in \Omega_{\kappa\lambda}$, if $\sigma \neq \tau$;
$(\langle 0, k+2, 5\rangle, p, q, \boldsymbol{\mu}, \mathrm{Sb}_0(p, q)) \in \Omega_{\kappa\lambda}$ for all $p, q \in \omega$;

(1) for any $k', b, c_0, \ldots, c_{k'-1} < \omega$ and any $\xi_0, \ldots, \xi_{k'-1} < \kappa$, if for all $i < k'$, $(c_i, \boldsymbol{\mu}, \xi_i) \in \Omega_{\kappa\lambda}$ and $(b, \boldsymbol{\xi}, \nu) \in \Omega_{\kappa\lambda}$, then $(\langle 1, k, b, c_0, \ldots, c_{k'-1}\rangle, \boldsymbol{\mu}, \nu) \in \Omega_{\kappa\lambda}$;

(2) for any b, if $(b, \boldsymbol{\mu}, \nu) \in \Omega_{\kappa\lambda}$, then

$$(\langle 2, k+1\rangle, b, \boldsymbol{\mu}, \nu) \in \Omega_{\kappa\lambda};$$

(3) if ν is the least ordinal such that $(\forall \pi < \rho)(\exists \xi < \nu).(b, \pi, \boldsymbol{\mu}, \xi) \in \Omega_{\kappa\lambda}$, then $(\langle 3, k+1, b\rangle, \rho, \boldsymbol{\mu}, \nu) \in \Omega_{\kappa\lambda}$;

(4) if $\nu < \lambda$, $(b, \nu, \boldsymbol{\mu}, 0) \in \Omega_{\kappa\lambda}$, and $(\forall \pi < \nu)(\exists \xi > 0).(b, \pi, \boldsymbol{\mu}, \xi) \in \Omega_{\kappa\lambda}$, then $(\langle 4, k, b\rangle, \boldsymbol{\mu}, \nu) \in \Omega_{\kappa\lambda}$.

As in previous cases, the definition of $\Omega_{\kappa\lambda}$ may be interpreted as a closure under certain functions. All of these functions have rank at most $\mathrm{Card}(\rho)$ for some $\rho < \kappa$ and it follows that the closure ordinal of the inductive definition is at most the least regular cardinal $\geqslant \kappa$ (this is what the reader was supposed to discover in Exercise I.3.10). We shall see that in general it is much smaller than this.

Since ν in clause (4) is initially restricted to κ, $\Omega_{\kappa\lambda} = \Omega_{\kappa\kappa}$ for all $\lambda \geq \kappa$. The hypothesis of clause (4) is never satisfied if $\lambda = 0$ so $\Omega_{\kappa 0}$ is in fact defined by clauses (0)–(3). If $\kappa \leq \kappa'$ and $\lambda \leq \lambda'$, then clearly $\Omega_{\kappa\lambda} \subseteq \Omega_{\kappa'\lambda'}$.

1.2 Lemma. *For all κ, λ, a, and $\boldsymbol{\mu}$, there is at most one ν such that $(a, \boldsymbol{\mu}, \nu) \in \Omega_{\kappa\lambda}$.*

Proof. Similar to that of Lemma II.2.2 with a transfinite induction replacing the simple induction. □

Thus we may set

$$\{a\}_{\kappa\lambda}(\boldsymbol{\mu}) \simeq \nu \leftrightarrow (a, \boldsymbol{\mu}, \nu) \in \Omega_{\kappa\lambda}.$$

In fact, we shall not be concerned with quite this degree of generality.

1.3 Definition. For all κ, λ, a, $\boldsymbol{\mu}$, and ν,

(i) $\{a\}_{\kappa}(\boldsymbol{\mu}) \simeq \nu \leftrightarrow (a, \boldsymbol{\mu}, \nu) \in \Omega_{\kappa\kappa}$;
(ii) $\{a\}_{\infty\lambda}(\boldsymbol{\mu}) \simeq \nu \leftrightarrow \exists\kappa\,.\,(a, \boldsymbol{\mu}, \nu) \in \Omega_{\kappa\lambda}$;
(iii) $\{a\}_{\infty}(\boldsymbol{\mu}) \simeq \nu \leftrightarrow \exists\kappa\,\exists\lambda\,.\,(a, \boldsymbol{\mu}, \nu) \in \Omega_{\kappa\lambda}$.

A partial function $F: {}^k\kappa \to \kappa$ is *κ-partial recursive* iff $F = \{a\}_\kappa$ for some $a \in \omega$ and *κ-recursive* iff it is κ-partial recursive and total ($\mathrm{Dm}\, F = {}^k\kappa$). A relation $R \subseteq {}^k\kappa$ is *κ-recursive* iff its characteristic function K_R is κ-recursive. R is *κ-semi-recursive* iff R is the domain of some κ-partial recursive function and *κ-co-semi-recursive* iff $({}^k\kappa \sim R)$ is κ-semi-recursive. There is a slight technical problem in making parallel definitions for the other two notions of Definition 1.3; for some a, $\{a\}_{\infty\lambda}$ and $\{a\}_\infty$ are not sets because their domains are proper classes of ordinals. We shall in practice ignore this point and proceed as if they were sets. The worried reader may either think of this chapter as formulated in a set theory which admits proper classes or verify that each use of a symbol $\{a\}_{\infty\lambda}$ or $\{a\}_\infty$ may be replaced by its definition in terms of sets. Thus we say that a partial function $F: {}^k\mathrm{Or} \to \mathrm{Or}$ is *(∞, λ)-partial recursive* [*∞-partial recursive*] iff $F = \{a\}_{\infty\lambda}$ [$F = \{a\}_\infty$] for some $a \in \omega$, etc.

In each of these cases we say that F is **-partial recursive in parameters* iff for some *-partial recursive function G and some $\boldsymbol{\rho}$, $F(\boldsymbol{\mu}) \simeq G(\boldsymbol{\mu}, \boldsymbol{\rho})$. If F (but not necessarily G) is total, then F is **-recursive in parameters.* When it is relevant, we say that F is **-(partial) recursive in the parameters $\boldsymbol{\rho}$.* Similar terminology applies to relations. Since indices are natural numbers, the constant function with value ρ is *-recursive for only denumerably many ρ, so that in general allowing parameters properly enlarges the collection of *-recursive functions (cf. Exercise 1.17).

The classes of κ-partial recursive functions and (∞, κ)-partial recursive functions are closed under κ-search and the class of ∞-partial recursive functions is closed under unbounded search. If $\kappa \leqslant \lambda$, then $\{a\}_\kappa \subseteq \{a\}_\lambda \subseteq \{a\}_{\infty\lambda} \subseteq \{a\}_\infty$, so each κ-partial recursive function F has natural extensions to λ-, (∞, λ)-, and ∞-partial recursive functions. Of course, different indices for F lead in general to different extensions. If $\{a\}_\kappa$ is total on ${}^k\kappa$, then no new computations with arguments less than κ are added in the extensions, but this is not in general true if $\{a\}_\kappa$ is partial. New computations with arguments and values from κ may be made possible by the availability to $\{a\}_\lambda$, $\{a\}_{\infty\lambda}$, and $\{a\}_\infty$ of intermediate arguments and values greater than κ. The $(\infty, 0)$-recursive functions will play somewhat the role here of the primitive recursive functions of ordinary recursion theory. They form a convenient class sufficient for many elementary calculations. Although it is not in general true that the restriction of an $(\infty, 0)$ recursive function to arguments from κ is κ-recursive, this will be true whenever κ is recursively regular (Definition 2.1).

1.4 Theorem. *For any k and any $F: {}^k\omega \to \omega$, F is ω-partial recursive iff F is $(\omega, 0)$-partial recursive iff F is (ordinary) partial recursive.*

Proof. Exercise 1.15. □

In the following we shall use only the obvious part of this theorem:if F is (ordinary) recursive, then F is $(\omega, 0)$-recursive and thus has an $(\infty, 0)$-partial recursive extension.

1.5 Theorem. *For all κ and λ, the classes of κ-partial recursive [(∞, λ)-partial recursive, ∞-partial recursive] functions and κ-recursive [(∞, λ)-recursive, ∞-recursive] relations are closed under definition by cases.*

Proof. We shall prove the result for κ-recursion. Suppose

$$F(\boldsymbol{\mu}) \simeq \begin{cases} G_i(\boldsymbol{\mu}), & \text{if } R_i(\boldsymbol{\mu}) \text{ for } i < k'; \\ G_{k'}(\boldsymbol{\mu}), & \text{otherwise;} \end{cases}$$

where the R_i $(i < k')$ are pairwise disjoint κ-recursive relations and the G_i $(i \leqslant k')$ are κ-partial recursive functions. We proceed by induction on k'. If $k' = 0$ there is nothing to prove, so suppose $k' > 0$ and the result is true for smaller values of k'. Then there exists a κ-partial recursive function H such that

$$H(\boldsymbol{\mu}) \simeq \begin{cases} G_i(\boldsymbol{\mu}), & \text{if } R_i(\boldsymbol{\mu}) \text{ for } 0 < i < k'; \\ G_{k'}(\boldsymbol{\mu}), & \text{otherwise;} \end{cases}$$

so that

$$F(\boldsymbol{\mu}) \simeq \begin{cases} G_0(\boldsymbol{\mu}), & \text{if } R_0(\boldsymbol{\mu}); \\ H(\boldsymbol{\mu}), & \text{otherwise.} \end{cases}$$

Let b and c be κ-indices of G_0 and H, respectively. Then

$$F(\boldsymbol{\mu}) \simeq \{\{\langle 0,4,4\rangle\}_\kappa(b, c, K_{R_0}(\boldsymbol{\mu}), 0)\}_\kappa(\boldsymbol{\mu})$$

and is thus κ-partial recursive. □

The equality relation on ordinals is $(\infty, 0)$-recursive, since

$$K_=(\sigma, \tau) = \{\langle 0,4,4\rangle\}_{\infty 0}(0, 1, \sigma, \tau).$$

Then the functions sg^+ and sg^- are $(\infty, 0)$-recursive and κ-recursive for all κ, and it follows as in § II.1 that the classes of κ-recursive, (∞, λ)-recursive, and ∞-recursive relations are Boolean Algebras (the arithmetic operations used there are applied only to natural numbers).

Suppose that R is κ-recursive,

$$P(\rho, \boldsymbol{\mu}) \leftrightarrow (\forall \pi < \rho) R(\pi, \boldsymbol{\mu}),$$

and

$$Q(\rho, \boldsymbol{\mu}) \leftrightarrow (\exists \pi < \rho) R(\pi, \boldsymbol{\mu}).$$

Then

$$K_P(\rho, \boldsymbol{\mu}) = (\sup\nolimits^+_{\pi<\rho} K_R(\pi, \boldsymbol{\mu})) \dot{-} 1$$

so P is also κ-recursive. That Q is κ-recursive follows by closure under complementation. The same applies to (∞, λ)- and ∞-recursion. We now have

$$\sigma < \tau \leftrightarrow (\exists \pi < \tau) \sigma = \pi$$

so that the relations, $<$, $\leqslant$, $>$, and $\geqslant$ are all $(\infty, 0)$-recursive.

1.6 Lemma. *The following relations are $(\infty, 0)$-recursive:*

(i) $\mathrm{Lm}(\rho) \leftrightarrow \rho$ *is a limit ordinal;*

(ii) $\mathrm{Suc}(\rho) \leftrightarrow \rho$ *is a successor ordinal;*

(iii) $\{\rho : \rho = \omega\}$;

(iv) $\{\rho : \rho < \omega\}$.

Proof. These are immediate from the preceding remarks and the following relationships:

$$\mathrm{Lm}(\rho) \leftrightarrow \rho \neq 0 \wedge (\forall \pi < \rho)\pi + 1 < \rho;$$

$$\mathrm{Suc}(\rho) \leftrightarrow (\exists \pi < \rho)\rho = \pi + 1;$$

$$\rho = \omega \leftrightarrow \mathrm{Lm}(\rho) \wedge (\forall \pi < \rho) \neg \mathrm{Lm}(\pi);$$

$$\rho < \omega \leftrightarrow \neg \mathrm{Lm}(\rho) \wedge (\forall \pi < \rho) \neg \mathrm{Lm}(\pi). \quad \square$$

Note that we are not asserting that the constant function with value ω is $(\infty, 0)$-recursive — it is not (Corollary 2.3). However, for any $\kappa > \omega$,

$$\omega = \kappa\text{-least } \pi . \mathrm{Lm}(\pi)$$

so this function is κ-recursive.

1.7 Lemma. *The predecessor function* Pd *is* $(\infty, 0)$-*recursive.*

Proof. Let

$$f(\pi, \rho) = \begin{cases} \pi, & \text{if } \quad \pi + 1 < \rho; \\ 0, & \text{otherwise.} \end{cases}$$

Then f is $(\infty, 0)$-recursive and $\mathrm{Pd}(\rho) = \sup^{+}_{\pi < \rho} f(\pi, \rho)$. $\square$

1.8 Recursion Theorem. *For any* κ *and any* κ-*partial recursive function* F, *there exists an* $\bar{e} < \omega$ *such that*

$$\{\bar{e}\}_\kappa(\boldsymbol{\mu}) \simeq F(\bar{e}, \boldsymbol{\mu}).$$

The same holds for (∞, λ)- *and* ∞-*recursion.*

Proof. As for Theorem II.2.6. $\square$

1.9 Theorem. *For any* κ *and* λ, *the classes of* κ-*partial recursive,* (∞, λ)-*partial recursive, and* ∞-*partial recursive functions are closed under primitive recursion.*

Proof. Suppose that F arises from κ-partial recursive functions G, H, and I by primitive recursion. Let f be defined by:

$$f(e, \rho, \boldsymbol{\mu}) \simeq \begin{cases} G(\boldsymbol{\mu}), & \text{if } \quad \rho = 0; \\ H(\{e\}_\kappa(\mathrm{Pd}(\rho), \boldsymbol{\mu}), \mathrm{Pd}(\rho), \boldsymbol{\mu}), & \text{if } \quad \mathrm{Suc}(\rho); \\ I(\sup^{+}_{\pi < \rho} \{e\}_\kappa(\pi, \boldsymbol{\mu}), \rho, \boldsymbol{\mu}), & \text{if } \quad \mathrm{Lm}(\rho). \end{cases}$$

Then f is κ-partial recursive so by the κ-Recursion Theorem there exists an index $\bar{e}$ such that $f(\bar{e}, \rho, \boldsymbol{\mu}) \simeq \{\bar{e}\}_\kappa(\rho, \boldsymbol{\mu})$. It is straightforward to prove by induction that $F = \{\bar{e}\}_\kappa$. □

1.10 Corollary. *The functions of ordinal addition, multiplication, and exponentiation are all* $(\infty, 0)$*-recursive.*

Proof. The ordinal addition function + arises from three $(\infty, 0)$-recursive functions G, H, and I by primitive recursion, where

$$G(\rho, \mu) = \mu, \qquad H(\nu, \rho, \mu) = \nu + 1, \quad \text{and} \quad I(\sigma, \rho, \mu) = \sigma.$$

It follows from Theorem 1.9 that + is $(\infty, 0)$-recursive. Similar arguments apply to multiplication and exponentiation. □

1.11 Theorem. *For all* κ *and* λ, *the classes of* κ*-partial recursive functions,* (∞, λ)*-partial recursive functions, and* ∞*-partial recursive functions are closed under* sup *and* inf.

Proof. Suppose H is κ-partial recursive,

$$F(\rho, \boldsymbol{\mu}) \simeq \sup_{\pi<\rho} H(\pi, \boldsymbol{\mu}),$$

and

$$G(\rho, \boldsymbol{\mu}) \simeq \inf_{\pi<\rho} H(\pi, \boldsymbol{\mu}).$$

Then $F(\rho, \boldsymbol{\mu}) \simeq \mathrm{Pd}(\overset{+}{\sup}_{\pi<\rho} H(\pi, \boldsymbol{\mu}))$ and $G(\rho, \boldsymbol{\mu}) \simeq H(I(\rho, \boldsymbol{\mu}), \boldsymbol{\mu})$, where I satisfies the primitive recursion:

$$I(0, \boldsymbol{\mu}) = 0$$

$$I(\rho + 1, \boldsymbol{\mu}) \simeq \begin{cases} \rho, & \text{if} \quad H(\rho, \boldsymbol{\mu}) < H(I(\rho, \boldsymbol{\mu}), \boldsymbol{\mu}); \\ I(\rho, \boldsymbol{\mu}), & \text{otherwise;} \end{cases}$$

$$I(\rho, \boldsymbol{\mu}) \simeq \sup_{\pi<\rho} I(\pi, \boldsymbol{\mu}), \quad \text{if} \quad \rho \text{ is a limit ordinal.} \quad \square$$

The κ-partial recursive functions are closed under κ-search and the ∞-partial recursive functions under unbounded search, so it follows that both classes are closed under bounded search. The next theorem shows that this is also true for the (∞, λ)-partial recursive functions, even when $\lambda = 0$.

1.12 Theorem. *For all* λ, *the class of* (∞, λ)*-partial recursive functions is closed under bounded search.*

Proof. The proof is an extension of that of II.1.5 (ii). If G is (∞, λ)-partial recursive and

$$F(\rho, \boldsymbol{\mu}) \simeq \text{“least”}\ \pi < \rho \,.\, G(\pi, \boldsymbol{\mu}) \simeq 0,$$

then F satisfies the primitive recursion:

$$F(0, \boldsymbol{\mu}) \simeq 0;$$

$$F(\rho + 1, \boldsymbol{\mu}) \simeq F(\rho, \boldsymbol{\mu}) + \mathrm{sg}^{+}(G(F(\rho, \boldsymbol{\mu}), \boldsymbol{\mu}));$$

$$F(\rho, \boldsymbol{\mu}) \simeq \sup_{\pi < \rho} F(\pi, \boldsymbol{\mu}), \quad \text{if} \quad \rho \text{ is a limit ordinal.} \quad \square$$

1.13 Corollary. *The ordinal subtraction function* $\dot{-}$ *and the sequence-manipulation functions and relation,* $\langle\ \ \rangle^{k}$, $(\ \ )$, $*$, *and* Sq *are all* $(\infty, 0)$-*recursive.*

Proof. Immediate from the definitions and preceding results. □

1.14–1.17 Exercises

1.14. Describe the computation trees for (κ, λ)-recursion.

1.15. Prove Theorem 1.4.

1.16. Compare the class of arithmetical relations on numbers with the class of relations on numbers which are $(\infty, 0)$-recursive in the parameter ω.

1.17. Show that there exists a countable ordinal κ such that properly more functions are κ-recursive in parameters than are simply κ-recursive.

1.18 Notes. The notion of recursiveness for functions on ordinals is a good example of an idea whose time had come and was discovered independently in one form or another by several people between 1960 and 1966. The idea was natural from several points of view, and this is reflected in the variety of motivations of those who made the key discoveries. From the side of set theory we have Takeuti's work of the 1950's in reformulating Gödel's constructibility in terms of a theory of ordinals. Takeuti [1960] and Kino–Takeuti [1962, – a] develop a theory of recursive ordinal functions essentially equivalent to ∞-recursiveness as described above and show that it behaves nicely when restricted to the ordinals less than any cardinal. Machover [1961] and Lévy [1963] arrived at essentially the same theory with the aim of using it to study the infinitary languages $\mathscr{L}_{\kappa\lambda}$.

Meanwhile Kreisel [1961] observed that in the analogy of Π_1^1 to Σ_1^0, the appropriate counterpart of Δ_1^1 is the class of *finite* sets, not the class of recursive sets. This is essentially because of the boundedness phenomenon — a Δ_1^1 subset of W is bounded (metafinite) whereas a recursive subset of ω is not bounded (finite). In a similar vein, Kreisel also proposed that the cardinality criteria, both in recursion theories and infinitary languages, should be replaced by more intrinsic definability criteria. These ideas led more or less directly to the metarecursion theory of Kreisel–Sacks [1965] in which the fundamental domain is the set of recursive ordinals as represented by a Π_1^1 set of unique notations for them. This is equivalent to ω_1-recursion as defined above.

The pieces were all fitted together independently by Kripke [1964, - a] and Platek [1966], who approached the area with the aim of generalizing ordinary recursion theory. Kripke expanded Kleene's equation calculus to serve for defining recursive functions over any initial segment of the ordinals, whereas Platek used an approach via primitive recursion and the search operator. These theories include all of the others and are equivalent to each other and in almost all respects to the one presented above.

Further discussion of some of these points may be found following Theorem 3.4 below and in the introduction to Barwise [1975]. The particular formulation of the theory that we have used is new here, but similar formulations have been used by several people in lectures.

2. Recursively Regular Ordinals

It is obvious that for many ordinals κ, the class of κ-partial recursive functions is not a very natural class. If κ is a successor ordinal, the successor function on κ is not total, hence not κ-recursive. Similarly, if κ is not closed under ordinal addition, multiplication, etc., then the restrictions of these functions to arguments from κ are only κ-partial recursive. Among other difficulties, this means that the sequence coding functions may not be everywhere defined. A natural "regularity" condition on κ is that it be closed under sufficiently many functions:

2.1 Definition. An ordinal κ is *recursively regular* iff κ is closed under all (∞, κ)-partial recursive functions.

An apparently stronger condition is
(a) for all all $a < \omega$, all $\boldsymbol{\mu} < \kappa$, and all ν

$$\{a\}_{\infty\kappa}(\boldsymbol{\mu}) \simeq \nu \leftrightarrow \{a\}_\kappa(\boldsymbol{\mu}) \simeq \nu.$$

Since the values of κ-partial recursive functions are by definition less than κ, (a) implies that κ is recursively regular. The converse is not obvious, however, since

$\{a\}_\kappa$ is restricted to using ordinals from κ in the middle of computations, while $\{a\}_{\infty\kappa}$ may use larger ordinals even to arrive at a value less than κ.

If we examine the schemas for computing (∞, κ)-partial recursive functions, it is intuitively clear that an (∞, κ) computation with arguments less than κ can arrive at a value $\geqslant \kappa$ only by means of the $\sup^+$ schema (3). The first time this happens, the function whose $\sup^+$ is $\geqslant \kappa$ is in fact κ-partial recursive, and we are led to the following condition:

(b) for all κ-partial recursive functions F and all $\rho, \boldsymbol{\mu} < \kappa$, if $F(\pi, \boldsymbol{\mu})$ is defined for all $\pi < \rho$, then $\sup^+_{\pi<\rho} F(\pi, \boldsymbol{\mu}) < \kappa$.

2.2 Theorem. *For all κ, κ is recursively regular iff* (a) *iff* (b).

Proof. Suppose first that κ is recursively regular and let F be a κ-partial recursive function with κ-index a. Then $\{a\}_{\infty\kappa}$ is an (∞, κ)-partial recursive function which extends F, and the function G defined by

$$G(\rho, \boldsymbol{\mu}) \simeq \sup^+_{\pi<\rho} \{a\}_{\infty\kappa}(\pi, \boldsymbol{\mu})$$

is likewise (∞, κ)-partial recursive. Then for $\rho, \boldsymbol{\mu} < \kappa$, since κ is closed under G, if for all $\pi < \rho$, $F(\pi, \boldsymbol{\mu})$ is defined, then

$$\sup^+_{\pi<\rho} F(\pi, \boldsymbol{\mu}) = G(\rho, \boldsymbol{\mu}) < \kappa.$$

We have thus deduced that condition (b) holds.

We observed above that (a) implies that κ is recursively regular, so it remains to prove that (b) implies (a). Given (b), we first observe that κ is a limit ordinal, since if $\mu < \kappa$ and $F(\pi, \mu) = \mu$ for all π, then

$$\mu + 1 = \sup^+_{\pi<1} F(\pi, \mu) < \kappa.$$

We shall prove for all τ by induction over $\Omega_{\tau\kappa}$ that for all a, $\boldsymbol{\mu}$, and ν,

$$(*) \qquad \boldsymbol{\mu} < \kappa \wedge (a, \boldsymbol{\mu}, \nu) \in \Omega_{\tau\kappa} \rightarrow (a, \boldsymbol{\mu}, \nu) \in \Omega_{\kappa\kappa}.$$

For $a = \langle 0, k, \dots \rangle$ this is immediate (that κ is a limit ordinal is needed for the successor function). Suppose that $a = \langle 1, k, b, c_0, \dots, c_{k'-1} \rangle$ and ξ_i $(i < k')$ are such that $(c_i, \boldsymbol{\mu}, \xi_i) \in \Omega_{\tau\kappa}$ and $(b, \boldsymbol{\xi}, \nu) \in \Omega_{\tau\kappa}$. Since $\boldsymbol{\mu} < \kappa$, the induction hypothesis yields that each $(c_i, \boldsymbol{\mu}, \xi_i) \in \Omega_{\kappa\kappa}$. Hence each $\xi_i < \kappa$, so again by the induction hypothesis, $(b, \boldsymbol{\xi}, \nu) \in \Omega_{\kappa\kappa}$. Thus also $(a, \boldsymbol{\mu}, \nu) \in \Omega_{\kappa\kappa}$.

The argument for clause (2) is even simpler and we omit it. Suppose that $a = \langle 3, k+1, b \rangle$, ρ and $\boldsymbol{\mu} < \kappa$, and $(a, \rho, \boldsymbol{\mu}, \nu) \in \Omega_{\tau\kappa}$. Then ν is the least ordinal such that for every $\pi < \rho$ there exists $\xi_\pi < \nu$ such that $(b, \pi, \boldsymbol{\mu}, \xi_\pi) \in \Omega_{\tau\kappa}$. By the induction hypothesis, for all $\pi < \rho$, $(b, \pi, \boldsymbol{\mu}, \xi_\pi) \in \Omega_{\kappa\kappa}$ — that is, $\{b\}_\kappa(\pi, \boldsymbol{\mu}) \simeq \xi_\pi$. Then by (b),

$$\nu = \sup^{+}\{\xi_\pi : \pi < \rho\} = \sup^{+}_{\pi<\rho}\{b\}_\kappa(\pi, \boldsymbol{\mu}) < \kappa.$$

It follows that $(a, \rho, \boldsymbol{\mu}, \nu) \in \Omega_{\kappa\kappa}$.

Finally, suppose that $a = \langle 4, k, b\rangle$, $\boldsymbol{\mu} < \kappa$, and $(a, \boldsymbol{\mu}, \nu) \in \Omega_{\tau\kappa}$. Then $\nu < \kappa$, $(b, \nu, \boldsymbol{\mu}, 0) \in \Omega_{\tau\kappa}$, and for all $\pi < \nu$, $(b, \pi, \boldsymbol{\mu}, \xi_\pi) \in \Omega_{\tau\kappa}$ for some $\xi_\pi > 0$. By the induction hypothesis (since if $\pi < \nu$, also $\pi < \kappa$), $(b, \nu, \boldsymbol{\mu}, 0) \in \Omega_{\kappa\kappa}$ and all $(b, \pi, \boldsymbol{\mu}, \xi_\pi) \in \Omega_{\kappa\kappa}$. Hence also $(a, \boldsymbol{\mu}, \nu) \in \Omega_{\kappa\kappa}$. $\square$

The term "recursively regular" is derived from condition (b). Indeed, if the modifier "κ-partial recursive" is removed from (b), it becomes just the condition that κ be a regular cardinal. In particular, every regular cardinal is recursively regular.

Of course, every recursively regular κ is closed under all $(\infty, 0)$-partial recursive functions and the restriction of any $(\infty, 0)$-partial recursive function to a recursively regular κ is κ-partial recursive.

2.3 Corollary. *The constant function with value ω is not $(\infty, 0)$-recursive. In fact, no constant function with value $\geqslant \omega$ is $(\infty, 0)$-recursive.*

Proof. Suppose that $F(\mu) = \nu \geqslant \omega$ for all μ and F is $(\infty, 0)$-recursive. ω is a regular cardinal, hence recursively regular, so as $0 < \omega$, also $F(0) < \omega$, a contradiction. $\square$

Although we have argued that some condition such as recursive regularity is *necessary* for an ordinal κ to support a reasonable recursion theory, we have not as yet given any reason to believe that such a condition is *sufficient*. To establish this, we shall show that there exists an $(\infty, 0)$-recursive relation T such that for all a, $\boldsymbol{\mu}$, and ν, and any recursively regular κ,

$$\{a\}_\kappa(\boldsymbol{\mu}) \simeq \nu \leftrightarrow (\exists \upsilon < \kappa)[T(a, \langle\boldsymbol{\mu}\rangle, \upsilon) \wedge (\upsilon)_0 = \nu].$$

In other words, for recursively regular κ, κ-recursion theory has a normal form very similar to that of § II.3 for ordinary recursion theory.

The proof is in two main steps. Let

$$T_0(a, \langle\boldsymbol{\mu}\rangle, \nu, \kappa, \lambda, \sigma) \leftrightarrow (a, \boldsymbol{\mu}, \nu) \in \Omega^\sigma_{\kappa\lambda},$$

where, as usual, $\Omega^\sigma_{\kappa\lambda}$ denotes the σ-th stage of the inductive definition of $\Omega_{\kappa\lambda}$. We show first that T_0 is $(\infty, 0)$-recursive. Next we prove that for recursively regular κ,

$$(a, \boldsymbol{\mu}, \nu) \in \Omega_{\kappa\kappa} \leftrightarrow (\exists \lambda < \kappa)(\exists \sigma < \kappa).(a, \boldsymbol{\mu}, \nu) \in \Omega^\sigma_{\lambda\lambda}.$$

It follows that if we set

$$T(a, \langle \boldsymbol{\mu} \rangle, \upsilon) \leftrightarrow T_0(a, \langle \boldsymbol{\mu} \rangle, (\upsilon)_0, (\upsilon)_1, (\upsilon)_2, (\upsilon)_3),$$

then T has the required property.

Let H be an $(\infty, 0)$-recursive function such that for all k, κ and all $\mu_0, \ldots, \mu_{k-1} < \kappa$, $\langle \boldsymbol{\mu} \rangle < H(k, \kappa)$, say $H(k, \kappa) = \langle \kappa_0, \ldots, \kappa_{k-1} \rangle$ with all $\kappa_i = \kappa$.

2.4 Lemma. *T_0 is $(\infty, 0)$-recursive.*

Proof. We shall define an $(\infty, 0)$-recursive function F and prove by induction that for all σ,

$$(*) \qquad F(a, \langle \boldsymbol{\mu} \rangle, \nu, \kappa, \lambda, \sigma) = \begin{cases} 0, & \text{if } (a, \boldsymbol{\mu}, \nu) \in \Omega^{\sigma}_{\kappa\lambda}; \\ 1, & \text{otherwise.} \end{cases}$$

If $\sigma = 0$, we set

$$(0) \quad G(a, \langle \boldsymbol{\mu} \rangle, \nu, \kappa, \lambda, \sigma) = \begin{cases} 0, & \text{if } a = \langle 0, k, \ldots \rangle \text{ and } \{a\}_{\kappa\lambda}(\boldsymbol{\mu}) \simeq \nu; \\ 1, & \text{otherwise.} \end{cases}$$

If $\sigma > 0$, then F should satisfy the following conditions:

(1) $F(\langle 1, k, b, c_0, \ldots, c_{k'-1} \rangle, \langle \boldsymbol{\mu} \rangle, \nu, \kappa, \lambda, \sigma) = 0$ iff for some $\xi < H(k', \kappa)$,
(i) $(\forall i < k')(\exists \tau < \sigma) . F(c_i, \langle \boldsymbol{\mu} \rangle, (\xi)_i, \kappa, \lambda, \tau) = 0$,
and
(ii) $(\exists \tau < \sigma) . F(b, \xi, \nu, \kappa, \lambda, \tau) = 0$;

(2) $F(\langle 2, k+1 \rangle, \langle b, \boldsymbol{\mu} \rangle, \nu, \kappa, \lambda, \sigma) = 0$ iff

$$(\exists \tau < \sigma) F(b, \langle \boldsymbol{\mu} \rangle, \nu, \kappa, \lambda, \tau) = 0;$$

(3) $F(\langle 3, k+1, b \rangle, \langle \rho, \boldsymbol{\mu} \rangle, \nu, \kappa, \lambda, \sigma) = 0$ iff $\nu, \rho < \kappa$,
(i) $(\forall \pi < \rho)(\exists \xi < \nu)(\exists \tau < \sigma) . F(b, \langle \pi, \boldsymbol{\mu} \rangle, \xi, \kappa, \lambda, \tau) = 0$,
and
(ii) $\neg(\exists \nu' < \nu)(\forall \pi < \rho)(\exists \xi < \nu')(\exists \tau < \sigma) . F(b, \langle \pi, \boldsymbol{\mu} \rangle, \xi, \kappa, \lambda, \tau) = 0$;

(4) $F(\langle 4, k, b \rangle, \langle \boldsymbol{\mu} \rangle, \nu, \kappa, \lambda, \sigma) = 0$ iff $\nu < \lambda$,
(i) $(\forall \pi < \nu)(\exists \xi < \kappa)(\exists \tau < \sigma)[\xi > 0 \wedge F(b, \langle \pi, \boldsymbol{\mu} \rangle, \xi, \kappa, \lambda, \tau) = 0]$,
and
(ii) $(\exists \tau < \sigma) . F(b, \langle \nu, \boldsymbol{\mu} \rangle, 0, \kappa, \lambda, \tau) = 0$;

(5) in all other cases, $F(a, \langle \boldsymbol{\mu} \rangle, \nu, \kappa, \lambda, \sigma) = 1$.

It is clear that $(*)$ holds if $\sigma = 0$. We assume as induction hypothesis that $\sigma > 0$ and $(*)$ with σ replaced by τ holds for all $\tau < \sigma$, and prove that $(*)$ itself holds. If $(a, \boldsymbol{\mu}, \nu) \notin \Omega^{\sigma}_{\kappa\lambda}$ by virtue of being of the wrong form, then $(*)$ holds by (5); otherwise one of (1)–(4) applies. Suppose first that $a = \langle 1, k, b, c_0, \ldots, c_{k'-1} \rangle$. Then by the induction hypothesis

$$F(a, \langle \boldsymbol{\mu} \rangle, \nu, \kappa, \lambda, \sigma) = 0 \leftrightarrow \text{for some } \xi_0, \ldots, \xi_{k'-1},$$

(i) $(\forall i < k')(\exists \tau < \sigma).(c_i, \boldsymbol{\mu}, \xi_i) \in \Omega^{\tau}_{\kappa\lambda}$,
and
(ii) $(\exists \tau < \sigma).(b, \boldsymbol{\xi}, \nu) \in \Omega^{\tau}_{\kappa\lambda}$.

These are exactly the conditions which guarantee that $(a, \boldsymbol{\mu}, \nu) \in \Omega^{\sigma}_{\kappa\lambda}$. The other cases are similar. □

2.5 Lemma. *For any recursively regular κ and all a, $\boldsymbol{\mu}$, and ν,*

$$(a, \boldsymbol{\mu}, \nu) \in \Omega_{\kappa\kappa} \leftrightarrow (\exists \lambda < \kappa)(\exists \sigma < \kappa).(a, \boldsymbol{\mu}, \nu) \in \Omega^{\sigma}_{\lambda\lambda}.$$

Proof. The implication $(\leftarrow)$ is immediate from the monotonicity of $\Omega_{\kappa\lambda}$ with respect to κ and λ. For the implication $(\rightarrow)$ we show that

$$\{(a, \boldsymbol{\mu}, \nu) : (\exists \lambda < \kappa)(\exists \sigma < \kappa).(a, \boldsymbol{\mu}, \nu) \in \Omega^{\sigma}_{\lambda\lambda}\}$$

is closed under the clauses (0)–(4) which define the monotone operator whose closure is $\Omega_{\kappa\kappa}$. Since κ is a limit ordinal, this is trivial for clause (0).

(1) Suppose that for all $i < k'$, $(c_i, \boldsymbol{\mu}, \xi_i) \in \Omega^{\sigma_i}_{\lambda_i\lambda_i}$ and $(b, \boldsymbol{\xi}, \nu) \in \Omega^{\sigma_{k'}}_{\lambda_{k'}\lambda_{k'}}$ with all λ_i, $\sigma_i < \kappa$. Since κ is a limit ordinal, there exist λ and $\sigma < \kappa$ such that for all $i \leq k'$, $\lambda_i < \lambda$ and $\sigma_i < \sigma$. But then clearly, $(\langle 1, k, b, c_0, \ldots, c_{k'-1}\rangle, \boldsymbol{\mu}, \nu) \in \Omega^{\sigma}_{\lambda\lambda}$.

(2) If for some $\lambda, \sigma < \kappa$, $(b, \boldsymbol{\mu}, \nu) \in \Omega^{\sigma}_{\lambda\lambda}$, then $(\langle 2, k+1\rangle, b, \boldsymbol{\mu}, \nu) \in \Omega^{\sigma+1}_{\lambda\lambda}$ and $\sigma + 1 < \kappa$.

(3) Suppose that $a = \langle 3, k+1, b\rangle$ and

$$(\forall \pi < \rho)(\exists \lambda < \kappa)(\exists \sigma < \kappa)(\exists \xi < \lambda).(b, \pi, \boldsymbol{\mu}, \xi) \in \Omega^{\sigma}_{\lambda\lambda}.$$

Then

$$(\forall \pi < \rho)(\exists \zeta < \kappa)(\exists \lambda < \zeta)(\exists \sigma < \zeta)(\exists \xi < \lambda)\, T_0(b, \langle \pi, \boldsymbol{\mu}\rangle, \xi, \lambda, \lambda, \sigma).$$

By Theorem 2.4, the relation R defined by

$$R(\zeta, \pi, b, \boldsymbol{\mu}) \leftrightarrow (\exists \lambda < \zeta)(\exists \sigma < \zeta)(\exists \xi < \lambda)\, T_0(b, \langle \pi, \boldsymbol{\mu}\rangle, \xi, \lambda, \lambda, \sigma)$$

is $(\infty, 0)$-recursive, hence κ-recursive, and we have

$$(\forall \pi < \rho)(\exists \zeta < \kappa) R(\zeta, \pi, b, \boldsymbol{\mu}).$$

Let F be the κ-recursive function such that

$$F(\pi, b, \boldsymbol{\mu}) = \kappa\text{-least } \zeta\,.\, R(\zeta, \pi, b, \boldsymbol{\mu}).$$

Because κ is recursively regular,

$$\tau = \sup^{+}_{\pi<\rho} F(\pi, b, \boldsymbol{\mu}) < \kappa.$$

It follows that

$$(\forall \pi < \rho)(\exists \lambda < \tau)(\exists \sigma < \tau)(\exists \xi < \lambda).(b, \pi, \boldsymbol{\mu}, \xi) \in \Omega^{\sigma}_{\lambda\lambda}$$

and thus

$$(\forall \pi < \rho)(\exists \lambda < \tau)(\exists \xi < \lambda).(b, \pi, \boldsymbol{\mu}, \xi) \in \Omega^{(\tau)}_{\tau\tau},$$

hence

$$(a, \rho, \boldsymbol{\mu}, \nu) \in \Omega^{\tau}_{\tau\tau},$$

where

$$\nu = \sup^{+}_{\pi<\rho} \cdot \{b\}_{\kappa}(\pi, \boldsymbol{\mu}).$$

Clause (4) is treated similarly. □

2.6 Theorem. *For any recursively regular* κ *and all* a, $\boldsymbol{\mu}$, *and* ν,

$$\{a\}_{\kappa}(\boldsymbol{\mu}) \simeq \nu \leftrightarrow (\exists \upsilon < \kappa)[T(a, \langle\boldsymbol{\mu}\rangle, \upsilon) \wedge (\upsilon)_0 = \nu].$$

Proof. The implication ($\leftarrow$) is immediate. Suppose that $\{a\}_{\kappa}(\boldsymbol{\mu}) \simeq \nu$. Then $\nu < \kappa$ and by Lemma 2.5 there are $\sigma, \lambda < \kappa$ such that $(a, \boldsymbol{\mu}, \nu) \in \Omega^{\sigma}_{\lambda\lambda}$. If $\upsilon = \langle \nu, \lambda, \lambda, \sigma \rangle$, then the recursive regularity of κ implies that $\upsilon < \kappa$ and clearly $T(a, \langle\boldsymbol{\mu}\rangle, \upsilon)$ and $(\upsilon)_0 = \nu$. □

We have a similar representation for (∞, λ)- and ∞-partial recursive functions:

2.7 Theorem. *For all* λ, a, $\boldsymbol{\mu}$, *and* ν,

(i) $\{a\}_{\infty\lambda}(\boldsymbol{\mu}) \simeq \nu \leftrightarrow \exists \upsilon\, [T(a, \langle\boldsymbol{\mu}\rangle, \upsilon) \wedge (\upsilon)_0 = \nu \wedge (\upsilon)_2 = \lambda]$;

(ii) $\{a\}_{\infty}(\boldsymbol{\mu}) \simeq \nu \leftrightarrow \exists \upsilon\, [T(a, \langle\boldsymbol{\mu}\rangle, \upsilon) \wedge (\upsilon)_0 = \nu]$. □

The theory of κ-(partial)-recursive functions and κ-(semi)-recursive relations can now be developed according to the pattern of § II.4. We list the facts below but omit proofs in most cases. For the next few paragraphs, κ is a fixed recursively regular ordinal.

2.8 Normal Form Corollary. *For any κ-partial recursive function F, there exists an $a \in \omega$ such that*

$$F(\boldsymbol{\mu}) \simeq (\kappa\text{-least } \upsilon\, .\, T(a, \langle \boldsymbol{\mu} \rangle, \upsilon))_0. \quad \square$$

From Theorem 2.6 it follows that a relation R is κ-semi-recursive iff for some $a \in \omega$

$$R(\boldsymbol{\mu}) \leftrightarrow (\exists \upsilon < \kappa)\, T(a, \langle \boldsymbol{\mu} \rangle, \upsilon).$$

Equivalently, R is κ-semi-recursive iff for some κ-recursive (or even $(\infty, 0)$-recursive) relation S,

$$R(\boldsymbol{\mu}) \leftrightarrow (\exists \sigma < \kappa)\, S(\sigma, \boldsymbol{\mu}).$$

One proves just as in § II.4 that the class of κ-semi-recursive relations properly includes the class of κ-recursive relations and is closed under finite intersection, bounded universal quantification, and relational composition with κ-partial recursive functions.

It is instructive to examine an alternative proof for closure under bounded quantification (which is also available in ordinary recursion theory). Suppose that R is κ-semi-recursive and S is a κ-recursive relation such that

$$R(\rho, \boldsymbol{\mu}) \leftrightarrow (\exists \sigma < \kappa)\, S(\sigma, \rho, \boldsymbol{\mu}).$$

Then

$$(\forall \pi < \rho)\, R(\pi, \boldsymbol{\mu}) \leftrightarrow (\forall \pi < \rho)(\exists \sigma < \kappa)\, S(\sigma, \pi, \boldsymbol{\mu}).$$

Let

$$V(\tau, \rho, \boldsymbol{\mu}) \leftrightarrow (\forall \pi < \rho)(\exists \sigma < \tau)\, S(\sigma, \pi, \boldsymbol{\mu}).$$

Then V is κ-recursive and

$$(\forall \pi < \rho)\, R(\pi, \boldsymbol{\mu}) \leftrightarrow (\exists \tau < \kappa)\, V(\tau, \rho, \boldsymbol{\mu}).$$

The implication $(\leftarrow)$ is obvious. On the other hand, if $(\forall \pi < \rho)\, R(\pi, \boldsymbol{\mu})$, let

$$F(\pi, \boldsymbol{\mu}) \simeq \kappa\text{-least } \sigma\, .\, S(\sigma, \pi, \boldsymbol{\mu})$$

and

$$\tau = \sup\nolimits^{+}_{\pi < \rho} F(\pi, \boldsymbol{\mu}).$$

By recursive regularity, $\tau < \kappa$ and clearly $V(\tau, \rho, \boldsymbol{\mu})$. Thus $(\forall \pi < \rho) R(\pi, \boldsymbol{\mu})$ defines a κ-semi-recursive relation. The reader versed in axiomatic set theory will observe here a similarity with the axiom of replacement. This analogy forms the basis for many of the results of § 7 (cf. Lemma 7.5).

2.9 Selection Theorem. *There exists a κ-partial recursive function* Sel_κ *such that for all a and $\boldsymbol{\mu}$,*

(i) $\exists \pi . \{a\}_\kappa(\pi, \boldsymbol{\mu}) \downarrow$ *iff* $\mathrm{Sel}_\kappa(a, \langle \boldsymbol{\mu} \rangle) \downarrow$, *and*

(ii) *if* $\exists \pi . \{a\}_\kappa(\pi, \boldsymbol{\mu}) \downarrow$, *then* $\{a\}_\kappa(\mathrm{Sel}_\kappa(a, \langle \boldsymbol{\mu} \rangle), \boldsymbol{\mu}) \downarrow$. □

2.10 Corollary. *For any κ-semi-recursive relation R, there exists a κ-partial recursive function* Sel_R *such that for all $\boldsymbol{\mu}$,*

$$\exists \pi R(\pi, \boldsymbol{\mu}) \leftrightarrow R(\mathrm{Sel}_R(\boldsymbol{\mu}), \boldsymbol{\mu}) \leftrightarrow \mathrm{Sel}_R(\boldsymbol{\mu}) \downarrow . \quad \square$$

2.11 Corollary. *The class of κ-semi-recursive relations is closed under bounded quantification, finite union, and $\exists^0_\kappa$ (existential quantification over κ).* □

2.12 Corollary. *The class of κ-partial recursive functions and κ-semi-recursive relations is closed under definition by positive cases.* □

2.13 Corollary. *A relation is κ-recursive iff it is both κ-semi-recursive and κ-co-semi-recursive (complement of a κ-semi-recursive relation).* □

2.14 Corollary. *For any partial function $F: {}^k\kappa \to \kappa$, F is κ-partial recursive iff its graph Gr_F is κ-semi-recursive and F is κ-recursive iff F is total and Gr_F is κ-recursive.* □

2.15 Corollary. *For any non-empty set $A \subseteq \kappa$, the following are equivalent:*

(i) *A is κ-semi-recursive;*

(ii) *$A = \mathrm{Im}(F)$ for some κ-recursive function F;*

(iii) *$A = \mathrm{Im}(F)$ for some κ-partial recursive function F.* □

2.16 Corollary. (i) *The class of κ-semi-recursive relations has the reduction property but not the separation property;*

(ii) *the class of κ-co-semi-recursive relations has the separation property but not the reduction property.* □

If $\lambda < \kappa$ is also recursively regular, one would in general expect that $\{a\}_\lambda \subsetneqq \{a\}_\kappa$. Hence, although every λ-partial recursive function has natural κ-partial recursive extensions, it is by no means obvious that every λ-partial recursive function is itself κ-partial recursive. Indeed, this cannot always be so. Since every regular cardinal is recursively regular, there exists a recursively regular ordinal κ such that there are uncountably many recursively regular $\lambda < \kappa$ (by the results of §3 this is true for $\kappa = \aleph_1$). For each such λ, let F_λ be the λ-recursive function with constant value 0 (and domain λ). Since there are uncountably many such F_λ's, not all can be κ-partial recursive.

2.17 Theorem. *For any recursively regular $\lambda < \kappa$, if $\{\sigma : \sigma < \lambda\}$ is κ-recursive, then every λ-partial recursive function is also κ-partial recursive and every λ-semi-recursive relation is κ-recursive.*

Proof. Under the condition given, for any λ-partial recursive G, if F is defined by:

$$F(\boldsymbol{\mu}) \simeq \begin{cases} G(\boldsymbol{\mu}), & \text{if } \mu_0, \ldots, \mu_{k-1} < \lambda; \\ \text{undefined}, & \text{otherwise}; \end{cases}$$

then $F = G$ and F is κ-partial recursive.

The constant function with value λ is κ-recursive since

$$\lambda = \kappa\text{-least } \tau \,.\, \tau \not< \lambda.$$

Then if R is any λ-semi-recursive relation there is an $a < \omega$ such that

$$R(\boldsymbol{\mu}) \leftrightarrow (\exists \upsilon < \lambda)\, T(a, \langle \boldsymbol{\mu} \rangle, \upsilon).$$

With respect to κ, the quantifier is bounded and the relation T is $(\infty, 0)$-recursive, so R is κ-recursive. □

We conclude this section with another characterization of the recursively regular ordinals.

2.18 Theorem. *For any κ, the following are equivalent:*

(i) *κ is recursively regular;*
(ii) $\Omega_{\kappa\kappa} = \bigcup\{\Omega^{\sigma}_{\lambda\lambda} : \lambda, \sigma < \kappa\}$;
(iii) $\Omega_{\kappa\kappa} = \Omega^{(\kappa)}_{\kappa\kappa}$.

Proof. That (i)→(ii) is Lemma 2.5. That (ii)→(iii) is obvious. Suppose that κ is *not* recursively regular. Then by Lemma 2.2 there exists a κ-partial recursive function F and ordinals ρ and $\boldsymbol{\mu} < \kappa$ such that $\sup^{+}_{\pi<\rho} F(\pi, \boldsymbol{\mu}) = \kappa$. Let G be the κ-partial recursive function defined by

$$G(e, \sigma) \simeq 0 \cdot \sup\nolimits^{+}_{\tau<\sigma} \{e\}_\kappa(\tau).$$

By the Recursion Theorem applied to a natural index for G, there exists an index $\bar{e}$ such that

$$G(\bar{e}, \sigma) \simeq \{\bar{e}\}_\kappa(\sigma)$$

and for all $\sigma < \kappa$, $(\bar{e}, \sigma, 0) \in \Omega_{\kappa\kappa} \sim \Omega_{\kappa\kappa}^{(\sigma)}$ — that is, $\{\bar{e}\}_\kappa(\sigma)$ requires at least σ steps to compute its value 0. Let a be a natural index such that

$$\{a\}_\kappa(\tau, \mu) \simeq 0 \cdot \sup^+_{\pi<\tau}\{\bar{e}\}_\kappa(F(\pi, \mu)).$$

Then $(a, \rho, \mu, 0) \in \Omega_{\kappa\kappa} \sim \Omega_{\kappa\kappa}^{(\kappa)}$, so condition (iii) fails. □

2.19 Corollary. *$\{\kappa : \kappa$ is recursively regular$\}$ is $(\infty, 0)$ recursive.*

Proof. Let H be the $(\infty, 0)$-recursive function defined just preceding Lemma 2.4 with the property that if $\mu \in {}^k\kappa$, then $\langle\mu\rangle < H(k, \kappa)$ and let $G(\kappa) = \sup^+_{k<\omega} H(k, \kappa)$. Then G is $(\infty, 0)$-recursive and for any κ, κ is recursively regular iff

$$(\forall a < \omega)(\forall \zeta < G(\kappa))(\forall \nu < \kappa)$$
$$[T_0(a, \zeta, \nu, \kappa, \kappa, \kappa) \leftrightarrow (\exists \sigma < \kappa)\, T_0(a, \zeta, \nu, \kappa, \kappa, \sigma)].$$ □

2.20–2.25 Exercises

2.20. Is condition (b) changed if the word "partial" is removed?

2.21. Show that for recursively regular κ, the class of κ-partial recursive functions is the smallest class of partial functions from κ to κ which contains the restriction of every $(\infty, 0)$-partial recursive function and is closed under composition and κ-search.

2.22. Show that if $\kappa > \omega$ is recursively regular and $R \subseteq {}^k\omega$ is κ-recursive, then R is also λ-recursive for some $\lambda < \kappa$.

2.23. Show that κ is recursively regular iff for every $F: \kappa \to \kappa$ which is κ-recursive in parameters, there exists a $\lambda < \kappa$ which is closed under F — that is, $(\forall \pi < \lambda) . F(\pi) < \lambda$.

2.24. A subset A of κ is κ-*finite* iff A is κ-recursive and for some $\rho < \kappa$, $A \subseteq \rho$. Show that κ is recursively regular iff the image of any κ-finite set under a κ-recursive function is κ-finite.

2.25. Under what conditions is the class of κ-(semi-) recursive relations closed under $\exists^0_\lambda$?

2.26 Notes. The term "recursively regular" is due to Platek [1966], who viewed the concept as a weakening of regularity. Kripke [1964] uses the term "admissible" for the same notion, presumably to indicate that the ordinals so named admit a recursion theory, and this term has become more widely used in the literature. We have used "recursively regular" both because it seems to us a more descriptive term and to save "admissible" for a different but closely related meaning in § 7.

The reader may be wondering why we have defined only recursive ordinal *functions* rather than *functionals.* Certainly there is no difficulty in adjusting the definition of $\Omega_{\kappa\lambda}$ to allow for functions from κ to κ as arguments and most of § 1 would go through unchanged. Things become more complicated, however, when we come to discuss recursive regularity. It is not hard to see that the only ordinals κ which would satisfy the condition corresponding to Definition 2.1:

$$\forall a\, (\forall \mu < \kappa)(\forall \mathbf{f} : \kappa \to \kappa)\forall \nu\, [\{a\}_{\infty\kappa}(\mu, \mathbf{f}) \simeq \nu \to \nu < \kappa]$$

are the regular cardinals. All would be well if the f were restricted to be κ-recursive, but then they could be replaced by their κ-indices, and it is not clear what would be gained. Recursive ordinal functionals are useful for some purposes but not for any in this book.

3. Ordinal Recursion and the Analytical Hierarchy

To this point our only examples of recursively regular ordinals are the infinite regular cardinals. The first goal of this section is to establish the recursive regularity of many countable ordinals (uncountably many, in fact). The existence of many recursively regular ordinals of each uncountable power will be established in § 5. Our other objective here is to explore the intimate connection between ω_1-recursion and the classes Π^1_1 and Δ^1_1, on the one hand, and between $\aleph_1$-recursion and the classes Σ^1_2 and Δ^1_2, on the other.

Recall that we observed following Lemma 1.6 that for any $\kappa > \omega$, the constant function with value ω is κ-recursive. It follows that operators such as $\overset{+}{\sup}_{\pi<\omega}$ and $(\exists \pi < \omega)$ may be freely used in defining κ-partial recursive functions and κ-recursive relations.

3.1 Lemma. *For any recursively regular* $\kappa > \omega$ *and any* $\gamma \in \mathrm{W}$, *if* γ *is* κ*-recursive, then* $\|\gamma\| < \kappa$.

Proof. Suppose $\kappa > \omega$ is recursively regular and there exists $\gamma \in \mathrm{W}$ such that γ is κ-recursive and $\|\gamma\| \geqslant \kappa$. Without loss of generality we may assume $\|\gamma\| = \kappa$. By use of the κ-Recursion Theorem we may define a κ-partial recursive function F such that for all $p \in \mathrm{Fld}(\gamma)$,

$$F(p) \simeq \kappa\text{-least } \pi . (\forall q < \omega)[q <_\gamma p \to F(q) < \pi],$$

and otherwise, $F(p) = 0$. For all $p \in \mathrm{Fld}(\gamma)$, $|p|_\gamma < \kappa$ and by induction on $|p|_\gamma$ it follows that $F(p)$ is defined for all $p < \omega$. But then by the recursive regularity of κ,

$$\kappa = \|\gamma\| = \sup^+\{|p|_\gamma : p \in \mathrm{Fld}(\gamma)\} = \sup^+_{p<\omega} F(p) < \kappa,$$

a contradiction. □

It follows immediately that no ordinal κ such that $\omega < \kappa < \omega_1$ is recursively regular, as for any such κ there exists an (ordinary) recursive function $\gamma \in \mathrm{W}$ such that $\|\gamma\| = \kappa$. γ is also κ-recursive and if κ were recursively regular, this would contradict Lemma 3.1.

We show next that ω_1 is recursively regular. We shall code ω_1-computations using elements of W as notations for the ordinals less than ω_1. The boundedness property (IV.2.1 (iii)) is the main tool for completing the proof.

Recall that for $m \in W$, $\|m\|$ is the order-type of the well-ordering $\leqslant_{\{m\}} = \{(p, q) : \{m\}(\langle p, q\rangle) = 0\}$, and for $m \notin W$, $\|m\| = \aleph_1$. There are Σ^1_1 relations $\leqslant_\Sigma$ and $<_\Sigma$ and Π^1_1 relations $\leqslant_\Pi$ and $<_\Pi$ such that whenever at least one of m and n belongs to W,

$$m \leqslant_\Sigma n \leftrightarrow \|m\| \leqslant \|n\| \leftrightarrow m \leqslant_\Pi n,$$

and

$$m <_\Sigma n \leftrightarrow \|m\| < \|n\| \leftrightarrow m <_\Pi n.$$

We shall write $\|\mathbf{m}\|$ for $(\|m_0\|, \ldots, \|m_{k-1}\|)$. Let

$$\Omega_W = \{\langle a, \mathbf{m}, n\rangle : \mathbf{m}, n \in W \wedge \{a\}_{\omega_1}(\|\mathbf{m}\|) \simeq \|n\|\}.$$

3.2 Lemma. $\Omega_W \in \Pi^1_1$.

Proof. We shall indicate how to give a direct Π^1_1 monotone inductive characterization of Ω_W — that is, we define a Π^1_1 monotone operator Γ such that $\bar{\Gamma} = \Omega_W$. It follows by Theorem II.3.1 that $\Omega_W \in \Pi^1_1$. The clauses which define Γ are translations of those which occur in the inductive definition of $\Omega_{\omega_1\omega_1}$ with ordinals less than ω_1 (the only ordinals that occur) replaced by their codes from W, and inequalities among these ordinals interpreted by $\leqslant_\Sigma$, etc.

Thus, Ω_W is the smallest set such that $\langle a, \mathbf{m}, n\rangle \notin \Omega_W$ unless $\mathbf{m}$ and n beong to W and for all k, n, $i < k$, and all $m_0, \ldots, m_{k-1}$, p, r, s, and $t \in W$,

(0) $\langle\langle 0, k, 0, n\rangle, \mathbf{m}, n'\rangle \in \Omega_W$ for all n' such that $\|n'\| = n$;

$\langle\langle 0, k, 1, i\rangle, \mathbf{m}, n\rangle \in \Omega_W$ for all n such that $n \preccurlyeq_\Pi m_i$ and $m_i \preccurlyeq_\Pi n$;
$\langle\langle 0, k, 2, i\rangle, \mathbf{m}, n\rangle \in \Omega_W$ for all n such that $m_i <_\Pi n$ and $\neg\exists q\,[m_i <_\Sigma q \wedge q <_\Sigma n]$;

etc.;

(1)
Similarly;
(2)

(3) if $\forall p\,[p <_\Sigma r \rightarrow \exists q\,(q <_\Pi n \wedge \langle b, p, \mathbf{m}, q\rangle \in \Omega_W)]$,
and $\neg\exists n'(n' <_\Sigma n \wedge \forall p\,[p <_\Pi r \rightarrow \exists q\,(q <_\Sigma n' \wedge \langle b, p, \mathbf{m}, q\rangle \in \Omega_W)])$,
then $\langle\langle 3, k+1, b\rangle, r, \mathbf{m}, n\rangle \in \Omega_W$;

(4) if $\forall p\,[p <_\Sigma n \rightarrow \exists q\,(\|q\| > 0 \wedge \langle b, p, \mathbf{m}, q\rangle \in \Omega_W)]$,
and $\exists q\,[\|q\| = 0 \wedge \langle b, n, \mathbf{m}, q\rangle \in \Omega_W$, then $\langle\langle 4, k, b\rangle, \mathbf{m}, n\rangle \in \Omega_W$.

We leave it to the reader to verify that this is in fact a Π^1_1 monotone inductive definition of Ω_W. □

3.3 Theorem. *ω_1 is the least recursively regular ordinal greater than ω.*

Proof. If remains to show that ω_1 is recursively regular; we use criterion (b) of Theorem 2.2. Let F be any ω_1-partial recursive function and ρ and $\boldsymbol{\mu} < \omega_1$ ordinals such that for all $\pi < \rho$, $F(\pi, \boldsymbol{\mu})$ is defined. We aim to show that $\sup^+_{\pi<\rho} F(\pi, \boldsymbol{\mu}) < \omega_1$. Let

$$A = \{n : (\exists \pi < \rho)\, F(\pi, \boldsymbol{\mu}) = \|n\|\}.$$

Since $\sup^+_{\pi<\rho} F(\pi, \boldsymbol{\mu}) = \sup^+\{\|n\| : n \in A\}$, it would suffice, by the Boundedness Theorem IV.2.1(iii) to show that $A \in \Sigma^1_1$. This is not in general true, but we shall find a Σ^1_1 subset $B \subseteq A$ such that

$$(*) \qquad \sup^+\{\|n\| : n \in A\} = \sup^+\{\|n\| : n \in B\}.$$

Let a be an ω_1-index for F and let r and $\mathbf{m}$ be fixed notations in W for ρ and $\boldsymbol{\mu}$, respectively. Let R be the Π^1_1 relation defined by

$$R(n, p) \leftrightarrow \langle a, p, \mathbf{m}, n\rangle \in \Omega_W.$$

Let Sel_R be the selection function with Π^1_1 graph given by Lemma IV.2.5 and set

$$B = \{n : \exists p\,[p <_\Sigma r \wedge \mathrm{Sel}_R(p) = n]\}.$$

Since by the hypothesis, for each $p <_\Sigma r$ there exist n such that $R(n, p)$, also

$$B = \{n : \exists p\,[p <_\Sigma r \wedge \forall n'(n' \neq n \rightarrow \mathrm{Sel}_R(p) \neq n')]\}.$$

Hence $B \in \Sigma_1^1$. That $(*)$ holds is clear; B contains a unique notation for each ordinal with a notation in A. □

3.4 Theorem. *For all k and all $R \subseteq {}^k\omega$,*
(i) *R is ω_1-semi-recursive* $\leftrightarrow R \in \Pi_1^1$;
(ii) *R is ω_1-recursive* $\leftrightarrow R \in \Delta_1^1$.

Proof. Suppose first that R is ω_1-semi-recursive, say with ω_1-semi-index a. It is easy to define a recursive function F such that for all m, $F(m) \in W$ and $\|F(m)\| = m$ (cf. Exercise IV.1.20). Then

$$R(\mathbf{m}) \leftrightarrow \exists n. \langle a, F(m_0), \ldots, F(m_{k-1}), n\rangle \in \Omega_W.$$

Hence by Lemma 3.2, $R \in \Pi_1^1$.

For the implication ($\leftarrow$) of (i), suppose that R is a Π_1^1 relation. By Theorem III.3.2 there exists a Π_1^0 monotone operator Γ such that $R \ll \bar{\Gamma}$. It suffices to show that $\bar{\Gamma}$ is ω_1-semi-recursive.

Let S be a recursive relation such that for all m and A,

$$m \in \Gamma(A) \leftrightarrow \forall p\, S(m, \bar{K}_A(p)).$$

Consider the following pair of recursively defined functions:

$$G_{(\ )}(m, 0) = 1; \quad \text{for} \quad \sigma > 0,$$

$$G_{(\ )}(m,\sigma) = \begin{cases} 0, & \text{if} \quad (\exists \pi < \sigma)\, G(m,\pi) = 0; \\ 1, & \text{otherwise;} \end{cases}$$

$$G(m,\sigma) = \begin{cases} 0, & \text{if} \quad (\forall p < \omega)\, S(m, \bar{G}_{(\ )}(p,\sigma)); \\ 1, & \text{otherwise.} \end{cases}$$

It is straightforward to prove by induction on σ that for all m,

$$G_{(\ )}(m,\sigma) = 0 \leftrightarrow m \in \Gamma^{(\sigma)},$$

and

$$G(m,\sigma) = 0 \leftrightarrow m \in \Gamma^{\sigma}.$$

We shall prove below that $G_{(\ )}$ and G are ω_1-recursive. Then since by Theorem IV.2.15, $|\Gamma| \leqslant \omega_1$, we have

$$m \in \bar{\Gamma} \leftrightarrow (\exists \sigma < \omega_1)\, G(m,\sigma) = 0.$$

By the remarks following Corollary 2.8, $\bar{\Gamma}$ is ω_1-semi-recursive.

To see that $G_{(\ )}$ and G are ω_1-recursive, let H be defined by:

$$H(e, \rho, m, \sigma) \simeq \begin{cases} 1, \text{ if } \rho = \sigma = 0; \\ \inf_{\pi<\sigma} \{e\}_{\omega_1}(1, m, \pi), \text{ if } \rho = 0 < \sigma; \\ \sup_{\pi<\omega} K_S(m, \langle\{e\}_{\omega_1}(0, 0, \sigma), \ldots, \{e\}_{\omega_1}(0, \pi - 1, \sigma)\rangle), \\ \hfill \text{if } \rho > 0. \end{cases}$$

Clearly H is ω_1-partial recursive, so by the Recursion Theorem there exists an $\bar{e}$ such that

$$\{\bar{e}\}_{\omega_1}(\rho, m, \sigma) \simeq H(\bar{e}, \rho, m, \sigma).$$

We leave to the reader the easy verification that

$$G_{(\ )}(m, \sigma) \simeq \{\bar{e}\}_{\omega_1}(0, m, \sigma),$$

and

$$G(m, \sigma) \simeq \{\bar{e}\}_{\omega_1}(1, m, \sigma).$$

For (ii), we have by Corollary 2.13, Theorem 3.3, and the fact that $\{\pi : \pi < \omega\}$ is ω_1-recursive:

$$\begin{aligned} R \text{ is } \omega_1\text{-recursve} &\leftrightarrow R \text{ and } {}^k\omega \sim R \text{ are } \omega_1\text{-semi-recursive} \\ &\leftrightarrow R \text{ and } {}^k\omega \sim R \text{ are } \Pi_1^1 \\ &\leftrightarrow R \in \Delta_1^1. \ \square \end{aligned}$$

It is interesting to compare the two recursion-theoretic characterizations of the classes of Π_1^1 and Δ_1^1 relations: (semi-) recursive in E and ω_1-(semi-) recursive. The image of any function $F : \omega \to \omega$ which is recursive in E is again recursive in E because Δ_1^1 is closed under existential number quantification ($\exists_\omega^0$). However, by Corollary 2.15 and the preceding theorem, every Π_1^1 set of numbers is the image of some ω_1-recursive function $G : \omega_1 \to \omega$. The difference, of course, lies in the domains of the functions; G has a "longer time" to enumerate its image. A Π_1^1 set is generable in ω_1 steps in any of several ways — via inductive definitions, reductions to recursive well-orderings, or ω_1-semi-recursion — and it is not surprising that it can be "effectively" enumerated in a sequence of length ω_1 but not, in general, in one of length ω. In fact, in any natural sense of effective enumeration, a set enumerable in fewer than ω_1 steps is necessarily Δ_1^1.

Another striking contrast arises in the theory of relative ω_1-recursion. As this topic is beyond the scope of this book, we shall merely sketch the facts and refer the reader to Shore [1977] or Simpson [1984?] for more details. Recall first the following two facts:

(1) (Friedberg–Mučnik Theorem (§ II.5)) there exist semi-recursive sets A and $B \subseteq \omega$ such that neither A is recursive in B nor B is recursive in A;

(2) (Theorem IV.2.12) for every two Π_1^1 sets A and $B \subseteq \omega$, either A is Δ_1^1 (hyperarithmetic) in B or B is Δ_1^1 in A.

Combining (2) with a relativized version of Corollary IV.1.9, we have also

(3) for every two sets A and $B \subseteq \omega$ which are semi-recursive in E, either A is recursive in B and E or B is recursive in A and E.

There are several possible generalizations of "recursive in" to "ω_1-recursive in", but we shall mention only two. The most natural in the present context arises by modifying the definition of $\Omega_{\omega_1\omega_1}$ to allow for the introduction of a parameter. For any $F: \omega_1 \to \omega_1$, let $\Omega_{\omega_1\omega_1}[F]$ be the smallest set satisfying clauses (0)–(4) as in the definition of $\Omega_{\omega_1\omega_1}$ and such that for all $\boldsymbol{\mu} \in {}^k\omega_1$,

$$(\langle 0, k, 3, i\rangle, \boldsymbol{\mu}, F(\mu_i)) \in \Omega_{\omega_1\omega_1}[F].$$

We then set $\{a\}_{\omega_1}^F(\boldsymbol{\mu}) \simeq \nu$ iff $(a, \boldsymbol{\mu}, \nu) \in \Omega_{\omega_1\omega_1}[F]$ and define the other notions accordingly. We say that A is *weakly ω_1-computable* from B iff for some a, $K_A = \{a\}_{\omega_1}^{K_B}$. From the foregoing it is not hard to deduce that for any A and $B \subseteq \omega$,

$$A \text{ is weakly } \omega_1\text{-computable from } B \leftrightarrow A \text{ is } \Delta_1^1 \text{ in } B.$$

Hence by (3) there is no pair of ω_1-semi-recursive sets A and $B \subseteq \omega$ such that neither is weakly ω_1-computable from the other.

In general, the inductive definition of $\Omega_{\omega_1\omega_1}[F]$ will require more than ω_1 stages, as there is no reason to expect that ω_1 will be recursively-in-F regular — that is, closed under all functions (∞, ω_1)-partial recursive in F. One natural restriction on the reduction of A to B is to require that each computation of $K_A(\mu)$ from B take fewer than ω_1 steps. Accordingly, we set

$$[a]_{\omega_1}^F(\boldsymbol{\mu}) \simeq \nu \quad \text{iff} \quad (a, \boldsymbol{\mu}, \nu) \in \Omega_{\omega_1\omega_1}^{(\omega_1)}[F]$$

and say that A is *ω_1-computable* from B iff for some a, $K_A = [a]_{\omega_1}^F$. It is proved in Sacks [1971] that

(4) there exist ω_1-semi-recursive sets A and $B \subseteq \omega$ such that neither A is ω_1-computable from B nor B is ω_1-computable from A.

The intuitive explanation for the contrast between (3) and (4) is an extension of that concerning enumerability discussed above. If $A \subseteq \omega$ is Π_1^1 but not Δ_1^1, then by Theorems IV.2.11–12, $\omega_1[A] > \omega_1$. Just as any set of numbers which is "effectively" enumerable in fewer than ω_1 steps is Δ_1^1, so any set enumerable in fewer than $\omega_1[A]$ steps is Δ_1^1 in A. Every Π_1^1 set is enumerable in ω_1 steps and is thus Δ_1^1 in A. The restriction in (4) to computations of length less than ω_1 eliminates this possibility and permits the construction of incomparable sets.

If one relaxes the requirement that A and B be subsets of ω, then the results are completely analogous to (1):

(5) there exist ω_1-semi-recursive sets A and $B \subseteq \omega_1$ such that neither A is weakly ω_1-computable from B nor B is weakly ω_1-computable from A. Analogous results hold for every recursively regular κ in place of ω_1. A proof may be found in Sacks–Simpson [1972].

Although we have formally established only the existence of one recursively regular ordinal between ω and $\aleph_1$, we need almost no further work to show that there are in fact uncountably many such ordinals, namely the ordinals $\omega_1[\beta]$ for arbitrary $\beta \in {}^\omega\omega$ (cf. Exercise 3.12). However it is not in general true that

$$R \text{ is } \omega_1[\beta]\text{-recursive} \leftrightarrow R \in \Delta_1^1[\beta].$$

To see this, note that there are only countably many β which are even ∞-recursive. If β is not ∞-recursive, then $\beta \in \Delta_1^1[\beta]$ but β is not $\omega_1[\beta]$-recursive. The failure of the corresponding fact for semi-recursive relations will be shown in § 5.

We turn now to the second level of the analytical hierarchy. The relevant ordinal κ here is not, as might be expected, the next recursively regular ordinal greater than ω_1, but $\aleph_1$, the first *regular* ordinal greater than ω. In fact, however, we shall see in § 5 that for relations and functions on numbers, the same results are obtained ith $\kappa = \delta_2^1$, the least ordinal not the order-type of a Δ_2^1 well-ordering of ω.

For any $\gamma \in \mathrm{W}$, let

$$\Omega_\gamma = \{\langle a, \mathbf{m}, n\rangle : \{a\}_{\|\gamma\|}(|m_0|_\gamma, \ldots, |m_{k-1}|_\gamma) \simeq |n|_\gamma\},$$

and

$$O(u, \gamma) \leftrightarrow \gamma \in \mathrm{W} \wedge u \in \Omega_\gamma.$$

3.5 Lemma. *O is Π_1^1.*

Proof. For each $\gamma \in \mathrm{W}$, Ω_γ is the closure $\bar{\Gamma}_\gamma$ of a monotone Π_1^1 operator. Γ_γ is defined as in the proof of Lemma 3.2 with the relations $\leqslant_\Sigma$ and $\leqslant_\Pi$ both replaced by $\leqslant_\gamma$. For $\gamma \notin \mathrm{W}$, set $\Gamma_\gamma(A) = \varnothing$ for all A. Then O is inductively defined by the family $\{\Gamma_\gamma : \gamma \in {}^\omega\omega\}$ and is thus Π_1^1 by Theorem III.3.15. □

3.6 Lemma. *For any R and S, if S is $\aleph_1$-recursive and for all ρ and $\boldsymbol{\mu}$,*

$$R(\rho, \boldsymbol{\mu}) \leftrightarrow \exists g\, (\forall p < \omega)[g(p) < \rho \wedge S(\bar{g}(p), \boldsymbol{\mu})],$$

then also R is $\aleph_1$-recursive.

Proof. For each ρ, let Γ_ρ be the monotone operator defined by:

$$\langle \boldsymbol{\mu}, \sigma\rangle \in \Gamma_\rho(B) \leftrightarrow \sigma < \bar{\rho} \wedge [\sim S(\sigma, \boldsymbol{\mu}) \vee (\forall \xi < \rho)\langle \boldsymbol{\mu}, \sigma * \langle \xi\rangle\rangle \in B],$$

where for each ρ, $\bar{\rho}$ is an ordinal larger than all $\langle \boldsymbol{\pi}\rangle$ with all $\pi_i < \rho$ ($\bar{\rho} = \overset{+}{\sup}_{k<\omega}\langle \rho_0, \ldots, \rho_k\rangle$ with all $\rho_j = \rho$). Exactly as in the proof of Theorem III.3.2 one may show that

$$\langle \boldsymbol{\mu}, \sigma\rangle \in \bar{\Gamma}_\rho \leftrightarrow \forall g\, (\exists p < \omega)[\rho \leq g(p) \vee \sim S(\sigma * \bar{g}(p), \boldsymbol{\mu})],$$

and hence that

$$R(\rho, \boldsymbol{\mu}) \leftrightarrow \langle \boldsymbol{\mu}, \langle\ \rangle\rangle \notin \bar{\Gamma}_\rho.$$

Let

$$P\,(\boldsymbol{\mu}, \sigma, \tau, \rho) \leftrightarrow \langle \boldsymbol{\mu}, \sigma\rangle \in \Gamma_\rho^\tau.$$

It is straightforward to verify as in the proof of Theorem 3.4 that P is $\aleph_1$-recursive, in fact, $(\infty, 0)$-recursive. Let

$$\begin{aligned} F(\rho, \boldsymbol{\mu}) &\simeq \aleph_1\text{-least } \tau(\forall \sigma < \bar{\rho})[\langle \boldsymbol{\mu}, \sigma\rangle \in \Gamma_\rho^\tau \leftrightarrow \langle \boldsymbol{\mu}, \sigma\rangle \in \Gamma_\rho^{(\tau)}] \\ &\simeq \aleph_1\text{-least } \tau(\forall \sigma < \bar{\rho})[P(\boldsymbol{\mu}, \sigma, \tau, \rho) \leftrightarrow (\exists \zeta < \tau) P(\boldsymbol{\mu}, \sigma, \zeta, \rho)]. \end{aligned}$$

Clearly F is $\aleph_1$-partial recursive. For a fixed $\boldsymbol{\mu}$ and ρ, Γ_ρ may be thought of as an operator over the countable set $\bar{\rho}$ and as such has a countable closure ordinal. In particular, for each $\sigma < \bar{\rho}$, there is a $\tau < \aleph_1$ such that

$$\langle \boldsymbol{\mu}, \sigma\rangle \in \Gamma_\rho^\tau \leftrightarrow \langle \boldsymbol{\mu}, \sigma\rangle \in \Gamma_\rho^{(\tau)}.$$

Since $\bar{\rho} < \aleph_1$ we have

$$F(\rho, \boldsymbol{\mu}) \simeq \sup_{\sigma < \bar{\rho}}\, (\aleph_1\text{-least } \tau)[\langle \boldsymbol{\mu}, \sigma\rangle \in \Gamma_\rho^\tau \leftrightarrow \langle \boldsymbol{\mu}, \sigma\rangle \in \Gamma_\rho^{(\tau)}]$$

and thus F is total, hence $\aleph_1$-recursive.

Finally,

$$\langle \boldsymbol{\mu}, \sigma\rangle \in \bar{\Gamma}_\rho \leftrightarrow \langle \boldsymbol{\mu}, \sigma\rangle \in \Gamma_\rho^{F(\rho, \boldsymbol{\mu})}$$

so that

$$R(\rho, \mu) \leftrightarrow \sim P(\mu, \langle\ \ \rangle, F(\rho, \mu), \rho).$$

Thus R is $\aleph_1$-recursive. □

3.7 Theorem. *For all k and all $R \subseteq {}^k\omega$,*
(i) *R is $\aleph_1$-semi-recursive $\leftrightarrow R \in \Sigma^1_2$;*
(ii) *R is $\aleph_1$-recursive $\leftrightarrow R \in \Delta^1_2$.*

Proof. Part (ii) follows immediately from (i) and the fact that $\aleph_1$ is recursively regular. Suppose first that R is $\aleph_1$-semi-recursive, say with $\aleph_1$-semi-index a. It is easy to define an arithmetical functional F such that for $\gamma \in \mathsf{W}$, $\mathsf{F}(m, \gamma)$ is defined for all m and $|\mathsf{F}(m, \gamma)|_\gamma = m$. Then

$$\begin{aligned}
R(\mathbf{m}) &\leftrightarrow \exists \nu . \{a\}_{\aleph_1}(\mathbf{m}) \simeq \nu \\
&\leftrightarrow \exists \nu\, (\exists \kappa < \aleph_1) . \{a\}_\kappa(\mathbf{m}) \simeq \nu \\
&\leftrightarrow \exists \nu\, (\exists \gamma \in \mathsf{W}) . \{a\}_{\|\gamma\|}(\mathbf{m}) \simeq \nu \\
&\leftrightarrow \exists \gamma\, [\gamma \in \mathsf{W} \wedge \exists n . \langle a, \mathsf{F}(m_0, \gamma), \ldots, \mathsf{F}(m_{k-1}, \gamma), n\rangle \in \Omega_\gamma].
\end{aligned}$$

The second equivalence follows from Lemma 2.5 and the recursive regularity of $\aleph_1$. R is Σ^1_2 by Lemma 3.5.

For the converse, suppose $R \in \Sigma^1_2$, say

$$R(\mathbf{m}) \leftrightarrow \exists \alpha\, \forall \beta\, \exists n\, P(\bar{\alpha}(n), \bar{\beta}(n), \mathbf{m})$$

for some recursive relation P. As in the proof of Theorem IV.1.1, for each $\mathbf{m}$ and α, there exists a linear ordering $\trianglelefteq^P_{\mathbf{m},\alpha}$ such that

$$R(\mathbf{m}) \leftrightarrow \exists \alpha\, [\trianglelefteq^P_{\mathbf{m},\alpha} \text{ is a well-ordering}].$$

Of course, $\trianglelefteq^P_{\mathbf{m},\alpha}$ is a countable ordering, hence is well-ordered just in case it is isomorphic to the usual $\leqslant$ on some ordinal $\pi < \aleph_1$. Hence

$$R(\mathbf{m}) \leftrightarrow \exists \alpha\, (\exists \pi < \aleph_1)\, \exists f\, \forall s\, \forall t\, [f(t) < \pi \wedge s \triangleleft^P_{\mathbf{m},\alpha} t \leftrightarrow f(s) < f(t)].$$

For s and t less than any given p, the relation in brackets depends only on values of α and f for arguments less than p. Hence, there exist $(\infty, 0)$-recursive relations U and V such that

$$\begin{aligned}
R(\mathbf{m}) &\leftrightarrow (\exists \pi < \aleph_1)\, \exists \alpha\, \exists f\, \forall p\, [f(p) < \pi \wedge \\
&\qquad\qquad (\forall s < p)(\forall t < p)\, U(\bar{\alpha}(p), \bar{f}(p), s, t, \mathbf{m})] \\
&\leftrightarrow (\exists \rho < \aleph_1)\, \exists g\, \forall p\, [g(p) < \rho \wedge V(\bar{g}(p), \mathbf{m})].
\end{aligned}$$

In the second implication ($\rightarrow$) we intend that $g(p) = \langle\alpha(p), f(p)\rangle$ and $\alpha = \langle\omega, \pi\rangle$. Then R is $\aleph_1$-semi-recursive by Lemma 3.6 and Corollary 2.11. $\square$

3.8–3.18 Exercises

3.8. Give another proof of Theorem 3.4 by showing directly that W is ω_1-semi-recursive.

3.9. Show that

(i) there exists an ω_1-partial recursive function F such that for all $c \in W$, $F(c) = \|c\|$;

(ii) there exists an ω_1-recursive function G such that for all $\rho < \omega_1$, $G(\rho) \in W$ and $\|G(\rho)\| = \rho$.

3.10. Let f be any ω_1-recursive function.

(i) Show that there exists a function g recursive in E such that for any $c \in W$, $g(c) \in W$ and $\|g(c)\| = f(\|c\|)$.

(ii) Show that there exists a primitive recursive function h such that for any $c \in W$, $h(c) \in W$ and $\|h(c)\| \geqslant f(\|c\|)$.

3.11. For any $R \subseteq {}^k\omega_1$, let

$$R_W(\mathbf{m}) \leftrightarrow m_0, \ldots, m_{k-1} \in W \quad \text{and} \quad R(\|\mathbf{m}\|).$$

Show that

(i) R is ω_1-semi-recursive $\leftrightarrow R_W \in \Pi_1^1$;

(ii) R is ω_1-finite $\leftrightarrow R_W \in \Delta_1^1$.

3.12. Show that for every $\beta \in {}^\omega\omega$, $\omega_1[\beta]$ is recursively regular, and if β is $\omega_1[\beta]$-recursive, then for any $R \subseteq {}^k\omega$,

$$R \text{ is } \omega_1[\beta]\text{-semi-recursive} \leftrightarrow R \in \Pi_1^1[\beta],$$

and

$$R \text{ is } \omega_1[\beta]\text{-recursive} \leftrightarrow R \in \Delta_1^1[\beta].$$

3.13. Characterize the least recursively regular ordinal $> \omega_1$.

3.14. Show that

(i) For all $\sigma < \omega_1$, the constant function with value σ is ω_1-recursive;

(ii) for all $\sigma < \delta_2^1$, the constant function with value σ is $\aleph_1$-recursive.

3.15. Show that if κ is recursively regular, α is κ-recursive and $R \subseteq {}^k\omega$ is $\Pi_1^1[\alpha]$, then R is κ-semi-recursive.

3.16. Suppose that κ is both recursively regular and a limit of recursively regular ordinals (recursively inaccessible — cf. § 6). Show

(i) if α is κ-recursive and $R \subseteq {}^k\omega$ is $\Pi_1^1[\alpha]$, then R is κ-recursive;

(ii) if $\alpha_0, \ldots, \alpha_{k-1}$ are κ-recursive and R is an arithmetical relation such that $\exists\beta\, \mathsf{R}(\mathbf{m}, \boldsymbol{\alpha}, \beta)$, then $\exists\beta\, [\beta$ is κ-recursive $\wedge\, \mathsf{R}(\mathbf{m}, \boldsymbol{\alpha}, \beta)]$ (use a relativized version of Theorem III.4.7);

(iii) if $S \subseteq {}^k\omega$ is defined by

$$S(\mathbf{m}) \leftrightarrow (\exists\kappa\text{-recursive } \alpha)(\forall\kappa\text{-recursive } \beta)\, \mathsf{R}(\mathbf{m}, \alpha, \beta)$$

with R arithmetical, then S is κ-semi-recursive;

(iv) $\{\alpha : \alpha$ is κ-recursive$\}$ is a model of the Δ_2^1-Comprehension schema.

3.17. Show that for any $R \subseteq {}^k\omega$,

$$R \text{ is } \aleph_1\text{-(semi-) recursive} \leftrightarrow R \text{ is } \delta_2^1\text{-(semi-) recursive}$$

(use the Basis Theorem IV.7.9).

3.18. Show that for any $A \subseteq \omega$ and any $R \subseteq {}^k\omega$, if A is $\aleph_1$-recursive in parameters and $R \in \Delta_2^1[A]$, then also R is $\aleph_1$-recursive in parameters (relativize the proof of Theorem 3.7).

3.19 Notes. The results here are first stated explicitly in Kripke [1964 a], but some of the ideas are present in Kino–Takeuti [1962, – a].

The various possible notions of relative ω_1- (or κ-) semi-recursiveness are discussed in Kreisel–Sacks [1965]. Simpson [1974] is another good survey of the (then) current state of knowledge about relative ordinal recursion.

4. *Ordinal Recursion and Type-2 Functionals*

We shall see in § 5 that the results of the preceding section cannot be extended to higher levels of the analytical hierarchy. They do, however, have some natural extensions to classes of relations (semi-) recursive in a type-2 functional I. In view of Theorems VI.1.8–9 and the fact that $\omega_1[\mathsf{E}] = \omega_1$, Theorem 3.4 may be phrased as: for all $R \subseteq {}^k\omega$,

$$R \text{ is } \omega_1[\mathsf{E}]\text{-(semi-) recursive} \leftrightarrow R \text{ is (semi-) recursive in } \mathsf{E}.$$

The primary goal of this section is to establish conditions under which this equivalence holds with E replaced by other functionals. In the process we shall also develop some machinery which will be useful in § 6.

It is easy to see that the equivalence cannot hold for arbitrary I, even assuming that E is recursive in I. There are only countably many ∞-recursive relations on ω and if R is any non-∞-recursive relation, there exist I such that R and E are recursive in I (take I to code R and E), but clearly R is not $\omega_1[\mathsf{I}]$-recursive.

For this section I is a fixed functional such that E is recursive in I. The proofs below depend heavily on the results and techniques of §§ VI.2–4 and the constructions are in many cases similar to ones which appear there. We shall omit much of the tedious detail and hope that the reader who has followed the proofs of these earlier sections will feel confident that he could fill in these details.

We first recall from § VI.4 that

$$U^{\mathsf{I}} = \{\langle a, \mathbf{m}\rangle : \{a\}^{\mathsf{I}}(\mathbf{m}) \text{ is defined}\}$$

and that there exist relations $\leqslant_+$ and $<_+$ semi-recursive in I and $\leqslant_-$ and $<_-$ co-semi-recursive in I such that whenever at least one of u and v belongs to U^{I},

$$u \leqslant_+ v \leftrightarrow |u|^{\mathsf{I}} \leqslant |v|^{\mathsf{I}} \leftrightarrow u \leqslant_- v,$$

and

$$u <_+ v \leftrightarrow |u|^{\mathsf{I}} < |v|^{\mathsf{I}} \leftrightarrow u <_- v.$$

By Theorem VI.4.17, the ordinals $|u|^{\mathsf{I}}$ for $u \in U^{\mathsf{I}}$ are exactly the ordinals less than $\omega_1[\mathsf{I}]$. We shall write $|\mathbf{m}|^{\mathsf{I}}$ for $(|m_0|^{\mathsf{I}}, \ldots, |m_{k-1}|^{\mathsf{I}})$ and omit the superscript in most cases.

4.1 Lemma. *There exist functions F, G, and H partial recursive in I such that for all u, m, p, and q,*

(*i*) $F(m) \in U^{\mathsf{I}}$ *and* $|F(m)| = m$;

(ii) *if* $m \in U^{\mathsf{I}}$, *then* $G(m) \in U^{\mathsf{I}}$ *and* $|G(m)| = |m| + 1$;

(iii) $H(p, q)$ *is defined iff* $|p|, |q| < \omega$, *and if so,*

$$|H(p, q)| = \mathrm{Sb}_0(|p|, |q|).$$

Proof. Exercise 4.17. □

4.2 Lemma. *There exists a primitive recursive function f such that for all a and all* $\mathbf{m} \in U^{\mathsf{I}}$,

(i) $\{f(a)\}^{\mathsf{I}}(\mathbf{m})$ *is defined iff* $\{a\}_{\omega_1[\mathsf{I}]}(|\mathbf{m}|)$ *is defined*;
(ii) $\{a\}_{\omega_1[\mathsf{I}]}(|\mathbf{m}|) \simeq |\{f(a)\}^{\mathsf{I}}(\mathbf{m})|$.

Proof. We define f via the Primitive Recursion Theorem to satisfy the following conditions:

(0) if $a = \langle 0, k, 0, n\rangle$, then $\{f(a)\}^{\mathsf{I}}(\mathbf{m}) \simeq F(n)$;
if $a = \langle 0, k, 1, i\rangle$, then $\{f(a)\}^{\mathsf{I}}(\mathbf{m}) \simeq m_i$;
if $a = \langle 0, k, 2, i\rangle$, then $\{f(a)\}^{\mathsf{I}}(\mathbf{m}) \simeq G(m_i)$;
if $a = \langle 0, k+4, 4\rangle$, then

$$\{f(a)\}^{\mathsf{I}}(p, q, r, s, \mathbf{m}) \simeq \begin{cases} p, & \text{if} \quad r \leqslant_+ s \quad \text{and} \quad s \leqslant_+ r; \\ q, & \text{if} \quad r <_+ s \quad \text{or} \quad s <_+ r; \end{cases}$$

if $a = \langle 0, k+2, 5\rangle$, then $\{f(a)\}^{\mathsf{I}}(p, q, m) \simeq H(p, q)$;

(1) if for some $k', b, c_0, \ldots, c_{k'-1}$, $a = \langle 1, k, b, c_0, \ldots, c_{k'-1}\rangle$, then $f(a) = \langle 1, k, 0, f(b), f(c_0), \ldots, f(c_{k'-1})\rangle$;

(2) if $a = \langle 2, k+1\rangle$, then $f(a)$ is defined as in case (2) of Lemma VI.2.2;

(3) if for some b, $a = \langle 3, k+1, b\rangle$, let R be the relation semi-recursive in I defined by:

$$\begin{aligned} R(n, r, \mathbf{m}) \leftrightarrow \forall p\, [p <_- r \rightarrow \exists q\, (q <_+ n \wedge \{f(b)\}^{\mathsf{I}}(p, \mathbf{m}) \simeq q)] \\ \wedge \neg \exists n' (n' <_- n \wedge \forall p\, [p <_+ r \rightarrow \exists q\, (q <_- n' \wedge \{f(b)\}^{\mathsf{I}}(p, \mathbf{m}) \simeq q)]), \end{aligned}$$

then choose $f(a)$ to be an index such that

(a) $\{f(a)\}^{\mathsf{I}}(r, \mathbf{m}) \simeq \mathrm{Sel}_R(r, \mathbf{m})$, and

(b) for any r, $\mathbf{m}$, and n such that $\{f(a)\}^{\mathsf{I}}(r, \mathbf{m}) \simeq n$, the set of subcomputations of $(f(a), r, \mathbf{m}, n)$ contains an element of the form $(f(b), p, \mathbf{m}, q)$ for each p such that $|p| < |r|$;

(4) if for some b, $a = \langle 4, k, b\rangle$, let S be the relation semi-recursive in I defined by:

$$\begin{aligned} S(n, \mathbf{m}) \leftrightarrow \forall p\, [p <_- n \rightarrow \exists q\, (|q| > 0 \wedge \{f(b)\}^{\mathsf{I}}(p, \mathbf{m}) \simeq q)] \\ \wedge \exists q\, (|q| = 0 \wedge \{f(b)\}^{\mathsf{I}}(n, \mathbf{m}) \simeq q), \end{aligned}$$

then choose $f(a)$ such that

(a) $\{f(a)\}^{\mathsf{I}}(\mathbf{m}) \simeq \mathrm{Sel}_S(\mathbf{m})$, and

(b) for any $\mathbf{m}$ and n such that $\{f(a)\}^{\mathsf{I}}(\mathbf{m}) \simeq n$, the set of subcomputations of $(f(a), \mathbf{m}, n)$ includes an element of the form $(f(b), p, \mathbf{m}, q)$ for each p such that $|p| \leqslant |n|$;

(5) if a is of none of these forms, $f(a) = 0$.

We leave to the reader the straightforward proof that f has the desired properties. The proof of the implication ($\leftarrow$) of (i) does not use clause (b) of cases (3) and (4) but for the implication ($\rightarrow$) these conditions are just what is needed to carry out an induction on subcomputations as in the proof of Theorem VI.2.8. □

4.3 Corollary. *For any k and any $R \subseteq {}^k\omega$,*

(i) *R is $\omega_1[\mathsf{I}]$-semi-recursive $\rightarrow R$ is semi-recursive in I;*

(ii) *R is $\omega_1[\mathsf{I}]$-recursive $\rightarrow R$ is recursive in I.*

Proof. For (i), let F and f be as in 4.1 and 4.2, respectively. Then if R is $\omega_1[\mathsf{I}]$-semi-recursive, say with $\omega_1[\mathsf{I}]$-semi-index a,

$$R(\mathbf{m}) \leftrightarrow \{f(a)\}^{\mathsf{I}}(F(m_0), \ldots, F(m_{k-1})) \text{ is defined}$$

so R is semi-recursive in I. (ii) follows immediately by Corollary VI.4.5. □

4.4 Theorem. *$\omega_1[\mathsf{I}]$ is recursively regular.*

Proof. The proof is similar to that of Theorem 3.3. Let F be any $\omega_1[\mathsf{I}]$-partial recursive function and ρ and $\boldsymbol{\mu}$ ordinals less than $\omega_1[\mathsf{I}]$ such that for all $\pi < \rho$, $F(\pi, \boldsymbol{\mu})$ is defined. We aim to show that $\sup^+_{\pi<\rho} F(\pi, \boldsymbol{\mu}) < \omega_1[\mathsf{I}]$. Let

$$A = \{n : (\exists \pi < \rho)\, F(\pi, \boldsymbol{\mu}) = |n|\}.$$

Choose r and $\mathbf{m}$, fixed elements of U^{I}, such that $|r| = \rho$ and each $|m_i| = \mu_i$, and set

$$R(n, p) \leftrightarrow F(|p|, |\mathbf{m}|) \simeq |n|.$$

It follows from Lemma 4.2 that R is semi-recursive in I. Let Sel_R be a selection function for R partial recursive in I (Corollary VI.4.2) and set

$$B = \{n : \exists p\, [p <_- r \wedge \forall n' (n' \neq n \rightarrow \mathrm{Sel}_R(p) \neq n')]\}.$$

Clearly $B \subseteq A$, B is co-semi-recursive in I, and

$$\sup{}^+\{|n| : n \in B\} = \sup{}^+\{|n| : n \in A\}.$$

Hence by the Boundedness Theorem VI.4.10 and Theorem VI.4.17,

$$\sup{}^+_{\pi<\rho} F(\pi, \boldsymbol{\mu}) = \sup{}^+\{|n| : n \in B\} < \kappa^{\mathsf{I}} = \omega_1[\mathsf{I}]. \quad \square$$

We consider next under what conditions the converse of Corollary 4.3 holds.

A natural requirement is suggested by Exercise 3.12. We want in some sense to require that I be $\omega_1[\mathsf{I}]$-recursive. Of course, this is not literally possible in our current formulation, but a weaker version of this condition, which asserts that I is $\omega_1[\mathsf{I}]$ computable *on* $\omega_1[\mathsf{I}]$-recursive functions, is both possible and sufficient.

We denote by $\{a\}_\kappa^\omega$ the restriction of $\{a\}_\kappa$ to arguments from ω.

4.5 Definition. For any κ, I is κ*-effective* iff there exists a κ-partial recursive function I such that for all a such that $\{a\}_\kappa^\omega \in {}^\omega\omega$,

$$\mathsf{I}(\{a\}_\kappa^\omega) = I(a).$$

A κ-index for I is also called a κ*-index* for I.

4.6 Example. For any $\kappa > \omega$, let

$$E(a) \simeq \inf_{\pi<\omega} \operatorname{sg}^+(\{a\}_\kappa(\pi)).$$

Then E is κ-partial recursive and if $\{a\}_\kappa^\omega \in {}^\omega\omega$, $E(a) \simeq \mathsf{E}(\{a\}_\kappa^\omega)$. Hence E is κ-effective for all $\kappa > \omega$.

4.7 Theorem. *There exists a primitve recursive function g such that for any d and any $\kappa > \omega$, if I is κ-effective with κ-index d, then for all a, $\mathbf{m} < \omega$,*

$$\{g(a, d)\}_\kappa(\mathbf{m}) \simeq \{a\}^{\mathsf{I}}(\mathbf{m}).$$

Proof. The proof is quite similar to that of Theorem VI.2.14. We define g via the Primitive Recursion Theorem to satisfy the following conditions:

(0) if $a = \langle 0, k, 0, \ldots \rangle$, then $g(a, d) = \langle 0, k, \ldots \rangle$;

(1) if for some $k', b, c_0, \ldots, c_{k'-1}$, $a = \langle 1, k, 0, b, c_0, \ldots, c_{k'-1} \rangle$, then $g(a, d) = \langle 1, k, g(b, d), g(c_0, d), \ldots, g(c_{k'-1}, d) \rangle$;

(2) if $a = \langle 2, k+1, 0 \rangle$, then $g(a, d)$ is defined as in the corresponding case of the proof of Lemma VI.2.2;

(3) if for some b, $a = \langle 3, k, 0, b \rangle$, then $g(a, d)$ is the natural index such that

$$\{g(a, d)\}_\kappa(\mathbf{m}) \simeq 0 \cdot \sup_{\pi<\omega} \{g(b, d)\}_\kappa(\pi, \mathbf{m}) + \{d\}_\kappa(c),$$

where c is an index such that $\{c\}_\kappa(\pi) \simeq \{g(b, d)\}_\kappa(\pi, \mathbf{m})$; such a c may be computed from $\mathbf{m}$ and an index for g;

(4) if a is of none of these forms, $g(a, d) = 0$.

The proof that g has the required property also goes much as in Theorem VI.2.14 and we make only a few remarks on the differences. In case (3) of the proof that

$$\{a\}^{\mathsf{I}}(\mathbf{m}) \simeq n \rightarrow \{g(a, d)\}_\kappa(\mathbf{m}) \simeq n,$$

we have as induction hypothesis that for some β and all p,

$$\{b\}^{\mathsf{I}}(p, \mathbf{m}) \simeq \{g(b, d)\}_\kappa(p, \mathbf{m}) \simeq \beta(p).$$

Then with c as in case (3) of the definition, $\{c\}^\omega_\kappa = \beta$, so

$$\begin{aligned}\{g(a, d)\}_\kappa(\mathbf{m}) &\simeq 0 \cdot \sup_{\pi<\omega}\{g(b, d)\}_\kappa(\pi, \mathbf{m}) + \{d\}_\kappa(c)\\ &\simeq 0 + \mathsf{I}(\{c\}^\omega_\kappa)\\ &\simeq \mathsf{I}(\beta) = n.\end{aligned}$$

For the converse implication we need to define a notion of subcomputation for $\Omega_{\kappa\kappa}$ and prove an analogue of Theorem VI.2.5. This presents no problems, technical or otherwise, and cases (0)–(2) of the proof proceed as before. In case (3), if $a = \langle 3, k, 0, b\rangle$ and $\{g(a, d)\}_\kappa(\mathbf{m}) \simeq n$, we have in particular that

$$\sup_{\pi<\omega} \{g(b, d)\}_\kappa(\pi, \mathbf{m}) \text{ is defined}$$

so that for some $\beta \in {}^\omega\omega$ and all p, $\{g(b, d)\}_\kappa(p, \mathbf{m}) \simeq \beta(p)$ and $\{c\}^\omega_\kappa = \beta$. Then by the induction hypothesis, $\beta = \lambda p.\{b\}^{\mathsf{I}}(p, \mathbf{m})$, so

$$\begin{aligned}\{a\}^{\mathsf{I}}(\mathbf{m}) &\simeq \mathsf{I}(\lambda p.\{b\}^{\mathsf{I}}(p, \mathbf{m}))\\ &\simeq \mathsf{I}(\{c\}^\omega_\kappa)\\ &\simeq \{d\}_\kappa(c) \simeq n. \quad \square\end{aligned}$$

4.8 Corollary. *For any κ such that I is κ-effective and any $R \subseteq {}^k\omega$,*
(i) *R is semi-recursive in $\mathsf{I} \to R$ is κ-semi-recursive;*
(ii) *R is recursive in $\mathsf{I} \to R$ is κ-recursive.* □

4.9 Corollary. *For any recursively regular κ, if I is κ-effective, then $\omega_1[\mathsf{I}] \leq \kappa$.*

Proof. Suppose that I is κ-effective and $\sigma < \omega_1[\mathsf{I}]$. Choose γ recursive in I such that $\|\gamma\| = \sigma$. By Corollary 4.8, γ is κ-recursive and thus by Lemma 3.1, $\sigma < \kappa$. □

4.10 Corollary. *If I is $\omega_1[\mathsf{I}]$-effective, then for all $R \subseteq {}^k\omega$,*
(i) *R is $\omega_1[\mathsf{I}]$-semi-recursive $\leftrightarrow R$ is semi-recursive in I;*
(ii) *R is $\omega_1[\mathsf{I}]$-recursive $\leftrightarrow R$ is recursive in I.*

Proof. Immediate from Corollaries 4.3 and 4.8. □

Of course, we already knew this for $\mathsf{I} = \mathsf{E}$ and to this point we have no other examples of functionals I which are $\omega_1[\mathsf{I}]$-effective. The rest of this section is devoted to finding some.

4.11 Definition. For any $d \in \omega$,

(i) $\mathrm{Om}[\mathsf{I}] = \{\omega_1[\mathsf{H}] : \mathsf{I}$ is recursive in $\mathsf{H}\}$;

(ii) $\mathrm{Ef}_d[\mathsf{I}] = \{\kappa : \kappa$ is recursively regular and I is κ-effective with index $d\}$;

(iii) I is *effective with index d* iff $\mathrm{Om}[\mathsf{I}] \subseteq \mathrm{Ef}_d[\mathsf{I}]$;

(iv) I is *effective* iff it is effective with some index.

Of course, E is effective and the conclusions of Corollary 4.10 hold for any effective I. Our main technique for finding other effective functionals is to show that if I is effective, so is its superjump I^{sJ}, which we recall is the functional such that

$$\mathsf{I}^{sJ}((\langle a, \mathbf{m}\rangle) * \alpha) = \begin{cases} 0, & \text{if } \{a\}^{\mathsf{I}}(\mathbf{m}, \alpha) \text{ is defined;} \\ 1, & \text{otherwise.} \end{cases}$$

Recall also that for any set A of ordinals,

$$\mathrm{Lim}(A) = \{\kappa : \kappa \in A \wedge (\forall \sigma < \kappa)(\exists \lambda \in A) . \sigma < \lambda < \kappa\}.$$

4.12 Lemma. $\mathrm{Om}[\mathsf{I}^{sJ}] \subseteq \mathrm{Lim}(\mathrm{Om}[\mathsf{I}])$.

Proof. Since I is recursive in I^{sJ}, $\mathrm{Om}[\mathsf{I}^{sJ}] \subseteq \mathrm{Om}[\mathsf{I}]$. Suppose $\kappa \in \mathrm{Om}[\mathsf{I}^{sJ}]$. It suffices to show that for any $\sigma < \kappa$ there exists a $\lambda \in \mathrm{Om}[\mathsf{I}]$ such that $\sigma < \lambda < \kappa$. By definition, $\kappa = \omega_1[\mathsf{H}]$ for some H such that I^{sJ} is recursive in H. Then for any $\sigma < \kappa$, there exists a γ recursive in H such that $\|\gamma\| = \sigma$. Then clearly,

$$\sigma < \omega_1[\gamma] \leqslant \omega_1[\mathsf{I}, \gamma].$$

Let G be a functional which codes I and γ, say $\mathsf{G}(\alpha) = \langle \gamma(\alpha(0)), \mathsf{I}(\alpha)\rangle$, and set

$$\lambda = \omega_1[\mathsf{I}, \gamma] = \omega_1[\mathsf{G}].$$

Then $\lambda \in \mathrm{Om}[\mathsf{I}]$ and $\sigma < \lambda$, so it remains to show that $\lambda < \kappa$.

By the techniques of Theorem VI.4.17 there exists a relation R semi-recursive in I and γ which well-orders a subset of U^{I}_{γ} in type λ. By Theorem VI.1.11, R is recursive in I^{sJ} and γ and thus recursive in H. Hence $\lambda < \omega_1[\mathsf{H}] = \kappa$. $\square$

We next need a slight generalization of Theorem 4.7:

4.13 Lemma. *There exists a primitive recursive function g such that for any d and* *any* $\kappa > \omega$, *if* I *is* κ*-effective with* κ*-index d, then for all a, e, and* $\mathbf{m} < \omega$ *such that* $\{e\}_\kappa^\omega \in {}^\omega\omega$,

$$\{g(a,d,e)\}_\kappa(\mathbf{m}) \simeq \{a\}^{\mathsf{I}}(\mathbf{m}, \{e\}_\kappa^\omega).$$

Proof. We need make only a few minor changes in the proof of Theorem 4.7. First, each case hypothesis should be changed from $a = \langle i, k, 0, \ldots\rangle$ to $a = \langle i, k, 1, \ldots\rangle$ to take account of the fact that $\{a\}^{\mathsf{I}}$ now has one function argument. The function g is defined exactly as before in cases (1)–(4) and in case (0) as long as $a \neq \langle 0, k, 1, 3, i, 0\rangle$ for some $i < k$. For such a we choose $g(a,d,e)$ to be an index such that

$$\{g(a,d,e)\}_\kappa(\mathbf{m}) \simeq \{e\}_\kappa(m_i) \simeq \{a\}^{\mathsf{I}}(\mathbf{m}, \{e\}_\kappa^\omega). \quad \square$$

4.14 Theorem. *For any d, there exists a d′ such that*

$$\mathrm{Lim}(\mathrm{Ef}_d[\mathsf{I}]) \subseteq \mathrm{Ef}_{d'}[\mathsf{I}^{\mathrm{sJ}}].$$

Proof. Suppose $\kappa \in \mathrm{Lim}(\mathrm{Ef}_d[\mathsf{I}])$. We shall need the following two facts:

(a) $\mathrm{Ef}_d[\mathsf{I}] \cap \kappa$ is κ-recursive;

(b) for any e, *if* $\{e\}_\kappa^\omega \in {}^\omega\omega$, then there exists a $\lambda \in \mathrm{Ef}_d[\mathsf{I}] \cap \kappa$ such that $\{e\}_\lambda^\omega = \{e\}_\kappa^\omega$.

Suppose for now that (a) and (b) are true. Note that under the conditions of (b), for any λ, $\{e\}_\lambda^\omega = \{e\}_\kappa^\omega$ just in case $\mathrm{Dm}\{e\}_\lambda^\omega = \omega$. Set

$$h(e) \simeq \kappa\text{-least } \lambda[\lambda \in \mathrm{Ef}_d[\mathsf{I}] \wedge (\forall p < \omega)(\exists \upsilon < \lambda)T(e, \langle p\rangle, \upsilon)],$$

where T is as defined in § 2. By (a), h is κ-partial recursive and by (b), if $\{e\}_\kappa^\omega \in {}^\omega\omega$, then $h(e)$ is defined, $h(e) \in \mathrm{Ef}_d[\mathsf{I}]$, and $\{e\}_{h(e)}^\omega = \{e\}_\kappa^\omega$.

Now if g is the function of Lemma 4.13, we have for any e such that $\{e\}_\kappa^\omega \in {}^\omega\omega$ and all a and $\mathbf{m}$,

$$\begin{aligned} \mathsf{I}^{\mathrm{sJ}}(\langle a, \mathbf{m}\rangle) * \{e\}_\kappa^\omega) \simeq 0 &\leftrightarrow \{a\}^{\mathsf{I}}(\mathbf{m}, \{e\}_\kappa^\omega) \text{ is defined} \\ &\leftrightarrow \{a\}^{\mathsf{I}}(\mathbf{m}, \{e\}_{h(e)}^\omega) \text{ is defined} \\ &\leftrightarrow \{g(a,d,e)\}_{h(e)}(\mathbf{m}) \text{ is defined.} \end{aligned}$$

Hence if

$$F(\langle a, \mathbf{m}\rangle, e) \simeq \begin{cases} 0, & \text{if } h(e)\downarrow \wedge (\exists \upsilon < h(e))T(g(a,d,e), \langle \mathbf{m}\rangle, \upsilon); \\ 1, & \text{if } h(e)\downarrow \wedge \neg(\exists \upsilon < h(e))T(g(a,d,e), \langle \mathbf{m}\rangle, \upsilon); \\ \text{undefined}, & \text{otherwise;} \end{cases}$$

and $\{e^+\}_\kappa(m) \simeq \{e\}_\kappa(m+1)$, then F is κ-partial recursive and

$$I^{sJ}(\{e\}_\kappa^\omega) \simeq I^{sJ}((\{e\}_\kappa(0)) * \{e^+\}_\kappa^\omega) \simeq F(\{e\}_\kappa(0), e^+).$$

Thus it suffices to choose d' so that

$$\{d'\}_\kappa(e) \simeq F(\{e\}_\kappa(0), e^+),$$

as then $\kappa \in \mathrm{Ef}_{d'}[\mathsf{I}^{sJ}]$.

To establish (a), note that for $\lambda < \kappa$,

$$\lambda \in \mathrm{Ef}_d[\mathsf{I}] \leftrightarrow \lambda \text{ is recursively regular } \wedge$$
$$(\forall a < \omega)(\{a\}_\lambda^\omega \in {}^\omega\omega \rightarrow \{d\}_\lambda(a) \text{ is defined}).$$

The implication $(\rightarrow)$ is immediate from the definition. Conversely, if the right-hand side holds and $\{a\}_\lambda^\omega \in {}^\omega\omega$, then $\{a\}_\lambda^\omega = \{a\}_\kappa^\omega$ and as $\{d\}_\lambda(a)$ is defined, we have

$$\{d\}_\lambda(a) \simeq \{d\}_\kappa(a) \simeq \mathsf{I}(\{a\}_\kappa^\omega) \simeq \mathsf{I}(\{a\}_\lambda^\omega).$$

Thus

$$\lambda \in \mathrm{Ef}_d[\mathsf{I}] \leftrightarrow \lambda \text{ is recursively regular } \wedge$$
$$(\forall a < \omega)((\forall p < \omega)(\exists \upsilon < \lambda)[T(a, \langle p\rangle, \upsilon) \wedge (\upsilon)_0 < \omega] \rightarrow (\exists \upsilon < \lambda)\, T(d, \langle a\rangle, \upsilon)),$$

which, with Corollary 2.19, implies (a).

For (b), suppose that $\{e\}_\kappa^\omega \in {}^\omega\omega$ and define a function G by

$$G(e, p) \simeq \kappa\text{-least } \upsilon\,.\, T(e, \langle p\rangle, \upsilon).$$

As κ is recursively regular and $\kappa > \omega$,

$$\sup_{p<\omega} G(e, p) < \kappa.$$

By the assumption that $\kappa \in \mathrm{Lim}(\mathrm{Ef}_d[\mathsf{I}])$, there exists a $\lambda \in \mathrm{Ef}_d[\mathsf{I}]$ such that $\sup_{p<\omega} G(e, p) < \lambda < \kappa$. Then $\{e\}_\lambda^\omega = \{e\}_\kappa^\omega$. □

4.15 Corollary. *If I is effective, then also I^{sJ} is effective.*

Proof. Immediate from 4.12 and 4.14. □

We recall the sequence of functionals defined following Theorem VI.6.11:

$$\mathsf{E}_0 = \mathsf{E} \quad \text{and} \quad \mathsf{E}_{r+1} = (\mathsf{E}_r)^{sJ}.$$

4.16 Corollary. *For all r and all $R \subseteq {}^k\omega$,*

(i) *R is $\omega_1[\mathsf{E}_r]$-semi-recursive $\leftrightarrow R$ is semi-recursive in E_r;*

(ii) *R is $\omega_1[\mathsf{E}_r]$-recursive $\leftrightarrow R$ is recursive in E_r.*

Proof. It is immediate by induction on r that each E_r is effective, hence $\omega_1[\mathsf{E}_r]$-effective. The result follows from Corollary 4.10. □

4.17–4.20 Exercises

4.17. Prove Lemma 4.1.

4.18. Show that if κ is recursively regular, α is κ-recursive, I is κ-effective, and $R \subseteq {}^k\omega$ is semi-recursive in I and α, then R is κ-semi-recursive. (This extends Exercise 3.15).

4.19 (Grilliot). Show that even if E is not recursive in I, $\omega_1[\mathsf{I}]$ is still recursively regular. (Use the result of Exercise VI.2.29.)

4.20. Show that if E is recursive in I and I is $\omega_1[\mathsf{I}]$-effective, then there exists an $\omega_1[\mathsf{I}]$-recursive function G such that for all $\rho < \omega_1[\mathsf{I}]$, $G(\rho) \in U^{\mathsf{I}}$ and $|G(\rho)|^{\mathsf{I}} = \rho$. (cf. Exercise 3.9)

4.21 Notes. The results of this section had the status of folklore already in the late 1960's and were probably known to Kripke and Platek even earlier, but we know of no published exposition prior to Aczel–Hinman [1974, § 2].

5. Stability

The fundamental definitions for this section are:

5.1 Definition. For any κ and λ,

(i) κ is *stable* iff κ is closed under all ∞-partial recursive functions;

(ii) κ is λ*-stable* iff κ is closed under all (∞, λ)-partial recursive functions;

(iii) κ is *weakly stable* iff κ is λ-stable for some $\lambda > \kappa$.

Note that κ is recursively regular iff κ is κ-stable. Then it is immediate from the definitions that

$$\kappa \text{ is stable} \leftrightarrow \kappa \text{ is } \lambda\text{-stable for all } \lambda \rightarrow \kappa \text{ is weakly stable} \rightarrow \kappa \text{ is recursively regular.}$$

We shall see that weak stability is a much stronger property than recursive regularity and that stability is also much stronger than weak stability.

The main result we prove in this section is that for every uncountable cardinal ρ, there are ρ stable ordinals less than ρ, hence also ρ recursively regular ordinals less than ρ. For $\rho = \aleph_1$, the last assertion follows from Exercise 3.12. The technique of the proof is an adaptation of the downward Löwenheim–Skolem Theorem of first-order logc. Suppose λ is recursively regular, $\sigma < \lambda$, and

$$A = \{F(\boldsymbol{\mu}) : \boldsymbol{\mu} \leq \sigma \wedge F \text{ is } (\infty, \lambda)\text{-partial recursive}\}.$$

As λ is recursively regular, $A \subseteq \lambda$. Furthermore, as the (∞, λ)-partial recursive functions are closed under composition, A is closed under all (∞, λ)-partial recursive functions. Hence if it should happen that A is in fact an ordinal κ, κ would be λ-stable. Since the projection functions are (∞, λ)-partial recursive, $\kappa > \sigma$. Of course, if $\kappa = \lambda$, this would be of no interest, but we shall see that if λ is chosen sufficiently large (for example, if $\mathrm{Card}(\lambda) > \mathrm{Card}(\sigma)$), then A will indeed turn out to be a λ-stable ordinal $\kappa < \lambda$. If we use all ∞-partial recursive functions to form A, then κ will be stable.

To show that A is an ordinal, that is, that A is transitive, we show that it cannot be collapsed. The reader should review that part of I.1.5 which deals with the collapsing function φ_A and its properties (1)–(5), which will be used in the following proof. We write $\varphi_A(\boldsymbol{\mu})$ for $(\varphi_A(\mu_0), \ldots, \varphi_A(\mu_{k-1}))$.

5.2 Lemma. *For any λ and any $A \subseteq \lambda$ [any $A \subseteq \mathrm{Or}$], if A is closed under all (∞, λ)-partial recursive functions [all ∞-partial recursive functions], then for all $\boldsymbol{\mu}$ and ν in A and all (∞, λ)-partial recursive functions F [all ∞-partial recursive functions F],*

$$F(\boldsymbol{\mu}) \simeq \nu \rightarrow F(\varphi_A(\boldsymbol{\mu})) \simeq \varphi_A(\nu).$$

Proof. Let A be a fixed subset of λ which is closed under all (∞, λ)-partial recursive functions. To improve legibility we shall write $\bar{\boldsymbol{\mu}}$ instead of $\varphi_A(\boldsymbol{\mu})$. Then we need to prove that for all $\boldsymbol{\mu}$ and $\nu \in A$ and all a,

$$\{a\}_{\infty\lambda}(\boldsymbol{\mu}) \simeq \nu \rightarrow \{a\}_{\infty\lambda}(\bar{\boldsymbol{\mu}}) \simeq \bar{\nu}.$$

Formally, this is a proof by induction over $\Omega_{\kappa\lambda}$, simultaneously for all $\kappa \geq \lambda$, that

$$(a, \boldsymbol{\mu}, \nu) \in \Omega_{\kappa\lambda} \rightarrow (a, \bar{\boldsymbol{\mu}}, \bar{\nu}) \in \Omega_{\kappa\lambda},$$

but we shall state matters directly in terms of (∞, λ)-computations.

We consider cases (0)–(4). If $A = \varnothing$, the result is vacuous and we assume otherwise.

(0) If $a = \langle 0, k, 0, n \rangle$ and $\{a\}_{\infty\lambda}(\boldsymbol{\mu}) \simeq n$, then $\{a\}_{\infty\lambda}(\bar{\boldsymbol{\mu}}) \simeq n = \bar{n}$ by (4) of I.1.5

($n \subseteq \omega \subseteq A$ because all constant functions with natural number values are (∞, λ)-recursive); if $a = \langle 0, k, 1, i\rangle$ and $\{a\}_{\infty\lambda}(\boldsymbol{\mu}) \simeq \mu_i$, then $\{a\}_{\infty\lambda}(\bar{\boldsymbol{\mu}}) \simeq \bar{\mu}_i$; if $a = \langle 0, k, 2, i\rangle$ and $\{a\}_{\infty\lambda}(\boldsymbol{\mu}) \simeq \mu_i + 1$, then $\{a\}_{\infty\lambda}(\bar{\boldsymbol{\mu}}) \simeq \bar{\mu}_i + 1 = \overline{\mu_i + 1}$ by (3) of I.1.5; if $a = \langle 0, k+4, 4\rangle$ and $\{a\}_{\infty\lambda}(\pi, \rho, \sigma, \tau, \boldsymbol{\mu}) \simeq \pi$, then $\sigma = \tau$ so $\bar{\sigma} = \bar{\tau}$ and $\{a\}_{\infty\lambda}(\bar{\pi}, \bar{\rho}, \bar{\sigma}, \bar{\tau}, \bar{\boldsymbol{\mu}}) \simeq \bar{\pi}$; if $a = \langle 0, k+4, 4\rangle$ and $\{a\}_{\infty\lambda}(\pi, \rho, \sigma, \tau, \boldsymbol{\mu}) \simeq \rho$, then $\sigma \neq \tau$, so by (1) of I.1.5, $\bar{\sigma} \neq \bar{\tau}$ and thus $\{a\}_{\infty\lambda}(\bar{\pi}, \bar{\rho}, \bar{\sigma}, \bar{\tau}, \bar{\boldsymbol{\mu}}) \simeq \bar{\rho}$; if $a = \langle 0, k+2, 5\rangle$ and $\{a\}_{\infty\lambda}(p, q, \boldsymbol{\mu}) \simeq \mathrm{Sb}_0(p, q)$, then $\{a\}_{\infty\lambda}(\bar{p}, \bar{q}, \bar{\boldsymbol{\mu}}) \simeq \{a\}_{\infty\lambda}(p, q, \bar{\boldsymbol{\mu}}) \simeq \mathrm{Sb}_0(p, q) \simeq \overline{\mathrm{Sb}_0(p, q)}$.

(1) If $a = \langle 1, k, b, c_0, \ldots, c_{k'-1}\rangle$ and $\{a\}_{\infty\lambda}(\boldsymbol{\mu}) \simeq \{b\}_{\infty\lambda}(\boldsymbol{\xi}) \simeq \nu$, where for $i < k'$, $\{c_i\}_{\infty\lambda}(\boldsymbol{\mu}) \simeq \xi_i$, then the closure property of A ensures that all $\xi_i \in A$, so by the induction hypothesis, $\{c_i\}_{\infty\lambda}(\bar{\boldsymbol{\mu}}) \simeq \bar{\xi}_i$ and $\{b\}_{\infty\lambda}(\bar{\boldsymbol{\xi}}) \simeq \bar{\nu}$. Hence

$$\{a\}_{\infty\lambda}(\bar{\boldsymbol{\mu}}) \simeq \{b\}_{\infty\lambda}(\bar{\boldsymbol{\xi}}) \simeq \bar{\nu}.$$

(2) If $a = \langle 2, k+1\rangle$ and $\{a\}_{\infty\lambda}(b, \boldsymbol{\mu}) \simeq \{b\}_{\infty\lambda}(\boldsymbol{\mu}) \simeq \nu$, then $\{a\}_{\infty\lambda}(\bar{b}, \bar{\boldsymbol{\mu}}) \simeq \{b\}_{\infty\lambda}(\bar{\boldsymbol{\mu}}) \simeq \bar{\nu}$ by the induction hypothesis and the fact that $\bar{b} = b$.

(3) If $a = \langle 3, k+1, b\rangle$ and $\{a\}_{\infty\lambda}(\rho, \boldsymbol{\mu}) \simeq \sup^+_{\pi<\rho}\{b\}_{\infty\lambda}(\pi, \boldsymbol{\mu}) \simeq \nu$, then

$$\begin{aligned}\{a\}_{\infty\lambda}(\bar{\rho}, \bar{\boldsymbol{\mu}}) &\simeq \sup\nolimits^+_{\sigma<\bar{\rho}}\{b\}_{\infty\lambda}(\sigma, \bar{\boldsymbol{\mu}})\\ &\simeq \sup\nolimits^+_{\pi<\rho \wedge \pi\in A}\{b\}_{\infty\lambda}(\bar{\pi}, \bar{\boldsymbol{\mu}})\\ &\simeq \sup\nolimits^+_{\pi<\rho \wedge \pi\in A}\overline{\{b\}_{\infty\lambda}(\pi, \boldsymbol{\mu})}.\end{aligned}$$

This second equality follows from (2) of I.1.5 and the third from the induction hypothesis. Let this value be ζ; we aim to show that $\zeta = \bar{\nu}$.

Since for $\pi < \rho$ and $\pi \in A$, $\{b\}_{\infty\lambda}(\pi, \boldsymbol{\mu}) \in A$ and $\{b\}_{\infty\lambda}(\rho, \boldsymbol{\mu}) < \nu$, also $\overline{\{b\}_{\infty\lambda}(\pi, \boldsymbol{\mu})} < \bar{\nu}$. Hence $\zeta \leqslant \bar{\nu}$. Suppose for a contradiction that $\zeta < \bar{\nu}$. Then by (2) of I.1.5, $\zeta = \bar{\tau}_0$ for some τ_0 such that $\tau_0 < \nu$ and $\tau_0 \in A$. Since $\tau_0 < \nu$, there exists a $\pi < \rho$ such that $\tau_0 \leqslant \{b\}_{\infty\lambda}(\pi, \boldsymbol{\mu})$. Let

$$h(\tau, \boldsymbol{\mu}) \simeq \lambda\text{-“least”}\ \pi[\tau \leqslant \{b\}_{\infty\lambda}(\pi, \boldsymbol{\mu})].$$

h is (∞, λ)-partial recursive and as $\lambda > \rho$, $h(\tau_0, \boldsymbol{\mu})$ is defined, say with value π_0. Since τ_0 and $\boldsymbol{\mu}$ are elements of A, also $\pi_0 \in A$ by the closure property of A. But

$$\tau_0 \leqslant \{b\}_{\infty\lambda}(\pi_0, \boldsymbol{\mu})$$

so

$$\bar{\tau}_0 \leqslant \overline{\{b\}_{\infty\lambda}(\pi_0, \boldsymbol{\mu})} < \zeta,$$

a contradiction.

(4) Suppose $a = \langle 4, k, b\rangle$ and

$$\{a\}_{\infty\lambda}(\boldsymbol{\mu}) \simeq (\lambda\text{-"least" } \pi . \{b\}_{\infty\lambda}(\pi, \boldsymbol{\mu}) \simeq 0) \simeq \nu.$$

For any $\sigma < \bar{\nu}$ there exists by (2) of I.1.5 a $\pi < \nu$ such that $\pi \in A$ and $\sigma = \bar{\pi}$. Then by the induction hypothesis,

$$\begin{aligned}\{b\}_{\infty\lambda}(\sigma, \bar{\boldsymbol{\mu}}) &\simeq \{b\}_{\infty\lambda}(\bar{\pi}, \bar{\boldsymbol{\mu}}) \\ &\simeq \overline{\{b\}_{\infty\lambda}(\pi, \boldsymbol{\mu})} > 0.\end{aligned}$$

On the other hand,

$$\{b\}_{\infty\lambda}(\bar{\nu}, \bar{\boldsymbol{\mu}}) \simeq \overline{\{b\}_{\infty\lambda}(\nu, \boldsymbol{\mu})} \simeq \bar{0} = 0.$$

Hence

$$\bar{\nu} \simeq \lambda\text{-"least" } \sigma[\{b\}_{\infty\lambda}(\sigma, \bar{\boldsymbol{\mu}}) \simeq 0] \simeq \{a\}_{\infty\lambda}(\bar{\boldsymbol{\mu}}).$$

The proof of the alternate version in brackets is exactly the same with all references to λ removed. □

5.3 Theorem. *For any* $\sigma \geqslant \omega$,

(i) *for any recursively regular* $\lambda > \sigma$, *there exists an ordinal* κ *such that* $\sigma < \kappa \leqslant \lambda$, $\mathrm{Card}(\kappa) = \mathrm{Card}(\sigma)$, *and* κ *is* λ*-stable*;

(ii) *there exists an ordinal* κ *such that* $\sigma < \kappa$, $\mathrm{Card}(\kappa) = \mathrm{Card}(\sigma)$, *and* κ *is stable.*

Proof. For (i), suppose λ is recursively regular and $\sigma < \lambda$, and set

$$A = \{F(\boldsymbol{\mu}) : \boldsymbol{\mu} \leqslant \sigma \wedge F \text{ is } (\infty, \lambda)\text{-partial recursive}\}.$$

As we noted above, A is closed under all (∞, λ)-partial recursive functions. Because λ is recursively regular, $A \subseteq \lambda$. Since the projection functions are (∞, λ)-recursive, $\sigma \subseteq A$ so by (4) of I.1.5, $\varphi_A(\boldsymbol{\mu}) = \boldsymbol{\mu}$ for all $\boldsymbol{\mu} \leqslant \sigma$.

For any $\nu \in A$ there exists $\boldsymbol{\mu} \leqslant \sigma$ and an (∞, λ)-partial recursive function F such that $F(\boldsymbol{\mu}) \simeq \nu$. By Lemma 5.2,

$$\varphi_A(\nu) \simeq F(\varphi_A(\boldsymbol{\mu})) \simeq F(\boldsymbol{\mu}) \simeq \nu.$$

It follows from (5) of I.1.5 that A is an ordinal $\kappa \leqslant \lambda$.

Clearly κ is λ-stable. Since there are only countably many (∞, λ)-partial recursive functions,

$$\mathrm{Card}(\kappa) = \mathrm{Card}(A) = \mathrm{Card}(\sigma) \cdot \aleph_0 = \mathrm{Card}(\sigma).$$

The proof of (ii) is obtained similarly from the bracketed portion of Lemma 5.2. □

5.4 Corollary. *For any cardinal $\rho > \omega$,*

$$\{\kappa : \kappa < \rho \wedge \kappa \text{ is stable}\} \quad \textit{and} \quad \{\kappa : \kappa < \rho \wedge \kappa \text{ is recursively regular}\}$$

both have cardinality ρ.

Proof. If ρ is an uncountable cardinal and $\sigma < \rho$, then by Theorem 5.3 there exists a stable κ such that $\mathrm{Card}(\kappa) = \mathrm{Card}(\sigma)$ and $\sigma < \kappa$, hence $\sigma < \kappa < \rho$. Thus the set of stable κ is cofinal in ρ. If ρ is regular, it follows that this set has cardinality ρ. The result for singular ρ is then immediate. Since a stable ordinal is recursively regular, there are also ρ recursively regular ordinals less than ρ. □

5.5 Lemma. *For any λ and ρ, if ρ is a limit of λ-stable ordinals [of stable ordinals], then ρ is λ-stable [stable].*

Proof. Suppose ρ is a limit of λ-stable ordinals. For any $\mu < \rho$ there exists a κ such that $\mu < \kappa < \rho$ and κ is λ-stable. Then for any (∞, λ)-partial recursive function F, $F(\mu) < \kappa < \rho$. Hence ρ is λ-stable. The proof for stability is identical. □

5.6 Corollary. *Every uncountable cardinal is stable and therefore recursively regular.*

Proof. It follows from Corollary 5.4 (or directly from 5.3) that any uncountable cardinal is a limit of stable ordinals, hence is stable by Lemma 5.5. □

We conclude this section by showing that the least stable ordinal is δ^1_2, the least ordinal not the order-type of a Δ^1_2 wellordering of ω. First we prove a generalization of Theorem 2.2.

5.7 Theorem. *For all κ and all $\lambda \geqslant \kappa$, the following are equivalent:*

(i) *κ is λ-stable;*

(ii) *for all $a \in \omega$, all $\mu < \kappa$, and all ν,*

$$\{a\}_{\infty\lambda}(\mu) \simeq \nu \leftrightarrow \{a\}_\kappa(\mu) \simeq \nu;$$

and if λ is recursively regular, also

(iii) *for all $a \in \omega$, all $\mu < \kappa$ and all ν,*

$$\{a\}_\lambda(\mu) \simeq \nu \leftrightarrow \{a\}_\kappa(\mu) \simeq \nu.$$

Proof. For recursively regular λ, the equivalence of (ii) and (iii) follows immediately from Theorem 2.2. From (ii) we may conclude that any (∞, λ)-partial recursive function applied to arguments from κ has the same value as some κ-partial recursive function, hence has value less than κ. Thus κ is λ-stable.

Suppose now that κ is λ-stable. Then an argument very similar to that used for the implication ((b)→(a)) of Theorem 2.2 will show that for all τ, a, $\boldsymbol{\mu}$, and ν,

$$\boldsymbol{\mu} < \kappa \wedge (a, \boldsymbol{\mu}, \nu) \in \Omega_{\tau\lambda} \rightarrow (a, \boldsymbol{\mu}, \nu) \in \Omega_{\kappa\kappa}.$$

Cases (0)–(2) are done in exactly the same way. For case (3) we use the fact that κ is recursively regular. For case (4), if $(a, \boldsymbol{\mu}, \nu) \in \Omega_{\tau\lambda}$, then $\{a\}_{\infty\lambda}(\boldsymbol{\mu}) \simeq \nu$ and because κ is closed under (∞, λ)-partial recursive functions, $\nu < \kappa$ and the proof proceeds as before. □

5.8 Theorem. *For all κ, the following are equivalent*:

(i) *κ is stable*;

(ii) *for all $a \in \omega$, all $\boldsymbol{\mu} < \kappa$, and all ν,*

$$\{a\}_\infty(\boldsymbol{\mu}) \simeq \nu \leftrightarrow \{a\}_\kappa(\boldsymbol{\mu}) \simeq \nu.$$

Proof. As for Theorem 5.7. □

5.9 Theorem. *δ_2^1 is the least stable ordinal.*

Proof. We show first that no $\kappa < \delta_2^1$ is stable. For any such κ there exists a $\gamma \in \mathrm{W} \cap \Delta_2^1$ such that $\|\gamma\| = \kappa$. By Theorem 3.7, γ is $\aleph_1$-recursive. If κ were stable, hence $\aleph_1$-stable, then by the preceding theorem, γ would be κ-recursive. This contradicts Lemma 3.1.

To prove that δ_2^1 is stable, suppose $\boldsymbol{\mu} < \delta_2^1$ and $\{a\}_\infty(\boldsymbol{\mu}) \simeq (\nu)$. It suffices to show that $\nu < \delta_2^1$. Since $\aleph_1$ is stable by Corollary 5.6, also $\{a\}_{\aleph_1}(\boldsymbol{\mu}) \simeq \nu$. Let $\boldsymbol{\varepsilon} = (\varepsilon_0, \ldots, \varepsilon_{k-1})$ be a sequence of elements of $\mathrm{W} \cap \Delta_2^1$ such that for $i < k$, $\|\varepsilon_i\| = \mu_i$. Then

$$\exists \gamma\, \exists \mathbf{m}\, \exists n\, [|\mathbf{m}|_\gamma = \|\boldsymbol{\varepsilon}\| \wedge O(\langle a, \mathbf{m}, n\rangle, \gamma)].$$

By Lemma 3.5, the expression in brackets defines a Δ_2^1 relation. Then by the Basis Theorem (IV.7.9).

$$(\exists \gamma \in \Delta_2^1)\, \exists \mathbf{m}\, \exists n\, [|\mathbf{m}|_\gamma = \|\boldsymbol{\varepsilon}\| \wedge O(\langle a, \mathbf{m}, n\rangle, \gamma)].$$

But then $\nu = |n|_\gamma$ for some $\gamma \in \mathrm{W} \cap \Delta_2^1$ so $\nu < \|\gamma\| < \delta_2^1$. □

5.10 Corollary. *For any $R \subseteq {}^k\omega$, the following are equivalent:*
(i) $R \in \Delta_2^1$ (Σ_2^1);
(ii) R *is* δ_2^1-*(semi-) recursive*;
(iii) R *is* κ-*(semi-) recursive for any* $\kappa \geqslant \delta_2^1$;
(iv) R *is* ∞-*(semi-) recursive.*

Proof. Immediate from Theorem 3.7 and the preceding two results. □

It follows, of course, that there is no analogue of Theorem 3.7 for higher levels of the analytical hierarchy. The equivalence of (i) and (ii) provides another hierarchy of length δ_2^1 for the Δ_2^1 relations on numbers (cf. V.6.8–9).

5.11 Corollary. *For any* ∞-*recursive set A of ordinals, and any stable ordinal* κ,
(i) *if* $A \neq \varnothing$, *then* $A \cap \delta_2^1 \neq \varnothing$;
(ii) *if* $A \not\subseteq \kappa$, *then* $A \cap \kappa$ *is cofinal in* κ;
(iii) *if* κ *is a cardinal and* $A \not\subseteq \kappa$, *then* $\mathrm{Card}(A \cap \kappa) = \kappa$.

Proof. Let A be ∞-recursive and set

$$F(\rho) \simeq \text{least } \mu \,.\, \mu \in A.$$

F is ∞-partial recursive and total if $A \neq \varnothing$. Since $0 < \delta_2^1$, $F(0) < \delta_2^1$ and $F(0) \in A$. Similarly, set

$$G(\rho, \sigma) \simeq \text{least } \mu[\mu \in A \wedge \mu > \sigma].$$

If $A \not\subseteq \kappa$, then for any $\sigma < \kappa$, $G(\rho, \sigma)$ is defined for all ρ. Hence $\sigma < G(0, \sigma) < \kappa$ and $G(0, \sigma) \in A$. (iii) is immediate from (ii). □

5.12–5.24 Exercises

5.12. Show that if $\kappa < \lambda$ and λ is stable, then κ is stable iff κ is λ-stable.

5.13. Show that ω_1 is not weakly stable.

5.14. Do there exist 0-stable ordinals κ such that $\omega < \kappa < \omega_1$?

5.15. Show that the least weakly stable ordinal is less than the least stable ordinal.

5.16. Suppose that both κ and $\lambda > \kappa$ are recursively regular. Show that κ is λ-stable iff for all λ-recursive relations R and all $\mu < \kappa$, $(\exists \pi < \lambda) R(\pi, \mu) \leftrightarrow (\exists \pi < \kappa) R(\pi, \mu)$.

5.17. Show that if κ is a limit of stable ordinals, then for any $R \subseteq {}^k\omega$, if R is κ-semi-recursive in parameters, then R is in fact κ-recursive in parameters.

5.18. Show that for any $A \subseteq \omega$, $\delta_2^1[A]$ is stable.

5.19. Show that for all $r \geqslant 3$, δ_r^1 is a limit of stable ordinals, hence is stable.

5.20. Show that for all $r \geqslant 1$ and all $R \subseteq {}^k\omega$, if R is δ_r^1-recursive in parameters, then $R \in \Delta_r^1$.

5.21. Prove the following extension of Corollary 5.10: for any $R \subseteq {}^k\omega$, R is ∞-(semi-) recursive in parameters iff R is $\delta_2^1[R]$-(semi-) recursive in parameters. (Suppose R is ∞-recursive in the parameters $\boldsymbol{\rho} = (\rho_0, \ldots, \rho_{n-1})$. Apply Theorem 5.2 to the set $A = \{F(\boldsymbol{\mu}) : \boldsymbol{\mu} \in \omega \cup \{\rho_0, \ldots, \rho_{n-1}\} \wedge F$ is ∞-partial recursive$\}$ to show that R is $\aleph_1$-recursive in parameters. Then apply a relativized version of the technique used in the proof of Theorem 5.9.)

5.22. Show that for any $A \subseteq \omega$, if A is ∞-recursive in parameters, then for all $R \subseteq {}^k\omega$,

(i) R is $\delta_2^1[A]$-semi-recursive in parameters $\leftrightarrow R \in \Sigma_2^1[A]$;

(ii) R is $\delta_2^1[A]$-recursive in parameters $\leftrightarrow R \in \Delta_2^1[A]$.

(Use the preceding exercise and Exercise 3.18.)

5.23. Suppose that every $R \subseteq {}^k\omega$ is ∞-recursive in parameters (this follows from the hypothesis V = L). Show that for all $r \geqslant 1$ and all $R \subseteq {}^k\omega$,

$$R \text{ is } \delta_r^1\text{-recursive in parameters} \leftrightarrow R \in \Delta_r^1.$$

5.24. Let Γ be an arithmetical inductive operator (not necessarily monotone) over ω. Show that $|\Gamma|$ is less than or equal to the least weakly stable ordinal.

5.25 Notes. The notion of stability is implicit already in Takeuti [1960] — he shows that uncountable cardinals are stable. Most of the other results are due to Kripke [1964] and Platek [1966].

6. *Recursively Large Ordinals*

The recursively regular ordinals are, of course, analogous to the regular cardinals of set theory. Furthermore, we know from Corollary 5.4 that every uncountable cardinal ρ is the ρ-th recursively regular ordinal. In particular, the recursively regular ordinals form a proper class. Let τ_ρ denote the ρ-th recursively regular ordinal — that is,

$$\tau_\rho = \text{least } \sigma[\sigma \text{ is recursively regular} \wedge (\forall \pi < \rho)\sigma > \tau_\pi].$$

Then $\tau_0 = \omega$, $\tau_1 = \omega_1$ and for all π and ρ, $\rho \leqslant \tau_\rho$ and $\pi < \rho \to \tau_\pi < \tau_\rho$. The fact mentioned above may now be stated: for any uncountable cardinal ρ, $\tau_\rho = \rho$. In set theory a fixed point in the enumeration of the regular cardinals is called (weakly) inaccessible. Accordingly, we say

6.1 Definition. For any κ, κ is *recursively inaccessible* iff $\kappa = \tau_\kappa$.

Thus every uncountable cardinal is recursively inaccessible, but these are by no means the only recursively inaccessible ordinals.

6.2 Lemma. *The function F such that $F(\rho) \simeq \tau_\rho$ is ∞-recursive. For any κ, the partial function F_κ such that $F_\kappa(\rho) \simeq \tau_\rho$ iff $\tau_\rho < \kappa$ is (∞, κ)-partial recursive and κ-recursive if κ is recursively regular.*

Proof. This follows from Theorem 1.9, Corollary 2.19, and the fact that the sequence τ_ρ is defined by the following primitive recursion:

$$\tau_0 = \omega;$$

$$\tau_{\rho+1} = \text{least } \sigma[\sigma \text{ is recursively regular} \wedge \sigma > \tau_\rho];$$

$$\tau_\rho = \text{least } \sigma[\sigma \text{ is recursively regular} \wedge \sigma \geqslant \sup^+_{\pi<\rho} \tau_\pi],$$

if ρ is a limit ordinal. □

6.3 Corollary. *For any κ, if κ is stable or even weakly stable, then κ is recursively inaccessible.*

Proof. Suppose first that κ is stable. Since the function F of the Lemma is ∞-recursive, κ is closed under F — that is, $\rho < \kappa \to \tau_\rho < \kappa$. Since κ is itself recursively regular, $\tau_\kappa \leqslant \kappa$. But always $\kappa \leqslant \tau_\kappa$ so in fact $\kappa = \tau_\kappa$.

If κ is λ-stable for some $\lambda > \kappa$, then κ is closed under F_λ. Hence, for all $\rho < \kappa$, if $F_\lambda(\rho)$ is defined (i.e., if $\tau_\rho < \lambda$), then $\tau_\rho < \kappa$. As in the previous case it suffices to show that $\rho < \kappa \to \tau_\rho < \kappa$. Suppose otherwise and let ρ_0 be the least $\rho < \kappa$ such that $\tau_{\rho_0} \geqslant \kappa$. Then as κ is itself recursively regular, $\tau_{\rho_0} = \kappa < \lambda$. But then $\tau_{\rho_0} < \kappa$, a contradiction. □

6.4 Corollary. *The least recursively inaccessible ordinal is less than the least stable ordinal.*

Proof. This is immediate from Corollary 6.3 and Exercise 5.15, but we also give a direct proof. Let τ^1_ρ denote the ρ-th recursively inaccessible ordinal:

$$\tau_\rho^1 = \text{least } \sigma[\tau_\sigma = \sigma \wedge (\forall \pi < \rho)\sigma > \tau_\pi^1].$$

By a proof similar to that of Lemma 6.2, the function F^1 such that $F^1(\rho) = \tau_\rho^1$ is ∞-recursive. Hence δ_2^1 is closed under F^1 so in particular, $F^1(0) < \delta_2^1$. □

The proof of this corollary suggests that these ideas may be carried much further. Again by analogy with set-theoretic terminology we call ordinals κ such that $\kappa = \tau_\kappa^1$ *recursively hyper-inaccessible.* Then every (weakly) stable κ is also recursively hyper-inaccessible and the least such ordinal is less than δ_2^1. If τ_ρ^2 enumerates the recursively hyper-inaccessible ordinals, then an ordinal κ such that $\kappa = \tau_\kappa^2$ is recursively *hyper-hyper-inaccessible,* etc. (cf. Exercise 6.23).

Clearly the first recursively regular ordinals $\tau_0, \tau_1, \ldots, \tau_\omega, \ldots, \tau_{\tau_\omega}, \ldots$ are all less than the least recursively inaccessible. The next two results give characterizations of this ordinal.

6.5 Theorem. *For all κ, κ is recursively inaccessible iff κ is recursively regular and a limit of recursively regular ordinals.*

Proof. Suppose first that κ is recursively inaccessible. Thus $\kappa = \tau_\kappa$ so κ is recursively regular. Furthermore, for any $\sigma < \kappa$, $\sigma \leqslant \tau_\sigma < \tau_\kappa = \kappa$, so κ is also a limit of recursively regular ordinals.

Conversely, suppose κ is recursively regular and a limit of recursively regular ordinals. Suppose further that κ is not recursively inaccessible so that $\kappa < \tau_\kappa$. The sequence τ_ρ is increasing so $\kappa = \tau_\rho$ for some $\rho < \kappa$. Then because κ is a limit of the recursively regular ordinals less than κ, $\kappa = \sup^+_{\pi<\rho} \tau_\pi$. Since the sequence τ_π for $\pi < \rho$ is κ-partial recursive, this contradicts the recursive regularity of κ. □

6.6 Theorem. *The least recursively inaccessible ordinal is $\omega_1[\mathsf{E}_1]$.*

Proof. By Lemma 4.12, Exercise VI.1.20, and the preceding theorem,

$$\begin{aligned}\omega_1[\mathsf{E}_1] \in \mathrm{Om}[\mathsf{E}_1] &\subseteq \mathrm{Lim}(\mathrm{Om}[\mathsf{E}]) \subseteq \mathrm{Lim}\{\kappa : \kappa \text{ is recursively regular}\}\\ &\subseteq \{\kappa : \kappa \text{ is recursively inaccessible}\}.\end{aligned}$$

Hence $\omega_1[\mathsf{E}_1]$ is recursively inaccessible. To see that it is the least such, let κ be any recursively inaccessible ordinal. By Example 4.6 there is an index d such that for every $\lambda > \omega$, E is λ-effective with index d. Hence by Theorem 4:14 and the preceding theorem, there is an index d' such that

$$\kappa \in \mathrm{Lim}\{\lambda : \lambda \text{ is recursively regular}\} \subseteq \mathrm{Lim}(\mathrm{Ef}_d[\mathsf{E}]) \subseteq \mathrm{Ef}_{d'}[\mathsf{E}_1].$$

In particular, E_1 is κ-effective so that by Corollary 4.9, $\omega_1[\mathsf{E}_1] \leqslant \kappa$. □

Next in the galaxy of "large cardinals" of set theory are the Mahlo cardinals; these also have their "recursive analogue".

6.7 Definition. For any κ, κ is *recursively Mahlo* iff for any $F:\kappa\to\kappa$ which is κ-recursive in parameters, there exists a recursively regular ordinal $\lambda<\kappa$ which is closed under F — that is, $(\forall\pi<\lambda)F(\pi)<\lambda$.

Compare Definition 6.7 with Exercise 2.23. Analogously as in set theory, we have

6.8 Lemma. *For any κ, if κ is recursively Mahlo, then κ is recursively regular, recursively inaccessible, recursively hyper-inaccessible, etc.*

Proof. Suppose that κ is recursively Mahlo. Let F be κ-partial recursive and let ρ, $\boldsymbol{\mu}<\kappa$ be such that $F(\pi,\boldsymbol{\mu})$ is defined for all $\pi<\rho$. Let G be the function κ-partial recursive in parameters defined by:

$$G(0)\simeq\rho,\quad\text{and for}\quad\pi>0,$$

$$G(\pi)\simeq\begin{cases}F(\pi,\boldsymbol{\mu}), & \text{if}\quad\pi<\rho;\\ 0, & \text{otherwise.}\end{cases}$$

There exists a recursively regular $\lambda<\kappa$ which is closed under G. In particular, $\rho=G(0)<\lambda$ so also

$$\sup^+_{\pi<\rho}F(\pi,\boldsymbol{\mu})=\sup^+_{\pi<\rho}G(\pi)<\lambda<\kappa.$$

Thus κ is recursively regular.

To show that κ is recursively inaccessible, it suffices by Theorem 6.5 to show that κ is a limit of recursively regular ordinals. For any $\sigma<\kappa$, set $H(\pi)=\sigma$ for all $\pi<\kappa$. Then H is κ-recursive in parameters so there exists a recursively regular ordinal $\lambda<\kappa$ which is closed under H. In particular, $\sigma=H(0)<\lambda$.

We leave to the reader (Exercise 6.24) the proof that κ is recursively hyper-inaccessible, etc. $\square$

6.9 Theorem. *For any κ, if κ is stable or even weakly stable, then κ is recursively Mahlo.*

Proof. Suppose first that κ is stable and F is κ-recursive in parameters, say $F(\pi)\simeq\{a\}_\kappa(\pi,\boldsymbol{\mu})$ with $\boldsymbol{\mu}<\kappa$. Since κ is recursively regular and $F(\pi)$ is defined for all $\pi<\kappa$, we have by Theorem 2.6

$$(\forall\pi<\kappa)(\exists\upsilon<\kappa)\,T(a,\langle\pi,\boldsymbol{\mu}\rangle,\upsilon).$$

Let G be the function defined by, for any $\boldsymbol{\rho}$,

$$G(\boldsymbol{\rho}) \simeq \text{least } \lambda[\lambda \text{ is recursively regular } \wedge \\ (\forall \pi < \lambda)(\exists \upsilon < \lambda)\, T(a, \langle \pi, \boldsymbol{\rho}\rangle, \upsilon)].$$

Clearly G is ∞-recursive and $G(\boldsymbol{\mu})$ is defined. Since κ is stable and $\boldsymbol{\mu} < \kappa$, also $G(\boldsymbol{\mu}) < \kappa$. Then $G(\boldsymbol{\mu})$ is a recursively regular ordinal less than κ which is closed under F.

The argument in case κ is only weakly stable is nearly identical. □

The least recursively Mahlo ordinal is $\omega_1[\mathbf{sJ}]$, but the proof of this fact is beyond the scope of this book (Harrington [1974]). It is not difficult to verify directly that this ordinal is less than δ^1_2; this also follows from Corollaries 6.12 and 6.19 below or from Corollary VII.1.9.

We next consider the notion of projectibility. Recursive regularity is an effective analogue of the set-theoretic property of being a regular cardinal. Nonprojectibility is in a sense an effective analogue of the property of being a cardinal, but the logical relationship between the notions is lost in the translation.

6.10 Definition. For any κ and λ,

(i) κ is *projectible to* λ iff there exists a one–one function F κ-recursive in parameters such that $\operatorname{Im} F \subseteq \lambda$;

(ii) κ^*, the *projectum* of κ, is the least λ such that κ is projectible to λ;

(iii) κ is *projectible* iff $\kappa^* < \kappa$; otherwise κ is *nonprojectible*.

We have already encountered numerous examples of ordinals which are projectible. In Exercise 3.9 is constructed an ω_1-recursive function G such that for all $\rho < \omega_1$,

$$G(\rho) \in W \quad \text{and} \quad \|G(\rho)\| = \rho.$$

Clearly G is one–one and thus it establishes that ω_1 is projectible to ω. Hence $\omega_1^* = \omega$. Similarly, the $\omega_1[\mathsf{I}]$-recursive function G of Exercise 4.20 such that for all $\rho < \omega_1[\mathsf{I}]$,

$$G(\rho) \in U^{\mathsf{I}} \quad \text{and} \quad |G(\rho)|^{\mathsf{I}} = \rho$$

shows that when E is recursive in I and I is $\omega_1[\mathsf{I}]$-effective, $\omega_1[\mathsf{I}]$ is projectible to ω. In particular, for all n, $\omega_1[\mathsf{E}_n]^* = \omega$. We could also show by similar means from the results at the end of § 3 that δ^1_2 is projectible to ω, but we shall have a simpler proof of this fact below (Corollary 6.15).

Of course, every infinite cardinal is nonprojectible, but there are many more nonprojectible ordinals.

6.11 Lemma. $\{\kappa : \kappa$ is recursively regular and nonprojectible$\}$ *is* $(\infty, 0)$-*recursive.*

Proof. By Corollary 2.19, $\{\kappa : \kappa$ is recursively regular$\}$ is $(\infty, 0)$-recursive. Let

$$R(a, \kappa, \lambda, \langle \boldsymbol{\mu} \rangle) \leftrightarrow \boldsymbol{\mu} < \kappa \wedge (\forall \rho < \kappa)(\exists \upsilon < \kappa)[T(a, \langle \rho, \boldsymbol{\mu} \rangle, \upsilon) \wedge (\upsilon)_0 < \lambda]$$

and

$$S(a, \kappa, \lambda, \langle \boldsymbol{\mu} \rangle) \leftrightarrow (\forall \rho\rho'\upsilon\upsilon' < \kappa)[T(a, \langle \rho, \boldsymbol{\mu} \rangle, \upsilon) \wedge T(a, \langle \rho', \boldsymbol{\mu} \rangle, \upsilon') \wedge (\upsilon)_0 = (\upsilon')_0 \rightarrow \rho = \rho'].$$

$R(a, \kappa, \lambda, \langle \boldsymbol{\mu} \rangle)$ holds whenever the index a defines a function F κ-recursive in parameters $\boldsymbol{\mu} < \kappa$ such that $\operatorname{Im} F \subseteq \lambda$. $S(a, \kappa, \lambda, \langle \boldsymbol{\mu} \rangle)$ holds in addition just in case this function is 1–1. Hence if κ is recursively regular,

$$\kappa \text{ is projectible} \leftrightarrow (\exists a < \omega)(\exists \lambda < \kappa)(\exists \boldsymbol{\mu} < \kappa)[R(a, \kappa, \lambda, \langle \boldsymbol{\mu} \rangle) \wedge S(a, \kappa, \lambda, \langle \boldsymbol{\mu} \rangle)].$$

Since R and S are obviously $(\infty, 0)$-recursive, this proves the Lemma. □

6.12 Corollary. *The least recursively regular nonprojectible ordinal is less than* δ_2^1. *There are* κ *recursively regular nonprojectible ordinals less than every uncountable cardinal* κ.

Proof. By Corollary 5.11. □

Although it is by no means obvious from the definitions, the least recursively regular nonprojectible ordinal is a quite large ordinal. We show this by use of techniques from § 5. For any σ and λ, let

$$\operatorname{st}_\lambda(\sigma) = \{F(\boldsymbol{\mu}) : \boldsymbol{\mu} \leqslant \sigma \wedge F \text{ is } (\infty, \lambda)\text{-partial recursive}\}$$

and

$$\operatorname{st}(\sigma) = \{F(\boldsymbol{\mu}) : \boldsymbol{\mu} \leqslant \sigma \wedge F \text{ is } \infty\text{-partial recursive}\}.$$

In the proof of Theorem 5.3, we essentially showed:

6.13 Lemma. *For any* $\sigma \geqslant \omega$ *and any recursively regular* λ,

(i) $\mathrm{st}_\lambda(\sigma)$ *is the least* λ*-stable ordinal greater than* σ; $\mathrm{st}_\lambda(\sigma) \leq \lambda$ *and* $\mathrm{Card}(\sigma) = \mathrm{Card}(\mathrm{st}_\lambda(\sigma))$;

(ii) $\mathrm{st}(\sigma)$ *is the least stable ordinal greater than* σ; $\mathrm{Card}(\sigma) = \mathrm{Card}(\mathrm{st}(\sigma))$. □

For example, $\mathrm{st}(\omega) = \delta_2^1$, the least stable ordinal. If κ is recursively regular, $\sigma < \kappa$, and there is no recursively regular λ such that $\sigma < \lambda < \kappa$, then $\mathrm{st}_\kappa(\sigma) = \kappa$. In particular, $\mathrm{st}_{\omega_1}(\omega) = \omega_1$.

We shall call an ordinal σ *sequence closed* iff for all $\boldsymbol{\mu} < \sigma$, also $\langle \boldsymbol{\mu} \rangle < \sigma$. For the most part we shall use only the obvious fact that if σ is either recursively regular or a limit of recursively regular ordinals, then σ is sequence closed.

6.14 Theorem. *For any recursively regular* λ *and any sequence closed ordinal* σ *such that* $\omega \leq \sigma < \lambda$, $\mathrm{st}_\lambda(\sigma)$ *and* $\mathrm{st}(\sigma)$ *are both projectible to* σ.

Proof. Let λ and σ be as stated and set $\kappa = \mathrm{st}_\lambda(\sigma)$. Since κ is λ-stable, for all $\boldsymbol{\mu} \leq \sigma$,

$$\{a\}_{\infty\lambda}(\boldsymbol{\mu}) \simeq \nu \leftrightarrow \{a\}_\kappa(\boldsymbol{\mu}) \simeq \nu.$$

Furthermore, if $\{a\}_{\infty\lambda}(\boldsymbol{\mu}) \simeq \nu$ with $\boldsymbol{\mu} \leq \sigma$, it is not hard to see that there exist a' and $\boldsymbol{\mu}'$ such that $\boldsymbol{\mu}' < \sigma$ and $\{a'\}_{\infty\lambda}(\boldsymbol{\mu}', \sigma) \simeq \nu$. Hence

$$\kappa = \{F(\boldsymbol{\mu}, \sigma) : \boldsymbol{\mu} < \sigma \wedge F \text{ is } \kappa\text{-partial recursive}\}.$$

By Theorem 2.6, for all ρ,

$$(*) \qquad \rho < \kappa \rightarrow (\exists a < \omega)(\exists \boldsymbol{\mu} < \sigma)(\exists \upsilon < \kappa)[T(a, \langle \boldsymbol{\mu}, \sigma \rangle, \upsilon) \wedge (\upsilon)_0 = \rho].$$

Let $G(\rho)$ be the least $((a, \boldsymbol{\mu}), \upsilon)$ for which the inside part holds — that is,

$$G(\rho) \simeq (\kappa\text{-least } \pi[(\pi)_0 < \sigma \wedge T((\pi)_{0,0}, (\pi)_{0,1} * \langle \sigma \rangle, (\pi)_1) \wedge (\pi)_{1,0} = \rho])_0.$$

Then G is κ-partial recursive in the parameter σ and by the definition of G, $\mathrm{Im}\, G \subseteq \sigma$. By $(*)$ and the assumption that σ is sequence closed, G is total. Furthermore, if $G(\rho) \simeq \nu$, then there exist π, a, $\boldsymbol{\mu}$, and ν such that $\pi = \langle \nu, \upsilon \rangle = \langle \langle a, \langle \boldsymbol{\mu} \rangle \rangle, \upsilon \rangle$ and $\{a\}_\kappa(\boldsymbol{\mu}, \sigma) \simeq \rho$. It follows that G is 1–1 and is thus a projection of κ to σ.

The proof for $\mathrm{st}(\sigma)$ is nearly identical. □

6.15 Corollary. δ_2^1 *is projectible to* ω. *For any* κ, *the least stable or recursively regular ordinal greater than* κ *is projectible to* κ.

Proof. We noted above that $\delta_2^1 = \mathrm{st}(\omega)$ and ω is sequence closed. Since a limit of sequence closed ordinals is sequence closed, for any κ there exists a largest sequence closed ordinal $\sigma \leqslant \kappa$. Then the least stable ordinal greater than κ is $\mathrm{st}(\sigma)$ and if λ is the least recursively regular ordinal greater than κ, then $\lambda = \mathrm{st}_\lambda(\kappa)$. By Theorem 6.14, $\mathrm{st}(\sigma)$ and $\mathrm{st}_\lambda(\sigma)$ are projectible to σ, hence also to κ. □

This result yields yet another construction of a Σ_2^1 well-ordering R of order-type δ_2^1 — namely, if F is a δ_2^1-recursive function which projects δ_2^1 to ω, let

$$R(m, n) \leftrightarrow m, n \in \mathrm{Im}(F) \wedge F^{-1}(m) < F^{-1}(n)$$

(cf. Exercises V.1.30, 33).

The key to the "largeness" properties of nonprojectible ordinals is:

6.16 Theorem. *For any recursively regular κ, if $\kappa^* > \omega$, then κ^* is a limit of κ-stable ordinals.*

Proof. Suppose κ is recursively regular and $\kappa^* > \omega$. Let

$$\sigma = \sup\{\lambda : \lambda < \kappa^* \wedge (\lambda \text{ is } \kappa\text{-stable} \vee \lambda = \omega)\}.$$

Clearly $\sigma \leqslant \kappa^*$. Suppose, for a contradiction, that $\sigma < \kappa^*$. Then either $\sigma = \omega$ or σ is the largest κ-stable ordinal less than κ^* (because a limit of κ-stable ordinals is itself κ-stable by Lemma 5.5). In either case σ is sequence closed so by Theorem 6.14, $\mathrm{st}_\kappa(\sigma)$ is projectible to σ. But $\mathrm{st}_\kappa(\sigma)$ is a κ-stable ordinal greater than σ so $\mathrm{st}_\kappa(\sigma) \geqslant \kappa^*$. Thus κ is projectible to $\mathrm{st}_\kappa(\sigma)$ and composing these projections yields a projection of κ to σ, a contradiction. Hence $\sigma = \kappa^*$, and since $\kappa^* > \omega$, κ^* is a limit of the κ-stable ordinals $\lambda < \kappa^*$. □

6.17 Corollary. *For any recursively regular κ, κ^* is also recursively regular, and if $\kappa^* > \omega$, then κ^* is κ-stable.*

Proof. If $\kappa^* = \omega$, then it is certainly recursively regular. Otherwise, κ^* is a limit of κ-stable ordinals, hence is itself κ-stable and recursively regular. □

6.18 Corollary. *For any recursively regular $\kappa > \omega$, if κ is nonprojectible, then κ is a limit of smaller κ-stable ordinals.* □

6.19 Corollary. *For any recursively regular $\kappa > \omega$, if κ is nonprojectible, then κ is recursively Mahlo, recursively inaccessible, recursively hyper-inaccessible, etc.*

Proof. Suppose that κ is recursively regular and nonprojectible, and let F be any

(∞, κ)-partial recursive function and $\mu < \kappa$ such that $(\forall \pi < \kappa) F(\pi, \mu)$ is defined. By Corollary 6.18 there exists $\lambda < \kappa$ such that $\mu < \lambda$ and λ is κ-stable. Then $(\forall \pi < \lambda) F(\pi, \mu) < \lambda$. Hence κ is recursively Mahlo. The other conclusions follow from Lemma 6.8. □

A concept closely related to projectibility is the following:

6.20 Definition. For any recursively regular κ and any λ and $\rho < \kappa$,

(i) λ is *κ-projectible to ρ* iff there exists a 1–1 function F which is κ-partial recursive in parameters such that $\mathrm{Dm}\, F = \lambda$ and $\mathrm{Im}\, F \subseteq \rho$;

(ii) λ is a *κ-cardinal* iff λ is not κ-projectible to any $\rho < \lambda$.

We shall show in the next section that λ is a κ-cardinal iff "λ is a cardinal" is true in L_κ, the class of sets constructible before κ. The proof of the following is left as Exercise 6.41.

6.21 Theorem. *For any recursively regular $\kappa > \omega$,*

(i) *ω is a κ-cardinal;*

(ii) *for any κ-cardinal $\lambda > \omega$, λ is a limit of κ-stable ordinals and is thus κ-stable;*

(iii) *if $\kappa^* < \kappa$, then κ^* is the greatest κ-cardinal.* □

6.22–6.42 Exercises

6.22. Let σ_ρ denote the ρ-th ordinal which is either recursively regular or a limit of recursively regular ordinals. Show that there exist ρ such that $\rho = \sigma_\rho$ but ρ is not recursively inaccessible.

6.23. Let $\tau_\rho^0 = \tau_\rho$, $\tau_\rho^{\nu+1} = \rho$-th ordinal κ such that $\tau_\kappa^\nu = \kappa$, and for limit ν, $\tau_\rho^\nu = \rho$-th ordinal κ such that $\tau_\kappa^\pi = \kappa$ for all $\pi < \nu$. $\lambda\rho . \tau_\rho^\nu$ enumerates the ordinals which are *ν-recursively inaccessible.* Show that

(i) if κ is recursively Mahlo and $\rho, \nu < \kappa$, then also $\tau_\rho^\nu < \kappa$;

(ii) for all $n < \omega$, $\tau_0^n = \omega_1[\mathrm{E}_n]$;

(iii) the least ordinal κ such that $\tau_0^\kappa = \kappa$ is less than the least recursively Mahlo ordinal.

6.24. Show that if κ is recursively Mahlo, then κ is recursively hyper-inaccessible.

6.25. For any recursively regular κ and any $\lambda < \kappa$ such that κ is projectible to λ,

$$\kappa = \sup{}^+\{\sigma : \sigma \text{ is the order-type of a } \kappa\text{-recursive well-ordering of } \lambda\}.$$

6.26. Show that for any recursively regular κ and any $\lambda \leqslant \kappa$, the following are equivalent:

(i) κ is projectible to λ;

(ii) there exists a set $A \subseteq \lambda$ and a function G κ-partial recursive in parameters such that $A \subseteq \mathrm{Dm}\, G$ and $G''A = \kappa$;

(iii) there exists a set $A \subseteq \lambda$ and a function H λ-partial recursive in parameters such that $A \subseteq \mathrm{Dm}\, H$ and $\sup^+ H''A = \kappa$;

(iv) there exists a set $B \subseteq \lambda$ such that B is κ-semi-recursive in parameters but not κ-recursive in parameters.

6.27. Show that if κ is recursively regular and not projectible to ω, then $\{\alpha : \alpha$ is κ-recursive$\}$ is a model of the Σ^1_2-Comprehension schema (use Exercise 3.16).

6.28. Suppose that $\mathsf{A} \subseteq {}^{\omega}\omega$ *has the following property*: *for any arithmetical relation* R *and any* $\alpha_0, \ldots, \alpha_{l-1} \in \mathsf{A}$,

$$\exists \beta\, \mathsf{R}(\mathbf{m}, \boldsymbol{\alpha}, \beta) \rightarrow (\exists \beta \in \mathsf{A})\, \mathsf{R}(\mathbf{m}, \boldsymbol{\alpha}, \beta);$$

(A is said to be a *β-model*). Let

$$\omega_1(\mathsf{A}) = \sup{}^+\{\|\gamma\| : \gamma \in \mathsf{A} \cap \mathsf{W}\}.$$

Show that

(i) if A is a model of Δ^1_0-Comprehension, then $\omega_1(\mathsf{A})$ is either recursively regular or a limit of recursively regular ordinals;

(ii) if A is a model of Π^1_1-Comprehension, then $\omega_1(\mathsf{A})$ is a limit of recursively regular ordinals;

(iii) if A is a model of Δ^1_2-Comprehension, then $\omega_1(\mathsf{A})$ is recursively inaccessible;

(iv) if A is a model of Σ^1_2-Comprehension, then $\omega_1(\mathsf{A})$ is not projectible to ω.
For (iii) use Lemma 3.5 and the result of Exercise IV.7.18 that Δ^1_2-Comprehension implies Σ^1_2-Choice.

6.29. In the notation of the preceding exercise, let $\omega_1^\kappa = \omega_1(\{\alpha : \alpha$ is κ-recursive in parameters$\})$. Show that for $\kappa > \omega$,

(i) if κ is recursively regular, then $\omega_1^\kappa \leqslant \kappa$;

(ii) there exist κ such that $\omega_1^\kappa < \kappa$;

(iii) if κ is recursively regular and projectible to ω, then $\omega_1^\kappa = \kappa$;

6.30. Show that

(i) if κ is projectible to ω and $\{\alpha : \alpha$ is κ-recursive in parameters$\}$ is a β-model and satisfies the Δ^1_2-Comprehension schema, then κ is recursively inaccessible;

(ii) if $\{\alpha : \alpha$ is κ-recursive in parameters$\}$ is a β-model and satisfies the Σ^1_2-Comprehension schema, then κ is not projectible to ω.

6.31. Show that for any recursively regular κ and any $R \subseteq {}^k\omega$, if R is κ-(semi-) recursive in parameters, then R is also ω_1^κ-(semi-) recursive in parameters. (Cf. Exercise 5.21. Use the same technique as there but now show that A collapses to some ordinal $\lambda \leqslant \omega_1^\kappa$ and that R is λ-recursive in parameters.)

6.32. Show that if κ is recursively regular, then ω_1^κ is recursively regular (cf. Exercise 6.28 (i)).

6.33. Show that for any κ, if E_1 is κ-effective, then ω_1^κ is recursively inaccessible, hence for κ projectible to ω, E_1 is κ-effective iff κ is recursively inaccessible.

6.34. Is δ_3^1 projectible?

6.35. Show that the class of sequence-closed ordinals is $(\infty, 0)$-recursive and if $\mathrm{Sqc}(\sigma)$ denotes the σ-th sequence-closed ordinal, then the function Sqc is $(\infty, 0)$-recursive.

6.36. Show that every stable ordinal is a limit of recursively regular nonprojectible ordinals.

6.37. Prove the converse of Corollary 6.18: if κ is a limit of smaller κ-stable ordinals, then κ is nonprojectible.

6.38. Say that κ is *projectible onto* λ iff there exists a one–one function F κ-recursive in parameters such that $\mathrm{Im}\, F = \lambda$ — that is, there is a one–one correspondence between κ and λ that is κ-recursive in parameters. κ is *strongly projectible* iff κ is projectible onto some $\lambda < \kappa$. Show

(i) if κ is nonprojectible or recursively regular, then κ is not strongly projectible;

(ii) if κ is a limit of sequence-closed ordinals but is not recursively regular, then κ is strongly projectible;

(iii) if κ is nonprojectible and a limit of sequence-closed ordinals, then κ is recursively regular.

6.39. Let $\kappa = \sup^+\{\tau_n : n < \omega\}$. What is κ^*?

6.40. Show that for any recursively regular κ and any $\lambda < \kappa$, λ is a κ-cardinal iff there is no one–one function F partial κ-recursive in parameters such that $\mathrm{Dm}\, F = \lambda$ and $\mathrm{Im}\, F$ is an ordinal $\rho < \lambda$.

6.41. Prove Theorem 6.21.

6.42. A κ-cardinal λ is *κ-regular* iff there is no $\rho < \lambda$ and no F which is

κ-partial recursive in parameters such that $\lambda = \sup_{\pi<\rho} F(\pi)$. Show that if λ is the least κ-cardinal greater than ω, then λ is κ-regular.

7. Ordinal Recursion and Constructible Sets

The perspective of this book has been recursion-theoretic. When several treatments of a topic were possible, we have chosen the one which seemed most in the spirit of recursion theory. This is particularly true of the present chapter where we have developed the theory of recursion on ordinals as a natural generalization of ordinary recursion theory and completely neglected an alternative and quite distinct set-theoretic description of the theory. In this section we present this description and prove its equivalence with ordinal recursion theory as given above.

In a nutshell, the point is that the theory of recursion on ordinals may equally well be viewed as a theory of definability over the universe L of constructible sets (introduced in § V.2). The foundation of this correspondence lies in the results of § III.5 where we see that ordinary recursion theory may be described in terms of definability over $\mathfrak{N}$, the standard model of arithmetic. If we add to this the facts that (1) there is a "simple" one–one correspondence between ω and L_ω, the class of sets constructible at finite levels, and (2) ordinary recursion theory is equivalent with ω-recursion theory (Theorem 1.4), we arrive at the conclusion that ω-recursion theory may be described in terms of definability over L_ω. The results below show that the same is true for any recursively regular ordinal κ in place of ω.

To provide ourselves with ample symbols, we shall need to adopt some new notational conventions for this section. In particular we shall use letters

$$a,\ b,\ c,\ d,\ t,\ u,\ v,\ w,\ x,\ y,\ z,\ M, \quad \text{and} \quad N \text{ for arbitrary sets}$$

and

$$f,\ g, \quad \text{and} \quad h \text{ for functions from } k\text{-tuples of sets to sets.}$$

These variables are used both in the metatheory and the formal theory. To the language $\mathscr{L}_{ZF}$ of set theory as described in § V.2 we add also $\vee$ (or) as a logical symbol.

7.1 Definition. The class of Δ_0 *formulas* of $\mathscr{L}_{ZF}$ is the smallest class X such that

(i) all atomic formulas of $\mathscr{L}_{ZF}$ and their negations belong to X;

(ii) for any $\mathfrak{A}$, $\mathfrak{B} \in X$, both $\mathfrak{A} \wedge \mathfrak{B}$ and $\mathfrak{A} \vee \mathfrak{B}$ belong to X;

(iii) for any $\mathfrak{A} \in X$ and any variables x and y, both $(\exists x \in y)\mathfrak{A}$ and $(\forall x \in y)\mathfrak{A}$ belong to X.

Of course, $(\exists x \in y)\mathfrak{A}$ is an abbreviation for $\exists x\,(x \in y \wedge \mathfrak{A})$, but we think of this *bounded quantifier* as a primitive symbol. The point of Δ_0 formulas is that no quantifiers occur without bounds. Note that the negation of a Δ_0 formula is logically equivalent to a Δ_0 formula.

Many of the elementary concepts of set theory are definable using only Δ_0 formulas. Among these are: $\varnothing$ (the empty set), the Boolean operations, ordered pair, cartesian product, relation and function, domain and range, ordinal, ordinal successor, finite ordinal, and supremum of a set of ordinals. As an example,

$$a = \mathrm{Dm}\, f \leftrightarrow (\forall x \in a)(\exists z \in f)(\exists w \in z)(\exists y \in w)\,.\, z = \langle x, y\rangle$$
$$\wedge (\forall z \in f)(\forall w \in z)(\forall x \in w)(\forall y \in w)\,.\, z = \langle x, y\rangle \rightarrow x \in a.$$

For a more detailed exposition of these facts see Barwise [1975].

The other important feature of Δ_0 formulas is that they are *absolute* for transitive $\in$-models. That is, for any transitive sets $M \subseteq N$, any Δ_0 formula $\mathfrak{A}$ with free variables included among $x_0, \ldots, x_{k-1}$ and any $u_0, \ldots, u_{k-1} \in M$,

$$(M, \in) \models \mathfrak{A}[\mathbf{u}] \quad \text{iff} \quad (N, \in) \models \mathfrak{A}[\mathbf{u}].$$

This is easily proved by induction on the class of Δ_0 formulas. In particular, this is true when N is the "real world" so that, for example, if $\mathfrak{D}$ is the Δ_0 formula given above that defines "domain," then for $a, f \in M$,

$$(M, \in) \models \mathfrak{D}[a, f] \quad \text{iff} \quad a = \mathrm{Dm}\, f.$$

Similarly,

$$(M, \in) \models \text{"}a \text{ is an ordinal"} \quad \text{iff} \quad a \text{ is an ordinal}.$$

These and similar facts are used repeatedly below without special mention.

7.2 Definition. The class of Σ *formulas* of $\mathscr{L}_{\mathrm{ZF}}$ is the smallest class X such that (i)–(iii) as in Definition 7.1 and

$(\mathrm{iv})_\Sigma$ for any $\mathfrak{A} \in X$ and any variable x, $\exists x\, \mathfrak{A}$ belongs to X.

The class of Π *formulas* of $\mathscr{L}_{\mathrm{ZF}}$ is the smallest class X such that (i)–(iii) as in Definition 7.1 and

$(\mathrm{iv})_\Pi$ for any $\mathfrak{A} \in X$ and any variable x, $\forall x\, \mathfrak{A}$ belongs to X.

Clearly every Δ_0 formula is both a Σ formula and a Π formula. The negation of a Σ formula is logically equivalent to a Π formula and vice versa.

For any set M, a relation $R \subseteq {}^k M$ is called Σ (Π)-*definable* over M iff for some Σ (Π) formula $\mathfrak{A}$ and some $u_0, \ldots, u_{l-1} \in M$,

$$R(v_0, \ldots, v_{k-1}) \leftrightarrow (M, \in) \models \mathfrak{A}[\mathbf{v}, \mathbf{u}].$$

R is Δ-*definable* over M iff R is both Σ-definable and Π-definable over M. Note that any total function $f: {}^k M \to M$ is Δ-definable over M iff it is Σ-definable over M because

$$f(\mathbf{x}) = y \leftrightarrow \forall z \, [z \neq y \to f(\mathbf{x}) \neq z].$$

To state the main results of this section we need one more concept, admissibility. The theory of admissible sets and its generalizations have proved of enormous importance in set theory, generalized recursion theory, and even model theory. We shall not be able here even to touch upon any aspect of this theory except its relationship to ordinal recursion theory, but we recommend most strongly that any reader of this section take time to read at least the beginning of Barwise [1975].

7.3 Definition. For any set M, M is *admissible* iff M is transitive and $(M, \in)$ satisfies the universal closures of the following formulas:

(i) (*Pair*) $\exists a \, (x \in a \wedge y \in a)$;

(ii) (*Union*) $\exists b \, (\forall y \in a)(\forall x \in y) . x \in b$;

(iii) (Δ_0-*Separation*) $\exists b \, \forall x \, (x \in b \leftrightarrow x \in a \wedge \mathfrak{A})$ for all Δ_0 formulas $\mathfrak{A}$ in which b does not occur free;

(iv) (Δ_0-*Collection*) $(\forall x \in a) \exists y \, \mathfrak{A} \to \exists b \, (\forall x \in a)(\exists y \in b) \mathfrak{A}$ for all Δ_0 formulas $\mathfrak{A}$ in which b does not occur free.

We shall prove that for any ordinal κ,

(1) L_κ is admissible iff κ is recursively regular;

(2) if L_κ is admissible, then for any relation $R \subseteq {}^k \kappa$,

(i) R is κ-semi-recursive in parameters iff R is Σ-definable over L_κ;

(ii) R is κ-recursive in parameters iff R is Δ-definable over L_κ;

We proceed first towards the implications ($\to$). We shall need a number of facts about admissible sets. To give proofs for them all would take us too far from our topic, so we shall merely state the results and again refer the reader to Barwise [1975].

7.4 Theorem. *For any admissible set M,*

(i) (Δ-*Separation*) *for any Δ-definable set $a \subseteq M$ and any $b \in M$, also $a \cap b \in M$;*

(ii) (Σ-*Collection*) *$(M, \in)$ satisfies the universal closure of*

$$(\forall x \in a) \exists y \, \mathfrak{A} \to \exists b \, (\forall x \in a)(\exists y \in b) \mathfrak{A}$$

for all Σ formulas $\mathfrak{A}$ in which b does not occur free;

(iii) (*Transitive Closure*) *for any $a \in M$,* TC(a), *the smallest transitive set b such that $a \in b$, also belongs to M;*

(iv) (*Σ-Recursion*) *for any function $g : {}^{k+2}M \to M$ which is Σ-definable over M, the* (unique) *function $f : {}^{k+1}M \to M$ such that for all $x_0, \ldots, x_{k-1}, y \in M$*

$$f(\mathbf{x}, y) = g(\mathbf{x}, y, f \upharpoonright \mathrm{TC}(y))$$

where

$$f \upharpoonright \mathrm{TC}(y) = \{(z, f(\mathbf{x}, z)) : z \in \mathrm{TC}(y)\},$$

is also Σ-definable over M. □

Let $o(M)$ denote the least ordinal not in M. If M is transitive, then $o(M)$ is exactly the set of ordinals in M. Note that $o(\mathrm{L}_\kappa) = \kappa$.

7.5 Lemma. *For any κ, any admissible set M with $\kappa = o(M)$, any ρ, $\boldsymbol{\mu} < \kappa$, and any partial function F from ${}^k\kappa$ into κ, if F is Σ-definable over M and $F(\pi, \boldsymbol{\mu})$ is defined for all $\pi < \rho$, then $\sup^+_{\pi<\rho} F(\pi, \boldsymbol{\mu}) < \kappa$.*

Proof. Let F be defined over M by the Σ formula $\mathfrak{A}$ and the parameters $\mathbf{u}$:

$$F(\pi, \boldsymbol{\mu}) \simeq \nu \leftrightarrow (M, \in) \models \mathfrak{A}[\pi, \boldsymbol{\mu}, \nu, \mathbf{u}].$$

Then by assumption we have

$$(M, \in) \models (\forall \pi \in \rho) \exists \nu\, \mathfrak{A}$$

and thus by Σ-Collection

$$(M, \in) \models \exists b\, (\forall \pi \in \rho)(\exists \nu \in b)\mathfrak{A}.$$

For such a, $b \in M$,

$$\sup^+_{\pi<\rho} F(\pi, \boldsymbol{\mu}) \leq \sup{}^+\{\nu : \nu \in b\} \leq \bigcup\{\nu : \nu \in b\} + 1 < \kappa. \quad \square$$

For any κ, let

$$T_\kappa(a, \langle \boldsymbol{\mu} \rangle, \nu, \kappa', \lambda, \sigma) \leftrightarrow \kappa', \lambda, \sigma < \kappa \wedge (a, \boldsymbol{\mu}, \nu) \in \Omega^\sigma_{\kappa'\lambda}.$$

7.6 Theorem. *For any κ and any admissible set M with $\kappa = o(M)$, T_κ is Δ-definable over M.*

Proof. The proof of Lemma 2.4 gives a recursive definition of the characteristic function F of T_κ in the form

$$F(a, \langle \boldsymbol{\mu} \rangle, \nu, \kappa', \lambda, \sigma) = G(a, \langle \boldsymbol{\mu} \rangle, \nu, \kappa', \lambda, \sigma, F \restriction \sigma).$$

If we establish that G is Σ-definable over M, then it follows by Σ-recursion that also F is Σ definable over M. Then since

$$\begin{aligned} T_\kappa(a, \langle \boldsymbol{\mu} \rangle, \nu, \kappa', \lambda, \sigma) &\leftrightarrow F(a, \langle \boldsymbol{\mu} \rangle, \nu, \kappa', \lambda, \sigma) = 0 \\ &\leftrightarrow F(a, \langle \boldsymbol{\mu} \rangle, \nu, \kappa', \lambda, \sigma) \neq 1, \end{aligned}$$

it follows that T_κ is Δ-definable over M.

The class of Σ-definable functions is clearly closed under definition by Σ cases. By inspection we see that the functions of ordinal arithmetic are introduced by Σ-recursion, and hence that their restrictions to ordinals $< \kappa$ are Σ-definable over M. The coding functions for finite sequences of natural numbers and ordinals are defined recursively from ordinal exponentiation, hence they are Σ-definable. The decoding functions use in addition the bounded search operation. If $f(\sigma, \mathbf{x}) = \text{least } \pi < \sigma \,.\, g(\pi, \mathbf{x}) = 0$, then

$$\begin{aligned} f(\sigma, \mathbf{x}) = y \leftrightarrow &[y \in \sigma \wedge g(y, \mathbf{x}) = 0 \wedge (\forall u \in y)(\exists z \neq 0) g(u, \mathbf{x}) = z] \\ &\vee [y = \sigma \wedge (\forall y \in \sigma)(\exists z \neq 0) \,.\, g(y, \mathbf{x}) = z]. \end{aligned}$$

Hence if g is Σ-definable over M, so is f. Finally, the equivalence,

$$f(\mathbf{x}, g(\mathbf{x})) = y \leftrightarrow \exists z \, [g(\mathbf{x}) = z \wedge f(\mathbf{x}, z) = y]$$

shows that the class of Σ-definable functions is closed under composition.

With these facts in mind, it is straightforward to check that G is Σ-definable. The cases depend on Σ conditions concerning the indices. The five subclauses of clause (0) which concern the constant, projection, and successor functions, and the characteristic function of equality are Δ_0 by the remarks following Definition 7.2; the last subclause involving the sequence coding functions is Σ by the preceding remarks. In the remaining clauses, all quantifiers are bounded and the conditions under which F has value 0 are thus Δ. In clause (1) we need to observe that the function H is defined by recursion from the sequence coding functions and is thus Σ-definable. It follows that the (complementary) conditions under which F has value 1 are also Δ. □

7.7 Theorem. *For any* κ, *any admissible set* M *with* $\kappa = o(M)$, *and all* a, $\boldsymbol{\mu}$, *and* ν,

$$(a, \boldsymbol{\mu}, \nu) \in \Omega_{\kappa\kappa} \leftrightarrow (\exists \lambda < \kappa)(\exists \sigma < \kappa) \,.\, (a, \boldsymbol{\mu}, \nu) \in \Omega^\sigma_{\lambda\lambda}.$$

Proof. We proceed almost exactly as in the proof of Lemma 2.5. The admissibility of M easily implies that κ is a limit ordinal. In clause (3), the relation R is Δ-definable over M by Theorem 7.6. Hence the function F is Σ-definable and Lemma 7.5 gives the inequality $\tau < \kappa$. The condition of recursive regularity may similarly be replaced by admissibility in clause (4). □

7.8 Corollary. *For any κ and any admissible set M with $\kappa = o(M)$, the relation*

$$R(a, \langle \mu \rangle, \nu) \leftrightarrow \{a\}_\kappa(\mu) \simeq \nu$$

is Σ-definable over M. Every function κ-partial recursive in parameters has graph Σ-definable over M. Every relation κ-semi-recursive in parameters is Σ-definable over M. Every relation Σ-recursive in parameters is Δ-definable over M.

Proof. These are all immediate from Theorems 7.6 and 7.7. □

7.9 Corollary. *For any κ, if there exists an admissible set M such that $\kappa = o(M)$, then κ is recursively regular.*

Proof. Immediate from Corollary 7.8, Lemma 7.5 and Theorem 2.2. □

The implications ($\rightarrow$) of (1) and (2) now follow by taking $M = \mathrm{L}_\kappa$ in Corollaries 7.8 and 7.9. To establish the converse implications we shall assign ordinals to the members of L_κ in such a way that Δ-definability over L_κ corresponds to recursiveness of the associated relations of ordinals. The assignment takes place in two steps: we first construct a language $\mathscr{L}$ in which every constructible set has a name and then "Gödel number" the language in such a way that whenever κ is recursively regular, names for elements of L_κ are assigned ordinals less than κ.

The symbols of $\mathscr{L}$ are $\dot{\in}$, $\dot{=}$, $\neg$, $\vee$, $\exists$, a variable v_i^σ for each ordinal σ and each $i < \omega$, and a constant symbol $\dot{\sigma}$ for each ordinal σ. The syntax of $\mathscr{L}$ is designed to imitate the process by which the levels of the hierarchy of constructible sets are defined. The variables v_i^σ intuitively range over L_σ. The terms of $\mathscr{L}$ include, in addition to the variables and constant symbols, *abstraction terms* $\hat{v}_i^\sigma \mathfrak{A}$. $\hat{v}_i^\sigma \mathfrak{A}$ is intended as a name for the element of $\mathrm{L}_{\sigma+1}$ defined over L_σ by the formula $\mathfrak{A}$. To make this work, $\mathfrak{A}$ should contain names only for elements of L_σ. This requirement leads us to define the classes of terms and formulas of $\mathscr{L}$ simultaneously by induction along with a notion of *degree*. The intention is that terms of degree $\leqslant \sigma$, other than variables, are names for elements of L_σ.

For all $i < \omega$ and all ordinals σ,

(i) v_i^σ is a term of degree σ;

(ii) $\dot{\sigma}$ is a term of degree $\sigma + 1$;

(iii) for any terms t and u, $t \dot{\in} u$ and $t \dot{=} u$ are formulas of degree $\max\{\deg(t), \deg(u)\}$;

(iv) for any formula $\mathfrak{A}$, $\neg\mathfrak{A}$ is a formula of degree $\deg(\mathfrak{A})$;

(v) for any formulas $\mathfrak{A}$ and $\mathfrak{B}$, $\mathfrak{A} \vee \mathfrak{B}$ is a formula of degree $\max\{\deg(\mathfrak{A}), \deg(\mathfrak{B})\}$;

(vi) for any formula $\mathfrak{A}$ of degree $\leqslant \sigma$, $\exists v_i^\sigma \mathfrak{A}$ is a formula of degree σ;

(vii) for any formula $\mathfrak{A}$ of degree $\leqslant \sigma$ with no free variables other than v_i^σ, $\hat{v}_i^\sigma \mathfrak{A}$ is a term of degree $\sigma + 1$;

(viii) an expression is a formula or term only if it is by virtue of (i)–(vii).

A term of $\mathscr{L}$ is called *closed* iff it is not a variable; a formula of $\mathscr{L}$ is *closed* iff it has no free variables. The intended interpretation of $\mathscr{L}$ is given by means of a function Val defined on closed terms and a predicate $\models$ defined on closed formulas. $\mathrm{Val}(t)$ is the set named by t and $\models \mathfrak{A}$ holds just in case $\mathfrak{A}$ is true under the interpretation. The definition is by a simultaneous recursion on degree with a subsidiary recursion on formulas at each level.

For all $i < \omega$ and all ordinals σ,

(i) $\mathrm{Val}(\dot{\sigma}) = \sigma$;

(ii) for any closed terms t and u, $\models t \mathrel{\dot{\in}} u$ iff $\mathrm{Val}(t) \in \mathrm{Val}(u)$ and $\models t \doteq u$ iff $\mathrm{Val}(t) = \mathrm{Val}(u)$;

(iii) for any closed formula $\mathfrak{A}$, $\models \neg\mathfrak{A}$ iff *not* $\models \mathfrak{A}$;

(iv) for any closed formulas $\mathfrak{A}$ and $\mathfrak{B}$, $\models \mathfrak{A} \vee \mathfrak{B}$ iff $\models \mathfrak{A}$ or $\models \mathfrak{B}$;

(v) for any closed formula $\exists v_i^\sigma \mathfrak{A}$, $\models \exists v_i^\sigma \mathfrak{A}$ iff for some closed term t of degree $\leqslant \sigma$, $\models \mathfrak{A}(t/v_i^\sigma)$;

(vi) for any term $\hat{v}_i^\sigma \mathfrak{A}$,

$$\mathrm{Val}(\hat{v}_i^\sigma \mathfrak{A}) = \{\mathrm{Val}(t) : t \text{ is a closed term of degree } \leqslant \sigma \text{ such that } \models \mathfrak{A}(t/v_i^\sigma)\}.$$

7.10 Theorem. *For all* κ,

$$\mathrm{L}_\kappa = \{\mathrm{Val}(t) : t \text{ is a closed term of degree } \leqslant \kappa\}.$$

Proof. We proceed by induction on κ. For $\kappa = 0$, there are no closed terms of degree 0 and both sides are empty. Assume as induction hypothesis that the result holds for any $\sigma < \kappa$ in place of κ. It κ is a limit ordinal, the result is immediate from the fact that the degree of a closed term is always a successor ordinal.

Suppose that $\kappa = \sigma + 1$. Clearly $\mathrm{Val}(\dot{\sigma}) = \sigma \in \mathrm{L}_\kappa$. We need to show that a subset of L_σ is definable over $(\mathrm{L}_\sigma, \in)$ (by a formula of $\mathscr{L}_{\mathrm{ZF}}$) just in case it is of the form $\mathrm{Val}(\hat{v}_i^\sigma \mathfrak{A})$ for some formula $\mathfrak{A}$ of $\mathscr{L}$ of degree $\leqslant \sigma$ with no free variables other than v_i^σ.

Let $\mathscr{L}_\sigma^*$ be a language obtained from $\mathscr{L}_{\mathrm{ZF}}$ by adding names $\dot{a}$ for each $a \in \mathrm{L}_\sigma$ and variables x_i^τ for each $i < \omega$ and $\tau < \sigma$. $(\mathrm{L}_\sigma, \in)^*$ denotes the corresponding expansion of $(\mathrm{L}_\sigma, \in)$ in which each name $\dot{a}$ is intepreted as a. Clearly the same subsets of L_σ are definable over $(\mathrm{L}_\sigma, \in)^*$ by formulas of $\mathscr{L}_\sigma^*$ as are definable over $(L_\sigma, \in)$ by formulas of $\mathscr{L}_{ZF}$. Note that over $(L_\sigma, \in)^*$ we have no need of parameters in the definitions as these are already built into the language $\mathscr{L}_\sigma^*$. To each term t and formula $\mathfrak{A}$ of $\mathscr{L}$ of degree $\leqslant \sigma$ we assign a term t^* and a formula $\mathfrak{A}^*$ of $\mathscr{L}_\sigma^*$, respectively, as follows: for all $i < \omega$ and all ordinals $\tau < \sigma$,

(i) $(v_i^\tau)^*$ is x_i^τ; $(v_i^\sigma)^*$ is x_i;
(ii) if $\mathrm{Val}(t) = a$, then $t^* = \dot{a}$;
(iii) $(t \dot{\in} u)^*$ is $(t^* \dot{\in} u^*)$ and $(t \doteq u)^*$ is $(t^* \doteq u^*)$;
(iv) $(\neg\mathfrak{A})^*$ is $(\neg\mathfrak{A}^*)$ and $(\mathfrak{A} \vee \mathfrak{B})^*$ is $\mathfrak{A}^* \vee \mathfrak{B}^*$;
(v) $(\exists v_i^\tau \mathfrak{A})^*$ is $(\exists x_i^\tau \in \dot{L}_\tau)\mathfrak{A}^*$;
(vi) $(\exists v_i^\sigma \mathfrak{A})^*$ is $\exists x_i\, \mathfrak{A}^*$.

We claim that for all closed formulas $\mathfrak{A}$ of $\mathscr{L}$ of degree $\leqslant \sigma$,

$$\vDash \mathfrak{A} \quad \text{iff} \quad (L_\sigma, \in)^* \vDash \mathfrak{A}^*.$$

The proof is by induction on closed formulas. For atomic formulas, the claim is obvious and the steps corresponding to $\neg$ and $\vee$ are trivial. We have then for $\tau < \sigma$,

$$\begin{aligned} \vDash \exists v_i^\tau \mathfrak{A} &\text{ iff for some closed term } t \text{ of degree } \leqslant \tau,\ \vDash \mathfrak{A}(t/v_i^\tau) \\ &\text{iff for some closed term } t \text{ of degree } \leqslant \tau,\ (L_\sigma, \in)^* \vDash \mathfrak{A}^*(\mathrm{V\dot{a}l}(t)/x_i^\tau) \\ &\text{iff for some } a \in L_\tau,\ (L_\sigma, \in)^* \vDash \mathfrak{A}^*(\dot{a}/x_i^\tau) \\ &\text{iff } (L_\sigma, \in)^* \vDash (\exists x_i^\tau \in \dot{L}_\tau)\mathfrak{A}^*. \end{aligned}$$

The second equivalence uses the easily proved fact that

$$\mathfrak{A}(t/v_i^\tau)^* = \mathfrak{A}^*(\mathrm{V\dot{a}l}(t)/x_i^\tau)$$

and the induction hypothesis on $\mathfrak{A}$. The third equivalence uses the induction hypothesis on τ. The step for $\exists v_i^\sigma \mathfrak{A}$ is similar.

Now we have for any closed formula $\mathfrak{A}$ of degree $\leqslant \sigma$ with at most v_i^σ free, and any $a \in L_\sigma$,

$$\begin{aligned} a \in \mathrm{Val}(\hat{v}_i^\sigma \mathfrak{A}) &\text{ iff for some closed term } t \text{ of degree } \leqslant \sigma, \\ &\qquad a = \mathrm{Val}(t) \quad \text{and} \quad \vDash \mathfrak{A}(t/v_i^\sigma) \\ &\text{iff } (L_\sigma, \in)^* \vDash \mathfrak{A}^*(\dot{a}/x_i). \end{aligned}$$

It follows that $\mathrm{Val}(\hat{v}_i^\sigma \mathfrak{A})$ is definable over $(L_\sigma, \in)^*$ and thus belongs to $L_{\sigma+1} = L_\kappa$.

For the converse inclusion suppose $b \in L_\kappa$, so for some closed formula $\mathfrak{B}$ of $\mathscr{L}_\sigma^*$,

$$b = \{a : (L_\sigma, \in)^* \vDash \mathfrak{B}(\dot{a}/x_i)\}.$$

We may clearly assume that neither the symbol $\wedge$ nor any of the variables x_i^τ occur in $\mathfrak{B}$. By the induction hypothesis, for every symbol $\dot{a}$ which occurs in $\mathfrak{B}$, there is a closed term t of $\mathscr{L}$ such that $t^* = \dot{a}$. It follows that there is a formula $\mathfrak{A}$ of $\mathscr{L}$ of degree $\leqslant \sigma$ with no free variables other than v_i^σ such that $\mathfrak{A}^* = \mathfrak{B}$. Thus $b = \mathrm{Val}(\hat{v}_i^\sigma \mathfrak{A})$. □

We now proceed to assign Gödel numbers to the terms and formulas of $\mathscr{L}$. The assignment is routine except for one trick in the choice of $\mathrm{gn}(v_i^\sigma)$. Let $\mathrm{Sqc}(\sigma)$ denote the σ-th sequence closed ordinal. Exercise 6.35 established that the function Sqc is $(\infty, 0)$-recursive. In particular, if κ is recursively regular and $\sigma < \kappa$, then also $\mathrm{Sqc}(\sigma) < \kappa$. We set

$$\mathrm{gn}(\dot{\sigma}) = \langle 0, \sigma \rangle;$$

$$\mathrm{gn}(v_i^\sigma) = \langle 1, i, \mathrm{Sqc}(\sigma) \rangle;$$

$$\mathrm{gn}(\hat{v}_i^\sigma \mathfrak{A}) = \langle 2, \mathrm{gn}(v_i^\sigma), \mathrm{gn}(\mathfrak{A}) \rangle;$$

$$\mathrm{gn}(t \doteq u) = \langle 3, \mathrm{gn}(t), \mathrm{gn}(u) \rangle;$$

$$\mathrm{gn}(t \mathrel{\dot{\in}} u) = \langle 4, \mathrm{gn}(t), \mathrm{gn}(u) \rangle;$$

$$\mathrm{gn}(\neg \mathfrak{A}) = \langle 5, \mathrm{gn}(\mathfrak{A}) \rangle;$$

$$\mathrm{gn}(\mathfrak{A} \vee \mathfrak{B}) = \langle 6, \mathrm{gn}(\mathfrak{A}), \mathrm{gn}(\mathfrak{B}) \rangle;$$

$$\mathrm{gn}(\exists v_i^\sigma \mathfrak{A}) = \langle 7, \mathrm{gn}(v_i^\sigma), \mathrm{gn}(\mathfrak{A}) \rangle.$$

7.11 Lemma. *For all* σ

(i) *for any formula* $\mathfrak{A}$ *of* $\mathscr{L}$ *of degree* $< \sigma$, $\mathrm{gn}(\mathfrak{A}) < \mathrm{Sqc}(\sigma)$;
(ii) *for any closed term* t *of* $\mathscr{L}$ *of degree* $\leqslant \sigma$, $\mathrm{gn}(t) < \mathrm{Sqc}(\sigma)$.

Proof. We proceed by induction on σ and assume as induction hypothesis that (i) and (ii) hold with τ in place of σ for all $\tau < \sigma$. First, if $\mathfrak{A}$ is of degree $< \sigma$, then any term t occurring in $\mathfrak{A}$ is of degree some $\tau < \sigma$. If t is closed, then $\mathrm{gn}(t) < \mathrm{Sqc}(\tau) < \mathrm{Sqc}(\sigma)$ by (ii) of the induction hypothesis. If t is v_i^τ, then $\mathrm{gn}(t) = \langle 1, i, \mathrm{Sqc}(\tau) \rangle < \mathrm{Sqc}(\sigma)$ because $\mathrm{Sqc}(\sigma)$ is sequence-closed. It follows in turn by induction on the subformulas of $\mathfrak{A}$ that $\mathrm{gn}(\mathfrak{A}) < \sigma$.

Now if t is a closed term of degree σ, then necessarily σ is a successor ordinal, say $\tau + 1$, and either $t = \dot{\tau}$ or for some i and some formula $\mathfrak{A}$ of degree $\leqslant \tau$ with at most v_i^τ free, $t = \hat{v}_i^\tau \mathfrak{A}$. In the first case, $\mathrm{gn}(t) = \langle 0, \tau \rangle < \mathrm{Sqc}(\sigma)$ because $\tau \leqslant \mathrm{Sqc}(\tau) < \mathrm{Sqc}(\sigma)$. In the second case, $\mathrm{gn}(\mathfrak{A}) < \mathrm{Sqc}(\sigma)$ by (i), so $\mathrm{gn}(t) = \langle 2, \langle 1, i, \mathrm{Sqc}(\tau) \rangle, \mathrm{gn}(\mathfrak{A}) \rangle < \mathrm{Sqc}(\sigma)$. □

7.12 Corollary. *For any* σ, *any formula* $\mathfrak{A}$ *of* $\mathscr{L}$ *of degree* $\leqslant \sigma$, *and any closed term* t *of degree* $\leqslant \sigma$,

(i) $\operatorname{gn}\mathfrak{A}(t/v_i^\sigma) < \operatorname{gn}(t \dot{\in} \hat{v}_i^\sigma \mathfrak{A}), \operatorname{gn}(\exists v_i^\sigma \mathfrak{A})$;

(ii) *for any* $a \in \operatorname{Val}(t)$, *there exists a term* u *such that* $a = \operatorname{Val}(u)$ *and* $\operatorname{gn}(u) < \operatorname{gn}(t)$.

Proof. (i) follows easily from the fact that under the given conditions, $\operatorname{gn}(t) < \operatorname{Sqc}(\sigma) < \operatorname{gn}(v_i^\sigma)$, so that $\operatorname{gn}\mathfrak{A}(t/v_i^\sigma) \leqslant \operatorname{gn}(\mathfrak{A})$. For (ii), if t is $\dot{\tau}$ with $\tau < \sigma$, then a is an ordinal $< \tau$ and $\operatorname{gn}(\dot{a}) = \langle 0, a\rangle < \langle 0, \tau\rangle = \operatorname{gn}(t)$. Otherwise, t is of the form $\hat{v}_i^\tau \mathfrak{A}$ with $\tau < \sigma$ and $\mathfrak{A}$ a formula of degree $\leqslant \tau$ with at most v_i^τ free. If $a \in \operatorname{Val}(t)$, then $a = \operatorname{Val}(u)$ for some closed term u of degree $\leqslant \tau$, and $\operatorname{gn}(u) < \operatorname{Sqc}(\tau) < \operatorname{gn}(v_i^\tau) < \operatorname{gn}(t)$. □

7.13 Lemma. *There exist* $(\infty, 0)$-*recursive functions and relations which satisfy the following conditions for any terms* t *and* u *and formula* $\mathfrak{A}$ *of* $\mathcal{L}$:

(i) $\operatorname{Ord}(\mu) \leftrightarrow \exists \sigma . \mu = \operatorname{gn}(\dot{\sigma})$;
(ii) $\operatorname{Var}(\mu) \leftrightarrow \exists \sigma \exists i . \mu = \operatorname{gn}(v_i^\sigma)$;
(iii) $\operatorname{Dg}(\operatorname{gn}(t)) = \deg(t)$;
$\operatorname{Dg}(\operatorname{gn}(\mathfrak{A})) = \deg(\mathfrak{A})$;
(iv) $\operatorname{Fr}(\operatorname{gn}(v_i^\sigma), \operatorname{gn}(\mathfrak{A})) \leftrightarrow v_i^\sigma$ *occurs free in* $\mathfrak{A}$;
(v) $\operatorname{Te}(\mu) \leftrightarrow \mu$ *is the Gödel number of some term of* $\mathcal{L}$;
(vi) $\operatorname{Fm}(\mu) \leftrightarrow \mu$ *is the Gödel number of some formula of* $\mathcal{L}$;
(vii) $\operatorname{Sb}(\operatorname{gn}(t), \operatorname{gn}(v_i^\sigma), \operatorname{gn}(\mathfrak{A})) = \operatorname{gn}(\mathfrak{A}(t/v_i^\sigma))$.

Proof. The calculations are routine and we leave them as Exercise 7.23. □

7.14 Theorem. *The relation* Tr *defined by*

$$\operatorname{Tr}(\mu) \leftrightarrow \mu \textit{ is the Gödel number of a closed formula } \mathfrak{A} \textit{ of } \mathcal{L} \textit{ such that } \models \mathfrak{A}$$

is $(\infty, 0)$-*recursive.*

Proof. We give a few of the conditions which Tr must satisfy and leave it to the reader to supply the remainder. The existence of Tr follows by application of the Recursion Theorem.

$$\operatorname{Tr}(\langle 4, \langle 0, \sigma\rangle, \langle 0, \tau\rangle\rangle) \leftrightarrow \sigma < \tau;$$

$$\operatorname{Tr}(\langle 4, \langle 0, \sigma\rangle, \operatorname{gn}(\hat{v}_j^\tau \mathfrak{B})\rangle) \leftrightarrow \sigma < \tau \wedge \operatorname{Tr}(\operatorname{gn}(\mathfrak{B}(\dot{\sigma}/v_j^\tau)));$$

$$\operatorname{Tr}(\langle 4, \operatorname{gn}(\hat{v}_i^\sigma \mathfrak{A}), \operatorname{gn}(\hat{v}_j^\tau \mathfrak{B})\rangle) \leftrightarrow (\exists \mu < \operatorname{Sqc}(\tau))$$

μ is the Gödel number of a closed term t of degree $\leqslant \tau$, $\operatorname{Tr}(\operatorname{gn}\mathfrak{B}(t/v_j^\tau))$, and $(\forall \nu < \operatorname{Sqc}(\sigma))$ if ν is the Gödel number of a closed term u of degree $\leqslant \sigma$, then $\operatorname{Tr}(\operatorname{gn}(\mathfrak{A}(u/v_i^\sigma)))$ iff $\operatorname{Tr}(\operatorname{gn}(u \dot{\in} t))$;

$$\mathrm{Tr}(\mathrm{gn}(\mathfrak{A} \vee \mathfrak{B})) \leftrightarrow \mathrm{Tr}(\mathrm{gn}(\mathfrak{A})) \text{ or } \mathrm{Tr}(\mathrm{gn}(\mathfrak{B}));$$

$$\mathrm{Tr}(\mathrm{gn}(\exists v_i^\sigma \mathfrak{A})) \leftrightarrow (\exists \mu < \mathrm{Sqc}(\sigma))\ \mu \text{ is the Gödel number of a closed term } t \text{ of degree } \leq \sigma \text{ and } \mathrm{Tr}(\mathfrak{A}(t/v_i^\sigma)).$$

In several cases we rely on Corollary 7.12 to ensure that the Gödel numbers on the right-hand side of each equivalence are less than those on the left. □

7.15 Corollary. *Every constructible set A of ordinals is $(\infty, 0)$-recursive in parameters. For every recursively regular κ, if $A \in \mathrm{L}_\kappa$, then A is κ-recursive in parameters.*

Proof. If $A \in \mathrm{L}_\kappa$, then by Theorem 7.10 there exists a closed term t of degree $\leq \kappa$ such that $A = \mathrm{Val}(t)$. Then for any σ,

$$\sigma \in A \leftrightarrow \models \dot{\sigma} \dot{\in} t \leftrightarrow \mathrm{Tr}(\langle 4, \langle 0, \sigma\rangle, \mathrm{gn}(t)\rangle).$$

Thus A is $(\infty, 0)$ recursive in the parameter $\mathrm{gn}(t)$. If κ is recursively regular, then $\mathrm{gn}(t) < \mathrm{Sqc}(\kappa) = \kappa$ and by Theorem 2.2, A is κ-recursive in the parameter $\mathrm{gn}(t)$. □

For any μ, let

$$^\circ\mu = \begin{cases} \mathrm{Val}(t), & \text{if } \mu \text{ is the Gödel number of a closed term } t; \\ \varnothing, & \text{otherwise.} \end{cases}$$

7.16 Lemma. *For any recursively regular κ,*

$$\mathrm{L}_\kappa = \{^\circ\mu : \mu < \kappa\}.$$

Proof. As in the preceding proof, if $A \in \mathrm{L}_\kappa$, then $A = \mathrm{Val}(t)$ for some t with $\mathrm{gn}(t) < \kappa$. Thus $A = {}^\circ\mathrm{gn}(t)$. On the other hand, it is easy to prove that for all terms, $\mathrm{dg}(t) \leq \mathrm{gn}(t)$ so that for any $\mu < \kappa$, if $\mu = \mathrm{gn}(t)$, then $\mathrm{dg}(t) < \kappa$ and by Theorem 7.10, $^\circ\mu = \mathrm{Val}(t) \in \mathrm{L}_\kappa$. □

For any formula $\mathfrak{A}$ of $\mathscr{L}_{\mathrm{ZF}}$ with free variables among $x_0, \ldots, x_{k-1}$ and any $\mu_0, \ldots, \mu_{k-1} < \kappa$, let

$$\mathrm{Sat}_{\mathfrak{A},\kappa}(\boldsymbol{\mu}) \leftrightarrow (\mathrm{L}_\kappa, \in) \vdash \mathfrak{A}[^\circ\boldsymbol{\mu}].$$

7.17 Theorem. *For any recursively regular κ and any Δ_0-formula $\mathfrak{A}$ of $\mathscr{L}_{\mathrm{ZF}}$, $\mathrm{Sat}_{\mathfrak{A},\kappa}$ is κ-recursive.*

Proof. We proceed by induction on Δ_0 formulas. For atomic formulas, the result is immediate from Theorem 7.14. The $\neg$, $\wedge$, and $\vee$ steps are trivial and we consider only bounded quantification. Suppose that $\mathfrak{A}$ is a Δ_0 formula and that $\mathrm{Sat}_{\mathfrak{A},\kappa}$ is κ-recursive. Then for ρ, $\boldsymbol{\mu} < \kappa$

$$\begin{aligned}(\mathrm{L}_\kappa, \in) \models (\exists x \in y)\,\mathfrak{A}[{}^\circ\rho, {}^\circ\boldsymbol{\mu}] \\ &\leftrightarrow (\exists a \in {}^\circ\rho).(\mathrm{L}_\kappa, \in) \models \mathfrak{A}[a, {}^\circ\boldsymbol{\mu}] \\ &\leftrightarrow (\exists \pi < \rho).\mathrm{Tr}(\langle 4, \pi, \rho\rangle) \wedge \mathrm{Sat}_{\mathfrak{A},\kappa}(\pi, \boldsymbol{\mu}).\end{aligned}$$

The second equivalence depends on Corollary 7.12 (ii). $\square$

7.18 Corollary. *For any recursively regular κ and any $R \subseteq {}^k\kappa$,*

(i) *if R is Σ-definable over L_κ, then R is κ-semi-recursive in parameters;*

(ii) *if R is Δ-definable over L_κ, then R is κ-recursive in parameters.*

Proof. It clearly suffices to prove (i), so suppose $\mathfrak{A}$ is a Δ_0 formula and R is defined by

$$R(\boldsymbol{\mu}) \leftrightarrow (\mathrm{L}_\kappa, \in) \models \exists x\, \mathfrak{A}[\boldsymbol{\mu}, \boldsymbol{b}].$$

Let $\rho_0, \ldots, \rho_{l-1}$ be ordinals $< \kappa$ such that $(\forall j < l).\ {}^\circ\rho_j = b_j$. Then

$$\begin{aligned}R(\boldsymbol{\mu}) &\leftrightarrow (\exists \nu < \kappa).(\mathrm{L}_\kappa, \in) \models \mathfrak{A}[{}^\circ\nu, {}^\circ\langle 0, \mu_0\rangle, \ldots, {}^\circ\langle 0, \mu_{k-1}\rangle, {}^\circ\rho_0, \ldots, {}^\circ\rho_{l-1}] \\ &\leftrightarrow (\exists \nu < \kappa)\mathrm{Sat}_{\mathfrak{A},\kappa}(\nu, \langle 0, \mu_0\rangle, \ldots, \langle 0, \mu_{k-1}\rangle, \rho_0, \ldots, \rho_{l-1}).\end{aligned}$$

Thus R is κ-semi-recursive in the parameters $\rho_0, \ldots, \rho_{l-1}$. $\square$

7.19 Corollary. *For any κ, if κ is recursively regular, then L_κ is admissible.*

Proof. It is trivial to check that for any limit ordinal κ, $(\mathrm{L}_\kappa, \in)$ satisfies the pair, union, and Δ_0-Separation axioms. Suppose that κ is recursively regular, that $\mathfrak{A}$ is a Δ_0 formula in which b does not occur free, and that for some $\boldsymbol{\mu}$, $\rho < \kappa$,

$$(\mathrm{L}_\kappa, \in) \models (\forall x \in a)\exists y\, \mathfrak{A}[{}^\circ\rho, {}^\circ\boldsymbol{\mu}].$$

Then

$$(\forall x \in {}^\circ\rho)(\exists \sigma < \kappa).(\mathrm{L}_\kappa, \in) \models \mathfrak{A}[{}^\circ\sigma, {}^\circ\boldsymbol{\mu}]$$

and

$$(\forall \pi < \rho)(\exists \sigma < \kappa)[\mathrm{Tr}(\langle 4, \pi, \rho\rangle) \rightarrow \mathrm{Sat}_{\mathfrak{A},\kappa}(\sigma, \boldsymbol{\mu})].$$

By the recursive regularity of κ and Theorem 7.14 and 7.17, if

$$\tau = \sup^{+}_{\pi<\rho}.\ \text{least}\ \sigma[\mathrm{Tr}(\langle 4, \pi, \rho\rangle) \to \mathrm{Sat}_{\mathfrak{A},\kappa}(\sigma, \boldsymbol{\mu})],$$

then $\tau < \kappa$, and if

$$\lambda = \sup^{+}_{\sigma<\tau} \mathrm{Dg}(\sigma),$$

then also $\lambda < \kappa$. Furthermore,

$$(\forall x \in {}^{\circ}\rho)\exists\sigma\, ({}^{\circ}\sigma \in \mathrm{L}_\lambda \text{ and } (\mathrm{L}_\kappa, \in) \models \mathfrak{A}[{}^{\circ}\sigma, {}^{\circ}\boldsymbol{\mu}]),$$

so

$$(\mathrm{L}_\kappa, \in) \models (\forall x \in a)(\exists y \in b)\mathfrak{A}[\mathrm{L}_\lambda, {}^{\circ}\rho, {}^{\circ}\boldsymbol{\mu}],$$

where b is interpreted as L_λ. Thus we have the conclusion of Δ_0-Collection. □

Finally we draw some interesting conclusions about relations on numbers.

7.20 Theorem. *For all $R \subseteq {}^k\omega$, the following are equivalent:*

(i) *R is ∞-recursive;*
(ii) *R is δ^1_2-recursive;*
(iii) *R is Δ^1_2;*
(iv) *$R \in \mathrm{L}_{\delta^1_2}$.*

Proof. The equivalence of (i), (ii), and (iii) is Corollary 5.10. If R is δ^1_2-recursive, then by Corollary 7.8, R is Δ-definable over $\mathrm{L}_{\delta^1_2}$. Since δ^1_2 is recursively regular, $\mathrm{L}_{\delta^1_2}$ is admissible and satisfies Δ-Separation. Hence $R \in \mathrm{L}_{\delta^1_2}$. On the other hand, if $R \in \mathrm{L}_{\delta^1_2}$, then by Corollary 7.15, R is δ^1_2-recursive in parameters. But by Exercise 3.14 (ii), R is then δ^1_2-recursive. □

7.21 Theorem. *For all $R \subseteq {}^k\omega$, the following are equivalent:*

(i) *R is ∞-recursive in parameters;*
(ii) *R is $\aleph_1$-recursive in parameters;*
(iii) *R is constructible;*
(iv) *$R \in \mathrm{L}_{\aleph_1}$.*

Proof. We prove that (ii)→(iv)→(iii)→(i)→(ii). If R is $\aleph_1$-recursive in parameters, then R is Δ-definable over $\mathrm{L}_{\aleph_1}$ and hence by Δ-Separation, $R \in \mathrm{L}_{\aleph_1}$. That (iv)→(iii) is trivial. That (iii)→(i) is immediate from Corollary 7.15. The implication (i)→(ii) is a weak version of Exercise 5.21. □

Note that using the full strength of Exercise 5.21 we could replace (ii) and (iv) by:

(ii)′ R is $\delta_2^1[R]$ recursive in parameters;
(iv)′ $R \in \mathrm{L}_{\delta_2^1[R]}$.

Note also that the equivalence of (ii) and (iii) implies that there are at most $\aleph_1$ constructible subsets of ω. Hence if $\mathrm{V} = \mathrm{L}$ then the Continuum Hypothesis holds.

7.22–7.29 Exercises

7.22. Prove Lemma 7.13.

7.23. Σ-*Separation* is the set of axioms

$$\exists b\, \forall x\, (x \in b \leftrightarrow x \in a \wedge \mathfrak{A})$$

for all Σ formulas $\mathfrak{A}$ in which b does not occur free. Show that for any admissible set M, $(M, \in)$ satisfies Σ-separation iff $(M, \in)$ satisfies the following schema of *strong* Δ_0-Collection:

$$\exists b\, (\forall x \in a)[\exists y\, \mathfrak{A} \rightarrow (\exists y \in b)\mathfrak{A}]$$

for all Δ_0 formulas $\mathfrak{A}$ in which b does not occur free.

7.24. Show that for any recursively regular κ, κ is non-projectible iff $(\mathrm{L}_\kappa, \in)$ satisfies Σ-Separation.

7.25. Show that for any recursively regular ordinal λ and any $\kappa \leqslant \lambda$, κ is λ-stable iff $(\mathrm{L}_\kappa, \in)$ is a Σ-*substructure* of $(\mathrm{L}_\lambda, \in)$ — that is, for any Σ formula $\mathfrak{A}$ with free variables among $x_0, \ldots, x_{k-1}$ and any $a_0, \ldots, a_{k-1} \in \mathrm{L}_\kappa$,

$$(\mathrm{L}_\kappa, \in) \models \mathfrak{A}[\boldsymbol{a}] \quad \text{iff} \quad (\mathrm{L}_\lambda, \in) \models \mathfrak{A}[\boldsymbol{a}]$$

(cf. Exercise 5.16).

7.26. Show that the special names $\dot{\sigma}$ in $\mathscr{L}$ are superfluous. Specifically, let $\mathscr{L}^-$ be the language obtained from $\mathscr{L}$ by omitting the names $\dot{\sigma}$. Show that for each ordinal σ there is a closed term t_σ of $\mathscr{L}^-$ of degree $\sigma + 1$ such that $\mathrm{Val}(t_\sigma) = \sigma$. Describe an assignment of Gödel numbers to $\mathscr{L}^-$ which makes $\lambda\sigma . \mathrm{gn}(t_\sigma)$ an $(\infty, 0)$-recursive function and check that all of the remaining results go through.

7.27. Define the classes of $\exists_r^0$ and $\forall_r^0$ formulas of $\mathscr{L}_{\mathrm{ZF}}$ as in Definition III.5.3, with $\exists_0^0 = \forall_0^0 = \Delta_0$. Note that if M is admissible, then every Σ formula is equivalent in M to an $\exists_1^0$ formula. Show that for any κ,

(i) κ is recursively regular iff L_κ is $\forall_2^0$-*reflecting* — that is, for any $\forall_2^0$ formula $\mathfrak{A}$ and any $\mathbf{u} \in \mathrm{L}_\kappa$, if $(\mathrm{L}_\kappa, \in) \models \mathfrak{A}[\mathbf{u}]$, then $(\exists \lambda < \kappa).(\mathrm{L}_\lambda, \in) \models \mathfrak{A}[\mathbf{u}]$;

(ii) κ is recursively Mahlo iff L_κ is $\forall_2^0$-*reflecting on the class of recursively regular ordinals* — that is, as above except that λ is required to be recursively regular.

7.28. Show that the Hypothesis of Constructibility is equivalent to: for all infinite cardinals κ, for any $\lambda < \kappa$, every subset of λ is κ-recursive in parameters.

7.29 (Shoenfield [1961]). Show that every Σ_2^1 relation $R \subseteq {}^k\omega$ is constructible.

7.30 Notes. The concepts and results of §§ 6 and 7 are again due to Kripke [1964] and Platek [1966]. Lévy [1965] first isolated the classes of Δ_0 and Σ formulas.

Epilogue

In contrast with those of a work of fiction, the lives of the characters of this book extend well beyond its covers. For the reader who has a yen to follow them a bit further, we discuss here briefly some of the more recent developments and offer some suggestions for further reading. In many cases the material is still very much in a state of development and we are attempting not to give a thorough and orderly presentation but rather to convey some of the flavor of current research. Some of the notational conventions of the preceding chapters do not apply here.

We begin with Inductive Definability, which has been either on stage or just in the wings in almost every section of the book. One of the most elementary questions which was not answered in the text is that of the closure ordinals and sets for most classes of non-monotone inductive operators over ω. To state these results concisely, we introduce some notation. For any class X of relations, let

$$|X| = \sup^{+}\{|\Gamma| : \Gamma \text{ is an inductive operator and } \Gamma \in X\};$$

$$X\text{-Ind} = \{R : R \text{ is reducible to } \bar{\Gamma} \text{ for some inductive operator } \Gamma \in X\};$$

$$X\text{-Hyp} = \{R : \text{both } R \text{ and } {\sim}R \text{ belong to } X\text{-Ind}\}.$$

Replacing "inductive" by "monotone" yields the definitions of $|X\text{-mon}|$, X-mon-Ind, and X-mon-Hyp.

For comparison we state first some of our earlier results in this notation:

(1) $|\Sigma^0_1| = \omega;$ (III.3.3)

$\Sigma^0_1\text{-mon-Ind} = \Sigma^0_1;$ (III.3.5)

$\Delta^0_{(\omega)} \subsetneqq \Sigma^0_1\text{-Ind} \subsetneqq \Delta^1_1;$ (III.3.6, 7)

(2) $|\Pi^0_1| = |\Pi^0_1\text{-mon}| = |\Pi^1_1\text{-mon}| = \omega_1;$ (IV.2.15, 16, 21)

$\Pi^0_1\text{-Ind} = \Pi^0_1\text{-mon-Ind} = \Pi^1_1\text{-mon-Ind} = \Pi^1_1;$ (III.3.1, 2, IV.2.17)

(3) $|\Sigma^1_1\text{-mon}| = \omega_1[\mathsf{E}^{\#}_1];$

$\Sigma^1_1\text{-mon-Ind} = \{R : R \text{ is semi-recursive in } \mathsf{E}^{\#}_1\};$ (VI.6.14)

(4) for all $r \geqslant 2$,

$|\Delta^1_r| = \delta^1_r$; (IV.7.10, III.3.34)

Δ^1_r-Ind $= \Delta^1_r$; (III.3.10)

Σ^1_r-mon-Ind $= \Sigma^1_r$; (III.3.14)

Π^1_r-mon-Ind $= \Pi^1_r$. (III.3.1)

Aanderaa [1974] proves the following:

(5) $|\Pi^1_1| < |\Sigma^1_1|$ and $|\Sigma^1_2| < |\Pi^1_2|$;

if $V = L$, then for all $r \geqslant 2$,

$$|\Delta^1_r| = |\Sigma^1_r\text{-mon}| < |\Sigma^1_r| < |\Pi^1_r|;$$

if PD, then as above for all even $r \geqslant 2$, while for all odd $r \geqslant 2$,

$$|\Delta^1_r| = |\Pi^1_r\text{-mon}| < |\Pi^1_r| < |\Sigma^1_r|.$$

The proof is based on the pre-wellordering property.

Recall from Exercise VIII.7.27 the notion κ is $\forall^0_r$-reflecting. In view of (i) of that exercise and Theorem VIII.3.4(i), the following result of Aczel–Richter [1974] is a generalization of (2):

(6) for all $r \geqslant 1$,

$|\Pi^0_r|$ = least $\kappa . L_\kappa$ is $\forall^0_{r+1}$-reflecting;

Π^0_r-Ind $= \{R : R$ is $|\Pi^0_r|$-semi-recursive in parameters$\}$.

Similarly, if we add to $\mathscr{L}_{ZF}$ second-order variables and define the classes of $\exists^1_r$ and $\forall^1_r$ formulas analogously, we have:

(7) $|\Sigma^1_1| = |\Sigma^1_1\text{-mon}|$ = least $\kappa . L_\kappa$ is $\exists^1_1$-reflecting;

Σ^1_1-Ind $= \Sigma^1_1$-mon-Ind $= \{R : R$ is $|\Sigma^1_1|$-semi-recursive in parameters$\}$;

(8) $|\Pi^1_1|$ = least $\kappa . L_\kappa$ is $\forall^1_1$-reflecting;

Π^1_1-Ind $= \{R : R$ is $|\Pi^1_1|$-semi-recursive in parameters$\}$.

Richter [1975] shows that

(9) $|\Pi^1_2|$ = least $\kappa . L_\kappa$ is $\forall^1_2$-reflecting;

Π^1_2-Hyp $= \{R : R$ is $|\Pi^1_2|$-recursive in parameters$\}$;

Π^1_2-Ind $\supsetneq \{R : R$ is $|\Pi^1_2|$-semi-recursive in parameters$\}$.

The direct analogues of (7)–(9) for Σ_2^1 or higher levels of the analytical hierarchy are not true, but Aczel–Richter [1974] have modified versions. They, and independently Cenzer [1974], obtained the interesting characterization

(10) $|\Pi_1^1| = \text{least } \kappa \,.\, \kappa \text{ is } \kappa^+ \text{ stable},$

where κ^+ denotes the least admissible ordinal greater than κ. Cenzer [1974] has similar characterizations for $|\Sigma_2^1|$ and for higher levels under the hypothesis $V = L$. Many of these facts are also proved in a more general form in Moschovakis [1974b] (see (31)–(33) below).

The results of the text concerning inductive operators over ${}^\omega\omega$, other than decomposable ones, are even more limited. We have only

(11) for all $r > 0$,

$\Pi_r^2\text{-mon-Ind} = \Pi_r^2$; (VI.7.10)

$\Sigma_r^2\text{-mon-Ind} = \Sigma_r^2$; (VI.7.15)

$\Delta_r^2\text{-Ind} = \Delta_r^2$. (VI.7.13)

Cenzer [1974b] announces (among others) the following results for operators over ${}^\omega\omega$:

(12) $|\Delta_1^0| = |\Pi_1^0\text{-mon}| = |\Pi_1^1\text{-mon}| = |\Sigma_1^1\text{-mon}| = \aleph_1$;

$\Pi_1^0\text{-mon-Ind} = \Pi_1^1\text{-mon-Ind} = \Pi_1^1$;

$\Delta_1^1\text{-Ind} \subseteq \Sigma_1^1\text{-mon-Ind} = \Sigma_2^1$;

(13) $|\Pi_1^1| = \omega_1[\mathbb{E}]$

$\Pi_1^1\text{-Ind} = \{\mathsf{R} : \mathsf{R}$ is semi-recursive in $\mathbb{E}$ (semi-hyperanalytical)$\}$;

(14) $|\Sigma_2^1| = |\Sigma_2^1\text{-mon}| = \omega_1[\mathbb{E}^{\#}]$;

$\Sigma_2^1\text{-Ind} = \Sigma_2^1\text{-mon-Ind} =$

$\{\mathsf{R} : \mathsf{R}$ is semi-recursive in $\mathbb{E}^{\#}$ (semi-hyperprojective)$\}$;

$|\Pi_2^1| > \omega_1[\mathbb{E}^{\#}]$.

The characterizations of Π_1^1-Ind and Σ_2^1-(mon)-Ind are essentially contained in Grilliot [1971a] read in conjunction with Hinman–Moschovakis [1971].

We turn next to an important unifying concept which is currently playing an important role in definability theory, that of a Spector class of relations over ω or over an arbitrary set M. A class X of relations over ω is called a *Spector class* iff it contains the recursive relations, is closed under $\cap$, $\cup$, $\exists^0$, $\forall^0$, and relational composition with recursive functions, is ω-indexable, and has the pre-wellordering property. This book is full of examples of Spector classes: Π^1_1, Σ^1_2, and the classes of relations semi-recursive in E, I for any type-2 functional I, in $\mathbb{E}$, $\mathbb{I}$ for any type-3 functional $\mathbb{I}$, or κ-semi-recursive for an admissible κ. The similarities in structure among these classes are largely explained by the fact that many of their properties are properties of any Spector class. Moschovakis [1974, 9A–C] gives a good exposition of the theory of Spector classes.

Call a type-2 functional I *normal* iff E is recursive in I. The 1-*envelope* and 1-*section* of I are the classes:

$$\text{1-env}(\mathsf{I}) = \{R : R \text{ is semi-recursive in } \mathsf{I}\};$$

$$\text{1-sc}(\mathsf{I}) = \{R : R \text{ is recursive in } \mathsf{I}\}.$$

For any normal I, 1-env(I) is a Spector class. The next two results characterize 1-env(I) and the collection of all 1-envelopes of normal type-2 functionals among all Spector classes. A class X of relations is called *closed under* I-*application* iff for any partial function $F \in X$ (that is, $\text{Gr}(F) \in X$), if $G(m) \simeq \mathsf{I}(\lambda p\,.\,F(p, m))$, then also $G \in X$. Moschovakis [1974a] shows:

(15) for any normal type-2 functional I, 1-env(I) is the smallest Spector class closed under I-application.

For any Spector class X, let $\Delta(X)$ denote the symmetric class $X \cap cX$ and let $\omega_1[X]$ denote the least ordinal not the order-type of a $\Delta(X)$ well-ordering of ω. For two Spector classes $X \subseteq Y$, set

$$X < Y \leftrightarrow X \subseteq \Delta(Y) \leftrightarrow \omega_1[X] < \omega_1[Y]$$

(the equivalence is a generalization of Theorem IV.2.14). A Spector class X is called *Mahlo* iff for any I such that X is closed under I-application, there exists a Spector class $Y < X$ also closed under I-application. Harrington–Kechris [1975] and independently Simpson show:

(16) A Spector class X is the 1-envelope of some normal type-2 functional iff X is not Mahlo.

These results have generalizations to higher types. For example, a type-3 functional $\mathbb{I}$ is called *normal* iff $\mathbb{E}$ is recursive in $\mathbb{I}$ and 2-env($\mathbb{I}$) and 2-sc($\mathbb{I}$) are the analogous classes of type-2 relations R. The criterion for X to be a Spector class over ${}^{\omega}\omega$ includes the conditions that X be closed under $\exists^1$ and $\forall^1$, so Corollary VII.2.11 implies that in general 1-env($\mathbb{I}$) is *not* a Spector class. It is, however, a *semi-Spector class*, defined by replacing closure under $\exists^1$ by the weaker closure under *deterministic* $\exists^1$: if P, Q $\in X$, P $\cap$ Q $= \varnothing$ and

$$\mathsf{R}(\mathbf{m}, \boldsymbol{\alpha}) \leftrightarrow \forall \beta\, [\mathsf{P}(\mathbf{m}, \boldsymbol{\alpha}, \beta) \vee \mathsf{Q}(\mathbf{m}, \boldsymbol{\alpha}, \beta)] \wedge \exists \beta\, \mathsf{P}(\mathbf{m}, \boldsymbol{\alpha}, \beta),$$

then also $\mathsf{R} \in X$. Moschovakis [1974a] proves

(17) for any normal type-3 functional $\mathbb{I}$, 2-env($\mathbb{I}$) is the smallest semi-Spector class over ${}^{\omega}\omega$ closed under $\mathbb{I}$-application,

and Kechris [1973a] shows

(18) a semi-Spector class X is the 2-envelope of some normal type-3 functional iff X is not Mahlo.

The extensions to types 4 and higher are now straightforward.

Sacks [1974] and [1978] has established parallel characterizations of the k-section of a normal type-n functional. For $k = 1$ and $n = 2$ this goes as follows. Let Code be the smallest class of subsets of ω such that $\{0\} \in \text{Code}$ and if for all n, $A_n \in \text{Code}$, then also $\{\langle m, n\rangle : m \in A_n\} \in \text{Code}$. To each $A \in \text{Code}$, assign a set by:

$$\text{set}(\{0\}) = \varnothing, \quad \text{and for} \quad A \neq \{0\},$$

$$\text{set}(A) = \{\text{set}(\{m : \langle m, n\rangle \in A\}) : n \in \omega\}.$$

The hereditarily countable sets are exactly those of the form set(A) for some $A \in \text{Code}$. Let As_1 (abstract 1-section) denote the theory in the language $\mathscr{L}_{\mathbf{ZF}}$ generated by the following axioms: Pair, Union, Δ_0-Separation (Definition VIII.7.3) and

(*Local Countability*) $\forall x.\ x$ is countable;

(Δ_0-DC) $\forall x\, \exists y\, \mathfrak{A} \rightarrow \exists f\, [f$ is a function with domain ω and $\forall n\, \mathfrak{A}(f(n)/x, f(n+1)/y)]$

for all Δ_0 formulas $\mathfrak{A}$.

Sacks then proves by a forcing argument:

(19) a countable class X of relations over ω closed under "recursive in" is the 1-section of some normal type-2 functional iff the structure $(\{\text{set}(A) : A \in X\}, \in)$ is a model of As_1.

In particular, this implies that all of the following classes of relations are 1-sections of normal type-2 functionals: Δ^1_r $(r \geq 1)$, $\{R : R \in \mathrm{L}_\kappa\}$ for any countable admissible κ, and the 1-section of any normal functional of type $\geqslant 2$. The extension of this last fact to higher types yields the following result known as the *Plus-One Theorem*:

(20) for any $k > 0$, the k-section of any normal functional of type $\geqslant (k+1)$ is also the k-section of some normal type-$(k+1)$ functional.

Moschovakis [1974a] shows that there is no corresponding result for envelopes — indeed:

(21) for any normal type-3 functional $\mathbb{I}$, 1-env($\mathbb{I}$) is not the 1-envelope of any normal type-2 functional.

The same technique shows that the classes Σ^1_r $(r \geqslant 1)$ and Π^1_r $(r \geqslant 2)$ are not 1-envelopes of normal type-2 functionals. However, Harrington [1973] proves the following *Plus-Two Theorem*:

(22) for any $k > 0$, the k-envelope of any normal functional of type $\geqslant (k+2)$ is also the k-envelope of some normal type-$(k+2)$ functional.

Before we leave the subject of recursion in higher types, we should mention the elegant Kechris–Moschovakis [1977]. Here is given an exposition of the basic notions and facts of this theory in the context of the general theory of inductive definability. Many of the difficult results are derived from simpler, more general results about inductive definitions.

One of the most surprising developments in recent years is that a large part of the theory developed for the type-structure over ω can be extended to the type-structure over an arbitrary set. The two basic tools for the development of this theory are inductive definability and admissible sets (with urelements). Indeed, many parts of the theory can be viewed from either point of view. The basic references here are Moschovakis [1974] and Barwise [1975]. The theory is quite complex by now and we can give here only a few examples of its scope.

Let $\mathfrak{M} = (M, R_0, \ldots, R_{n-1})$ be a relational structure for some first-order language $\mathscr{L}$. Add to $\mathscr{L}$ names for the elements of M and second-order variables $U, V, W, \ldots$ ranging over relations. A formula $\mathfrak{A}$ is called ***first-order positive*** iff all occurrences of second-order variables are free and positive. If $\mathfrak{A}$ is first-order positive with $x_0, \ldots, x_{k-1}$, U free, then $\mathfrak{A}$ defines a monotone operator $\Gamma_{\mathfrak{A}}$ over ${}^kM$: for $R \subseteq {}^kM$,

$$\Gamma_{\mathfrak{A}}(R) = \{\mathbf{a} : \mathfrak{M} \models \mathfrak{A}[\mathbf{a}, R]\}.$$

(Monotonicity follows from the fact that U occurs only positively.) A relation

$S \subseteq {}^kM$ is $\mathfrak{M}$*-positive-Inductive* ($\mathfrak{M}$-pos-Ind) iff for some operator $\Gamma_{\mathfrak{A}}$ over ${}^{k+l}M$ and some $\mathbf{b} \in {}^lM$, for all $\mathbf{a} \in {}^kM$,

$$S(\mathbf{a}) \leftrightarrow (\mathbf{a}, \mathbf{b}) \in \bar{\Gamma}_{\mathfrak{A}}.$$

S is $\mathfrak{M}$*-positive-Hyperelementary* ($\mathfrak{M}$-pos-Hyp) iff both R and ${}^kM \sim R$ are $\mathfrak{M}$-pos-Ind. Much of the general theory is motivated by the examples of the two structures:

$$\mathfrak{N}_0 = (\omega, R_+, R_\cdot) = \text{standard (relational) structure for arithmetic;}$$

$$\mathfrak{N}_1 = (\omega \cup {}^\omega\omega, \omega, R_+, R_\cdot, \text{Apl}) = \text{standard (relational) structure for analysis,}$$

where R_+ and $R_\cdot$ are the graphs of the functions $+$ and $\cdot$ and $\text{Apl}(m, n, \alpha) \leftrightarrow \alpha(m) = n$. For these we have

(23) $\quad \mathfrak{N}_0\text{-pos-Ind} = \Pi_1^1 = 1\text{-env}(\mathsf{E});$

$$\mathfrak{N}_1\text{-pos-Ind} = \bigcup\{2\text{-env}(\mathsf{E}^{\#}, \beta) : \beta \in {}^\omega\omega\}.$$

This second equality should be contrasted with (14); here arbitrary parameters from ${}^\omega\omega$ may be introduced in the reduction.

For any structure $\mathfrak{M}$ we call a class X of relations over M an $\mathfrak{M}$*-Spector-class* iff it contains the $\mathfrak{M}$-elementary (first-order definable over $\mathfrak{M}$) relations, is closed under $\cap$, $\cup$, $\exists^M$, $\forall^M$, and relational composition with $\mathfrak{M}$-elementary functions, is M-indexable (contains a universal relation) and has the pre-wellordering property. Moschovakis [1974] shows that for an arbitrary $\mathfrak{M}$, the class $\mathfrak{M}$-pos-Ind has all of these properties except perhaps M-indexability. Note that in the absence of a universal relation the pre-wellordering property may be established in the form of Exercise V.1.20(ii); the proof is along the lines of Exercise III.3.33. To secure indexability he restricts to the class of acceptable structures, which have the requisite coding ability. $\mathfrak{M}$ is *acceptable* iff there exist $N \subseteq M$, $\preccurlyeq \subseteq N \times N$, and a one–one (coding) function $\langle\cdot\rangle$ from the set of all finite sequences from M into M such that $\preccurlyeq$ orders N in type ω and N, $\preccurlyeq$, and the following relations and functions are all $\mathfrak{M}$-elementary:

$$\text{Seq}(a) \leftrightarrow a \text{ belongs to the range of } \langle\cdot\rangle;$$

$$\begin{aligned} \text{lh}(a) &= \dot{0}, \quad \text{if} \quad \neg\text{Seq}(a); \\ &= \dot{n}, \quad \text{if} \quad a = \langle a_1, \ldots, a_n\rangle; \end{aligned}$$

$$\begin{aligned} \text{q}(a, \dot{m}) &= a_m, \quad \text{if} \quad a = \langle a_1, \ldots, a_n\rangle \quad \text{and} \quad 1 \leq m \leq n; \\ &= \dot{0}, \quad \text{otherwise;} \end{aligned}$$

where $\dot{n}$ denotes the n-th element of N in the ordering $\preccurlyeq$.

For an acceptable structure $\mathfrak{M}$ the proof that $\mathfrak{M}$-pos-Ind is M-indexable proceeds by showing that this is exactly the class of relations definable over $\mathfrak{M}$ by a *game formula*:

$$\forall b_0 \exists b_1 \forall b_2 \exists b_3 \cdots \bigvee_{n \in \omega} \mathfrak{A}(\langle b_0, \ldots, b_{2n-1} \rangle, \mathbf{a})$$

with $\mathfrak{A}$ elementary. This fact is a generalization of Exercise III.3.22. Then Moschovakis [1974] shows:

(24) for any acceptable structure $\mathfrak{M}$, $\mathfrak{M}$-pos-Ind is the smallest $\mathfrak{M}$-Spector class.

One of the most interesting results of the theory is a very broad generalization of (23). A relation $S \subseteq {}^kM$ is called $\mathfrak{M} - \Pi_1^1$ iff it is definable over $\mathfrak{M}$ by a second-order formula with parameters in which the second-order variables are quantified only universally. Then Moschovakis [1974, 8A.1] proves

(25) for any countable acceptable structure $\mathfrak{M}$,

$$\mathfrak{M}\text{-pos-Ind} = \mathfrak{M} - \Pi_1^1.$$

Note that the assumption of countability is really necessary here since $\mathfrak{N}_1$-pos-Ind is a proper subset of $\mathbf{\Pi}_1^2 = \mathfrak{N}_1 - \Pi_1^1$ (cf. discussion following VII.2.11). Barwise [1975, VI.5.2] shows that the hypothesis of acceptability in (25) (and elsewhere) may be replaced by the weaker requirement that M admit an M-pos-Ind pairing function.

We consider now the approach via admissible sets. For the moment we restrict attention to transitive $\in$-structures — that is, structures of the form $\mathfrak{M} = (M, \in, R_0, \ldots, R_{n-1})$ where M is a transitive set and $\in$ is the restriction of the membership relation to M. Below we mention how this restriction may be removed. For such $\mathfrak{M}$, set

$$\mathrm{HYP}(\mathfrak{M}) = \bigcap \{N : N \text{ is an admissible set and } M, R_0, \ldots, R_{n-1} \in N\}.$$

It turns out that $\mathrm{HYP}(\mathfrak{M})$ is itself admissible and is thus the smallest admissible set containing $M, R_0, \ldots, R_{n-1}$. Then Barwise [1975, IV.3.1, 3 and VI.5.1] shows

(26) for any countable transitive $\in$-structure $\mathfrak{M}$ and any $S \subseteq {}^kM$, S is $\mathfrak{M} - \Pi_1^1$ iff S is Σ definable over $(\mathrm{HYP}(\mathfrak{M}), \in)$;

(27) for any transitive $\in$-structure $\mathfrak{M}$ which admits an $\mathfrak{M}$-pos-Ind pairing function and any $S \subseteq {}^kM$, S is $\mathfrak{M}$-pos-Ind iff S is Σ definable over $(\mathrm{HYP}(\mathfrak{M}), \in)$.

Note that in (27) since $M \in \mathrm{HYP}(\mathfrak{M})$, it follows by Δ-Separation that S is $\mathfrak{M}$-pos-Hyp iff $S \in \mathrm{HYP}(\mathfrak{M})$. In view of Barwise [1975, VI.5.8], (27) provides an alternative method for proving that $\mathfrak{M}$-pos-Ind is a Spector class.

Applying (26) to the case $\mathfrak{M} = (\mathrm{L}_\kappa, \in)$, we obtain the following pretty generalization of Theorem VIII.3.4:

(28) for any countable admissible ordinal κ, and any $R \subseteq {}^k\kappa$,
(i) R is κ^+-semi-recursive iff R is $\kappa - \Pi^1_1$;
(ii) R is κ^+-recursive iff R is $\kappa - \Delta^1_1$;

where κ^+ is the least admissible ordinal greater than κ and R is $\kappa - \Pi^1_1$ iff for some κ-recursive relation $\mathsf{P} \subseteq {}^{k,1}\kappa$ (cf. VIII.2.26),

$$R(\mu) \leftrightarrow (\forall f : \kappa \to \kappa)(\exists \pi < \kappa)\mathsf{P}(\pi, \mu, f).$$

The characterization (27) of a particular $\mathfrak{M}$-Spector class in terms of definability over an admissible structure can be extended to give a representation for arbitrary $\mathfrak{M}$-Spector classes via the notion of a companion. For any $\mathfrak{M}$-Spector class X, let

$$\mathrm{Cmp}(X) = \bigcap\{N : N \text{ is an admissible set and } \Delta(X) \subseteq N\}.$$

$\mathrm{Cmp}(X)$ is called the *Companion Set* of X. By way of abbreviation, for any $C \subseteq \mathrm{Cmp}(X)$, let us write $\Sigma[C]$ ($\Delta[C]$) for the class of all relations on $\mathrm{Cmp}(X)$ which are Σ-definable (Δ-definable) over the structure $(\mathrm{Cmp}(X), \in, C)$. A structure of this type is called admissible iff it satisfies the axioms and schemata Pair, Union, Δ_0-Separation, and Δ_0-Collection formulated in the appropriate language, $\mathscr{L}_{\mathrm{ZF}}$ with an additional unary relation symbol. The *Companion Theorem*, Moschovakis [1974, 9E.1] asserts

(29) for any $\mathfrak{M}$-Spector class there exists a set $C_X \subseteq \mathrm{Cmp}(X)$ such that
(i) $(\mathrm{Cmp}(X), \in, C_X)$ is admissible;
(ii) $\mathrm{Cmp}(X)$ is C_X-projectible on M (that is, there exists a $\Delta[C_X]$ function f which maps a subset of M onto $\mathrm{Cmp}(X)$);
(iii) $\mathrm{Cmp}(X)$ is C_X-resolvable (that is, there exists a $\Delta[C_X]$ function $g : o(\mathrm{Cmp}(X)) \to \mathrm{Cmp}(X)$ such that $\mathrm{Cmp}(X) = \bigcup\{g(\sigma) : \sigma < o(\mathrm{Cmp}(X))\}$).
(iv) for any $S \subseteq {}^kM$, $S \in X$ iff $S \in \Sigma[C_X]$.

Furthermore, if C'_X is any other set satisfying (i)–(iv), then $\Sigma[C_X] = \Sigma[C'_X]$.

The class $\Sigma[C_X]$ is called the *Companion Class* of X and the pair $(\mathrm{Cmp}(X), \Sigma[C_X])$ is the *Companion* of X. Result (27) may now be restated as

(30) for any transitive $\in$-structure $\mathfrak{M}$ which admits an $\mathfrak{M}$-pos-Ind pairing function, the Companion of $\mathfrak{M}$-pos-Ind is $(\mathrm{HYP}(\mathfrak{M}), \Sigma[\varnothing])$.

Moschovakis [1974] and [1974b] computes the companions of many other Spector classes. We shall discuss here just one further class of examples which generalizes (6). A structure $\mathfrak{M}$ is *almost acceptable* iff there exist relations and functions N, $\preccurlyeq$, Seq, lh, and q as before which are $\mathfrak{M}$-pos-Hyp (instead of $\mathfrak{M}$-elementary). Let $\mathscr{L}^*$ be the first-order language for the expanded structure $\mathfrak{M}^* = (M, R_0, \ldots, R_{n-1}, \preccurlyeq, \mathrm{Seq}, \mathrm{lh}, \mathrm{q})$ together with names for the elements of M and second-order variables $U, V, W, \ldots$. The classes of $\exists^0_r$ and $\forall^0_r$ formulas of $\mathscr{L}^*$ are defined as in III.5.3 with $\preccurlyeq$ playing the role of $\leqslant$ for bounded quantifiers. If $\mathfrak{A}$ is a formula of $\mathscr{L}^*$ with $x_0, \ldots, x_{k-1}$, U free, then $\mathfrak{A}$ defines an inductive operator $\Gamma_{\mathfrak{A}}$ over ${}^kM$ by: for $R \subseteq {}^kM$,

$$\Gamma_{\mathfrak{A}}(R) = R \cup \{\mathbf{a} : \mathfrak{M}^* \models \mathfrak{A}[\mathbf{a}, R]\}.$$

Of course, $\Gamma_{\mathfrak{A}}$ is in general no longer monotone. A relation $S \subseteq {}^kM$ is called $\mathfrak{M}$-Π^0_r-*Inductive* ($\mathfrak{M}$-Σ^0_r-*Inductive*) iff for some $\forall^0_r$ ($\exists^0_r$) formula $\mathfrak{A}$ and some $\mathbf{b} \in {}^lM$, for all $\mathbf{a} \in {}^kM$,

$$S(\mathbf{a}) \leftrightarrow (\mathbf{a}, \mathbf{b}) \in \bar{\Gamma}_{\mathfrak{A}}.$$

Theorem 11 of Moschovakis [1974b] verifies that this is independent of the particular N, $\preccurlyeq$, Seq, lh, and q chosen. He further shows:

(31) for any almost acceptable structure $\mathfrak{M}$,
(i) $\mathfrak{M}$-Σ^0_2-Ind $=$ $\mathfrak{M}$-pos-Ind;
(ii) for all $r \geqslant 2$, $\mathfrak{M}$-Π^0_r-Ind $=$ $\mathfrak{M}$-Σ^0_{r+1}-Ind and is an $\mathfrak{M}$-Spector class.

In fact, $\mathfrak{M}$-Π^0_r-Ind is chatacterized as the smallest $\mathfrak{M}$-Spector class satisfying either one of two natural additional conditions. Note that the restriction to $r \geqslant 2$ is again necessary because by (13),

$$\mathfrak{N}_1\text{-}\Pi^0_1\text{-Ind} = \mathbf{\Pi}^1_1\text{-Ind} = \bigcup\{2\text{-env}(\mathbb{E}, \beta) : \beta \in {}^{\omega}\omega\}$$

is not a Spector class by VII.2.11. However by Moschovakis [1974b, Theorem 29]

(32) for any acceptable structure $\mathfrak{M}$, $\mathfrak{M}$-Π^0_1-Ind is the smallest $\mathfrak{M}$-semi-Spector class.

As for companions, Moschovakis [1974b, Theorem 23] shows

(33) for any almost acceptable transitive $\in$-structure $\mathfrak{M}$ and any $r \geq 2$, the companion of $\mathfrak{M}$-Π^0_r-Ind is $(\mathrm{HYP}(\mathfrak{M}, \Pi^0_r), \Sigma[\varnothing])$, where

$$\mathrm{HYP}(\mathfrak{M}, \Pi^0_r) = \bigcap\{N : N \text{ is a } \forall^0_{r+1}\text{-reflecting admissible set and } M, R_0, \ldots R_{n-1} \in N\}.$$

Similar but more complicated characterizations are given for $\mathfrak{M}$-Σ^m_r-Ind and $\mathfrak{M}$-Π^m_r-Ind for all m, $r \geq 1$.

At any given level of definability the class of monotone operators is intermediate between the class of positive operators and the class of arbitrary inductive operators. In many cases there are strictly more relations definable by arbitrary operators than by positive operators. For arithmetical operators over ω the relations definable by positive or by monotone operators coincide; a general version of this follows easily from (25):

(34) for any countable acceptable structure $\mathfrak{M}$, $\mathfrak{M}$-pos-Ind $=$ $\mathfrak{M}$-mon-Ind.

On the other hand it was known from (7) (and earlier) that over ω, Σ^1_1-Ind $=$ Σ^1_1-mon-Ind. Harrington–Kechris [1976] show that this situation is not at all infrequent. They call a class Z of inductive operators over an almost acceptable structure $\mathfrak{M}$ *adequate* if it satisfies certain mild definability conditions, which in particular are satisfied by all of the classes $\mathfrak{M}$-Σ^k_r and $\mathfrak{M}$-Π^k_r $(k + r \geq 1)$. We say *well-foundedness is cZ over* $\mathfrak{M}$ iff the second-order relation "S is not well-founded" belongs to Z. Then Harrington–Kechris prove

(35) for any almost acceptable structure $\mathfrak{M}$ and any adequate class Z, if well-foundedness is cZ over $\mathfrak{M}$ and

$$cZ \subseteq \mathfrak{M}\text{-}Z\text{mon-Ind},$$

then

$$\mathfrak{M}\text{-}Z\text{-mon-Ind} = \mathfrak{M}\text{-}Z\text{-Ind}.$$

In particular, it follows that if well-foundedness is Π^0_1 over $\mathfrak{M}$, as it is for $\mathfrak{N}_1$, then for all $r \geq 2$, $\mathfrak{M}$-Π^0_r-Ind $=$ $\mathfrak{M}$-Π^0_r-mon-Ind, and $\mathfrak{M}$-mon-Ind $=$ $\mathfrak{M}$-Ind. As another application, note that for any almost acceptable $\mathfrak{M}$, well-foundedness is Π^1_1 over $\mathfrak{M}$. Thus if $\mathfrak{M}$ is also countable, it follows from (24) and (25) that $\mathfrak{M}$-Σ^1_1-mon-Ind $=$ $\mathfrak{M}$-Σ^1_1-Ind. A similar result holds for uncountable almost acceptable $\mathfrak{M}$ if we replace 'Π^1_1' by '$\mathfrak{M}$-pos-Ind'.

To extend the methods involving admissible sets to arbitrary relational structures $\mathfrak{M}$, Barwise [1975] develops the theory of admissible sets with urelements. The elements of M are treated as initially given objects and a class V_M of sets over M is defined recursively by:

$$V_M(0) = \varnothing;$$
$$V_M(\sigma + 1) = \mathbf{P}(M \cup V_M(\sigma));$$
$$V_M(\sigma) = \bigcup\{V_M(\tau) : \tau < \sigma\} \text{ for limit } \sigma.$$

Let $\mathscr{L}^*$ be a first-order language which includes the language of $\mathfrak{M}$, unary relation symbols U ("is an urelement") and S ("is a set"), and the membership symbol $\in$. Any set $A \in V_M$ determines a structure

$$\mathbb{A}_{\mathfrak{M}} = (M \cup A, R_0, \ldots, R_{n-1}, M, A, \in)$$

for $\mathscr{L}^*$. Such an A is called *admissible over* $\mathfrak{M}$ iff $M \cup A$ is transitive and $\mathbb{A}_{\mathfrak{M}}$ satisfies the axioms and schemata Pair, Union, Δ_0-Separation, and Δ_0-Collection formulated in $\mathscr{L}^*$. If in addition the set M of urelements belongs to A, then A is called *admissible above* $\mathfrak{M}$. Much of the theory of admissible sets carries over to admissible sets with urelements. As an example, define

$$\mathrm{HYP}_{\mathfrak{M}} = \bigcap\{A : A \text{ is admissible above } \mathfrak{M}\}.$$

Then $\mathrm{HYP}_{\mathfrak{M}}$ is itself admissible above $\mathfrak{M}$ and the following extension of (27) holds:

(36) for any structure $\mathfrak{M}$ which admits an $\mathfrak{M}$-pos-Ind pairing function and any $S \subseteq {}^kM$, S is $\mathfrak{M}$-pos-Ind iff S is Σ-definable over $\mathrm{HYP}_{\mathfrak{M}}$.

Still another approach to the study of definability over arbitrary structures is through abstract and axiomatic computation theories. Moschovakis [1969] develops a theory of computations over a structure $\mathfrak{M}$ by working in a larger set M^* obtained by closing $M \cup \{0\}$ under a pairing function. Since M^* may not have a natural wellordering, he is forced to deal with multiple-valued functions. A relation $\{e\}(\mathbf{x}) \rightarrow y$ is defined inductively and the functions so indexed are called *$\mathfrak{M}$-search computable* because of the inclusion of the following clause in the inductive definition:

$$\text{if } \{b\}(y, \mathbf{x}) \rightarrow 0, \quad \text{then } \{\langle 9, k, b\rangle\}(\mathbf{x}) \rightarrow y.$$

Gordon [1970] shows that $\mathfrak{M}$-search computability coincides with Σ-definability over $\mathrm{HF}_{\mathfrak{M}}$, the structure over $\mathfrak{M}$ whose sets are the hereditarily finite sets in V_M. If a schema of the form

$$\text{if } \forall y\, \exists u . \{b\}(y, \mathbf{x}) \rightarrow u \quad \text{and} \quad \exists y . \{b\}(y, \mathbf{x}) \rightarrow 0,$$
$$\text{then } \{\langle 10, k, b\rangle\}(\mathbf{x}) \rightarrow 0;$$
$$\text{if } \forall y\, \exists u\, [u \neq 0 \wedge \{b\}(y, \mathbf{x}) \rightarrow u], \quad \text{then } \{\langle 10, k, b\rangle\}(\mathbf{x}) \rightarrow 1;$$

is added to those for search computability, the resulting functions are called $\mathfrak{M}$-*search computable* in E. Combining the results of Moschovakis [1969] and [1974] and Barwise [1975]

(37) for any structure $\mathfrak{M}$ which admits an $\mathfrak{M}$-pos-Ind pairing function and any $S \subseteq {}^kM$, S is $\mathfrak{M}$-semi-search computable in E iff S is Σ-definable over $\mathrm{HYP}_{\mathfrak{M}}$ iff S is $\mathfrak{M}$-pos-Ind.

If the E-schema is left in but the search schema is removed, the resulting functions are called $\mathfrak{M}$-*prime computable in* E. For $\mathfrak{N}_1$ this is equivalent to recursion in $\mathbb{E}$ (as opposed to $\mathbb{E}^{\#}$ for search computability in E) and in general Grilliot [1971a] shows essentially

(38) for any acceptable structure $\mathfrak{M}$, $\mathfrak{M}$-Π_1^0-Ind is the class of relations $\mathfrak{M}$-semi-prime computable in E.

Axiomatic computation theories have been studied by several people. Wagner [1969] and Strong [1968] developed much of the elementary theory. Moschovakis [1971] introduced axioms concerning the (ordinal) length of a computation and Fenstad [1974] adds the relation "is a subcomputation of". There is a natural notion of a Spector computation theory such that a class of relations is a Spector class iff it is the envelope of a Spector theory. Many of the deepest results of recursion theory on higher types such as the Grilliot Selection Theorem and the Plus-One and Plus-Two Theorems have been generalized to the axiomatic setting. The current state of the theory is surveyed in Fenstad [1975] and treated in detail in Moldestad [1977] and Fenstad [1980?].

References

Aanderaa, S.
1974 *Inductive definitions and their closure ordinals*, in Fenstad–Hinman [1974], 207–220.

Ackermann, W.
1928 *Zum Hilbertschen Aufbau der reelen Zahlen*, **Math. Annalen** 99 (1928) 118–133.

Aczel, P.
1970 *Representability in some systems of second order arithmetic*, **Israel Jour. Math.** (1970) 309–328.

1975 *Quantifiers, games, and inductive definitions*, in Kanger [1975], 1–14.

Aczel, P. and Hinman, P.G.
1974 *Recursion in the superjump*, in Fenstad–Hinman [1974], 3–41.

Aczel P. and Richter, W.H.
1974 *Inductive definitions and reflecting properties of admissible ordinals*, in Fenstad–Hinman [1974], 301–381.

Addison, J.W. Jr.
1954 **On Some Points of the Theory of Recursive Functions,** Ph.D. thesis; University of Wisconsin, Madison, 1954, 92 pp.
1955 *Analogies in the Borel, Lusin, and Kleene hierarchies*, I, II (abstracts), **Bull. Amer. Math. Soc.** 61 (1955) 75, 171–172.
1959 *Separation principles in the hierarchies of classical and effective descriptive set theory*, **Fund. Math.** 46 (1959) 123–135.
1959a *Some consequences of the axiom of constructibility*, **Fund. Math.** 46 (1959) 337–357.
1965 *The undefinability of the definable* (abstract), **Notices Amer. Math. Soc.** 12 (1965) 347.

Addison, J.W. Jr. and Moschovakis, Y.N.
1968 *Some consequences of the axiom of definable determinateness*, **Proc. Nat. Acad. Sci. USA** 59 (1968) 708–712.

Amstislavskij, V.I.
1970 *Effective R-sets and transfinite extensions of recursive hierarchies*, **Fund. Math.** 68 (1970) 61–68.
1970a *On the decomposition of a field of sets obtained by an R-operation over recursive sets*, **Soviet Mathematics** 11 (1970) 419–422.

Anonymous
1961 **Infinitistic Methods** (Proceedings of the Symposium on Foundations of Mathematics, Warsaw, 1959), Pergamon Press, Oxford, 1961, 362 pp.

Bar-Hillel, Y.
1965 **Logic, Methodology, and Philosophy of Science II** (Proceedings of the International Congress, Jerusalem, 1964, Editor), North-Holland, Amsterdam, 1965, 440 pp.

1970 **Mathematical Logic and Foundations of Set Theory** (Proceedings of an international colloquium, Jerusalem, 1968, Editor), North-Holland, Amsterdam, 1970, 145 pp.

Bar-Hillel, Y., Poznanski, E.I.J., Rabin, M.O., and Robinson, A.

1961 **Essays on the Foundations of Mathematics** (Dedicated to A.A. Fraenkel on his seventieth anniversary, Editors), Magnus Press, Hebrew University, Jerusalem, 1961, 351 pp.

Barnes, R.F.

1965 *The classification of recursive sets of number-theoretic functions* (abstract), **Notices Amer. Math. Soc.** 12 (1965) 622–623.

Barwise, K.J.

1975 **Admissible Sets and Structures,** Perspectives in Mathematical Logic 1, Springer-Verlag, Berlin–Heidelberg–New York, 1975, 394 pp.

1977 **Handbook of Mathematical Logic,** North-Holland, Amsterdam, 1977.

Barwise, K.J. and Fisher, E.

1970 *The Shoenfield absoluteness lemma,* **Israel Jour. Math.** 8 (1970) 329–339.

Blackwell, D.

1967 *Infinite games and analytic sets,* **Proc. Nat. Acad. Sci. U.S.A.** 58 (1967) 1836–1837.

Bulloff, J.J., Hahn, S.W., and Holyoke, T.C.

1969 **Foundations of Mathematics** (Symposium papers commemorating the sixtieth birthday of Kurt Gödel, Columbus, Ohio, 1966, Editors), Springer-Verlag, Berlin–Heidelberg–New York, 1969, 195 pp.

Cenzer, D.

1974 *Ordinal recursion and inductive definitions,* in Fenstad–Hinman [1974], 221–264.

1974a *Analytic inductive definitions,* **Jour. Symb. Log.** 39 (1974) 310–312.

1974b *Inductively defined sets of reals,* **Bull. Amer. Math. Soc.** 80 (1974) 485–487.

Chang, C.C. and Moschovakis, Y.N.

1970 *The Suslin–Kleene theorem for V_κ with cofinality $(\kappa) = \omega$,* **Pacific Jour. Math.** 35 (1970) 565–569.

Clarke, D.A.

1964 **Hierarchies of Predicates of Finite Types,** Memoirs of the Amer. Math. Soc. 51, American Mathematical Society, Providence, R.I., 1964, 95 pp.

Cohen, P.J.

1963/64 *The independence of the continuum hypothesis,* I, II, **Proc. Nat. Acad. Sci. U.S.A.** 50 (1963) 1143–1148, 51 (1964) 105–110.

Crossley, J.N.

1967 **Sets, Models, and Recursion Theory** (Proceedings of the Summer School in Mathematical Logic and Logic Colloquium, Leicester, England, 1965, Editor), North-Holland, Amsterdam, 1967, 331 pp.

Davis, M.

1964 *Infinite games of perfect information,* **Advances in Game Theory,** Annals of Math. Studies 52 (1964) 85–101.

Dekker, J.C.E.

1962 **Recursive Function Theory** (Proceedings of the Fifth Symposium in Pure Mathematics of the Amer. Math. Soc., New York, 1961, Editor), American Mathematical Society, Providence, R.I., 1962, 247 pp.

Devlin, K.J.

1973 **Aspects of Constructibility,** Lecture Notes in Mathematics 354, Springer-Verlag, Berlin–Heidelberg–New York, 1973, 240 pp.

1974 *An introduction to the fine structure of the constructible hierarchy (results of Ronald Jensen),* in Fenstad–Hinman [1974], 123–163.

Enderton, H.B.
1964 *Hierarchies in recursive function theory*, **Trans. Amer. Math. Soc.** 111 (1964) 457–471.
1970 *The unique existential quantifier*, **Archiv. f. math. Logik und Grundl.** 13/1–2 (1970) 52–54.
1972 **A Mathematical Introduction to Logic,** Academic Press, New York, 1972, 293 pp.

Feferman, S.
1964/65 *Some applications of the notions of forcing and generic sets*, **Fund. Math.** 56 (1964/65) 325–345.

Fenstad, J.E.
1971 **Proceedings of the Second Scandanavian Logic Symposium** (Oslo, 1970, Editor), North-Holland, Amsterdam, 1971, 413 pp.
1971a *The axiom of determinateness*, in Fenstad [1971], 41–61.
1974 *On axiomatizing recursion theory*, in Fenstad–Hinman [1974], 385–404.
1975 *Computation theories: an axiomatic approach to recursion on general structures*, in Müller *et al.* [1975], 143–168.
1980? **Computation Theories: an Axiomatic Approach to Recursion on General Structures,** Perspectives in Mathematical Logic, Springer-Verlag, Berlin–Heidelberg–New York, to appear.

Fenstad, J.E. and Hinman, P.G.
1974 **Generalized Recursion Theory** (Proceedings of the 1972 Oslo Symposium, Editors), North-Holland, Amsterdam, 1974, 456 pp.

Fenstad, J.E. and Normann, D.
1974 *On absolutely measurable sets*, **Fund. Math.** 81 (1974) 91–98.

Friedman, H.M.
1971 *Determinateness in the low projective hierarchy*, **Fund. Math.** 72 (1971) 79–95.

Gale, D. and Stewart, F.M.
1953 *Infinite games with perfect information*, **Contributions to the Theory of Games II,** Ann. of Math. Studies 28 (1953) 245–266.

Gandy, R.O.
1960 *Proof of Mostowski's conjecture*, **Bull. Acad. Polon. Sci.** 8 (1960) 571–575.
1967 *Computable functionals of finite type* I, in Crossley [1967], 202–242.
1967a *General recursive functionals of finite type and hierarchies of functions*, **Ann. Fac. Sci. Univ. Clermont-Ferrand** 35 (1967) 5–24.

Gandy, R.O. and Yates, C.E.M.
1971 **Logic Colloquium '69** (Proceedings of the Summer School and Colloquium in Mathematical Logic, Manchester, 1969, Editors), North-Holland, Amsterdam, 1971, 451 pp.

Gödel, K.
1931 *Über formal unentscheidbare Sätze der Principia Mathematica und verwandter Systeme*, I, **Monatshefte für Mathematik und Physik** 38 (1931) 173–198.
1939 *Consistency-proof for the generalized continuum-hypothesis*, **Proc. Nat. Acad. Sci. U.S.A.** 25 (1939) 220–224.

Gordon, C.
1970 *Comparisons between some generalizations of recursion theory*, **Compositio Math.** 22 (1970) 333–346.

Grilliot, T.J.
1969 *Selection functions for recursive functionals*, **Notre Dame J. of Formal Logic** 10 (1969) 225–234.
1969a *Hierarchies based on objects of finite types*, **Jour. Symb. Log.** 34 (1969) 177–182.
1971 *On effectively discontinuous type-2 objects*, **Jour. Symb. Log.** 36 (1971) 245–248.
1971a *Inductive definitions and computability*, **Trans. Amer. Math. Soc.** 158 (1971) 309–317.

Harrington, L.A.

1973 **Contributions to Recursion Theory on Higher Types (or, a Proof of Harrington's Conjecture)**, Ph.D. thesis, MIT, 1973.

1974 *The superjump and the first recursively Mahlo ordinal*, in Fenstad–Hinman [1974], 43–52.

1975 *Π^1_2 sets and Π^1_2 singletons*, **Proc. Amer. Math. Soc.** 52 (1975) 356–360.

1978 *Consistency and independence results in descriptive set theory*, I, II, to appear in **Ann. of Math.**

Harrington, L.A. and Mac Queen, D.B.

1976 *Selection in abstract recursion theory*, **Jour. Symb. Log.** 41 (1976) 153–158.

Harrington, L.A. and Kechris, A.S.

1975 *On characterizing Spector classes*, **Jour. Symb. Log.** 40 (1975) 19–24.

1976 *On monotone vs. non-monotone induction*, **Bull. Amer. Math. Soc.** 82 (1976) 888–890.

Hinman, P.G.

1964 *Some theorems on relative recursiveness of higher type objects (abstract)*, **Notices Amer. Math. Soc.** 11 (1964) 677.

1966 **Ad Astra per Aspera: Hierarchy Schemata in Recursive Function Theory**, Ph.D. thesis, University of California, Berkeley, 1966, 133 pp.

1969 *Hierarchies of effective descriptive set theory*, **Trans. Amer. Math. Soc.** 142 (1969) 111–140.

1969a *Some applications of forcing to hierarchy problems in arithmetic*, **Zeitschr. f. math. Logik u. Grundlagen d. Math.** 15 (1969) 341–352.

1973 *The finite levels of the hierarchy of effective R-sets*, **Fund. Math.** 74 (1973) 1–10.

Hinman, P.G. and Moschovakis, Y.N.

1971 *Computability over the continuum*, in Gandy–Yates [1971], 77–105.

Jensen, R.B. and Karp, C.

1971 *Primitive recursive set functions*, in Scott [1971], 143–176.

Kanger, S.

1975 **Proceedings of the Third Scandinavian Logic Symposium** (Uppsala, Sweden, 1973, Editor), North-Holland, Amsterdam, 1975, 214 pp.

Kantorovitch L. and Livenson, E.

1932/33 *Memoir on the analytical operations and projective sets*, I and II, **Fund. Math.** 18 (1932) 214–279 and 20 (1933) 54–97.

Kechris, A.S.

1973 *Measure and category in effective descriptive set theory*, **Ann. Math. Logic** 5 (1973) 337–384.

1973a *The structure of envelopes: a survey of recursion theory in higher types*, MIT Logic Seminar notes, 1973, 28 pp.

1974 *On projective ordinals*, **Jour. Symb. Log.** 39 (1974) 269–282.

Kechris, A.S. and Moschovakis, Y.N.

1977 *Recursion in higher types*, in Barwise [1977], 681–737.

Kino, A. and Takeuti, G.

1962 *On hierarchies of predicates of ordinal numbers*, **Jour. Math. Soc. Japan** 14 (1962) 199–232.

1962a *A note on predicates of ordinal numbers*, **Jour. Math. Soc. Japan** 14 (1962) 367–378.

Kleene, S.C.

1935/36 *General recursive functions of natural numbers*, **Math. Ann.** 112 (1935/36) 727–742.

1943 *Recursive predicates and quantifiers*, **Trans. Amer. Math. Soc.** 53 (1943) 41–73.

1950 *A symmetric form of Gödel's theorem*, **Proceedings of the Koninklijke Nederlandse Akademie van Wetenschappen** 53 (1950) 800–802 and **Indag. Math.** 12 (1950) 244–246.

1952 *Recursive functions and intuitionistic mathematics*, **Proceedings of the International Congress of Mathematicians**, Cambridge, Mass., 1950, 679–685, American Mathematical Society, Providence, R.I., 1952.

1955 *On the forms of the predicates in the theory of constructive ordinals* (second paper), **Amer. J. Math.** 77 (1955) 405–428.

1955a *Arithmetic predicates and function quantifiers,* **Trans. Amer. Math. Soc.** 79 (1955) 312–340.

1955b *Hierarchies of number-theoretic predicates,* **Bull. Amer. Math. Soc.** 61 (1955) 193–213.

1959 *Recursive functionals and quantifiers of finite types* I, **Trans. Amer. Math. Soc.** 91 (1959) 1–52.

1959a *Quantification of number-theoretic functions,* **Compos. Math.** 14 (1959) 23–41.

1963 *Recursive functionals and quantifiers of finite types* II, **Trans. Amer. Math. Soc.** 108 (1963) 106–142.

Kondô, M.

1938 *Sur l'uniformization des complémentaires analytiques et les ensembles projectifs de la seconde classe,* **Japanese Jour. of Math.** 15 (1938) 197–230.

Kreisel, G.

1961 *Set theoretic methods suggested by the notion of potential totality,* in Anonymous [1961], 103–140.

1962 *The axiom of choice and the class of hyperarithmetic functions,* **Indag. Math.** 24 (1962) 307–319.

1971 *Some reasons for generalizing recursion theory,* in Gandy–Yates [1971], 139–198.

Kreisel, G. and Sacks, G.E.

1965 *Metarecursive sets,* **Jour. Symb. Log.** 30 (1965) 318–338.

Kripke, S.

1964 *Transfinite recursions on admissible ordinals,* I, II (abstracts), **Jour. Symb. Log.** 29 (1964) 161–162.

1964a *Admissible ordinals and the analytic hierarchy* (abstract), **Jour. Symb. Log.** 29 (1964) 162.

Kuratowski, K.

1966 **Topology I,** Academic Press, New York, 1966, 560 pp.

Lerman, M.

198? **The Uppersemilattice of Degrees of Unsolvability,** Perspectives in Mathematical Logic, Springer-Verlag, Berlin–Heidelberg–New York, to appear.

Lévy, A.

1963 *Transfinite computability* (abstract), **Notices Amer. Math. Soc.** 10 (1963) 286.

1965 **A Hierarchy of Formulas in Set Theory,** Memoirs of the Amer. Math. Soc. 57, American Mathematical Society, Providence, R.I., 1965, 76 pp.

1965a *Definability in axiomatic set theory* I, in Bar-Hillel [1965], 127–151.

1978 **An Introduction to Set Theory,** Perspectives in Mathematical Logic 3, Springer-Verlag, Berlin–Heidelberg–New York, to appear.

Ljapunov, A.A.

1949 *On δs-operations preserving measurability and the property of Baire* (Russian), **Mat. Sbornik N.S.** 24 (66) (1949) 119–127.

1953 *R*-Sets (Russian), **Trudy Mat. Inst. Steklov.** 40, Izdat Akad. Nauk SSSR, Moscow, 1953, 67 pp.

1953a *On the classification of R-sets* (Russian), **Mat. Sbornik N.S.** 32 (74) (1953) 255–262.

Ljapunov, A.A., Stschegolkow, E.A., and Arsenin, W.J.

1955 **Arbeiten zur Deskriptiven Mengenlehre,** Deutscher Verlag der Wissenschaften, Berlin, 1955, 108 pp.

Lorenzen, P. and Myhill, J.

1959 *Constructive definition of certain analytic sets of numbers,* **Jour. Symb. Log.** 24 (1959) 37–49.

Luzin, N.

1930 **Lecons sur les Ensembles Analytiques et leurs Applications,** Gauthier-Villars, Paris, 1930, 328 pp.

Machover, M.

1961 *The theory of transfinite recursion,* **Bull. Amer. Math. Soc.** 67 (1961) 575–578.

Martin, D.A.

1968 *The axiom of determinateness and reduction principles in the analytical hierarchy*, **Bull. Amer. Math. Soc.** 74 (1968) 687–689.

1970 *Measurable cardinals and analytic games*, **Fund. Math.** 66 (1970) 287–291.

1975 *Borel determinacy*, **Annals of Math.** 102 (1975) 363–371.

1977 *Projective sets and cardinal numbers: some questions related to the continuum problem*, mimeographed, to appear revised in **Jour. Symb. Log.**

1979? **Borel and Projective Games,** Perspectives in Mathematical Logic, Springer-Verlag, Berlin–Heidelberg–New York, to appear.

Moldestad, J.

1977 **Computations in Higher Types,** Lecture Notes in Mathematics 574, Springer-Verlag, Berlin–Heidelberg–New York, 1977, 203 pp.

Moschovakis, Y.N.

1967 *Hyperanalytic predicates*, **Trans. Amer. Math. Soc.** 129 (1967) 249–282.

1969 *Abstract first-order computability* I, **Trans. Amer. Math. Soc.** 138 (1969) 427–464, II 138 (1969) 465–504.

1969a *The lack of hierarchies on the second projective class* (*preliminary report*) (abstract), **Jour. Symb. Log.** 34 (1969) 155–156.

1970 *Determinacy and prewellorderings of the continuum*, in Bar-Hillel [1970], 24–63.

1970a *The Suslin–Kleene theorem for countable structures*, **Duke Jour. Math.** 37 (1970) 344–352.

1971 *Axioms for computation theories — first draft*, in Gandy–Yates [1971], 199–255.

1971a *Uniformization in a playful universe*, **Bull. Amer. Math. Soc.** 77 (1971) 731–736.

1972 *The game quantifier*, **Proc. Amer. Math. Soc.** 31 (1972) 245–250.

1973 *Analytic definability in a playful universe*, in Suppes *et al.* [1973], 77–85.

1974 **Elementary Induction on Abstract Structures,** North-Holland, Amsterdam, 1974, 218 pp.

1974a *Structural characterizations of classes of relations*, in Fenstad–Hinman [1974], 53–79.

1974b *On non-monotone inductive definability*, **Fund. Math.** 82 (1974) 39–83.

1978 *On the basic notions in the theory of induction*, to appear in the Proceedings of the Fifth International Congress for Logic, Methodology, and Philosophy of Science, London, Ontario, Canada, 1975.

1979? **Descriptive Set Theory: A Foundational Approach,** North-Holland, Amsterdam, to appear.

Mostowski, A.M.

1946 *On definable sets of positive integers*, **Fund. Math.** 34 (1946) 81–112.

1955 *Contributions to the theory of definable sets and functions*, **Fund. Math.** 42 (1955) 271–275.

1969 **Constructible Sets with Applications,** North-Holland, Amsterdam 1969, 269 pp.

Müller, G.H., Oberschelp, A., and Potthoff, K.

1975 **Logic Conference Kiel, 1974** (Proceedings of the International Summer Institute and Logic Colloquium, Editors), Lecture Notes in Mathematics 499, Springer-Verlag, Berlin–Heidelberg–New York, 1975, 651 pp.

Mycielski, J.

1964 *On the axiom of determinateness*, **Fund. Math.** 53 (1964) 205–224.

1971 *On some consequences of the axiom of determinateness*, in Scott [1971], 265–266.

Mycielski, J. and Steinhaus, H.

1962 *A mathematical axiom contradicting the axiom of choice*, **Bull. Acad. Polon. des Sciences** 10 (1962) 1–3.

Mycielski, J. and Swierczkowski, S.

1964 *On the Lebesgue measurability and the axiom of determinateness*, **Fund. Math.** 54 (1964) 67–71.

Myhill, J.

1961 *Category methods in recursion theory*, **Pacific Jour. Math.** 11 (1961) 1479–1486.

Oxtoby, J.C.
1971 **Measure and Category; A Survey of the Analogies between Topological and Measure Spaces,** Springer-Verlag, Berlin–Heidelberg–New York, 1971, 95 pp.

Paris, J.B.
1972 $\mathrm{ZF} \vdash \Sigma^0_4$ *determinateness*, **Jour. Symb. Log.** 37 (1972) 661–667.

Péter, R.
1935 *Konstruktion nichtrekursiver Funktionen*, **Math. Annalen** 111 (1935) 42–60.
1967 **Recursive Functions,** Academic Press, New York, 1967, 300 pp.

Platek, R.A.
1966 **Foundations of Recursion Theory,** Ph.D. thesis and supplement, Stanford University, 1966.

Post, E.L.
1944 *Recursively enumerable sets of positive integers and their decision problems*, **Bull. Amer. Math. Soc.** 50 (1944) 284–316.

Putnam, H.
1964 *On hierarchies and systems of notations*, **Proc. Amer. Math. Soc.** 15 (1964) 44–50.

Richter, W.H.
1968 *Constructively accessible ordinal numbers*, **Jour. Symb. Log.** 33 (1968) 43–55.
1971 *Recursively Mahlo ordinals and inductive definitions*, in Gandy–Yates [1971], 273–288.
1975 *The least* Σ^1_2 *and* Π^1_2 *reflecting ordinals*, in Müller *et al.* [1975], 568–578.

Rogers, H., Jr.
1959 *Recursive functions over well-ordered partial orderings*, **Proc. Amer. Math. Soc.** 10 (1959) 847–853.
1967 **Theory of Recursive Functions and Effective Computability,** McGraw-Hill, New York, 1967, 482 pp.

Sacks, G.E.
1963 **Degrees of Unsolvability,** Annals of Math. Studies 55, Princeton Univ. Press, Princeton, 1963, 174 pp.
1966 *Metarecursively enumerable sets and admissible ordinals*, **Bull. Amer. Math. Soc.** 72 (1966) 59–64.
1966a *Post's problem, admissible ordinals, and regularity*, **Trans. Amer. Math. Soc.** 124 (1966) 1–23.
1967 *Metarecursion theory*, in Crossley [1967], 243–263.
1969 *Measure-theoretic uniformity in recursion theory and set theory*, **Trans. Amer. Math. Soc.** 142 (1969) 381–420.
1971 *On the reducibility of* Π^1_1 *sets*, **Advances in Math.** 7 (1971) 57–82.
1974 *The 1-section of a type-n object*, in Fenstad–Hinman [1974], 81–93.
1978 *The k-section of a type-n object*, **Amer. J. Math.,** to appear.

Sacks, G.E. and Simpson, S.G.
1972 *The α-finite injury method*, **Ann. Math. Log.** 4 (1972) 343–367.

Scott, D.S.
1971 **Axiomatic Set Theory I** (Proceedings of the Thirteenth Symposium in Pure Mathematics of the Amer. Math. Soc., Los Angeles, 1967, Editor), American Mathematical Society, Providence, R.I., 1971, 474 pp.

Shoenfield, J.R.
1958 *The class of recursive functions*, **Proc. Amer. Math. Soc.** 9 (1958) 690–692.
1961 *The problem of predicativity*, in Bar-Hillel *et. al.* [1961], 132–139.
1962 *The form of the negation of a predicate*, in Dekker [1962], 131–134.
1967 **Mathematical Logic,** Addison-Wesley, Reading, Mass., 1967, 344 pp.
1968 *A hierarchy based on a type-2 object*, **Trans. Amer. Math. Soc.** 134 (1968) 103–108.
1971 *Unramified forcing*, in Scott [1971], 357–381.

1971a **Degrees of Unsolvability,** North-Holland, Amsterdam, 1971, 111 pp.
1971b *Measurable cardinals,* in Gandy–Yates [1971], 19–49.

Shore, R.A.
1977 *α-Recursion theory,* in Barwise [1977].

Simpson, S.G.
1974 *Degree theory on admissible ordinals,* in Fenstad–Hinman [1974], 165–193.
1984? **Admissible Recursion Theory,** Perspectives in Mathematical Logic, Springer-Verlag, Berlin–Heidelberg–New York, to appear.

Solovay, R.M.
1969 *On the cardinality of Σ^1_2 sets of reals,* in Bulloff *et al.* [1969], 58–73.

Spector, C.
1955 *Recursive well-orderings,* **Jour. Symb. Log.** 20 (1955) 151–163.
1958 *Measure-theoretic construction of incomparable hyperdegrees,* **Jour. Symb. Log.** 23 (1958) 280–288.
1959 *Hyperarithmetic quantifiers,* **Fund. Math.** 48 (1959) 313–320.
1961 *Inductively defined sets of natural numbers,* in Anonymous [1961], 97–102.

Strong, H.R.
1968 *Algebraically generalized recursive function theory,* **IBM Jour. Res. Devel.** 12 (1968) 465–475.

Suppes, P., Henkin, L., Joja A., and Moisil, Gr. C.
1973 **Logic, Methodology, and Philosophy of Science IV** (Proceedings of the Fourth International Congress for Logic, Methodology, and Philosophy of Science, Bucharest, Romania, 1971, Editors) North-Holland, Amsterdam, 1973, 981 pp.

Suzuki, Y.
1964 *A complete classification of the Δ^1_2-functions,* **Bull. Amer. Math. Soc.** 70 (1964) 246–253.

Takeuti, G.
1960 *On the recursive functions of ordinal numbers,* **Jour. Math. Soc. of Japan** 12 (1960) 119–128.
1965 *Recursive functions and arithmetic functions of ordinal numbers,* in Bar-Hillel [1965], 179–196.

Tanaka, H.
1967 *Some results in the effective descriptive set theory,* **Publ. Res. Inst. Math. Sci. Kyoto Univ.,** Ser. A, 3 (1967) 11–52.
1968 *A basis result for Π^1_1 sets of positive measure,* **Comment. Math. Univ. St. Paul** 16 (1968) 115–127.
1970 *On analytic well-orderings,* **Jour. Symb. Log.** 35 (1970) 198–203.
1970a *On a Π^0_1 set of positive measure,* **Nagoya Math. J.** 38 (1970) 139–144.

Tugué, T.
1960 *Predicates recursive in a type-2 object and Kleene hierarchies,* **Comment. Math. Univ. St. Paul** 8 (1960) 97–117.

Turing, A.M.
1936 *On computable numbers with an application to the Entscheidungsproblem,* **Proc. London Math. Soc.,** Series 2, 42 (1936) 230–265 and 43 (1936) 544–546.

Wagner, E.G.
1969 *Uniform reflexive structures: on the nature of Gödelizations and relative computability,* **Trans. Amer. Math. Soc.** 144 (1969) 1–41.

Zuckerman, M.M.
1974 **Sets and Transfinite Numbers,** Macmillan, New York, 1974, 423 pp.

Global Notational Conventions

Objects of type 0:

$a, b, \ldots, e, i, j, \ldots, w$	natural numbers = elements of ω

Objects of type 1:

$\alpha, \beta, \gamma, \delta, \varepsilon$	total unary functions $\omega \to \omega$ = elements of ${}^{\omega}\omega$
$f, g, h, E, F, \ldots, K$	partial functions ${}^{k}\omega \to \omega$
A, B, C, D, M, N, O, W	sets of natural numbers = subsets of ω
$P, Q, \ldots, V$	k-ary relations on ω = subsets of ${}^{k}\omega$

Objects of type 2:

$\mathsf{E}, \mathsf{F}, \ldots, \mathsf{K}$	partial functionals ${}^{k,l}\omega \to \omega$
$\mathsf{A}, \mathsf{B}, \mathsf{C}, \mathsf{D}, \mathsf{M}, \mathsf{N}, \mathsf{W}$	sets of (total unary) functions = subsets of ${}^{\omega}\omega$
$\mathsf{P}, \mathsf{Q}, \ldots, \mathsf{V}$	(k, l)-ary relations = subsets of ${}^{k,l}\omega$

Objects of type 3:

$\mathbb{E}, \mathbb{F}, \ldots, \mathbb{K}$	partial functionals ${}^{k,l,l'}\omega \to \omega$
$\mathbb{A}, \mathbb{B}, \mathbb{C}, \mathbb{D}, \mathbb{M}, \mathbb{N}, \mathbb{W}$	sets of total unary functionals = subsets of ${}^{({}^{\omega}\omega)}\omega$
$\mathbb{P}, \mathbb{Q}, \ldots, \mathbb{V}$	(k, l, l')-ary relations = subsets of ${}^{k,l,l'}\omega$

Objects of type 4:

$\mathcal{E}, \mathcal{F}, \ldots, \mathcal{K}$	partial functionals ${}^{k,l,l',l''}\omega \to \omega$
$\mathcal{P}, \mathcal{Q}, \ldots, \mathcal{V}$	(k, l, l', l'')-ary relations = subsets of ${}^{k,l,l',l''}\omega$

Other:

$\kappa, \lambda, \mu, \nu, \pi, \rho, \sigma, \tau, \upsilon$	ordinal numbers
Γ, Λ	inductive operators
$\mathcal{L}$	formal language
$\mathcal{T}$	formal theory
$\mathfrak{A}, \mathfrak{B}, \mathfrak{C}$	formulas of a formal language
$\mathfrak{M}, \mathfrak{N}, \mathfrak{U}$	structures, models
x, y, z	arbitrary objects
X, Y, Z	arbitrary sets
$\varphi, \psi, \chi, \theta$	arbitrary functions

The last three categories are often subject to local conventions

In most instances a bold-face letter denotes a finite sequence of objects of the type denoted by the light-face letter, Exceptions: $\mathbf{\Sigma}, \mathbf{\Pi}, \mathbf{\Delta}, \mathbf{\nabla}, \boldsymbol{\kappa}, \boldsymbol{\omega}_1[-]$.

Special Notations

Chapter I

Chapter II

Chapter III

Chapter IV

Chapter V

Epilogue

Subject Index

For EU product safety concerns, contact us at Calle de José Abascal, 56–1°,
28003 Madrid, Spain or eugpsr@cambridge.org.

www.ingramcontent.com/pod-product-compliance
Lightning Source LLC
LaVergne TN
LVHW102010210926
848483LV00010B/242
9781107168244